*Introductory
Mathematical Analysis*

Introductory Mathematical Analysis

Sixth Edition

Edgar D. Eaves
University of Tennessee

J. Harvey Carruth
University of Tennessee

Allyn and Bacon, Inc.
Boston   London   Sydney   Toronto

Copyright © 1985, 1978, 1974, 1969, 1964, 1958 by Allyn and Bacon, Inc., 7 Wells Avenue, Newton, Massachusetts 02159. All rights reserved. No part of the material protected by this copyright notice may be reproduced or utilized in any form or by any means, electronic or mechanical, including photocopying, recording, or by any information storage and retrieval system, without written permission from the copyright owner.

Library of Congress Cataloging-in-Publication Data

Eaves, Edgar D.
   Introductory mathematical analysis.
   Includes index.
   1. Mathematical analysis. I. Carruth, J. Harvey
(James Harvey), 1938– . II. Title.
QA300.E3 1985   515   85-7507
ISBN 0-205-08259-9

Printed in the United States of America

10 9 8 7 6 5 4 3 2 1   89 88 87 86 85

To Our Wives

DOROTHEA AND KAYLA

Contents

Preface   xiii

1
Number Systems   1

    1.1 Real Numbers   1
    1.2 The Real Number System   7
    1.3 Some Basic Properties of Real Numbers   14
    1.4 Fractions   23
    1.5 The Order Relations and Absolute Value   42
    1.6 Coordinate Systems   48
    1.7 Complex Numbers   55

2
Elementary Algebraic Processes   61

    2.1 Algebraic Expressions   61
    2.2 Exponents   68
    2.3 Radicals   78
    2.4 Polynomials in One Variable   82
    2.5 Polynomials in Two or More Variables   92
    2.6 Special Products and Factoring   97
    2.7 Algebraic Fractions   106
    2.8 The Binomial Theorem   113

3

Polynomials, Equations, and Inequalities   117

- 3.1 Polynomials in One Variable   117
- 3.2 Equations   119
- 3.3 Linear Equations in One Variable   127
- 3.4 Quadratic Equations in One Variable   130
- 3.5 The Method of Factoring   133
- 3.6 Solving the Equation $[f(x)]^2 = k^2$ 136
- 3.7 The Method of Completing the Square   139
- 3.8 The Quadratic Formula Method   145
- 3.9 Facts About the Roots of the Quadratic Equation $ax^2 + bx + c = 0$ 151
- 3.10 Equations in Quadratic Form   153
- 3.11 Polynomials: Two Major Problems   154
- 3.12 Facts Concerning the Roots of a Polynomial Equation   162
- 3.13 Rational Roots   170
- *3.14 Irrational Roots   174
- 3.15 Equations Containing Fractions   177
- 3.16 Equations Containing Radicals   182
- 3.17 Inequalities in One Variable   184
- 3.18 Higher-Degree Polynomial Inequalities   189
- 3.19 Inequalities Involving Quotients of Polynomials   193

4

Relations, Functions, and Graphs   196

- 4.1 Relations and Functions   197
- 4.2 Graphs of Relations and Functions   209
- 4.3 An Introduction to Analytic Geometry   217
- 4.4 Equations of Sets of Points in the Plane   222
- 4.5 Straight Lines   224
- 4.6 Equations of Straight Lines   230
- 4.7 Circles and Their Equations   238
- 4.8 Parabolas and Their Equations   243
- 4.9 Ellipses and Their Equations   248
- 4.10 Hyperbolas and Their Equations   252
- 4.11 Graphs of Linear Inequalities in Two Variables   256

5

Some Basic Concepts of Elementary Calculus   259

- 5.1 Average Rates   259
- 5.2 Average Rates Over a Flexible Interval   266
- 5.3 The Limit Concept and Some Limit Theorems   269
- 5.4 Continuous Functions   276
- 5.5 Instantaneous Rates   281

Contents

    5.6 The Derivative of a Function   286
    5.7 Rules for Differentiation   289
    5.8 Derivatives of Products, Quotients, and Powers   295
    5.9 Implicit Differentiation   303

6

Applications of the Derivative   308

    6.1 Interpretation of a Derivative   308
    6.2 Successive Differentiation   314
    6.3 Some Important Theorems   315
    6.4 Increasing and Decreasing Functions   318
    6.5 Maxima and Minima   324
    6.6 Applied Problems in Maxima and Minima   331
    6.7 The Chain Rule for Differentiation and Related Rates   337

7

Integration   342

    7.1 Antiderivatives and Integration Formulas   343
    7.2 Applications of Integration   346
    7.3 Equations of Curves   353
    7.4 Integrals and Differentials   354
    7.5 Areas Found by Integration   363
*7.6 Volumes of Solids   373
    7.7 Integration: A Process of Summation   375

8

An Introduction to Linear Algebra   381

    8.1 Systems of Two Linear Equations in Two Variables   381
    8.2 The Method of Determinants   390
    8.3 The Concept of a Matrix   401
    8.4 Matrix Algebra   408
    8.5 The Inverse of a Square Matrix   416
    8.6 Systems of Linear Equations Written as a Single Matrix Equation   420
*8.7 Systems of Linear Inequalities   423
*8.8 Linear Programming   427

9

Simple Interest, Compound Interest, and Annuities   437

    9.1 Percentage   437
    9.2 Simple Interest   441

9.3 Present Value at Simple Interest 449
9.4 Compound Interest 457
9.5 Present Value at Compound Interest 464
9.6 Equations of Value 469
9.7 Annuities 473
9.8 Present Value of an Annuity 478
9.9 Amortization of a Debt 483
9.10 Finding the Value of an Annuity on any Conversion Date 489

10

Exponential and Logarithmic Functions 494

10.1 Compounding Continuously and Exponential Functions 495
10.2 Natural Logarithms Defined 504
10.3 The Laws of Natural Logarithms 507
10.4 The Derivatives of $e^{f(x)}$ and $\ln f(x)$ 513
10.5 The Integrals $\int e^{f(x)} f'(x)\,dx$ and $\int \dfrac{f'(x)}{f(x)}\,dx$ 517
*10.6 Logarithms to Other Bases

11

Trigonometric Functions 525

11.1 Angles and Their Measures 525
11.2 Triangles 529
11.3 The Trigonometric Functions 532
11.4 Related Angles 537
11.5 The Trigonometric Functions of Acute Angles 541
11.6 Tables of Trigonometric Functions 547
11.7 Solution of Right Triangles 550
11.8 Angles of Elevation and Angles of Depression 552
11.9 Projections, Vectors, and Components of Vectors 555
11.10 Oblique Triangles 560
11.11 Basic Trigonometric Identities 565
*11.12 Functions of $(A + B)$, $(A - B)$, and $2A$ 570
*11.13 Graphs of the Trigonometric Functions 578
11.14 Differentiation and Integration of Trigonometric Functions 582

12

Probability and Statistics 593

12.1 Probability 593
12.2 Permutations and Combinations 596
12.3 The Binomial Expansion; Probability Rules 601
12.4 Mathematical Expectation 605
12.5 Probability Distributions 607

Contents

    12.6 Mean Value, Variance, and Standard Deviation   613
    12.7 Populations and Samples   617
    12.8 Normal Probability Distributions   620

Appendix A Logic and Sets 625

    A.1 Logic, Propositions, and Connectives   626
    A.2 Bi-implications and Theorems   637
    A.3 Sets, Elements, and Membership   640
    A.4 Subsets   646
    A.5 Set Operations   652
    A.6 The Cartesian Product of Two Sets   658
  *A.7 Equipollent Sets   662
  *A.8 Countable Sets   665
  *A.9 Open Statements and Quantifiers   668

Appendix B Tables 673

    Table I Natural Trigonometric Functions   674
    Table II Compound Amount   697
    Table III Present Value   703
    Table IV Exponential Functions   709
    Table V Natural Logarithms   715
    Table VI Amount of an Annuity of 1 per Annum   717
    Table VII Present Value of 1 per Annum   723

Answers to the Even-Numbered Problems 729

Index 801

Preface

Like its five predecessors, this text is designed to treat, in an introductory fashion, most of the mathematical topics commonly taught at the first-year college and university level. The book gives a modern, integrated treatment of the basic concepts of algebra, trigonometry, analytic geometry, calculus, linear algebra, and mathematics of finance. Primarily, the book is designed to give those students who take only one year of college mathematics an opportunity to obtain as broad a background in mathematics as is possible in the one school year allotted to the subject. The text is not designed for a full-year course in calculus and analytic geometry, although calculus of polynomials, exponentials, logarithms, and trigonometric functions are treated rather thoroughly.

The text contains more material than can normally be covered in an academic year course meeting three regular fifty-minute periods each week. However, the book is written in such a way that much picking and choosing of topics is possible. For example, any of the sections marked with an asterisk may be omitted without loss of continuity. Chapters 1 through 3 can be used for a one-quarter course in college algebra (or Chapters 1 through 4 with topics from Appendix A for a one-semester course). Chapters 5 through 7, with topics from Chapters 4, 10, and 11, can be used for a one-quarter or a one-semester course on differential and integral calculus. Any combination of topics chosen from Chapters 8 through 12 can be used for a quarter or semester "additional topics" course. Chapter 9 includes enough material to be used for an entire quarter course on mathematics of finance. Each of Chapters 8 through 12 is essentially self-contained, which gives the instructor a rather wide variety of choices regarding topics to omit if pushed for time. Finally, Appendix A can be used for a course on set theory and logic.

This new edition follows essentially the same general outline as the fifth edition and covers, for the most part, the same major topics. Some of the most significant features of the text are:

1. The use of hand calculators is encouraged throughout the text, particularly in the material on college algebra and mathematics of finance. Access to a hand calculator is *not* required for the treatment of any section in the text. Whenever the use of a calculator is suggested, a parallel noncalculator treatment is always given. This emphasis on the use of hand calculators allows the inclusion of many more (optional) examples and exercises which involve real-life problems. In particular, this text is the first to our knowledge which could be used to teach a course in mathematics of finance without the use of tables. Again, tables are included in the text and a completely satisfactory course can be taught on this material without the use of calculators.
2. The material is organized in several natural "packages":
   a. College Algebra: Chapters 1 through 3
   b. Functions, Relations, Equations, and Graphs: Chapter 4
   c. Calculus of Polynomials: Chapters 5 through 7
   d. Linear Algebra: Chapter 8
   e. Mathematics of Finance: Chapter 9
   f. Special Functions (Exponential, Logarithmic, and Trigonometric) Including Derivatives and Integrals: Chapters 10 and 11
   g. Probability and Statistics: Chapter 12
   h. Logic and Sets: Appendix A

   These "packages" make for ready adaptation to semester, quarter, or other sequences or to individual courses in college freshman-level mathematics.
3. A large number of concrete examples are used to motivate concepts prior to their treatment, and then additional examples are given following their treatment for reinforcement purposes.
4. The results which are most important in problem solving are collected and isolated in tabular form immediately preceding the appropriate problem lists.
5. Comprehensive problem lists are included. These are graded from routine to challenging. Answers to even-numbered problems are provided at the end of the text and answers to odd-numbered problems are provided in an instructor's supplement. Many changes have been made in the problem lists in this edition of the text.
6. In addition to the many illustrative examples, solutions to representative problems are often given within the problem lists themselves (especially in the material on college algebra, where such ready access and review seem most needed by students).
7. Rules for differentiation are introduced earlier than in previous editions. Somewhat less emphasis is given to motivating the differentiation process in this edition.
8. A section has been added which treats definite integrals as limits of sums. Previous editions used only the antiderivative approach.

Preface

9. A chapter on probability and statistics has been reinstated in this edition.
10. Differentiation and integration of most of the basic functions is included (polynomial, exponential, logarithmic, and trigonometric).
11. Material on logic and set theory is included in Appendix A, but little emphasis is placed on the language of set theory within the text. Set theoretic notation for intervals of real numbers is used and explained fully within the text itself.
12. Certain sections and certain problems are marked with an asterisk (*). These are more difficult than other materials in the text and their omission does not destroy continuity in any way.

The authors are deeply grateful for the valuable suggestions and criticisms of their colleagues C. G. Doss, Donald Dessart, and H. T. Mathews. Special thanks go to Cindy Blair for her able preparation, typing, and proofreading of the manuscript. Thanks are also due to Jose DeLuna and Ed Branding for their assistance in checking the answers to problems in the problem lists. Also, thanks are due to the members of the office staff of the Department of Mathematics, Barbara Bateman, Cindy Blair, Mary Ann Bledsoe, Jessica Daugherty, Jane Garron, Rose Nelson, Sandy Rucker, and Rosanna Templin, for their friendly assistance and to the many teachers in all parts of the country who have used previous editions for their many helpful suggestions.

Finally, it is a pleasure to acknowledge the valuable assistance, complete cooperation, and considerable patience of the staff of Allyn and Bacon.

E. D. E.
J. H. C.

1
Number Systems

Introduction

It is the purpose of this chapter to study in some detail the characteristic properties of the real number system and the complex number system. We shall exhibit the definitions and axioms which determine these two systems, and use these definitions and axioms in proving some theorems that are basic in the study of elementary mathematics. For some, much of this chapter may be a review. If so, it is nevertheless well for the student not to take the review lightly, but rather to be certain that all the topics covered are clear to him or her, for these topics furnish the background for the material that is to follow. For those who are unfamiliar with the various set theoretic notations which will be used herein, explanations and definitions may be found in Appendix A.

1.1 REAL NUMBERS

The concept of numbers is basic in mathematics and every student has studied about and worked with numbers since grade school days. In fact, the mathematical objects called numbers are so commonplace in all our lives that most people are apt to take them for granted without really knowing or understanding their characteristic properties. Since a knowledge and a clear understanding of these properties are so essential as background for further study of mathematics, we feel it worthwhile to give a brief discussion of such properties at this point.

Our experience in working with numbers started in early childhood when we learned to count. In the counting process we became acquainted with the set of *positive integers* {1, 2, 3, 4, ...}. The set of positive integers is also known as the set of *natural numbers* and as the set of *counting numbers*. For convenience, we shall denote the set of positive integers by the capital letter N; that is, $N = \{1, 2, 3, 4, ...\}$. We shall at times use small letters such as a, b, c, etc., to represent arbitrary positive integers.

In grade school we learned to add, subtract, multiply, and divide positive integers. We soon discovered that we can add any two positive integers or multiply any two positive integers and the result of either operation is always a positive integer. However, with only positive integers to work with, the operations of subtraction and division are not always possible. For example, it is not possible to subtract 5 from 2 or to divide 7 by 4 if our arsenal of numbers contains only positive integers. To remedy these shortcomings, it was necessary to enlarge our set of numbers to include some new elements and to extend the operations of addition, subtraction, multiplication, and division to the larger set.

First, we included *zero* (0) and then the *negatives* of the positive integers $\{-1, -2, -3, -4, ...\}$, which together with the set of positive integers form the set of *integers* $\{..., -4, -3, -2, -1, 0, 1, 2, 3, 4, ...\}$. Next we included such elements as $\frac{1}{2}, \frac{2}{3}, -\frac{4}{5}$, called *fractions*.

DEFINITION 1.1. Any numbers that can be expressed in the form $\frac{a}{b}$, where a and b both are integers and $b \neq 0$, is called a *rational number*.

Notice that every integer n is a rational number since it can be expressed as $n = \frac{n}{1}$. Thus the set of rational numbers consists of all the integers together with all the fractions, $\frac{a}{b}$, which do not represent integers, but a and b are both integers with $b \neq 0$.

DEFINITION 1.2. An integer b is called a *factor* (or a *divisor*) of an integer a if there is an integer c such that a is the product of b and c; that is, $a = bc$. In this case, c is likewise a factor of a.

Since 1 is a factor of every integer, it follows that each integer has at least two factors, itself and 1.

DEFINITION 1.3. A positive integer is called a *prime* if it is greater than 1 and has only itself and 1 as positive factors. A positive integer greater than 1 and not a prime is called a *composite*.

Example 1   Numbers such as 2, 3, 5, 7, 11, etc., are prime, while numbers such as 6, 9, 12, 14, etc., are composites.

We next state, without proof, two very useful theorems, the second of which is a consequence of the first.

THEOREM 1.1. (UNIQUE FACTORIZATION THEOREM). *A positive integer greater than 1 can be expressed as the product of factors, which are primes, in one and only one way except for the order in which the factors are written in the product.*

THEOREM 1.2. *Suppose that the letters a and p represent positive integers and p is a prime. If p is a factor of a^2, then p is a factor of a.*

DEFINITION 1.4. If two positive integers a and b have the same positive integer c as a factor, then c is called a common factor of a and b.

Example 2 2 is a common factor of 4 and 8; 2, 3, 5, 6, 10, and 15 are common factors of 30 and 60.

DEFINITION 1.5. A fraction such as a/b, where a and b are integers, is said to be in *simplest* or *lowest form* if a and b have no common factors other than 1.

Example 3 The fractions $\frac{2}{3}, \frac{4}{9}, \frac{8}{15}$, and $\frac{10}{21}$ are in simplest form, while $\frac{4}{6}, \frac{9}{18}$, and $\frac{14}{21}$ are not in simplest form.

In the study of arithmetic, we learned that each rational number is expressible as an infinite decimal. For example,

$$2 = 2.00000\ldots$$

$$\frac{1}{2} = 0.50000\ldots$$

$$\frac{1}{3} = 0.33333\ldots$$

$$\frac{7}{22} = 0.3181818\ldots$$

$$\frac{2}{7} = 0.285714285714\ldots$$

In each of these decimal representations, it is seen that a fixed digit or a fixed set of digits repeats indefinitely; for this reason such decimals are called *repeating decimals*. It can be shown that if a rational number is expressed as an infinite decimal it will be a repeating decimal and, conversely, that every repeating infinite decimal is the expansion of some rational number. It is easy to exhibit infinite decimals that are not repeating decimals and hence are not rational numbers, for example, 0.25303003000300003 ….

DEFINITION 1.6. A number whose infinite decimal expansion is not a repeating decimal is called an *irrational number*.

It is obvious from the definition of an irrational number that the set of rational numbers and the set of irrational number are disjoint. That is, no number can be both rational and irrational.

THEOREM 1.3. $\sqrt{3}$ *is not a rational number.*

We shall give an indirect proof. That is, we shall assume the theorem false and show that this assumption leads to a contradiction.

Proof. Assume $\sqrt{3}$ is a rational number. Then it follows from the definition of a rational number that there must exist two integers a and b, which we shall assume to have no common factors other than 1, such that

(1) $$\frac{a}{b} = \sqrt{3}$$

Hence

(2) $$a^2 = 3b^2$$

This last equation implies that the prime number 3 is a factor of a^2 and thus 3 is a factor of a (Theorem 1.2). If 3 is a factor of a, then $a = 3c$, or $a^2 = 9c^2$, where c is some positive integer. Substituting for a^2 in Equation (2), we have

$$3b^2 = 9c^2$$

or $$b^2 = 3c^2$$

which shows that 3 is a factor of b^2 and hence of b. But this is impossible since the numbers a and b are assumed to have no common factors other than 1. Thus our assumption that $\sqrt{3}$ is a rational number must be false. We conclude that $\sqrt{3}$ is not a rational number.

It is interesting to note that if a hand calculator is used to "evaluate" $\sqrt{3}$, the number obtained is 1.732050808075. What the calculator is indicating is an approximation to $\sqrt{3}$, which is accurate to 12 decimal places. Hence, problems solved by use of a hand calculator are often not exact solutions. However, they are ordinarily accurate enough for all practical purposes.

DEFINITION 1.7. A *real* number is a number that is either a *rational* number or an *irrational* number. The set of *real* numbers is the set consisting of all the *rational* numbers and all the *irrational* numbers. Every real number can be expressed as an infinite decimal.

SEC. 1.1   Real Numbers

It must be mentioned at this point that the process of "adding" two infinite decimals is quite a complicated one and is beyond the scope of this book. The process of "multiplying" two infinite decimals is even more complicated. Also, one must even equate certain different infinite decimals (for example, one must equate 0.99999... to 1.00000...). Suffice it to say that the well-known operations of addition and multiplication of fractions (rational numbers) can be extended to the set of real numbers. If one is interested in adding $\sqrt{2}$ to $\sqrt{3}$, then about the best that can be done is to write down $\sqrt{2}+\sqrt{3}$. Of course, $\sqrt{2}$ and $\sqrt{3}$ can be approximated (accurate to any desired number of decimal places) and these approximations can be added, yielding an approximation to $\sqrt{2}+\sqrt{3}$. This is what one does when using a calculator to "evaluate" $\sqrt{2}+\sqrt{3}$. The result obtained might be

$$\sqrt{2}+\sqrt{3} = 1.414213562373 + 1.732050807568$$

Again we emphasize that this is only an approximation to $\sqrt{2}+\sqrt{3}$ and not the exact value. A rather mild challenge is afforded in the seeking of the entire 13 significant digits in each of the above numbers since the calculator displays at most 10 digits. This exercise is left to the interested reader. In the remaining sections of this chapter we shall discuss the characteristic properties of the real and complex number systems.

Exercises 1.1

1. Make a list of all the positive integers which are primes and less than 100.

2. Express each of the given numbers as a product of primes.
   a. 42   **b.** 57   **c.** 154   **d.** 2,737

3. Find the common prime factors of each given pair of numbers.
   a. 18 and 30   **b.** 70 and 120

4. Find all the factors which are integers of each given number.
   a. 6   **b.** -12 **c.** 75   **d.** 24   **e.** -27

5. Express each given number as an infinite decimal.
   a. $\dfrac{8}{5}$ **b.** $\dfrac{3}{8}$ **c.** $\dfrac{22}{7}$ **d.** $\dfrac{11}{3}$ **e.** $\dfrac{728}{25}$

6. Determine which of the given numbers are rational. Give a reason for your answer in each case.
   a. 1.24   **b.** 1.414
   c. 0   **d.** 3.1416
   e. 3.142857142857142857...   **f.** 1.3201001000100001...

7. Use a method similar to that used in the proof of Theorem 1.3 to prove that each of the following numbers is not rational.
   a. $\sqrt{2}$ b. $\sqrt{5}$ c. $\sqrt[3]{2}$

8. Every real number belongs to one or more of the following sets: positive integers, integers, rational numbers, and irrational numbers. For example, the number 6 is a positive integer, it is an integer, and it is a rational number; hence it belongs to three of the four above-mentioned sets. On the other hand, the number -5 belongs to just two of these sets, while the number $\frac{2}{3}$ belongs only to the set of rational numbers, and the number $\sqrt{3}$ belongs only to the set of irrational numbers. Prepare a table as illustrated below and place each of the following numberrs in the set or sets to which it belongs: 7, -4, -378, $\sqrt{9}$, $\frac{\pi}{3}$, -2, 46, 0, $\frac{3}{7}$, $\sqrt[3]{8}$, $\frac{6.4}{0.32}$, $-\frac{5}{3}$, $\sqrt[3]{2}$, 34,796, $\frac{0}{3}$, $\sqrt{4}$, $4+\sqrt{3}$.

Positive Integers	Integers	Rational Numbers	Irrational Numbers
6	6	6	$\sqrt{3}$
	-5	-5	
		$\frac{2}{3}$	

9. An integer a is an even integer if and only if $a=2n$, where n is some integer. An integer b is an odd integer if and only if it can be expressed in the form $b=2n+1$ or $b=2m-1$, where n and m are integers. Assume also that the sum of any two integers is an integer and that the product of any two integers is an integer.
   a. Prove that the square of an even integer is an even integer.
   b. Prove that the square of an odd integer is an odd integer.
   c. Prove that the sum of any two odd integers is an even integer.
   d. Prove that the product of any two odd integers is an odd integer.
   e. Prove that an integer is an even integer if its square is an even integer.
   f. Prove that an integer is an odd integer if its square is an odd integer.

10. It was proved in Theorem 1.3 that $\sqrt{3}$ is not rational. Prove that $2+\sqrt{3}$ is not rational.

11. Show by an example that:
    a. The sum of two irrational numbers may be rational.
    b. The product of two irrational numbers may be rational.
    c. The sum of two primes may be a prime. How many such examples can you find?

1.2 THE REAL NUMBER SYSTEM

Mathematical Systems in General

In general, we think of a mathematical system as consisting of a set of elements, a set of definitions, one or more operations that can be performed upon the elements, a set of axioms, and some theorems which can be proved from the definitions and axioms. Thus the expression "the system of natural numbers" means much more than just the numbers themselves; they are only the elements involved.

In this section we shall consider a very important mathematical system, the system of real numbers. Its importance lies in the fact that the study of arithmetic and elementary algebra is nothing more than the study of the properties of the real number system. In fact, in the remainder of the text we shall be concerned chiefly with combinations and manipulations of expressions that represent real numbers.

The Real Number System

As was pointed out in our previous remarks, one of the chief ingredients of any mathematical system is a set of elements. In the case of the real number system, as you no doubt have already guessed, the set of elements involved is the set of real numbers. Let us call this set R and write $R = \{a, b, c, \ldots\}$, where it is to be understood that the lowercase letters a, b, c, etc., represent real numbers.

One of the most important relations of arithmetic is that of equality, and one of the most important symbols is the symbol for equals, $=$.

DEFINITION 1.8. Two letters, a and b, representing real numbers are said to be *equal* if and only if they represent the same real number. To indicate that a and b are equal, we write $a = b$, which is read "a equals b" or "a is equal to b."

We shall assume that the following properties of equality hold.

PROPERTIES OF EQUALITY

If a, b, and c represent any real numbers, then

PROPERTY 1. $a = a$ (reflexive property)

PROPERTY 2.   If $a = b$, then $b = a$ (symmetric property)

PROPERTY 3. If $a=b$ and $b=c$, then $a=c$ (transitive property)

Property 3 justifies the following principle, which we shall state as Property 4.

PROPERTY 4. If $a=b$, then b may be substituted for a in any expression, equality, or statement involving a without changing the value of the given expression or the truth or falsity of the given equality or statement (substitution principle).

Property 2 states that the order in which an equality is written does not affect the equality.

Example 1 | If $15=3a$, then $3a=15$

Example 2 | If $7a-3=b^2+2c$, then $b^2+2c=7a-3$

Property 3 may be roughly translated: "If each of two real numbers is equal to the same third real number, then the numbers are equal to each other."

The student probably remembers Property 4, the substitution principle, in the form "equals may be substituted for equals." The substitution principle is a most useful tool in simplifying or changing the form of mathematical expressions or equalities.

Example 3 | $5+(3+4)=12$ is a true statement; $3+4=7$ is a true statement. Hence we can substitute (7) for (3+4) in the first equality and write $5+7=12$, which is also a true statement.

In mathematics, one frequently refers to a particular set A of elements as being closed under a particular operation, or we may say that set A possesses the closure property with respect to a particular operation. To see what is meant by such a statement, let us consider the set $A=\{-1, 0, 1\}$ with the operation multiplication. It is obvious in this case that if we multiply any element of A by itself or by another element of A the resulting product is always one of the three elements of A. We say that A is closed under multiplication. Notice, however, that if we use the operation addition instead of multiplication, set A is not closed under addition, since $1+1=2$ and 2 is not in A.

DEFINITION 1.9. A set A is said to be closed under an operation if the result of this operation on any two elements of A also produces an element of A, whether the two elements used are the same element or are two different elements.

Consider the set of positive integers $N=\{1, 2, 3, 4, 5, ...\}$. It is a well-known fact that the sum of two positive integers is a positive integer, and the

SEC. 1.2 The Real Number System

product of two positive integers is a positive integer. Thus, according to the above definition, we can say that the set N is closed under addition and under multiplication. It is obvious that N is not closed under subtraction or under division.

We take, as the two basic operations of our system, the well-known operations of addition and multiplication (see Section 1.1). If two real numbers are represented by a and b, then their sum is represented by $a+b$ and their product is represented by either $a \cdot b$ or $(a)(b)$. It will be convenient (though inaccurate) henceforth to write "Let a, b, and c be any three real numbers," instead of "Let a, b, and c represent any three real numbers." Where no confusion is likely, products such as $a \cdot b$ or $(a)(b)$ will be expressed as ab (confusion would, of course, result from omitting the dot in a product such as $2 \cdot 3$, which would give 23 and not 2 times 3).

The following axioms, Axioms 1 to 11, are basic in all operations involving expressions that represent real numbers. Throughout, a, b, c, and d will be real numbers.

THE CLOSURE LAWS

AXIOM 1. $a+b$ is a unique real number

AXIOM 2. $a \cdot b$ is a unique real number

Axioms 1 and 2 state that if any two real numbers are added or multiplied the result will always be a unique real number. A brief way of saying this is to say, "The set of real numbers is closed under the operations of addition and multiplication."

THE COMMUTATIVE LAWS

AXIOM 3. $a+b=b+a$

AXIOM 4. $ab=ba$

These two axioms state that the order in which two numbers are added or multiplied does not affect the result of that addition or multiplication.

Example 4 $3+5=5+3=8$

Example 5 $(3)(5)=(5)(3)=15$

THE ASSOCIATIVE LAWS

AXIOM 5. $(a+b)+c=a+(b+c)$

AXIOM 6. $(ab)c = a(bc)$

It should be noted that addition and multiplication are binary operations. That is, they are operations which we perform with two numbers. For example, if we wish to add three numbers, we must first add two of them and then add their sum to the third. In fact, the expression $a+b+c$ or the expression abc has no meaning unless we define it. So we define $a+b+c$ and abc by the following definitions.

DEFINITION 1.10. $a+b+c = (a+b)+c$

DEFINITION 1.11. $abc = (ab)c$

Now that we have given meaning to an expression involving three numbers, we can extend these definitions to an expression involving four numbers. In fact, we define $a+b+c+d$ and $abcd$ by

$$a+b+c+d = (a+b+c)+d \quad \text{and} \quad abcd = (abc)d$$

This procedure obviously could be extended to define an expression involving any number of real numbers. It follows from Axioms 3 and 5 that, although to add three or more real numbers we must first add two of them and then add this sum to a third, and so on, the way to select the first two to be added, or the way to select each subsequent number to be added to the sum already obtained, does not affect the final result. A similar remark holds for Axioms 4 and 6 in relation to finding the product of three or more real numbers.

Example 6
$3+5+2 = (3+5)+2 = 8+2 = 10$
$ = 3+(5+2) = 3+7 = 10$
$ = (3+2)+5 = 5+5 = 10$

Example 7
$(3)(5)(2) = [(3)(5)](2) = (15)(2) = 30$
$ = (3)[(5)(2)] = (3)(10) = 30$
$ = [(3)(2)](5) = (6)(5) = 30$

Example 8
$(a)(b)(a) = a^2 b$

THE DISTRIBUTIVE LAW

AXIOM 7. $a(b+c) = ab + ac$. In words, this relation states that the operation multiplication is distributive with respect to addition.

Example 9
$5(4+3) = (5)(4) + (5)(3) = 20 + 15 = 35$
Check: $(5)(7) = 35$

Example 10
$5(2a - 3b) = 10a - 15b$

SEC. 1.2   The Real Number System

THE AXIOMS DEFINING THE IDENTITY ELEMENTS

AXIOM 8.   *There is a unique real number zero, denoted by 0, which has the special property that, if a is any real number, then $a+0=0+a=a$. Zero is the only real number having this property and, because of this property, it is said to be the additive identity element of the set of real numbers.*

AXIOM 9.   *There is a unique real number one, denoted by 1, which is different from zero and has the special property that if a is any real number then $a \cdot 1 = 1 \cdot a = a$. Because of this property, 1 is called the* multiplicative identity element *of the set of real numbers.*

THE AXIOMS DEFINING INVERSE ELEMENTS

AXIOM 10.   *For each real number a, there exists one and only one real number, which we denote by $(-a)$, such that $a + (-a) = (-a) + a = 0$. The real number $(-a)$ is called the additive inverse of the real number a.*

AXIOM 11.   *If a is any real number other than zero, there exists one and only one real number, denoted by $\frac{1}{a}$, such that $(a)\left(\frac{1}{a}\right) = \left(\frac{1}{a}\right)(a) = 1$. The number $\frac{1}{a}$ is called the multiplicative inverse of a.*

A comment about notation. For clarity and consistency, the authors would prefer to use the built-up fractional notation $\frac{1}{a}$ entirely. However, another common notation for the multiplicative inverse of a is $1/a$. This latter form is considerably easier and less expensive to set in type and, consequently, its use will be prevalent in the text.

DEFINITION 1.12.   The real number b is said to be the *reciprocal* of the real number a if and only if $a \cdot b = 1$.

Since, by Axiom 4, $a \cdot b = b \cdot a$, it follows from Definition 1.11 that, if the real number b is the reciprocal of the real number a, then a is also the reciprocal of b. From Axiom 11 it follows that if the real number b is the reciprocal of the real number a, then $b = 1/a$; that is, the reciprocal of a number is its multiplicative inverse, and conversely. It is important to note that zero is the only real number that does not have a reciprocal. The reciprocal of zero is not defined; in other words, division by zero is not defined.

The previous paragraph points out one of the most important rules of operations when dealing with real numbers. *Never divide by zero!* Three other very common errors which occur in algebra are attempts to distribute exponents, radicals, and reciprocals with respect to addition. (We shall treat exponents and radicals in detail in Chapter 2.)

Example 11 $(2+3)^3 \neq 2^3 + 3^3$, for
$(2+3)^3 = 5^3 = 125$, whereas
$2^3 + 3^3 = 8 + 27 = 35$

Example 12 $\sqrt{9+16} \neq \sqrt{9} + \sqrt{16}$, for
$\sqrt{9+16} = \sqrt{25} = 5$, whereas
$\sqrt{9} + \sqrt{16} = 3 + 4 = 7$

Example 13 $\dfrac{1}{2+2} \neq \dfrac{1}{2} + \dfrac{1}{2}$, for
$\dfrac{1}{2+2} = \dfrac{1}{4}$, whereas
$\dfrac{1}{2} + \dfrac{1}{2} = 1$

These errors occur so frequently that we set them apart here in a Warning Box. These can be thought of as rules involving *things **not** to do*.

Warning

1. Never divide by zero. The expression $\dfrac{a}{0}$ has no meaning
2. $(a+b)^n \neq a^n + b^n$ in general
3. $\sqrt{a+b} \neq \sqrt{a} + \sqrt{b}$ in general
4. $1/(a+b) \neq 1/a + 1/b$ for any real numbers a and b

In algebra we are more interested in rules involving *things that we **are** allowed to do*. Some of these rules are embodied in Axioms 1 to 11, and we isolate them here for convenience.

Let a, b, and c be any three real numbers. Then:

AXIOM 1. $a+b$ is a unique real number
AXIOM 2. $a \cdot b$ is a unique real number
AXIOM 3. $a+b=b+a$
AXIOM 4. $ab=ba$
AXIOM 5. $(a+b)+c=a+(b+c)$
AXIOM 6. $(ab)c=a(bc)$
AXIOM 7. $a(b+c)=ab+ac$

SEC. 1.2 The Real Number System

> AXIOM 8. $a + 0 = 0 + a = a$
> AXIOM 9. $a \cdot 1 = 1 \cdot a = a$
> AXIOM 10. $a + (-a) = (-a) + a = 0$
> AXIOM 11. $(a)\left(\dfrac{1}{a}\right) = \left(\dfrac{1}{a}\right) a = 1$ provided $a \neq 0$

In more advanced courses in mathematics these same 11 axioms are taken as the axioms of an *abstract* mathematical system called a *field*. That is, any set of objects with two operations satisfying these axioms is *by definition* a field. Thus the real number system is a field.

Exercises 1.2

In the following exercises the letters a, b, c, ... represent real numbers unless otherwise specified.

1. By using Definition 1.10 and any axioms which apply, prove
$$a + b + c = c + b + a$$

2. Prove $(a + b + c) + d = (a + b) + (c + d)$, stating each definition or axiom used.

3. Assuming $2 + 3 = 5$, what axiom (or axioms) justify the statement
$$2a + 3a = 5a$$

4. **a.** Write the additive inverse for each of the following numbers: $8, \sqrt{2}, -4, -\dfrac{1}{2}, \dfrac{3}{8}, 0, 4.25$.
   b. Find the reciprocal of each of the following numbers that has a reciprocal: $\dfrac{2}{3}, 0.4, -\dfrac{7}{8}, -\dfrac{4}{3}, \pi/3, \dfrac{0}{2}, -0.25, \sqrt{5}, 1.2/4.8, \dfrac{1}{3}$.

5. Prove: $ab + de + ac + df = a(b + c) + d(e + f)$, stating each definition or axiom used.

6. **a.** What axiom justifies the statement $(2 \cdot 3)5 = 2(3 \cdot 5)$?
   b. Prove $abc = cba$. Justify each step with a definition or axiom.

7. Define the product $abcde$ and use your definition plus any axioms needed to prove $abcde = (ab)(cde)$.

8. Prove that the set consisting of all the even positive integers is closed under addition and also under multiplication. You may assume that the set of positive integers is closed under addition and under multiplication.

9. Find a proper subset of R (the set of real numbers), not identical with the set of even positive integers, that is closed under (**a**) addition, (**b**) subtraction, (**c**) multiplication, (**d**) division, (**e**) both addition and subtraction, (**f**) addition and multiplication, (**g**) addition, subtraction, and multiplication.

10. Show that the set $S = \{5, 10, 15, \ldots, 5n, \ldots\}$ is closed under addition and multiplication.

11. Suppose we invent, just for fun, a new operation and call it "multivation." Let us denote this new operation by a small rectangle, $\square$, and define it by the relation $a \square b = ab + 2$.
    a. Find $3 \square 4$, $-2 \square 5$, $(-1) \square (-2)$, and $0 \square 7$.
    b. Is the set of positive integers closed under the operation of "multivation"?
    c. Is the statement "$a \square (b+c) = a \square b + a \square c$ for all a, b, c in R" true or false?

12. It follows from the distributive law (Axiom 7) that $(9 \cdot 43) + (9 \cdot 57) = 9(43 + 57) = 9 \cdot 100 = 900$. Use the distributive law (Axiom 7) to evaluate each of the following:
    a. $(7 \cdot 27) + (7 \cdot 13)$ b. $(17 \cdot 63) + (17 \cdot 37)$
    c. $(92 \cdot 58) + (92 \cdot 42)$ d. $(375 \cdot 850) + (375 \cdot 150)$

13. Sometimes Axiom 7 (the distributive law) can be used to express a sum as a product. For example, $ab + ac = a(b + c)$. In such a case we say that we have factored the sum. Use the distributive law to factor each of the following.
    a. $a^2 + ab$ b. $de + df$
    c. $ab + ac + db + dc$ d. $42 + 54$
    e. $56 + 88$ f. $27 + 45$

14. Justify each of the following statements by stating an axiom as the reason for each step you make:
    a. $0 + 7 = 7$ b. $(-a) + (a + b) = b$
    c. $1 \cdot (a + b) + 0 = a + b$ d. $-(ab) + a(a + b) = a^2$

15. Is the set consisting of all the rational numbers closed under division?

1.3 SOME BASIC PROPERTIES OF REAL NUMBERS

Symbols of Grouping

In the previous section we defined $a + b + c$ to be $(a + b) + c$. This is an example of what one might call *grouping*, and the symbols involved are parentheses. The parentheses about $a + b$ indicate that we are to consider $a + b$ as a single number. Hence, if one was asked to calculate $(2 + 5) + 3$, the parentheses about $2 + 5$ indicate that the 2 should be added to the 5 *first* and then that result should be added to 3. It so happens that the order in which any string of real numbers is added is immaterial. However, it is often desirable to indicate that certain terms of an algebraic expression are to be grouped together and considered as a single number. Symbols of grouping in common use are *parentheses* ( ), *brackets* [ ], and *braces* { }. Since, by the distributive law,

$$k(a + b - c) = ka + kb - kc$$

SEC. 1.3   Some Basic Properties of Real Numbers

it follows that, for $k = +1$,
$$(+1)(a+b-c) = (+1)(a) + (+1)(b) + (+1)(-c)$$
$$= a+b-c$$

But
$$(+1)(a+b-c) = +(a+b-c)$$
$$= (a+b-c)$$

Hence
$$(a+b-c) = a+b-c$$

Similarly, for $k = -1$, it follows that
$$(-1)(a+b-c) = (-1)(a) + (-1)(b) + (-1)(-c)$$
$$= -a-b+c$$

Also,
$$(-1)(a+b-c) = -(a+b-c)$$

Therefore,
$$-(a+b-c) = -a-b+c$$

This shows that to remove parentheses preceded by a plus sign, actual or implied, the terms involved are not changed. But to remove parentheses preceded by a minus sign, the sign of each term inside the parentheses must be changed. Hence, to insert parentheses preceded by a plus sign, no changes in the terms are to be made, while to insert parentheses preceded by a minus sign, it is necessary to change the sign of each term placed within the parentheses.

Example 1
$$3a + 2bc + b - c = 3a + (2bc + b) - c$$
$$= 3a + 2bc + (b - c)$$
$$= (3a + 2bc) + (b - c)$$

When removing symbols of grouping where one set of symbols is contained within another, the following rule is suggested: first remove the innermost set of symbols and collect like terms; then continue this process until all symbols of grouping have been removed.

Example 2
$$8a - \{6a - [3 - (a - 3)]\} = 8a - \{6a - [3 - a + 3]\}$$
$$= 8a - \{6a - [6 - a]\}$$
$$= 8a - \{6a - 6 + a\}$$
$$= 8a - \{7a - 6\}$$
$$= 8a - 7a + 6$$
$$= a + 6$$

Some Basic Properties of Addition and Multiplication

We shall now state several theorems we shall need later. We prove only one of these theorems and leave the proofs of the remaining results to the interested reader. The student will probably recognize these results as rules of operation he or she once memorized. Throughout the section, a, b, and c will be real numbers.

THEOREM 1.4. If $a=b$, then $a+c=b+c$.

Proof: $a+c=a+c$ (Property 1 of Section 1.2)
$a=b$ (Given)
$a+c=b+c$ (Property 4 of Section 1.2)

For the sake of brevity and convenience, let us agree to write $-a$ for $(-a)$ when no ambiguity can result. Similarly, we agree that $+(+a)=+(a)=(a)=a$. We can now formulate the following definition.

DEFINITION 1.13. The symbol $a-b$ is defined to mean $a+(-b)$. Recall (Axiom 10) that $-b$ is the additive inverse of b. The process of adding $-b$ to a to obtain $a-b$ is called subtraction. The expression "subtract b from a" means "add $-b$ to a."

It should be noted in this connection that we are now using the minus sign in two different ways. It is being used to indicate the additive inverse of a number and also to indicate the operation of subtraction. This, however, should not lead to confusion, for the sense in which it is being used should always be apparent.

Theorem 1.4 basically says that we can add the same number to both sides of an equation and obtain another equation. If, instead of adding c to both sides, we add $-c$ to both sides, we obtain the result that you can also subtract the same number from both sides of an equation and obtain another equation.

THEOREM 1.5. If $a=b$, then $a-c=b-c$.

Theorems 1.4 and 1.5 are very useful in solving equations, as illustrated in the next two examples.

Example 3   Solve the equation $a-3=8$.

Solution: $a-3=8$ (Given)
$a-3+3=8+3$ (Theorem 1.4, adding 3 to both sides)
$a+0=11$ (Axiom 10 and addition)
$a=11$ (Axiom 8)

Example 4   Solve the equation $a+3=8$.

Solution: $a+3=8$ (Given)
$a+3-3=8-3$ (Theorem 1.5, subtracting 3 from both sides)
$a+0=5$ (Axiom 10 and subtraction)
$a=5$ (Axiom 8)

THEOREM 1.6. If $a=b$, then $ac=bc$.

SEC. 1.3   Some Basic Properties of Real Numbers

DEFINITION 1.14. *If $b \neq 0$, the symbol $\dfrac{a}{b}$ is defined to mean $(a)\left(\dfrac{1}{b}\right)$. We call $\dfrac{a}{b}$ the quotient of a and b. The process of finding the quotient $\dfrac{a}{b}$ is called division.*

The symbol $\dfrac{a}{b}$ is called a *fraction* and, if a and b are both integers, it is called a *simple fraction*. In the fraction $\dfrac{a}{b}$, a is called the *numerator* of the fraction and b is called the *denominator* of the fraction.

THEOREM 1.7.   *If $a = b$ and $c \neq 0$, then $\dfrac{a}{c} = \dfrac{b}{c}$.*

Theorem 1.6 says that we can multiply both sides of an equation by the same number and obtain another equation. Theorem 1.7 says that we can divide both sides of an equation by the same *nonzero* number and obtain another equation. Again, these two theorems are very useful in solving equations, as the next two examples illustrate.

Example 5   Solve the equation $\dfrac{a}{3} = 15$.

Solution: $\dfrac{a}{3} = 15$ *(Given)*

$\left(\dfrac{a}{3}\right)(3) = (15)(3)$ *(Theorem 1.6, multiplying both sides by 3)*

$(a)\left(\dfrac{1}{3}\right)(3) = 45$ *(Definition 1.13 and multiplication)*

$(a)(1) = 45$ *(Axiom 11)*
$a = 45$ *(Axiom 9)*

Example 6   Solve the equation $3a = 15$.

Solution: $3a = 15$ *(Given)*

$\dfrac{3a}{3} = \dfrac{15}{3}$ *(Theorem 1.7, dividing both sides by 3)*

$(3a)\left(\dfrac{1}{3}\right) = 5$ *(Definition 1.6 and division)*

$$(a3)\left(\frac{1}{3}\right) = 5 \quad \text{(Axiom 4)}$$

$$a\left[(3)\left(\frac{1}{3}\right)\right] = 5 \quad \text{(Axiom 6)}$$

$$a \cdot 1 = 5 \quad \text{(Axiom 11)}$$

$$a = 5 \quad \text{(Axiom 9)}$$

THEOREM 1.8. $-(-a) = a.$

THEOREM 1.9. $(a)(-b) = (-a)(b) = -(ab).$

THEOREM 1.10. $(-a)(-b) = ab.$

By setting $a = 1$ in Theorem 1.9, we obtain the following important corollary. (A corollary is simply a fact which is an immediate consequence of the theorem.)

COROLLARY 1. $(-1)b = -b = -(b)$

Now, by using Corollary 1 and Axiom 7, we obtain a second corollary to Theorem 1.9.

COROLLARY 2. $-a - b = -a + (-b) = (-1)a + (-1)b = (-1)(a+b) = -(a+b)$

The following examples illustrate Theorems 1.8, 1.9, and 1.10 and Corollaries 1 and 2.

Example 7 $-(-9) = 9, \quad -[-(-9) = -(9)] = -9$

Example 8 $(3)(-2) = (-3)(2) = -(3 \cdot 2) = -(6) = -6$

Example 9 $(-5)(-3) = (5)(3) = 15$

Example 10 $(-1)\left(\frac{3}{4}\right) = -\left(\frac{3}{4}\right) = -\frac{3}{4}$

Example 11 $-11 - 6 = -(11+6) = -17$

Example 12 Subtract 35 from 20.

Solution: $20 - 35 = -35 + 20$
$= -(35 - 20)$
$= -(15)$
$= -15$

SEC. 1.3   Some Basic Properties of Real Numbers   19

Example 13 Subtract 27 from -32.

$$\begin{aligned}\text{Solution:}\quad -32-27 &= -(32+27)\\ &= -(59)\\ &= -59\end{aligned}$$

Example 14 Subtract -45 from 30.

$$\begin{aligned}\text{Solution:}\quad 30-(-45) &= 30+45\\ &= 75\end{aligned}$$

Example 15 Subtract -40 from -60.

$$\begin{aligned}\text{Solution:}\quad -60-(-40) &= -[60+(-40)]\\ &= -(60-40)\\ &= -(20)\\ &= -20\end{aligned}$$

Some Products and Quotients Involving Zero

The number zero is a special sort of number in many ways. In Axiom 8 we learned that if we add zero to any number a the result is a; for example, $3+0=3$, $-5+0=-5$, $\sqrt{2}+0=\sqrt{2}$, etc. Earlier in this section we defined division for all real numbers except zero. So, according to our definition, division by zero is not a permissible operation. The fact that division by zero is undefined actually stems from Axiom 11, which provides for a reciprocal for every real number except zero. It is not surprising, therefore, that students often have trouble with operations involving zero.

We shall now state some theorems which, if completely understood by the student, should help make clear how to handle any situation in which zero is involved.

THEOREM 1.11. *If a is any real number, then $a \cdot 0 = 0$.*

THEOREM 1.12. *If $a \cdot b = 0$, then $a = 0$ or $b = 0$.*

Combining Theorems 1.11 and 1.12, we obtain:

THEOREM 1.13. *$a \cdot b = 0$ if and only if $a = 0$ or $b = 0$.*

Example 16 If $(a-2)(a+3) = 0$, what can you say about a?

Solution: Since $(a-2)(a+3) = 0$, we conclude that $a-2 = 0$ or $a+3 = 0$ (Theorem 1.12). Hence (adding 2 to both sides of the first equation and subtracting 3 from both sides of the second equation) $a = 2$ or $a = -3$. So we can say that $a = 2$ or $a = -3$.

THEOREM 1.14. *If $b \neq 0$, then $0/b = 0$.*

Example 17 $\quad \left| \dfrac{0}{3} = 0, \quad \dfrac{0}{(-7)} = 0, \quad \text{etc.} \right.$

THEOREM 1.15. *If $b \neq 0$ and $a/b = 0$, then $a = 0$.*

Example 18 $\quad \left| \text{ If } \dfrac{a}{3} = 0, \text{ then } a = 0. \right.$

Combining Theorems 1.14 and 1.15, we obtain:

THEOREM 1.16. *If $b \neq 0$, then $a/b = 0$ if and only if $a = 0$.*

Axioms 7 to 11 and the results of Section 1.3 are so important for computation in mathematics that we set them apart and distinguish them with the title *Rules of Operation*.

Rules of Operations

Let $a, b, c, d, e, \ldots$ be any real numbers.

RULE 1.1. $\quad +(a+b-c+d-e \ldots) = (a+b-c+d-e \ldots)$
$\qquad\qquad\qquad = a+b-c+d-e \ldots$
RULE 1.2. $\quad -(a+b-c+d-e \ldots) = -a-b+c-d+e \ldots$
RULE 1.3. When removing symbols of grouping, remove the innermost set of symbols first, collect like terms, and continue this process until all symbols have been removed.
RULE 1.4. $\quad a(b+c) = ab + ac$
RULE 1.5. $\quad a + 0 = 0 + a = a$
RULE 1.6. $\quad a \cdot 1 = 1 \cdot a = a$
RULE 1.7. $\quad a + (-a) = (-a) + a = a - a = -a + a = 0$
RULE 1.8. $\quad (a)\left(\dfrac{1}{a}\right) = \left(\dfrac{1}{a}\right)(a) = 1 \quad$ (provided $a \neq 0$)
RULE 1.9. If $a = b$, then $a + c = b + c$
RULE 1.10. If $a = b$, then $a - c = b - c$
RULE 1.11. If $a = b$, then $ac = bc$
RULE 1.12. If $a = b$ and $c \neq 0$, then $\dfrac{a}{c} = \dfrac{b}{c}$
RULE 1.13. $\quad -(-a) = a$
RULE 1.14. $\quad (a)(-b) = (-a)(b) = -(ab)$
RULE 1.15. $\quad (-a)(-b) = ab$

SEC. 1.3 Some Basic Properties of Real Numbers

RULE 1.16. $(-1)(b) = -b = -(b)$
RULE 1.17. $-a-b = -(a+b)$
RULE 1.18. $a \cdot 0 = 0 \cdot a = 0$
RULE 1.19. $a \cdot b = 0$ if and only if $a = 0$ or $b = 0$
RULE 1.20. $\dfrac{0}{b} = 0$ (provided $b \neq 0$)
RULE 1.21. $\dfrac{a}{b} = 0$ if and only if $a = 0$ (provided $b \neq 0$)

Exercises 1.3

1. Perform each indicated subtraction.
   a. 273 from 428
   b. 341 from 286
   c. -50 from 120
   d. 27 from -60
   e. -200 from -125
   f. 23 from 0
   g. -4.3 from 0
   h. 0.582 from 0.0379
   i. -4.24 from 9.659

2. In each of the statements replace the letter x, where possible, by a real number which will make the statement true.
   a. $\dfrac{15}{5} = x$
   b. $\dfrac{51}{x} = 3$
   c. $\dfrac{x}{18} = -4$
   d. $\dfrac{0}{3} = x$
   e. $\dfrac{72.9}{2.43} = x$
   f. $\dfrac{x}{6} = 0$
   g. $\dfrac{20}{x} = 0$
   h. $(1.37)(0) = x(4.5)$
   i. $\dfrac{x}{7} = 0$
   j. $(5)(x) = (6)(10)$

3. Simplify each given expression by writing it as a single number with the proper sign.
   a. $(4)(-5)$
   b. $(-3)(-7)$
   c. $-(4+8)$
   d. $-(6)(-3)$
   e. $-[(-2)-(-8)]$
   f. $-(8-10)$
   g. $-[6-(6)]$
   h. $(6-2)(4-7)$
   i. $(3-4)(5-9)$

4. In each of the following expressions, remove all symbols of grouping and combine terms.
   a. $(8-3)-(4-7+1)$
   b. $(-6)(4)+(-3)(-7)$
   c. $(-9)(-2)-(-4)(-3)$
   d. $-3(-5)+[-(-2)]$
   e. $\{-[5-(6-8)+(7-5)]-(4-2)\}$
   f. $2+\{6a-[4+a-(2a+1)]\}$
   g. $2a-[b-(7+3b)+4a]-5+8b$

5. Write each given expression as an equal expression with the last two terms enclosed in parentheses preceded by a plus sign.
   a. $a-b+c$
   b. $a+b-c+d$
   c. $a-b+c-d$
   d. $a+b-c-d$

6. Write each given expression as an equal expression with the last two terms enclosed in parentheses preceded by a minus sign.
   a. $a+b-c$ b. $a-b+c$
   c. $a-b-c+d$ d. $a+b-c-d$

7. If each lot in a certain residential section measures 66 feet across the front and if there are 14 lots in a certain block, how long is that block? How many such lots would there be in a distance of 1 mile (5,280 feet)?

8. A certain airplane weighs 216,000 pounds at take-off. One sixth of this weight is fuel. If the fuel weighs 7 pounds per gallon, how many gallons of fuel does the plane carry at take-off?

9. Solve each of the following equations giving a reason for each step in the solution. The letters a, b, and c in the equations represent real numbers.
   a. $c+5=-1$ b. $a-11=-4$ c. $b+9=3$ d. $-3-c=7$
   e. $2a-1=3$ f. $-3b+2=5$ g. $2a-1=a+3$ h. $5c+7-2c=c-3$

To illustrate what these instructions mean, a solution to part **a** follows.

Solution of **a**:

$c+5=-1$	(Given)
$c+5-5=-1-5$	(Rule 1.10, subtracting 5 from both sides)
$c+0=-1-5$	(Rule 1.7)
$c=-1-5$	(Rule 1.5)
$c=-(1+5)$	(Rule 1.17)
$c=-(6)$	(Addition)
$c=-6$	(Rule 1.16)

After doing several problems carefully, as above, the student should find it possible to simply give the reason(s) for each step in his or her head. The important thing is that each manipulation must be based on one of the *Rules of Operation* or on simple addition, subtraction, multiplication, or division of real numbers. For example, a brief form of the solution to part **a** might be:

Brief Solution of **a**:

$c+5=-1$	
$c+5-5=-1-5$	(Subtracting 5 from both sides)
$c+0=-(1+5)$	(Rule 1.17)
$c=-6$	

10. Is subtraction commutative? That is, is $a-b=b-a$ for any two real numbers a and b?

11. Is subtraction associative? That is, is $(a-b)-c=a-(b-c)$ for all a, b, and c?

12. Is multiplication distributive with respect to subtraction? That is, is $a(b-c)=ab-ac$ for all a, b, and c?

13. Prove $-(a-b)=-a+b$, and justify each step in your proof.

14. Prove $(a+b)(c+d)=ac+bc+ad+bd$, and justify each step in your proof.

1.4 FRACTIONS

Experience has taught us that many entering college freshmen need considerable review when it comes to computations involving fractions. For this reason we give a somewhat laborious treatment of fractions in this section. We beg the indulgence of those students who find the treatment too elementary and promise those students more challenging topics later in the text. We would also observe that the rules of operation for fractions are essentially identical to rules we shall develop in Chapter 2 for quotients of algebraic expressions.

We saw in Section 1.3 (Definition 1.14) that if a and b are real numbers and $b \neq 0$, then a/b is called a fraction. Also, if both a and b are integers, then a/b is called a simple fraction. Since fractions are only defined when the denominator is nonzero, *throughout the remainder of this chapter when we write "Let a/b be a fraction," or when we simply write a/b, it will be understood that a and b are real numbers and $b \neq 0$*.

A comment on notation. As was the case with reciprocals, it is less expensive and easier to use the notation a/b. Consequently, a/b will be the prevalent notation throughout the text. However, the two symbols a/b and $\dfrac{a}{b}$ may, and will, be used interchangeably.

Simple fractions arise often in everyday life. For example, if a person sets out to make one-half of a recipe, then he or she must be able to multiply several numbers by $\dfrac{1}{2}$ (or divide several numbers by 2, depending on how you prefer to look at it). It is especially important that multiplication of fractions be understood if the recipe calls for $\dfrac{3}{4}$ cup of flour, for in this case one-half recipe calls for $\left(\dfrac{1}{2}\right)\left(\dfrac{3}{4}\right) = \dfrac{3}{8}$ cup of flour. Another example where fractions might arise is in computing grades. Suppose that, in a given course, the final grade is based on regular examinations, the final examination, and homework. If $\dfrac{1}{2}$ of the grade is based on the regular examinations and $\dfrac{1}{3}$ of the grade is based on the final examination, what portion of the grade is based on homework? Here one must be able to add $\dfrac{1}{2}$ to $\dfrac{1}{3}$, obtaining $\dfrac{5}{6}$, and then subtract $\dfrac{5}{6}$ from 1 to obtain $\dfrac{1}{6}$, the desired answer.

In view of the importance of computations involving fractions, the remainder of this section will deal with multiplication, division, addition, and subtraction of fractions. Before treating these topics, however, it is important that we make several observations in regard to fractions.

Equality of Fractions

The reader is probably aware that $\frac{1}{2}, \frac{2}{4}, \frac{3}{6}, \frac{4}{8}$, and $\frac{100}{200}$ all represent the same real number. The characteristic property shared by each pair of these fractions turns out to be the following. If one of these fractions is a/b and another is c/d then $ad = bc$. For example, if $a/b = \frac{2}{4}$ and $c/d = \frac{100}{200}$, then $ad = (2)(200) = 400 = (4)(100) = bc$. It can be shown that if a/b and c/d are any two fractions, then a/b and c/d represent the same real number if and only if $ad = bc$. In this case we say that $a/b = c/d$.

DEFINITION 1.15. (EQUALITY OF FRACTIONS). Let a/b and c/d be any two fractions. Then $a/b = c/d$ if and only if $ad = bc$, that is, if and only if the numerator of the first times the denominator of the second is equal to the denominator of the first times the numerator of the second.

Example 1 | Show that $\frac{9}{21} = \frac{3}{7}$.

Solution: $(9)(7) = 63 = (21)(3)$, and so $\frac{9}{21} = \frac{3}{7}$.

Some easy consequences of Definition 1.15 are the following theorems, which are extremely useful in computations involving fractions.

THEOREM 1.17. If $c \neq 0$, then $\frac{ac}{bc} = \frac{a}{b}$.

Proof: $(ac)b = acb$
$= bca$
$= (bc)a$

and so $\frac{ac}{bc} = \frac{a}{b}$

THEOREM 1.18. If $c \neq 0$, then $\frac{(a/b)}{(c/b)} = \frac{a}{c}$.

Proof: Left to the interested reader.

Theorems 1.17 and 1.18 can be summarized by saying that *the value of a fraction is not changed when the numerator and denominator are both multiplied or divided by the same number, provided they are not multiplied or divided by zero.*

SEC. 1.4 Fractions

This fact will be isolated as a rule of operation at the end of this section. Theorem 1.17 will be used primarily to find common denominators of fractions, whereas Theorem 1.18 will be used primarily to reduce fractions to simplest form. (Both of these topics will be treated later in this section.)

Example 2
$$\frac{1}{3} = \frac{1 \cdot 2}{3 \cdot 2} = \frac{2}{6}$$

Example 3
$$\frac{15}{35} = \frac{3 \cdot 5}{7 \cdot 5} = \frac{3}{7}$$

Signs Associated with a Fraction

Another important property of fractions deals with certain signs (+ or −) associated with various parts of fractions. If $\frac{a}{b}$ is a fraction, and a and b are positive real numbers, then so are $\frac{a}{-b}$, $\frac{-a}{b}$, $\frac{-a}{-b}$, $-\left(\frac{a}{b}\right)$, $-\left(\frac{a}{-b}\right)$, $-\left(\frac{-a}{b}\right)$, and $-\left(\frac{-a}{-b}\right)$. One sees that there are two possibilities, + or −, for each of:

(i) the sign of the entire fraction
(ii) the sign of the numerator
(iii) the sign of the denominator

These are called *the signs associated with the fraction*. The sign of the entire fraction is placed on the same level as the *line of division*. The signs of the numerator and denominator are placed, respectively, above and below the line of division. If no sign appears in one of the three possible positions, then the sign of that *position* is implied to be +.

Example 4 What are the signs associated with $\frac{4}{-7}$?

Solution: The sign of the entire fraction is +.
The sign of the numerator is +.
The sign of the denominator is −.

Example 5 What are the signs associated with $-\frac{-4}{7}$?

Solution: The sign of the entire fraction is −.
The sign of the numerator is −.
The sign of the denominator is +.

When we say either "Change the sign ..." or "The sign may be changed ..." we mean:

(a) If the sign was $+$ to begin with, it becomes $-$.
(b) If the sign was $-$ to begin with, it becomes $+$.

The next theorem is extremely important in computations involving fractions. The proof is left to the interested reader as an exercise.

THEOREM 1.19. *Any two of the three signs associated with a fraction may be changed without changing the value of the fraction.*

Example 6 $\quad -\dfrac{-1}{5} = \dfrac{1}{5} = \dfrac{-1}{-5} = -\dfrac{1}{-5}$

Example 7 $\quad \dfrac{3}{-7} = \dfrac{-3}{7} = -\dfrac{3}{7} = -\dfrac{-3}{-7}$

Example 8 $\quad \dfrac{a}{b} = \dfrac{-a}{-b} = -\dfrac{a}{-b} = -\dfrac{-a}{b}$

Example 9 $\quad -\dfrac{a}{b} = \dfrac{-a}{b} = \dfrac{a}{-b} = -\dfrac{-a}{-b}$

Since the denominator b of a fraction a/b is never zero, the reader is probably familiar with the fact that b must be either a *positive* real number or a *negative* real number (these ideas will be discussed at some length in Section 1.5). In the case that b is negative, Theorem 1.19 allows us to change the sign of the denominator and one of the other signs associated with the fraction without changing the value of the fraction. This leads to the next corollary.

COROLLARY. *Any fraction may be expressed as (i.e., is equal to) a fraction which has a positive denominator.*

Example 10 $\quad -\dfrac{-2}{17}$ already has a positive denominator.

Example 11 $\quad \dfrac{3}{-5} = -\dfrac{3}{5}$ and $-\dfrac{3}{5}$ has a positive denominator.

Expressing Fractions in Simplest Form

In Section 1.1 we discussed the notion of one natural number being a factor of or a divisor of another natural number. This concept can be extended to dealing with integers in general, whether they are positive, negative, or zero.

SEC. 1.4   Fractions

DEFINITION 1.16.   An integer b is called a *factor* or a divisor of an integer a if there is an integer c such that $a = bc$.

We point out that if b is a nonzero integer, then b is a factor of the integer a if and only if a/b is an integer. Hence, in testing to see whether or not a given integer b is a factor of another given integer a, we simply check to see whether or not "b divides a with no remainder."

Example 12   3 is a factor of 6 since $\frac{6}{3} = 2$.

Example 13   4 is not a factor of 35 since $\frac{35}{4} = 8.75$, which is not an integer.

DEFINITION 1.17.   If two integers a and b have the same integer c as a factor, then c is called a *common factor* of a and b.

It should be observed at this point that both 1 and -1 are factors of any integer a since $\frac{a}{1} = a$ and $\frac{a}{-1} = -a$, both of which are integers. Hence every pair of integers has both 1 and -1 as common factors. Theorem 1.18 tells us that, if the numerator and denominator of a fraction have a common factor c (note that $c \neq 0$ since the denominator is not zero), then we can *cancel* the c. If c is neither 1 nor -1, this cancellation has the effect of *reducing the size* of both numerator and denominator. What we do then is to simply keep cancelling common factors out of the numerator and denominator until the only common factors left are 1 and -1. This process is called reducing the fraction to *simplest form*.

DEFINITION 1.18.   A simple fraction with positive denominator is said to be in *simplest form* when the numerator and denominator have no common factors other than 1 and -1.

Example 14 $\frac{3}{7}$ is in simplest form since 3 and 7 have no common factors other than 1 and -1.

Example 15 $\frac{6}{21}$ is not in simplest form since 6 and 21 have 3 as a common factor.

Example 16   Reduce $\frac{6}{21}$ to simplest form.

Solution: $\frac{6}{21} = \frac{2 \cdot 3}{7 \cdot 3} = \frac{2}{7}$, which is in simplest form.

Example 16 illustrates a method for reducing any fraction to simplest form.

METHOD FOR REDUCING FRACTIONS TO SIMPLEST FORM: Express both numerator and denominator as products of primes and then cancel as many of these primes (*in pairs*) as possible.

Example 17 | Reduce $\dfrac{660}{1{,}610}$ to simplest form.

Solution: $\dfrac{660}{1{,}610} = \dfrac{2\cdot 2\cdot 3\cdot 5\cdot 11}{2\cdot 5\cdot 7\cdot 23} = \dfrac{2\cdot 3\cdot 11}{7\cdot 23} = \dfrac{66}{161}$

Common Denominators

Suppose we begin with two fractions $\dfrac{a}{b}$ and $\dfrac{c}{d}$. Since $d \neq 0$, we can use Theorem 1.17 to obtain the equation

$$\frac{a}{b} = \frac{ad}{bd}$$

Similarly, since $b \neq 0$, we obtain the equation

$$\frac{c}{d} = \frac{bc}{bd}$$

Notice that $\dfrac{ad}{bd}$ and $\dfrac{bc}{bd}$ have the same denominator bd. What has been accomplished is a verification of the following theorem:

THEOREM 1.20. *Given any two fractions, it is possible to write two new fractions that are, respectively, equal to the two given fractions such that the two new fractions have the same denominator. (We say that two fractions that have the same denominator have a* common denominator.*)*

Example 18 | Express $\dfrac{2}{3}$ and $\dfrac{4}{5}$ as fractions having a common denominator.

Solution: $\dfrac{2}{3} = \dfrac{2\cdot 5}{3\cdot 5} = \dfrac{10}{15}$ and $\dfrac{4}{5} = \dfrac{4\cdot 3}{5\cdot 3} = \dfrac{12}{15}$

The process outlined above can be extended easily to any number of fractions, as the next example illustrates:

SEC. 1.4   Fractions

Example 19 Express $\dfrac{2}{3}, \dfrac{4}{5}$, and $\dfrac{7}{10}$ as fractions having a common denominator.

Solution: We use the product $3 \cdot 5 \cdot 10 = 150$ of the three denominators as the common denominator. Hence

$$\frac{2}{3} = \frac{2 \cdot 50}{3 \cdot 50} = \frac{100}{150}, \quad \frac{4}{5} = \frac{4 \cdot 30}{5 \cdot 30} = \frac{120}{150}, \quad \frac{7}{10} = \frac{7 \cdot 15}{10 \cdot 15} = \frac{105}{150}.$$

In Example 19 it turns out that many numbers, other than 150, would have served as common denominators of the three fractions $\dfrac{2}{3}, \dfrac{4}{5}$, and $\dfrac{7}{10}$. Examples of such numbers are 30, 60, 90, and 120. It should be noted that each of the numbers that could be used as a common denominator of the three given fractions is a number that is exactly divisible by each denominator of the three fractions. This gives us a criterion for determining a common denominator of any group of fractions. This criterion may be stated as follows: *A common denominator of a set of fractions must contain as a factor the denominator of each fraction of the set.* For any given set of fractions, many numbers meeting the condition of the criterion can be found. Ordinarily, the preferred common denominator is the smallest positive integer which is a common denominator. This number is called the *least common denominator*.

DEFINITION 1.19. The *least common denominator* (abbreviated LCD) of a set of simple fractions is the smallest positive integer d with the property that each denominator in the set of fractions is a factor of d.

METHOD FOR FINDING LEAST COMMON DENOMINATORS: Find the prime factors of the denominator of each given fraction, take the product of all the different prime factors thus found, and include in this product each different factor as many times as that factor occurs in any one of the denominators of the given fractions. This product is the desired LCD.

Example 20 Find the LCD of the fractions $\dfrac{4}{25}$ and $\dfrac{7}{10}$.

Solution: The first denominator $25 = 5 \cdot 5$ has 5 as its only prime factor; however, since 5 occurs twice as a factor of this denominator, 5 must also occur twice as a factor of the LCD. The second denominator $10 = 2 \cdot 5$ has 2 and 5 as its prime factors, and each occurs as a factor only once. Now, the LCD already has the factor 5 twice because of the first denominator, and so the factor 5 in the second denominator can be ignored. The factor 2 of the second denominator must be a factor of the LCD, however. Therefore, the LCD is $5 \cdot 5 \cdot 2 = 50$.

It is not really enough to find the LCD of a set of simple fractions. One must also be able to express each of the original fractions as fractions having this

LCD as a denominator. The technique for accomplishing this task is simply to take each fraction $\dfrac{a}{b}$ in the set and multiply both numerator and denominator by the appropriate integer c to make the new denominator equal to the LCD. This integer c is rather easily seen to be $\dfrac{\text{LCD}}{b}$, since $(b)\left(\dfrac{\text{LCD}}{b}\right) = \text{LCD}$. This technique is illustrated in the next example.

Example 21   Find the LCD of the fractions $\dfrac{1}{6}, \dfrac{3}{8}$, and $\dfrac{7}{10}$ and express each fraction as a fraction whose denominator is this LCD.

Solution: Factoring each denominator gives $6 = 2\cdot 3$, $8 = 2\cdot 2\cdot 2$, and $10 = 2\cdot 5$. Thus the LCD is $2\cdot 2\cdot 2\cdot 3\cdot 5 = 120$. Now

$$\frac{1}{6} = \frac{1\cdot 20}{6\cdot 20} = \frac{20}{120}, \quad \frac{3}{8} = \frac{3\cdot 15}{8\cdot 15} = \frac{45}{120}, \quad \frac{7}{10} = \frac{7\cdot 12}{10\cdot 12} = \frac{84}{120}$$

Addition and Subtraction of Fractions

Having discussed the various properties of fractions, we proceed now to the computational rules for dealing with fractions. First we must decide how to add, subtract, multiply, and divide fractions. Curiously, addition and subtraction are, in a sense, more complicated than multiplication and division when dealing with fractions. However, since the more natural order is to treat addition and subtraction first, this is the route we shall take. Several methods will be given for adding fractions. The student should not be dismayed by thinking that he or she must become proficient in the use of each method. The important thing is to be able to correctly add fractions using any valid method (or rule). We shall discuss the methods we feel are most commonly used.

Among the first things we learn about fractions are the facts that 2 halves make a whole $\left(\text{that is, } \dfrac{1}{2} + \dfrac{1}{2} = 1\right)$, 3 thirds make a whole $\left(\text{that is, } \dfrac{1}{3} + \dfrac{1}{3} + \dfrac{1}{3} = 1\right)$, and so on. Another early observation is that 1 third taken together with another third yields 2 thirds $\left(\text{that is, } \dfrac{1}{3} + \dfrac{1}{3} = \dfrac{2}{3}\right)$. These facts illustrate the method for adding several fractions having a common denominator. The method is actually a theorem, however; we state it simply as a "method" and leave its proof to the interested reader.

METHOD FOR ADDING FRACTIONS WITH A COMMON DENOMINATOR: Add the numerators and leave the denominator alone.

SEC. 1.4   Fractions

Example 22 $\dfrac{2}{7} + \dfrac{3}{7} = \dfrac{2+3}{7} = \dfrac{5}{7}$

Example 23 $\dfrac{1}{5} + \dfrac{3}{5} + \dfrac{2}{5} = \dfrac{1+3+2}{5} = \dfrac{6}{5}$

Example 24 $\dfrac{2}{13} + \dfrac{3}{13} + \dfrac{5}{13} + \dfrac{1}{13} = \dfrac{2+3+5+1}{13} = \dfrac{11}{13}$

We have observed that, given any set of fractions, there are many common denominators for the set and that, given any such common denominator d, each fraction can be expressed as a fraction having denominator d. This fact, together with the method given above for adding fractions having a common denominator, yields one method for adding any set of fractions.

FIRST METHOD FOR ADDING FRACTIONS

1. Find a common denominator d.
2. Express each fraction as a fraction with denominator d.
3. Add the numerators of these new fractions and use d as denominator.

Example 25 Add $\dfrac{3}{4}$ to $\dfrac{2}{5}$.

Solution: 1. $d = 4 \cdot 5 = 20$ is a common denominator

2. $\dfrac{3}{4} = \dfrac{3 \cdot 5}{4 \cdot 5} = \dfrac{15}{20}$ and $\dfrac{2}{5} = \dfrac{2 \cdot 4}{5 \cdot 4} = \dfrac{8}{20}$

3. $\dfrac{15 + 8}{20} = \dfrac{23}{20}$

Hence

$$\dfrac{3}{4} + \dfrac{2}{5} = \dfrac{3 \cdot 5}{4 \cdot 5} + \dfrac{2 \cdot 4}{5 \cdot 4} = \dfrac{15}{20} + \dfrac{8}{20} = \dfrac{15 + 8}{20} = \dfrac{23}{20}$$

Example 26 Find the sum $\dfrac{2}{3} + \dfrac{1}{4} + \dfrac{3}{5}$.

Solution: 1. $d = 3 \cdot 4 \cdot 5 = 60$ is a common denominator

2. $\dfrac{2}{3} = \dfrac{2 \cdot 20}{3 \cdot 20} = \dfrac{40}{60}$, $\dfrac{1}{4} = \dfrac{1 \cdot 15}{4 \cdot 15} = \dfrac{15}{60}$, and $\dfrac{3}{5} = \dfrac{3 \cdot 12}{5 \cdot 12} = \dfrac{36}{60}$

3. $\dfrac{40 + 15 + 36}{60} = \dfrac{91}{60}$

Hence

$$\frac{2}{3} + \frac{1}{4} + \frac{3}{5} = \frac{2 \cdot 20}{3 \cdot 20} + \frac{1 \cdot 15}{4 \cdot 15} + \frac{3 \cdot 12}{5 \cdot 12}$$

$$= \frac{40}{60} + \frac{15}{60} + \frac{36}{60}$$

$$= \frac{40 + 15 + 36}{60}$$

$$= \frac{91}{60}$$

After solving a few problems as illustrated in Examples 25 and 26, the student will probably realize that the method illustrated above can be condensed by writing on your paper only the equations appearing after the word "Hence" in Examples 25 and 26.

Yet another reduction in the amount of writing required to solve such problems can be achieved as illustrated in the following solution to Example 26.

Another Solution to Example 26: First, mentally observe that $3 \cdot 4 \cdot 5 = 60$ is a common denominator for the three fractions. Then write

$$\frac{2}{3} + \frac{1}{4} + \frac{3}{5} = \frac{}{60}$$

Next say to yourself, "3 goes into 60 twenty times," and write $2 \cdot 20$ above the 60. Then say to yourself, "4 goes into 60 fifteen times," and write $+1 \cdot 15$ above the 60. Finally, say to yourself, "5 goes into 60 twelve times," and write $+3 \cdot 12$ above the 60. Once these things are done, all that appears on your paper is

$$\frac{2}{3} + \frac{1}{4} + \frac{3}{5} = \frac{2 \cdot 20 + 3 \cdot 15 + 3 \cdot 12}{60}$$

To complete the solution, simply write

$$= \frac{40 + 15 + 36}{60}$$

$$= \frac{91}{60}$$

So the complete solution would appear on your paper as

$$\frac{2}{3} + \frac{1}{4} + \frac{3}{5} = \frac{2 \cdot 20 + 1 \cdot 15 + 3 \cdot 12}{60}$$

$$= \frac{40 + 15 + 36}{60}$$

$$= \frac{91}{60}$$

SEC. 1.4 Fractions

This alternative solution is the method many students use, and so we rephrase the first method for adding fractions to reflect this technique.

ALTERNATIVE FIRST METHOD FOR ADDING FRACTIONS

1. Find a common denominator d.
2. Take each fraction (in order), divide the denominator of this fraction into d, and multiply this result by the numerator.
3. Add the numbers obtained in step 2 and use this sum as the numerator.

Example 27   Find the sum $\dfrac{3}{4} + \dfrac{5}{6} + \dfrac{11}{12}$ and express the result in simplest form.

Solution: Let us select the common denominator $d = 24$ (for no special reason except to illustrate that any common denominator can be used).

$$\begin{aligned}
\frac{3}{4} + \frac{5}{6} + \frac{11}{12} &= \frac{3\cdot 6 + 5\cdot 4 + 11\cdot 2}{24} \\
&= \frac{18 + 20 + 22}{24} \\
&= \frac{60}{24} \\
&= \frac{5\cdot 12}{2\cdot 12} \\
&= \frac{5}{2}
\end{aligned}$$

The reader has probably observed that in most examples so far we have used the product of all denominators as our common denominator. This method will always work since the product of all denominators is a common denominator of a collection of fractions. In fact, this is a method which is used by many students, and so we set it apart as a separate "method" even though it is a special case of the first method for adding fractions.

SECOND METHOD FOR ADDING FRACTIONS

1. Use the product of all denominators as a common denominator d.
2. Take each fraction (in order), divide the denominator into d (obtaining the product of the *other* denominators in this case), and multiply this result by the numerator.
3. Add the numbers obtained in step 2 and use this sum as the numerator.

Example 28 | Find the sum $\frac{1}{4} + \frac{2}{3} + \frac{7}{10}$.

Solution:
$$\frac{1}{4} + \frac{2}{3} + \frac{7}{10} = \frac{1(3 \cdot 10) + 2(4 \cdot 10) + 7(4 \cdot 3)}{120}$$
$$= \frac{1(30) + 2(40) + 7(12)}{120}$$
$$= \frac{30 + 80 + 84}{120}$$
$$= \frac{194}{120} = \frac{97 \cdot 2}{60 \cdot 2} = \frac{97}{60}$$

It so happens that 120 is not the LCD of the fractions in Example 28. If we had used the LCD, $4 \cdot 3 \cdot 5 = 60$, then the numbers we dealt with would have been smaller. Since the first method for adding fractions says to find *any* common denominator, the LCD is an acceptable choice. Again, many students use this method, and so we set it apart as a separate method, even though it is a special case of the first method.

THIRD METHOD FOR ADDING FRACTIONS

1. Use the LCD as a common denominator.
2. Take each fraction (in order), divide the denominator into the LCD, and multiply this result by the numerator.
3. Add the numbers obtained in step 2 and use this sum as the numerator.

Solution to Example 28, using the third method:
$$\frac{1}{4} + \frac{2}{3} + \frac{7}{10} = \frac{1(15) + 2(20) + 7(6)}{60}$$
$$= \frac{15 + 40 + 42}{60}$$
$$= \frac{97}{60}$$

All our examples to this point have dealt with addition of fractions whose numerators and denominators were positive. We now indicate, through the solution of two examples, how one adds and subtracts fractions some of whose signs are negative.

Example 29 | Subtract $\frac{2}{3}$ from $\frac{5}{7}$.

SEC. 1.4   Fractions   35

$$\text{Solution:} \quad \frac{5}{7} - \frac{2}{3} = \frac{5}{7} + \frac{(-2)}{3} = \frac{5 \cdot 3 + (-2) \cdot 7}{21}$$
$$= \frac{15 + (-14)}{21}$$
$$= \frac{15 - 14}{21}$$
$$= \frac{1}{21}$$

$$\text{Alternate Solution:} \quad \frac{5}{7} - \frac{2}{3} = \frac{5 \cdot 3 - 2 \cdot 7}{21}$$
$$= \frac{15 - 14}{21}$$
$$= \frac{1}{21}$$

Example 30   Add $\dfrac{7}{-9}$ to $\dfrac{-1}{3}$.

$$\text{Solution:} \quad \frac{7}{-9} + \frac{-1}{3} = -\frac{7}{9} - \frac{1}{3}$$
$$= -\left(\frac{7}{9} + \frac{1}{3}\right)$$
$$= -\frac{7 \cdot 1 + 1 \cdot 3}{9}$$
$$= -\frac{7 + 3}{9}$$
$$= -\frac{10}{9}$$

Examples 29 and 30 only illustrate how we might go about adding and subtracting fractions with various signs involved. *There is no substitute for practice when it comes to gaining proficiency in such computations.*

Improper and Mixed Fractions

In several of the examples of this section the answers have had the numerator larger than the denominator. One such was Example 26 for which the answer was $\dfrac{91}{60}$. For the most part we shall leave answers in this form. Fractions having the numerator larger than the denominator are sometimes called *improper fractions*. Actually, they are in no way "improper." The simple fact is that in describing physical quantities we seldom use improper fractions. For example, one would be likely to say, "we have three and one-half miles to go," rather than "we have seven-halves of a mile to go." Hence, when dealing with physical quantities we shall ordinarily convert improper fractions to what are called *mixed fractions*. A *mixed fraction* is simply the sum of an integer and a fraction. Also, instead of writing $3 + \frac{1}{2}$, we write $3\frac{1}{2}$.

Example 31 Express the improper fraction $\frac{23}{3}$ as a mixed fraction.

Solution:
$$\frac{23}{3} = \frac{21 + 2}{3}$$
$$= \frac{21}{3} + \frac{2}{3}$$
$$= 7 + \frac{2}{3}$$
$$= 7\frac{2}{3}$$

Notice that we could obtain the same result by dividing 23 by 3, obtaining 7 with a remainder of 2, and writing $7\frac{2}{3}$.

Example 32 Express the mixed fraction $5\frac{2}{3}$ as an improper fraction.

Solution:
$$5\frac{2}{3} = 5 + \frac{2}{3}$$
$$= \frac{5}{1} + \frac{2}{3}$$
$$= \frac{5 \cdot 3 + 2 \cdot 1}{3}$$
$$= \frac{15 + 2}{3}$$
$$= \frac{17}{3}$$

Notice that we could obtain the same result by saying "3 times 5 equals 15 plus 2 equals 17" to obtain the numerator of the improper fraction.

The alternative approaches to the solutions of Examples 31 and 32 (that is, the approaches following the word "Notice" in each case) are perfectly acceptable, *if applied correctly*, and should be used by those students who feel more comfortable in doing so.

Multiplication and Division of Fractions

Now that the notation, terminology, and other machinery have been developed for fractions, the rules for multiplying and dividing fractions are quite straightforward. One fact that should be emphasized at this point is that the word "of" ordinarily means "multiply." For example, if you are asked to "take two-fifths of five-sixths," then you should multiply $\frac{2}{5}$ by $\frac{5}{6}$.

RULE FOR MULTIPLYING FRACTIONS

1. Multiply all numerators and use this product as the numerator.
2. Multiply all denominators and use this product as the denominator.

Example 33
$$\frac{2}{7} \cdot \frac{5}{11} = \frac{2 \cdot 5}{7 \cdot 11} = \frac{10}{77}$$

Example 34
$$\frac{3}{4} \cdot \frac{5}{6} \cdot \frac{7}{23} = \frac{3 \cdot 5 \cdot 7}{4 \cdot 6 \cdot 23} = \frac{105}{552}$$

Example 35
$$\left(-\frac{1}{3}\right)\left(\frac{-2}{7}\right)\left(\frac{11}{-2}\right) = \left(-\frac{1}{3}\right)\left(-\frac{2}{7}\right)\left(-\frac{11}{2}\right)$$
$$= -\left(\frac{1}{3}\right)\left(\frac{2}{7}\right)\left(\frac{11}{2}\right)$$
$$= -\frac{1 \cdot 2 \cdot 11}{3 \cdot 7 \cdot 2}$$
$$= -\frac{1 \cdot 11}{3 \cdot 7}$$
$$= -\frac{11}{21}$$

At least two mildly interesting maneuvers took place in the solution to Example 35. One was in the second step, where we used the fact that $(-a)(-b)(-c) = -(a)(b)(c)$ (see Section 1.3). The other was in the fourth step, where the 2 was "cancelled." Actually, we used Theorem 1.18 and divided both numerator and denominator by 2. Often this process is indicated by drawing "slash marks" through both 2's. If this is done, the fourth step would appear as

$$-\frac{1 \cdot \not{2} \cdot 11}{3 \cdot 7 \cdot \not{2}} = \frac{1 \cdot 11}{3 \cdot 7}$$

A similar procedure may be adopted in order to *reduce* other fractions as illustrated in the next example. *We will ordinarily reduce each fraction to simplest* form and express each fraction in such a way that at most one of the signs associated with it is minus.

Example 36

$$\left(\frac{3}{4}\right)\left(\frac{-2}{9}\right)\left(\frac{6}{-7}\right) = \left(\frac{3}{4}\right)\left(-\frac{2}{9}\right)\left(-\frac{6}{7}\right)$$

$$= \left(\frac{3}{4}\right)\left(\frac{2}{9}\right)\left(\frac{6}{7}\right)$$

$$= \frac{3 \cdot 2 \cdot 6}{4 \cdot 9 \cdot 7}$$

$$= \frac{3 \cdot 6}{2 \cdot 9 \cdot 7} = \frac{\overset{2}{6}}{2 \cdot 3 \cdot 7}$$

$$= \frac{\overset{1}{2}}{2 \cdot 7} = \frac{1}{7}$$

Many students will combine several of the cancellations into one step, and a portion of the solution might appear on paper as follows:

$$\frac{3 \cdot \overset{2}{2} \cdot \overset{2}{6}}{\overset{}{4} \cdot \overset{}{9} \cdot 7} = \frac{1}{7}$$

or perhaps as

$$\left(\frac{3}{4}\right)\left(\frac{2}{9}\right)\left(\frac{\overset{3}{6}}{7}\right) = \frac{1}{7}$$

Students taking this approach should be forewarned that mistakes happen easily in such a procedure, and it is sometimes difficult for a teacher to isolate the exact point at which a mistake is made. Hence the warning is repeated: *Whatever computations you make, make them carefully and be sure that they are justified by one of the valid rules of operation.*

While it is not uncommon to multiply three or more fractions, we ordinarily only attempt to divide one fraction by another.

RULE FOR DIVIDING FRACTIONS

$$\frac{a/b}{c/d} = \frac{a}{b} \cdot \frac{d}{c}$$

SEC. 1.4 Fractions

That is, to divide $\frac{a}{b}$ by $\frac{c}{d}$, we *invert* $\frac{c}{d}$ and multiply $\frac{a}{b}$ by the result; hence the time-honored phrase, "To divide by a fraction, invert it and multiply."

Example 37 Divide $\frac{3}{5}$ by $\frac{7}{17}$.

Solution: $\dfrac{3/5}{2/17} = \dfrac{3}{5} \cdot \dfrac{17}{2} = \dfrac{51}{10}$

Example 38 Divide $\dfrac{21}{-35}$ by $\dfrac{-4}{-5}$ and express your answer in simplest form with no more than one minus sign.

Solution: $\dfrac{21/-35}{-4/-5} = \dfrac{-21/35}{4/5}$

$= -\dfrac{21/35}{4/5}$

$= -\left(\dfrac{21}{35}\right)\left(\dfrac{5}{4}\right) = -\dfrac{21}{7 \cdot 4} = -\dfrac{3}{4}$

A final word of caution is in order. Among the most prevalent of all errors in algebraic computation is "wrong cancellation." We illustrate what is meant by this statement by several examples:

Example 39 $\dfrac{2+3}{2+4}$ is neither equal to $\dfrac{3}{4}$, nor is it equal to $\dfrac{4}{5}$. The first number is often obtained by incorrectly "canceling the 2's" and treating the "canceled 2's" as zeros. Thus the incorrect solution would appear on a paper as

$\dfrac{\cancel{2}+3}{\cancel{2}+4} = \dfrac{3}{4}$ WRONG!

The second number is often obtained by incorrectly "canceling the 2's" and treating the "canceled 2's" as ones. Thus the incorrect solution would appear on a paper as

$\dfrac{\cancel{2}+3}{\cancel{2}+4} = \dfrac{4}{5}$ WRONG!

Example 40 $\dfrac{3+\cancel{2}}{\cancel{4}} = \dfrac{3}{2}$ WRONG!

Example 41
$$\frac{3+\cancel{2}^{1}}{\cancel{4}_{2}} = \frac{4}{2} \qquad \text{WRONG!}$$

The key to avoiding such errors is to always be careful to divide the *entire numerator* and the *entire denominator* by the same number, not just part of the numerator or part of the denominator. The temptation is to insert a Warning Box at this point; however, the number of possible types of "wrong cancellation" is so great that it would be impossible to make such a table comprehensive. *Suffice it to say that great care must be taken in the process of cancellation.*

The several rules of operation for dealing with fractions are gathered together here in tabular form.

Rules of Operation

Let a, b, c, and d be real numbers.

RULE 1.22. $\quad \dfrac{a}{b} = \dfrac{c}{d}$ if and only if $ad = bc$

RULE 1.23. $\quad \dfrac{ac}{bc} = \dfrac{a}{b}$, provided $c \neq 0$

RULE 1.24. $\quad \dfrac{\frac{a}{c}}{\frac{b}{c}} = \dfrac{a}{b}$, provided $c \neq 0$

RULE 1.25. $\quad \dfrac{a}{b} = \dfrac{-a}{-b} = -\dfrac{-a}{b} = -\dfrac{a}{-b}$

RULE 1.26. $\quad -\dfrac{a}{b} = \dfrac{-a}{b} = \dfrac{a}{-b} = -\dfrac{-a}{-b}$

RULE 1.27. $\quad \dfrac{a}{b} + \dfrac{c}{d} = \dfrac{(D/b)a + (D/d)c}{D}$, where D is any common denominator for $\dfrac{a}{b}$ and $\dfrac{c}{d}$

RULE 1.28. $\quad \dfrac{a}{b} \cdot \dfrac{c}{d} = \dfrac{a \cdot c}{b \cdot d}$

RULE 1.29. $\quad \dfrac{\frac{a}{b}}{\frac{c}{d}} = \dfrac{a}{b} \cdot \dfrac{d}{c}$, provided $\dfrac{c}{d} \neq 0$

SEC. 1.4 Fractions

Exercises 1.4

1. Which of the following pairs of fractions are equal?
   a. $\dfrac{3}{5}, \dfrac{7}{9}$ b. $\dfrac{2}{3}, \dfrac{14}{21}$ c. $\dfrac{4}{7}, \dfrac{16}{28}$ d. $\dfrac{4}{5}, \dfrac{20}{25}$ e. $\dfrac{7}{15}, \dfrac{42}{90}$ f. $\dfrac{1}{9}, \dfrac{8}{72}$

2. Express each of the following factors in four different forms by "changing signs."
   a. $-\dfrac{3}{7}$ b. $\dfrac{-4}{5}$ c. $\dfrac{2}{-9}$ d. $\dfrac{5}{6}$ e. $\dfrac{-8}{-15}$ f. $-\dfrac{-3}{-4}$

3. Express each of the following fractions with a positive denominator.
   a. $\dfrac{2}{-3}$ b. $\dfrac{-4}{5}$ c. $\dfrac{-8}{-15}$ d. $-\dfrac{-5}{-6}$ e. $\dfrac{3}{4}$ f. $-\dfrac{-11}{17}$

4. Express each of the following fractions in simplest form.
   a. $\dfrac{7}{35}$ b. $\dfrac{9}{63}$ c. $\dfrac{25}{85}$ d. $\dfrac{-32}{-96}$ e. $\dfrac{240}{560}$ f. $\dfrac{10}{4,235}$

5. Find the LCD of each of the following sets of fractions.
   a. $\dfrac{1}{8}, \dfrac{5}{12}$ b. $\dfrac{3}{26}, \dfrac{4}{39}$ c. $\dfrac{4}{21}, \dfrac{7}{18}$ d. $\dfrac{1}{8}, \dfrac{1}{3}, \dfrac{1}{5}$ e. $\dfrac{3}{8}, \dfrac{2}{4}, \dfrac{7}{35}$ f. $\dfrac{4}{15}, \dfrac{11}{35}, \dfrac{17}{21}$

6. Express each of the following fractions with no minus signs where possible and, in any event, with as few minus signs as possible. [In Parts e and f, a, b, and c represent real numbers.]
   a. $-\dfrac{3}{-8}$ b. $-\dfrac{-6}{-11}$ c. $\dfrac{-12}{-3}$ d. $\dfrac{(-5)(-2)}{4-9}$ e. $-\dfrac{-a-b}{a-b}$ f. $\dfrac{a-c}{-a-b}$

7. Express each of the following fractions as a fraction in simplest form that has both a positive numerator and a positive denominator.
   a. $\dfrac{6+4}{3-5}$ b. $-\dfrac{8-2}{4-6}$ c. $\dfrac{-12-6}{5+4}$ d. $\dfrac{-7-2}{3-6}$ e. $\dfrac{8+7}{3+2}$ f. $\dfrac{13+7}{9-5}$

8. Combine each given expression into a single fraction reduced to simplest form.
   a. $\dfrac{4}{9} + \dfrac{7}{18}$ b. $\dfrac{3}{5} + 2/3$ c. $\dfrac{7}{12} - \dfrac{4}{3}$ d. $\dfrac{3}{2} + \dfrac{1}{6} + \dfrac{5}{8}$
   e. $\dfrac{1}{2} - \dfrac{5}{12} + 1/3$ f. $\dfrac{5}{6} - \dfrac{1}{8} - \dfrac{7}{12}$ g. $\dfrac{13}{15} - \dfrac{11}{35} + \dfrac{17}{21}$ h. $\dfrac{5}{18} + \dfrac{7}{12} - 5/8$

9. Find each indicated product and reduce to simplest form.

a. $\dfrac{3}{2} \cdot \dfrac{5}{4}$ **b.** $\left(-\dfrac{1}{3}\right)\left(\dfrac{6}{5}\right)$ **c.** $\left(-\dfrac{2}{5}\right)\left(-\dfrac{9}{8}\right)$ **d.** $\left(-\dfrac{5}{6}\right)\left(\dfrac{7}{4}\right)\left(\dfrac{8}{15}\right)$

e. $\left(\dfrac{3}{8}\right)\left(-\dfrac{5}{3}\right)\left(-\dfrac{4}{3}\right)$ **f.** $\left(-\dfrac{4}{9}\right)\left(\dfrac{-3}{9}\right)\left(\dfrac{7}{-2}\right)$ **g.** $\left(-\dfrac{9}{15}\right)\left(\dfrac{-5}{-21}\right)\left(\dfrac{7}{-4}\right)$

10. Perform each indicated division.

a. $5/6 \div \dfrac{2}{3}$ **b.** $-\dfrac{3}{14} \div \dfrac{5}{7}$ **c.** $-\dfrac{4}{11} \div \dfrac{2}{-33}$ **d.** $\dfrac{24}{15} \div \dfrac{-6}{5}$

e. $0 \div \dfrac{2}{7}$ **f.** $\dfrac{-2}{-3} \div \dfrac{9}{-4}$ **g.** $\dfrac{-3/5}{7/20}$ **h.** $\dfrac{\frac{4}{-5}}{-2/3}$ **i.** $\dfrac{\frac{14}{3}}{7}$

11. Simplify each of the following complex fractions.

a. $\dfrac{1 - 1/6}{\dfrac{1}{2} + 1/3}$ **b.** $\dfrac{\dfrac{1}{5} - 3/4}{\dfrac{3}{5} + 7/20}$ **c.** $\dfrac{\dfrac{3}{4} - 8/5}{\dfrac{8}{15} - \dfrac{4}{5}}$

d. $\dfrac{\dfrac{5}{7} \div \dfrac{3}{14}}{\dfrac{2}{3} \div \dfrac{5}{9}}$ **e.** $\dfrac{\dfrac{3}{8} + \dfrac{9}{5} - \dfrac{7}{26}}{\dfrac{7}{4} - \dfrac{9}{20} + 1}$ **f.** $\dfrac{\left(-\dfrac{1}{3}\right) \div \left(\dfrac{2}{-9}\right)}{\dfrac{3}{2} + \left(-\dfrac{1}{6}\right)}$

Solution to **a**: $\dfrac{1 - \dfrac{1}{6}}{\dfrac{1}{2} + \dfrac{1}{3}} = \dfrac{\dfrac{6-1}{6}}{\dfrac{3+2}{6}} = \dfrac{5/6}{5/6} = \dfrac{5}{6} \cdot \dfrac{6}{5} = 1$

Alternate Solution: $\dfrac{1 - \dfrac{1}{6}}{\dfrac{1}{2} + \dfrac{1}{3}} = \dfrac{\left(1 - \dfrac{1}{6}\right)6}{\left(\dfrac{1}{2} + \dfrac{1}{3}\right)6} = \dfrac{6-1}{3+2} = \dfrac{5}{5} = 1$

1.5 THE ORDER RELATIONS AND ABSOLUTE VALUE

It is often necessary to compare two real numbers. Unfortunately, their ability to be compared does not follow from the axioms and theorems already stated. To

SEC. 1.5   The Order Relations and Absolute Value   43

compare real numbers, we need the following order axioms. The relation "is less than," $<$, is defined for the real numbers by these axioms.

The Order Axioms

AXIOM 12.   *If a and b are any two real numbers, then one and only one of the following statements is true: $a < b$, $a = b$, $b < a$.*

AXIOM 13.   *If a, b, and c are real numbers such that $a < b$ and $b < c$, then $a < c$.*

AXIOM 14.   *If a, b, and c are real numbers such that $a < b$, then $a + c < b + c$.*

AXIOM 15.   *If a, b, and c are real numbers such that $a < b$ and $0 < c$, then $ac < bc$.*

AXIOM 16.   *If a, b, and c are real numbers such that $a < b$ and $c < 0$, then $bc < ac$.*

We often find it desirable to say that one number "is greater than" another number. To make this possible, we define the statement $a > b$, which is read "'a is greater than b," as meaning $b < a$. Hence $a > b$ and $b < a$ are equivalent statements. Statements such as $a > b$ or $a < b$, involving "greater than" or "less than" symbols, are called *inequalities*.

Axiom 10 of Section 1.2 guarantees the existence of the additive inverse $-a$ for each real number a. Although this additive inverse is often referred to as the negative of a, this does not mean that a is necessarily a positive number and $-a$ is a negative number. What then is a positive real number? What is a negative real number? Both of these questions are answered in the following definition.

DEFINITION 1.20.   A real number a is said to be *positive* if $a > 0$, and it is said to be *negative* if $a < 0$.

It follows from Axiom 12 that the real number 0 is neither positive nor negative. Also, Axiom 15 can now be stated, if $a < b$ and c is positive, then $ac < bc$. Similarly, Axiom 16 can be stated, if $a < b$ and c is negative, then $bc < ac$.

Example 1   Let $a = -5$, $b = -2$, and $c = 4$; then $-5 < -2$ and $(-5)(4) < (-2)(4)$, or $-20 < -8$.

Example 2   Let $a = -5$, $b = -2$, and $c = -4$; then $-5 < -2$, but $(-5)(-4) > (-2)(-4)$, or $20 > 8$.

In applying Axiom 15 in Example 1, we say that *the direction of the inequality was retained*. Similarly, in applying Axiom 16 in Example 2, we say that *the direction of the inequality was reversed*. Hence Axioms 15 and 16 can be summarized in the following way:

1. *Multiplication of both sides of an equality by a positive number retains the direction of the inequality.*
2. *Multiplication of both sides of an inequality by a negative number reverses the direction of the inequality.*

These observations illustrate only a few of the many facts about inequalities. A number of the more useful properties of inequalities are collected in the table of Rules of Operation at the end of this section. Illustrations of how they are used are given in the exercises.

Finally, it is often useful to know and state that one number is "less than or equal to" another number. If a is either less than b or equal to b, then we write $a \leq b$. If $a \leq b$, then we shall sometimes write $b \geq a$ and say b is "greater than or equal to" a.

The Absolute Value of a Real Number

In operating with positive and negative numbers, it is convenient to introduce what is called the *absolute value* of a real number a, denoted by the symbol $|a|$ and defined as follows.

DEFINITION 1.21. $|a| = \begin{cases} a, & \text{if } a \text{ is positive or zero} \\ -a, & \text{if } a \text{ is negative} \end{cases}$

Thus the absolute value of a nonzero number is always positive.

Example 3 $|0| = 0$, $|6| = 6$, $|-7| = -(-7) = 7$. In the last case, $a = -7$, and hence $-a = -(-7) = 7$.

Let us now use the concept of absolute value to summarize the rules for adding, subtracting, multiplying, and dividing positive and negative numbers.

Addition: To add two real numbers having like signs, add their absolute values and prefix their common sign. To add two real numbers having unlike signs, find the difference of their absolute values and prefix to it the sign of the number having the larger absolute value. (In either case, the result of adding is called the *algebraic sum* of the two numbers.)

Subtraction: To subtract a real number a from a real number b is to find a real number c, such that $a + c = b$. This operation can be accomplished by changing the sign of *a* and adding the result to *b* in accordance with the rules for addition.

Multiplication: To multiply two real numbers, neither of which is zero, multiply their absolute values; prefix the plus sign to the product if the two numbers have like signs, and prefix the minus sign if they have unlike signs.

Division: To divide a real number *a* by a nonzero real number *b*, find a number *c* such that $a=bc$. In carrying out this operation, find the quotient of the absolute value of *a* divided by the absolute value of *b*; prefix to it the plus sign if *a* and *b* have like signs, and prefix the minus sign if *a* and *b* have unlike signs. (*Remember*: Division by zero is not defined. To understand this fact better, one may consider that, if *a* is not zero and *b* is zero, it is impossible to find a number *c* such that $a=bc$, but that, on the other hand, if $a=0$ and $b=0$, any real number *c* will satisfy the relation $a=bc$.)

The most important relationships between absolute value and inequality are isolated as Rules 1.45 through 1.50 in the following table. Again, illustrations of their use are contained in the exercises.

Rules of Operations

Let *a*, *b*, *c*, and *d* be real numbers.

RULE 1.30. If $a<b$ and $b<c$, then $a<c$
RULE 1.31. If $a<b$, then $a+c<b+c$
RULE 1.32. If $a<b$ and $c>0$, then $ac<bc$
RULE 1.33. If $a<b$ and $c>0$, then $\dfrac{a}{c}<\dfrac{b}{c}$
RULE 1.34. If $a<b$ and $c<0$, then $ac>bc$
RULE 1.35. If $a<b$ and $c<0$, then $\dfrac{a}{c}>\dfrac{b}{c}$
RULE 1.36. If $a>0$, then $-a<0$
RULE 1.37. If $a<b$, then $-a>-b$
RULE 1.38. $a<b$ if and only if $b-a>0$
RULE 1.39. $ab>0$ if, and only if, either:
    (i) $a>0$ and $b>0$
    or
    (ii) $a<0$ an $b<0$
RULE 1.40. $ab<0$ if, and only if, either:
    (i) $a>0$ and $b<0$
    or
    (ii) $a<0$ and $b>0$
RULE 1.41. If $a>0$, then $1/a>0$

RULE 1.42. If $a < 0$, then $1/a < 0$
RULE 1.43. If $a < b$ and $ab > 0$, then $1/a > 1/b$
RULE 1.44. $a/b < c/d$ if, and only if, $ad < bc$, provided $b > 0$ and $d > 0$
RULE 1.45. $|a| \geq 0$
RULE 1.46. If $a \neq 0$, then $|a| > 0$
RULE 1.47. $|a| < b$ if, and only if, $-b < a < b$
RULE 1.48. $|a| \leq b$ if, and only if, $-b \leq a \leq b$
RULE 1.49. If $b > 0$, then $|a| > b$ if, and only if, either:
 (i) $a > b$
 or
 (ii) $a < -b$
RULE 1.50. If $b > 0$, then $|a| \geq b$ if, and only if, either:
 (i) $a \geq b$
 or
 (ii) $a \leq -b$

Exercises 1.5

1. Indicate the order relation between the numbers in each of the following pairs.
  a. 3, 4 **b.** $-3, 2$ **c.** $0, -1$ **d.** $-2, 1$ **e.** $-4, -5$
  f. $12.3, 12\frac{1}{3}$ **g.** $-1, -2$ **h.** $-8, 0$ **i.** $-423, 1$

Solution to **a**: $3 < 4$

Solution to **g**: $-1 > -2$. To see this, we use Rule 1.37 and the fact that $1 < 2$. We could obtain the same result by multiplying both sides of $1 < 2$ by the negative number -1 and applying Rule 1.34.

2. Arrange the following numbers in ascending order (that is, smallest one first, etc.).

$$-1, 3, 2, 0, \sqrt{3}, -4, -6, -\frac{1}{2}, \sqrt{7}, 13, \sqrt{100}, -6\frac{1}{4}, 21$$

3. Indicate the order relation between ac and bc in each of the following cases.
  a. $a=2, b=5, c=4$ **b.** $a=2, b=-5, c=4$
  c. $a=-2, b=-5, c=4$ **d.** $a=2, b=5, c=-4$
  e. $a=2, b=-5, c=-4$ **f.** $a=-2, b=-5, c=-4$

Solution to **b**: $ac = 2 \cdot 4 = 8$ and $bc = (-5) \cdot 4 = -20$, and so $ac > bc$. We could have solved the problem by observing that $a = 2 > -5 = b$ and then multiplying both sides by 4. Since 4 is positive, the direction of the inequality is retained, and we have $ac > bc$.

4. Use Rule 1.44 to select the larger of the two fractions in each of the following pairs of fractions.

  a. $\frac{2}{3}, \frac{4}{7}$ **b.** $\frac{11}{17}, \frac{20}{31}$ **c.** $\frac{371}{586}, \frac{728}{1,107}$ **d.** $\frac{40,783}{31,426}, \frac{81,565}{62,752}$

SEC. 1.5 The Order Relations and Absolute Value

Solution to **b**: $11 \cdot 31 = 341 > 340 = 17 \cdot 20$ and so $\dfrac{11}{17} > \dfrac{20}{31}$.

5. Find the absolute value of each of the following numbers.

    a. 11   **b.** $7\dfrac{1}{3}$ **c.** -6.1 **d.** $-\sqrt{2}$ **e.** $3-7$ **f.** $\dfrac{2-1}{3-5}$ **g.** $-\dfrac{-2}{-3}$

6. Find the absolute value of each of the following numbers: $5, -8, -\dfrac{3}{4}, \sqrt{2}, -189, 17, -2.5$.

7. Find the value of each of the following expressions; write in simplest form free of absolute value symbols.

    a. $|7-5|$ **b.** $|3-8|$ **c.** $\dfrac{|5-9|}{|6-4|}$ **d.** $|-3| \cdot |8|$ **e.** $\dfrac{4-8}{|-2|}$ **f.** $\dfrac{|5-5|}{|-7|}$

8. Label each of the following statements as T (true) or F (false).

    a. $|4| = 4$ **b.** $|-5| = -5$
    c. $|2+3| = |2| + |3|$ **d.** $|-7| > 1$
    e. If $|a| < 5$, then $-5 < a < 5$ **f.** $|3-\pi| = \pi - 3$
    g. $|2-7| = |7-2|$ **h.** $|5-8| \leq |5| - |8|$

9. a. Find four different numbers each of which lies between 17/29 and 18/29.

    b. Which is the larger, $\dfrac{7}{-11}$ or $\dfrac{10}{-17}$?

10. If $|a| < 2$, what can be said about a?

    Solution: Using Rule 1.47, $-2 < a < 2$.

11. If $|a| < -2$, what can be said about a?

    Solution: Using Rule 1.47, $-(-2) < a < -2$ or, equivalently, $2 < a < -2$. However, there are no real numbers which are greater than 2 and less than -2 at the same time. Hence we conclude that there is no real number a such that $|a| < -2$.

12. If $|a| \geq 5$, what can be said about a?

13. If $|a| \geq -2$, what can be said about a?

14. If $|a| \leq 3$, what can be said about a?

15. If $|a| \leq 0$, what can be said about a?

16. Prove: **a.** If $a > 0$ and $b > 0$, then $a + b > 0$.
               b. If $a < 0$ and $b < 0$, then $a + b < 0$.

    Solution to **a**: Since $a > 0$, we use Rule 1.31 to obtain $a + b > 0 + b = b$. Now, using the fact that $b > 0$, we have both $a + b > b$ and $b > 0$. Hence Rule 1.30 may be applied to obtain $a + b > 0$.

17. Prove: If $a<b$, then $a<(a+b)/2<b$. It is useful to observe that $(a+b)/2$ is the midpoint of the line segment joining a to b.
18. By examining the four cases (1) a and b both positive, (2) a and b both negative, (3) one positive and the other negative, (4) one is zero, prove that $|a|\cdot|b|=|ab|$.
19. By examining cases similar to those used in Exercise 18, prove that if a and b are real numbers, then $|a+b|<|a|+|b|$ or $|a+b|=|a|+|b|$, and state when the equality sign holds. This proves what is known as the *triangle inequality*: $|a+b|\leq|a|+|b|$.

1.6 COORDINATE SYSTEMS

The Number Line

In working with real numbers it is often helpful to be able to think of them as being placed on a *number line*; that is, each number of the set of real numbers corresponds to a point on a line. Such a correspondence can be established in the following way.

Consider, for convenience, a horizontal line L extending indefinitely in both directions (Figure 1.1). Choose a point 0 on L and let this point correspond to the number zero. We shall call this point the *zero point*, or the *origin*. Choose a second point A, distinct from 0 and to the right of it; let the point correspond to the number $+1$. Taking the distance $0A$ as the unit of distance, the number scale is now fixed if we consider distances measured to the right as positive and those to the left as negative. If a is any positive real number, the point on the number line at a distance a units to the right of the origin shall be taken as the point which corresponds to the number a. If a is a negative real number, the point on the number line at a distance $|a|$ units to the left of the origin shall be taken as the point corresponding to the negative number a. Thus the point that corresponds to 5 is the point on the number line five units to the right of the origin, while the point that corresponds to -4 is the point four units to the left of the origin. Conversely, a point P to the right of the origin, whose distance from the origin is d units, shall have as its corresponding real number the positive real number d and, similarly, if Q is a point d units to the left of the origin, then Q shall have as its corresponding real number the negative real number $-d$. This device provides a correspondence such that to each real number there is one and only one corresponding point on the number line. We shall assume that, conversely, to each point on the number line there corresponds one and only one real number.

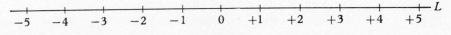

Figure 1.1

SEC. 1.6   Coordinate Systems

It follows, from the method used in setting up the correspondence between the set of real numbers and the points on the number line and from our definition of "less than," that the real number a is less than the real number b if and only if the point on the number line that corresponds to a lies to the left of the point that corresponds to b.

Example 1   Draw a number line and indicate on this line the points which correspond to the real numbers $-3, -2, -1, 0, 1, 2,$ and 3. Also indicate the points which correspond to $2\frac{4}{7}$ and to $\sqrt{2}$.

Solution: See Figure 1.2.

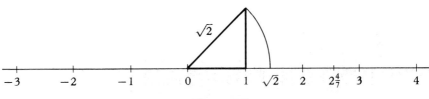

Figure 1.2

It is conventional to call the real number associated with a particular point on a number line the coordinate of that point and hence to refer to a number line as a *one-dimensional coordinate system*. We have already observed that such a coordinate system gives us a graphical interpretation of the relative magnitudes of numbers. For example, the algebraic fact $4 < 7$ corresponds to the geometric fact that the point with coordinate 4 lies to the left of the point with coordinate 7. We say that a number line is the graph of the set of all real numbers. The *graph* of a set S of real numbers consists of the set of all the points on a number line whose coordinates are the elements of S. Thus, if S contains only one number, then the graph of S consists of just one point; or, in general, if S contains n distinct numbers, then the graph of S consists of the n points on a number line whose coordinates are the n numbers in S. It may happen that a set S consists of the numbers a and b and all the numbers between a and b, or perhaps the set contains the numbers between a and b, but neither a nor b. Other possibilities are that S contains all numbers between a and b and contains a but not b, or contains b but not a. In each of these cases the graph consists of a segment of a number line which may contain both, one, or neither of its endpoints. In discussing such cases the following notation will be helpful.

Notation: Let a and b be any two real numbers, such that $a < b$. Now consider a number line and the two fixed points on this line which represent the numbers a and b, respectively. For convenience we shall refer to the point corresponding to a as the point a and the point corresponding to b as the point *b.* The set of points on the number line consisting of the point *a*, the point *b*, and all the points between *a* and *b* is called a *closed interval* and is denoted by the symbol [*a, b*]. The graph of this set of points is shown in Figure 1.3(a). Note that we indicate that *a* and *b* are included in the set by the black dots at their points.

The set of points consisting only of the points between *a* and *b*, not including either *a* or *b*, is called an *open interval* and is denoted by the symbol (*a, b*). The graph of (*a, b*) is shown in Figure 1.3(b). Note the small open circles at points *a* and *b*, indicating that they do not belong to the set.

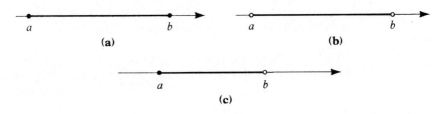

Figure 1.3

If we wish to include in our set the point *a* but not the point *b*, we denote such a set by the symbol [*a, b*) and represent it graphically as in Figure 1.3(c). Similarly, the symbol (*a, b*] denotes all the points on the number line between *a* and *b* together with the point *b*, but not including the point *a*. Such a set of points is often referred to as a *half-open* interval. In a similar manner, we shall denote

(i) the set of all numbers greater than *a* by the symbol (a, ∞)
(ii) the set of all numbers greater than or equal to *a* by the symbol $[a, \infty)$
(iii) the set of all numbers less than *b* by the symbol $(-\infty, b)$
(iv) the set of all numbers less than or equal to *b* by the symbol $(-\infty, b]$

It should be observed that

1. $[a, b] = \{x | x \in R \text{ and } a \leq x \leq b\}$
2. $[a, b) = \{x | x \in R \text{ and } a \leq x < b\}$
3. $(a, b] = \{x | x \in R \text{ and } a < x \leq b\}$
4. $(a, b) = \{x | x \in R \text{ and } a < x < b\}$
5. $(a, \infty) = \{x | x \in R \text{ and } a < x\}$
6. $(-\infty, b) = \{x | x \in R \text{ and } x < b\}$

Rectangular Coordinate System

In the previous subsection we were able to set up a one-to-one correspondence between the points on a line and the real numbers. We shall now use the same

SEC. 1.6   Coordinate Systems

general idea to set up a one-to-one correspondence between the points in a plane and ordered pairs of real numbers. Ordered pairs are briefly described in Appendix A in preparation for the definition of the Cartesian product of two sets. Since the concept of ordered pairs of real numbers is of great importance in this and later sections, we state the following definitions.

DEFINITION 1.22.   An *ordered pair* of real numbers is a pair of real numbers, denoted symbolically as (a, b), where a is designated as the *first element or component* of the pair and b is designated as the *second element or component* of the pair. The first element is always listed first in the parentheses. Furthermore, two ordered pairs (a, b) and (c, d) are equal if and only if $a=c$ and $b=d$. In particular, $(a, b)=(b, a)$ if and only if $a=b$.

NOTE: We have used the same notation for the ordered pair with the first element a and second element b as we used in our previous remarks for the open interval between a and b. This is a rather standard practice, and the context in which the notation is used should prevent any ambiguity.

Suppose that X is some set of real numbers and that Y is also a set of real numbers. In Appendix A the Cartesian product $X \times Y$ is defined as the set of all possible ordered pairs (x, y) whose first element, x, belongs to X and whose second element, y, belongs to Y; x and y may or may not be identical. Stated in symbols.

$$X \times Y = \{(x, y) | x \in X \text{ and } y \in Y\}$$

Example 2   If $X = \{1, 2, 3\}$ and $Y = \{5, 7\}$, then

$$X \times Y = \{(1, 5)(1, 7), (2, 5), (2, 7), (3, 5), (3, 7)\}$$

Example 3   If $X = \{1, 2, 3\}$ and $Y = \{1, 2, 3\}$, then

$$X \times Y = \{(1, 1), (1, 2), (1, 3), (2, 1), (2, 2), (2, 3), (3, 1), (3, 2), (3, 3)\}$$

In mathematics we are often concerned with sets whose elements are ordered pairs of real numbers and thus have a need for some method of representing such a set graphically. That is, we need a system which produces an association between a point in the plane and an ordered pair of real numbers. The system we are about to describe does just that. In fact, this system establishes a one-to-one correspondence between the points in a plane and the ordered pairs (x, y) of the set $X \times Y$ of ordered pairs of real numbers, where $X = Y = R$ and R is the set of all real numbers. The set $X \times Y$ is thus the set of all possible ordered pairs of real numbers.

To establish a one-to-one correspondence between the points in the plane and the ordered pairs of the set $X \times Y$, we first construct in the plane two perpendicular lines $x'x$ and $y'y$, which we shall call the *x axis* and the *y axis*, respectively. For purposes of discussion, it is convenient to think of the *x* axis as being horizontal and the y axis as being vertical. Their point of intersection we shall call the *origin*. We next set up a one-dimensional coordinate system on each of the two axes. On each axis, we make the origin correspond to the real number 0. On the x axis the points to the right of the origin are made to correspond to the positive real numbers, and the points to the left of the origin are made to correspond to the negative real numbers. The unit of distance used on each axis is arbitrary and may be different for the two axes. In fact, it is often desirable to use different units of measurement, particularly when the units on the two are to be interpreted as representing different physical units of measurement which are not comparable (temperature, distance, speed, etc.).

The framework just described, the x and y axes together with their respectively arbitrary scales, is called a *rectangular coordinate system*. The word "rectangle" here refers to the fact that the axes are mutually perpendicular.

We can now use this system to associate with each point P in the plane an ordered pair of real numbers. First, suppose that P is a point on the x axis. Since a one-to-one correspondence between the points on the x axis and the real numbers has already been established, each point on the x axis has one and only one coordinate in this one-dimensional system. Thus P corresponds to some real number a; we associate with P the ordered pair $(a, 0)$, which signifies a on the x axis and 0 on the y axis. Similarly, if P is on the y axis, it has a coordinate, b, in the one-dimensional coordinate system already established on that axis. In this case, associate with the point P the ordered pair $(0, b)$. Finally, let P be a point that is not on either of the axes. Through it, draw a line parallel to the y axis and intersecting the x axis in some point M. Draw also a line through P parallel to the x axis and intersecting the y axis in some point N (see Figure 1.4). The point M, being on the x axis, will correspond to some real number a, and the point N, being on the y axis, will correspond to some real number b. We shall associate with the point P the ordered pair (a, b) of real numbers (Figure 1.4).

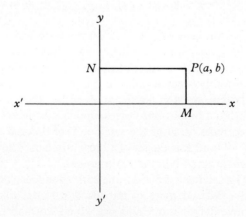

Figure 1.4

SEC. 1.6 Coordinate Systems

Thus, by means of a rectangular coordinate system, we have made correspond to each point P in the plane an ordered pair (a, b) of real numbers. Conversely, we agree to associate with each ordered pair of real numbers the point determined by it. For example, suppose we are given the ordered pair (a, b). The real number a corresponds to a point M on the x axis, and the number b corresponds to a point N on the y axis. The line drawn through M parallel to the y axis and the line drawn through N parallel to the x axis will intersect in the desired point P.

We have now established a one-to-one correspondence between the points of a plane and the ordered pairs (a, b) of real numbers. When a rectangular coordinate system of x and y axes has been established in a plane, we often call the plane the xy plane.

DEFINITION 1.23. The real numbers a and b in the ordered pair (a, b) associated with the point P are called the *coordinates* of the point P. The first number, a, of the pair is called the *x coordinate* of P, and the second number, b, is called the *y coordinate* of P.

We shall often refer to a point with coordinates (a, b) as the point (a, b). We shall also say $P(a, b)$, which is read "the point P with coordinates (a, b)." The x coordinate of a point is also known as the *abscissa* of the point and the y coordinate as the *ordinate* of the point. If a point lies to the right of the y axis its abscissa is a positive number; if it lies to the left, the abscissa is negative. Similarly, the ordinate of a point is positive if the point lies above the x axis, and it is negative if it lies below. We therefore interpret the abscissa and ordinate of a point as being the directed distances of the point from the x and y axes, respectively. That is, the abscissa and ordinate of a point each exhibit two kinds of information about the point, direction and distance. For example, if the abscissa of a point P is -3, the minus sign gives the direction of the point from the y axis and $|-3|$ gives the distance from the point to the y axis in terms of the units used in the scale on the x axis. So the point P is on a line parallel to the y axis and 3 units to the left.

Example 4 | The point $(3, -2)$ is a point that is on a line parallel to the y axis and 3 units to the right. It is also on a line parallel to the x axis and 2 units below it. Hence it is the point of intersection of these two lines. See Figure 1.5.

Let us now point out some rather obvious but yet very important details in connection with a rectangular coordinate system. The coordinates of the origin are obviously $(0, 0)$. The x and y axes divide the plane into four portions called *quadrants*. These quadrants are numbered I, II, III, and IV, beginning with the one in the upper-right corner and going counterclockwise. See Figure 1.6. If a point lies in quadrant I, its coordinates are both positive, and conversely. In quadrant II the abscissa is negative and the ordinate is positive. In quadrant III

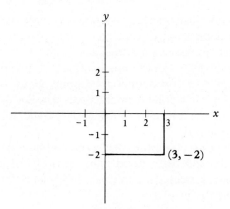

Figure 1.5

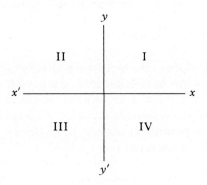

Figure 1.6

both abscissa and ordinate are negative. In quadrant IV the abscissa is positive and the ordinate is negative. The abscissa of every point on the y axis is zero, and the ordinate of every point on the x axis is zero. Locating a point in the plane having a given pair of real numbers as its coordinates is called *plotting the point*. In Figure 1.5 we plotted the point whose coordinates were $(3, -2)$.

We shall represent the coordinates of a general point in the plane as (x, y), but if we want to talk about a specific fixed point in the plane without using its actual numerical coordinates, we shall either use subscripts, such as (x_1, y_1) or (x_2, y_2), or we shall use letters at the beginning of the alphabet, as (a, b) or (c, d). Two points whose ordinates are the same lie on a line parallel to the x axis, and two points whose abscissas are the same lie on a line parallel to the y axis. We may define *distance* between two such points as follows.

DEFINITION 1.24. The distance between the points $P_1(x_1, y_1)$ and $P_2(x_2, y_1)$, denoted by $d(P_1, P_2)$, is defined as $d(P_1, P_2) = |x_2 - x_1|$. The distance between the points $Q_1(x_1, y_1)$ and $Q_2(x_1, y_2)$ is defined as $d(Q_1, Q_2) = |y_2 - y_1|$.

SEC. 1.7   Complex Numbers

The distance between two points not on a line parallel to one of the coordinate axes is meaningless except when the same scale is used on both axes.

Exercises 1.6

1. Draw the graph of each interval.
   a. $[1, 5]$ b. $[-3, 2]$ c. $[-5, 5]$ d. $(1, 4)$ e. $(-1, 3)$ f. $(-2, 1]$
   g. $[3, 5)$ h. $(-\infty, 2)$ i. $(-3, \infty)$

2. Draw a number line and label the points corresponding to 0, 1, 3, -2, $\frac{3}{2}$, 2.4, $-\frac{5}{6}$, $\sqrt{5}$, and $\sqrt{8}$.

3. Find the distance between the given pairs of points.
   a. $P_1(3, 2)$ and $P_2(10, 2)$ b. $Q_1(5, -2)$ and $Q_2(5, -9)$
   c. $P_3(-4, 3)$ and $P_4(2, 3)$ d. $Q_3(-1, 7)$ and $Q_4(-1, 5)$

4. Plot the following points.
   a. $P_1(5, 3)$ b. $P_2(-2, 1)$ c. $P_3(4, -7)$ d. $P_4(-6, -3)$ e. $P_5(0, 4)$
   f. $P_6(-5, 0)$

5. Using the same scale on both the x and y axes, plot the following points: $(-5, 7)$, $(-3, 5)$, $(0, 2)$, $(5, -3)$, $(8, -6)$. Do these points seem to have any property in common?

6. Where must a point lie if its abscissa is 3? If -3? Where must a point lie if its ordinate is 5? If -5?

7. Assuming that the same scale is used on both axes, where would all the points lie whose abscissas and ordinates are equal? Where would they lie if the scale units are different on the two axes?

8. Using the same scale on each of the axes, plot the following points: $(5, 0)$, $(4, 3)$, $(3, 4)$, $(0, 5)$, $(-3, 4)$, $(-4, 3)$, $(-5, 0)$, $(-4, -3)$, $(-3, -4)$, $(0, -5)$, $(3, -4)$, $(4, -3)$. Connect these points by a smooth curve. What conclusion do you come to?

1.7 COMPLEX NUMBERS

Although our concern in this text is chiefly with real numbers, problems will arise in which the real number system is not adequate for their solution. These problems will occur primarily in solving polynomial equations in Chapter 3. Typically, we shall be interested in solving equations of the form $a^2 = -b$ for a, where b is a positive real number. Since $-b$ is negative whenever b is positive and since $a^2 \geq 0$ for all real numbers a, there is no real number a with the property that $a^2 = -b$ when b is positive. What is done to overcome this difficulty is, basically, to introduce a new number i, called an *imaginary number*, which has the property that $i^2 = -1$. (In electrical engineering, the symbol j is used.)

DEFINITION 1.25. The symbol *i* will be used to denote the *imaginary unit*, a number whose square is equal to -1. For emphasis we sometimes write $i = \sqrt{-1}$ and say, "*i* is the square root of -1." *Pure imaginary numbers* are obtained by writing real numbers immediately to the left or to the right of the symbol *i*. Hence each pure imaginary number is of the form bi for some real number b. *Complex numbers* are symbols of the form $a + bi$, where a and b are real numbers. We agree that, if $b = 0$, then $a + bi$ is simply the real number a, and that if $a = 0$, then $a + bi$ is the pure imaginary number bi. The *conjugate* of the complex number $a + bi$ is the complex number $a - bi$.

For purposes, the following properties of complex numbers suffice. A more detailed treatment of the operations on the set of complex numbers follows in the subsection which concludes this section.

PROPERTY 1. If $a^2 = -b$, where b is a positive real number, then either $a = i\sqrt{b}$ or $a = -i\sqrt{b}$ (the definition and properties of $\sqrt{b}$ will be treated in Chapter 2).

PROPERTY 2. $a + bi = c + di$ if and only if $a = c$ and $b = d$. In particular, $a + bi = 0$ if and only if $a = 0$ and $b = 0$.

PROPERTY 3. Addition of Complex Numbers

$$(a + bi) + (c + di) = (a + c) + (b + d)i$$

PROPERTY 4. Multiplication of Complex Numbers

$$(a + bi)(c + di) = (ac - bd) + (ad + bc)i$$

Example 1 If $a^2 = -4$, what can be said about a?

Solution: Either $a = 2i$ or $a = -2i$.

Example 2 If $a^2 = -2$, what can be said about a?

Solution: Either $a = i\sqrt{2}$ or $a = -i\sqrt{2}$.

Example 3 If $(a - 1)^2 = -9$, what can be said about a?

Solution: Either $a - 1 = 3i$ or $a - 1 = -3i$. Hence either $a = 1 + 3i$ or $a = 1 - 3i$. This situation is usually abbreviated by writing $a = 1 \pm 3i$ and is read "*a* equals one plus or minus three *i*."

Problems similar to that of Example 3 will be extremely important in solving quadratic equations in Chapter 3.

SEC. 1.7   Complex Numbers   57

Example 4   Use Property 2 to determine real values for x and y for which the equation $3x+2yi = 9-8i$ is a true statement.

Solution: According to Property 2, $3x+2yi = 9-8i$ if and only if $3x = 9$ and $2y = -8$. Hence $x = 3$ and $y = -4$.

Example 5   Find the indicated sum: $(7+5i)+(4+3i)$.

Solution: $(7+5i)+(4+3i) = 7+5i+4+3i = 7+4+(5+3)i = 11+8i$.

It should be pointed out that to find the product of two complex numbers it is not necessary or desirable to use Property 4 as a formula. The best procedure in finding the product of two complex numbers is to use the rules of algebra for finding the product of two algebraic expressions and replace i^2 by -1 wherever it occurs.

Example 6   Find the indicated product: $(2-5i)(4+3i)$.

Solution: $(2-5i)(4+3i) = 8+6i-20i-15i^2$
$= 8-14i+15$
$= 23-14i$

The Complex Number System

The very informal treatment given above to complex numbers, while sufficient for the remainder of this text, is not adequate for many purposes. For this reason, we present a more formal treatment of the complex number system. In view of Property 2 ($a+bi = c+di$ if and only if $a = c$ and $b = d$) and the corresponding property of ordered pairs [$(a, b) = (c, d)$ if and only if $a = c$ and $b = d$], we shall formally define complex numbers as ordered pairs of real numbers.

DEFINITION 1.26.   A *complex number* is an ordered pair, (a, b), of real numbers. The complex number (a, b) is usually represented symbolically in the form $a+bi$ or $a+ib$, these symbols being assumed to be equal. We call a the *real part* and b the *imaginary part* of the complex number. The *conjugate* of the complex number (a, b) is the complex number $(a, -b)$.

When a new class of numbers is defined, in order to obtain a mathematical system, it is necessary to define laws of operation for those numbers. The operations of addition and multiplication on complex numbers are defined as follows (a, b, c, and d are real numbers).

DEFINITION 1.27.   Addition of Complex Numbers

$$(a+bi)+(c+di) = (a+c)+(b+d)i$$

DEFINITION 1.28. Multiplication of Complex Numbers
$$(a+bi)(c+di) = (ac-bd) + (ad+bc)i$$

The following facts can be demonstrated with little difficulty and are left as exercises for the student.

FACT 1: The set of real numbers can be "identified" with the subset of the complex numbers consisting of those with imaginary part 0 (identifying the real number a with the complex number $a+0i$). If this is done, then the operations of addition and multiplication of complex numbers reduce precisely to addition and multiplication of real numbers.

FACT 2: $(a+bi)-(c+di) = (a-c)+(b-d)i$

FACT 3: $\dfrac{1}{c+di} = \left(\dfrac{c}{c^2+d^2}\right) - \left(\dfrac{d}{c^2+d^2}\right)i \quad$ if $c+di \neq 0$

FACT 4: $\dfrac{a+bi}{c+di} = \left(\dfrac{ac+bd}{c^2+d^2}\right) + \left(\dfrac{bc-ad}{c^2+d^2}\right)i \quad$ if $c+di \neq 0$

FACT 5: The complex number system, that is, the set of complex numbers $C = \{a+bi \mid a \in R \text{ and } b \in R\}$ with the operations of addition and multiplication defined above satisfies the 11 field axioms of Section 1.2. Hence the complex number system is a field.

PROOF OF FACT 4

We multiply the numerator and denominator of the fraction $\dfrac{a+bi}{c+di}$ by the conjugate of the denominator, $c-di$, as follows:

$$\frac{a+bi}{c+di} = \left(\frac{a+bi}{c+di}\right)\left(\frac{c-di}{c-di}\right) = \frac{ac-adi+bci-bdi^2}{c^2-cdi+cdi-d^2i^2}$$

Now replace i^2 by -1:

$$= \frac{ac-adi+bci+bd}{c^2+d^2}$$

$$= \frac{(ac+bd)+(bc-ad)i}{c^2+d^2}$$

$$= \left(\frac{ac+bd}{c^2+d^2}\right) + \left(\frac{bc-ad}{c^2+d^2}\right)i = x+yi$$

It should be observed that we have just proved that any fraction whose denominator is a complex number, $c+di \neq 0$, can be expressed in the form $x+yi$

SEC. 1.7 Complex Numbers

by multiplying the numerator and denominator of the given fraction by the conjugate of the denominator, replacing i^2 by -1, and simplifying.

Example 7 Express the fraction $\dfrac{6+7i}{4+3i}$ in the form $x+yi$.

Solution:
$$\dfrac{6+7i}{4+3i} = \left(\dfrac{6+7i}{4+3i}\right)\left(\dfrac{4-3i}{4-3i}\right) = \dfrac{24-18i+28i-21i^2}{16-12i+12i-9i^2}$$
$$= \dfrac{24+10i+21}{16+9}$$
$$= \dfrac{45+10i}{25}$$
$$= \dfrac{9}{5}+\dfrac{2}{5}i$$

Exercises 1.7

1. Find each indicated sum.
   a. $(7+3i)+(2+11i)$ b. $(5-6i)+(-3+4i)$
   c. $(-1+5i)+(-2+i)$ d. $(-6-12i)+(-8-11i)$
   e. $(2+3i)+(2-3i)$ f. $5+(4-6i)$
   g. $(-7-3i)+9$ h. $(23-17i)+(65+43i)$

2. Find each indicated difference.
   a. $(7+6i)-(4+3i)$ b. $(13-5i)-(8-2i)$
   c. $(-5-4i)-(3-2i)$ d. $(5-i)-(7-3i)$
   e. $(8-5i)-(4-7i)$ f. $(-6+2i)-(-12-15i)$

3. Find each indicated product.
   a. $(3+2i)(5+4i)$ b. $(7-i)(-3-3i)$
   c. $(1-i)(1+i)$ d. $(1+i)(3-2i)$
   e. $(1+i)(1+i)$ f. $5i(2i-7)$
   g. $(3-5i)(6-2i)$ h. $(1-i)(1-i)$

4. Use the method used in the proof of Fact 4 to express each given fraction in the form $x+yi$.
   a. $\dfrac{1+i}{1-i}$ b. $\dfrac{4}{3+i}$ c. $\dfrac{4+3i}{2+5i}$ d. $\dfrac{5-2i}{i}$ e. $\dfrac{25i}{4-3i}$ f. $\dfrac{-4+2i}{-2-3i}$

5. Verify that each given equation is a true statement.
   a. $(2+3i)^2-4(2+3i)+13=0$ b. $(2-3i)^2-4(2-3i)+13=0$
   c. $i^4+5i^2+4=0$ d. $(-1+2i)^2+2(-1+2i)+5=0$
   e. $(-1-2i)^2+2(-1-2i)+5=0$ f. $4\left(\dfrac{3+5i}{2}\right)^2-12\left(\dfrac{3+5i}{2}\right)+34=0$

6. Use the definition of equality of two complex numbers to determine real values for x and y for which each given equation is a true statement.
   a. $2x + 5yi = 4 - 10i$
   b. $3x - 2i = 9 - yi$
   c. $i(y + xi) + 4 - 7i = 0$
   d. $(x + y + 2) + (x - y - 4)i = 0$

7. Suppose $z_1 = a + bi$, $z_2 = c + di$, and $z_3 = e + fi$, where a, b, c, d, e, and f are real numbers. Show that:
   a. $z_1 + z_2 = z_2 + z_1$
   b. $z_1 z_2 = z_2 z_1$
   c. $(z_1 + z_2) + z_3 = z_1 + (z_2 + z_3)$
   d. $(z_1 z_2) z_3 = z_1 (z_2 z_3)$
   e. $z_1(z_2 + z_3) = z_1 z_2 + z_1 z_3$

2

Elementary Algebraic Processes

2.1 ALGEBRAIC EXPRESSIONS

Elementary algebra may be described as a generalized arithmetic in which letters are used to represent numbers. For example, the arithmetic facts that $5+5+5=(3)(5)$, $6+6+6=(3)(6)$, and $(-4)(-4)(-4)=(3)(-4)$ are all special cases of the more general fact that $a+a+a=3a$ for every real number a. In this chapter we shall introduce several types of "algebraic expressions" and treat the rules of operation which are used in computations involving these expressions. The ability to combine, manipulate, and simplify various sorts of algebraic expressions (for example, $3x$, $2x^2-3x+1$, x^2y+xy^2, $\sqrt{x^2+y^2}$, $\frac{3x^{1/2}}{5x^{3/2}}$, etc.) turns out to be extremely important in many situations. Before giving an imprecise definition of an algebraic expression (no attempt will be made to give a precise definition), it is essential to acquaint ourselves with certain terminology.

Letters near the end of the alphabet (such as u, v, w, x, y, z), when used to represent numbers, are called *variables*. (Later in the text we shall also allow variables to represent algebraic expressions; see Appendix A for a more detailed treatment.) It is often desirable in mathematics to use a letter to represent a fixed number. When this is done, we ordinarily use a letter near the beginning of the alphabet (such as a, b, c, d, e, f) and refer to such letters as *arbitrary constants*. Arbitrary constants enable us to represent by a single expression a whole set of other expressions. For example, each of the statements

(1) x^2+2x+1 is a perfect square
(2) $4x^2+12x+9$ is a perfect square

is a special case of the more general statement:
(3) $a^2x^2 + 2abx + b^2$ is a perfect square
where a and b are used to represent arbitrary constants. In (1), $a=1$ and $b=1$, whereas in (2), $a=2$ and $b=3$.

Roughly speaking, an *algebraic expression* is any expression involving specific numbers, variables, arbitrary constants, sums, differences, products, quotients, exponents, and radicals (exponents and radicals will be discussed later in this chapter).

Our approach in this chapter will be to treat elementary types of algebraic expressions first and then treat increasingly more complicated types as the chapter progresses. We shall attempt to develop techniques necessary to perform the basic operations of addition, subtraction, multiplication, and division when dealing with each type of algebraic expression. Also of interest will be factoring of certain types of algebraic expressions into products of simpler expressions (primarily polynomials in one or two variables).

A Beginning

In Section 1.3, twenty-one rules of operation were listed for dealing with real numbers. Each of these rules has a counterpart in algebraic expressions. We illustrate this fact by considering two of the rules individually.

Example 1 Consider Rule 1.4, which states that $a(b+c) = ab + ac$ for all real numbers a, b, and c. This would suggest that we should accept $x(y+z) = xy + xz$ as a rule for multiplying the algebraic expression x by the algebraic expression $y+z$. This is, in fact, what we do. Hence, if we were listing rules of operation for algebraic expressions, we might list

Rule 2.4. $x(y+z) = xy + xz$

Example 2 Consider Rule 1.7, which states that $a + (-a) = (-a) + a = a - a = -a + a = 0$ for each real number a. This would suggest that we should accept $x + (-x) = (-x) + x = x - x = -x + x = 0$ as a rule for adding the algebraic expression x to the algebraic expression $-x$.

Example 3 Returning to Example 1, we would also agree that $a(x+y) = ax + ay$, that $3(x+y) = 3x + 3y$, and that $3(x+1) = 3x + 3$.

For those who regard the contents of this section to this point as so much "mumbo jumbo," we would hasten to assure you that there is indeed a difference between the rules.

RULE 1.4: $a(b+c) = ab + ac$ for all real numbers a, b, and c.

SEC. 2.1 Algebraic Expressions

RULE 2.4: The algebraic expression $x(y+z)$ is to be identified with (i.e., set equal to) the algebraic expression $xy+xz$.

Rule 1.4 states a property of multiplication and addition of real numbers, whereas Rule 2.4 gives an *admissible operation* that one is allowed to perform on the algebraic expression $x(y+z)$. Without Rule 2.4 we would have no reason to write $x(y+z) = xy+xz$. As algebraic expressions, they are certainly not the same. The first contains only one x and the second contains two x's. Mathematicians treat the algebraic expression $x(y+z)$ as a purely formal expression having no particular connection with the real number system. However, the dismayed student should not dispair, for in the end nothing is lost by simply thinking of the variables in an algebraic expression as representing real numbers. The reason for this is the following statement, which underlies the development of rules of operation for algebraic expressions.

We define our rules of operation for algebraic expressions in such a way that, if any real numbers are substituted for the variables in the expressions, the resulting equation is valid.

Example 4 To add the algebraic expressions $2x, (-3y), 5x, 2y, 4a$, and 3, we proceed as follows:

$$(2x)+(-3y)+(5x)+(2y)+(4a)+(3) = 2x-3y+5x+2y+4a+3$$
$$= 2x+5x-3y+2y+4a+3$$
$$= 7x-y+4a+3$$

For the first equality we used an analog of Rule 1.1 concerning removal of parentheses. For the second equality we used an analog of Axiom 3 allowing us to change the order of "terms." The final equality is obtained by applying Rule 2.4 as developed in Example 1.

In the next 11 examples, we shall indicate solutions to problems involving addition, subtraction, multiplication, and division of some basic types of algebraic expressions. In each case we use unwritten rules of operation which are analogous to rules of operations for real numbers (as in Example 4). We take the attitude that a comprehensive list of rules of operation for algebraic expressions is unnecessary, since the operations performed are completely analogous to those studied in Chapter 1.

Example 5 Subtract $4ax$ from $10ax$.

Solution: $10ax - 4ax = (10-4)ax = 6ax$

Example 6 Subtract $10ax$ from $4ax$.

Solution: $4ax - 10ax = (4-10)ax = -6ax$

Example 7 Subtract $(-4ax)$ from $10ax$.

Solution: $10ax - (-4ax) = 10ax + 4ax = (10+4)ax = 14ax$

Example 8 Add $3x+2y$, $x-4y$, and $-2x+7y$.

Solution:
$$\begin{aligned}(3x+2y)+(x-4y)+(-2x+7y) &= 3x+2y+x-4y-2x+7y\\ &= 3x+x-2x+2y-4y+7y\\ &= (3+1-2)x+(2-4+7)y\\ &= 2x+5y\end{aligned}$$

Example 9 Subtract $3x-2y+5$ from $8x-7y+2$.

Solution:
$$\begin{aligned}(8x-7y+2)-(3x-2y+5) &= 8x-7y+2-3x+2y-5\\ &= 8x-3x-7y+2y+2-5\\ &= (8-3)x+(-7+2)y+(2-5)\\ &= 5x-5y-3\end{aligned}$$

Example 10 Add $\frac{2}{3}x - \frac{3}{4}xy + \frac{1}{9}y - \frac{2}{15}$ to $-\frac{2}{3}y + \frac{2}{3} - \frac{5}{3}xy + \frac{1}{6}x$.

Solution:
$$\left(\frac{2}{3}x - \frac{3}{4}xy + \frac{1}{9}y - \frac{2}{15}\right) + \left(-\frac{2}{3}y + \frac{2}{3} - \frac{5}{3}xy + \frac{1}{6}x\right)$$
$$= \frac{2}{3}x - \frac{3}{4}xy + \frac{1}{9}y - \frac{2}{15} - \frac{2}{3}y + \frac{2}{3} - \frac{5}{3}xy + \frac{1}{6}x$$
$$= \frac{2}{3}x + \frac{1}{6}x - \frac{3}{4}xy - \frac{5}{3}xy + \frac{1}{9}y - \frac{2}{3}y - \frac{2}{15} + \frac{2}{3}$$
$$= \left(\frac{2}{3} + \frac{1}{6}\right)x - \left(\frac{3}{4} + \frac{5}{3}\right)xy + \left(\frac{1}{9} - \frac{2}{3}\right)y + \left(\frac{2}{3} - \frac{2}{15}\right)$$
$$= \left(\frac{4+1}{6}\right)x - \left(\frac{9+20}{12}\right)xy + \left(\frac{1-6}{9}\right)y + \left(\frac{10-2}{15}\right)$$
$$= \frac{5}{6}x - \frac{29}{12}xy - \frac{5}{9}y + \frac{8}{15}$$

Notice that in each of these examples we have "collected like terms" and added (or subtracted, as the case may be) numerical parts. Roughly speaking, the *terms* of an algebraic expression are the largest portion of the expression which contains neither a plus sign nor a minus sign. Two terms are *alike* if they can be written so as to contain the same variables (and perhaps the same arbitrary constants) to the same powers, and in the same order. If a term is written so that it contains only one fixed number, this number is called the *numerical coefficient of the term*. This terminology is illustrated in the next example.

SEC. 2.1 Algebraic Expressions

Example 11 The terms of the algebraic expression $5y+3x-2y+5xy+6$ are $5y$, $3x$, $2y$, $5xy$, and 6. The first, $(5y)$, and the third, $(2y)$, are the only like terms. Hence, if we wanted to "simplify" this algebraic expression, we would collect the two like terms:

$$5y+3x-2y+5xy+6 = 3x+5y-2y+5xy+6$$
$$= 3x+(5-2)y+5xy+6$$
$$= 3x+3y+5xy+6$$

Finally, the numerical coefficient of the term $5y$ is 5, of $3x$ is 3, of $-2y$ is -2, of $5xy$ is 5, and of 6 is 6.

The terminology introduced above will be useful as we treat increasingly more complicated algebraic expressions.

Example 12 Multiply $x+1$ by $y+2$.

Solution: $(x+1)(y+2) = x(y+2)+1(y+2)$
$\qquad\qquad\qquad\quad = xy+2x+y+2$

Example 13 Multiply $2x-3$ by $-4y+5$.

Solution: $(2x-3)(-4y+5) = 2x(-4y+5)-3(-4y+5)$
$\qquad\qquad\qquad\qquad\quad = -8xy+10x+12y-15$

Example 14 Divide $(x+1)(x-3)$ by $(x+1)(2x+5)$.

Solution: $\dfrac{(x+1)(x-3)}{(x+1)(2x+5)} = \dfrac{x-3}{2x+5}$

Here we have divided both numerator and denominator by $x+1$. If we were interested in applying this "result" to a situation involving real numbers, we would need to be sure that $x+1 \neq 0$, that is, $x \neq -1$. Consequently, we usually write

$$\dfrac{(x+1)(x-3)}{(x+1)(2x+5)} = \dfrac{x-3}{2x+5} \quad \text{if } x \neq -1$$

Example 15 Divide $x+3$ by $x+2$.

Solution: There is not much that can be done to modify the expression $\dfrac{x+3}{x+2}$. Occasionally, it is useful to perform the following operations on this expression:

$$\dfrac{x+3}{x+2} = \dfrac{x+2+1}{x+2} = \dfrac{x+2}{x+2} + \dfrac{1}{x+2} = 1 + \dfrac{1}{x+2}$$

The student is perhaps more familiar with long division as a process for obtaining the same result:

$$x+2 \overline{)\begin{array}{c} 1 \\ x+3 \\ \underline{x+2} \\ 1 \end{array}} \quad \text{so} \quad \frac{x+3}{x+2} = 1 + \frac{1}{x+2}$$

in much the same way that

$$5 \overline{)\begin{array}{c} 3 \\ 17 \\ \underline{15} \\ 2 \end{array}} \quad \text{so} \quad \frac{17}{5} = 3 + \frac{2}{5} = 3\frac{2}{5}$$

WARNING: Be sure not to cancel the x in this problem.

$$\frac{\cancel{x}+3}{\cancel{x}+2} = \frac{3}{2} \quad \text{IS WRONG!}$$

$$\frac{\overset{1}{\cancel{x}}+3}{\underset{1}{\cancel{x}}+2} = \frac{1+3}{1+2} = \frac{4}{3} \quad \text{IS WRONG!}$$

If you want to divide both numerator and denominator by an expression, you must divide the *entire* numerator and the *entire* denominator by the expression. See Examples 39, 40, and 41 in Section 1.4 for a treatment of wrong cancellation.

Exercises 2.1

1. Find the sums of the expressions.
   a. $3xy, \; -5xy, \; 9xy, \; -2xy, \; xy$
   b. $2x+y, \; x-3y, \; 7x-4y, \; -6x+5y$
   c. $5x-3y-4z, \; -x-2y+3z, \; -2x+3y-z$
   d. $4xy-3x, \; 4x-5xy, \; -2x+7xy$
   e. $2x+y-3, \; 3x+2y-1, \; 4x-3y+6, \; -6x+y+5$

Solution to **d**:

$$\begin{aligned}
(4xy-3x)&+(4x-5xy)+(-2x+7xy) \\
&= 4xy-3x+4x-5xy-2x+7xy \quad \text{(Removing parentheses)} \\
&= 4xy-5xy+7xy-3x+4x-2x \quad \text{(Collecting like terms)} \\
&= (4-5+7)xy+(-3+4-2)x \\
&= 6xy-x
\end{aligned}$$

SEC. 2.1 Algebraic Expressions

2. Subtract the first expression from the second expression.
   a. $3x - 4y$, $6x - 3y + 5$ b. $5x - 7$, $3x - 4$
   c. $7x - 3y$, $-4x + 6y$ d. $3x - 4xy + 2y$, $2x - xy - 8y$
   e. $2x - 5y + 3z$, $ax + by + z$

   Solution to **a**:

   $$(6x - 3y + 5) - (3x - 4y) = 6x - 3y + 5 - 3x + 4y \quad \text{(Removing parentheses)}$$
   $$= 6x - 3x - 3y + 4y + 5 \quad \text{(Collecting like terms)}$$
   $$= (6 - 3)x + (-3 + 4)y + 5$$
   $$= 3x + y + 5$$

3. Add $\frac{2}{5}x + \frac{5}{9}y - \frac{1}{2}z$ to $\frac{3}{10}x - \frac{2}{3}y + \frac{1}{4}z$.

4. Subtract $\frac{3}{4}x + \frac{2}{7}xy - \frac{5}{8}y$ from $\frac{5}{12}x - \frac{3}{2}y - \frac{3}{14}xy$.

5. In each of the following, remove all symbols of grouping and collect like terms where possible.
   a. $(2x - 3y + 5) - (4x - 7y + 2) - (3x - 2y - 1)$
   b. $[(7x + 2y) - y] - [(5x - 6) - (3y - x)] - (2x - y + 4)$
   c. $\{[4x - (5xy - 2y) + 3xy] - (2x - 3y)\} - (8xy + y)$
   d. $[3x - (2y + 5) - (4x - 3y + 1)] - (3x - 2y - 4)$
   e. $(3a - 2b) - [(2a + 3b) - (9a + b)]$
   f. $4x - 2y - [3x - (5x + 7y)]$
   g. $(7x - 3) - 5x - [(x - 4) - (3x - 1)]$
   h. $\{[(2x + 8y) - 4y] - (-6x - 5y)\}$

   Solution to **d**:
   $$[3x - (2y + 5) - (4x - 3y + 1)] - (3x - 2y - 4)$$
   $$= [3x - 2y - 5 - 4x + 3y - 1] - 3x + 2y + 4$$
   $$= 3x - 2y - 5 - 4x + 3y - 1] - 3x + 2y + 4$$
   $$= 3x - 4x - 3x - 2y + 3y + 2y - 5 - 1 + 4$$
   $$= (3 - 4 - 3)x + (-2 + 3 + 2)y + (-5 - 1 + 4)$$
   $$= -4x + 3y - 2$$

6. Find the product of each given pair of expressions.
   a. $x + 1$, $y - 3$ b. $3x - 2$, $4y + 7$
   c. $\frac{1}{2}x + \frac{2}{3}y - 1$, $-\frac{4}{5}z + \frac{2}{7}a - \frac{3}{4}$ d. $(x - 2y + 3)$, $(-2x + 5y)$
   e. $3ax - 4by + 2z - 8$, $5c + \frac{3}{2}w - 10$

   Solution to **b**:
   $$(3x - 2)(4y + 7) = 3x(4y + 7) - 2(4y + 7)$$
   $$= 12xy + 21x - 8y - 14$$

7. Divide the first expression by the second expression in each of the following.
   a. $2x+6y$, $4x+2y$
   b. $(x-3)(y+5)$, $(y+5)(x-3)$
   c. $(x+2y)(x+2y)$, $(x+2y)(x-y)$
   d. $(3x-6y)(4x+12y)$, $(x-2y)(x+3y)$
   e. $(2x+8y)(3x+y)$, $(x-4y)(6x+2y)$

2.2 EXPONENTS

In Section 1.1 we used the notation a^2 and b^2 in proving that $\sqrt{3}$ is an irrational number. The assumption was made that the reader was familiar with the rudiments of powers and roots of real numbers. In this section and the next we treat these topics in some detail.

DEFINITION 2.1. Let a be a real number. We define $a^1 = a$, $a^2 = a \cdot a$, $a^3 = a \cdot a \cdot a$ and so on. So, if m is a positive integer, we define a^m as the product of m factors, each of which is equal to a.

Example 1
$$2^3 = 2 \cdot 2 \cdot 2 = 8, \quad 3^2 = 3 \cdot 3 = 9, \quad 3^4 = 3 \cdot 3 \cdot 3 \cdot 3 = 81$$

$$(-2)^3 = (-2)(-2)(-2) = -8, \quad \left(\frac{1}{2}\right)^3 = \left(\frac{1}{2}\right)\left(\frac{1}{2}\right)\left(\frac{1}{2}\right) = \frac{1}{8}$$

$$\left(-\frac{2}{3}\right)^4 = \left(-\frac{2}{3}\right)\left(-\frac{2}{3}\right)\left(-\frac{2}{3}\right)\left(-\frac{2}{3}\right) = \frac{16}{81}$$

We refer to a^m as the mth *power of a*; a is called the *base* and m is called the *exponent* of a^m.

In the next two subsections we develop five properties of exponents. In each case we motivate the property with examples prior to stating the property and then illustrate the property with additional examples. Throughout each subsection, a and b will be real numbers and m and n will be positive integers.

The Same Base and Different Exponents

Example 2
$$2^3 \cdot 2^2 = (2 \cdot 2 \cdot 2)(2 \cdot 2) = 2 \cdot 2 \cdot 2 \cdot 2 \cdot 2 = 2^5$$

$$\left(\frac{1}{3}\right)\left(\frac{1}{3}\right)^3 = \left(\frac{1}{3}\right)^1\left(\frac{1}{3}\right)^3 = \left(\frac{1}{3}\right)\left(\frac{1}{3} \cdot \frac{1}{3} \cdot \frac{1}{3}\right) = \frac{1}{3} \cdot \frac{1}{3} \cdot \frac{1}{3} \cdot \frac{1}{3} = \left(\frac{1}{3}\right)^4$$

$$\left(-\frac{1}{5}\right)^2\left(-\frac{1}{5}\right)^4 = \left[\left(-\frac{1}{5}\right)\left(-\frac{1}{5}\right)\right]\left[\left(-\frac{1}{5}\right)\left(-\frac{1}{5}\right)\left(-\frac{1}{5}\right)\left(-\frac{1}{5}\right)\right]$$

SEC. 2.2 Exponents

$$= \left(-\frac{1}{5}\right)\left(-\frac{1}{5}\right)\left(-\frac{1}{5}\right)\left(-\frac{1}{5}\right)\left(-\frac{1}{5}\right)\left(-\frac{1}{5}\right) = \left(-\frac{1}{5}\right)^6$$

Notice that $2^3 \cdot 2^2 = 2^{3+2}$, $\left(\frac{1}{3}\right)^1 \left(\frac{1}{3}\right)^3 = \left(\frac{1}{3}\right)^{1+3}$, and $\left(-\frac{1}{5}\right)^2 \left(-\frac{1}{5}\right)^4 = \left(-\frac{1}{5}\right)^{2+4}$.

These facts illustrate the fact that in order to *multiply* two powers of the same base, one simply adds the exponents.

PROPERTY 1. $a^m \cdot a^n = a^{m+n}$.

Example 3

$(51)^{11} \cdot (51)^{14} = (51)^{11+14} = (51)^{25}$

$\left(-\frac{1}{15}\right)^9 \left(-\frac{1}{15}\right)^3 = \left(-\frac{1}{15}\right)^{9+3} = \left(-\frac{1}{15}\right)^{12}$

$\left(17\frac{1}{2}\right)\left(17\frac{1}{2}\right)^3 = \left(17\frac{1}{2}\right)^{1+3} = \left(17\frac{1}{2}\right)^4$

$a^5 \cdot a^7 = a^{5+7} = a^{12}$

$b^m \cdot b^3 = b^{m+3}$

QUOTIENTS

Example 4

$\dfrac{2^3}{2^2} = \dfrac{2 \cdot 2 \cdot 2}{2 \cdot 2} = 2 = 2^1$

$\dfrac{(-5)^2}{(-5)^5} = \dfrac{(-5)(-5)}{(-5)(-5)(-5)(-5)(-5)} = \dfrac{1}{(-5)(-5)(-5)} = \dfrac{1}{(-5)^3}$

$\dfrac{(7/5)^2}{(7/5)} = \dfrac{(7/5)(7/5)}{7/5} = 7/5 = (7/5)^1$

$\dfrac{(23)^{10}}{(23)^{10}} = 1$

Notice that

$$\dfrac{2^3}{2^2} = 2^{3-2}, \quad \dfrac{(-5)^2}{(-5)^5} = \dfrac{1}{(-5)^{5-2}}, \quad \dfrac{(7/5)^2}{(7/5)^1} = (7/5)^{2-1}$$

and

$$\dfrac{(23)^{10}}{(23)^{10}} = (23)^{10-10}$$

So, in order to *divide* one power of a by another power of a, we first determine which power has the larger exponent. If the numerator has the larger exponent, then we obtain a to the difference of the exponents. If the denominator has the larger exponent, then we obtain the reciprocal of a to the difference of the exponents. If the exponents are the same, we obtain simply 1. More precisely:

PROPERTY 2. If $a \neq 0$, $\dfrac{a^m}{a^n} = \begin{cases} a^{m-n} & \text{if } m > n \\ \dfrac{1}{a^{n-m}} & \text{if } m < n \\ 1 & \text{if } m = n \end{cases}$

Example 5

$\dfrac{2^5}{2^2} = 2^{5-2} = 2^3 = 8$

$\dfrac{2^2}{2^5} = \dfrac{1}{2^{5-2}} = \dfrac{1}{2^3} = \dfrac{1}{8}$

$\dfrac{(-1/2)^7}{(-1/2)^4} = \left(-\dfrac{1}{2}\right)^{7-4} = \left(-\dfrac{1}{2}\right)^3 = -\dfrac{1}{8}$

$\dfrac{(-1/2)^4}{(-1/2)^7} = \dfrac{1}{(-1/2)^{7-4}} = \dfrac{1}{(-1/2)^3} = \dfrac{1}{-1/8} = -8$

$\dfrac{a^{250}}{a^{221}} = a^{250-221} = a^{29}$

POWERS

Example 6

$(3^2)^3 = (3^2)(3^2)(3^2) = (3 \cdot 3)(3 \cdot 3)(3 \cdot 3)$
$= 3 \cdot 3 \cdot 3 \cdot 3 \cdot 3 \cdot 3 = 3^6$

$[(-7)^3]^4 = [(-7)^3][(-7)^3][(-7)^3][(-7)^3]$
$= [(-7)(-7)(-7)][(-7)(-7)(-7)][(-7)(-7)(-7)][(-7)(-7)(-7)]$
$= (-7)^{12}$

Notice that $(3^2)^3 = 3^{2 \cdot 3}$ and $[(-7)^3]^4 = (-7)^{3 \cdot 4}$. These facts illustrate that, in order to take a *power* of a power, one multiplies the exponents. More precisely:

PROPERTY 3. $(a^m)^n = a^{m \cdot n}$

Example 7

$(5^3)^7 = 5^{3 \cdot 7} = 5^{21}$

$\left[\left(-\dfrac{3}{2}\right)^{51}\right]^{11} = \left(-\dfrac{3}{2}\right)^{51 \cdot 11} = \left(-\dfrac{3}{2}\right)^{561}$

$(a^5)^{21} = a^{5 \cdot 21} = a^{105}$

$(b^5)^n = b^{5 \cdot n} = b^{5n}$

SEC. 2.2   Exponents

Different Bases and the Same Exponent

PRODUCT

Example 8
$$2^2 \cdot 3^2 = (2 \cdot 2)(3 \cdot 3) = (2 \cdot 3)(2 \cdot 3) = (2 \cdot 3)^2$$
$$5^3 \cdot 4^3 = (5 \cdot 5 \cdot 5)(4 \cdot 4 \cdot 4) = (5 \cdot 4)(5 \cdot 4)(5 \cdot 4) = (5 \cdot 4)^3$$

That $2^2 \cdot 3^2 = (2 \cdot 3)^2$ and $5^3 \cdot 4^3 = (5 \cdot 4)^3$ illustrate the fact that, to *multiply* two powers which have the same exponent, we multiply the bases and use the same exponent. More precisely:

PROPERTY 4. $a^m \cdot b^m = (a \cdot b)^m$

Example 9
$$11^{10} \cdot 9^{10} = (11 \cdot 9)^{10} = (99)^{10}$$
$$(-8)^4 \cdot (-7)^4 = [(-8)(-7)]^4 = 56^4$$
$$a^{17} \cdot 6^{17} = (a \cdot 6)^{17} = (6a)^{17}$$
$$(156^m) \cdot (2^m) = (156 \cdot 2)^m = 312^m$$

QUOTIENTS

Example 10
$$\frac{6^3}{5^3} = \frac{6 \cdot 6 \cdot 6}{5 \cdot 5 \cdot 5} = \left(\frac{6}{5}\right)^3$$
$$\frac{(-3)^4}{4^4} = \frac{(-3)(-3)(-3)(-3)}{4 \cdot 4 \cdot 4 \cdot 4} = \frac{-3}{4} \cdot \frac{-3}{4} \cdot \frac{-3}{4} \cdot \frac{-3}{4} = \left(\frac{-3}{4}\right)^4$$

That $\frac{6^3}{5^3} = \left(\frac{6}{5}\right)^3$ and $\frac{(-3)^4}{4^4} = \left(\frac{-3}{4}\right)^4$ illustrate the fact that, to *divide* one power by another power which has the same exponent, we divide the bases and use the same exponent. More precisely:

PROPERTY 5.   If $b \neq 0$, $\dfrac{a^m}{b^m} = \left(\dfrac{a}{b}\right)^m$

Example 11
$$\frac{12^5}{6^5} = \left(\frac{12}{6}\right)^5 = 2^5 = 32$$
$$\frac{(-8)^3}{2^3} = \left(\frac{-8}{2}\right)^3 = (-4)^3 = -64$$
$$\frac{456^{23}}{249^{23}} = \left(\frac{456}{249}\right)^{23}$$

$$\frac{a^5}{b^5} = \left(\frac{a}{b}\right)^5$$

$$\frac{a^{39}}{5^{39}} = \left(\frac{a}{5}\right)^{39}$$

The five properties of exponents developed to this point are really the most important properties powers of real numbers possess. Unfortunately, these properties are only expressed for exponents which are positive integers. The remainder of this section will be devoted to extending the notion of powers to include arbitrary real numbers as exponents. To make this extension, however, we must restrict our attention to positive bases. The extension process goes through the steps: negative and zero exponents, fractional exponents, and, finally, arbitrary exponents. The final subsection of this section will deal with powers of algebraic expressions.

Negative and Zero Exponents

Property 2 listed above is somewhat cumbersome since one must consider three possibilities. We proceed now to give definitions for a^0 and a^{-m}, where a is a real number different from 0 and m is a positive integer. These definitions will allow us to state Property 2 more concisely, and they will enable us to perform computations on expressions involving various combinations of positive, negative, and zero exponents.

DEFINITION 2.2. If a is a real number and $a \neq 0$, then a^0 is defined by the equation
$$a^0 = 1$$

The symbol 0^0 is not defined.

DEFINITION 2.3. If a is a real number, $a \neq 0$, and m is a positive integer, then a^{-m} is defined by the equation
$$a^{-m} = \frac{1}{a^m}$$

Example 12 $\quad 5^0 = 1, \quad (-17)^0 = 1, \quad (1{,}000)^0 = 1, \quad \left(\frac{21}{97}\right)^0 = 1$

Example 13 $\quad 5^{-2} = \frac{1}{5^2} = \frac{1}{25}$

$(-17)^{-1} = \frac{1}{(-17)^1} = \frac{1}{-17} = -\frac{1}{17}$

SEC. 2.2 Exponents

$$(56)^{-5} = \frac{1}{(56)^5}$$

$$a^{-29} = \frac{1}{a^{29}} \quad \text{if } a \neq 0$$

In light of Definitions 2.2 and 2.3, Property 2 may be restated as

$$\frac{a^m}{a^n} = a^{m-n} \quad \text{if } a \neq 0$$

Example 14

$$\frac{3^2}{3^5} = 3^{2-5} = 3^{-3}$$

$$\frac{-6}{(-6)^8} = (-6)^{1-8} = (-6)^{-7}$$

$$\frac{a^9}{a^{13}} = a^{9-13} = a^{-4}$$

Fractional Exponents

What meaning should be given to a fractional power of a real number a? For example, what should $a^{1/2}$ be? If such a power is to satisfy Property 1, then we must have

$$(a^{1/2})(a^{1/2}) = a^{1/2 + 1/2} = a^1 = a$$

Looking at the matter from another direction, if such a power is to satisfy Property 3, then we must have

$$(a^{1/2})^2 = a^{(1/2) \cdot 2} = a^1 = a$$

Hence we would like for $a^{1/2}$ to be a number whose square is a. It also happens that, for each positive real number a, there is a positive real number b with the property that $b^2 = a$. Notice also that if $b^2 = a$, then also $(-b)^2 = b^2 = a$. These observations lead to the next definition.

DEFINITION 2.4. Let a be a nonnegative real number. The real number b is said to be a square root of a if and only if $b^2 = a$. Each positive real number has two square roots, equal in absolute value and opposite in sign. The number 0 has only one square root, 0 itself.

Example 15

The square roots of 4 are 2 and -2.
The square roots of 100 are 10 and -10.
The square roots of 1 are 1 and -1.
The only square root of 0 is 0.

Important Note. Negative real numbers do not have square roots within the real number system. Recall that in Section 1.7 we observed that square roots of negative real numbers are imaginary numbers. It is possible to make sense out of certain fractional powers of negative numbers, but the complications involved make the project hardly worth the effort in an elementary text such as this one. It is also possible to treat positive fractional powers of 0, but then one must add disclaimers when dealing with arbitrary fractional powers. Because of these problems, we shall only define fractional powers of positive real numbers. Finally, when we extend the rules of operation to arbitrary powers of variables and algebraic expressions, we shall insist that all bases be positive in any applications of computational results.

DEFINITION 2.5. If a is a positive real number, then $a^{1/2}$ is defined to be the *positive real number whose square is a.* (For those who are more familiar with the "radical" notation $\sqrt{a}$, it will be treated in the next section.)

Example 16

$4^{1/2} = 2$ since $2^2 = 4$

$100^{1/2} = 10$ since $10^2 = 100$

$\left(\dfrac{1}{9}\right)^{1/2} = \dfrac{1}{3}$ since $\left(\dfrac{1}{3}\right)^2 = \dfrac{1}{9}$

$-(4^{1/2}) = -2$

$(-4)^{1/2}$ is not defined

NOTE: It is not true that $4^{1/2} = \pm 2$, even though both $+2$ and -2 are square roots of 4. We have defined $4^{1/2}$ to be the *positive* real number whose square is 4. Hence $4^{1/2} = 2$, and that's that!

To generalize the notion of a root, if m is a positive integer, b is an *m*th *root of a* if $b^m = a$. It can be shown that every positive real number a has exactly one positive *m*th root. If m is even, then a does not have a negative real *m*th root (since an odd power of a negative number is negative). In view of these facts, we make the following definition.

DEFINITION 2.6. Let a be a positive real number, and let m be a positive integer. The *principal* mth *root of a*, denoted $a^{1/m}$, is the positive real number whose mth power is a. Hence we have the important equation $(a^{1/m})^m = a$.

Example 17

$(16)^{1/4} = 2$ since $(2)^4 = 16$

$(64)^{1/3} = 4$ since $(4)^3 = 64$

$(100)^{1/2} = 10$ since $(10)^2 = 100$

SEC. 2.2   Exponents

$$\left(\frac{1}{27}\right)^{1/3} = \frac{1}{3} \quad \text{since } \left(\frac{1}{3}\right)^3 = \frac{1}{27}$$

$$(1)^{1/10} = 1 \quad \text{since } (1)^{10} = 1$$

We now extend our definition to include all fractions (with integers as numerator and denominator) as exponents.

DEFINITION 2.7. Let m and n be positive integers and let a be a positive real number. We define $a^{m/n}$ by the equation

$$a^{m/n} = (a^{1/n})^m$$

We also define $a^{-m/n}$ by the equation

$$a^{-m/n} = \frac{1}{a^{m/n}}$$

Example 18

$$16^{3/4} = (16^{1/4})^3 = 2^3 = 8$$

$$(27)^{2/3} = (27^{1/3})^2 = 3^2 = 9$$

$$\left(\frac{1}{25}\right)^{3/2} = \left[\left(\frac{1}{25}\right)^{1/2}\right]^3 = \left(\frac{1}{5}\right)^3 = \frac{1}{125}$$

$$(8)^{-2/3} = \frac{1}{8^{2/3}} = \frac{1}{(8^{1/3})^2} = \frac{1}{2^2} = \frac{1}{4}$$

It so happens that $(a^{1/n})^m = (a^m)^{1/n}$. Hence we can calculate $a^{m/n}$ in two different ways. The approach using $(a^{1/n})^m$ is almost always easier since one deals with smaller numbers. This fact is illustrated in the next example.

Example 19   Evaluate $(125)^{4/3}$.

Solution: If we first raise 125 to the fourth power, we obtain 244,140,625, and then extract the cube root, we obtain 625. In symbols

$$(125)^{4/3} = (125^4)^{1/3} = (244,140,625)^{1/3} = 625$$

Whereas, if we use Definition 2.7, we have

$$(125)^{4/3} = (125^{1/3})^4 = 5^4 = 625$$

Yet another solution would be

$$(125)^{4/3} = (125^{1/3})^4 = (125^{1/3})^3(125^{1/3}) = (125)(5) = 625$$

If the student uses a hand calculator to evaluate $(125)^{4/3}$, he or she might simply divide 4 by 3 and raise 125 to this power using the function y^x.

Arbitrary Exponents

Now that we have introduced the definitions of various powers of positive real numbers up to and including fractional (rational) powers, we simply mention that the notion can be extended to enable us to take powers such as $5^{\sqrt{2}}$ and $11^{-\sqrt{3}}$. That is, we can take a positive real number a and raise it to any real-number power we like. For those with a hand calculator having the functions $\sqrt{x}$ and y^x, you might calculate $5^{\sqrt{2}} = 6.70499$ (rounded to five decimal places) and $11^{-\sqrt{3}} = 0.01571$ (rounded to five decimal places). We emphasize that these are only approximations.

Finally, if the notion of powers is extended so that we can take arbitrary powers of positive real numbers, then the five properties of exponents remain valid.

Example 20

$$3^{\sqrt{2}} \cdot 3^4 = 3^{\sqrt{2}+4}$$

$$\frac{a^{\sqrt{11}}}{a^3} = a^{\sqrt{11}-3}$$

$$(15^{\sqrt{2}})^{\sqrt{2}} = 15^{\sqrt{2}\cdot\sqrt{2}} = 15^2 = 225$$

$$9^{\sqrt{3}} \cdot \sqrt{2}^{\sqrt{3}} = (9 \cdot \sqrt{2})^{\sqrt{3}}$$

$$\frac{a^{\sqrt{11}}}{b^{\sqrt{11}}} = \left(\frac{a}{b}\right)^{\sqrt{11}}$$

Exponents for Algebraic Expressions

As one might expect, the properties of exponents can be used to develop rules of operation for powers of algebraic expressions.

Example 21

$$(x+1)^{1/2}(x+1)^{3/2} = (x+1)^{1/2+3/2} = (x+1)^2$$

$$\frac{(3x-4)^{5/2}}{(3x-4)^{9/2}} = (3x-4)^{5/2-9/2} = (3x-4)^{-2} = \frac{1}{(3x-4)^2}$$

$$[(x+2y)^{\sqrt{2}}]^3 = (x+2y)^{3\sqrt{2}}$$

$$(5x-2xy+y)^5(7x+16)^5 = [(5x-2xy+y)(7x+16)]^5$$

$$\frac{(11x-15y)^{1/2}}{(3x-y)^{1/2}} = \left(\frac{11x-15y}{3x-y}\right)^{1/2}$$

We summarize this entire section with the following table of rules of operation.

SEC. 2.2 Exponents

Rules of Operation for Exponents

> Let A and B be positive real numbers, variables, or algebraic expressions, and let r and s be real numbers.
>
> RULE 2.1. $A^r \cdot A^s = A^{r+s}$
>
> RULE 2.2. $\dfrac{A^r}{A^s} = A^{r-s}$
>
> RULE 2.3. $(A^r)^s = A^{r \cdot s}$
>
> RULE 2.4. $A^r \cdot B^r = (A \cdot B)^r$
>
> RULE 2.5. $\dfrac{A^r}{B^r} = \left(\dfrac{A}{B}\right)^r$
>
> To apply these rules to real numbers, we insist that $A > 0$ and $B > 0$ when the specific values are substituted for the variables.

Exercises 2.2

1. Evaluate:

a. 5^{-2}
b. $(3.4)^0$
c. $\left(\dfrac{2}{3}\right)^{-1}$
d. $\left(\dfrac{1}{3}\right)^{-2}$
e. $(-2)^{-4}$

f. $(3^{-1} - 4^{-1})^{-1}$
g. -4^2
h. $16^{1/2}$
i. $8^{1/3}$
j. $8^{-2/3}$

k. $(25)^{3/2}$
l. $(-37)^0$
m. $\left(\dfrac{1}{32}\right)^{3/5}$
n. $(64)^{5/3}$

Solution to **c**: $\left(\dfrac{2}{3}\right)^{-1} = \dfrac{1}{\left(\dfrac{2}{3}\right)^1} = \dfrac{1}{\dfrac{2}{3}} = \dfrac{3}{2}$

2. Making use of Rules 2.1 through 2.5 and definitions, perform the indicated operations and leave all answers free of negative exponents.

a. $3^2 \cdot 3^4$
b. $x^3 \cdot x^5$
c. $2^3 \cdot 2^{-5} \cdot 2^6$
d. $a^2 \cdot a^7 \cdot a^{-4}$

e. $\dfrac{x^4 \cdot x^{-3}}{x^{-2}}$
f. $7^{-3} \cdot 7^3$
g. $(2^2)^3$
h. $(a^{-1} b^2)^{-3}$

i. $(x^{-1} \cdot y^{3/2})^{-2}$ j. $(2 \cdot 3)^3 (2 \cdot 1)^3$ k. $\dfrac{(63)^{1/2}}{(7)^{1/2}}$ l. $\left(\dfrac{x^2 y^{-1}}{2^5}\right)\left(\dfrac{x^{-3} z^{-1}}{y}\right)$

3. Express each of the following in the simplest form you can and free of negative exponents.

   a. $2^3 \cdot 3^{-2}$ b. $4x^{-3}$ c. $\dfrac{5}{2x^{-2}}$ d. $\dfrac{3x^{-3} y^4}{2x^{-2} y^{-2}}$

   e. $\dfrac{5^{-1} \cdot 3^{-2}}{4^{-2} + 7}$ f. $[(9)^{-2}(3^3)]^2$

2.3 RADICALS

A very common notion which is used in dealing with (principal) roots of real numbers is *radical notation*. Again, we deal only with positive real numbers.

DEFINITION 2.8. Let a be a positive real number. We define $\sqrt{a}$ by the equation

$$\sqrt{a} = a^{1/2}$$

and, for a positive integer n, we define $\sqrt[n]{a}$ by the equation

$$\sqrt[n]{a} = a^{1/n}$$

The symbol $\sqrt[n]{a}$ is called a *radical*; n is called the *index* and a is called the *radicand*.

We read $\sqrt{a}$ as "the square root of a" and $\sqrt[n]{a}$ as "the nth root of a." We mean, of course, the principal square root and the principle nth root, respectively. We could always use $\sqrt[2]{a}$ in place of $\sqrt{a}$, but mathematical tradition causes us to follow the time-honored practice of treating square roots as being somehow special.

Example 1 $\sqrt{25} = 5$, $\sqrt[4]{16} = 2$, $\sqrt[3]{27} = 3$, $\sqrt{\dfrac{1}{4}} = \dfrac{1}{2}$

Some of the properties of radicals are isolated as follows:

PROPERTY 1. $(\sqrt[n]{a})^n = a$

SEC. 2.3 Radicals

Example 2 $(\sqrt[3]{2})^3 = 2, \quad \left(\sqrt[11]{\dfrac{3}{4}}\right)^{11} = \dfrac{3}{4}, \quad (\sqrt{2{,}155})^2 = 2{,}155$

PROPERTY 2. $\sqrt[n]{a}\,\sqrt[n]{b} = \sqrt[n]{ab}$

Example 3
$\sqrt[2]{3}\,\sqrt[2]{3} = \sqrt[2]{9} = 3$

$\sqrt[3]{2}\,\sqrt[3]{4} = \sqrt[3]{8} = 2$

$\sqrt[5]{11}\,\sqrt[5]{6} = \sqrt[5]{66}$

$\sqrt[11]{22} = \sqrt[11]{(2)(11)} = \sqrt[11]{2}\,\sqrt[11]{11}$

$\sqrt[4]{-3}\,\sqrt[4]{5}$ is not defined since $\sqrt[4]{-3}$ is not defined

PROPERTY 3. $\dfrac{\sqrt[n]{a}}{\sqrt[n]{b}} = \sqrt[n]{\dfrac{a}{b}}$

Example 4
$\dfrac{\sqrt{8}}{\sqrt{2}} = \sqrt{\dfrac{8}{2}} = \sqrt{4} = 2$

$\dfrac{\sqrt[5]{11}}{\sqrt[5]{5}} = \sqrt[5]{\dfrac{11}{5}}$

$\sqrt[3]{\dfrac{5}{8}} = \dfrac{\sqrt[3]{5}}{\sqrt[3]{8}} = \dfrac{\sqrt[3]{5}}{2}$

$\dfrac{\sqrt[6]{-2}}{\sqrt[6]{-4}}$ is not defined

As in the case of exponents, we develop rules of operation for algebraic expressions involving radicals by using the properties of radicals.

Example 5 $(\sqrt{x + 2xy - 3})^2 = x + 2xy - 3$

$\sqrt[3]{3x - 2y + a}\,\sqrt[3]{4 - z} = \sqrt[3]{(3x - 2y + a)(4 - z)}$

$\sqrt[5]{\dfrac{ax - by + z}{3xy + 4yz}} = \dfrac{\sqrt[5]{ax - by + z}}{\sqrt[5]{3xy + 4yz}}$

Rules of Operation for Radicals

Let A and B be positive real numbers, variables, or algebraic expressions and let n be a positive integer.

RULE 2.6. $(\sqrt[n]{A})^n = A$

RULE 2.7. $\sqrt[n]{A}\sqrt[n]{B} = \sqrt[n]{AB}$

RULE 2.8. $\dfrac{\sqrt[n]{A}}{\sqrt[n]{B}} = \sqrt[n]{\dfrac{A}{B}}$

RULE 2.9. $\sqrt[n]{A^m} = (\sqrt[n]{A})^m$

To apply these rules to real numbers, we insist that $A > 0$ and $B > 0$ when the specific values are substituted for the variables.

Exercises 2.3

1. Evaluate:
   a. $\sqrt{49}$
   b. $\sqrt[5]{32}$
   c. $\sqrt{256}$
   d. $\sqrt{(-5)^2}$
   e. $\sqrt{3^2 + 4^2}$
   f. $\sqrt{\dfrac{4}{9}}$
   g. $\sqrt[3]{(27)^2}$
   h. $\sqrt{(17+21)^2}$
   i. $\sqrt[3]{\dfrac{1}{8}}$
   j. $\sqrt{7^4}$
   k. $\sqrt{4^7}$
   l. $\sqrt{(7-11)^2}$
   m. $\sqrt[4]{81}$
   n. $\sqrt{121}$

   Solution to g: If we use the fact that $\sqrt[3]{(27)^2} = [(27)^2]^{1/3}$, then we obtain $\sqrt[3]{(27)^2} = [(27)^2]^{1/3} = (27^{1/3})^2 = 3^2 = 9$.

2. By writing each number under a radical as the product of two factors and then applying Rule 2.7, simplify each of the following.
   a. $\sqrt{18}$
   b. $\sqrt{32}$
   c. $\sqrt[3]{16}$
   d. $\sqrt[3]{54}$
   e. $\sqrt{200}$
   f. $\sqrt{75}$
   g. $\sqrt{160}$
   h. $\sqrt[3]{128}$
   i. $\sqrt[3]{24}$
   j. $\sqrt{98}$
   k. $\sqrt{68}$
   l. $\sqrt{800}$

   Solution to a: $\sqrt{18} = \sqrt{9\cdot 2} = \sqrt{9}\sqrt{2} = 3\sqrt{2}$

3. Reduce each of the following to as simple a form as you can:
   a. $\sqrt{3}\sqrt{27}$
   b. $\sqrt{2}\sqrt{3}\sqrt{6}$
   c. $\sqrt{3}\sqrt{15}\sqrt{5}$
   d. $\sqrt{6}\sqrt{18}$
   e. $\sqrt{3} + \sqrt{12}$
   f. $\sqrt{20} + \sqrt{45}$
   g. $\sqrt{27} + \sqrt{48} - \sqrt{75}$
   h. $\sqrt{32} - \sqrt{128} + \sqrt{72}$

SEC. 2.3 Radicals

Solution to e:
$$\sqrt{3}+\sqrt{12} = \sqrt{3}+\sqrt{4\cdot 3}$$
$$= \sqrt{3}+\sqrt{4}\sqrt{3}$$
$$= \sqrt{3}+2\sqrt{3}$$
$$= (1+2)\sqrt{3}$$
$$= 3\sqrt{3}$$

4. Simplify each of the following, leaving the denominators free of radicals.

a. $\dfrac{\sqrt{8}}{\sqrt{2}}$ b. $\dfrac{\sqrt{18}}{\sqrt{2}}$ c. $\dfrac{1}{\sqrt[3]{2}}$ d. $\sqrt{\dfrac{7}{16}}$ e. $\sqrt{\dfrac{18}{25}}$

f. $\sqrt{\dfrac{3}{8}}$ g. $\dfrac{\sqrt{x^7}}{\sqrt[3]{x^3}}$ h. $\dfrac{\sqrt[3]{16}}{\sqrt[3]{32}}$ i. $\dfrac{\sqrt{6}+\sqrt{24}}{\sqrt{2}}$

Solution to d: $\sqrt{\dfrac{7}{16}} = \dfrac{\sqrt{7}}{\sqrt{16}} = \dfrac{\sqrt{7}}{4}$

5. Express each of the following in equivalent form using radicals instead of fractional exponents.

a. $(3x)^{1/2}$ b. $x^{3/2}$ c. $x^{4/7}$ d. $x^{-2/3}$ e. $(x-1)^{2/5}$ f. $(4+x^2)^{-1/2}$

Solution to d: $x^{-2/3} = \dfrac{1}{x^{2/3}} = \dfrac{1}{(x^{1/3})^2} = \dfrac{1}{(\sqrt[3]{x})^2}$

or $\quad x^{-2/3} = \dfrac{1}{x^{2/3}} = \dfrac{1}{(x^2)^{1/3}} = \dfrac{1}{\sqrt[3]{x^2}}$

Either solution is correct.

6. Rewrite each of the following using fractional exponents instead of radicals.

a. $\sqrt{x^3}$ b. $4/\sqrt{x}$ c. $\sqrt[4]{x^3}$ d. $\dfrac{4}{\sqrt[3]{x}}$ e. $\sqrt[3]{(4+x)^2}$ f. $\dfrac{1}{\sqrt[5]{x+y^2}}$

Solution to a: $\sqrt{x^3} = (x^3)^{1/2} = x^{3/2}$

7. a. Show that $\sqrt{\dfrac{1}{2}} = \dfrac{\sqrt{2}}{2}$. $\left(\text{Hint: } \sqrt{\dfrac{1}{2}} = \sqrt{\dfrac{1\cdot 2}{2\cdot 2}} = \sqrt{\dfrac{2}{4}}\right)$

b. By multiplying numerator and denominator by $\sqrt{2}+1$, show that

$$\dfrac{1}{\sqrt{2}-1} = \sqrt{2}+1$$

c. In the same manner as for part **b**, show that

$$\frac{4}{3-\sqrt{5}} = 3 + \sqrt{5}$$

2.4 POLYNOMIALS IN ONE VARIABLE

Among the most common types of algebraic expressions are the polynomials in one variable. Example 1 includes several such algebraic expressions.

Example 1
$$3x^2 - 2x + 1, \quad x^3 = 1, \quad 5x^3 + 4x^2 - 3x$$
$$\frac{2}{3}x^5 - 2x^4 + \frac{11}{15}x^2 - \frac{7}{9}$$

If a particular power of the variable is missing, we may always insert it with a coefficient of zero.

Example 2
$$x^3 - 1 = x^3 + 0 \cdot x^2 + 0 \cdot x - 1$$
$$5x^3 + 4x^2 - 3x = 5x^3 + 4x^2 - 3x + 0$$
$$\frac{2}{3}x^5 - \sqrt{2x^4} + \frac{11}{15}x^2 - \frac{7}{9} = \frac{2}{3}x^5 - \sqrt{2x^4} + 0 \cdot x^3 + \frac{11}{15}x^2 + 0 \cdot x - \frac{7}{9}$$

These examples motivate our next definition.

DEFINITION 2.9. A *polynomial in the variable x* is an algebraic expression which can be written in the form

$$A_n x^n + A_{n-1} x^{n-1} + \cdots + A_1 x + A_0$$

where the A's are numbers belonging to some number system (ordinarily the real number system) and n is a positive integer or zero.

If $A_n \neq 0$, the polynomial $A_n x^n + A_{n-1} x^{n-1} + \cdots + A_1 x + A_0$ is said to be of *degree n*, and we say that it is an *nth-degree polynomial*. If $n = 0$ and $A_0 \neq 0$, the polynomial consists of the number A_0 and its degree is 0. The polynomial consisting of just the number 0 does not have a degree, and it is called the *zero polynomial*. The A's are called *coefficients*; A_i is called the *ith coefficient* for each i between 0 and n; A_n is called the *leading coefficient*; and A_0 is called the *constant term*. Each $(A_i x^i)$ is called a *term* of the polynomial.

SEC. 2.4 Polynomials in One Variable

Example 3 For the polynomial $3x^2 - 2x + 1$:

(a) The degree is 2 and we call it a second-degree polynomial or a *quadratic polynomial*.
(b) The coefficients are $A_2 = 3$, $A_1 = -2$, and $A_0 = 1$.
(c) The leading coefficient is 3.
(d) The constant term is 1.
(e) The terms are $3x^2$, $-2x$, and 1.

Example 4 For the polynomial $x^3 - 1$:

(a) The degree is 3 and we call it a third-degree polynomial or a *cubic polynomial*.
(b) The coefficients are $A_3 = 1$, $A_2 = 0$, $A_1 = 0$, and $A_0 = -1$.
(c) The leading coefficient is 1.
(d) The constant term is -1.
(e) The terms are x^3 and -1 (or we could include the other "zero" terms $0 \cdot x^2$ and $0 \cdot x$).

Example 5 For the polynomial $\frac{2}{3}x^7 + \frac{5}{9}x^3 - \frac{1}{7}$:

(a) The degree is 7 and we call it a seventh-degree polynomial.
(b) The coefficients are $A_7 = \frac{2}{3}$, $A_6 = 0$, $A_5 = 0$, $A_4 = 0$, $A_3 = \frac{5}{9}$, $A_2 = 0$, $A_1 = 0$, and $A_0 = -\frac{1}{7}$.
(c) The leading coefficient is $\frac{2}{3}$.
(d) The constant term is $-\frac{1}{7}$.
(e) The terms are $\frac{2}{3}x^7$, $\frac{5}{9}x^3$, and $-\frac{1}{7}$ (or we could include the other "zero" terms $0 \cdot x^6$, $0 \cdot x^5$, etc.).

Addition and Subtraction of Polynomials

The observations and computations made in Section 2.1 make the processes of addition and subtraction of polynomials rather straightforward. Basically, one simply collects like terms and then adds or subtracts coefficients. We illustrate by examples.

Example 6
$$\begin{aligned}(x^2 - 2x + 1) + (2x^2 + x - 5) &= x^2 - 2x + 1 + 2x^2 + x - 5 \\ &= x^2 + 2x^2 - 2x + x + 1 - 5 \\ &= (1 + 2)x^2 + (-2 + 1)x + (1 - 5) \\ &= 3x^2 - x - 4\end{aligned}$$

Another acceptable procedure for adding polynomials is what one might call long-hand addition.

$$\begin{array}{r} x^2 - 2x + 1 \\ +2x^2 + x - 5 \\ \hline 3x^2 - x - 4 \end{array}$$

Example 7

$$\begin{aligned} (3x^2 - x + 4) - (2x^2 - 1) &= 3x^2 - x + 4 - 2x^2 + 1 \\ &= 3x^2 - 2x^2 - x + 4 + 1 \\ &= (3 - 2)x^2 - x + 5 \\ &= x^2 - x + 5 \end{aligned}$$

Or, using long-hand subtraction,

$$\begin{array}{r} 3x^2 - x + 4 \\ -2x^2 - 1 \\ \hline x^2 - x + 5 \end{array}$$

Example 8 Add $x^2 - 5x$, $-3x^2 + 1$, $4x - 3$, and $4x^2 - 8x + 2$.

Solution

$$\begin{aligned} (x^2 - 5x) + (-3x^2 + 1) &+ (4x - 3) + (4x^2 - 8x + 2) \\ &= x^2 - 5x - 3x^2 + 1 + 4x - 3 + 4x^2 - 8x + 2 \\ &= x^2 - 3x^2 + 4x^2 - 5x + 4x - 8x + 1 - 3 + 2 \\ &= (1 - 3 + 4)x^2 + (-5 + 4 - 8)x + (1 - 3 + 2) \\ &= 2x^2 - 9x \end{aligned}$$

Or, using tabular addition,

$$\begin{array}{r} x^2 - 5x \\ -3x^2 + 1 \\ 4x - 3 \\ 4x^2 - 8x + 2 \\ \hline 2x^2 - 9x \end{array}$$

Example 9 Subtract $\dfrac{2}{3}x^2 - \dfrac{4}{5}x + \dfrac{1}{2}$ from $\dfrac{9}{10}x^3 - \dfrac{3}{2}x^2 - \dfrac{1}{10}x + \dfrac{3}{4}$.

Solution

$$\left(\frac{9}{10}x^3 - \frac{3}{2}x^2 - \frac{1}{10}x + \frac{3}{4}\right) - \left(\frac{2}{3}x^2 - \frac{4}{5}x + \frac{1}{2}\right)$$

SEC. 2.4 Polynomials in One Variable

$$= \frac{9}{10}x^3 - \frac{3}{2}x^2 - \frac{1}{10}x + \frac{3}{4} - \frac{2}{3}x^2 + \frac{4}{5}x - \frac{1}{2}$$

$$= \frac{9}{10}x^3 + \left(-\frac{3}{2} - \frac{2}{3}\right)x^2 + \left(-\frac{1}{10} + \frac{4}{5}\right)x + \left(\frac{3}{4} - \frac{1}{2}\right)$$

$$= \frac{9}{10}x^3 + \left(\frac{-9-4}{6}\right)x^2 + \left(\frac{-1+8}{10}\right)x + \left(\frac{3-2}{4}\right)$$

$$= \frac{9}{10}x^3 - \frac{13}{6}x^2 + \frac{7}{10}x + \frac{1}{4}$$

Or, using long-hand subtraction,

$$\begin{array}{cccc} \frac{9}{10}x^3 - \frac{3}{2}x^2 & -\frac{1}{10}x & +\frac{3}{4} \\ -\quad\quad\quad \frac{2}{3}x^2 & -\frac{4}{5}x & +\frac{1}{2} \\ \hline \frac{9}{10}x^3 + \left(-\frac{3}{2}-\frac{2}{3}\right)x^2 & +\left(-\frac{1}{10}+\frac{4}{5}\right)x & +\left(\frac{3}{4}-\frac{1}{2}\right) \end{array}$$

and proceed as above to reduce the coefficients.

Multiplication of Polynomials

Once again, the observations and computations made in Section 2.1, together with the rules for multiplying powers, make the process of multiplication of polynomials rather straightforward. Basically, one multiplies each term of the first polynomial by each term of the second polynomial. We illustrate by examples.

Example 10

$(x + 1)(x - 1) = x(x - 1) + 1(x - 1)$
$\qquad\qquad\quad = x^2 - x + x - 1$
$\qquad\qquad\quad = x^2 - 1$

Or, using long-hand multiplication,

$$\begin{array}{r} x + 1 \\ x - 1 \\ \hline x^2 + x \\ -x - 1 \\ \hline x^2 \quad\quad -1 \end{array}$$

Still another approach is as follows: Multiply the first term of the first expression (that is, x) by each term in the second expression (x and -1), and then multiply the second term in the first expression (that is, 1) by each term in the second expression (again x and -1). The sum of all of these products is the desired result. This process can be abbreviated as follows:

$$(x + 1)(x - 1) = x^2 - x + x - 1$$

where the numbers accompanying the arrows indicate the order in which the products are taken: x times x first, x times -1 second, 1 times x third, and 1 times -1 fourth.

Example 11

$$(x^2 - 2x + 1)(x + 3) = x^2(x + 3) - 2x(x + 3) + 1(x + 3)$$
$$= x^3 + 3x^2 - 2x^2 - 6x + x + 3$$
$$= x^3 + (3 - 2)x^2 + (-6 + 1)x + 3$$
$$= x^3 + x^2 - 5x + 3$$

Or, using long-hand multiplication,

$$\begin{array}{r} x^2 - 2x + 1 \\ x + 3 \\ \hline x^3 - 2x^2 + x \\ + 3x^2 - 6x + 3 \\ \hline x^3 + x^2 - 5x + 3 \end{array}$$

or,

$$(x^2 - 2x + 1)(x + 3) = x^3 + 3x^2 - 2x^2 - 6x + x + 3$$

Example 12

Multiply $x^3 + 2x - 5$ by $3x^2 - 7x + 4$.

Solution

$$(x^3 + 2x - 5)(3x^2 - 7x + 4)$$
$$= x^3(3x^2 - 7x + 4) + 2x(3x^2 - 7x + 4) - 5(3x^2 - 7x + 4)$$

SEC. 2.4 Polynomials in One Variable

$$= 3x^5 - 7x^4 + 4x^3 + 6x^3 - 14x^2 + 8x - 15x^2 + 35x - 20$$
$$= 3x^5 - 7x^4 + (4 + 6)x^3 + (-14 - 15)x^2 + (8 + 35)x - 20$$
$$= 3x^5 - 7x^4 + 10x^3 - 29x^2 + 43x - 20$$

Or, using long-hand multiplication,

$$
\begin{array}{r}
x^3 + 2x - 5 \\
3x^2 - 7x + 4 \\
\hline
3x^5 + 6x^3 - 15x^2 \\
-7x^4 - 14x^2 + 35x \\
4x^3 + 8x - 20 \\
\hline
3x^5 - 7x^4 + 10x^3 - 29x^3 + 43x - 20
\end{array}
$$

Sometimes, usually in multiplication, you may want to "check your answer." One technique for doing this is simply to substitute some real number for x in each polynomial and see if an equality is obtained. For computational purposes the easiest number to substitute is 1, since one obtains simply the sum of the coefficients when 1 is substituted for x in a polynomial in x. Although this checking technique is not foolproof, it usually works.

Example 13

To check our answer for Example 12, we might write as follows:

Check: When $x = 1$,

$$x^3 + 2x - 5 = 1 + 2 - 5 = -2$$
$$3x^2 - 7x + 4 = 3 - 7 + 4 = 0$$

and so

$$(x^3 + 2x - 5)(3x^2 - 7x + 4) = (-2)(0) = 0$$

Also

$$3x^5 - 7x^4 + 10x^3 - 29x^2 + 43x - 20 = 3 - 7 + 10 - 29 + 43 - 20 = 0$$

Therefore, the answer is probably correct.

Division of Polynomials

In attempting to divide one polynomial P by another polynomial D, we seek a third polynomial Q such that $P = D \cdot Q$, for then we have the equation

$$\frac{P}{D} = Q$$

if we divide both sides by D.

Example 14 $\quad \left| \quad \dfrac{x^2 - 1}{x + 1} = x - 1 \quad$ since $x^2 - 1 = (x + 1)(x - 1)$

Example 15 $\quad \left| \quad \dfrac{10x^2 + 29x - 21}{2x + 7} = 5x - 3 \quad$ since $10x^2 + 29x - 21 = (2x + 7)(5x - 3)$

In examples such as Example 15, $10x^2 + 29x - 21$ is called the *dividend*, $2x + 7$ is called the *divisor*, and $5x - 3$ is called the *quotient*. It is desirable to have a method of finding the quotient directly from the dividend and the divisor. Such a method, suggested by the method of long division in arithmetic, is illustrated in the following example:

Example 16 $\quad \Big|\quad$ Divide $10x^2 + 29x - 21$ by $2x + 7$.

Solution: The dividend and divisor, having been written in the order of descending powers of x, are arranged in a manner similar to that used in long division in arithmetic:

$$
\begin{array}{r}
5x \;-\; 3 \\
2x + 7 \,\overline{\big)\, 10x^2 + 29x - 21} \\
10x^2 + 35x \\ \hline
-\,6x - 21 \\
-\,6x - 21 \\ \hline
0x + 0
\end{array}
$$

The first term of the quotient is obtained by dividing the first term of the dividend by the first term of the divisor: $10x^2/2x = 5x$. Next, multiply the entire divisor by this first term of the quotient, placing the resulting terms under the like terms of the dividend, and subtract to obtain the new dividend. The second term of the quotient is now obtained by dividing the first term of the new dividend by the first term of the divisor, as before: $(-6x)/2x = -3$. Multiply the divisor by this second term of the quotient and place it under the like terms of the new dividend. Now subtract to obtain a second new dividend. Since in our example this second new dividend is zero, our solution is complete.

A similar procedure can be used in dividing any polynomial in x by another polynomial. (In arranging the dividend in the order of descending powers of some letter, it may be helpful to supply any missing powers by using zero coefficients for them.) In performing divisions in arithmetic we often have a remainder. Thus $\dfrac{17}{7} = 2 + \dfrac{3}{7}$. Here the quotient is 2 and the remainder is 3. Similarly, in dividing polynomials in x we often have a remainder. In fact, this is always the case unless the divisor is a factor of the dividend, in which case we say the remainder is zero.

SEC. 2.4　Polynomials in One Variable

Example 17　Divide $8x^3 - 16x + 13$ by $2x - 3$.

Solution

$$
\begin{array}{r}
4x^2 + 6x + 1 \\
2x-3 \overline{\smash{\big)}\, 8x^3 + 0x^2 - 16x + 13} \\
\underline{8x^3 - 12x^2} \\
12x^2 - 16x + 13 \\
\underline{12x^2 - 18x} \\
2x + 13 \\
\underline{2x - 3} \\
+16
\end{array}
$$

The quotient is $4x^2 + 6x + 1$ and the remainder is $+16$. This result can be expressed in the form

$$\frac{8x^3 - 16x + 13}{2x - 3} = 4x^2 + 16x + 1 + \frac{16}{2x - 3}$$

We check our result by multiplying the quotient by the divisor and adding the remainder to obtain the dividend:

$$(4x^2 + 6x + 1)(2x - 3) + 16 = 8x^3 - 16x + 13$$

Exercises 2.4

1. In each part, find the sum of the polynomials.
   a. $3x^2 - 4x + 7$, $2x^2 - 5x + 2$
   b. $3x - 2$, $4x + 5$, $5x - 3$
   c. $x^4 + 1$, $x^3 - 1$
   d. $5x^2 - 4x + 3$, $2x^3 - 4$
   e. $11x^7 - 6x^4 + x^3 - 1$, $29x^5 + 47x^2$
   f. $\frac{3}{4}x^3 - \frac{1}{7}x + \frac{2}{5}, \frac{4}{9}x^2 + \frac{2}{3}x - \frac{1}{8}$
   g. $\frac{1}{4}x^4 + \frac{1}{3}x^3 + \frac{1}{2}x^2 + x + 1, \frac{1}{4}x^3 + \frac{1}{3}x^2 + \frac{1}{2}x + 1$
   h. $\sqrt{2x^5} - \sqrt{3x^3}, 3\sqrt{3x^3} + \frac{2}{5}x, \frac{3}{4}x - 7$

   Solution to a: $(3x^2 - 4x + 7) + (2x^2 - 5x + 2)$
   $$= (3 + 2)x^2 + (-4 - 5)x + (7 + 2)$$
   $$= 5x^2 - 9x + 9$$

Or, alternatively,
$$3x^2 - 4x + 7$$
$$+2x^2 - 5x + 2$$
$$\overline{5x^2 - 9x + 9}$$

2. In each part, subtract the first polynomial from the second.
   a. $x^2 - 2x + 1, \ x^2 + 3x - 4$
   b. $x^2 + 3x - 4, \ x^2 - 2x + 1$
   c. $5x^2 + 4x + 3, \ 6x^2 - 12x - 5$
   d. $11x^5 - 6x^9 + 5, \ -5x^7 + 11x^5 - 2$
   e. $9x - 4, \ -5x + 2$
   f. $\frac{1}{2}x + \frac{3}{2}x^2 - 1, \ x^2 - x + 2$
   g. $-\frac{1}{3}x^3 + \frac{3}{5}x - \frac{1}{7}, \frac{5}{9}x - \frac{3}{7}x^3 + \frac{11}{23}x^2 - \frac{17}{19}$
   h. $\sqrt{7x^4} - \frac{6}{7}x^3 + \sqrt{3x} - 2, \ \sqrt{2x^4} + \frac{2}{9}x^3 - \sqrt{3x} + \frac{10}{3}$

Solution to a: $(x^2 + 3x - 4) - (x^2 - 2x + 1)$
$$= x^2 + 3x - 4 - x^2 + 2x - 1$$
$$= (1 - 1)x^2 + (3 + 2)x + (-4 - 1)$$
$$= 0x^2 + 5x - 5$$
$$= 5x - 5$$

Or, alternatively,
$$x^2 + 3x - 4$$
$$-x^2 - 2x + 1$$
$$\overline{5x - 5}$$

3. In each part, find the product and express your answer as a polynomial with descending powers of x.
   a. $(x - 2)(x + 1)$
   b. $(x - 2)(x + 1)(2x - 3)$
   c. $(5x - 3)^2$
   d. $(2x^2 - 3x + 5)(x^2 - 1)$
   e. $(x + 1)^3$
   f. $(x - 1)^3$
   g. $(x + 2)(x^2 - 2x + 4)$
   h. $(x^2 + x + 1)(2x^2 + 4x + 8)$
   i. $\left(\frac{1}{2}x^2 + \frac{2}{3}x - \frac{3}{4}\right)\left(\frac{1}{7}x^2 + \frac{5}{6}x - \frac{2}{3}\right)$
   j. $(x^3 - 4)(x^4 - 5)$

Solution to a: $(x - 2)(x + 1) = x(x + 1) - 2(x + 1)$
$$= x^2 + x - 2x - 2$$
$$= x^2 + (1 - 2)x - 2$$
$$= x^2 - x - 2$$

SEC. 2.4 Polynomials in One Variable

Or, alternatively,

$$
\begin{array}{r}
x - 2 \\
x + 1 \\
\hline
x^2 - 2x \\
x - 2 \\
\hline
x^2 - x - 2
\end{array}
$$

4. Divide, finding quotient and remainder, and leave the result in the form

$$\frac{\text{dividend}}{\text{divisor}} = \text{quotient} + \frac{\text{remainder}}{\text{divisor}}$$

   a. $x^2 - 4x - 5$ by $x + 1$
   b. $8x^2 - 4x$ by $4x$
   c. $3x^2 + 13x - 10$ by $3x - 2$
   d. $2x^2 + 8x - 5$ by $x - 1$
   e. $2x^3 - 3x^2 - 11x + 6$ by $2x - 1$
   f. $x^4 - 16x^3 + 86x^2 - 176x + 105$ by $x - 7$
   g. $-6x^3 + 5x^2 - 4$ by $-2x^2 + 3x - 2$
   h. $2x^3 + 7x^2 + 10x + 6$ by $2x + 3$
   i. $2x^3 + 3x^2 - 2x - 3$ by $x - 2$
   j. $x^4 + 6x^3 - 16x^2 - 150x - 225$ by $x - 5$
   k. $2x^3 + 5x^2 - 3x - 4$ by $2x + 1$
   l. $3x^5 - 7x^4 + 10x^3 - 29x^2 + 43x - 20$ by $3x^2 - 7x + 4$
   m. $8x^3 - 12x^2 + 6x - 65$ by $2x - 5$
   n. $x^5 - 32$ by $x - 2$
   o. $3x^3 - 4x^2 + 2x - 1$ by $2x - 1$
   p. $x^4 + 1$ by $x + 1$

Solution to **d**:

$$
\begin{array}{r}
2x + 10 \\
x - 1 \overline{\smash{)}\,2x^2 + 8x - 5} \\
2x^2 - 2x \\
\hline
10x - 5 \\
10x - 10 \\
\hline
5
\end{array}
$$

So

$$\frac{2x^2 + 8x - 5}{x - 1} = 2x + 10 + \frac{5}{x - 1}$$

2.5 POLYNOMIALS IN TWO OR MORE VARIABLES

By combining the techniques of Sections 2.2 and 2.4, we are able to perform calculations with algebraic expressions which involve sums of various products of powers. These can involve any number of variables and arbitrary constants. Rather than giving a formal definition of polynomials in several variables and formally giving rules of operation for performing calculations with these, we shall illustrate these ideas with several examples.

Addition and Subtraction

Example 1

$$(x + 3y - 2) + (2x + y - 4) = x + 2x + 3y + y - 2 - 4$$
$$= (1 + 2)x + (3 + 1)y + (-2 - 4)$$
$$= 3x + 4y - 6$$

Or, using long-hand addition,

$$\begin{array}{r} x + 3y - 2 \\ +2x + y - 4 \\ \hline 3x + 4y - 6 \end{array}$$

Example 2 Add $x^2 - 5xy$, $-3x^2 + y^2$, $4xy - 3y^2$, and $4x^2 - 8xy$.

Solution

$$(x^2 - 5xy) + (-3x^2 + y^2) + (4xy - 3y^2) + (4x^2 - 8xy)$$
$$= x^2 - 3x^2 + 4x^2 - 5xy + 4xy - 8xy + y^2 - 3y^2$$
$$= (1 - 3 + 4)x^2 + (-5 + 4 - 8)xy + (1 - 3)y^2$$
$$= 2x^2 - 9xy - 2y^2$$

Or, using tabular addition,

$$\begin{array}{r} x^2 - 5xy \\ -3x^2 + y^2 \\ 4xy - 3y^2 \\ 4x^2 - 8xy \\ \hline 2x^2 - 9xy - 2y^2 \end{array}$$

Example 3 Subtract $\sqrt{3}x^2yz^3 + 5xy^2z - 3xz^2$ from $11xy^2z - 2x^2yz^2 + \sqrt{11}xz^2 + 8x^3yz^3 - 25$.

Solution

$$(11xy^2z - 2x^2yz^2 + \sqrt{11}xz^2 + 8x^3yz^3 - 25) - (\sqrt{3}x^2yz^3 + 5xy^2z - 3xz^2)$$

SEC. 2.5   Polynomials in Two or More Variables

$$= 11xy^2z - 2x^2yz^2 + \sqrt{11}xz^2 + 8x^3yz^3 - 25 - \sqrt{3}x^2yz^3 - 5xy^2z + 3xz^2$$
$$= (11-5)xy^2z - 2x^2yz^2 + (\sqrt{11}+3)xz^2 + 8x^3yz^3 - 25 - \sqrt{3}x^2yz^3$$
$$= 6xy^2z - 2x^2yz^2 + (\sqrt{11}+3)xz^2 + 8x^3yz^3 - \sqrt{3}x^2yz^3 - 25$$

Or, using long-hand subtraction,

$$\begin{array}{l} 11xy^2z - 2x^2yz^2 + \sqrt{11}xz^2 \qquad\quad + 8x^3yz^3 - 25 \\ -\;\;5xy^2z \qquad\qquad -\quad 3xz^2 \qquad\qquad\qquad\qquad\qquad + \sqrt{3}x^2yz^3 \\ \hline 6xy^2z - 2x^2yz^2 + (\sqrt{11}+3)xz^2 + 8x^3yz^3 - 25 - \sqrt{3}x^2yz^3 \end{array}$$

Notice that in each case we have collected like terms and added or subtracted their numerical coefficients. Basically, this is the procedure for adding or subtracting such algebraic expressions.

The reader has almost certainly noticed that the long-hand process is really more of a short-hand process. That is, it simply involves less writing. The main purpose in giving the parenthetical or one-line-at-a-time solution in each case is to illustrate the fact that we are simply doing with algebraic expressions what we have done earlier with real numbers. Also, in dealing with factoring in the next section, the parenthetical notation is preferable to the long-hand notation. That is, we prefer to write

$$(x-y)(x+y) = x^2 - y^2$$

rather than

$$\begin{array}{r} x - y \\ \times\; x + y \\ \hline x^2 - y^2 \end{array}$$

From now on we shall use one form or the other in each example, but not both.

Multiplication

Example 4
$$(x-y)(x+y) = x(x+y) - y(x+y)$$
$$= x^2 + xy - yx - y^2$$
$$= x^2 - y^2$$

Example 5   Multiply $3ax - 5y + 7$ by $2xy$.

Solution: $2xy(3ax - 5y + 7) = 6ax^2y - 10xy^2 + 14xy$

In general, to multiply two algebraic expressions, multiply each term of one of the expressions by each term of the other and add the resulting terms.

Example 6 Multiply $5x^2y - 3xy^2$ by $2x + 7xy$.

Solution
$$
\begin{array}{r}
5x^2y - 3xy^2 \\
2x + 7xy \\ \hline
10x^3y - 6x^2y^2 \\
35x^3y^2 - 21x^2y^3 \\ \hline
10x^3y - 6x^2y^2 + 35x^3y^2 - 21x^2y^3
\end{array}
$$

Ordinarily, we would *factor out* the largest algebraic expression we can by using an analog of the distributive law (see Rule 1.4). That is, we might write the answer as

$$x^2y(10x - 6y + 35xy - 21y^2)$$

Division

Division of one polynomial involving several variables by another can be a tedious undertaking. However, in many cases things work out nicely. Again we illustrate with examples.

Example 7 Divide $x^2 - 2xy + y^2$ by $x - y$.

Solution
$$
\begin{array}{r}
x - y \\
x-y\overline{\smash{)}x^2 - 2xy + y^2} \\
\underline{x^2 - xy} \\
-xy + y^2 \\
\underline{-xy + y^2} \\
0
\end{array}
$$

Hence
$$\frac{x^2 - 2xy + y^2}{x - y} = x - y$$

Example 8 Divide $5x^3y - 10x^2 + 15xy^2 - x^2y^2 + 2xy - 3y^3$ by $5x - y$.

Solution
$$
\begin{array}{r}
x^2y - 2x + 3y^2 \\
5x-y\overline{\smash{)}5x^3y - 10x^2 + 15xy^2 - x^2y^2 + 2xy - 3y^3} \\
\underline{5x^3y - x^2y^2} \\
-10x^2 + 15xy^2 + 2xy - 3y^3 \\
\underline{-10x^2}+ 2xy \\
15xy^2 - 3y^3 \\
\underline{15xy^2 - 3y^3} \\
0
\end{array}
$$

SEC. 2.5 Polynomials in Two or More Variables

So the quotient is
$$x^2y - 2x + 3y^2$$

Example 9 Divide $10x^2 - 5xy + 7y^3$ by $x + y$.

Solution

$$
\begin{array}{r}
10x - 15y \\
x+y \overline{\smash{\big)} 10x^2 - 5xy + 7y^3} \\
\underline{10x^2 + 10xy} \\
-15xy + 7y^3 \\
\underline{-15xy - 15y^2} \\
7y^3 + 15y^2
\end{array}
$$

So
$$\frac{10x^2 - 5xy + 7y^3}{x+y} = 10x - 15y + \frac{7y^3 + 15y^2}{x+y}$$

(See Example 15 in Section 2.1, Example 17 in Section 2.4, and Exercise 4 in the Exercises.) An equally correct "solution" would be

$$
\begin{array}{r}
7y^2 - 7xy - 5x + 7x^2 \\
y+x \overline{\smash{\big)} 7y^3 - 5xy + 10x^2 } \\
\underline{7y^3 + 7xy^2} \\
-5xy + 10x^2 - 7xy^2 \\
\underline{-7xy^2 - 7x^2y} \\
-5xy + 10x^2 + 7x^2y \\
\underline{-5xy - 5x^2} \\
15x^2 + 7x^2y \\
\underline{7x^2y + 7x^3} \\
15x^2 - 7x^3
\end{array}
$$

So
$$\frac{10x^2 - 5xy + 7y^3}{x+y} = 7y^2 - 7xy - 5x + 7x^2 + \frac{15x^2 - 7x^3}{x+y}$$

We see that two entirely different approaches yield two seemingly different answers. The key is that in each case the product of the divisor and the answer is the dividend. Checking this fact is a nice exercise for the reader in the two solutions for Example 9.

Exercises 2.5

1. In each part, find the sum of the given expressions.
   a. $5x^2y, -8x^2y, 7x^2y, -4x^2y, 2x^2y$
   b. $3x + 5y, 4x - 74, -6x + 2y, 8x - 3y$
   c. $2x - 3y + 6z, -4x - 5y - 8z, -7x + 2y + z$
   d. $x^2 - 3xy + 2y^2, 4x^2 + xy - 5y^2, 3x^2 + 2xy + y^2$
   e. $4x - 3y + 2, 7x + 5y - 6, -8x - 10y + 7, x + 2y + 1$
   f. $\sqrt{2}x^2 - \frac{2}{3}xy + \frac{3}{2}y^2, 3x^2 - \frac{3}{4}y^2 + \frac{11}{12}y$

2. Subtract the first expression from the second in each part.
   a. $5x - 4y, -2x - 7y$
   b. $4x^2 - 3xy, x^2 + 2xy - y^2$
   c. $7xyz + 3xy - 6yx, 2yz - 6xz + 10xyz$
   d. $4x^2 - \sqrt{3}y^2 + 5, \frac{3}{2}y^2 + \frac{1}{3}x^2 - 75$

3. In each part, remove all symbols of grouping and collect like terms where possible.
   a. $[(5x - 3y) + 2y] - [(4x - 3) - (2y - x)] + (x - y - 3)$
   b. $(2x - 5y) - [(3x + 2y) - (7x - 4y)]$
   c. $\{[(x + y) - 3y] - (-2x - 7y)\}$
   d. $\{[2x^2 - (3xy - 5y^2) + xy] - (x + 3y^2)\} - (6xy - y^2)$

4. In each part, perform the indicated multiplication and simplify your answer as much as you can.
   a. $3x^2y(2x^3 - 5xy^2 + 3y^2)$
   b. $(x - y)(x^2 + 5xy - 4y^2)$
   c. $(x + 2y)(x^2 - 3xy + y^2)$
   d. $(4x + 3y)(5x - 2y)$
   e. $(2x + 3y)(3x - 2y)$
   f. $(4x + 7y)(2x - 5y)$
   g. $(x + y)(x^2 - xy + y^2)$
   h. $(x + y)^3$
   i. $[(x + y) - 1][(x + y) + 1]$
   j. $(2x - y)(4x^2 + 2xy + y^2)$
   k. $(5x + 7y)(5x - 7y)$
   l. $(a + b + c)^2$
   m. $(x + 2y)(x - y)(3x + 2y)$
   n. $(2x + y)(x - y)(3x - 2y)$
   o. $(6x + 7y)^2$
   p. $(x - 2y + 3)(2x + y + z)$

5. In each part, perform the indicated division and thereby find the quotient $q(x)$ and the remainder R.
   a. Divide $x^2 + 6xy + 9y^2$ by $x + 3y$
   b. Divide $2x^2 - 3xy + y^2$ by $2x - y$
   c. Divide $x^3 + 6x^2y + 12xy^2 + 8y^3$ by $x + y$
   d. Divide $8x^3 + 12x^2y - 6xy^2 - y^3$ by $2x + y$
   e. Divide $x^3 - 8y^3$ by $x - 2y$
   f. Divide $2x^2 - xy - 6y^2 + x + 19y - 15$ by $2x + 3y - 5$
   g. Divide $x^3 + 2x^2y^2 - y^3$ by $x^2 + y$

2.6 SPECIAL PRODUCTS AND FACTORING

Certain products arise often enough in algebraic computation to justify our giving them special attention. For example $(x + y)(x - y) = x^2 - y^2$ is probably recognized, from high school algebra, as a *recurring equality*. The student probably even remembers describing this equality by saying "the difference of two squares is the product of their sum with their difference." The phrasing of this observation brings up another point. We learn to factor the expression $x^2 - y^2$ into a product of simpler expressions (that is, write $x^2 - y^2$ as the product of the two simpler expressions $x + y$ and $x - y$). This can be useful in dealing with various algebraic manipulations. Let us illustrate with examples.

Example 1 | Add $x - y$ to $x^2 - y^2$ and then factor.

Solution:
$$(x - y) + (x^2 - y^2) = (x - y) + (x + y)(x - y)$$
$$= [1 + (x + y)](x - y)$$
$$= (1 + x + y)(x - y)$$

Of course, $x^2 + x - y^2 - y$ is also an acceptable answer, but we can see that it might be easier to evaluate $(1 + x + y)(x - y)$ for particular values of x and y than it would be to evaluate $x^2 + x - y^2 - y$.

Example 2 | Multiply $x - y$ by $x^2 - y^2$.

Solution:
$$(x - y)(x^2 - y^2) = (x - y)(x + y)(x - y)$$
$$= (x + y)(x - y)^2$$

Again, this solution to Example 2 has no conceptual advantage over the methods used in the previous section where the solution might have been

$$(x - y)(x^2 - y^2) = x^3 - xy^2 - yx^2 + y^3$$

However, occasionally it is very helpful to know several expressions which are equal to a given expression and then to use the expression that does the most good. This is particularly evident when dealing with certain algebraic fractions, as the next two examples illustrate. (Also, Section 2.7 will be devoted entirely to algebraic fractions.)

Example 3 | Divide $x^2 - y^2$ by $x - y$.

Solution: $\dfrac{x^2 - y^2}{x - y} = \dfrac{(x + y)(x - y)}{x - y} = x + y$ if $x - y \neq 0$

Example 4 | Simplify the algebraic fraction $\dfrac{x-y}{x^2-y^2}$.

Solution: $\dfrac{x-y}{x^2-y^2} = \dfrac{x-y}{(x+y)(x-y)} = \dfrac{1}{x+y}$ if $x - y \neq 0$

We proceed to calculate several additional *special products*. We shall illustrate the use of each special product in calculating more complicated products and in factoring more complicated expressions. In every case but the last (Rule 2.15), simple multiplication justifies the result and the justifications are left as exercises (see Exercise 6).

Throughout this section A and B will represent either real numbers, arbitrary constants, or algebraic expressions.

RULE 2.10: $(A + B)^2 = A^2 + 2AB + B^2$. That is, the square of a sum of two terms equals the square of the first term *plus* twice the product of the two terms, plus the square of the second term.

Example 5 | $(3x + 2y)^2 = (3x)^2 + 2(3x)(2y) + (2y)^2$
$= 9x^2 + 12xy + 4y^2$

Example 6 | Factor $x^2 + 4x + 4$.

Solution: $x^2 + 4x + 4 = x^2 + 2(2)(x) + 2^2$
$= (x + 2)^2$

With practice, we gain proficiency in organizing and calculating such special products and in recognizing and factoring such special polynomials.

Example 7 | $[(x + 1) + (2y + 3)]^2 = (x + 1)^2 + 2(x + 1)(2y + 3) + (2y + 3)^2$
$= x^2 + 2x + 1 + 2(2xy + 3x + 2y + 3) + 4y^2 + 12y + 9$
$= x^2 + 2x + 6x + 4xy + 4y^2 + 4y + 12y + 1 + 6 + 9$
$= x^2 + 8x + 4xy + 4y^2 + 16y + 16$

Example 8 | Factor $16x^2 + 40xy + 25y^2$.

Solution: $16x^2 + 40xy + 25y^2 = (4x)^2 + 2(4x)(5y) + (5y)^2$
$= (4x + 5y)^2$

RULE 2.11: $(A - B)^2 = A^2 - 2AB + B^2$. That is, the square of the difference of two terms equals the square of the first term *minus* twice the product of the two terms plus the square of the second term.

SEC. 2.6 Special Products and Factoring

Example 9

$$(\sqrt{2}x - \sqrt{3}y)^2 = (\sqrt{2}x)^2 - 2(\sqrt{2}x)(\sqrt{3}y) + (\sqrt{3}y)^2$$
$$= 2x^2 - 2\sqrt{2}\sqrt{3}xy + 3y^2$$
$$= 2x^2 - 2\sqrt{6}xy + 3y^2$$

Example 10

Factor $16x^2 - 48xy + 36y^2$.

Solution: $16x^2 - 48xy + 36y^2 = (4x)^2 - 2(4x)(6y) + (6y)^2$
$$= (4x - 6y)^2$$

Alternate Solution: $16x^2 - 48xy + 36y^2 = 4(4x^2 - 12xy + 9y^2)$
$$= 4[(2x)^2 - 2(2x)(3y) + (3y)^2]$$
$$= 4(2x - 3y)^2$$

To see that the first answer is the same as the second,

$$(4x - 6y)^2 = [2(2x - 3y)]^2 = 2^2(2x - 3y)^2$$
$$= 4(2x - 3y)^2$$

RULE 2.12: $(A + B)(A - B) = A^2 - B^2$. That is, the product of the sum of two terms with the difference of the same two terms equals the difference of the squares of the two terms.

Example 11

$(x^2 + 2x)(x^2 - 2x) = (x^2)^2 - (2x)^2 = x^4 - 4x^2$

Example 12

Factor $3x^2 - 4y^2$.

Solution: $3x^2 - 4y^2 = (\sqrt{3}x)^2 - (2y)^2$
$$= (\sqrt{3}x + 2y)(\sqrt{3}x - 2y)$$

RULE 2.13: $(A + B)(A^2 - AB + B^2) = A^3 + B^3$. That is (reading from right to left), the sum of the cubes of two terms equals the *sum* of the two terms times the square of the first term *minus* the product of the two terms plus the square of the second term.

Example 13

$(x + 2y)(x^2 - 2xy + 4y^2) = (x + 2y)[x^2 - (x)(2y) + (2y)^2]$
$$= x^3 + (2y)^3$$
$$= x^3 + 8y^3$$

Example 14 | Factor $27x^4 + 64xy^3$.

Solution: First we factor out the common x.
$$27x^4 + 64xy^3 = x(27x^3 + 64y^3)$$
$$= x[(3x)^3 + (4y)^3]$$
$$= x(3x + 4y)[(3x)^2 - (3x)(4y) + (4y)^2]$$
$$= x(3x + 4y)(9x^2 - 12xy + 16y^2)$$

The solution to Example 14 and the alternate solution to Example 10 illustrate an important technique. *When factoring, first factor out the largest algebraic expression common to each term.* We shall continue to use this technique in later examples.

RULE 2.14: $(A - B)(A^2 + AB + B^2) = A^3 - B^3$. That is (reading from right to left), the difference of the cubes of two terms equals the *difference* of the two terms times the square of the first term *plus* the product of the two terms plus the square of the second term.

Example 15 | $(x - 2y)(x^2 + 2xy + 4y^2) = (x - 2y)[x^2 + (x)(2y) + (2y)^2]$
$$= x^3 - (2y)^3$$
$$= x^3 - 8y^3$$

Example 16 | Factor $21y^3 - 7x^3$.

Solution: $21y^3 - 7x^3 = 7(3y^3 - x^3)$
$$= 7[(\sqrt[3]{3}y)^3 - x^3]$$
$$= 7(\sqrt[3]{3}y - x)[(\sqrt[3]{3}y)^2 + \sqrt[3]{3}yx + x^2]$$
$$= 7(\sqrt[3]{3}y - x)(\sqrt[3]{9}y^2 + \sqrt[3]{3}xy + x^2)$$

RULE 2.15: If n is a positive integer, then
$$(A - B)(A^{n-1} + A^{n-2}B + A^{n-3}B^2 + \cdots + A^2B^{n-3} + AB^{n-2} + B^{n-1}) = A^n - B^n$$

This seemingly complicated special product arises quite naturally in many situations. If $n = 1$, it gives nothing more than $A - B = A - B$. If $n = 2$, it gives $(A - B)(A + B) = A^2 - B^2$, that is, Rule 2.12. If $n = 3$, it gives $(A - B)(A^2 + AB + B^2) = A^3 - B^3$, that is, Rule 2.14. To indicate why Rule 2.15 is true, let us use long-hand multiplication.

SEC. 2.6   Special Products and Factoring

$$A^{n-1} + A^{n-2}B + A^{n-3}B^2 + \cdots + A^2B^{n-3} + AB^{n-2} + B^{n-1}$$
$$\times\ A - B$$

$$\begin{array}{l} A^n + A^{n-1}B + A^{n-2}B^2 \quad + \cdots + A^3B^{n-3} + A^2B^{n-2} + AB^{n-1} \\ \quad -\ A^{n-1}B - A^{n-2}B^2 \quad -\cdots - A^3B^{n-3} - A^2B^{n-2} - AB^{n-1} - B^n \end{array}$$

$$A^n \hspace{8cm} - B^n$$

So, the middle terms cancel each other out, yielding the desired equality. Actually, to give a careful proof, a process known as mathematical induction must be employed.

Example 17 $(x - y)(x^3 + x^2y + xy^2 + y^3) = x^4 - y^4$

Example 18 Factor $64y^5 - 2x^5$.

Solution

$$\begin{aligned} 64y^5 - 2x^5 &= 2(32y^5 - x^5) \\ &= 2[(2y)^5 - x^5] \\ &= 2(2y - x)[(2y)^4 + (2y)^3x + (2y)^2x^2 + (2y)x^3 + x^4] \\ &= 2(2y - x)(16y^4 + 8y^3x + 4y^2x^2 + 2yx^3 + x^4) \end{aligned}$$

Factoring Expressions of the Form $ax^2 + bxy \times cy^2$

Among the most common types of polynomials that one is asked to factor in algebra are the polynomials in x and y of the form $ax^2 + bxy + cy^2$. We provide several examples which illustrate techniques applicable to this sort of problem.

Example 19 Factor $4x^2 + 20xy + 25y^2$.

Solution: If we recall Rule 2.10, we might notice that

$$4x^2 + 20xy + 25y^2 = (2x)^2 + 2(2x)(5y) + (5y)^2$$
$$= (2x + 5y)^2$$

However, even if we do not notice that $4x^2 + 20xy + 25y^2$ is a perfect square, we can still solve the problem as follows:

First write

$$4x^2 + 20xy + 25y^2 = (\quad x \quad y)(\quad x \quad y)$$

leaving room for coefficients of x and y and also leaving off the signs between x and y on the right.

Next: Notice that the coefficient of y^2 is positive, and so the signs in the factorization must be the same. Otherwise, their product would be negative. So insert + signs in each factor so that you now have the following on your paper:

$$4x^2 + 20xy + 25y^2 = (\quad x + \quad y)(\quad x + \quad y)$$

Continue to leave room for coefficients of x and y.

Third: Think about the possible factorizations of 4 (the coefficient of x^2), choose one such factorization, and use the factors as coefficients for x. Let's say you get lucky and choose $4 = 2 \cdot 2$ as the first factorization to try. Using 2 as a coefficient of each x, you then have on your paper

$$4x^2 + 20xy + 25y^2 = (2x + \quad y)(2x + \quad y)$$

Notice again that we have left room for coefficients of y.

Finally: Think about the possible factorizations of 25 (the coefficient of y^2), choose one such factorization, and use the factors as coefficients for y. Once again, suppose you get lucky and choose $25 = 5 \cdot 5$ as the first factorization you try. Then you would use 5 as a coefficient of each y and mentally check to see if an equality is obtained. So you would now have on your paper

$$4x^2 + 20xy + 25y^2 = (2x + 5y)(2x + 5y)$$

However, don't just leave it at that until you mentally check that the equality does hold. In this case it does.

In Example 19, if we had chosen one of the other factorizations of 4, that is, $4 = 1 \cdot 4$ or $4 = 4 \cdot 1$, we would have found in the final step that no factorization of 25 yielded an equality. Using $4 = 1 \cdot 4$, for example,

$$4x^2 + 20xy + 25y^2 \neq (x + y)(4x + 25y)$$

$$4x^2 + 20xy + 25y^2 \neq (x + 5y)(4x + 5y)$$

$$4x^2 + 20xy + 25y^2 \neq (x + 25y)(4x + y)$$

and $25 = 1 \cdot 25$, $25 = 5 \cdot 5$, and $25 = 25 \cdot 1$ are the only factorizations of 25. So we would simply have gone back to the third step and used a different factorization of 4. This illustrates the fact that

Factoring is a trial and error process.

Example 20

Factor $x^2 + 7x + 12$.

Solution: In this case your first step is to write

$$x^2 + 7x + 12 = (x \quad)(x \quad)$$

since no y's appear. Now the only factorization of 1 as a product of positive integers is $1 = 1 \cdot 1$, so you do not need to place any coefficients in front of x. You

SEC. 2.6 Special Products and Factoring

should recognize that the constant terms will be both plus or both minus since the constant term of $x^2 + 7x + 12$ *is* $+12$. Next observe that these constant terms must be positive, for otherwise you would obtain a negative coefficient of x in the product and this coefficient is $+7$. So enter a plus after each x and have on your paper

$$x^2 + 7x + 12 = (x + \quad)(x + \quad)$$

Finally, consider possible factorizations of 12, this time keeping in mind that you want the sum of the factors to be 7. The possible factorizations of 12 are $12 = 1 \cdot 12, 12 = 2 \cdot 6, 12 = 3 \cdot 4, 12 = 4 \cdot 3, 12 = 6 \cdot 2,$ and $12 = 12 \cdot 1$. Clearly, the only factorizations whose sum is 7 are $12 = 3 \cdot 4$ and $12 = 4 \cdot 3$, and it really doesn't matter which we choose. So your final step is to enter the constant terms 3 and 4. After all of this thinking, plotting, and scheming, all that appears on your paper is the correct solution:

$$x^2 + 7x + 12 = (x + 3)(x + 4)$$

Example 21 Factor $x^2 - 4x - 21$.

Solution: First write $x^2 - 4x - 21 = (x \quad)(x \quad)$. Next notice that the constant terms must differ in sign in order for their product to be -21. So enter either $+$ first and $-$ second or the other way around. On your paper would appear

$$x^2 - 4x - 21 = (x + \quad)(x - \quad)$$

Finally, think about the factorizations of 21 with integers as factors, that is, $21 = 1 \cdot 21, 21 = 3 \cdot 7, 21 = 7 \cdot 3,$ and $21 = 21 \cdot 1$. Mentally substitute each possibility, checking to see if the coefficient of x in the product is -4. It turns out that the second factorization, $21 = 3 \cdot 7$, does the job and so, again, all that appears on your paper is

$$x^2 - 4x - 21 = (x + 3)(x - 7)$$

In the next two examples, we abbreviate the steps by indicating, one step at a time, how the solution might be developed.

Example 22 Factor $3x^2 - 8xy + 4y^2$.

Solution: Step 1. $3x^2 - 8xy + 4y^2 = (x \quad y)(3x \quad y)$
Step 2. $3x^2 - 8xy + 4y^2 = (x - y)(3x - y)$
Step 3. $3x^2 - 8xy + 4y^2 = (x - 2y)(3x - 2y)$

Again, however, all that needs to appear on your paper is

$$3x^2 - 8xy + 4y^2 = (x - 2y)(3x - 2y)$$

Example 23 Factor $6x^2 + 7xy - 20y^2$.

Solution: Step 1. $6x^2 + 7xy - 20y^2 = (\ x \quad y)(x \quad y)$
Step 2. $6x^2 + 7xy - 20y^2 = (6x \quad y)(x \quad y)$

Step 3. Whoops. No combination of factors of 20 with one plus and the other minus will work for the factorization of 6 as 6·1. So we go back to Step 2 and use a different factorization of 6.
Step 2 again. $\quad 6x^2 + 7xy - 20y^2 = (2x \quad y)(3x \quad y)$
Step 3. $\quad 6x^2 + 7xy - 20y^2 = (2x + 5y)(3x - 4y)$
Once again, all that needs to appear on your paper is
$$6x^2 + 7xy - 20y^2 = (2x + 5y)(3x - 4y)$$

Example 24 Factor $40x^2 - 41x - 21$.

Solution: $40x^2 - 41x - 21 = (8x + 3)(5x - 7)$
(See Examples 19 through 23 for explanations.)

Example 25 Factor $4x^2 + 5xy + 12y^2$.

Solution: By checking cases, you will find that it is not possible to factor this polynomial into "first-degree" factors with integer coefficients.

Example 26 Factor $20x^2 + 26xy - 6y^2$.

Solution: $\quad 20x^2 + 26xy - 6y^2 = 2(10x^2 + 13xy - 3y^2)$
$\qquad\qquad\qquad\qquad\quad = 2(5x - y)(2x + 3y)$

For ease of reference we list the special product rules in a table here.

Special Products

Let A and B be real numbers, arbitrary constants, or algebraic expressions.

RULE 2.10. $(A + B)^2 = A^2 + 2AB + B^2$
RULE 2.11. $(A - B)^2 = A^2 - 2AB + B^2$
RULE 2.12. $(A + B)(A - B) = A^2 - B^2$
RULE 2.13. $(A + B)(A^2 - AB + B^2) = A^3 + B^3$
RULE 2.14. $(A - B)(A^2 + AB + B^2) = A^3 - B^3$
RULE 2.15. $(A - B)(A^{n-1} + A^{n-2}B + \cdots + AB^{n-2} + B^{n-1}) = A^n - B^n$ for any positive integer n

---

Exercises 2.6

1. Memorize Rules 2.10, 2.11, and 2.12 and use them to perform the following multiplications by inspection.
   a. $(x + 3y)^2$ **b.** $(4x - 7y)^2$ **c.** $(5x + 2y)^2$

SEC. 2.6 Special Products and Factoring

d. $(3x - 2y)(3x + 2y)$ e. $(7x - 5y)^2$ f. $(8x - 3y)^2$
g. $(12x - 9y)^2$ h. $(2x + 8y)(2x - 8y)$ i. $(2x^3y + 3y)^2$
j. $(x + y)(x - y)$ k. $(11x - 4y)(11x + 4y)$ l. $[(x + y) + 3]^2$

2. Factor by removing the largest expression common to each term of the following expressions.
   a. $5a^2x^3 - 10axy$ b. $24x^3y^2 + 6x^2y^3$
   c. $6a^3b^2 + 8a^2b^3 - 4z^4b^2$ d. $15 + 25x - 30y$
   e. $6x^3y^3z^3 - 12x^2y^2z^2$ f. $2(x - 3)(x + 1)^2 + 2(x - 3)^2(x + 1)$
   g. $4x(2x - y) - 7(2x - y)$ h. $8x(2x + 1)^3(x^2 - 1)^3 + 6(2x + 1)^2(x^2 - 1)^4$
   Solution to a: $5a^2x^3 - 10axy = 5ax(ax^2 - 2y)$

3. Factor each of the following without introducing imaginary numbers as coefficients.
   a. $x^2 - 9y^2$ b. $4z^2 - 25b^2$ c. $x^3 - 4x$
   d. $8x^2 - 18y^2$ e. $0.04x^2 - 1.69y^2$ f. $x^3 - 1$
   g. $27x^3 - 8y^3$ h. $64x^3 + 27y^3$ i. $16a^4 - 2a$
   j. $x^6 - y^6$ k. $x^2 - 5$ l. $(x + y)^2 - 9$
   m. $81x^4 - 256$ n. $27a^3 - 8$ o. $x^4 - y^4$
   p. $a^2b^2 - b^4$ q. $45x^2 - 80y^2$ r. $(x + y)^2 - (a - b)^2$

4. Using Rules 2.10 and 2.11, express each of the following expressions as a perfect square.
   a. $x^2 + 6x + 9$ b. $x^2 - 8x + 16$ c. $4x^2 - 20x + 25$
   d. $9x^2 - 24xy + 16y^2$ e. $36x^2 - 84xy + 49y^2$ f. $64x^2 + 80xy + 25y^2$
   g. $2x^2 + 6\sqrt{2}xy + 9y^2$ h. $x + 4y\sqrt{x} + 4y^2$ i. $9x - 6\sqrt{xy} + y$
   j. $0.16x^2 + 0.96x + 1.44$

5. Find, if possible, the factors with integer coefficients of the following expressions:
   a. $x^2 + 5x + 6$ b. $x^2 - 2x - 3$ c. $x^2 - 5x - 24$
   d. $2x^2 - 5x + 2$ e. $3x^2 - x - 6$ f. $10x^2 - 11x - 6$
   g. $x^2 - 7x + 12$ h. $x^2 + 7x + 12$ i. $10x^2 - 14x - 12$
   j. $6x^2 - 10x + 10$ k. $15x^2 + 28x - 32$ l. $18x^2 - 21x + 5$
   m. $2x^2 - 5xy - 12y^2$ n. $21x^2 + 41x + 10$ o. $12x^2 + 11x - 15$
   p. $14x^2 + 29xy - 15y^2$ q. $6x^2 - 31x + 60$ r. $8x^2 + 22x - 21$
   Solution to k: $15x^2 + 28x - 32 = (5x - 4)(3x + 8)$

6. By performing the indicated multiplication in each case, verify Rules 2.10 through 2.14.
   Example: Rule 2.10, $(A + B)^2 = A^2 + 2AB + B^2$
   Solution: $(A + B)^2 = (A + B)(A + B) = A(A + B) + B(A + B)$
   $$= A^2 + AB + AB + B^2$$
   $$= A^2 + 2AB + B^2$$

2.7 ALGEBRAIC FRACTIONS

In algebra, just as in arithmetic, the indicated quotient of two expressions is called a fraction; for example, $\dfrac{2x}{3y}$, $\dfrac{2a+7b}{x+y}$, and $\dfrac{x^2-2xy+7y^2}{x^2+3xy+2y^2}$ are fractions. The dividend in an indicated division is called the *numerator* of the fraction, and the divisor is called the *denominator*. In the fraction $\dfrac{2a+7b}{x+y}$, the numerator is $2a+7b$ and the denominator is $x+y$. The rules in algebra for operating with fractions are entirely analogous to those developed in Section 1.4. Hence the student would do well at this point to review that section.

First, we agree that there are three signs associated with each algebraic fraction: the sign of the numerator, the sign of the denominator, and the sign of the entire fraction. Any two of the signs associated with a given fraction can be changed without changing the fraction.

Example 1
$$\frac{x+2}{y-3} = -\frac{x+2}{-(y-3)} = -\frac{-(x+2)}{y-3} = \frac{-(x+2)}{-(y-3)}$$

Example 2
$$-\frac{3a-x}{2y} = \frac{3a-x}{-2y} = \frac{-(3a-x)}{2y} = -\frac{-(3a-x)}{-2y}$$

Recall that we defined a simple fraction to be in simplest form (Definition 1.17) if the numerator and denominator have no common factors other than 1 and -1. For the analogous concept in algebraic fractions, we need to understand the notion of a *simple factor*. Basically, a *simple factor* is an algebraic expression which cannot be expressed as a product of two or more factors. An algebraic fraction is said to be in *simplest form* if the numerator and denominator have no common factors. Hence, to reduce an algebraic fraction to simplest form, factor the numerator and denominator into simple factors. Then divide both the numerator and denominator by each factor common to both. The resulting fraction is the original fraction reduced to simplest form.

Example 3
Reduce $\dfrac{2x^2-2xy-4y^2}{2x^2-8y^2}$ to simplest form.

Solution:
$$\frac{2x^2-2xy-4y^2}{2x^2-8y^2} = \frac{2(x-2y)(x+y)}{2(x-2y)(x+2y)}$$
$$= \frac{x+y}{x+2y} \quad \text{if } x-2y \neq 0$$

Here, as always, we are only allowed to divide both numerator and denominator by a quantity if that quantity is not zero. This is the reason for the qualification "if $x-2y \neq 0$" in the solution to Example 3.

SEC. 2.7 Algebraic Fractions

The definition of a common denominator of several fractions and methods for finding common denominators were extremely important in adding and subtracting fractions. For algebraic fractions the situation is much the same. First, any two algebraic fractions may be rewritten in such a way that they have a *common denominator*, that is, the same denominator. Basically, the least common denominator of a set of algebraic fractions is the "smallest" algebraic expression containing each denominator as a factor. A method for finding the LCD of a set of algebraic fractions is to find the simple factors of the denominators of all given fractions and take the product of all the factors that are different from each other, raising each such simple factor to the highest power which appears in any denominator.

Example 4 Find the LCD of the fractions $\dfrac{3}{x^2+2x+1}, \dfrac{4x}{x^2-1}, \dfrac{1+2x}{x^2-x-2}$, and write each fraction as a fraction whose denominator is the LCD.

Solution: If we factor each denominator, we get $(x^2+2x+1)=(x+1)^2$, $x^2-1=(x+1)(x-1)$, $x^2-x-2=(x-2)(x+1)$. The LCD of the three given fractions is therefore:

$$(x+1)^2(x-1)(x-2)$$

Our fractions can now be written:

$$\frac{3}{x^2+2x+1} = \frac{3(x-1)(x-2)}{(x+1)^2(x-1)(x-2)}$$

$$\frac{4x}{x^2-1} = \frac{4x(x+1)(x-2)}{(x+1)^2(x-1)(x-2)}$$

$$\frac{1+2x}{x^2-x-2} = \frac{(1+2x)(x+1)(x-1)}{(x+1)^2(x-1)(x-2)}$$

As in the case of fractions involving only numbers, the LCD is only one of many common denominators. For example, the product of all given denominators is a common denominator.

Addition and Subtraction of Algebraic Fractions

Recall in Section 1.4 we observed that, to add (or subtract) fractions with a common denominator, one simply adds (or subtracts) the numerators and leaves the denominator alone. The exact same process is applied to adding (or subtracting) algebraic fractions which have the same denominator. We illustrate by example.

Example 5 | Find the sum of the three algebraic fractions

$$\frac{4x}{x^2 - y^2}, \quad \frac{x + 2y}{x^2 - y^2}, \quad \text{and} \quad -\frac{3x}{x^2 - y^2}$$

and reduce your answer to simplest form.

Solution:
$$\frac{4x}{x^2 - y^2} + \frac{x + 2y}{x^2 - y^2} - \frac{3x}{x^2 - y^2} = \frac{4x + x + 2y - 3x}{x^2 - y^2}$$

$$= \frac{2x + 2y}{x^2 - y^2}$$

$$= \frac{2(x + y)}{(x - y)(x + y)}$$

$$= \frac{2}{x - y} \quad \text{if } x + y \neq 0$$

Example 6 | Subtract $\dfrac{x - 2y}{x^2 + 2xy + y^2}$ from $\dfrac{2x - y}{x^2 + 2xy + y^2}$ and reduce your answer to simplest form.

Solution:
$$\frac{2x - y}{x^2 + 2xy + y^2} - \frac{x - 2y}{x^2 + 2xy + y^2} = \frac{(2x - y) - (x - 2y)}{x^2 + 2xy + y^2}$$

$$= \frac{2x - x - y + 2y}{x^2 + 2xy + y^2}$$

$$= \frac{x + y}{x^2 + 2xy + y^2}$$

$$= \frac{x + y}{(x + y)^2} = \frac{1}{x + y} \quad \text{if } x + y \neq 0$$

The techniques for adding algebraic fractions which do not have the same denominator are again completely analogous to those developed in Section 1.4. There a total of four methods were given. We select only one to phrase for algebraic fractions.

RULE FOR ADDING ALGEBRAIC FRACTIONS

1. Find a common denominator D.
2. Take each fraction (in order), divide the denominator of this fraction into D, and multiply this result by the numerator.
3. Add the numerators obtained in step 2 and use this sum as the numerator.

SEC. 2.7 Algebraic Fractions

Example 7   Add $\dfrac{y}{x}$ to $\dfrac{x}{y}$.

Solution: $\dfrac{y}{x} + \dfrac{x}{y} = \dfrac{y^2 + x^2}{xy}$

Example 8   Find the sum of the three algebraic fractions

$$\dfrac{1}{x+1},\quad \dfrac{3}{x-2},\quad \text{and}\quad \dfrac{2}{x+3}$$

Solution

$$\dfrac{1}{x+1} + \dfrac{3}{x-2} + \dfrac{2}{x+3} = \dfrac{(1)(x-2)(x+3) + (3)(x+1)(x+3) + 2(x+1)(x-2)}{(x+1)(x-2)(x+3)}$$

$$= \dfrac{(x^2 + x - 6) + 3(x^2 + 4x + 3) + 2(x^2 - x - 2)}{(x+1)(x-2)(x+3)}$$

$$= \dfrac{x^2 + 3x^2 + 2x^2 + x + 12x - 2x - 6 + 9 - 4}{(x+1)(x-2)(x+3)}$$

$$= \dfrac{6x^2 + 11x - 1}{(x+1)(x-2)(x+3)}$$

Example 9   Combine into a single fraction and reduce your answer to simplest form:

$$\dfrac{2x}{x+1} + \dfrac{x}{1-x} - \dfrac{4}{x^2-1}$$

Solution: $\dfrac{2x}{x+1} + \dfrac{x}{1-x} - \dfrac{4}{x^2-1} = \dfrac{2x}{x+1} - \dfrac{x}{x-1} - \dfrac{4}{(x+1)(x-1)}$

$$= \dfrac{(2x)(x-1) - (x)(x+1) - 4}{(x+1)(x-1)}$$

$$= \dfrac{2x^2 - 2x - x^2 - x - 4}{(x+1)(x-1)}$$

$$= \dfrac{x^2 - 3x - 4}{(x+1)(x-1)}$$

$$= \dfrac{(x+1)(x-4)}{(x+1)(x-1)}$$

$$= \dfrac{x-4}{x-1}\quad \text{if } x+1 \neq 0$$

Notice that we have used the LCD as our common denominator in this example. This is ordinarily the most convenient denominator to use.

Multiplication of Algebraic Fractions

For multiplication, we simply repeat the rule developed in Section 1.4.

RULE FOR MULTIPLYING ALGEBRAIC FRACTIONS

1. Multiply all numerators and use this product as the numerator.
2. Multiply all denominators and use this product as the denominator.

Example 10 Multiply $\dfrac{5}{x^2 - y^2}$ by $x + y$.

Solution: $\dfrac{5}{x^2 - y^2} \cdot \dfrac{x + y}{1} = \dfrac{5(x + y)}{(x - y)(x + y)} = \dfrac{5}{x - y}$ if $x + y \neq 0$

Example 11 Multiply $\dfrac{(x + y)}{(x - y)}$ by $\dfrac{(x^2 + xy - 2y^2)}{(x^2 + 2xy + y^2)}$.

Solution: $\dfrac{x + y}{x - y} \cdot \dfrac{x^2 + xy - 2y^2}{x^2 + 2xy + y^2} = \dfrac{(x + y)(x + 2y)(x - y)}{(x - y)(x + y)(x + y)} = \dfrac{x + 2y}{x + y}$

if neither $x - y = 0$ nor $x + y = 0$.

Division of Algebraic Fractions

Again, we simply repeat the rule developed in Section 1.4.

RULE FOR DIVIDING ALGEBRAIC FRACTIONS

$$\dfrac{A/B}{C/D} = \dfrac{A}{B} \cdot \dfrac{D}{C}$$

So: "In order to divide by a fraction, invert it and multiply."

Example 12 $\dfrac{y}{2x^2} \div \dfrac{x - 3}{6x} = \dfrac{y}{2x^2} \cdot \dfrac{6x}{x - 3} = \dfrac{6xy}{2x^2(x - 3)} = \dfrac{3y}{x(x - 3)}$ if $x \neq 0$

Complex Fractions

A complex fraction is a fraction whose numerator, denominator, or both, contain fractions such as

$$\dfrac{\frac{2}{3}}{\frac{5}{7}}, \quad \dfrac{2 + \frac{1}{3}}{\frac{3}{4} - 1}, \quad \dfrac{\frac{3}{x + 2} + \frac{1}{x - 3}}{\dfrac{2x + 5}{x^2 - x - 6}}, \quad \text{etc.}$$

SEC. 2.7 Algebraic Fractions

A *simple fraction* is a fraction which has no fractions in either the denominator or numerator.

One method of reducing a complex fraction to a simple fraction is to multiply the numerator and denominator of the complex fraction by the LCD of all the fractions in both the numerator and denominator of the complex fraction.

Example 13 Write as a simple fraction: $\dfrac{\dfrac{1}{2}+\dfrac{3}{4}}{\dfrac{2}{3}-\dfrac{1}{6}}$.

Solution: The LCD of the four fractions in the numerator and denominator is 12. Multiply numerator and denominator of the complex fraction by 12:

$$\frac{\left(\dfrac{1}{2}+\dfrac{3}{4}\right)12}{\left(\dfrac{2}{3}-\dfrac{1}{6}\right)12} = \frac{6+9}{8-2} = \frac{15}{6} = \frac{5}{2}$$

Example 14 Reduce to a simple fraction: $\dfrac{\dfrac{2}{x}-\dfrac{3}{y}}{\dfrac{4}{x^2}-\dfrac{9}{y^2}}$

Solution: The LCD of the four fractions in the numerator and denominator of the given complex fraction is $x^2 y^2$. Multiply the numerator and denominator of the complex fraction by $x^2 y^2$ and obtain

$$\frac{\left(\dfrac{2}{x}-\dfrac{3}{y}\right)x^2y^2}{\left(\dfrac{4}{x^2}-\dfrac{9}{y^2}\right)x^2y^2} = \frac{2xy^2-3x^2y}{4y^2-9x^2} = \frac{xy(2y-3x)}{(2y-3x)(2y+3x)} = \frac{xy}{3x+2y}$$

if $2y - 3x \neq 0$.

Exercises 2.7

Simplify, where possible, each of the following fractions.

1. $\dfrac{32}{48}$ 2. $\dfrac{72}{45}$ 3. $\dfrac{5+3}{11+5}$

4. $\dfrac{\dfrac{3}{2}+\dfrac{5}{6}}{\dfrac{2}{3}-\dfrac{1}{2}}$
5. $\dfrac{ab+ac}{b+c}$
6. $\dfrac{x^2-5x}{x^2-25}$

7. $\dfrac{4x-2y}{2y}$
8. $\dfrac{3x^2+2x-5}{2x-2}$
9. $\dfrac{x^2-2x-3}{x^2-9}$

10. $\dfrac{4x^2+4xy+y^2}{4x^2+2xy}$
11. $\dfrac{4x^2+2x-12}{2x^2-13x+15}$
12. $\dfrac{x^3-x^2y+xy^2}{x^3+y^3}$

13. $\dfrac{x^3-y^3}{x^2+xy+y^2}$
14. $\dfrac{x^2+xy-12y^2}{x^2-6xy+y^2}$
15. $\dfrac{2x^2+13xy+15y^2}{2x^2+7xy-15y^2}$

16. $\dfrac{\dfrac{1}{x}-\dfrac{1}{y}}{\dfrac{1}{x}+\dfrac{1}{y}}$
17. $\dfrac{9x^2-4y^2}{6x^2-13xy+6y^2}$
18. $\dfrac{x^4-y^4}{x^4-2x^2y^2+y^4}$

Perform the indicated operations in each of the following problems and leave the answer as a simple fraction reduced to simplest form.

19. $\dfrac{7}{15}-\dfrac{2}{5}+\dfrac{13}{30}$
20. $\dfrac{3}{5}-\dfrac{11}{25}+\dfrac{7}{20}$

21. $\dfrac{\dfrac{3}{4}-\dfrac{7}{10}+\dfrac{5}{8}}{\dfrac{7}{8}+\dfrac{3}{5}-\dfrac{9}{20}}$
22. $\dfrac{7}{12}+\dfrac{13}{20}-\dfrac{17}{15}$

23. $\dfrac{x-y}{6}-\dfrac{x+y}{3}$
24. $\dfrac{x+3}{8}-\dfrac{x-5}{4}+\dfrac{3x+7}{2}$

25. $\dfrac{x-3}{7x}-\dfrac{4}{x}$
26. $\dfrac{2x-1}{6}-\dfrac{3x+2}{4}+\dfrac{x-2}{3}$

27. $\dfrac{2x-2}{3}-\dfrac{x+1}{6}+\dfrac{x-1}{4}$
28. $\dfrac{5}{2x-5}-\dfrac{4}{2x+5}$

29. $\dfrac{2x}{x+1}-\dfrac{x}{x-1}+\dfrac{1}{x+1}$
30. $\dfrac{3x+2}{3x}-\dfrac{2x^2+x}{2x^2}$

31. $\dfrac{2x}{x^2-1}-\dfrac{1}{x-1}+\dfrac{1}{x+1}$
32. $\dfrac{2x-1}{x^2-4}+\dfrac{3}{x-2}-\dfrac{2}{x+2}$

33. $\dfrac{3}{x^2} - \dfrac{x-y}{xy^2} + \dfrac{1}{y^2}$

34. $\dfrac{x+1}{x^2-9} - \dfrac{2}{x+3} + \dfrac{1}{x-3}$

35. $\dfrac{4-x}{x^2-2x+4} + \dfrac{1}{x+2}$

36. $\dfrac{2}{(x-1)^2} - \dfrac{1}{1-x} + \dfrac{1}{x}$

37. $\dfrac{x+2}{x^2-9} \cdot \dfrac{x-3}{x^2-x} \cdot \dfrac{x^3+3x^2}{5x+10}$

38. $\dfrac{x^2-3x+2}{x^3-25x} \cdot \dfrac{x^2-6x+5}{x^2-4x+4}$

39. $\dfrac{2x^2-5x+2}{3x+5} \div 2x-1$

40. $\left(\dfrac{x^2}{9} - \dfrac{4}{x^2}\right) \div \left(\dfrac{x}{3} + \dfrac{2}{x}\right)$

41. $\dfrac{7}{x^2-x} + \dfrac{2}{x^2+x} - \dfrac{4}{x^2-1}$

42. $\left(\dfrac{2}{x} + \dfrac{x}{2}\right) \div \left(\dfrac{2}{x} - \dfrac{x}{2}\right)$

43. $\dfrac{xy}{x+y} - \dfrac{x^2}{x-y}$

44. $\dfrac{\dfrac{m+1}{m-1} + \dfrac{m-1}{m+1}}{\dfrac{m-1}{m+1} - \dfrac{m+1}{m-1}}$

45. $\dfrac{x^2+3x}{x^2+x-6} + \dfrac{5x}{x^2-9} - \dfrac{4}{3x^2-12}$

46. $\dfrac{x-4}{x^2-4} - \dfrac{x+2}{x^2} + \dfrac{1}{x^2-2x}$

47. $\dfrac{3 + \dfrac{x-1}{x+1}}{3 - \dfrac{x+1}{x-1}}$

48. $\dfrac{\dfrac{1}{x+h+3} - \dfrac{1}{x+3}}{h}$

49. $\dfrac{\dfrac{x+h+2}{x+h+3} - \dfrac{x+2}{x+3}}{h}$

50. $\dfrac{x^2+3x+2}{x^2-x-2} \cdot \dfrac{x^2+x-2}{x^2+4x+4} \div \dfrac{x^2-x}{x^2+x-6}$

2.8 THE BINOMIAL THEOREM

We have already given (in Section 2.6) rules for expanding the second and third powers of the binomial $x + y$. (A *binomial* is simply an algebraic expression with two terms.) It is desirable to have a simple rule for expanding the general nth power of the binomial $x + y$, where n is a positive integer. To find such a rule, let us examine the following expansions, each of which is easily verified by direct multiplication.

$$(x + y)^1 = x + y$$
$$(x + y)^2 = x^2 + 2xy + y^2$$
$$(x + y)^3 = x^3 + 3x^2y + 3xy^2 + y^3$$
$$(x + y)^4 = x^4 + 4x^3y + 6x^2y^2 + 4xy^3 + y^4$$

Careful inspection of the above expansions suggests these facts regarding the expansion of $(x + y)^n$:

1. The first term is x^n, the last term is y^n, and the total number of terms is $n + 1$.
2. The symbol x occurs in each term except the last, and the exponent of x decreases by 1 in each succeeding term. The symbol y occurs in each term except the first; its exponent is 1 in the second term and increases by 1 in each succeeding term. In each term the sum of the exponents of x and y is n.
3. The coefficient of the second term is n. From any given term, the coefficient of the succeeding term may be obtained by multiplying the coefficient of the given term by the exponent of x in that term and then dividing this product by the number of the given term.

The above observations enable us to write a general formula, called *The Binomial Theorem*, for the expansion of the general nth power of the binomial $x + y$, where n is a positive integer.

THEOREM 2.1 (THE BINOMIAL THEOREM). *For any positive integer n,*

$$(x + y)^n = x^n + nx^{n-1}y + \frac{n(n-1)}{2}x^{n-2}y^2$$
$$+ \frac{n(n-1)(n-2)}{2 \cdot 3}x^{n-3}y^3 + \cdots + y^n$$

A rigorous proof of this theorem is beyond the scope of this text and will be omitted here.

To be able to use the binomial theorem effectively in expanding binomials, the student should carefully memorize the rule for determining from a given term the coefficient of the succeeding term. (See Fact 3 in the preceding list.)

Example 1 Expand $(x + y)^6$.

Solution: The first two terms of the expansion are obviously $x^6 + 6x^5y$.

Third term of expansion: $\quad \frac{6 \cdot 5}{2}x^4y^2 = 15x^4y^2$

Fourth term of expansion: $\quad \frac{15 \cdot 4}{3}x^3y^3 = 20x^3y^3$

SEC. 2.8   The Binomial Theorem   115

Fifth term of expansion: $\dfrac{20\cdot 3}{4}x^2y^4 = 15x^2y^4$

Sixth term of expansion: $\dfrac{15\cdot 2}{5}xy^5 = 6xy^5$

Seventh term of expansion: $\dfrac{6\cdot 1}{6}x^0y^6 = y^6$

Therefore,

$$(x+y)^6 = x^6 + 6x^5y + 15x^4y^2 + 20x^3y^3 + 15x^2y^4 + 6xy^5 + y^6$$

Another observation is useful at this point. Notice that if one writes the expansions of $x + y$ in the following form an unusual thing happens, if the coefficients are placed in a similar pyramid form:

$(x+y)^0 = 1$ 1

$(x+y)^1 = x + y$ 1   1

$(x+y)^2 = x^2 + 2xy + y^2$ 1   2   1

$(x+y)^3 = x^3 + 3x^2y + 3xy^2 + y^3$ 1   3   3   1

$(x+y)^4 = x^4 + 4x^3y + 6x^2y^2 + 4xy^3 + y^4$ 1   4   6   4   1

⋮   ⋮

Notice that each number in this pyramid is the sum of the two numbers above it, to the left and to the right. This is one rather neat way to obtain the expansion of $(x + y)^n$ for small values of n. The number pyramid is called Pascal's pyramid or Pascal's triangle.

Example 2   Expand $(x + y)^6$ using Pascal's triangle.

Solution:

$n = 0$						1						
$n = 1$					1		1					
$n = 2$				1		2		1				
$n = 3$			1		3		3		1			
$n = 4$		1		4		6		4		1		
$n = 5$	1		5		10		10		5		1	
$n = 6$	1	6		15		20		15		6		1

Hence

$$(x+y)^6 = x^6 + 6x^5y + 15x^4y^2 + 20x^3y^3 + 15x^2y^4 + 6xy^5 + y^6$$

It often happens that it is necessary to expand a binomial in which one or both of its terms contain two or more factors. In such a case it is best to enclose each term in parentheses, use the rule to expand the binomial, and then simplify by removing all parentheses. The following example illustrates how this is done.

Example 3

Expand $(2x - y^2)^5$.

Solution: $(2x - y^2)^5 = [(2x) + (-y^2)]^5$

$= (2x)^5 + 5(2x)^4(-y^2) + 10(2x)^3(-y^2)^2$

$+ 10(2x)^2(-y^2)^3 + 5(2x)(-y^2)^4 + (-y^2)^5$

Simplifying,

$(2x - y^2)^5 = 32x^5 - 80x^4y^2 + 80x^3y^4 - 40x^2y^6 + 10xy^8 - y^{10}$

Exercises 2.8

In Exercises 1 through 12, expand each given expression and simplify where possible. (Use Facts 1, 2, and 3 listed at the beginning of this section.)

1. $(x + h)^3$
2. $(x + 2)^4$
3. $(x - 2)^5$
4. $(x + 2y)^5$
5. $(2x + y)^6$
6. $(x - 2y)^5$
7. $(x + y)^7$
8. $(3x + 2y)^4$
9. $(x^2 + 2y)^4$
10. $\left(x - \dfrac{1}{x}\right)^5$
11. $(x - y)^7$
12. $\left(x + \dfrac{1}{\sqrt{x}}\right)^8$

Find and simplify the first four terms in the expansion of Exercises 13 through 16.

13. $(x + y)^{20}$
14. $(a - b)^{40}$
15. $(x + 2y)^{10}$
16. $(1 - x^2)^{12}$

17. The $(r + 1)$st term in the expansion of $(x + y)^n$ is $\dfrac{(r + 1)(r + 2) \ldots (n)}{(1)(2)(3) \ldots (n - r)} x^{n-r} y^r$, where $0 \leq r \leq n$. Use this fact to find the indicated term in the expansion of each given expression.

   a. Sixth term of $(x + y)^{10}$
   b. Fourth term of $(x - y)^8$
   c. Tenth term of $(x - 2y)^{14}$
   d. Twentieth term of $(2x - y)^{30}$

   Solution to c: Find tenth term of $(x - 2y)^{14}$
   $r + 1 = 10$, hence $r = 9$, $n = 14$, $n - r = 5$

   Tenth term: $\dfrac{10 \cdot 11 \cdot 12 \cdot 13 \cdot 14}{1 \cdot 2 \cdot 3 \cdot 4 \cdot 5} x^5 (-2y)^9 = -1{,}025{,}024 x^5 y^9$

18. By writing $(1.01)^5$ as $(1 + 0.01)^5$, expand and thus evaluate, rounding off your answer correct to five decimal places.

3

Polynomials, Equations, and Inequalities

Introduction

Probably the most common problem encountered in elementary mathematics is that of solving an equation or inequality of some type. It is our purpose in this chapter to develop methods and procedures for solving the following:

1. Polynomial equations in one variable.
2. Polynomial inequalities in one variable.
3. Equations and inequalities whose solutions may be reduced to the problem of solving a polynomial equation or inequality.

3.1 POLYNOMIALS IN ONE VARIABLE

Recall that in Section 2.4 (Definition 2.9) we defined a polynomial in the variable x to be an algebraic expression which can be written in the form

$$A_n x^n + A_{n-1} x^{n-1} + \cdots + A_1 x + A_0$$

where the A's are numbers belonging to some number system (usually the A's are real numbers) and n is a positive integer or zero. It would be useful if the student would review the terminology developed in Section 2.4 before proceeding further. We shall deal almost entirely with polynomials having real numbers as coefficients. Occasionally, complex numbers will appear as coefficients in a polynomial. Typical examples of polynomials are:

(a) $3x^2 - 7x + 5$
(b) $5x - 8$
(c) $x^3 + 2ix - 4$
(d) $2x^7 + 5x^3 + 1$
(e) $\frac{1}{2}x^4 - 9$
(f) $0x + 6$

For convenience, we shall at times use a symbol such as $P(x)$, read "P of x," to denote a polynomial involving the variable x. In such a case we may write

$$P(x) = A_n x^n + A_{n-1} x^{n-1} + \cdots + A_1 x + A_0$$

By $P(c)$ we shall mean the number obtained by substituting the number c for x in the polynomial $P(x)$.

Example 1 If $P(x) = 2x^3 - 5x^2 + 7x + 4$, find $P(2)$ and $P(-1)$.

Solution:
$P(2) = 2(2)^3 - 5(2)^2 + 7(2) + 4$
$= 16 - 20 + 14 + 4 = 14$
$P(-1) = 2(-1)^3 - 5(-1)^2 + 7(-1) + 4$
$= -2 - 5 - 7 + 4 = -10$

Notice that we have no difficulty substituting any real number for x in a polynomial, and the result of the computation is a real number in each case. This is not true of algebraic expressions in general. For example, we are not allowed to substitute 1 for x in the expression $\dfrac{1}{x-1}$, nor are we allowed to substitute -3 for x in the expression $\sqrt{x+2}$ (unless we are interested in complex numbers at the time). In future discussions we shall use symbols such as $f(x), g(x), u(x), v(x)$, etc., read "f of x," "g of x," "u of x," "v of x," etc., to denote algebraic expressions which may or may not be polynomials. Whenever such a symbol is used to denote a polynomial it will be so stated.

In view of the definition of a polynomial, none of the following algebraic expressions are polynomials:

$$3x^2 + 7\sqrt{x} + 2$$

$$x^3 + 5x - \frac{7}{x} + 3$$

$$\frac{x^4 - 3x^2 + 10}{x^2}$$

Exercises 3.1

1. State whether each of the following expressions is or is not a polynomial. For each polynomial, give its degree n, its leading coefficient A_n, and its constant term A_0.

    a. $2x^4 - 5x^3 - 4x^2 + 7x + 10$
    b. $x^3 - 4x + 2\sqrt{x} - 5$
    c. $8 + 5x - 4x^3 - 6x^2$
    d. 8

SEC. 3.2   Equations   119

    e. $(3x-2)^2 + 4(2x+5) - 6$ f. $\dfrac{x-1}{x+2} + \dfrac{3x}{x-5} + 8$

    g. $\sqrt{x+1} + 5x - 4$ h. $2x^3 + \sqrt{5x^2 - 4x + 2}$

2. If $f(x) = x^3 - 4x^2 + 5x + 3$, find $f(0)$, $f(-2)$, $f(3)$, and $f(a)$.
3. Let $f(x) = 3x^4 - 2x^3 + 8x + 7$ and $g(x) = 4x^3 + 5x^2 - 6$. Find:
    a. $f(x) + g(x)$ b. $f(x) - g(x)$
4. Let $f(x) = x^5 - 4x^3 + 2$ and $g(x) = 3x^2 - 1$. Find:
    a. $f(x)g(x)$ b. $[g(x)]^2$
5. Let $f(x) = 3x - 1$, $g(x) = 2x + 5$, and $h(x) = 6 - 5x$. Find:
    a. $f(x) + g(x) - h(x)$ b. $f(x)g(x) + [h(x)]^2$

3.2 EQUATIONS

It is common practice to call any statement involving equality relation, that is, involving the equality symbol, =, an *equation*. In fact, the student is quite accustomed to seeing statements of the form

$$2 + 3 = 5$$

$$5 + 7 = 10$$

$$x + 2 = 7$$

$$2x + 5x = 7x$$

$$x^2 + 3x + 2 = 5 + 2x - x^2$$

and has learned to refer to such statements as equations. We shall use the same terminology in the remainder of this text. A very important fact concerning equations is that an equation may be a true statement, it may be false, or it may be neither true nor false. Consider, for example, the equation $2 + 3 = 5$. This is certainly a true statement, since $2 + 3$ and 5 are just different symbols for the same number. The equation $5 + 7 = 10$ is obviously false since $5 + 7$ and 10 are symbols which represent two entirely different numbers. We could change this equation into a true statement by writing $5 + 7 \neq 10$, the negation of the original statement. On the other hand, consider the equation $x + 2 = 7$. This equation is an open statement in the sense that, until some particular number is substituted for x, it is neither true nor false. It is obvious, however, that if we substitute the number 5 for x, the equation becomes a true statement because $5 + 2 = 7$ is true. It is equally obvious that if we substitute any number other than 5 for x the equation becomes a false statement.

    Recall that in Chapter 2 we used letters near the beginning of the alphabet to represent fixed numbers and called them *arbitrary constants*. We also adopted the convention of using letters near the end of the alphabet to denote *variables*. Arbitrary constants enable us to represent by a single equation a set of particular equations. For example, $3x - 5 = 0$ and $2x + 7 = 0$ are particular first-degree equations in the variable x. However, if we make the statement $ax + b = 0$, we are making a statement about these equations and, in fact, we are making a statement about all possible first-degree equations in the variable x; that is, no matter what numbers are substituted for a and b in the equation $ax + b = 0$, provided $a \neq 0$, the result is always a first-degree equation in the variable x.

It will be convenient for our purposes to think of equations as being of two kinds, numerical equations and equations involving variables. By a numerical equation we mean an equation that makes a statement about one or more specific numbers, $2 + 3 = 5$, $4 + 2 = 7 - 1$, and $5 + 3 = 8$. As has been pointed out above, a numerical equation is a true statement or a false statement, but not both. We shall call a numerical equation that makes a true statement a *numerical identity*. The equations just cited are numerical identities.

Equations involving variables include all equations that contain one or more letters representing arbitrary numbers from some set of numbers. Some typical equations of this kind are $3x + 2 = 10$, $2x + 3x = 5x$, $x^2 + 3x + 8 = 0$, $x^2 - 5x = 3x^2 + 7$, $y = 4 - x^2$, $x^2 + y^2 - 5x + 2y = 9$, $ax + by + c = 0$, etc.

As we have seen, an equation involving a variable is neither true nor false until some specific number is substituted for the variable in the equation. If this is done, however, the equation becomes either true or false, but not both. Equations that are true for all values of their variables, such as $2x + 3x = 5x$ or $(x + y)^2 = x^2 + 2xy + y^2$, also are called *identities*. Equations that are true for some values of their variables and false for others, such as $3x = 15$ or $x + y = 5$, are called *conditional equations*.

The expression to the left of the equality sign in an equation is often referred to as the *left side* of the equation, and the expression to the right of the equality sign as the *right side*.

DEFINITION 3.1. An equation involving one or more variables is said to be an *identity* if the equation is true for all values of its variables for which both the left side and the right side are defined.

DEFINITION 3.2. An equation involving one or more variables is said to be a *conditional equation* if it is true for some values of its variables and false for others.

Since in an identity the left side has the same value as the right side for all possible values of their variables, either side may be substituted for the other in any expression, statement, or equation, wherever one side may occur, without changing the possible values of the expression or the truth or falsity of the statement or equation. This application, of substituting one side of an identity

SEC. 3.2 Equations

for the other, is one of the most important uses of identities and is certainly the one that occurs most often. In fact, each time we change from the expanded form of an expression to the factored form or combine terms in an expression or, in general, substitute equals for equals, we are making use of one or more identities.

Solutions and Solution Sets

One of the basic problems in elementary mathematics is that of "solving conditional equations." In this and succeeding sections, we shall be concerned with discussing methods and procedures for solving a particular type of conditional equation called a *polynomial equation in one variable*.

Typical examples of such equations are:

(a) $2x^3 + 4x - 1 = x^2 + 6x + 7$
(b) $(x+2)(x-3) = x + 9$
(c) $x^2 - 3x = 10$
(d) $x^4 - 2x^3 - 4x^2 + x + 4 = 0$

DEFINITION 3.3. A *polynomial equation in the variable x* is an equation of the form $f(x) = g(x)$, where $f(x)$ and $g(x)$ are polynomials in x.

Let us begin our discussion of ways and means of solving conditional equations in one variable by familiarizing ourselves with some terminology and notation that may or may not be new to you. First, we want to state precisely what is meant by a *solution* of an equation involving one variable. As was pointed out in our previous remarks, an equation involving a variable is an open statement and as such is neither true nor false. However, when a specific number a is substituted for the variable, our equation becomes a proposition (see Appendix A) and is either true or false, but not both.

If, when we substitute a for x in a particular equation, the resulting equation is true (that is, the resulting equation is a numerical identity), then a is called a *solution*, or a *root*, of the equation. For example, the number 2 is a solution of the equation

$$2x^3 + 4x - 1 = x^2 + 6x + 7$$

because, when 2 is substituted for x in the left side of the equation, we obtain

$$2(2)^3 + 4(2) - 1 = 16 + 8 - 1 = 23$$

and when 2 is substituted for x in the right side of the equation we obtain

$$(2)^2 + 6(2) + 7 = 4 + 12 + 7 = 23$$

Therefore, when 2 is substituted for x in the equation, we obtain the numerical identity

$$2(2)^3 + 4(2) - 1 = (2)^2 + 6(2) + 7$$

and so 2 is a solution of the equation

$$2x^3 + 4x - 1 = x^2 + 6x + 7$$

A particularly nice form in which to write the verification that 2 is a solution is

(1)
$$2(2)^3 + 4(2) - 1 = 16 + 8 - 1 = 23$$
$$\parallel$$
$$(2)^2 + 6(2) + 7 = 4 + 12 + 7 = 23$$

where the lines are written left to right, in order, and the last (vertical) equals sign is written only after both 23's have been obtained. By reading left to right on the first line and then reading right to left on the second line, we see that the following sequence of equalities holds:

(2)
$$2(2)^3 + 4(2) - 1 = 16 + 8 - 1$$
$$= 23$$
$$= 4 + 12 + 7$$
$$= (2)^2 + 6(2) + 7$$

This is perhaps an even more preferable presentation of the verification that 2 is a solution. However, one generally begins at both ends and works toward the middle and, in this form, it is difficult to decide how far apart to place $2(2)^3 + 4(2) - 1$ and $(2)^2 + 6(2) + 7$. For this reason, we shall use the form (1) rather than form (2) when checking answers within the text. This form can be used equally well to show that 3 is not a solution of the equation, as follows:

$$2(3)^3 + 4(3) - 1 = 54 + 12 - 1 = 65$$
$$\not\parallel$$
$$(3)^2 + 6(3) + 7 = 9 + 18 + 7 = 34$$

Example 1 Show that 5 is a solution of the equation $x^2 - 3x = 10$.

Solution: $(5)^2 - 3(5) = 25 - 15 = 10$

Example 2 Show that -3 is a solution of the equation $(x+2)(x-3) = x + 9$.

Solution: $(-3+2)(-3-3) = (-1)(-6) = 6$
$$\parallel$$
$$-3 + 9 \qquad\qquad = 6$$

Example 3 Show that 1 is not a root of the equation

$$2x^2 - 3x + 11 = x^4 + 5x - 7$$

SEC. 3.2 Equations

Solution: $2(1)^2 - 3(1) + 11 = 2 - 3 + 11 = 10$
$(1)^4 + 5(1) - 7 = 1 + 5 - 7 = -1$

We summarize the above discussion in the following definitions.

DEFINITION 3.4. A *solution*, or *root*, of an equation in one variable is a number which, when substituted for the variable in the equation, reduces the equation to a numerical identity. In particular, the number a is a solution, or root, of the equation $f(x) = g(x)$ if and only if $f(a) = g(a)$, where $f(x)$ and $g(x)$ are algebraic expressions in x.

The instruction "solve the equation $f(x) = g(x)$" means "find all the roots of the equation $f(x) = g(x)$." We shall find the next definition useful.

DEFINITION 3.5. The set S, consisting of all the roots of a given equation in one variable, is called the *solution set* of the equation. Thus, to solve an equation is to find its solution set.

It should be observed that Definitions 3.4 and 3.5 apply to all equations in one variable and not just to polynomial equations.

That two or more equations may have the same solution set is easily verified. For example, consider the equations

$$5x - 2 = 2x + 4, \qquad 3x = 6, \qquad x = 2$$

Although no two of these equations are the same, a simple check will show that each has the same solution set, $S = \{2\}$. In such a case we say the equations are equivalent. To be more specific, we make the following definition.

DEFINITION 3.6. Two equations in one variable are said to be *equivalent* if and only if their solution sets are the same, that is, are identical.

To indicate that two equations are equivalent, we use the symbol $\leftrightarrow$ (which is also used for bi-implications in Appendix A). Thus, to indicate that the equation $f(x) = g(x)$ is equivalent to the equation $u(x) = v(x)$, we write

$$f(x) = g(x) \leftrightarrow u(x) = v(x)$$

which is read, "The equation $f(x) = g(x)$ is equivalent to the equation $u(x) = v(x)$." For example, referring to the three equivalent equations listed above, we may write $5x - 2 = 2x + 4 \leftrightarrow 3x = 6$, $3x = 6 \leftrightarrow x = 2$, and hence $5x - 2 = 2x + 4 \leftrightarrow x = 2$; or, more simply,

$$5x - 2 = 2x + 4 \leftrightarrow 3x = 6 \leftrightarrow x = 2$$

Warning

> The student is cautioned that the equals symbol, $=$, should never be used to indicate the equivalence of two equations. The equations $3x = 6$ and $x = 2$ would not be called "equal." In fact, we do not even define "equality of equations." So it is permissible (and correct) to write $3x = 6 \leftrightarrow x = 2$, whereas it is not permissible to write $3x = 6 = x = 2$.

The importance of the concept of equivalent equations lies in the fact that it is the basis of a procedure for solving equations. To see that this statement is true, let us say that an equation obtained from another equation by any process whatever is called a *derived equation*. In such a case we say that the second equation was derived from the first. Now suppose we wish to solve a given equation. If through one or more equivalent equations, each derived from the preceding, we finally arrive at an equation whose solution set is obvious, we have thus found the solution of the given equation. To carry out such a procedure, we must know what operations produce derived equations equivalent to the given equation. It can be shown that the following operations always produce a derived equation, which is equivalent to the equation from which it was derived.

1. *Changing the order of the terms in one side of the equation (or in both sides separately).*

Example 4 $3 - 5x^2 + 2x + x^3 = 0 \leftrightarrow x^3 - 5x^2 + 2x + 3 = 0$

Example 5 $x + 5x^3 - 3 = -11x + 5 - 8x^2 \leftrightarrow 5x^3 + x - 3 = -8x^2 - 11x + 5$

2. *Combining like terms in one side of the equation (or in both sides separately).*

Example 6 $5x - 3x + x = 10 - 4 \leftrightarrow 3x = 6$

Example 7 $3x^2 - 2x + 5x - 1 = x^2 - 4x^2 - 8x + 7 - 5$
$\leftrightarrow 3x^2 + 3x - 1 = -3x^2 - 8x + 2$

3. *Using an identity to change one side of the equation (or using two identities to change both sides simultaneously).*

Example 8 $x^2 + 2x + 1 = 0 \leftrightarrow (x + 1)^2 = 0$

Example 9 $x^2 - 2x + 1 = x^2 - 1 \leftrightarrow (x - 1)^2 = (x - 1)(x + 1)$

SEC. 3.2 Equations

4. *Multiplying both sides of the equation by the same nonzero constant k.*

Example 10
$$\frac{3}{4}x + 1 = \frac{1}{2}x - 3 \leftrightarrow 4\left(\frac{3}{4}x + 1\right) = 4\left(\frac{1}{2}x - 3\right)$$
$$\leftrightarrow 3x + 4 = 2x - 12$$

Here $k = 4$.

5. *Dividing both sides of the equation by the same nonzero constant c.*

Example 11
$$3x^2 - 6x - 9 = 0 \leftrightarrow \frac{1}{3}(3x^2 - 6x - 9) = 0$$
$$\leftrightarrow x^2 - 2x - 3 = 0$$

Here $c = 3$.

6. *Adding or subtracting the same polynomial to or from both sides of the equation.*

Example 12
$$5x - 2 = 2x + 4 \leftrightarrow 5x - 2 + (2 - 2x) = 2x + 4 + (2 - 2x)$$
$$\leftrightarrow 5x - 2x = 4 + 2 \leftrightarrow 3x = 6$$

Here we have added the polynomial $2 - 2x$ to both sides.

Example 13
$$3x^2 - 2x + 7 = x^2 - x + 2$$
$$\leftrightarrow (3x^2 - 2x + 7) - (x^2 - x + 2) = (x^2 - x + 2) - (x^2 - x + 2)$$
$$\leftrightarrow x^2 - x + 5 = 0$$

Here we have subtracted the polynomial $x^2 - x + 2$ from both sides.

Example 13 illustrates the fact that, if we subtract the right-hand side of a polynomial equation from both sides, we obtain an equivalent equation with 0 on the right-hand side. Symbolically,

$$f(x) = g(x) \leftrightarrow f(x) - g(x) = 0$$

Hence, if we so desire, we can always reduce a polynomial equation to an equivalent equation of the form $F(x) = 0$. Such polynomial equations [those in the form $F(x) = 0$] are said to be in *standard form*.

Example 14 Write the equation $(x - 1)(2x + 3) = (x + 2)(5 - x^2)$ in standard form.

Solution: Performing the indicated multiplication in each side and applying one or more of the six operations listed above, we obtain:

$$2x^2 + x - 3 = 5x - x^3 + 10 - 2x^2$$
$$\leftrightarrow 2x^2 + x - 3 - (5x - x^3 + 10 - 2x^2) = 0$$
$$\leftrightarrow x^3 + 4x^2 - 4x - 13 = 0$$

Reducing a polynomial equation to standard form is one of the first steps we shall take in solving such equations in almost every case. One notable exception is the case of linear, or first-degree, equations. These will be treated in the next section. We need the following definitions to accurately describe what we mean by first-degree equations.

DEFINITION 3.7. The *degree* of a polynomial equation in one variable, $F(x) = 0$ (written in standard form), is the same as the degree of the polynomial $F(x)$.

Example 15 The degree of the equation $3x^3 - 2x^2 - 11x + 8 = 0$ is 3.

Example 16 What is the degree of the equation
$$5x^2 - 3x + 2x^2 - 1 = 10x^2 - 5x - 3x^2 + 15$$

Solution: First we reduce the equation to standard form:
$$5x^2 - 3x + 2x^2 - 1 = 10x^2 - 5x - 3x^2 + 15$$
$$\leftrightarrow 7x^2 - 3x - 1 = 7x^2 - 5x + 15$$
$$\leftrightarrow 2x - 16 = 0$$

Now we observe that the degree of the equation is 1 (even though the original form of the equation had some second powers of x involved).

Example 16 illustrates the fact that, to determine the degree of a polynomial equation in one variable, one first writes the equation in standard form and then applies Definition 3.7.

Exercises 3.2

In each exercise, write the given polynomial equation in standard form and give its degree.

1. $x^2 + 2x - 3 = (x - 1)^2 + 7$
2. $(2x + 3)(5x - 7) = 0$
3. $(x + 2)^3 = 10$
4. $5x - 4 = 3x + 6$
5. $(2x + 3)^2 - 5(2x + 3) = 6$
6. $2x^2 + 4x + 3 = (2x - 1)(x + 2)$
7. $(\sqrt{2}x + 1)(\sqrt{2}x - 1) = 3 - x - x^2$
8. $4x^3 - x^2 + x + 1 = x^3 - 3x^2 + 7x - 2$
9. $\dfrac{2}{3}x^2 - \dfrac{1}{7}x + \dfrac{5}{17} = \dfrac{1}{6}x^2 + \dfrac{3}{5}x - \dfrac{5}{9}$
10. $x - \dfrac{1}{6}x^2 + \sqrt{2} = \dfrac{2}{5}x^2 - \sqrt{3} + \sqrt{7}x$

3.3 LINEAR EQUATIONS IN ONE VARIABLE

Perhaps the simplest polynomial equations are those of the type:

(a) $2x = 3$ (b) $-\frac{2}{3}x + 2 = 0$

(c) $\sqrt{3}x = \sqrt{2}x - 1$ (d) $x + 3 = 5x + 1$

Notice that the largest power of x that appears is the first power (that is, each of these polynomial equations has degree 1). Such polynomial equations are called *linear equations in one variable*.

DEFINITION 3.8. A *linear equation in one variable* is a polynomial equation whose degree is 1.

Linear equations in one variable can be solved rather easily, as the following example illustrates:

Example 1 Solve each of the linear equations (a) to (d) listed above.

Solution: (a) $2x = 3 \leftrightarrow x = \frac{3}{2}$ (dividing both sides by 2). Now the solution set S of the derived equation $x = \frac{3}{2}$ is easily seen to be $S = \left\{\frac{3}{2}\right\}$. Hence the solution set of the original equation $2x = 3$ is also $S = \left\{\frac{3}{2}\right\}$.

(b) $-\frac{2}{3}x + 2 = 0 \leftrightarrow -\frac{2}{3}x = -2$ (Subtracting 2 from both sides)

$\leftrightarrow x = \frac{(-2)}{-\left(\frac{2}{3}\right)}$ $\left(\text{Dividing both sides by } -\frac{2}{3}\right)$

$\leftrightarrow x = 3$

Again, the solution set S of the derived equation $x = 3$ is easily seen to be $S = \{3\}$, and so this is also the solution set of the original equation $-\frac{2}{3}x + 2 = 0$.

(c) $\sqrt{3}x = \sqrt{2}x - 1$

$\leftrightarrow \sqrt{3}x - \sqrt{2}x = -1$ (Subtracting $\sqrt{2}x$ from both sides)

$\leftrightarrow (\sqrt{3} - \sqrt{2})x = -1$

$\leftrightarrow x = \frac{(-1)}{\sqrt{3} - \sqrt{2}}$ (Dividing both sides by $\sqrt{3} - \sqrt{2}$)

Once again we see that $S = \left\{\dfrac{(-1)}{\sqrt{3}-\sqrt{2}}\right\}$. Various algebraic manipulations might be performed on this "answer" to obtain other "answers," which are just as correct. For example,

$$S = \left\{-\dfrac{1}{\sqrt{3}-\sqrt{2}}\right\}, \quad S = \left\{\dfrac{1}{\sqrt{2}-\sqrt{3}}\right\}, \quad \text{and} \quad S = \{-\sqrt{3}-\sqrt{2}\}$$

are all correct (the last is obtained by multiplying numerator and denominator by $\sqrt{3}+\sqrt{2}$, a process known as rationalizing the denominator).

(d) $x + 3 = 5x + 1 \leftrightarrow x - 5x = 1 - 3$. Here we have collected like terms on opposite sides by first subtracting $5x$ from both sides and then subtracting 3 from both sides (so we have skipped a step). Actually, we could obtain the same result by the one step of subtracting $5x + 3$ from both sides. To complete the solution,

$$x - 5x = 1 - 3 \leftrightarrow -4x = -2$$
$$\leftrightarrow x = \dfrac{-2}{-4}$$
$$\leftrightarrow x = \dfrac{1}{2}$$

Hence the solution set is $S = \left\{\dfrac{1}{2}\right\}$.

The procedure followed in solving each of the above linear equations can be summarized as follows:

Step 1. Collect all terms involving x on the left and all constant terms on the right.
Step 2. Combine like terms on both sides.
Step 3. Divide both sides by the coefficient of x.

We illustrate the use of this procedure in two additional examples.

Example 2   Find the solution set S of the equation $3x - 2 = x + 8$.

Solution: $3x - 2 = x + 8 \leftrightarrow 3x - x = 8 + 2$ (*Step 1*)
$\leftrightarrow 2x = 10$ (*Step 2*)
$\leftrightarrow x = \dfrac{10}{2} = 5$ (*Step 3*)

Hence $S = \{5\}$.

If the original equation involves fractions, it is often useful to *clear the equation of fractions* before beginning with Step 1. This technique is illustrated in the next example.

SEC. 3.3 Linear Equations in One Variable

Example 3 Solve the equation $\dfrac{x+1}{4} - \dfrac{2x-9}{10} = \dfrac{3}{2}$.

Solution: We start by clearing the equation of fractional coefficients. We do this by multiplying both sides by 20, the least common denominator of all fractions involved. Our solution is as follows:

$$\dfrac{x+1}{4} - \dfrac{2x-9}{10} = \dfrac{3}{2}$$
$$\leftrightarrow 20\left(\dfrac{x+1}{4} - \dfrac{2x-9}{10}\right) = 20\left(\dfrac{3}{2}\right)$$
$$\leftrightarrow 5(x+1) - 2(2x-9) = 30$$
$$\leftrightarrow 5x + 5 - 4x + 18 = 30$$
$$\leftrightarrow 5x - 4x = 30 - 5 - 18$$
$$\leftrightarrow x = 7$$

Therefore, $S = \{7\}$.

If a linear equation is written in standard form, it appears as $A_1 x + A_0 = 0$ with $A_1 \neq 0$. It is a fairly standard practice to use the notation $ax + b = 0$ with $a \neq 0$ to represent a general linear equation. If we go through the steps of solving such a general equation, we obtain

$$ax + b = 0 \leftrightarrow ax = -b \qquad \text{(Step 1)}$$
$$\leftrightarrow x = \dfrac{(-b)}{a} = -\dfrac{b}{a} \qquad \text{(Step 3; Step 2 is not needed)}$$

Hence we find that the solution set of the general linear equation $ax + b = 0$ with $a \neq 0$ is $S = \left\{-\dfrac{b}{a}\right\}$. It follows from these observations that a linear equation in one variable has one and only one root. It will be shown in a later section that the graph of a polynomial function defined by an equation of the form $f(x) = ax + b$ (with $a \neq 0$) is a straight line. This fact is the motivation for calling equations that can be reduced to the form $ax + b = 0$ (with $a \neq 0$) linear equations.

Exercises 3.3

Solve each of the following equations.

1. $7x = 5x + 6$
2. $3x - 10 = 2x - 12$
3. $4x - 8 = 6x - 13$
4. $7(3x - 2) - 9(2x - 1) = 10$
5. $4(3x + 2) - 4x = 3(2 + x) + 17$
6. $(y - 2)(y + 1) = y(y - 3) - 6$
7. $3x^2 - x(2x - 3) = x^2 + x - 6$
8. $\dfrac{6x - 5}{4} = \dfrac{10 + 15}{2}$

9. $\dfrac{3-2x}{4} - \dfrac{5}{3} = \dfrac{3x-1}{6}$

10. $\dfrac{1}{2}\left(\dfrac{3}{4}x - 4\right) = \dfrac{5}{6}\left(4 - \dfrac{2}{3}x\right)$

11. $\dfrac{4}{3}\left(\dfrac{1}{2} - \dfrac{4}{5}x\right) = \dfrac{1}{2}\left(\dfrac{5}{3}x + \dfrac{1}{15}\right)$

12. $\dfrac{x}{2} - \dfrac{x}{5} = 3$

13. $\dfrac{2}{3}x - \dfrac{3}{5}x = \dfrac{4}{15}$

14. $\dfrac{3x}{4} + \dfrac{5x}{8} = \dfrac{x}{2} + \dfrac{7}{14}$

15. $\dfrac{3-2x}{4} = \dfrac{5}{3} + \dfrac{3x-1}{6}$

16. $\dfrac{5}{6} - y + 2 = \dfrac{4+3y}{3}$

17. $\dfrac{2x-3}{6} + \dfrac{12x+1}{9} = 2 - \dfrac{3x-7}{3}$

In each of the following exercises, solve for x in terms of the various constants involved. Then use your hand calculator to approximate the value of x rounded to two decimal places.

18. $4.316x - 18.24 = -6.441x + 8.46$

19. $\sqrt{7}x - 3.4 = \sqrt{3}x + 10.68$

20. $\dfrac{x}{5.34} - \dfrac{1}{2} = 2.74x + \sqrt{8}$

3.4 QUADRATIC EQUATIONS IN ONE VARIABLE

It was shown in Section 3.2 that every polynomial equation in one variable can be written in the standard form

$$A_n x^n + A_{n-1} x^{n-1} + \cdots + A_1 x + A_0 = 0$$

Hence every polynomial equation in one variable and of degree 2 can be written in the standard form $ax^2 + bx + c = 0$, with $a \neq 0$. Such equations are called *quadratic equations*.

DEFINITION 3.9. A *quadratic equation in one variable* is a polynomial in one variable whose degree is 2.

It will be sufficient for our purposes to assume that the coefficients a, b, and c in the standard form $ax^2 + bx + c = 0$ are real numbers, unless otherwise stated. If $a = 0$, the degree of the equation is not 2 and so the equation is not a quadratic equation.

In Section 3.3 we learned to solve any linear equation. We now proceed to

SEC. 3.4 Quadratic Equations in One Variable

give the techniques which will allow us to solve any quadratic equation. These are the method of factoring (which only works for certain special equations), the method of completing the square (which works for all quadratic equations), and the quadratic formula method (which works for all quadratic equations and is easier to apply than the method of completing the square). These methods will be treated in Sections 3.5, 3.7, and 3.8, respectively. In the remainder of this section we shall develop certain facts and processes which are extremely useful in solving both quadratic equations and higher-degree polynomial equations.

Solving Factored Polynomial Equations

We have found that the solution set of the equation $x - 1 = 0$ is $S_1 = \{1\}$, and the solution set of the equation $x - 5 = 0$ is $S_2 = \{5\}$. We can use these two facts to solve the equation $(x - 1)(x - 5) = 0$ as follows:

Example 1 Solve equation $(x - 1)(x - 5) = 0$.

Solution: According to Theorem 1.13 (or Rule 1.19), if a and b are real numbers, then $a \cdot b = 0$ if and only if $a = 0$ or $b = 0$. The counterpart of this fact for algebraic expressions is: If A and B are algebraic expressions, then $A \cdot B = 0$ if and only if $A = 0$ or $B = 0$. Applying this with $A = x - 1$ and $B = x - 5$, we see that

$$(x - 1)(x - 5) = 0 \quad \text{if and only if} \quad (x - 1 = 0 \text{ or } x - 5 = 0)$$

Now, $x - 1 = 0$ if and only if $x = 1$, and $x - 5 = 0$ if and only if $x = 5$. Hence

$$(x - 1 = 0 \text{ or } x - 5 = 0) \quad \text{if and only if} \quad (x = 1 \text{ or } x = 5)$$

Therefore, $(x - 1)(x - 5) = 0$ if and only if $x = 1$ or $x = 5$. It follows from the definition of a solution set that the solution set for $(x - 1)(x - 5) = 0$ is $S = \{1, 5\}$, the set consisting of the two numbers 1 and 5. We shall adopt the following shorthand approach to the solution of this problem:

$$(x - 1)(x - 5) = 0 \leftrightarrow x - 1 = 0 \quad \text{or} \quad x - 5 = 0$$
$$\leftrightarrow x = 1 \quad \text{or} \quad x = 5$$

Hence $S = \{1, 5\}$.

Notice that we have used the $\leftrightarrow$ equivalence symbol to assert that the original equation is equivalent to the *disjunction* of two equations (see Appendix A). This is a new twist but it should cause no confusion. One final observation is that the solution set S of the equation $(x - 1)(x - 5) = 0$ is the *union* of the solution sets $S_1 = \{1\}$ of the equation $x - 1 = 0$ and $S_2 = \{5\}$ of the equation $x - 5 = 0$ (see Appendix A). That is, S consists of those numbers which are in S_1 or in S_2 or both. (The union of two sets S_1 and S_2 is denoted $S_1 \cup S_2$.) This fact motivates the next theorem.

THEOREM 3.1. *Suppose $f(x)$ and $g(x)$ are polynomials in x. The solution set S of the equation $f(x)g(x) = 0$ is the union of the solution set T_1 of $f(x) = 0$ and the solution set T_2 of $g(x) = 0$, that is, $S = T_1 \cup T_2$.*

The next theorem is an easily obtained generalization of Theorem 3.1.

THEOREM 3.2. *Let $f_1(x)$, $f_2(x)$, ..., $f_n(x)$ be polynomials in x and let $F(x) = f_1(x)f_2(x) \ldots f_n(x)$. The solution set of the equation $F(x) = 0$ is the union of the solution sets of the equations $f_1(x) = 0$, $f_2(x) = 0$, ..., $f_n(x) = 0$.*

Example 2

Find the solution set S of the equation $(x-1)(x-5)(x-8) = 0$.

First Solution: Let T_1 be the solution set of $x - 1 = 0$, T_2 the solution set of $x - 5 = 0$, and T_3 the solution set of $x - 8 = 0$. Then since

$$x - 1 = 0 \leftrightarrow x = 1, \quad x - 5 = 0 \leftrightarrow x = 5, \quad \text{and} \quad x - 8 = 0 \leftrightarrow x = 8$$

it follows that $T_1 = \{1\}$, $T_2 = \{5\}$, and $T_3 = \{8\}$. Therefore, applying Theorem 3.2, we have

$$S = T_1 \cup T_2 \cup T_3 = \{1, 5, 8\}$$

Alternate Solution:

$$(x-1)(x-5)(x-8) = 0 \leftrightarrow x - 1 = 0 \quad \text{or} \quad x - 5 = 0 \quad \text{or} \quad x - 8 = 0$$
$$\leftrightarrow x = 1 \quad \text{or} \quad x = 5 \quad \text{or} \quad x = 8$$

Hence $S = \{1, 5, 8\}$.

Example 3

Solve the equation $\left(\dfrac{5}{11}x - \sqrt{3}\right)\left(-\sqrt{2}x - \dfrac{15}{16}\right) = 0$.

Solution

$$\left(\frac{5}{11}x - \sqrt{3}\right)\left(-\sqrt{2}x - \frac{15}{16}\right) = 0 \leftrightarrow \frac{5}{11}x - \sqrt{3} = 0 \quad \text{or} \quad -\sqrt{2}x - \frac{15}{16} = 0$$

$$\leftrightarrow \frac{5}{11}x = \sqrt{3} \quad \text{or} \quad -\sqrt{2}x = \frac{15}{16}$$

$$\leftrightarrow x = \frac{11\sqrt{3}}{5} \quad \text{or} \quad x = -\frac{15}{16\sqrt{2}}$$

Hence $S = \left\{\dfrac{11\sqrt{3}}{5}, -\dfrac{15}{16\sqrt{2}}\right\}$. We could use a hand calculator to approximate these two numbers and obtain $x \cong 3.81$ or $x \cong -0.66$ (rounded to two decimal places), where $\cong$ means "is approximately equal to."

Exercises 3.4

Solve each of the following equations.

1. $(x-5)(3x+1) = 0$
2. $(x-5)(x+3)(x-4)(x-7) = 0$
3. $x(3x-2) = 0$
4. $(x - \sqrt{17})(5x - 3) = 0$
5. $\left(\dfrac{2}{3}x - 3\right)\left(\dfrac{2}{7}x - \dfrac{14}{3}\right) = 0$
6. $(\sqrt{3}x - 2)(4x - \sqrt{7})\left(\dfrac{2}{9}x + \dfrac{1}{7}\right) = 0$
7. $(4x - 7)^2 = 0$ (see note following Example 2 of Section 3.5).

In each of the following exercises, first express x in terms of the various constants involved and then use a hand calculator to approximate each solution, rounded to two decimal places.

8. $\left(\dfrac{3}{2}x - \dfrac{1}{5}\right)^3 \left(-\dfrac{3}{8}x + 2\right)^2 (3x - \sqrt{27})^5 = 0$
9. $(12.375x - 7.846)(-3.241x - 6.832) = 0$
10. $(\sqrt{5}x - \sqrt{3})(-\sqrt{3}x + 7) = 0$

3.5 THE METHOD OF FACTORING

Section 2.6 was devoted to special products and factoring of polynomial expressions. In many instances it is possible to write a quadratic equation in standard form and then factor the left side. When this is possible, we can use the techniques of Section 3.4 to solve the equation. We illustrate with seven examples.

Example 1   Solve the equation $x^2 - 1 = 0$.

Solution: $x^2 - 1 = 0 \leftrightarrow (x - 1)(x + 1) = 0$
$\leftrightarrow x - 1 = 0$ or $x + 1 = 0$
$\leftrightarrow x = 1$ or $x = -1$

Hence $S = \{1, -1\} = \{\pm 1\}$

Example 2 | Solve the equation $4x^2 - 4x + 1 = 0$.

Solution: $4x^2 - 4x + 1 = 0 \leftrightarrow (2x-1)(2x-1) = 0$
$\leftrightarrow 2x - 1 = 0$
$\leftrightarrow 2x = 1$
$\leftrightarrow x = \dfrac{1}{2}$

Hence $S = \left\{\dfrac{1}{2}\right\}$

NOTE. Example 2 illustrates that a quadratic equation may have only one root. Actually $2x - 1$ is a factor of $4x^2 - 4x + 1$ twice, and we say that $\tfrac{1}{2}$ is a *double root* of the equation. Therefore, we might use an abuse of language and say that $4x^2 - 4x + 1 = 0$ has two roots, but each of them is $\tfrac{1}{2}$. We sometimes indicate this by writing $S = \{\tfrac{1}{2}, \tfrac{1}{2}\}$.

Example 3 | Solve the equation $22x^2 + 38x - 12 = 0$.

Solution: $22x^2 + 38x - 12 = 0 \leftrightarrow (11x - 3)(2x + 4) = 0$
$\leftrightarrow 11x - 3 = 0$ or $2x + 4 = 0$
$\leftrightarrow 11x = 3$ or $2x = -4$
$\leftrightarrow x = \dfrac{3}{11}$ or $x = -2$

Hence $S = \left\{\dfrac{3}{11}, -2\right\}$

Example 4 | Solve the equation $x^2 - 3x = 2x - 6$.

Solution: $x^2 - 3x = 2x - 6 \leftrightarrow x^2 - 5x + 6 = 0$
$\leftrightarrow (x-2)(x-3) = 0$
$\leftrightarrow x - 2 = 0$ or $x - 3 = 0$
$\leftrightarrow x = 2$ or $x = 3$

Hence $S = \{2, 3\}$

Example 5 | Solve the equation $(5x - 1)(2x - 1) = 13$.

Solution: $(5x - 1)(2x - 1) = 13 \leftrightarrow 10x^2 - 7x + 1 = 13$
$\leftrightarrow 10x^2 - 7x - 12 = 0$
$\leftrightarrow (5x + 4)(2x - 3) = 0$
$\leftrightarrow 5x + 4 = 0$ or $2x - 3 = 0$
$\leftrightarrow 5x = -4$ or $2x = 3$
$\leftrightarrow x = -\dfrac{4}{5}$ or $x = \dfrac{3}{2}$

Hence $S = \left\{-\dfrac{4}{5}, \dfrac{3}{2}\right\}$

SEC. 3.5 The Method of Factoring

Example 6 Solve the equation $9x^2 - 15x + 1 = 15 - 26x - 6x^2$.

Solution: $9x^2 - 15x + 1 = 15 - 26x - 6x^2 \leftrightarrow 15x^2 + 11x - 14 = 0$
$$\leftrightarrow (3x - 2)(5x + 7) = 0$$
$$\leftrightarrow 3x - 2 = 0 \quad \text{or}$$
$$5x + 7 = 0$$
$$\leftrightarrow 3x = 2 \quad \text{or} \quad 5x = -7$$
$$\leftrightarrow x = \frac{2}{3} \quad \text{or} \quad x = -\frac{7}{5}$$

Hence $S = \left\{ \frac{2}{3}, -\frac{7}{5} \right\}$

Example 7 Solve the equation $(2x - 1)(3x + 1) = (2x - 1)(x - 2)$.

Solution: $(2x - 1)(3x + 1) = (2x - 1)(x - 2) \leftrightarrow 6x^2 - x - 1$
$$= 2x^2 - 5x + 2$$
$$\leftrightarrow 4x^2 + 4x - 3 = 0$$
$$\leftrightarrow (2x - 1)(2x + 3) = 0$$
$$\leftrightarrow 2x - 1 = 0 \quad \text{or}$$
$$2x + 3 = 0$$
$$\leftrightarrow 2x = 1 \quad \text{or} \quad 2x = -3$$
$$\leftrightarrow x = \frac{1}{2} \quad \text{or} \quad x = -\frac{3}{2}$$

Hence $S = \left\{ \frac{1}{2}, -\frac{3}{2} \right\}$

Alternate Solution: Some sharp-eyed students might have recognized the common factor $2x - 1$ on both sides of the original equation. This allows a solution in which the sides need not be expanded.

$$(2x - 1)(3x + 1) = (2x - 1)(x - 2) \leftrightarrow (2x - 1)(3x + 1)$$
$$- (2x - 1)(x - 2) = 0$$
$$\leftrightarrow (2x - 1)[(3x + 1) - (x - 2)] = 0$$
$$\leftrightarrow (2x - 1)(2x + 3) = 0$$

and the remainder of the solution is as above.

Exercises 3.5

Solve each of the following quadratic equations by factoring.

1. $x^2 - 5x - 6 = 0$

Solution: $x^2 - 5x - 6 = 0 \leftrightarrow (x + 1)(x - 6) = 0$
$$\leftrightarrow (x + 1) = 0 \quad \text{or} \quad x - 6 = 0$$
$$\leftrightarrow x = -1 \quad \text{or} \quad x = 6$$

Hence $S = \{-1, 6\}$

2. $x^2 - 3x - 10 = 0$
3. $2x^2 = 3x$
4. $x^2 - 2x - 3 = 0$
5. $x^2 - 25 = 0$
6. $x^2 = 5$
7. $9x^2 - 12x + 4 = 0$
8. $2x^2 + 7x - 4 = 0$
9. $x^2 - 3x + 2 = 0$
10. $2x^2 + 7x - 15 = 0$
11. $(x + 3)(x + 4) = 2$
12. $(x + 1)(x + 2) = 0$
13. $(3x - 2)(x - 3) = -2$
14. $x(x - 5) = -4$
15. $3x^2 + 5x + 1 = x^2 - 2x - 4$
16. $(x - 3)(x - 1) = 2x + 3$
17. $3x^2 - 7x = 4x - 10$
18. $y^2 + 4y + 4 = 0$
19. $5x^2 - 2x - 24 = 0$
20. $2x^2 + 9x + 1 = 2x - 4$
21. $x(x - 2) = 3$
22. $15x^2 - 23x + 4 = 0$
23. $14x^2 + 11x - 15 = 0$
24. $2x^2 - 11x + 14 = 0$
25. $6x^2 + 13x - 28 = 0$
26. $8x^2 - 14x - 15 = 0$
27. $6x^2 - 31x - 60 = 0$
28. $x^2 - ax = ab - bx$
29. $2n^2 - 3n = 2n - 2$
30. $x^2 + 3bx = 2ax + 6ab$
31. $(2x - 3)^2 = 4$
32. Find a quadratic equation whose solution set is the given set

   a. $\{3, -2\}$ b. $\left\{\dfrac{3}{2}, \dfrac{1}{2}\right\}$ c. $\{-3, -3\}$ d. $\left\{\dfrac{3}{5}, -\dfrac{4}{7}\right\}$

   Solution to **a**: 3 is a solution of $x - 3 = 0$, and -2 is a solution of $x + 2 = 0$. Hence $(x - 3)(x + 2) = 0$ has the solution set $\{3, -2\}$. Written in standard form, this equation becomes $x^2 - x - 6 = 0$.

3.6 SOLVING THE EQUATION $[f(x)]^2 = k^2$

The method of factoring developed in Section 3.5 can be used to solve equations of the form $[f(x)]^2 = k^2$, where k is a constant. This modified procedure will be extremely important in the method of completing the square (Section 3.7), and we illustrate it by four examples.

Example 1 Find the solution set of the equation $x^2 = 49$.

Solution: $x^2 = 49 \leftrightarrow x^2 - 49 = 0$
$\leftrightarrow (x - 7)(x + 7) = 0$
$\leftrightarrow x - 7 = 0 \text{ or } x + 7 = 0$
$\leftrightarrow x = 7 \text{ or } x = -7$

Hence $S = \{7, -7\}$

SEC. 3.6   Solving the Equation $[f(x)]^2 = k^2$

Example 2   Solve the equation $(2x - 3)^2 = 10$.

Solution:
$$(2x - 3)^2 = 10 \leftrightarrow (2x - 3)^2 - 10 = 0$$
$$\leftrightarrow [(2x - 3) - \sqrt{10}][(2x - 3) + \sqrt{10}] = 0$$
$$\leftrightarrow (2x - 3) - \sqrt{10} = 0 \quad \text{or} \quad (2x - 3) + \sqrt{10} = 0$$
$$\leftrightarrow 2x = 3 + \sqrt{10} \quad \text{or} \quad 2x = 3 - \sqrt{10}$$
$$\leftrightarrow x = \frac{3 + \sqrt{10}}{2} \quad \text{or} \quad x = \frac{3 - \sqrt{10}}{2}$$

Hence
$$S = \left\{ \frac{3 + \sqrt{10}}{2}, \frac{3 - \sqrt{10}}{2} \right\}$$

NOTE: That $x = \dfrac{3 + \sqrt{10}}{2}$ or $x = \dfrac{3 - \sqrt{10}}{2}$ is often indicated by the shorthand $x = \dfrac{3 \pm \sqrt{10}}{2}$, which is read x equals 3 plus or minus $\sqrt{10}$ divided by 2. Hence we might write $S = \left\{ \dfrac{3 \pm \sqrt{10}}{2} \right\}$.

Example 3   Solve the equation $x^2 = -4$.

Solution: Referring back to Property 1 of complex numbers in Section 1.7, we find that
$$x^2 = -4 \leftrightarrow x = 2i \quad \text{or} \quad x = -2i$$

We could also take the following approach:
$$x^2 = -4 \leftrightarrow x^2 + 4 = 0$$
$$\leftrightarrow x^2 + (-2i)^2 = 0$$
$$\leftrightarrow [x - (-2i)][x + (-2i)] = 0$$
$$\leftrightarrow (x + 2i)(x - 2i) = 0$$
$$\leftrightarrow x + 2i = 0 \quad \text{or} \quad x - 2i = 0$$
$$\leftrightarrow x = -2i \quad \text{or} \quad x = 2i$$

Clearly, the first approach is shorter and it is the one we shall use henceforth. The solution set is $S = \{\pm 2i\}$.

Example 4   Solve the equation $(4x - 2)^2 = -8$.

Solution:
$$(4x - 2)^2 = -8 \leftrightarrow 4x - 2 = \sqrt{8}i \quad \text{or} \quad 4x - 2 = -\sqrt{8}i$$
$$\leftrightarrow 4x = 2 + \sqrt{8}i \quad \text{or} \quad 4x = 2 - \sqrt{8}i$$
$$\leftrightarrow x = \frac{2 + \sqrt{8}i}{4} \quad \text{or} \quad x = \frac{2 - \sqrt{8}i}{4}$$

Even this solution can be shortened by adopting the following shorthand:

$$(4x - 2)^2 = -8 \leftrightarrow 4x - 2 = \pm\sqrt{8}i$$
$$\leftrightarrow 4x = 2 \pm \sqrt{8}i$$
$$\leftrightarrow x = \frac{2 \pm \sqrt{8}i}{4}$$

This is a perfectly acceptable solution. One final point, however, is that in algebra we usually attempt to write answers in as simple a form as possible. Hence we would probably carry the solution a couple of steps further, as follows:

$$x = \frac{2 \pm \sqrt{8}i}{4} \leftrightarrow x = \frac{2 \pm 2\sqrt{2}i}{4} \quad \text{Using properties of radicals (Section 2.3)}$$

$$\leftrightarrow x = \frac{1 \pm \sqrt{2}i}{2} \quad \text{Dividing numerator and denominator by 2}$$

Hence, if we were giving an answer to this exercise, we would probably give it in the form $S = \left\{\dfrac{1 \pm \sqrt{2}i}{2}\right\}$.

The techniques illustrated in Examples 1 through 4 can be summarized, and generalized, in the next theorem:

THEOREM 3.3. *Let k^2 be a real number which may be positive, negative, or zero. The solution set S of the equation $[f(x)]^2 = k^2$ is the union of the solution set T_1 of the equation $f(x) = k$ and the solution set T_2 of the equation $f(x) = -k$.*

Exercises 3.6

Solve each of the following quadratic equations using the method described in this section:

1. $x^2 = 7$

Solution: $x^2 = 7 \leftrightarrow x = \pm\sqrt{7}$

Hence $\qquad\qquad\qquad S = \{\pm\sqrt{7}\}$

2. $x^2 + 16 = 0$.

Solution: $x^2 + 16 = 0 \leftrightarrow x^2 = -16$
$\leftrightarrow x = \pm 4i$

Hence $\qquad\qquad\qquad S = \{\pm 4i\}$

3. $(x - 1)^2 = 6$ **4.** $(x + 2)^2 = -9$

5. $(2x - 3)^2 = 9$ **6.** $(x + 5)^2 = 16$

7. $(5x - 2)^2 = -9$ **8.** $(3x + 2)^2 = -5$

SEC. 3.7   The Method of Completing the Square

9. $\left(\dfrac{2}{3}x - \dfrac{11}{5}\right)^2 = \dfrac{16}{25}$ 10. $\left(\dfrac{1}{7}y + \dfrac{3}{17}\right)^2 = -\dfrac{5}{4}$

In each of Exercises 11 through 14, first solve for x in terms of the various constants involved and then use a hand calculator to approximate each solution (rounded to three decimal places).

11. $(\sqrt{15x} - 3)^2 = 12$ 12. $(-2x + \sqrt{3})^2 = 16$

13. $(\sqrt{5x} + 2)^2 = -21$

Solution: $(\sqrt{5x} + 2)^2 = -21 \leftrightarrow \sqrt{5x} + 2 = \pm\sqrt{21}\,i$
$$\leftrightarrow \sqrt{5x} = -2 \pm \sqrt{21}\,i$$
$$\leftrightarrow x = \dfrac{-2 \pm \sqrt{21}\,i}{\sqrt{5}}$$
$$\leftrightarrow x = -\dfrac{2}{\sqrt{5}} \pm \dfrac{\sqrt{21}}{\sqrt{5}}\,i$$

Hence $\quad S = \left\{-\dfrac{2}{\sqrt{5}} \pm \dfrac{\sqrt{21}}{\sqrt{5}}\,i\right\}$

Using a hand calculator and rounding to three decimal places,
$$x \cong -0.894 \pm 2.049i$$

14. $(-\sqrt{19x} - 2\sqrt{2})^2 = -513.14$

3.7   THE METHOD OF COMPLETING THE SQUARE

We saw in the previous section that equations of the form $[f(x)]^2 = k^2$ could be readily solved, provided $f(x)$ was a first-degree (linear) polynomial. For example, the equation $(x + 1)^2 = 4$ is equivalent to $x + 1 = \pm 2$, from which we immediately find $S = \{1, -3\}$. The idea in the method of completing the square is to take a given quadratic equation and find an equivalent equation which has the form $[f(x)]^2 = k^2$. Then we simply use the technique of the last section to complete the solution. As a background for this method, let us recall the form of certain special products:

$$(x + 1)^2 = x^2 + 2x + 1$$
$$(x + 2)^2 = x^2 + 4x + 4$$
$$(x + 3)^2 = x^2 + 6x + 9$$
$$(x + n)^2 = x^2 + 2nx + n^2$$

Each such equation is an algebraic identity, and hence either side may be substituted for the other in any expression or equation in which that side occurs.

Inspection of the right side in each of these identities shows, first, that the coefficient of x^2 is 1 and, second, that the last term is the square of one-half the coefficient of x in the middle term. Hence, if given two terms of the form $x^2 + ax$, one can always form a perfect square simply by adding a third term found by squaring one-half the coefficient of x: $x^2 + ax + (\frac{1}{2}a)^2$. This expression is, in fact, the square of $(x + \frac{1}{2}a)$, for

$$\left(x + \frac{1}{2}a\right)^2 = x^2 + ax + \frac{1}{4}a^2$$

We illustrate how to use these observations to solve quadratic equations in the next five examples. It should be observed at this point that, once the student learns to use the quadratic formula (in Section 3.8), he or she will probably never again use completing the square to solve a quadratic equation. However, we shall illustrate in Examples 7 and 8 that the ability to "complete the square" can be useful for other purposes as well.

Example 1 Solve the equation $x^2 + 4x - 6 = 0$.

Solution: If we were to attempt to use factoring, we would find that the left side does not factor with integer coefficients. So we take the following approach:

$$x^2 + 4x - 6 = 0 \leftrightarrow x^2 + 4x = 6$$
$$\leftrightarrow x^2 + 4x + 4 = 6 + 4$$

Here we have taken half of 4, squared it to obtain 4, and added this 4 to both sides of the equation. The result is that we have *completed the square* on the left, and so we can continue the equivalences as follows:

$$\leftrightarrow (x + 2)^2 = 10$$
$$\leftrightarrow x + 2 = \pm\sqrt{10}$$
$$\leftrightarrow x = -2 \pm \sqrt{10}$$

Hence $S = \{-2 \pm \sqrt{10}\}$

Notice that, from the equation $(x + 2)^2 = 10$ on, we are doing exactly what was done in Section 3.6.

Example 2 Solve the equation $3x^2 - 24x - 45 = 0$.

Solution: Again, the left side will not factor with integer coefficients, so we proceed as in Example 1 with one minor modification. First we divide both sides by 3 so that the leading coefficient becomes 1.

$$3x^2 - 24x - 45 = 0 \leftrightarrow x^2 - 8x - 15 = 0$$
$$\leftrightarrow x^2 - 8x = 15$$
$$\leftrightarrow x^2 - 8x + 16 = 15 + 16$$

SEC. 3.7 The Method of Completing the Square 141

(Again, taking half of -8, squaring it to obtain 16, and adding 16 to both sides *completes the square* on the left.) Continuing the equivalences,

$$\leftrightarrow (x-4)^2 = 31$$
$$\leftrightarrow x - 4 = \pm\sqrt{31}$$
$$\leftrightarrow x = 4 \pm \sqrt{31}$$

Hence $\qquad S = \{4 \pm \sqrt{31}\}$

Example 3 Solve the equation $3x^2 - 4x = 2$.

Solution: $3x^2 - 4x = 2 \leftrightarrow x^2 - \dfrac{4}{3}x = \dfrac{2}{3}$

$$\leftrightarrow x^2 - \dfrac{4}{3}x + \dfrac{4}{9} = \dfrac{2}{3} + \dfrac{4}{9}$$

$\left(\text{Half of } -\dfrac{4}{3} \text{ is } -\dfrac{2}{3} \text{ whose square is } \dfrac{4}{9}.\right)$

$$\leftrightarrow \left(x - \dfrac{2}{3}\right)^2 = \dfrac{(2)(3) + (4)(1)}{9}$$
$$\leftrightarrow \left(x - \dfrac{2}{3}\right)^2 = \dfrac{10}{9}$$
$$\leftrightarrow x - \dfrac{2}{3} = \pm\dfrac{\sqrt{10}}{3}$$
$$\leftrightarrow x = \dfrac{2}{3} \pm \dfrac{\sqrt{10}}{3}$$

Hence, $S = \left\{\dfrac{2}{3} \pm \dfrac{\sqrt{10}}{3}\right\}$. We could use a hand calculator to approximate each solution and find that $\dfrac{2}{3} + \dfrac{\sqrt{10}}{3} \cong 1.721$ and $\dfrac{2}{3} - \dfrac{\sqrt{10}}{3} \cong -0.387$, each approximation being rounded to three decimal places.

Example 4 Solve the equation $\dfrac{1}{2}x^2 - 2x + \dfrac{9}{2} = 0$.

Solution: $\dfrac{1}{2}x^2 - 2x + \dfrac{9}{2} = 0 \leftrightarrow x^2 - 4x + 9 = 0$

$$\leftrightarrow x^2 - 4x = -9$$
$$\leftrightarrow x^2 - 4x + 4 = -9 + 4$$

(Half of -4 is -2 whose square is 4.)

$$\leftrightarrow (x-2)^2 = -5$$
$$\leftrightarrow x - 2 = \pm\sqrt{5}i$$
$$\leftrightarrow x = 2 \pm \sqrt{5}i$$

Hence $\qquad S = \{2 \pm \sqrt{5}i\}$

Example 5 | Solve the equation $-x^2 + \sqrt{3}x + \frac{5}{6} = 0$.

Solution: $-x^2 + \sqrt{3}x + \frac{5}{6} = 0 \leftrightarrow x^2 - \sqrt{3}x - \frac{5}{6} = 0$

$$\leftrightarrow x^2 - \sqrt{3}x = \frac{5}{6}$$

$$\leftrightarrow x^2 - \sqrt{3}x + \left(\frac{\sqrt{3}}{2}\right)^2 = \frac{5}{6} + \left(\frac{\sqrt{3}}{2}\right)^2$$

$$\leftrightarrow \left(x - \frac{\sqrt{3}}{2}\right)^2 = \frac{5}{6} + \frac{3}{4}$$

$$\leftrightarrow \left(x - \frac{\sqrt{3}}{2}\right)^2 = \frac{(5)(2) + (3)(3)}{12}$$

$$\leftrightarrow \left(x - \frac{\sqrt{3}}{2}\right)^2 = \frac{19}{12}$$

$$\leftrightarrow x - \frac{\sqrt{3}}{2} = \pm\sqrt{\frac{19}{12}}$$

$$\leftrightarrow x = \frac{\sqrt{3}}{2} \pm \sqrt{\frac{19}{12}}$$

Hence $S = \left\{\frac{\sqrt{3}}{2} \pm \sqrt{\frac{19}{12}}\right\}$

Here one might modify the answer in several ways; for example,

$$\frac{\sqrt{3}}{2} \pm \sqrt{\frac{19}{12}} = \frac{\sqrt{3}}{2} \pm \frac{\sqrt{19}}{\sqrt{12}}$$

$$= \frac{\sqrt{3}}{2} \pm \frac{\sqrt{19}}{2\sqrt{3}} = \frac{\sqrt{3}}{2} + \frac{\sqrt{19}\sqrt{3}}{2 \cdot 3}$$

$$= \frac{\sqrt{3}}{2} \pm \frac{\sqrt{57}}{6} = \frac{3\sqrt{3} \pm \sqrt{57}}{6}$$

However, the first answer is correct, sufficient, and as easy to approximate with a hand calculator as any of the others.

The method of completing the square, illustrated in Examples 1 through 5, may be used to solve any quadratic equation with real coefficients. For convenience, we list one sequence of steps, which could be called:

THE METHOD OF COMPLETING THE SQUARE

Step 1. Collect terms involving x^2 and x on the left and constant terms on the right.

Step 2. Combine like terms on both sides.

SEC. 3.7 The Method of Completing the Square

Step 3. Make the coefficient of x^2 equal to 1 (if it is not already 1) by dividing both sides by the coefficient of x^2.

Step 4. Take half of the coefficient of x, square it, and add this square to both sides.

Step 5. Write the left side as a perfect square and combine the constants on the right side.

Step 6. Use the method for solving equations of the form $[f(x)]^2 = k^2$ (Section 3.6) to solve the resulting equation.

Since many students have difficulty with the method of completing the square, we give an additional example, labeling each step as it is taken.

Example 6 Solve the equation $-2x^2 - 5x - 9 = x^2 - 2x + 1$.

Solution

$$-2x^2 - 5x - 9 = x^2 - 2x + 1 \leftrightarrow -2x^2 - x^2 - 5x + 2x = 9 + 1 \quad \text{(Step 1)}$$
$$\leftrightarrow -3x^2 - 3x = 10 \quad \text{(Step 2)}$$
$$\leftrightarrow x^2 + x = -\frac{10}{3} \quad \text{(Step 3)}$$
$$\leftrightarrow x^2 + x + \frac{1}{4} = -\frac{10}{3} + \frac{1}{4} \quad \text{(Step 4)}$$
$$\leftrightarrow \left(x + \frac{1}{2}\right)^2 = -\frac{37}{12} \quad \text{(Step 5)}$$
$$\left.\begin{array}{l}\leftrightarrow x + \dfrac{1}{2} = \pm\sqrt{\dfrac{37}{12}}\,i \\[2mm] \leftrightarrow x = -\dfrac{1}{2} \pm \sqrt{\dfrac{37}{12}}\,i\end{array}\right\} \quad \text{(Step 6)}$$

Hence $$S = \left\{-\frac{1}{2} \pm \sqrt{\frac{37}{12}}\,i\right\}$$

Finally, we give two examples to show that the ability to complete the square can be useful to find smallest (in Example 7) or largest (in Example 8) values of second-degree polynomials. The reader is referred to Chapter 6 for an expanded treatment of maximum and minimum values of expressions.

Example 7 Find the smallest value of the polynomial $f(x) = x^2 - 2x - 3$.

Solution: We go through the motions of completing the square as follows:
$$x^2 - 2x - 3 = x^2 - 2x \quad - 3$$
$$= x^2 - 2x + 1 - 3 - 1$$

(Here we have completed the square on $x^2 - 2x$. To do this, we added 1. But in order not to change the expression, we must also subtract 1.)

$$= (x - 1)^2 - 4$$

Now, for any real number x, $(x-1)^2 \geq 0$, and so, for any real number x, $(x-1)^2 - 4 \geq -4$. Moreover, by taking $x = 1$, $(x-1)^2 - 4 = -4$. Therefore, -4 is the smallest value of $f(x) = x^2 - 2x - 3$, and this value occurs when $x = 1$.

Example 8 Find the largest value of the polynomial $f(x) = -2x^2 - 3x + 5$.

Solution:
$$-2x^2 - 3x + 5 = -2\left(x^2 + \frac{3}{2}x\right) + 5$$

$$= -2\left(x^2 + \frac{3}{2}x + \frac{9}{16}\right) + 5 + \frac{9}{8}$$

$\left[\text{Here we have completed the square on } x^2 + \frac{3}{2}x. \text{ This results in adding } (-2)\left(\frac{9}{16}\right)\right.$
$= -\frac{9}{8}$, that is, subtracting $\frac{9}{8}$. In order not to change the expression, we must also add $\left.\frac{9}{8}.\right]$

$$= -2\left(x + \frac{3}{4}\right)^2 + \frac{49}{8}$$

In a manner similar to that used in Example 7, $-2\left(x + \frac{3}{4}\right)^2 \leq 0$ for all real numbers x, and so $-2\left(x + \frac{3}{4}\right)^2 + \frac{49}{8} \leq \frac{49}{8}$ for all real numbers x. Also, if $x = -\frac{3}{4}$, then $-2\left(x + \frac{3}{4}\right)^2 + \frac{49}{8} = \frac{49}{8}$. Therefore, the largest value of the polynomial $f(x) = -2x^2 - 3x + 5$ is $\frac{49}{8}$, and this value occurs when $x = -\frac{3}{4}$.

Exercises 3.7

Determine the number which must be added to each of the following expressions so that the resulting expression is a perfect square. Write the resulting expression as a perfect square.

1. $x^2 + 4x$

    Solution: Half of 4 is 2, whose square is 4. Hence 4 must be added to $x^2 + 4x$ in order to make it a perfect square. Finally,
    $$x^2 + 4x + 4 = (x + 2)^2$$

2. $t^2 - 12t$ 3. $y^2 - \frac{4}{3}y$ 4. $x^2 - \frac{3}{2}x$ 5. $v^2 + \frac{5}{3}v$

6. $y^2 - \frac{7}{5}y$ 7. $t^2 + at$ 8. $x^2 + (m/n)x$

Solve each of the following equations by use of the method of completing the square.

9. $x^2 + 4x - 2 = 0$

   Solution: $x^2 + 4x - 2 = 0 \leftrightarrow x^2 + 4x = 2$
   $\leftrightarrow x^2 + 4x + 4 = 2 + 4$
   $\leftrightarrow (x+2)^2 = 6$
   $\leftrightarrow x + 2 = \pm\sqrt{6}$
   $\leftrightarrow x = -2 \pm \sqrt{6}$

   Hence $S = \{-2 \pm \sqrt{6}\}$

10. $t^2 - 12t + 12 = 0$
11. $3y^2 - 4y - 1 = 0$
12. $2x^2 - 3x - 7 = 0$
13. $3v^2 + 5v - 3 = 0$
14. $5y^2 - 7y - 4 = 0$
15. $x^2 + 3x - 3 = 0$
16. $5x^2 + 2x = 24$
17. $x^2 + 2x - 1 = 0$
18. $2x^2 - 4x - 5 = 0$
19. $t^2 + at - a^2 = 0$
20. $nx^2 + mx + p = 0$
21. $3x^2 - 11x + 5 = 0$
22. $2x^2 + 5x + 2 = 0$
23. $x^2 - 2x - 5 = 0$
24. $x^2 - 3x + 4 = 0$

25. In each of Exercises 17 and 18, use a hand calculator to approximate each solution, rounding to three decimal places.

26. Find the largest value of the polynomial $f(x) = -x^2 + 3x - 2$.

27. Find the smallest value of the polynomial $f(x) = \frac{3}{4}x^2 - \frac{1}{2}x + 2$.

3.8 THE QUADRATIC FORMULA METHOD

We found in the previous section that any quadratic equation with real coefficients could be solved by the method of completing the square. At the end of Section 3.3 we observed that the general linear equation $ax + b = 0$ (with $a \neq 0$) has as its solution set $S = \left\{-\dfrac{b}{a}\right\}$. We proceed now to use the method of completing the square to solve the general quadratic equation $ax^2 + bx + c = 0$ (with $a \neq 0$). The result can be expressed in any of the following forms:

(1) $\qquad x = \dfrac{-b + \sqrt{b^2 - 4ac}}{2a} \quad \text{or} \quad x = \dfrac{-b - \sqrt{b^2 - 4ac}}{2a}$

(2) $\qquad x = \dfrac{-b \pm \sqrt{b^2 - 4ac}}{2a}$

(3) $\qquad S = \left\{\dfrac{-b + \sqrt{b^2 - 4ac}}{2a}, \dfrac{-b - \sqrt{b^2 - 4ac}}{2a}\right\}$

(4) $\qquad S = \left\{\dfrac{-b \pm \sqrt{b^2 - 4ac}}{2a}\right\}$

(5) $$S = \left\{ -\frac{b}{2a} \pm \sqrt{\left(-\frac{b}{2a}\right)^2 + \left(\frac{c}{a}\right)} \right\}$$ *This form is useful for calculator computation.*

The net effect of this fact is that we need only perform the method of completing the square once and then simply apply this result to solve any quadratic equation we like. Before deriving the quadratic formula, we give three examples illustrating its use.

Example 1 Use the quadratic formula to find the solution set of the equation

$$2x^2 + x = 2x + 15$$

Solution: Writing the given equation in standard form, we have

$$2x^2 - x - 15 = 0$$

If we compare this equation with the general equation $ax^2 + bx + c = 0$, we see that $a = 2$, $b = -1$, and $c = -15$. Now substitute these values for a, b, and c in the quadratic formula. This gives

$$x = \frac{-(-1) \pm \sqrt{(-1)^2 - 4(2)(-15)}}{2(2)}$$
$$= \frac{1 \pm \sqrt{1 + 120}}{4} = \frac{1 \pm \sqrt{121}}{4}$$
$$= \frac{1 \pm 11}{4}$$

Therefore,

$$x = \frac{1 + 11}{4} \quad \text{or} \quad x = \frac{1 - 11}{4}$$

So

$$x = 3 \quad \text{or} \quad x = -\frac{10}{4} = -\frac{5}{2}$$

Hence the solution set is $S = \left\{ 3, -\frac{5}{2} \right\}$.

Example 2 Find the solution set S of the equation $x^2 - 4x + 13 = 0$.

Solution: Here we have $a = 1$, $b = -4$, and $c = 13$. Substituting these values for a, b, and c in the quadratic formula, we get

$$x = \frac{4 \pm \sqrt{16 - 52}}{2} = \frac{4 \pm \sqrt{-36}}{2} = \frac{4 \pm 6i}{2} = 2 \pm 3i$$

Hence $$S = \{2 + 3i, 2 - 3i\} = \{2 \pm 3i\}$$

The quadratic formula is a most important tool and should be carefully memorized. It is useful not only for solving second-degree polynomials in one

SEC. 3.8 The Quadratic Formula Method

variable but is very effective in solving second-degree polynomial equations in two variables for one of the variables, as the following example will illustrate.

Example 3

Solve the equation $3x^2 - 4xy + y^2 - 6x + 2y = 0$ for y in terms of x.

Solution: Rewrite the given equation in the form

$$y^2 - (4x - 2)y + (3x^2 - 6x) = 0$$

Now we treat this as a quadratic equation in y with $a = 1$, $b = -(4x - 2)$, and $c = 3x^2 - 6x$. Substituting these values for a, b, and c in the quadratic formula, we obtain

$$y = \frac{(4x - 2) \pm \sqrt{(4x - 2)^2 - 4(3x^2 - 6x)}}{2}$$

$$= \frac{(4x - 2) \pm \sqrt{16x^2 - 16x + 4 - 12x^2 + 24x}}{2}$$

$$= \frac{(4x - 2) \pm \sqrt{4x^2 + 8x + 4}}{2}$$

$$= (2x - 1) \pm \sqrt{x^2 + 2x + 1}$$

$$= (2x - 1) \pm \sqrt{(x + 1)^2}$$

$$= (2x - 1) \pm |x + 1|$$

Let us derive the quadratic formula. First, if $a < 0$, we could multiply both sides by -1 and obtain an equivalent equation whose leading coefficient is positive. Therefore, we shall assume that $a > 0$ to begin with (see Exercise 33). Following the procedure for completing the square,

$$ax^2 + bx + c = 0 \leftrightarrow ax^2 + bx = -c \quad \text{(Step 2)}$$

$$\leftrightarrow x^2 + \frac{b}{a}x = -\frac{c}{a} \quad \text{(Step 3)}$$

$$\leftrightarrow x^2 + \frac{b}{a}x + \left(\frac{b}{2a}\right)^2 = -\frac{c}{a} + \left(\frac{b}{2a}\right)^2 \quad \text{(Step 4)}$$

$$\left.\begin{array}{l} \leftrightarrow \left(x + \dfrac{b}{2a}\right)^2 = -\dfrac{c}{a} + \left(\dfrac{b}{2a}\right)^2 \\[2mm] \leftrightarrow \left(x + \dfrac{b}{2a}\right)^2 = \dfrac{b^2 - 4ac}{4a^2} \end{array}\right\} \quad \text{(Step 5)}$$

$$\leftrightarrow x + \frac{b}{2a} = \pm\sqrt{\frac{b^2 - 4ac}{4a^2}}$$

$$\leftrightarrow x + \frac{b}{2a} = \pm\frac{\sqrt{b^2 - 4ac}}{2a} \quad \text{(We have used } a > 0 \text{ here.)}$$

$$\leftrightarrow x = \frac{-b \pm \sqrt{b^2 - 4ac}}{2a}$$

Now that the quadratic formula has been derived, we shall apply it to solve each of the five examples from Section 3.7. It will be seen that sometimes the computations are easier using the quadratic formula and sometimes they are not. Either method, completing the square or the quadratic formula, is entirely acceptable. The quadratic formula is easiest to apply if you are not interested in simplifying your answers, or if you plan to use a calculator to approximate the solutions. Also, one only needs to remember a formula, rather than a method, when applying the quadratic formula.

Example 4

(See Example 1 of Section 3.7.) Solve the equation $x^2 + 4x - 6 = 0$.

Solution: By formula (2)

$$x = \frac{-4 \pm \sqrt{(4)^2 - (4)(1)(-6)}}{(2)(1)}$$

$$= \frac{-4 \pm \sqrt{16 + 24}}{2}$$

$$= \frac{-4 \pm \sqrt{40}}{2}$$

$$= \frac{-4 \pm 2\sqrt{10}}{2}$$

$$= -2 \pm \sqrt{10}$$

Hence $S = \{-2 \pm \sqrt{10}\}$

Example 5

(See Example 2 of Section 3.7.) Solve the equation $3x^2 - 24x - 45 = 0$.

Solution: First we divide both sides by 3 and obtain the equivalent equation $x^2 - 8x - 15 = 0$. Then, using formula (2), we obtain

$$x = \frac{-(-8) \pm \sqrt{(-8)^2 - (4)(1)(-15)}}{(2)(1)}$$

$$= \frac{8 \pm \sqrt{64 + 60}}{2}$$

$$= \frac{8 \pm \sqrt{124}}{2}$$

$$= \frac{8 \pm 2\sqrt{31}}{2}$$

$$= 4 \pm \sqrt{31}$$

Hence $S = \{4 \pm \sqrt{31}\}$. In this case, it is perhaps easier to use completing the square if one wants the answer in this form. However, if one is interested in approximating to within three decimal places, a hand calculator can be used to go straight from the first step to the approximate solutions:

$$x \cong 9.568 \quad \text{or} \quad x \cong -1.568$$

SEC. 3.8   The Quadratic Formula Method

Example 6 (See Example 3 of Section 3.7.) Solve the equation $3x^2 - 4x = 2$.

Solution: First the equation is written in standard form, $3x^2 - 4x - 2 = 0$. Then formula (2) is applied to obtain

$$x = \frac{-(-4) \pm \sqrt{(-4)^2 - (4)(3)(-2)}}{(2)(3)}$$

$$= \frac{4 \pm \sqrt{16 + 24}}{6}$$

$$= \frac{4 \pm \sqrt{40}}{6}$$

$$= \frac{4 \pm 2\sqrt{10}}{6}$$

$$= \frac{2 \pm \sqrt{10}}{3}$$

Hence $S = \left\{ \dfrac{2 \pm \sqrt{10}}{3} \right\}$

Example 7 (See Example 4 of Section 3.7.) Solve the equation $\dfrac{1}{2}x^2 - 2x + \dfrac{9}{2} = 0$.

Solution: First we multiply both sides by 2 (in order to clear fractions) and obtain

$$x^2 - 4x + 9 = 0$$

Then, use formula (2) to obtain

$$x = \frac{-(-4) \pm \sqrt{(-4)^2 - (4)(1)(9)}}{(2)(1)}$$

$$= \frac{4 \pm \sqrt{16 - 36}}{2}$$

$$= \frac{4 \pm \sqrt{-20}}{2}$$

$$= \frac{4 \pm 2\sqrt{5}i}{2}$$

$$= 2 \pm \sqrt{5}i$$

Hence $S = \{2 \pm \sqrt{5}i\}$

Example 8 (See Example 5 of Section 3.7.) Solve the equation

$$-x^2 + \sqrt{3}x + \frac{5}{6} = 0$$

Solution: Clearing fractions, we obtain

$$-6x^2 + 6\sqrt{3}x + 5 = 0$$

$$x = \frac{-6\sqrt{3} \pm \sqrt{(6\sqrt{3})^2 - (4)(-6)(5)}}{(2)(-6)}$$

$$= \frac{-6\sqrt{3} \pm \sqrt{108 + 120}}{-12}$$

$$= \frac{-6\sqrt{3} \pm \sqrt{228}}{-12}$$

$$= \frac{\sqrt{3}}{2} \mp \frac{\sqrt{228}}{12}$$

This means that

$$\frac{-6\sqrt{3} + \sqrt{228}}{-12} = \frac{\sqrt{3}}{2} - \frac{\sqrt{228}}{12}$$

and

$$\frac{-6\sqrt{3} - \sqrt{228}}{-12} = \frac{\sqrt{3}}{2} + \frac{\sqrt{228}}{12}$$

Hence

$$S = \left\{ \frac{\sqrt{3}}{2} \pm \frac{\sqrt{228}}{12} \right\}$$

This answer is simply another form of the answers given to Example 5 in Section 3.7.

Exercises 3.8

Use the quadratic formula to solve each of the following equations.

1. $x^2 - 4x + 2 = 0$
2. $x^2 + 4x - 3 = 0$
3. $x^2 - 2x + 3 = 0$
4. $6x^2 - 29x + 35 = 0$
5. $x^2 - 10x + 12 = 0$
6. $5x^2 - 8x + 2 = 0$
7. $7x^2 - 19x + 24 = 2x^2 - 15x + 25$
8. $11x^2 - 10x + 2 = 0$
9. $mx^2 + nx + p = 0$
10. $10x^2 + 7x - 3 = 0$
11. $3.6x^2 - 3.27x + 0.13 = 0$
12. $1{,}000x^2 - 605x - 864 = 0$
13. $5x^2 - 12x + 15 = 2x^2 - 4x - 5$
14. $2x^2 - 5x + 12 = 0$
15. $12x^2 - 8ax + a^2 = 0$
16. $6x^2 - 18x - 24 = 0$
17. $10x^2 + 29x + 10 = 0$
18. $3x^2 + 8x - 3 = 0$
19. $6x^2 - 8x + 7 = 0$
20. $5x^2 - 14mx + 9m^2 = 0$
21. $4x^2 + 12x - 15 = 0$
22. $x^2 - 6x + 13 = 0$
23. $3x^2 + 5x + 8 = 0$
24. $4x^2 + 25 = 0$

SEC. 3.9   Facts About the Roots of the Quadratic Equation $ax^2 + bx + c = 0$ 151

Solve each of the following equations for y in terms of x.

25. $x^2 - 3xy + 2y^2 - 3x + 3y = 0$
26. $x^2 - 2xy + y^2 - 25 = 0$
27. $2x^2 + 3xy + y^2 - 2x - y = 0$
28. $4x^2 + 4xy + y^2 + 2x + y - 2 = 0$
29. $2x^2 - 3xy + y^2 + 11x - 8y + 15 = 0$
30. $3x^2 - 5xy + 2y^2 + 8x - 5y - 3 = 0$
31. $2x^2 + xy - y^2 - x + 5y - 6 = 0$
32. $2x^2 - 2xy + y^2 + 8x - 12y + 36 = 0$

33.* Show that the quadratic formula also holds if $a < 0$. That is, go through the derivation of the quadratic formula and make modifications where necessary in the case that $a < 0$.

3.9 FACTS ABOUT THE ROOTS OF THE QUADRATIC EQUATION $ax^2 + bx + c = 0$

Often it is desirable to have information about the nature of the roots of a certain quadratic equation. That is, we may need to know whether the roots are real, whether they are equal, whether they are imaginary, etc. A careful look at the quadratic formula will suggest a method for obtaining such information without actually solving the equation.

Let r_1 and r_2 denote the roots of the general quadratic equation $ax^2 + bx + c = 0$. Then, by the quadratic formula,

$$r_1 = \frac{-b + \sqrt{b^2 - 4ac}}{2a} \quad \text{and} \quad r_2 = \frac{-b - \sqrt{b^2 - 4ac}}{2a}$$

The expression $(b^2 - 4ac)$, called the *discriminant*, which appears under the radical sign in the above formula, determines the nature of the roots r_1 and r_2 of any given quadratic equation. Thus, if a, b, and c are real numbers, the following possible cases result:

Case 1. $b^2 - 4ac > 0$; then r_1 and r_2 are real and unequal.
Case 2. $b^2 - 4ac = 0$; then r_1 and r_2 are real and equal.
Case 3. $b^2 - 4ac < 0$; then r_1 and r_2 are unequal and neither is real.

If we agree to count r_1 and r_2 as two roots even though $r_1 = r_2$, then the above discussion justifies the theorem.

THEOREM 3.4.   *Every polynomial equation, in one variable, of degree 2 has exactly two roots.*

To aid us in proving a very useful theorem, we next find the sum and product of the two roots r_1 and r_2.

$$r_1 + r_2 = \frac{-b + \sqrt{b^2 - 4ac}}{2a} + \frac{-b - \sqrt{b^2 - 4ac}}{2a} = -\frac{b}{a}$$

$$r_1 r_2 = \left(\frac{-b + \sqrt{b^2 - 4ac}}{2a}\right)\left(\frac{-b - \sqrt{b^2 - 4ac}}{2a}\right)$$

$$= \frac{(-b)^2 - (b^2 - 4ac)}{4a^2} = \frac{4ac}{4a^2} = \frac{c}{a}$$

THEOREM 3.5. *If r_1 and r_2 are the roots of the quadratic equation $ax^2 + bx + c = 0$, then*

$$ax^2 + bx + c = a(x - r_1)(x - r_2)$$

Proof:
$$a(x - r_1)(x - r_2) = a[x^2 - (r_1 + r_2)x + r_1 r_2]$$
$$= a\left[x^2 - \left(-\frac{b}{a}\right)x + \frac{c}{a}\right]$$
$$= ax^2 + bx + c$$

The importance of this theorem lies in the fact that it gives us a method for factoring any quadratic $ax^2 + bx + c$.

Example 1   Factor the quadratic $3x^2 - 2x - 7$.

Solution: First we find the roots r_1 and r_2 of the equation $3x^2 - 2x - 7 = 0$. Using the quadratic formula,

$$x = \frac{2 \pm \sqrt{4 + 84}}{6} = \frac{2 \pm 2\sqrt{22}}{6} = \frac{1 \pm \sqrt{22}}{3}$$

Thus
$$r_1 = \frac{1 + \sqrt{22}}{3} \quad \text{and} \quad r_2 = \frac{1 - \sqrt{22}}{3}$$

Applying Theorem 3.5, we obtain

$$3x^2 - 2x - 7 = 3\left(x - \frac{1 + \sqrt{22}}{3}\right)\left(x - \frac{1 - \sqrt{22}}{3}\right)$$

Exercises 3.9

Compute the value of the discriminant $b^2 - 4ac$, and thus determine the nature of the roots of each of the following equations.

1. $x^2 - 4x + 3 = 0$
2. $3x^2 - 7x - 5 = 0$
3. $x^2 - 8x + 16 = 0$
4. $5x^2 - 60x + 39 = 0$
5. $2x^2 - 5x + 7 = 0$
6. $3x^2 - 4x + 5 = 0$
7. $4x^2 - 2x - 3 = 0$
8. $5 - 4x = 3x^2$
9. $x^2 - 2x - 6 = 0$
10. $x^2 - 5x + 1 = 0$
11. $3x^2 + 6x - 2 = 0$
12. $2x^2 - 3x - 8 = 0$

13. $x^2 - 4\sqrt{3}x + 12 = 0$ **14.** $4x^2 - 12x - 9 = 0$ **15.** $9x^2 - 4x + 1 = 1$
16. $3x^2 - 4x + 4 = 0$

Determine the value or values of the constant k for which each of the following equations has equal roots.

17. $x^2 + kx + 9 = 0$ **18.** $x^2 + kx + 2k = 0$
19. $4x^2 + 4kx + 9 = 0$ **20.** $x^2 + 2(k+3)x + 12k = 0$
21. $5x^2 + 2(k-1)x + 3 = k(2x^2 - 1)$ **22.** $7y^2 - 2y + 1 = ky(y-2)$

Without solving the equations, find the sum and product of the roots for each of the following equations (use Theorem 3.4).

23. $2x^2 - 8x + 5 = 0$ **24.** $4x^2 - 3x + 8 = 0$ **25.** $13x^2 - 65x + 39 = 0$
26. $7x^2 - 9x - 6 = 0$ **27.** $6x - 2x^2 + 5 = 0$ **28.** $8x^2 = 3x - 5$
29. $3x^2 = 5x$ **30.** $5x - 4 = 2x - 8x^2$

3.10 EQUATIONS IN QUADRATIC FORM

An equation in one variable is said to be in *quadratic form* if by a suitable substitution of a new variable it can be expressed as a quadratic equation in the new variable. The following table exhibits some equations in quadratic form and a suitable substitution that could be used in each case.

Original Equation	Suitable Substitution	Quadratic Equation in y
$x^4 - 10x^2 + 9 = 0$	$y = x^2$	$y^2 - 10y + 9 = 0$
$x - \sqrt{x} - 2 = 0$	$y = \sqrt{x}$	$y^2 - y - 2 = 0$
$(x^2 - 2x)^2 - 11(x^2 - 2x) + 24 = 0$	$y = x^2 - 2x$	$y^2 - 11y + 24 = 0$
$4x^{-4} - 5x^{-2} + 1 = 0$	$y = x^{-2}$	$4y^2 - 5y + 1 = 0$

In the following examples we illustrate methods of solving equations in quadratic form.

Example 1   Solve the equation $x^4 - 10x^2 + 9 = 0$.

Solution: Let $y = x^2$; then our equation becomes

$$y^2 - 10y + 9 = 0 \leftrightarrow (y-1)(y-9) = 0$$
$$\leftrightarrow y - 1 = 0 \quad \text{or} \quad y - 9 = 0$$
$$\leftrightarrow y = 1 \quad \text{or} \quad y = 9$$
$$\leftrightarrow x^2 = 1 \quad \text{or} \quad x^2 = 9$$
$$\leftrightarrow x = \pm 1 \quad \text{or} \quad x = \pm 3$$

The solution set of the given equation is the set $\{\pm 1, \pm 3\}$.

Example 2 — Solve the equation $x - \sqrt{x} - 2 = 0$.

Solution: Let $y = \sqrt{x}$. This means y cannot be a negative number. We now have
$$y^2 - y - 2 = 0 \leftrightarrow (y+1)(y-2) = 0$$
$$\leftrightarrow y + 1 = 0 \quad \text{or} \quad y - 2 = 0$$
$$\leftrightarrow y = -1 \quad \text{or} \quad y = 2$$

The possibility $y = -1$ must be rejected since y cannot be negative. Therefore,
$$\sqrt{x} = 2 \quad \text{or} \quad x = 4$$

Hence $x = 4$ is the only solution of the given equation.

Example 3 — Solve the equation $(x^2 - 2x)^2 - 11(x^2 - 2x) + 24 = 0$.

Solution: Let $y = x^2 - 2x$; then
$$(x^2 - 2x)^2 - 11(x^2 - 2x) + 24 = 0$$
$$\leftrightarrow y^2 - 11y + 24 = 0$$
$$\leftrightarrow (y - 8)(y - 3) = 0$$
$$\leftrightarrow y - 8 = 0 \quad \text{or} \quad y - 3 = 0$$
$$\leftrightarrow y = 8 \quad \text{or} \quad y = 3$$
$$\leftrightarrow x^2 - 2x = 8 \quad \text{or} \quad x^2 - 2x = 3$$
$$\leftrightarrow x^2 - 2x - 8 = 0 \quad \text{or} \quad x^2 - 2x - 3 = 0$$
$$\leftrightarrow (x - 4)(x + 2) = 0 \quad \text{or} \quad (x - 3)(x + 1) = 0$$
$$\leftrightarrow x - 4 = 0 \quad \text{or} \quad x + 2 = 0 \quad \text{or} \quad x - 3 = 0 \quad \text{or} \quad x + 1 = 0$$
$$\leftrightarrow x = 4 \quad \text{or} \quad x = -2 \quad \text{or} \quad x = 3 \quad \text{or} \quad x = -1$$

Therefore, the solution set of the given equation is $\{-1, -2, 3, 4\}$.

Exercises 3.10

Solve each of the following equations.

1. $x^4 - 13x^2 + 36 = 0$
2. $(x^2 - 3x)^2 - 2(x^2 - 3x) - 8 = 0$
3. $x - 2\sqrt{x} - 3 = 0$
4. $x^4 - 10x^2 + 9 = 0$
5. $2x - 13\sqrt{x} + 20 = 0$
6. $36x^{-4} - 13x^{-2} + 1 = 0$
7. $6x^{-2} - 19x^{-1} + 10 = 0$
8. $x^4 - 7x^2 - 8 = 0$

3.11 POLYNOMIALS: TWO MAJOR PROBLEMS

In Section 3.1 a polynomial in one variable, say x, was defined as an algebraic expression that could be written in the form
$$A_n x^n + A_{n-1} x^{n-1} + \cdots + A_1 x + A_0$$

SEC. 3.11 Polynomials: Two Major Problems

where n is a positive integer or zero and the coefficients $A_n, A_{n-1}, \ldots, A_1, A_0$ are arbitrary constants. If $A_n \neq 0$, the polynomial is of degree n. The coefficients are assumed to be elements of the complex number system unless otherwise restricted. The universal set U to which the variable x belongs will also be assumed to consist of the set of complex numbers.

In the study of such polynomials, two major problems are of interest: first, to find the value of the polynomial for a particular value of the variable x and, second, to find the value or values of x for which the polynomial will have some given value. For convenience, the symbol $f(x)$ will be used throughout the remainder of this chapter to represent the general polynomial given above; that is,

$$f(x) = A_n x^n + A_{n-1} x^{n-1} + \cdots + A_1 x + A_0$$

We can now restate our two problems symbolically in the following manner.

1. Given that $f(x)$ is a polynomial and a is a number, find $f(a)$.
2. Given that $f(x)$ is a polynomial and M is a number, find the solution set of the polynomial equation $f(x) = M$.

Synthetic Substitution

Problem 1 is nothing more than a problem of substitution, and as such it is more laborious than difficult. For example, suppose that

$$f(x) = 2x^5 - 3x^3 + 7x^2 - 2x + 10$$

and our problem is to find $f(7)$. By direct substitution we get

$$f(7) = 2(7)^5 - 3(7)^3 + 7(7)^2 - 2(7) + 10$$

To simplify this expression, it is necessary to evaluate $(7)^2$, $(7)^3$, and $(7)^5$, which, although not difficult, is time consuming. Performing these operations, we finally obtain

$$f(7) = 2(16{,}807) - 3(343) + 7(49) - 14 + 10 = 32{,}924$$

If the degree of the polynomial had been higher or the number substituted for x had been larger, the labor involved would have been even more arduous. It is our purpose in this section to give a simple scheme for performing such a substitution. This scheme is called *synthetic substitution* and will be illustrated by using the polynomial just discussed.

Step 1. Write on the first line the coefficients of the polynomial in the order of descending powers of x, supplying zero coefficients for missing powers of x; see the systematic arrangement below.

Step 2. Rewrite the first coefficient, 2, on the third line, immediately below its position on the first line; see below.

Step 3. Multiply this first coefficient by the given value of x, in this case 7, and enter the product, 14, on the second line directly below the second coefficient. Now add and write the sum, 14, below on the third line. Similarly, we enter the product of 14 and 7 under the third coefficient, -3, add, and write the sum below, etc. The final number in the third line, 32,924, is the value of the polynomial when $x = 7$, showing that

$$f(7) = 32{,}924$$

First line: $\quad 2 + 0 - 3 + 7 - 2 + 10\,\underline{|\,7}$
Second line: $14 + 98 + 665 + 4{,}704 + 32{,}914$

Third line: $\quad 2 + 14 + 95 + 672 + 4{,}702 + 32{,}924$

That this scheme will work for any polynomial $f(x)$ and for any value of x can be seen from the following. Starting with a number A_n, multiplying it by any value of x, and adding A_{n-1} gives $A_n x + A_{n-1}$. Multiplying this sum by x and adding A_{n-2} gives $A_n x^2 + A_{n-1} x + A_{n-2}$, etc. Repeating this operation n times, we finally arrive at

$$A_n x^n + A_{n-1} x^{n-1} + \cdots + A_1 x + A_0$$

which is the value of the polynomial.

Example 1 If $f(x) = 3x^4 + 7x^3 - 6x + 8$, what is $f(-2)$?

Solution: $\quad 3 + 7 + 0 - 6 + 8\,\underline{|-2}$
$\, -6 - 2 + 4 + 4$

$ 3 + 1 - 2 - 2 + 12$

Hence $f(-2) = 12$. Let us now verify this by direct substitution:

$$f(-2) = 3(-2)^4 + 7(-2)^3 - 6(-2) + 8 = 48 - 56 + 12 + 8 = 12$$

Example 2 If $f(x) = 6x^4 + x^3 + 4x^2 + x - 2$, what is $f(-\tfrac{2}{3})$?

Solution: $\quad 6 + 1 + 4 + 1 - 2\,\underline{|-\tfrac{2}{3}}$
$\, -4 + 2 - 4 + 2$

$ 6 - 3 + 6 - 3 + 0$

The value of the polynomial when $x = -\tfrac{2}{3}$ is thus seen to be zero, or $f(-\tfrac{2}{3}) = 0$.

SEC. 3.11   Polynomials: Two Major Problems

The Zeros of a Polynomial

Our second problem concerning polynomials is a little more difficult: Given that $f(x)$ is a polynomial and M is a number, find the set of numbers a for which $f(a) = M$. In other words, this is a problem of solving the equation

(1) $$f(x) = M$$

where by a solution of the equation we mean, of course, a number a such that

$$f(a) = M$$

Let $$F(x) = f(x) - M$$

it follows that if $$f(a) = M$$

then $$F(a) = M - M = 0$$

Hence any solution of   (2) $F(x) = 0$

is also a solution of $f(x) = M$

and conversely. Therefore, Equations (1) and (2) are equivalent.

The problem at hand is thus the problem of finding the zeros, or the roots, of an equation

$$F(x) = 0$$

where $F(x)$ is a polynomial in x. Recall (Section 3.2) that such an equation is a polynomial equation written in standard form.

Polynomial equations of the first and second degree were discussed in sufficient detail earlier and hence need not be considered here except to recall the quadratic formula and to emphasize again that with the aid of this formula the roots of any polynomial equation of the second degree can be written down immediately in terms of the coefficients a, b, and c when our equation has been written in the standard form

$$ax^2 + bx + c = 0$$

which is the most general polynomial equation of the second degree. It will be recalled that, by completing the square, we found the roots to be

$$x = \frac{-b \pm \sqrt{b^2 - 4ac}}{2a}$$

which is called the *quadratic formula*. The quadratic formula, which expresses the roots of the general polynomial equation of the second degree in terms of its coefficients, might lead one to suppose that a formula could be found for expressing the roots of a polynomial equation of any degree in terms of its coefficients. Unfortunately, however, this is not the case. In fact, a famous Norwegian mathematician, N. H. Abel, proved in the early part of the nineteenth century that polynomial equations of higher degree than the fourth cannot, in general, be solved algebraically. Although we will not be able to find general methods for solving polynomial equations, we can develop a certain technique that will aid us in finding the roots of many such equations. In particular, we shall develop a technique for finding the rational roots of any polynomial equation. To do this, we shall need the help of certain theorems regarding polynomials and polynomial equations, and these we shall now consider.

The Remainder and Factor Theorems

If a polynomial $f(x)$ of degree n is divided by $x - a$, one obtains a quotient, say $q(x)$, which is also a polynomial, and a remainder R, which will be a constant if the division is carried to completion as described in Section 2.5. Moreover, the relation

(3) $$f(x) = (x - a)q(x) + R$$

as was pointed out in Chapter 2, is a numerical identity for all values of x. We usually wrote Equation (3) in the form

$$\frac{f(x)}{x - a} = q(x) + \frac{R}{x - a}$$

in Chapter 2. Now, since Equation (3) is an identity, it must hold when $x = a$; hence

$$f(a) = (a - a)q(a) + R = 0 \cdot q(a) + R = R$$

We now state this result as a theorem.

THE REMAINDER THEOREM

If a polynomial $f(x)$ is divided by a binomial $x - a$ until a remainder R not containing x is obtained, this remainder is equal to the value of the polynomial when $x = a$; that is, $R = f(a)$.

In certain special cases, R in Equation (3) is zero, and then this equation becomes

(4) $$f(x) = (x - a)q(x)$$

which shows that in such a case $x - a$ is a factor of $f(x)$. Now, since $f(a) = R$ and R is zero, it follows that $f(a) = 0$, which establishes the following theorem.

THE FACTOR THEOREM

A polynomial $f(x)$ contains $x - a$ as a factor if and only if $f(a) = 0$.

SEC. 3.11 Polynomials: Two Major Problems

In other words, if a is a root of the equation $f(x) = 0$, then $x - a$ is a factor of $f(x)$ and, conversely, if $x - a$ is a factor of $f(x)$, then a is a root of $f(x) = 0$.

Example 3

Find the remainder when $2x^4 - 3x^3 + 7x$ is divided by $x + 2$.

Solution: Here $f(x) = 2x^4 - 3x^3 + 7x$ and, since $x - a = x + 2$, it follows that $a = -2$. Thus

$$R = f(-2) = 32 + 24 - 14 = 42$$

Let us check this result by long division:

$$
\begin{array}{r}
2x^3 - 7x^2 + 14x - 21 \\
x+2 \overline{\smash{\big)}\, 2x^4 - 3x^3 + 0x^2 + 7x + 0} \\
\underline{2x^4 + 4x^3} \\
-7x^3 + 0x^2 + 7x + 0 \\
\underline{-7x^3 - 14x^2} \\
14x^2 + 7x + 0 \\
\underline{14x^2 + 28x} \\
-21x + 0 \\
\underline{-21x - 42} \\
+42
\end{array}
$$

This division shows not only that the remainder is 42 but also that the quotient $q(x)$ is $q(x) = 2x^3 - 7x^2 + 14x - 21$.

This same information can be obtained by synthetic substitution. We now demonstrate this method, using this same example:

$$
\begin{array}{r}
2 - 3 + 0 + 7 + 0 \,\underline{|-2} \\
4 + 14 - 28 + 42 \\
\hline
2 - 7 + 14 - 21 + 42
\end{array}
$$

The last number in the third line, as we have already learned, gives us $f(-2)$, or the remainder. Now compare the other numbers in line three with the coefficients in $q(x)$: they are identical. The reason for this will become obvious when we notice that in the substitution at each step we multiply by -2 and add, whereas in the division we multiply by $+2$ and subtract, and these are equivalent operations. Hence synthetic substitution has provided us with both the remainder and the quotient, and so the process is also called *synthetic division*.

Example 4 Find the quotient and remainder when $2x^3 - 3x^2 + 4$ is divided by $(x - 2)$.

Solution: Here $x - a = x - 2$, and hence $a = 2$. Using synthetic division, we obtain

$$\begin{array}{r} 2 - 3 + 0 + 4 \underline{|2} \\ + 4 + 2 + 4 \\ \hline 2 + 1 + 2 + 8 \end{array}$$

This shows that $q(x) = 2x^2 + x + 2$ and $R = 8$. The student may check this result by long division.

Example 5 Determine whether $x - 2$ is a factor of $x^6 - 64$.

First Solution: Here $f(x) = x^6 - 64$ and $a = 2$. Substitution of 2 for x gives $f(2) = 2^6 - 64 = 0$; therefore, $x - 2$ is a factor of $x^6 - 64$.

Second Solution: Let us use synthetic division:

$$\begin{array}{r} 1 + 0 + 0 + 0 + 0 + 0 - 64\underline{|2} \\ + 2 + 4 + 8 + 16 + 32 + 64 \\ \hline 1 + 2 + 4 + 8 + 16 + 32 + 0 \end{array}$$

Hence $x - 2$ is a factor of $x^6 - 64$ and the second factor is

$$x^5 + 2x^4 + 4x^3 + 8x^2 + 16x + 32$$

That is,

$$x^6 - 64 = (x - 2)(x^5 + 2x^4 + 4x^3 + 8x^2 + 16x + 32)$$

This second solution has the obvious advantage of providing us with the second factor as well.

Example 6 Determine whether $x - 1$ is a factor of $x^3 + 7x^2 - 3x - 4$.

Solution: Synthetic division gives

$$\begin{array}{r} 1 + 7 - 3 - 4\underline{|1} \\ + 1 + 8 + 5 \\ \hline 1 + 8 + 5 + 1 \end{array}$$

Since $f(1) \neq 0$, it follows that $x - 1$ is not a factor of the given polynomial.

Exercises 3.11

Use the Remainder Theorem to find the remainder R in each of the following, given the first expression divided by the second.

1. $x^3 - 5x^2 + 4x + 8;\ x - 2$
2. $x^4 - 3x^2 + 5x + 3;\ x + 1$

SEC. 3.11 Polynomials: Two Major Problems

3. $x^3 - 5;\ x + 2$
4. $2x^3 - 5x^2 + 7x;\ x - 3$
5. $3x^5 - 4x^3 + 6x^2 + 2;\ x - 2$
6. $4x^{15} - 8x^{13} + 7x^6 - 5x^2 + 12;\ x$

Use the Factor Theorem to determine whether the first expression is a factor of the second.

7. $x - 1;\ 7x^3 - 11x + 4$
8. $x + 2;\ x^6 - 64$
9. $x - 2;\ x^4 - 8x^2 + 16$
10. $x + 2;\ x^7 - 128$
11. $x;\ x^5 + 3x^2 + 5$
12. $x - \frac{3}{2};\ 4x^3 - 12x^2 + 9x$

13. $2x - 1;\ 4x^3 + 4x^2 + 11x - 7$ (*Hint:* If $x - \frac{1}{2}$ is a factor, does this give any indication concerning $2x - 1$?)

14. Find all the factors of $x^3 + 2x^2 - 5x - 6$ given that $x - 2$ is one factor.

In each of the Exercises 15 through 24, use synthetic division to find the quotient $q(x)$ and the remainder R when the first expression is divided by the second. Use the Remainder Theorem as a check.

15. $x^3 + 5x^2 + 6x - 7;\ x - 1$
16. $4x^3 + 6x^2 - 7;\ x + 3$
17. $x^4 + 2x^2 + 3;\ x + 2$
18. $6x^5 - 8x^3 + 4x^2 - 2x;\ x - \frac{1}{2}$
19. $14x^5 + 4x^4 + 6x^3 + 8x + 2;\ x + \frac{1}{2}$
20. $x^5 + 2x^4 - 2x^3 + 3x^2 + x - 5;\ x - 2$
21. $x^4 + 2;\ x + 1$
22. $9x^3 - 4x + 6;\ x + \frac{1}{3}$
23. $2x^3 - 4x^2 - 2x + 7;\ x - 3$
24. $4x^2 - 2x^3 + 6x - 8;\ x + 4$

25. If n is a positive integer, show that
   a. $x - a$ is a factor of $x^n - a^n$.
   b. $x + a$ is a factor of $x^n - a^n$ if n is even.
   c. $x + a$ is a factor of $x^n + a^n$ if n is odd.

26. Use the Remainder Theorem to find the value of k such that if $2x^3 - 5x^2 + 7x + k$ is divided by $x - 2$ the remainder is 10.

27. Find the value of k for which $x - 3$ is a factor of $3x^4 - 8x^3 + kx^2 - 9x + 9$.

28. Determine whether or not $x + 3$ is a factor of $5x^4 + 17x^3 + 6x^2 + 9x + 27$.

29. Determine whether or not $2x - 3$ is a factor of $2x^3 - 9x^2 - 11x + 30$.

30. If $q(x) = A_n x^n + A_{n-1} x^{n-1} + \cdots + A_1 x + A_0$, show that $x - 1$ is a factor of $q(x)$ if the sum of the coefficients of $q(x)$ is zero.

3.12 FACTS CONCERNING THE ROOTS OF A POLYNOMIAL EQUATION

Complex Roots of a Polynomial Equation

Two complex numbers that differ only in the sign of the imaginary part are called *conjugate complex numbers*. Each is said to be the conjugate of the other. For example, $2 + 3i$ and $2 - 3i$ are conjugates or, in more general terms, $a + bi$ and $a - bi$ are conjugates. We wish now to prove a very useful theorem regarding conjugate complex numbers as roots of polynomial equations.

THEOREM 3.6. *If a polynomial equation with real coefficients has the complex number $a + bi$ as a root, where $b \neq 0$, it has also the conjugate complex number $a - bi$ as a root.*

Proof: Let $f(x) = 0$ be a polynomial equation having $a + bi$, where $b \neq 0$, as a root. We form the product

$$\begin{aligned} P(x) &= [x - (a + bi)][x - (a - bi)] \\ &= [(x - a) - bi][(x - a) + bi] \\ &= (x - a)^2 - (bi)^2 \\ &= (x - a)^2 + b^2 \\ &= x^2 - 2ax + a^2 + b^2 \end{aligned}$$

Now divide $f(x)$ by $P(x)$ and carry out the division until a remainder is reached that is not of higher degree in x than the first. This remainder can be written in the form $Rx + S$, in which R and S are real numbers since the coefficients of $f(x)$ and $P(x)$ are real. If the quotient obtained in the division just described is represented by $q(x)$, we can write the following identity:

$$f(x) = q(x)P(x) + Rx + S$$

We wish next to show that R and S are both zero and that, hence, the product $P(x)$ is a factor of $f(x)$. Since an identity holds for all values of x, we can write a second identity:

$$f(a + bi) = q(a + bi)P(a + bi) + R \cdot (a + bi) + S$$

By hypothesis, $a + bi$ is a root of $f(x) = 0$; hence

$$f(a + bi) = 0$$

Furthermore, it follows from the definition of $P(x)$ that

$$P(a + bi) = 0$$

The second identity is therefore reduced to

$$R \cdot (a + bi) + S = 0$$

$$(Ra + S) + Rbi = 0 + 0i$$

In view of the definition of equality for complex numbers, we have

$$Ra + S = 0 \quad \text{and} \quad Rb = 0$$

But $b \neq 0$ according to the statement of our theorem; therefore, $R = 0$ and, consequently, $S = 0$. This gives the identity

$$f(x) = q(x)P(x)$$

From the definition of $P(x)$, it is easily seen that

$$P(a - bi) = 0$$

and hence

$$f(a - bi) = 0$$

This establishes our theorem.

The theorem just proved is sometimes stated thus: *Complex roots of polynomial equations with real coefficients occur in conjugate pairs.*

Number of Roots

We have learned from previous discussions that every polynomial equation of the first degree has one and only one root and that an equation of the second degree, a quadratic equation, has two roots. From these results one might suspect that the number of roots of a polynomial equation of degree n is n. This is true provided each multiple root is counted as many times as the root occurs. For instance, the equation

$$x^3 - 3x^2 + 3x - 1 = 0$$

can be factored as follows:

$$(x - 1)(x - 1)(x - 1) = 0$$

Setting each factor equal to zero and solving for x, we get three roots,

$$x = 1, 1, 1$$

In this case all three roots are equal, and we say that 1 is a multiple root of multiplicity 3 of the given equation. Strictly speaking, r is a root of multiplicity k of the polynomial equation $f(x) = 0$ if and only if $(x - r)^k$ is a factor of $f(x)$ and $(x - r)^{k+1}$ is not a factor of $f(x)$. When the roots of a polynomial equation are counted, each multiple root of multiplicity k is counted k times. We shall now state two important theorems regarding the roots of a polynomial equation.

THEOREM 3.7 (FUNDAMENTAL THEOREM OF ALGEBRA). *Every polynomial equation of degree $n > 0$ has at least one root. Moreover, this root may be a real number, a pure imaginary number, or a general complex number.*

We shall accept this theorem without proof, since any proof that could be given is beyond the scope of our work here.

THEOREM 3.8. *Every polynomial equation of degree n has exactly n roots, each root of multiplicity k being counted as k roots.*

Proof: Let $f(x) = 0$ be a polynomial equation of degree n. By the fundamental theorem of algebra, this equation has at least one root, which we shall denote by r_1. It follows from the factor theorem that $x - r_1$ is a factor of $f(x)$, and so the given equation can be written

$$f(x) = (x - r_1)q_1(x) = 0$$

The quotient $q_1(x)$ is a polynomial, and hence the equation $q_1(x) = 0$ must have at least one root, say r_2. By the factor theorem,

$$q_1(x) = (x - r_2)q_2(x)$$

Consequently,

$$f(x) = (x - r_1)(x - r_2)q_2(x) = 0$$

This same type of argument can be repeated n times to give

(1) $$f(x) = q_n(x)(x - r_1)(x - r_2) \cdots (x - r_n) = 0$$

Since $f(x)$ is of degree n, it follows that $q_1(x)$ must be of degree $n - 1$, $q_2(x)$ of degree $n - 2$, $q_3(x)$ of degree $n - 3$, etc., and $q_n(x)$ of degree $n - n$, or zero. Thus $q_n(x)$ is a constant. In particular,

if $$f(x) = A_n x^n + A_{n-1} x^{n-1} + \cdots + A_1 x + A_0 = 0$$

then $$q_n(x) = A_n$$

and

(2) $$f(x) = A_n(x - r_1)(x - r_2) \cdots (x - r_n) = 0$$

This last equation shows that every polynomial equation of degree n has at least n roots, $r_1, r_2, r_3, \ldots$, and r_n, which may or may not all be distinct. However, in the sense of our theorem, all are counted whether they are distinct or not.

To prove that the given equation, $f(x) = 0$, cannot have more than n roots, let us suppose that any polynomial equation at all, $P(x) = 0$, has m roots with $m > n$. Then, if $r_1, \ldots, r_m$ are the m roots mentioned, $(x - r_1)(x - r_2) \cdots (x - r_m)$ is a factor of $P(x)$ (by successive applications of the factor theorem). However, $(x - r_1)(x - r_2) \cdots (x - r_m)$ is easily seen to have degree m, and so $P(x)$ has degree at least m. But $f(x)$ has degree n and $n < m$ so that $f(x)$ cannot have m roots with $m > n$. Our conclusion is that Equation (2) has exactly n roots, counting each root of multiplicity k as k roots.

SEC. 3.12   Facts Concerning the Roots of a Polynomial Equation

Upper and Lower Limits of the Real Roots of a Polynomial Equation

A real number U is an *upper limit* of the real roots of a polynomial equation if no root of the equation is larger than U. Similarly, a real number L is a *lower limit* of the real roots of a polynomial equation if no root of the equation is smaller than L. We shall find it useful to recognize that a certain number is an upper limit (or lower limit) of the real roots of an equation. This observation will preclude the necessity of testing any larger (or smaller) numbers as possible roots and, hence, will be a time saver.

Example 1   Show that $U = 3$ is an upper limit of the real roots of the equation

$$f(x) = 2x^3 - 5x^2 - 3x + 7 = 0$$

Solution: Synthetically dividing $f(x)$ by $x - 3$ [or, saying this another way, synthetically dividing the coefficients of $f(x)$ by 3], we obtain

$$\begin{array}{r} 2 - 5 - 3 + 7 \underline{\lfloor 3} \\ + 6 + 3 + 0 \\ \hline 2 + 1 + 0 + 7 \end{array}$$

Notice that all the numbers in the third line are either positive or zero. If we use a number larger than 3 for our synthetic division, all the numbers in the third line will be larger than those found above (except the first, which remains the same). So the remainder when synthetically dividing by a number larger than 3 will be larger than 7. In particular, the remainder will not be zero, and, by the remainder theorem, this number larger than 3 cannot be a root.

The solution to Example 1 illustrates the following fact, which we isolate as a *Test for Upper Limits.*

TEST FOR UPPER LIMITS OF THE REAL ROOTS OF A POLYNOMIAL EQUATION

The *positive* real number U is an upper limit of the real roots of the polynomial equation $f(x) = 0$ if, when $f(x)$ is synthetically divided by $x - U$ [that is, the coefficients of $f(x)$ are synthetically divided by U], all numbers in the third line are either positive or zero.

Caution: The test above applies only to *positive* numbers U. Similarly, the test for lower limits will apply only to *negative* numbers L. Moreover, both tests give sufficient conditions only. It is possible, for example, to have a positive real number U which is an upper limit for the real roots of a polynomial equation $f(x) = 0$, and yet the third line in the synthetic division contains a negative number. The next example illustrates this fact.

Example 2

For the equation $x^2 - 4x + 4 = 0$ we have $S = \{2\}$, since

$$x^2 - 4x + 4 = 0 \leftrightarrow (x-2)^2 = 0$$
$$\leftrightarrow x - 2 = 0$$
$$\leftrightarrow x = 2$$

Therefore, 3 is an upper limit of the real roots of this equation. However, when $x^2 - 4x + 4$ is synthetically divided by $x - 3$, we obtain

$$\begin{array}{r}1 - 4 + 4\,\underline{|3} \\ +3 - 3 \\ \hline 1 - 1 + 1\end{array}$$

which has a negative number in the third line. Hence, our test for upper limits does not show that 3 is an upper limit, and yet 3 is, in fact, an upper limit of the real roots.

Now we turn to a test for lower limits. Again, the test is illustrated and motivated by an example.

Example 3

Show that $L = -2$ is a lower limit of the real roots of the equation

$$f(x) = 2x^3 - 5x^2 - 3x + 7 = 0$$

Solution: Synthetically dividing $f(x)$ by $x + 2$ [or, saying this another way, synthetically dividing the coefficients of $f(x)$ by -2], we obtain

$$\begin{array}{r}2 - 5 - 3 + 7\,\underline{|-2} \\ -4 + 18 - 30 \\ \hline 2 - 9 + 15 - 23\end{array}$$

Notice that the numbers in the third line alternate in sign. If we use a number smaller than -2 for our synthetic division (*Note:* Numbers smaller than -2 must have absolute value larger than 2. So -3 is smaller than -2 but -1 is not), then the numbers in the third line will still alternate in sign, and they will be larger in absolute value than those found above (except the first, which remains the same). In particular, the remainder will not be zero, and, by the remainder theorem, this number smaller than -2 cannot be a root.

The solution to Example 3 illustrates the following fact, which we isolate as a *Test for Lower Limits.*

TEST FOR LOWER LIMITS OF THE REAL ROOTS OF A POLYNOMIAL EQUATION

The *negative* real number L is a lower limit of the real roots of the polynomial equation $f(x) = 0$ if, when $f(x)$ is synthetically divided by $x - L$ [that is, the coefficients of $f(x)$ are synthetically divided by L], the numbers in the third line

SEC. 3.12   Facts Concerning the Roots of a Polynomial Equation   167

are alternately positive and negative (again allowing zeros to occur and counting them as either positive or negative).

The caution preceding Example 2 should be reemphasized here. These tests yield only sufficient conditions in order that a number be an upper limit or a lower limit. The next, and final, example of this section illustrates that a negative number L can be a lower limit of the real roots of a polynomial equation without having alternately positive and negative signs in the third line of the synthetic division.

Example 4

For the equation $x^2 + 4x + 4 = 0$, we have $S = \{-2\}$, since

$$x^2 + 4x + 4 = 0 \leftrightarrow (x+2)^2 = 0$$
$$\leftrightarrow x + 2 = 0$$
$$\leftrightarrow x = -2$$

Therefore, -3 is a lower limit of the real roots of this equation. However, when $x^2 + 4x + 4$ is synthetically divided by $x + 3$, we obtain

$$\begin{array}{r} 1 + 4 + 4 \underline{\lvert -3} \\ -3 - 3 \\ \hline 1 + 1 + 1 \end{array}$$

which does not have alternating signs in the third line. Again, our test for lower limits does not show that -3 is a lower limit, and yet -3 is, in fact, a lower limit of the real roots.

Descartes' Rule of Signs

Before attempting to solve a polynomial equation of degree higher than the second, it is often helpful to have as much information about the roots of the equation as can be readily obtained. In the preceding subsections we have discussed methods and theorems which will be helpful in obtaining such information. Descartes' rule of signs, which we shall state but not prove, will help us to gain additional information about the real roots of an equation.

If a polynomial is arranged in descending powers of x, there is said to be a variation in sign whenever two successive terms have opposite signs. Missing powers are to be ignored. For example,

$$f(x) = x^4 - 2x^2 + 3x - 5$$

has three variations in sign, one between the first and second terms, another between the second and third terms, and another between the third and fourth terms.

Let $f(-x)$ denote the polynomial obtained by replacing x by $-x$ in the polynomial $f(x)$. Since $(-x)^2 = x^2, (-x)^3 = -x^3, (-x)^4 = x^4$, etc., it is obvious that $f(-x)$ can be obtained by changing the sign of the coefficient of each odd power in $f(x)$. Thus

if
$$f(x) = x^4 + 2x^3 - x^2 - 7x + 5$$
then
$$f(-x) = x^4 - 2x^3 - x^2 + 7x + 5$$

If a is a root of $f(x) = 0$, then $-a$ is a root of $f(-x) = 0$; hence the number of negative roots of $f(x) = 0$ is precisely the same as the number of positive roots of $f(-x) = 0$.

DESCARTES' RULE OF SIGNS

A polynomial equation $f(x) = 0$ with real coefficients has either as many positive real roots as $f(x)$ has variations in sign or less than that by an even integer; the equation $f(x) = 0$ has either as many negative real roots as $f(-x)$ has variations in sign or less than that by an even integer.

A proof of Descartes' rule can be found in any standard text on the theory of equations.

Example 5   What information does Descartes' rule of signs give regarding the roots of the equation $f(x) = 2x^3 + 5x^2 - 3x + 4 = 0$?

Solution: Since $f(x)$ has two variations in sign, $f(x) = 0$ has either two positive real roots or zero positive real roots. Now consider
$$f(-x) = -2x^3 + 5x^2 + 3x + 4 = 0$$
Since $f(-x)$ has only one variation in sign, $f(x) = 0$ must have exactly one negative real root, for the first integer which is less than 1 by an even integer is -1, and the number of roots cannot be -1. Therefore, the given equation has either two positive real roots and one negative real root, or it has two imaginary roots and one negative root. It must have exactly three roots.

Exercises 3.12

In Exercises 1 through 8, determine the smallest positive integer which the upper limit test shows is an upper limit of the real roots. Also, determine the largest negative integer which the lower limit test shows is a lower limit of the real roots.

1. $x^3 - 2x^2 - x + 2 = 0$

    Solution: For upper limits we begin with 1:

    $$\begin{array}{r} 1 - 2 - 1 + 2 \underline{|1} \\ 1 \\ \hline 1 - 1 \end{array}$$

    We can stop here because a negative number has already appeared

SEC. 3.12   Facts Concerning the Roots of a Polynomial Equation   169

$$\begin{array}{r}1-2-1+2\,\underline{|2}\\2\quad\ \ 0\end{array}$$

$$1\quad 0-1 \qquad \text{Again, stop}$$

$$\begin{array}{r}1-2-1+2\,\underline{|3}\\+3+3+6\end{array}$$

$$1+1+2+8$$

So 3 is the answer to the first part. For the second part, begin with -1 and continue to -2, and so on, until the signs in the third line alternate. The answer in this case happens to be -1 since

$$\begin{array}{r}1-2-1+2\,\underline{|-1}\\-1+3-2\end{array}$$

$$1-3+2-0$$

Notice that we have considered the zero in the third line to have a negative sign.

2. $x^3 - 3x + 5 = 0$ 3. $x^3 - 3x^2 + 2x + 3 = 0$

4. $x^4 + 2x^3 + 3x^2 + 2x + 1 = 0$ 5. $x^4 + x^3 - 3x^2 - 4x + 2 = 0$

6. $2x^4 - 3x^2 + 2x^2 - 5x - 2 = 0$ 7. $2x^3 - 3x^2 - 7x + 8 = 0$

Using Descarte's Rule of Signs and any other pertinent material discussed thus far in this chapter, obtain as much information as you can about the roots of each of the following equations.

8. $x^3 - 2x^2 - x - 4 = 0$

Solution: $f(x) = x^3 - 2x^2 - x - 4$ has only one variation in sign and so the equation has exactly one positive real root. $f(-x) = -x^3 - 2x^2 + x - 4$ has two variations in sign and so the equation has either two negative real roots or no negative real roots.

  Using the techniques applied in Exercises 1 through 7, we find that the positive real root is less than 3. Also, if the equation has any negative real roots, both must be greater than -1. Finally, if there are no negative real roots, then the equation has two imaginary roots.

  A suitable form for the answer would be: One positive real root less than 3 and either two negative real roots greater than -1 or two imaginary roots.

9. $2x^4 - x^3 - 3x^2 + 2x + 3 = 0$ 10. $x^4 - 5x^2 + 4 = 0$

11. $x^4 + x^3 + 3x^2 + 4 = 0$ 12. $3x^4 - 5x^3 + 6x^2 - 3x + 1 = 0$

13. $4x^5 - 9 = 0$ 14. $4x^5 + 9 = 0$

15. $x^3 + 3x^2 + 9 = 0$ 16. $x^4 - 2x^3 + 3x^2 - 5x - 6 = 0$

3.13 RATIONAL ROOTS

Recall that a rational number is a real number that can be expressed in the form $\frac{a}{b}$, where a and b are integers (with $b \neq 0$). A root of an equation that is a rational number is called a *rational root* of the equation. The next theorem will allow us to find all rational roots of any polynomial equation which has integers as coefficients. The process is one of trial and error, and the facts developed in previous sections of this chapter will be helpful in minimizing the number of trials required.

THEOREM 3.9. *If the coefficients $A_n, A_{n-1}, \ldots, A_1, A_0$ of a polynomial equation*

(1) $$f(x) = A_n x^n + A_{n-1} x^{n-1} + \cdots + A_1 x + A_0 = 0$$

are integers, and if $\frac{c}{d}$ is a rational number in lowest terms that satisfies the equation, then c is a factor of A_0 and d is a factor of A_n.

Proof: Since $\frac{c}{d}$ is a rational number in lowest terms, c and d are integers that have no common factor. By hypothesis, $\frac{c}{d}$ is a root of $f(x) = 0$; hence

$$A_n \left(\frac{c}{d}\right)^n + A_{n-1} \left(\frac{c}{d}\right)^{n-1} + \cdots + A_1 \left(\frac{c}{d}\right) + A_0 = 0$$

Simplifying and multiplying through by d^n, we get

(2) $$A_n c^n + A_{n-1} c^{n-1} d + \cdots + A_1 c d^{n-1} + A_0 d^n = 0$$

Now transpose the last term of the left side of this equation to the right and factor the left side thus:

$$c(A_n c^{n-1} + A_{n-1} c^{n-2} d + \cdots + A_1 d^{n-1}) = -A_0 d^n$$

Since each letter in the last equation represents an integer and since sums and products of integers all give integers, the quantity inside the parentheses in the left side is an integer, and so the left side is the product of two integers. That is,

$$c \cdot (\text{integer}) = -A_0 d^n$$

If two numbers are equal, a factor of one must be a factor of the other. Hence c is a factor of $A_0 d^n$. But since c has no factor in common with d, it has no factor in common with d^n and therefore it is a factor of A_0.

In a similar manner, we can write Equation (2) as

$$d(A_{n-1} c^{n-1} + A_{n-2} c^{n-2} d + \cdots + A_1 c d^{n-2} + A_0 d^{n-1}) = -A_n c^n$$

SEC. 3.13 Rational Roots

from which it follows that d is a factor of A_n, and the proof of our theorem is complete.

COROLLARY. Any rational root of a polynomial equation of the form
$$x^n + A_{n-1}x^{n-1} + \cdots + A_1 x + A_0 = 0$$
where $A_{n-1}, \ldots, A_1, A_0$ are integers, is itself an integer and is a factor of A_0. (Note that $A_n = 1$.)

Example 1   Find the rational roots of the equation $f(x) = 2x^3 - 9x^2 + 7x + 6 = 0$.

Solution: By the above theorem, if $\dfrac{c}{d}$ is a rational root of the given equation, then c is a factor of 6 and d is a factor of 2. The possible rational roots in this case then are $\pm\frac{1}{2}, \pm\frac{3}{2}, \pm 1, \pm 2, \pm 3, \pm 6$.

To determine whether any of these numbers are roots, we shall use synthetic division, recalling that a is a root of $f(x) = 0$ if the remainder is zero after division of $f(x)$ by $x - a$, and that if the remainder is not zero, then a is not a root.

Let us first try the number $x = \frac{1}{2}$.

$$\begin{array}{r} 2 - 9 + 7 + 6 \,\underline{|\tfrac{1}{2}} \\ +1 - 4 \\ \hline 2 - 8 + 3 \end{array}$$

The division need not be continued because $(3)(\tfrac{1}{2})$ is not an integer, and once a fraction, and not an integer, is introduced in the third line of our synthetic substitution process, there is no need to carry the division further, for such fractions will persist and prevent a final zero. This will be true in general if the coefficients of the given equations are integers. We next try 1, $\tfrac{3}{2}$, and then 2.

$$\begin{array}{r} 2 - 9 + 7 + 6\,\underline{|1} \\ +2 - 7 + 0 \\ \hline 2 - 7 + 0 + 6 \end{array} \qquad \begin{array}{r} 2 - 9 + 7 + 6\,\underline{|\tfrac{3}{2}} \\ +3 - 9 - 3 \\ \hline 2 - 6 - 2 + 3 \end{array} \qquad \begin{array}{r} 2 - 9 + 7 + 6\,\underline{|2} \\ +4 - 10 - 6 \\ \hline 2 - 5 - 3 + 0 \end{array}$$

These divisions show that 1 and $\tfrac{3}{2}$ are not roots but that 2 is a root. By the factor theorem, $x - 2$ is a factor, and the given equation can be written
$$(x - 2)(2x^2 - 5x - 3) = 0$$
The new equation, $2x^2 - 5x - 3 = 0$, obtained by removing the factor $x - 2$, is called the *depressed equation*. Now this equation can be solved as a quadratic to yield the remaining two roots. Factoring the depressed equation, we have
$$(2x + 1)(x - 3) = 0$$
Solving for x, we get $x = -\tfrac{1}{2}$ and $x = 3$. The roots of the given equation are $x = 2, 3, -\tfrac{1}{2}$.

Check: These may be checked by direct substitution or by forming the product $(x - 2)(x - 3)(2x + 1) = 2x^3 - 9x^2 + 7x + 6$.

It should be emphasized that a given number may be a root of an equation more than once; that is, it may be a multiple root. For this reason, one should work with the depressed equation at each step. There is no other way of discovering that the number is a multiple root. Furthermore, if a number has been tried and found not to be a root of the original equation, it will not be a root of the depressed equation at any step and need not be tried again. It should also be noted that the above procedure will enable us to find all the roots of a polynomial equation if not more than two of the roots are irrational or imaginary. This follows from the fact that, when all the rational roots have been found, the depressed equation will be a quadratic and can thus be solved by the quadratic formula.

Example 2 Find all the roots of the equation $f(x) = x^6 - 9x^4 + 16x^3 - 9x^2 + 1 = 0$.

Solution: By Descartes' rule of signs, the given equation has either four, two, or no positive real roots. Since the equation

$$f(-x) = x^6 - 9x^4 - 16x^3 - 9x^2 + 1 = 0$$

has two variations in sign, the original equation has either two or no negative real roots. It must have six roots. The possible rational roots are ± 1. We try $x = 1$ as follows:

$$\begin{array}{r}1 + 0 - 9 + 16 - 9 + 0 + 1 \underline{|1} \\ +1 + 1 - 8 + 8 - 1 - 1 \\ \hline 1 + 1 - 8 + 8 - 1 - 1 + 0 \end{array}$$

and find that 1 is a root. Let us keep trying $x = 1$ in the depressed equation at each step until it fails to give a zero remainder:

$$\begin{array}{r}1 + 1 - 8 + 8 - 1 - 1 \underline{|1} \\ +1 + 2 - 6 + 2 + 1 \\ \hline 1 + 2 - 6 + 2 + 1 \end{array}$$

$$\begin{array}{r}1 + 2 - 6 + 2 + 1 \\ +1 + 3 - 3 - 1 \\ \hline 1 + 3 - 3 - 1 \end{array}$$

$$\begin{array}{r}1 + 3 - 3 - 1 \\ +1 + 4 + 1 \\ \hline 1 + 4 + 1 + 0 \end{array}$$

The final depressed equation is the quadratic $x^2 + 4x + 1 = 0$, which we can solve by the quadratic formula, getting

$$x = \frac{-4 \pm \sqrt{16 - 4}}{2} = -2 \pm \sqrt{3}$$

The roots of the original equation are 1, 1, 1, 1, $-2 \pm \sqrt{3}$.

SEC. 3.13 Rational Roots

Exercises 3.13

Find all the rational roots of each of the following equations. Where possible find all the roots of the equation.

1. $x^3 - 6x^2 + 11x - 6 = 0$
2. $x^3 - 2x^2 - 5x + 6 = 0$
3. $3x^3 - 2x^2 + 6x - 4 = 0$
4. $x^3 - 5x^2 + 7x - 3 = 0$
5. $x^3 + 3x^2 - 25x - 75 = 0$
6. $x^3 - 4x^2 + 9x - 10 = 0$
7. $x^4 - 3x^3 - 11x^2 + 3x + 10 = 0$
8. $6x^4 - x^3 - 39x^2 - 6x + 40 = 0$
9. $8x^4 + 30x^3 + 21x^2 - 20x - 12 = 0$
10. $x^4 - 6x^2 + 4x = 0$
11. $2x^3 - 3x^2 - 8x + 12 = 0$
12. $x^4 - 18x^2 + 81 = 0$
13. $4x^4 - 8x^3 + 39x^2 + 2x - 10 = 0$
14. $x^4 - x^3 - 2x^2 - 4x - 24 = 0$
15. $x^4 - x^3 - 10x^2 - 8x = 0$
16. $4x^3 - 8x^2 + 13x + 25 = 0$
17. $x^3 - x^2 - 8x + 12 = 0$
18. $x^4 - 2x^3 + 2x^2 - 8x + 8 = 0$
19. $x^4 - 9x^3 + 30x^2 - 44x + 24 = 0$
20. $2x^3 - 9x^2 + 7x + 6 = 0$
21. $2x^3 - 45x^2 + 2{,}000 = 0$
22. $x^5 - 5x^4 + 6x^3 + 11x^2 - 43x + 30 = 0$
23. $x^5 - 5x^4 + 3x^3 + 13x^2 - 8x - 12 = 0$
24. $8x^3 - 36x^2 + 54x - 27 = 0$
25. $2x^4 - 3x^3 - 3x^2 + 7x - 3 = 0$
26. $x^4 - 10x^2 + 9 = 0$
27. $4x^3 - 36x^2 + 83x - 66 = 0$
28. $x^4 + x^3 - 11x^2 - 9x + 18 = 0$
29. $120x^3 - 91x^2 + 54x + 40 = 0$
30. $2x^3 - 11x^2 + x + 35 = 0$

Factor each of the following polynomials as products of polynomials having integers as coefficients. Carry the factorization as far as possible.

31. $x^3 - 5x^2 + 2x + 8$
32. $x^4 + 3x^3 - 3x^2 - 7x + 6$
33. $4x^4 + 4x^3 + 5x^2 + 8x - 6$

Solution: By using the techniques applied in Exercises 1 through 30, we find that the roots are $\frac{1}{2}, -\frac{3}{2},$ and $\pm\sqrt{2}i$. Hence, by the factor theorem,

$$4x^4 + 4x^3 + 5x^2 + 8x - 6 = 4\left(x - \frac{1}{2}\right)\left(x + \frac{3}{2}\right)(x + \sqrt{2}i)(x - \sqrt{2}i)$$

$$= 2\left(x - \frac{1}{2}\right)2\left(x + \frac{3}{2}\right)(x^2 + 2)$$

$$= (2x - 1)(2x + 3)(x^2 + 2)$$

The last two factors were combined in order to obtain integer coefficients.

34. $4x^4 + x^2 - 5x$
35. $4x^3 + 5x^2 + 13x + 3$

36. $x^4 + 9x^3 + 30x^2 + 44x + 24$ 37. $x^6 - 64$

38. $x^5 + 5x^4 - 4x^3 - 46x^2 - 8x + 120$ 39. $8x^3 - 1$

40. $12x^3 - 28x^2 - 7x + 5$

Form equations with integral coefficients having the following solution sets.

41. $\{1, 2, 3\}$ 42. $\{-1, -1, 2\}$ 43. $\{2, \pm\sqrt{3}\}$

Solution: $(x - 2)(x - \sqrt{3})(x + \sqrt{3}) = 0$ is an equation having the given set as a solution set. Expanding, we obtain

$$(x - 2)(x^2 - 3) = 0$$

and finally

$$x^3 - 2x^2 - 3x + 6 = 0$$

This is an equation of the desired type.

3.14* IRRATIONAL ROOTS

In general, irrational roots of a polynomial equation of degree $n > 2$ cannot be found exactly. However, it is always possible to approximate any such root as closely as we please. There are many well-known methods for approximating the irrational roots of a polynomial equation, only the simplest of which will be considered here. Other methods, such as Horner's, Graeffe's, and Newton's, are discussed in most texts on the theory of equations.

Basically, the method we shall describe depends only on the following fact:

LOCATION PRINCIPLE: If $f(x)$ is a polynomial with real coefficients and if $f(a)$ and $f(b)$ are opposite in sign, then $f(x) = 0$ has at least one root between a and b.

Example 1 Consider the polynomial equation $f(x) = x^3 - 7x + 3 = 0$. This equation has the property that $f(-x) = -x^3 + 7x + 3$, and so it has exactly one negative real root [since there is one variation in sign among the coefficients of $f(-x)$]. There are two variations in sign among the coefficients of $f(x)$, so the equation has either two positive real roots or no positive real roots. The possible rational roots are ± 1, ± 3. Using synthetic substitutions, we obtain the following table:

x	-3	-2	-1	0	1	2	3
$f(x)$	-3	9	9	3	-3	-3	9

Notice that none of the possible rational roots are roots, and so any real roots of the equation $f(x) = 0$ must be irrational.

SEC. 3.14* Irrational Roots

Using the location principle, we observe that $f(x) = 0$ must have at least one root between -3 and -2 [since $f(-3) = -3$ and $f(-2) = 9$ are opposite in sign]; it must have at least one root between 0 and 1 [since $f(0) = 3$ and $f(1) = -3$ are opposite in sign]; and it must have at least one root between 2 and 3 [since $f(2) = -3$ and $f(3) = 9$ are opposite in sign]. Now, the degree of $f(x)$ is three, and so the equation $f(x) = 0$ has exactly three roots. It follows that there is exactly one root between -3 and -2, exactly one root between 0 and 1, exactly one real root between 2 and 3, and all these are irrational roots.

The above observations show that we have made a first approximation to the three roots of this equation. Now we want to make the approximations within one decimal place of the exact answers. To do this, we (for example) evaluate $f(x)$ at numbers of the form $-2.9, -2.8, \ldots, -2.2, -2.1$ until we find the place where the values are opposite in sign. Here $f(-2.9) = -0.933$ and $f(-2.8) = 0.648$. Therefore, the root which lies between -3 and -2 must, in fact, lie between -2.9 and -2.8. Hence either -2.9 or -2.8 is an approximation to one root of the equation which is within one decimal place of the exact value of the root. To approximate this root to within two decimal places, we evaluate $f(x)$ at points of the form $-2.89, -2.88, \ldots, -2.82, -2.81$ until we find the place where the values are opposite in sign. Here $f(-2.84) = -0.026304$ and $f(-2.83) = 0.144813$. Therefore, the root which lies between -2.9 and -2.8 must, in fact, lie between -2.84 and -2.83. Again, either -2.84 or -2.83 is an approximation to one root of the equation which is within two decimal places of the exact value of the root. By continuing this process of "dividing successive intervals into ten parts," we can obtain approximations to this root which are within any desired number of decimal places of the exact value of the root.

Finally, the same process can be applied to approximate the other two roots to any desired degree of accuracy.

Even using synthetic substitution, the process described in Example 1 can involve tedious computations. The computations are considerably easier using a hand calculator. In fact, by using a programmable hand calculator, a simple program can be written which allows one to find the value of $f(x)$ for any value of x by simply punching the value of x and then starting the program. If properly written, the calculator will compute the value of $f(x)$, display it, and stop. Then another value of x can be punched in, the program can be started, and the value of $f(x)$ will be displayed. This can be extremely helpful in curve sketching (see Section 3.15 and Chapter 4) as it makes the construction of a table of values rather trivial. A block diagram for a program could look something like the diagram shown on p. 176 [for the polynomial $f(x) = x^3 - 7x + 3$].

The previous paragraph indicated that a programmable calculator could be used to quickly evaluate $f(x)$ for any value of x. It so happens that such a calculator can be programmed to do all the work for you in approximating irrational roots of polynomial equations. For example, one can be programmed to approximate the root of $x^3 - 7x + 3 = 0$ which lies between -3 and -2 to within (say) 5 decimal places. Basically, the program is written so as to evaluate $f(x)$ at -2.5 (the midpoint of the line between -3 and -2). Then the program

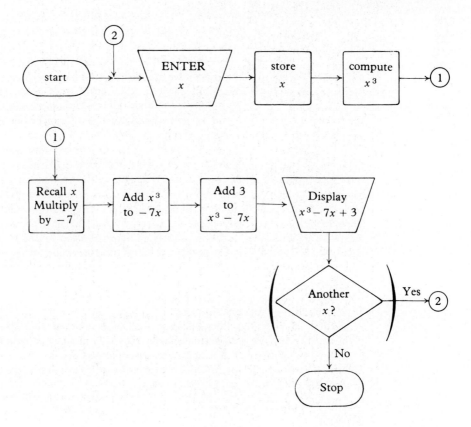

"tests" this value to see whether -2.5 is to the left or to the right of the root. If -2.5 is to the left, then the program will evaluate $f(x)$ at the midpoint of the line between -2.5 and -2. If -2.5 is to the right, then the program will evaluate $f(x)$ at the midpoint of the line between -3 and -2.5. All this is done automatically (provided the program is properly written), and the process continues through enough steps so that the last interval has length less than 0.00001. This is to ensure that the approximation is within 5 decimal places of the exact value of the root. It so happens that such a program would go through this "bisection" process 17 times before arriving at the desired approximation. However, such a process is performed very rapidly. Solving such problems on a programmable calculator can serve as an introduction to computer programming.

Exercises 3.14

Use the method described in Example 1 to approximate, to one decimal place, the irrational roots of each of the following equations.

1. $2x^3 + x^2 - 10x - 4 = 0$ 2. $x^3 - 3x - 4 = 0$

SEC. 3.15 Equations Containing Fractions

3. $x^4 - 2x^3 - 4x^2 + 4x + 4 = 0$ **4.** $x^4 - x^3 - 5x^2 + 7x - 2 = 0$

5. $x^3 + 3x - 2 = 0$ **6.** $x^3 - 4x + 1 = 0$

7. Use the method discussed in the last section to approximate to three decimal places the positive root of the equation in Exercise 1.

8. Find, to three decimal places, the least positive root of the equation $x^3 - 3x + 1 = 0$.

Find (accurate to three decimal places) the indicated principal roots.

9. $\sqrt[3]{4}$ **10.** $\sqrt[3]{10}$ **11.** $\sqrt[4]{6}$ **12.** $\sqrt[5]{8}$

3.15 EQUATIONS CONTAINING FRACTIONS

In the preceding sections of this chapter, methods were given for finding all the rational roots of a polynomial equation. In fact, we are now able to find the solution set of any polynomial equation provided at most two of its roots are not rational numbers. This suggests that if the problem of solving a given equation can be reduced to the problem of solving an equivalent polynomial equation, then our chances for finding the solution set of the given equation should be good.

In this section we shall consider a type of equation containing fractions whose solution always leads to the solution of a polynomial equation. These are equations of the form $\dfrac{P(x)}{Q(x)} = 0$, where $P(x)$ and $Q(x)$ are polynomials. Some typical examples of such equations are

$$\frac{3x + 7}{2 - x} = 0 \quad \text{and} \quad \frac{x^3 - 7x + 3}{x^3 + x - 1} = 0$$

Our method for solving such equations is based on the fact that a fraction $\dfrac{A}{B} = 0$ if and only if $A = 0$ and $B \neq 0$. We illustrate the technique by way of examples.

Example 1   Find the solution set of the equation

$$\frac{(x-1)(x+2)}{(x+3)(x-5)} = 0$$

Solution: Using the observation made above, the given equation holds if and only if

$$(x - 1)(x + 2) = 0 \quad \text{and} \quad (x + 3)(x - 5) \neq 0$$

Hence we proceed to solve the equation

$$(x - 1)(x + 2) = 0$$

and then check to see if any of these roots make $(x + 3)(x - 5) = 0$.

$$(x - 1)(x + 2) = 0 \leftrightarrow x - 1 = 0 \quad \text{or} \quad x + 2 = 0$$
$$\leftrightarrow x = 1 \quad \text{or} \quad x = -2$$

Now, neither 1 nor -2 is a solution to $(x + 3)(x - 5)$. Therefore, the solution set of the original equation is $S = \{1, -2\}$.

Example 2 Find the solution set of the equation

$$\frac{x^2 - 1}{x^2 - 2x + 1} = 0$$

Solution: Again, the given equation holds if and only if $x^2 - 1 = 0$ and $x^2 - 2x + 1 \neq 0$. Solving the equation $x^2 - 1 = 0$, we obtain

$$x^2 - 1 = 0 \leftrightarrow (x - 1)(x + 1) = 0$$
$$\leftrightarrow x - 1 = 0 \quad \text{or} \quad x + 1 = 0$$
$$\leftrightarrow x = 1 \quad \text{or} \quad x = -1$$

Now, if $x = 1$, then $x^2 - 2x + 1 = 0$, and so 1 cannot be a solution to the original equation (it makes the denominator zero). Hence 1 must be discarded. However, if $x = -1$, then $x^2 - 2x + 1 = 4 \neq 0$, and so -1 is a solution to the original equation. It follows that the solution set of the original equation is $S = \{-1\}$.

In Section 3.2 we observed that we could take a polynomial equation of the form $f(x) = g(x)$ and change it to the equivalent equation $f(x) - g(x) = 0$. The same is true of equations involving fractions or other algebraic expressions. We isolate this fact as a theorem.

THEOREM 3.10. *If $f(x)$ and $g(x)$ are algebraic expressions, then $f(x) = g(x) \leftrightarrow f(x) - g(x) = 0$.*

This theorem gives the necessary equipment to solve additional equations involving fractions. Again, we illustrate with two examples.

Example 3 Solve the equation $\dfrac{2}{(x - 1)(x + 1)} = \dfrac{1}{(x - 1)} - 2$.

Solution

$$\frac{2}{(x - 1)(x + 1)} = \frac{1}{x - 1} - 2 \leftrightarrow \frac{2}{(x - 1)(x + 1)} - \frac{1}{x - 1} + 2 = 0$$

(by Theorem 3.10)

$$\leftrightarrow \frac{2 - (x + 1) + 2(x - 1)(x + 1)}{(x - 1)(x + 1)} = 0$$

$$\leftrightarrow \frac{2x^2 - x - 1}{(x - 1)(x + 1)} = 0$$

SEC. 3.15 Equations Containing Fractions

Now, the last equation is in the form $\dfrac{P(x)}{Q(x)} = 0$ with $P(x)$ and $Q(x)$ polynomials, so we use the technique of Examples 1 and 2 to solve it.

$$2x^2 - x - 1 = 0 \leftrightarrow (2x+1)(x-1) = 0$$
$$\leftrightarrow 2x + 1 = 0 \quad \text{or} \quad x - 1 = 0$$
$$\leftrightarrow 2x = -1 \quad \text{or} \quad x = 1$$
$$\leftrightarrow x = -\frac{1}{2} \quad \text{or} \quad x = 1$$

If $x = -\dfrac{1}{2}$, $(x-1)(x+1) = \left(\dfrac{1}{2} - 1\right)\left(\dfrac{1}{2} + 1\right) = \left(-\dfrac{1}{2}\right)\left(\dfrac{3}{2}\right) \neq 0$, and so $-\dfrac{1}{2}$ is a solution of the original equation. However, if $x = 1$, $(x-1)(x+1) = 0$, and so $x = 1$ must be discarded. It follows that the solution set of the original equation is
$$S = \left\{-\frac{1}{2}\right\}.$$

Example 4   Find the solution set of the equation

$$\frac{2}{x-1} + \frac{5}{x+1} = \frac{4}{x^2-1}$$

Solution

$$\frac{2}{x-1} + \frac{5}{x+1} = \frac{4}{x^2-1} \leftrightarrow \frac{2}{x-1} + \frac{5}{x+1} - \frac{4}{x^2-1} = 0$$
(by Theorem 3.10)
$$\leftrightarrow \frac{2(x+1) + 5(x-1) - 4}{(x-1)(x+1)} = 0$$
$$\leftrightarrow \frac{7x - 7}{(x-1)(x+1)} = 0$$

Now
$$7x - 7 = 0 \leftrightarrow 7x = 7$$
$$\leftrightarrow x = 1$$

But if $x = 1$, then $(x-1)(x+1) = 0$. Therefore, there are no solutions to the original equation and we have $S = \{\ \}$.

The Method of Clearing of Fractions

Another procedure commonly used for solving equations involving fractions is the following. Suppose we wish to find the solution set S of a given equation involving quotients of polynomials.

PROCEDURE:

Step 1. Find the LCD of all the fractions involved in the given equation. It will always be a polynomial.

Step 2. Multiply both sides of the given equation by the LCD found in Step 1. This step is sometimes called *clearing the equation of fractions.* The resulting derived equation will always be a polynomial equation.

Step 3. Find the solution set T of the polynomial equation derived in Step 2. The solution set S of the given equation is a subset of T; that is, every number in S is also in T. However, there may be one or more numbers in T that do not belong to S. A number which belongs to T but does not belong to S is often called an *extraneous root.*

Step 4. Check, by actual substitution into the given equation, each number in T to determine whether or not it is a root of that equation.

Step 5. S is the set of all the numbers in T which are roots of the given equation. Any number in T which is a zero of the LCD in Step 1 cannot be a root of the given equation.

We shall illustrate the method just described by using it to solve the equation of Example 3.

Example 5

Find the solution set S of the equation

$$\frac{2}{(x-1)(x+1)} = \frac{1}{x-1} - 2$$

Solution: We follow the steps outlined above.

Step 1. The LCD of all the fractions involved is $(x-1)(x+1)$.
Step 2. Multiply both sides of the given equation by $(x-1)(x+1)$.

$$(x-1)(x+1)\left[\frac{2}{(x-1)(x+1)}\right] = (x-1)(x+1)\left(\frac{1}{x-1} - 2\right)$$

Simplify each side and obtain

$$2 = x + 1 - 2(x-1)(x+1)$$

Writing the last equation in standard form, we have

$$2x^2 - x - 1 = 0$$

Step 3. Find the solution set T of the equation

$$2x^2 - x - 1 = 0$$

or in factored form

$$(2x+1)(x-1) = 0$$

The solution set of this equation is

$$T = \left\{-\frac{1}{2}, 1\right\}$$

SEC. 3.15 Equations Containing Fractions

Step 4. Checking $-\frac{1}{2}$ in the given equation shows that it is a root. Substituting 1 for x makes the denominator of two of the fractions in the given equation zero; hence 1 is not a root. Therefore,

Step 5. $S = \left\{-\frac{1}{2}\right\}$

This result shows why Step 4 is absolutely necessary when this method is used.

Remark: Before attempting to solve an equation involving fractions, one should list, either mentally or on paper, each of the numbers which when substituted for x would reduce to zero any denominator of a fraction involved in the equation. Any such number cannot possibly be a root of that equation. These numbers are usually easily found by inspection.

Exercises 3.15

Solve each of the following examples.

1. $\dfrac{(x-3)(x+2)}{(x+1)(x-3)} = 0$

2. $\dfrac{x^2 - 2x - 3}{x(x-5)} = 0$

3. $\dfrac{3x+1}{x} = 4$

4. $\dfrac{2}{x+2} - \dfrac{x}{x+1} = 1$

5. $\dfrac{3x}{4} - \dfrac{5x}{2} = -7$

6. $\dfrac{5}{2x} + \dfrac{4}{3x} = \dfrac{23}{12}$

7. $\dfrac{2}{x} - \dfrac{10+x}{6x} = -\dfrac{5}{6}$

8. $\dfrac{4}{x-2} = \dfrac{3}{x-1}$

9. $\dfrac{2}{x-1} + \dfrac{5}{x+1} = \dfrac{4}{x^2-1}$

10. $\dfrac{3x+2}{x-2} = 3$

11. $\dfrac{1}{x-2} + \dfrac{1}{x+2} = \dfrac{4}{x^2-4}$

12. $\dfrac{3x+4}{6x-5} = \dfrac{2x+5}{4x-1}$

13. $\dfrac{2}{x+1} - \dfrac{4x+6}{x-1} = 3$

14. $\dfrac{6x+9}{x^2+x-2} + \dfrac{2x+3}{x+2} - \dfrac{2x+5}{x-1} = 0$

15. $\dfrac{5}{x-1} - \dfrac{7}{x-2} + \dfrac{2}{x-3} = 0$

16. $\dfrac{6x}{2x+1} + \dfrac{6}{2x-1} = \dfrac{2x+3}{2x-1} + 2$

17. $\dfrac{3}{x-1} - \dfrac{4}{x} = 1$

18. $\dfrac{2x}{x+1} + \dfrac{x}{1-x} = \dfrac{4}{x^2-1}$

19. $\dfrac{2x+5}{x-1} - \dfrac{2x+3}{x-1} = \dfrac{6x+9}{x^2+x-2}$

20. $\dfrac{x-2}{x+3} + \dfrac{x+3}{x-2} = \dfrac{7-6x}{x^2+x-6}$

21. $\dfrac{x-4}{x^2-4} + \dfrac{1}{x^2-2x} = \dfrac{x+2}{x^2}$

22. $\dfrac{x}{x+3} + \dfrac{5x^2}{x^2-9} = 0$

3.16 EQUATIONS CONTAINING RADICALS

Certain equations in one variable where the variable occurs under one or more square root radicals can often be reduced to a polynomial equation by squaring both sides of the equation until the resulting equation no longer involves radicals. Such a squaring operation may lead to a new equation that is not equivalent to the original equation. However, the solution set of the given equation will always be a subset of the solution set of the new equation.

We illustrate how such equations can be solved by way of examples.

Example 1 Consider the equation $x = 1$. The solution set of this equation is clearly $S = \{1\}$. However, if we square both sides, we obtain $x^2 = 1$, which has $T = \{-1, 1\}$ as its solution set. Therefore, if in the process of solving an equation we square both sides and ultimately arrive at a new equation with solution set T, *it is absolutely necessary* to check each member of T by substituting it in the original equation in order to find S, the solution set of the original equation. As in the previous section, the root -1 to the equation $x^2 = 1$ is called an *extraneous root* (introduced by squaring both sides of the equation $x = 1$).

Example 2 Solve the equation $\sqrt{x-5} = 2$.

Solution: If $\sqrt{x-5} = 2$, then $(\sqrt{x-5})^2 = 2^2$, and we use a single arrow to denote this fact as follows:

$$\sqrt{x-5} = 2 \to x - 5 = 4 \qquad \text{Squaring both sides}$$
$$\leftrightarrow x = 9$$

Checking $x = 9$ in the original equation, we have $\sqrt{9-5} = \sqrt{4} = 2$, and so 9 is, in fact, a solution to the original equation (and it is the only solution). Hence $S = \{9\}$.

Example 3 Solve the equation $\sqrt{x+7} - \sqrt{2x-3} = 2$.

Solution: $\sqrt{x+7} - \sqrt{2x-3} = 2 \leftrightarrow \sqrt{x+7} = \sqrt{2x-3} + 2$
Squaring both sides $\to (\sqrt{x+7})^2 = (\sqrt{2x-3}+2)^2$
Expanding (Don't forget the $\leftrightarrow x + 7 = 2x - 3 + 4\sqrt{2x-3} + 4$
  middle term!)

SEC. 3.16 Equations Containing Radicals

Simplifying	$\leftrightarrow 6 - x = 4\sqrt{2x-3}$
Square again	$\rightarrow (6-x)^2 = (4\sqrt{2x-3})^2$
Expanding	$\leftrightarrow 36 - 12x + x^2 = 16(2x-3)$
Simplifying	$\leftrightarrow x^2 - 44x + 84 = 0$
Factoring	$\leftrightarrow (x-2)(x-42) = 0$
	$\leftrightarrow x - 2 = 0 \quad \text{or} \quad x - 42 = 0$
	$\leftrightarrow x = 2 \quad \text{or} \quad x = 42$
Test $x = 2$	

$$\sqrt{2+7} - \sqrt{4-3} = \sqrt{9} - \sqrt{1}$$
$$= 3 - 1 = 2$$
$$2 \in S$$

Test $x = 42$

$$\sqrt{42+7} - \sqrt{84-3} = \sqrt{49} - \sqrt{81}$$
$$= 7 - 9 \neq 2$$
$$42 \notin S$$

Hence $S = \{2\}$

These examples illustrate the following fact, which is the basic tool in solving equations involving radicals.

THEOREM 3.11. *Let $f(x)$ and $g(x)$ represent arbitrary algebraic expressions which may or may not involve radicals. If S is the solution set of $f(x) = g(x)$ and T is the solution set of $[f(x)]^2 = [g(x)]^2$, then $S \subseteq T$. We denote this symbolically as*

$$f(x) = g(x) \rightarrow [f(x)]^2 = [g(x)]^2$$

(*Note the single arrow in this case.*)

Exercises 3.16

Find the solution set of each of the following equations.

1. $\sqrt{x+2} = 2$
2. $\sqrt{2x+1} = 7 - x$
3. $\sqrt{2x+5} + \sqrt{x+2} = 5$
4. $\sqrt{x} + \sqrt{2x+1} = 5$
5. $x + 2\sqrt{x-1} = 9$
6. $x - \sqrt{x^2-9} = 1$
7. $x + \sqrt{x-7} = 9$
8. $\sqrt{7x-6} - \sqrt{2x+3} = 3$
9. $\sqrt{3x+7} - \sqrt{2x-2} = 2$
10. $\sqrt{2x-5} + x - 3 = 1$
11. $\sqrt{5x+1} - \sqrt{x+1} = 2$
12. $\sqrt{5x-1} = \sqrt{x+1}$

13. $\sqrt{5x+1} - \sqrt{x+1} = 2$ 14. $\sqrt{5x-1} = \sqrt{x+1}$

15. $\sqrt{2x+1} - \sqrt{x-3} = \sqrt{x}$ 16. $\sqrt{2x+7} - \sqrt{2-x} = \sqrt{x+3}$

17. $\sqrt{5x-6} + \sqrt{x+1} = \sqrt{7x+4}$ 18. $\sqrt{3x+4} - 2\sqrt{x-3} = 2$

3.17 INEQUALITIES IN ONE VARIABLE

Inequalities were defined and treated in Section 1.5. They are statements such as $3 < 7, 6 - 5 < 4 - 1, 5 > 2$, etc. We wish to extend the definition to include open statements such as $x - 3 < 0$, $x - 2 < 7$, $3x + 4 \leq x + 10$, $x^2 - 3x - 4 > 0$, or in fact any open statement of the form $f(x) < g(x)$, where $f(x)$ and $g(x)$ are polynomials with real numbers as coefficients. Just as was the case with equations, an inequality may be a true statement, or it may be a false statement, or, if it involves variables, it may be neither true nor false. For example, $7 > 3$ is a true statement, $8 - 2 < 4 + 1$ is a false statement, but $x + 2 < 3$ is neither true nor false until a specific number is substituted for x.

Inequalities involving one variable are of three types.

1. *Conditional inequalities:* Those inequalities which yield true statements for some values of the variable and false statements for other values of the variable. Example: $2x > 3$ is true if $x = 2$ and false if $x = 1$.
2. *Absolute inequalities:* Those which yield true statements for all values of the variable. Example: $x + 1 > x$.
3. *Contradictory inequalities:* Those which yield false statements for all values of the variable. Example: $x + 1 < x$.

The real number a is said to be a *solution* of a given inequality in one variable if, and only if, the inequality becomes a true statement when a is substituted for that variable in the inequality. For example, 2 is a solution of $3x + 1 < 10$, because $3(2) + 1 = 7$, and $7 < 10$ is a true statement. Similarly, 1 is a solution, and so is -1. In fact, any real number less than 3 is a solution. However, 4 is not a solution, since $3(4) + 1 = 13$, and $13 < 10$ is a false statement. Clearly, 3 is not a solution, nor is any number greater than 3 a solution.

The set of all the solutions of an inequality is called the *solution set* of the inequality. The solution set S of the above inequality is $S = \{x | x < 3\}$, which is read "S is the set of all real numbers x such that x is a number less than 3." In this case the solution set contains an infinite number of elements. This fact is rather typical of inequalities, for most inequalities commonly encountered do have infinitely many solutions. The graph of $S = \{x | x < 3\}$ is given in Figure 3.1. Here we indicate that 3 is not in the graph by placing an open circle at its spot.

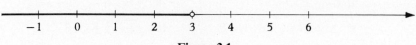

Figure 3.1

SEC. 3.17 Inequalities in One Variable

To solve an inequality is to find its solution set. The methods of solving inequalities are very similar to those of solving equations. For example, we say two inequalities are *equivalent* if, and only if, they have the same solution set. Also, the usual procedure in solving an inequality is to attempt to reduce the inequality to an equivalent inequality whose solution set can be found by inspection. Seven basic facts are needed for solving the types of inequalities in which we shall be interested. These are set apart as rules of operation as follows:

RULE 3.1. Any algebraic expression which appears in an inequality can be added to or subtracted from both sides to obtain an equivalent inequality.

Example 1 $3x - 5 < 0 \leftrightarrow 3x < 5$ Adding 5 to both sides

Example 2 $3x + 5 < 5x + 7 \leftrightarrow 3x - 5x < 7 - 5$ Subtracting 5x and 5 from both sides

RULE 3.2. Multiplying or dividing both sides of an inequality by a positive real number and retaining the direction of the inequality yields an equivalent inequality.

Example 3 $\frac{2}{3}x < \frac{1}{6} \leftrightarrow 4x < 1$ Multiplying both sides by 6

Example 4 $3x < 12 \leftrightarrow x < 4$ Dividing both sides by 3

RULE 3.3. Multiplying or dividing both sides of an inequality by a negative real number and reversing the direction of the inequality yields an equivalent inequality.

Example 5 $-2x < 4 \leftrightarrow x > -2$ Dividing both sides by -2

Example 6 $-\frac{1}{5}x > 7 \leftrightarrow x < -35$ Multiplying both sides by -5

RULE 3.4. If $f(x)$ is an algebraic expression and $a \geq 0$, then

$$|f(x)| < a \leftrightarrow -a < f(x) < a \quad \text{and} \quad |f(x)| \leq a \leftrightarrow -a \leq f(x) \leq a$$

Example 7 $|x - 2| < 3 \leftrightarrow -3 < x - 2 < 3$

Example 8 $|3x^2 - 2x + 7| \leq \sqrt{5} \leftrightarrow -\sqrt{5} \leq 3x^2 - 2x + 7 \leq \sqrt{5}$

RULE 3.5. If $f(x)$ is an algebraic expression and $a > 0$, then

$$|f(x)| > a \leftrightarrow f(x) > a \quad \text{or} \quad f(x) < -a$$

and

$$|f(x)| \geq a \leftrightarrow f(x) \geq a \quad \text{or} \quad f(x) \leq -a$$

Example 9 $\quad |x| \geq 10 \leftrightarrow x \geq 10 \quad \text{or} \quad x \leq -10$

Example 10 $\quad |3x-5| > 2 \leftrightarrow 3x-5 > 2 \quad \text{or} \quad 3x-5 < -2$

RULE 3.6. If $f(x)$ and $g(x)$ are algebraic expressions, then

$$f(x) \cdot g(x) > 0 \leftrightarrow \begin{cases} f(x) > 0 \quad \text{and} \quad g(x) > 0 \\ \quad \text{or} \\ f(x) < 0 \quad \text{and} \quad g(x) < 0 \end{cases}$$

Example 11 $\quad (x-2)(x+1) > 0 \leftrightarrow \begin{cases} x-2 > 0 \quad \text{and} \quad x+1 > 0 \\ \quad \text{or} \\ x-2 < 0 \quad \text{and} \quad x+1 < 0 \end{cases}$

RULE 3.7. If $f(x)$ and $g(x)$ are algebraic expressions, then

$$f(x) \cdot g(x) < 0 \leftrightarrow \begin{cases} f(x) > 0 \quad \text{and} \quad g(x) < 0 \\ \quad \text{or} \\ f(x) < 0 \quad \text{and} \quad g(x) > 0 \end{cases}$$

Example 12 $\quad (2x+5)(x-3) < 0 \leftrightarrow \begin{cases} 2x+5 > 0 \quad \text{and} \quad x-3 < 0 \\ \quad \text{or} \\ 2x+5 < 0 \quad \text{and} \quad x-3 > 0 \end{cases}$

We now isolate the seven rules of operation for inequalities in tabular form and then proceed to illustrate how they are used to solve various types of inequalities.

Rules of Operation for Inequalities

RULE 3.1. Any algebraic expression which appears in an inequality can be added to or subtracted from both sides to obtain an equivalent inequality.

RULE 3.2. Multiplying or dividing both sides of an inequality by a positive real number and retaining the direction of the inequality yields an equivalent inequality.

RULE 3.3. Multiplying or dividing both sides of an inequality by a negative real number and reversing the direction of the inequality yields an equivalent inequality.

RULE 3.4. If $f(x)$ is an algebraic expression and $a \geq 0$, then

$$|f(x)| < a \leftrightarrow -a < f(x) < a$$

and

$$|f(x)| \leq a \leftrightarrow -a \leq f(x) \leq a$$

SEC. 3.17 Inequalities in One Variable

RULE 3.5. If $f(x)$ is an algebraic expression and $a \geq 0$, then
$$|f(x)| > a \leftrightarrow f(x) > a \quad \text{or} \quad f(x) < -a$$
and
$$|f(x)| \geq a \leftrightarrow f(x) \geq a \quad \text{or} \quad f(x) \leq -a$$

RULE 3.6. If $f(x)$ and $g(x)$ are algebraic expressions, then
$$f(x) \cdot g(x) > 0 \leftrightarrow \begin{cases} f(x) > 0 \text{ and } g(x) > 0 \\ \text{or} \\ f(x) < 0 \text{ and } g(x) < 0 \end{cases}$$

RULE 3.7. If $f(x)$ and $g(x)$ are algebraic expressions, then
$$f(x) \cdot g(x) < 0 \leftrightarrow \begin{cases} f(x) > 0 \text{ and } g(x) < 0 \\ \text{or} \\ f(x) < 0 \text{ and } g(x) > 0 \end{cases}$$

Linear Inequalities in One Variable

The simplest polynomial inequalities in one variable are those which have, or may be reduced to, one of the following forms: $ax + b < 0$, $ax + b \leq 0$, $ax + b > 0$, $ax + b \geq 0$. Such inequalities are called linear inequalities in one variable. Procedures for solving such inequalities are illustrated in the following examples:

Example 13 Solve the inequality $3x - 5 < 0$.

Solution: $3x - 5 < 0 \leftrightarrow 3x < 5$ Adding 5 to both sides
$\leftrightarrow x < \dfrac{5}{3}$ Dividing both sides by 3

Thus the solution set S of $3x - 5 < 0$ is $S = \{x | x < \tfrac{5}{3}\}$. The graph of S is shown in Figure 3.2. Observe that the solution set T of $3x - 5 \leq 0$ is identical with the solution set of $3x - 5 < 0$ except T contains one more number than S, that is, $\tfrac{5}{3}$. The graph of T is shown in Figure 3.3.

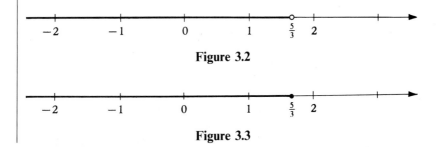

Figure 3.2

Figure 3.3

Example 14   Find the solution set S of the inequality $3x + 5 < 5x + 7$.

Solution: $3x + 5 < 5x + 7 \leftrightarrow 3x - 5x < 7 - 5$
$\leftrightarrow -2x < 2$ *Subtracting $5x$ and 5 from both sides*
$\leftrightarrow x > -1$ *Dividing both sides by -2*

Therefore, $S = \{x | x > -1\}$. The graph of S is shown in Figure 3.4.

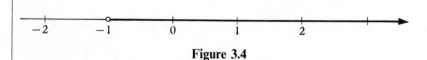

Figure 3.4

Inequalities Involving Absolute Values

It is sometimes necessary to solve linear inequalities which involve absolute values. For example, inequalities such as $|x| > 3$, $|2x - 1| < 5$, $|3 - 2x| \leq 1$, etc. To solve such inequalities, we use Rules 3.4 and 3.5.

Example 15   Solve the inequality $|3 - 2x| < 5$.

Solution: $|3 - 2x| < 5 \leftrightarrow -5 < 3 - 2x < 5$ *Rule 3.4*
$\leftrightarrow -8 < -2x < 2$ *Two applications of Rule 3.1*
$\leftrightarrow 4 > x > -1$ *Two applications of Rule 3.3*

Hence $\{x | |3 - 2x| < 5\} = (-1, 4)$. See Figure 3.5 for the graph.

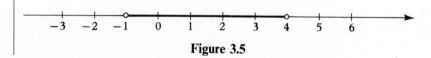

Figure 3.5

Note: In Example 15, $-5 < 3 - 2x < 5$ means that both $-5 < 3 - 2x$ and $3 - 2x < 5$ must hold simultaneously. Hence we need to subtract 3 from both sides of two inequalities to obtain the next line. This is why "Two applications of Rule 3.1" is given as a reason for this step.

Example 16   Solve the inequality $|5x + 2| \geq 3$.

Solution: $|5x + 2| \geq 3 \leftrightarrow 5x + 2 \geq 3$ or $5x + 2 \leq -3$ *Rule 3.5*
$\leftrightarrow 5x \geq 1$ or $5x \leq -5$ *Two applications of Rule 3.1*
$\leftrightarrow x \geq \dfrac{1}{5}$ or $x \leq -1$ *Two applications of Rule 3.2*

SEC. 3.18 Higher-Degree Polynomial Inequalities

Hence the solution set S consists of all points x which satisfy either $x \geq \frac{1}{5}$ or $x \leq -1$. This set can be written as the *union* of the two sets $\{x \mid x \leq -1\} = (-\infty, -1]$ and $\{x \mid x \geq \frac{1}{5}\} = [\frac{1}{5}, \infty)$ (see Appendix A). Hence $S = (-\infty, -1] \cup [\frac{1}{5}, \infty)$, and the graph of this set is shown in Figure 3.6.

Figure 3.6

Exercises 3.17

Solve each of the following inequalities.

1. $x + 4 < 7$
2. $x - 2 \geq 5$
3. $2x - 3 \leq x + 1$
4. $5x - 3 < 0$
5. $5 - 2x > 9$
6. $4x + 3 \leq 2x + 7$
7. $3x + 5 < 5x - 3$
8. $4x - 10 < 5x - 3$
9. $2x - 8 \geq 6x + 5$
10. $|x| > 5$
11. $|x| \leq 3$
12. $|x - 2| < 3$
13. $|x + 1| > 5$
14. $|x + 3| < 4$
15. $|x - 2| \geq 3$
16. $|4x - 5| > 3$
17. $|3x - 2| \geq 7$
18. $|2x - 3| \leq 5$
19. $|3x + 8| \leq 11$
20. $|x - 1| < 0.01$

21. Write an inequality which states that the distance from x to 3 is less than 2.
22. Write an inequality which expresses the fact that the distance from x to 2 is greater than 3.

Solve the following inequalities.

23. $|x - 1| < 2x$ [*Hint:* Check the intervals $(-\infty, 1)$ and $(1, \infty)$]
24. $|x - 3| > 2x + 1$
25. $|x - 5| < |x + 1|$ [Check the intervals $(-\infty, -1)$, $(-1, 5)$, and $(5, \infty)$]
26. $|2x - 1| < |x - 4|$

3.18 HIGHER-DEGREE POLYNOMIAL INEQUALITIES

In this section we shall develop a procedure for solving any polynomial inequality which it might be necessary to solve in this text. We illustrate the procedure by way of example.

Example 1 Find the solution set of the inequality
$$f(x) = (x+2)(x-1)(x-3)(x-5) < 0$$

Solution

Step 1. Find all the *zeros* of the polynomial $f(x)$; that is, find the roots of the equation $f(x) = 0$. [Note that if r is a root of the equation $f(x) = 0$ then $f(r) = 0$. This is the motivation for calling r a *zero* of the polynomial $f(x)$.] In this case we can see by inspection that the zeros are $-2, 1, 3,$ and 5.

Step 2. Draw a number line and use these four points to divide it into five intervals (intervals 1 and 5 are infinite in length), as shown in Figure 3.7.

```
Interval 1      Interval 2      Interval 3      Interval 4      Interval 5
―――――――――●――――――――――――●――――――――――――●――――――――――――●―――――――――――
        -2            1            3            5
```

Figure 3.7

Step 3. Select a number in each of the five intervals and determine whether $f(x)$ is positive or negative when that number is substituted for x. In this case we pick -3 in interval 1, 0 in interval 2, 2 in interval 3, 4 in interval 4, and 6 in interval 5, and find that $f(-3) > 0$, $f(0) < 0$, $f(2) > 0$, $f(4) < 0$, and $f(6) > 0$. Now, it turns out that the value of $f(x)$ does not change sign in any given interval, and we can indicate this fact by placing $+$ signs above each interval where a positive value is obtained and $-$ signs above each interval where a negative value is obtained (see Figure 3.8). Also, since $f(x) = 0$ at each of $-2, 1, 3,$ and 5, we place a zero above each.

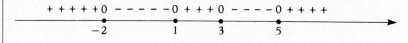

Figure 3.8

Step 4. By inspection, simply write down the set consisting of those points for which $f(x) < 0$. This set is
$$S = \{x \mid -2 < x < 1 \quad \text{or} \quad 3 < x < 5\} = (-2, 1) \cup (3, 5)$$

Example 2 Use Figure 3.8 to solve each of the following inequalities.

(a) $(x+2)(x-1)(x-3)(x-5) \leq 0$
(b) $(x+2)(x-1)(x-3)(x-5) > 0$
(c) $(x+2)(x-1)(x-3)(x-5) \geq 0$

SEC. 3.18 Higher-Degree Polynomial Inequalities

Solution: By simply inspecting Figure 3.8, we find:

(a) $S = \{x \mid -2 \leq x \leq 1 \text{ or } 3 \leq x \leq 5\} = [-2, 1] \cup [3, 5]$
(b) $S = \{x \mid -\infty < x < -2 \text{ or } 1 < x < 3 \text{ or } 5 < x\} = (-\infty, -2) \cup (1, 3) \cup (5, \infty)$
(c) $S = \{x \mid -\infty < x \leq -2 \text{ or } 1 \leq x \leq 3 \text{ or } 5 \leq x < \infty\} = (-\infty, -2] \cup [1, 3] \cup [5, \infty)$

If the equation $f(x) = 0$ has no real roots at all, then either $f(x) > 0$ for all values of x or $f(x) < 0$ for all values of x. This fact enables us to solve inequalities of the type treated in the next two examples.

Example 3 Solve the inequality $f(x) = 3x^2 - x + 4 > 0$.

Solution: Since $b^2 - 4ac = (-1)^2 - (4)(3)(4) = 1 - 48 = -47 < 0$, this equation has no real roots. Therefore, it suffices to evaluate $f(x)$ at one particular real number to find out whether all values are positive or all values are negative. Zero is always the easiest number to use since $f(0)$ is always the constant term of the polynomial $f(x)$, 4 in this case. Now $4 > 0$, and so $f(x) > 0$ for all real numbers x. Hence $S = (-\infty, \infty)$, the set of all real numbers.

Example 4 Solve the inequality $f(x) = 3x^2 - x + 4 \leq 0$.

Solution: As in Example 3, $f(x) = 0$ has no real roots, and so we found that $f(x) > 0$ for all real numbers x. Therefore, there are no real numbers x for which $f(x) \leq 0$, and we conclude $S = \{\ \}$, the empty set.

We solve one additional example and then give a step-by-step description of the technique used. We note that the procedure can also be used to solve linear inequalities which were treated in the previous section.

Example 5 Solve the inequality $f(x) = (x + 2)(x - 1)(2x^2 - 3x + 7) \geq 0$.

Solution

Step 1. The zeros of the polynomial $f(x)$ are readily found to be -2 and 1 (the factor $2x^2 - 3x + 7$ has no real roots).
Step 2. Drawing a number line and dividing it into three intervals, we obtain Figure 3.9.

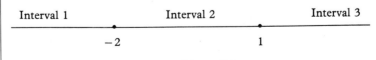

Figure 3.9

Step 3. Select -3 in interval 1, 0 in interval 2, and 2 in interval 3, and obtain $f(-3) > 0$, $f(0) < 0$, and $f(2) > 0$. Now draw the number line shown in Figure 3.10.

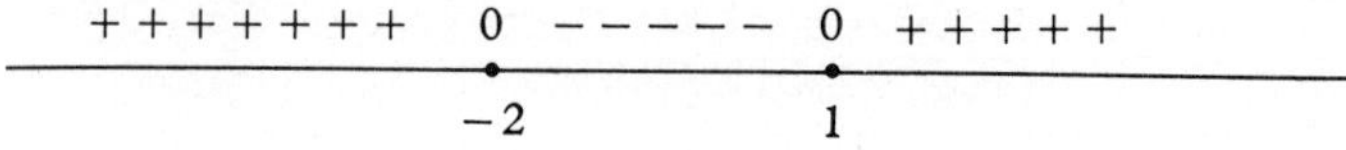

```
+ + + + + + +   0  - - - - -  0  + + + + +
─────────────────●───────────●──────────────
              -2            1
```

Figure 3.10

Step 4. By inspection, write down the solution set:
$$S = \{x \mid -\infty < x \leq -2 \text{ or } 1 \leq x < \infty\} = (-\infty, -2] \cup [1, \infty)$$

Procedure for Solving Polynomial Inequalities

Let $f(x) = A_n x^n + A_{n-1} x^{n-1} + \cdots + A_1 x + A_0$ be a polynomial with real coefficients and consider an inequality of any of the forms:

(a) $f(x) < 0$ (b) $f(x) \leq 0$
(c) $f(x) > 0$ (d) $f(x) \geq 0$

Step 1. Find all distinct real roots of the equation $f(x) = 0$; that is, find all zeros of the polynomial $f(x)$.
Step 2. Draw a number line and divide it into intervals by labeling the roots below the appropriate points on the number line (there is always one more interval than there are distinct real roots).
Step 3. Select a number in each interval and evaluate $f(x)$ at each such number to determine what sign should be placed above each interval. Place the appropriate sign above each interval and place a zero above each root.
Step 4. By inspection, write down the solution to the inequality in question.

Exercises 3.18

Solve each of the following inequalities.

1. $x(x + 1)(x - 2) < 0$
2. $x(x - 1)(x - 4) \leq 0$
3. $x(x + 1)(x - 2) > 0$
4. $x(x - 1)(x - 4) \geq 0$
5. $(x + 3)(x - 1)(x - 4) \leq 0$
6. $(x + 1)(x - 1)(x - 3) \geq 0$

SEC. 3.19   Inequalities Involving Quotients of Polynomials

7. $(x-2)(x-4)(x-6) \leq 0$
8. $(2x+1)(3x-1)(x-2) < 0$
9. $(x+2)(x-1)(x-3)(x-6) > 0$
10. $2x^3 - 9x^2 + 7x + 6 > 0$
11. $x^3 + 11x > 6x^2 + 6$
12. $(x-1)^3(x-3)(x-5)^2 < 0$
13. $(x+2)(x-2)(x^2+1) \leq 0$
14. $(x^2-9)(3x^2-2x+1) < 0$
15. $(x+1)^5(x-1)^3(x-3)^7 < 0$
16. $(x^2+2x-3)(-2x^2+x-1) > 0$
17. $(x+4)(x+2)(x-1)(x-3)(x-5) > 0$
18. $2x^4 - 5x^3 < 11x^2 - 20x - 12$

3.19 INEQUALITIES INVOLVING QUOTIENTS OF POLYNOMIALS

Certain inequalities involving quotients of polynomials can be solved by adapting the process described in the previous section. We shall illustrate how this is done by way of example.

Example 1   Solve the inequality $\dfrac{x+3}{x(x-1)} < 0$.

Solution: Since $\dfrac{1}{x(x-1)}$ and $x(x-1)$ have the same sign as long as both are defined (i.e., as long as $x \neq 0$ and $x \neq 1$), $\dfrac{x+3}{x(x-1)}$ and $(x+3)(x)(x-1)$ have the same sign (again, as long as $x \neq 0$ and $x \neq 1$). Therefore, it suffices to solve the inequality

$$(x+3)(x)(x-1) < 0$$

and then omit $x = 0$ and/or $x = 1$ in case they turn out to be solutions. Using the process described in Section 3.18, we obtain

$$T = \{x \mid x < -3 \ \text{ or } \ 0 < x < 1\} = (-\infty, -3) \cup (0, 1)$$

as the solution set for $(x+3)(x)(x-1) < 0$. Since neither 0 nor 1 is in this set, no elements need be discarded. Therefore, the solution set to the original equation is

$$S = T = (-\infty, -3) \cup (0, 1)$$

Example 2   Solve the inequality $\dfrac{(x+2)(x-5)}{x-3} \geq 0$.

Solution: As in Example 1, the solution set of this inequality is obtained by finding the solution set T of the inequality

$$(x+2)(x-5)(x-3) \geq 0$$

and then discarding 3 if it happens to be in T. Using the procedure of Section 3.18, we obtain

$$T = \{x \mid -2 \le x \le 3 \quad \text{or} \quad 5 \le x < \infty\} = [-2, 3] \cup [5, \infty)$$

Discarding 3, we find that the solution set of the original inequality is

$$S = \{x \mid -2 \le x < 3 \quad \text{or} \quad 5 \le x < \infty\} = [-2, 3) \cup [5, \infty)$$

Example 3 Find the solution set S of the inequality

$$\frac{2x}{x+1} - \frac{x}{x-1} \ge \frac{4}{x^2 - 1}$$

Solution

$$\frac{2x}{x+1} - \frac{x}{x-1} \ge \frac{4}{x^2-1} \leftrightarrow \frac{2x}{x+1} - \frac{x}{x-1} - \frac{4}{x^2-1} \ge 0$$

$$\leftrightarrow \frac{x^2 - 3x - 4}{x^2 - 1} \ge 0$$

$$\leftrightarrow \frac{(x+1)(x-4)}{(x+1)(x-1)} \ge 0$$

We solve this last inequality as we did Examples 1 and 2. The solution set T of the inequality $(x+1)(x-4)(x+1)(x-1) \ge 0$ is

$$T = \{x \mid -\infty < x \le 1 \quad \text{or} \quad 4 \le x < \infty\} = (-\infty, 1] \cup [4, \infty)$$

Discarding -1 and 1, we find that the solution set of the original inequality is

$$S = \{x \mid -\infty < x < -1 \quad \text{or} \quad -1 < x < 1 \quad \text{or} \quad 4 \le x < \infty\}$$
$$= (-\infty, -1) \cup (-1, 1) \cup [4, \infty)$$

Exercises 3.19

Solve each of the following inequalities.

1. $\dfrac{x+1}{(x-1)(x-3)} < 0$

2. $\dfrac{x+2}{x(x-4)} > 0$

3. $\dfrac{1}{(x+1)(x-1)(x-4)} < 0$

4. $\dfrac{(x+3)(x-2)}{x-4} > 0$

5. $\dfrac{(x-1)(x-5)}{(x+1)(x-3)} < 0$

6. $\dfrac{(3x+2)(x-2)}{2x+3} \le 0$

7. $\dfrac{(x+1)(x-5)}{(x-1)(x-3)} > 0$

8. $\dfrac{(2x+1)(x^2 - 5x + 8)}{(3x-5)(2x^2 + 3x + 2)} > 0$

SEC. 3.19 Inequalities Involving Quotients of Polynomials

9. $\dfrac{2x-1}{x^2-4} + \dfrac{3}{x-2} < \dfrac{2}{x+2}$

10. $\dfrac{4}{x-3} < \dfrac{3}{2x-5}$

11. $x - \dfrac{2}{x} < 3$

12. $\dfrac{2x-3}{x-2} < \dfrac{3}{2x-5}$

13. $\dfrac{2}{x} + 1 < \dfrac{9}{x+2}$

14. $\dfrac{3x}{x-3} + \dfrac{x-7}{x+2} < \dfrac{4x^2-5}{x^2-x-6}$

4

Relations, Functions, and Graphs

A Preview

From our daily experiences and from our study of elementary and high school mathematics, each of us has learned of and become familiar with many physical objects, or quantities, which are measurable. A physical object is said to be measurable if it is common practice to associate with the object a real number which represents the size or magnitude of the object in terms of well-known units of measure. Areas, volumes, lengths, distances, speed, mass, force, time, etc., are examples of measurable physical quantities. Most of us are also familiar with pairs of such objects, where one member of the pair is related to the other member in the sense that the numerical measure of the one depends on, or is determined by, the numerical measure of the other, and a change in the numerical measure of one member of the pair provides a change in the corresponding numerical value of the other. For example, the area (A square units) of a square is determined by the length (x units) of a side of the square.

Other familiar examples of pairs of physical objects that are related include the following. The distance (D miles) you can drive a car in five hours depends on the average speed (V miles per hour) at which you drive the car. Many cities have a sales tax. In each of these cities the sales tax is a certain percent of the amount of the purchase. Thus the sales tax on a particular item purchased depends on the cost of the item. The amount (number of dollars) of an electric bill each month depends on the number of kilowatt hours of electricity used during the month. The heat loss through the roof of a house on a cold day depends on the thickness of the insulation between the ceiling and the roof.

The pairs of physical quantities that are related could obviously be extended to any length. However, a large number of such examples have to do with cases which are so complicated and so complex that no elementary means of analyzing their modes of variation exists. On the other hand, so many cases of importance can be studied with the proper mathematical tools that it is worthwhile to investigate the nature of such tools and to master the use of a few of the basic ones. The subject which provides these tools is called *calculus*.

Elementary calculus is concerned with two major problems. The first is that of finding the rate at which the numerical measure of one of two related physical quantities changes per unit change in the numerical measure of the other when the specific relation between the numerical measures of the two quantities is known. This problem is called the *rate problem*. The process of finding such a rate is called *differentiation*, and a rate found by differentiation is called a *derivative*. The second major problem with which calculus is concerned is the reverse of the rate problem, that is, given the rate at which the numerical measure of one quantity is changing with respect to the numerical measure of a second quantity, find the relation between the numerical measures of the two quantities. The process of finding this relation from the given rate is called *integration*. A mastery of the processes of differentiation and integration will enable us to solve with ease and precision a number of problems that would be most difficult if not impossible otherwise. Chapters 5, 6, and 7 will be devoted to the study of elementary calculus. First, however, we need to provide a bit more background material. The purpose of this chapter is to study in some detail the general concept of a relation and to give particular attention to a special kind of relation called a *function*. This chapter also contains a brief introduction to analytic geometry, which will be most helpful in our study of calculus.

4.1 RELATIONS AND FUNCTIONS

In the preview we talked about the numerical measures of two physical quantities being related but did not say precisely what is meant by such a statement. In this section we shall attempt to remedy that situation by stating some definitions that should make clear our meaning in the above statements.

First, it is important to understand that a physical quantity of the type mentioned in the preview has not just one numerical measure but the totality of its possible numerical measures is a subset of R, the set of all real numbers. For example, if A is the numerical measure of the area of a square, then the number A is in the set R. Similarly, if S represents the length of a side of a square, the number S is also in the set R.

DEFINITION 4.1. Suppose X and Y are subsets of R. Let x represent a numerical measure of one physical quantity, and let y represent a numerical measure of a second physical quantity. Suppose also that $x \in X$ and $y \in Y$. A *relation* between X and Y is a rule of correspondence which assigns, or makes correspond, to each value of x in X one or more values of y in Y. Such a rule of correspondence pairs with each real number x in X one or more values of y in Y, and hence generates a set of ordered pairs (x, y) of real numbers.

A correspondence, or relation, between the set X and the set Y can be expressed by means of a verbal statement, a formula, a list of pairs, or by one of many other means. But no matter how this relation is expressed, the end result is always a set of ordered pairs of real numbers. It seems quite natural, therefore, to make the following general definition.

DEFINITION 4.2. A *relation* is any set of ordered pairs of real numbers; that is, a *relation* is any subset of $R \times R$, where R is the set of all real numbers.

The set X consisting of all the first elements x of the ordered pairs (x, y) of a relation is called the *domain of definition of the relation*. The set Y consisting of all the second elements y of this set of ordered pairs (x, y) is called the *range of the relation*.

Example 1 Consider the relation $A = \{(-1, 0), (1, 2), (1, -1), (2, 3), (2, -2), (3, 5)\}$. The domain of definition of A is the set $\{-1, 1, 2, 3\}$. The range of A is the set $\{-2, -1, 0, 2, 3, 5\}$.

Example 2 Suppose $X = \{2, 3, 4\}$ and $Y = \{1, 2, 3\}$. List all the elements of the relation B if $B = \{(x, y) | y < x, x \in X, \text{ and } y \in Y\}$.

Solution: $B = \{(2, 1), (3, 1), (3, 2), (4, 1), (4, 2), (4, 3)\}$

The Function Concept

One of the most important concepts in mathematics is that of a special type of relation called a *function*. Let us first consider some specific relations that are of the type we have in mind.

If one drives a car at the average speed of 50 miles per hour, the distance D (miles) traveled in t hours is related to t by the rule

(1) $$D = 50t$$

Since the distance D traveled in t hours depends on the value of t, we shall usually write

(1') $$D(t) = 50t$$

and the expression $D(t)$ can be read "the distance traveled in t hours." We say that Formula (1') defines D as a function of t, meaning that the formula generates a set of ordered pairs $(t, D(t))$.

The area A (square units) of a square having a side of length x (units) is related to x by the rule

(2) $$A = x^2$$

or

(2') $$A(x) = x^2$$

In this case, we say that Formula (2') defines A as a function of x.

An object falls vertically downward from a rest position. The distance s (feet) the body falls in the first t seconds, neglecting air resistance, is related to t by the formula

(3) $$s(t) = 16t^2$$

We say that Formula (3) defines s as a function of t.

The examples given above have two important features in common, which furnish the basis for the general concept of a function. The first and most striking feature is that in each example we have a pairing of things and that the numerical value of a certain one of the things is determined by the numerical value of the other. Each of the three formulas generates a set of ordered pairs of real numbers, which suggests that a function is a relation. A close inspection shows that the relations determined by Equations (1'), (2'), and (3) have a second common feature; that is, in each relation no two distinct ordered pairs have the same first element. Relations in general do not have this common property. This leads us to the following equivalent definitions.

DEFINITION 4.3. A *function* is a definitely specified set of ordered pairs of real numbers such that if we represent an arbitrary pair of the set by (x, y) there is one and only one value of y corresponding to each permissible value of x.

DEFINITION 4.3'. A *function* is a relation with the property that no two distinct ordered pairs of the relation have the same first element.

In Section 4.2 we shall discuss a method of associating with any set X of ordered pairs of real numbers a set of points in the plane called the graph of X. An interesting and distinguishing feature of the graph of a function is the fact that no two distinct points of its graph lie on the same vertical line.

It follows from the definition of a function that every function is a relation but that not every relation is a function. In fact, a relation is a function if and only if any two of its ordered pairs with equal first elements have second elements which are also equal. For example, neither the relation A nor the relation B of Examples 1 and 2 is a function. It should be noted, however, that a function may possess two or more ordered pairs with unequal first elements, even though their second elements are equal.

Example 3 Let $C = \{(-1, 2), (1, 2), (2, 2), (3, 4)\}$. Relation C is a function.

Example 4 Let $D = \{(-1, 2), (1, 2), (1, 3), (2, 2), (3, 4)\}$. Relation D is not a function, because it possesses the two pairs $(1, 2)$ and $(1, 3)$ with equal first elements but unequal second elements.

As in the case of relations in general, the set X consisting of all the values of the first variable, x, in the ordered pairs (x, y) which define the function, is called the *domain of definition of the function*. The set Y consisting of all the values of the second variable, y, is called the *range of the function*. The variable x, which represents the numbers in the domain of definition of a function, is called the *independent variable*; the variable y, which represents the numbers in the range, is called the *dependent variable*. The dependent variable y is also called the *value of the function at x*.

It is important to recognize that, to determine a function, all that is really needed are the domain of the independent variable x and a rule which pairs with each number x in the domain one and only one number y in the range.

Example 5 Let us define a function by first declaring that its domain is the set $\{-1, 0, 1, 2, 3\}$. Now, suppose to each value of x we assign the number y and we apply the rule "square and add 3." The function thus defined is $\{(-1, 4), (0, 3), (1, 4), (2, 7), (3, 12)\}$. The range of this function is the set $\{3, 4, 7, 12\}$. The pair $(0, 3)$ is one member of the function; it is said, also, to belong to the function. The value of the function at $x = 0$ is 3, and the value of the function at $x = 2$ is 7.

Functional Notation

In our discussion of sets in Appendix A we use capital letters to designate the sets, being unmindful of whether the elements of the set were ordered pairs or not. However, it is conventional to use lowercase letters to designate functions, and so in general we shall follow this practice and only on occasion use capital letters. A very convenient form of designating a function is represented by the symbol $f: (x, y)$, which is read "the function f whose ordered pairs are (x, y)." The phrase "y is a function of x" implies the same thing. Now, it was mentioned in the previous subsection that the dependent variable y is called the value of the function at x. It is customary also to represent the value of a function $f: (x, y)$ at x by the symbol $f(x)$, which is read "the value of f at x" or, more simply, "f of x." Since y and $f(x)$ are different names for the value of f at x, we can write $y = f(x)$, which is read "y equals f of x" or, more precisely, "y is the value of the function f at x." We must emphasize here that $f(x)$ does not mean f times x and is never so read.

In this connection, we must warn the student of the double usage of the symbol $f(x)$, prevalent in mathematical literature. The notation $f(x)$ is used to symbolize the function f and also to symbolize the value of the function. It is

SEC. 4.1 Relations and Functions

common practice to write "the function $f(x)$" as an abbreviation for "the function f whose value at x is $f(x)$" and to write "the function $f(x) = 2x + 1$" as an abbreviation for "the function f defined by the equation $f(x) = 2x + 1$." We shall be guilty of using such abbreviations on occasion, but it should cause no confusion if the student clearly understands that $f(x)$ is not the function but is the value of the function at x. The function is f, a set of ordered pairs. The equation $f(x) = 2x + 1$ is not the function but a rule for obtaining the value of the function f for each value of x in the domain. Hence, if $x = 2$, then $f(2) = 2(2) + 1 = 5$; that is, $f(2)$ represents the value of the function f at $x = 2$. Similarly, we would represent the value of f at $x = a$ by $f(a)$, and we would evaluate $f(a)$ by replacing x by a thus: $f(a) = 2a + 1$.

It will be recalled that the symbols $f(x)$, $g(x)$, $u(x)$, and $v(x)$ were introduced in Section 3.1 to represent polynomials. Since each polynomial in one variable, say x, can be interpreted as the value of a function at x, we were then just using functional notation. The following two examples illustrate the force and convenience of functional notation.

Example 6 Suppose that $f: (x, y)$ is defined by the equation $f(x) = x^2 - 5x + 2$. Then

$$f(2) = (2)^2 - 5(2) + 2 = -4$$

$$f(0) = (0)^2 - 5(0) + 2 = 2$$

$$f(-3) = (-3)^2 - 5(-3) + 2 = 26$$

$$f(4 + h) = (4 + h)^2 - 5(4 + h) + 2$$

$$= 16 + 8h + h^2 - 20 - 5h + 2$$

$$= h^2 + 3h - 2$$

Example 7 If $f(x) = \dfrac{x}{x-1}$, what are $f(2), f(-1), f(x^2)$, and $f\left(\dfrac{1}{x}\right)$?

Solution:

$$f(2) = \frac{2}{2-1} = 2$$

$$f(-1) = \frac{-1}{-1-1} = \frac{1}{2}$$

$$f(x^2) = \frac{x^2}{x^2 - 1}$$

$$f\left(\frac{1}{x}\right) = \frac{\dfrac{1}{x}}{\dfrac{1}{x} - 1} = \frac{1}{1-x}$$

In our discussion we have used the letter f to represent the function, the letter x to represent the independent variable, and the letter y to represent the dependent variable. These are the letters customarily used to represent functions and variables in general: "a function," "the function," "any function," etc. Any other letters would have served equally well, and many others will be used at various times throughout the remainder of this text, especially when we are speaking of some particular variable or particular function.

It is quite possible for two or more quantities under discussion to be functions of the same independent variable. In such a case it would be necessary, to avoid confusion, to represent each different function and each different quantity by different letters. For example, suppose you leave a certain place at a certain time with a full tank of gas in your car. Suppose, further, that you drive at a certain rate of speed. The distance traveled, $d(t)$ miles, and the amount of gasoline, $n(t)$ gallons, remaining in your tank after t hours are both functions of t. Here it is necessary to use different letters d and n because the two functions are different. That is, we would not use the symbol f to denote each of these functions.

So far we have restricted our definition of a function to that of one independent variable. However, it is easy to see how our definition and also our notation could be generalized so as to describe functions of any number of variables. Also, it is easy to cite examples of quantities that are functions of more than one independent variable. For example, the area A of a triangle is a function of its base b and altitude h or, expressed in functional notation, $A(b, h)$; the volume V of a rectangular box is a function of its length l, width w, and height h or, in symbols, $V(l, w, h)$. In this text we shall be concerned primarily with functions of a single independent variable. We turn next to a discussion of some common methods of expressing the rules which define such functions.

Rules Used in Defining Functions

It has already been mentioned that a function is usually defined by giving the domain of definition and the rule of correspondence (see Example 5). A rule may be expressed in many ways, but the following are the four most common.

A verbal statement. Consider the sentence, "Now is the time for all good men to come to the aid of their party." Note that this sentence contains 16 words. We now define the function $f: (x, y)$ as follows: The domain of definition is the set $X = \{x | 1 \leq x \leq 16\}$. To each x in X the number y which corresponds to x is the number of letters in the xth word of the given sentence. For example, $f(1) = 3$ because the first word of the given sentence contains three letters.

A formula or an equation. This is usually the most desirable means of stating the rule which defines a function because of its simplicity and convenience. This will be the method used to define most of the functions studied in the remainder of this text.

A table of values. The rule here is to "look it up in the table." Often, the only practical means of defining certain functions is by means of a table of values.

A graph. The rule here is "read it from the graph." This rule and the previous two are, for our purposes, the most important ways of stating a rule which defines a function, and each will be discussed in more detail in the following subsections.

Functions Defined by a Formula or an Equation

It should be emphasized at this point that no matter what method is used for stating the rule of correspondence, a function is not completely determined until we know precisely its domain of definition. Once the domain of definition of a function is known and the rule defining the function is given, the range can be determined. However, to carefully determine the range of such a function can be a tedious task, and we do not stress this process in the text (see Exercise 18). It is conventional not to mention the domain when a function is defined by an equation. This is not mathematically legitimate; however, it is really time saving, provided we have a definite agreement in advance as to what assumptions are being made concerning the domain in such a case.

For our purposes in this text we make the following agreement:

> *When a function is defined by means of an equation and no mention is made of its domain, we shall assume that the domain is the largest subset of R, the set of real numbers, for which the equation makes sense.*

In other words, the domain is assumed to be the set of all possible real numbers, each of which when substituted for x in the defining equation yields a *real number* as a value of y. Consider, for example, the function defined by the equation $y = \sqrt{x}$. For y to be a real number, x cannot be a negative real number, whereas, if $x \geq 0$, then $y = \sqrt{x}$ is a real number. Hence the domain is the set $X = \{x | x \geq 0\}$. The following examples may help to further clarify the procedure for finding the domain of a function defined by an equation.

Example 8   Let $f: (x, y)$ be defined by the equation $f(x) = 2x + 1$. We are assuming here that the domain of definition X is the set of all the real numbers.

Example 9   Suppose $f:(x, y)$ is defined by $y = \sqrt{x - 3}$. Since by agreement y must be real, it follows that x is in the domain if an only if $x - 3 \geq 0$, that is, if and only if $x \geq 3$. Hence the domain of this function is $X = \{x | x \geq 3\}$.

Example 10 If $y = f(x) = \dfrac{1}{x-1}$, what is the domain of f?

Solution: The only value of x for which the function is not defined is $x = 1$, for $\dfrac{1}{1-1} = \dfrac{1}{0}$, and division by zero is not defined. For every other value of x, $f(x)$ is a well-defined real number. Hence the domain is the set $X = \{x | x \neq 1\}$.

Example 11 If $y = f(x) = x^2 - 2x$, what is the domain of f?

Solution: Since $x^2 - 2x$ is a real number for every real number x, the domain of this function is the entire set of real numbers R.

Example 12 Consider the function $f: (x, y)$ defined by the equation $y = 4$. Since the domain is not explicitly stated, it is, according to our agreement, the set consisting of all the real numbers. The range is the set $Y = \{4\}$. In other words, the function f has the same value for every x in the domain. We call such a function a *constant* function.

DEFINITION 4.4. A function $f: (x, y)$ is called a *constant function* if its range Y consists of a single element, that is, of one and only one real number.

Functions Defined by a Table of Values

A very simple way to show how one quantity varies with another is by means of a table of values. In fact, this is such a forceful and convenient way of conveying certain types of information that we see it in use every day. To cite a few such examples: the daily paper may have a table showing the hourly temperature for the preceding 24 hours; each United States Series E Bond has on it a table, showing its redemption value each half year until maturity; the parcel post clerk consults a table to see how much postage to put on your package; the cashier in a store refers to a table to determine the amount of sales tax due on a purchase; federal income tax forms carry tables from which one can read the income tax due on a given adjusted gross income. The table mentioned in each of the above examples obviously defines a function.

For examples of tables of values which define functions and which are of more immediate interest to us, we may turn to Appendix B. Detailed discussions of these tables will be given later in the text. Although the procedures used in compiling these tables are complicated and somewhat involved, it is unnecessary for us to concern ourselves with these procedures. We shall be content to take as our rule when working with a function defined by one of these tables the simple command, "look it up in the table."

When a function is defined by means of a table of values, it is very important to know whether the table gives the complete rule or whether the domain of definition of the function actually consists of all the numbers in some interval while the table gives only values of the function at certain selected points

SEC. 4.1 Relations and Functions

of the interval. The tables in the appendix are of the latter type. With such tables it is possible to find an approximate value of the function at intermediate points not given in the table by a method called *interpolation*, to be described later. Examples of the two types of tables are given below.

Example 13 The following table shows the number of hotel rooms, N, that are let at a given rate, P dollars per day.

P	6	7	8	10	12	14
N	40	80	100	200	160	120

This table defines completely the function, in the sense that intermediate values are of no significance. The domain of definition is the set $\{6, 7, 8, 10, 12, 14\}$, and the range is the set $\{40, 80, 100, 120, 160, 200\}$.

Example 14 A liquid was heated to 90°C and allowed to cool to room temperature. The temperature, $T°$, of the liquid after t minutes of cooling was that shown below.

t	0	2	4	6	8	10	12	14	16	18	20
$T°$	90	74	62	54	49	44	40	37	34	32	30

Since the liquid obviously was cooling through the interval of time $t = 0$ to $t = 20$, it seems reasonable to assume that the domain of t consists of at least all the numbers in the closed interval [0, 20]. This being assumed, one may then reasonably ask such questions as, "What was the temperature of the liquid when t was 3.2?" or "How many minutes after cooling started was the temperature 50°?"

Interpolation by proportional parts. If we can assume that the domain of an independent variable includes not only the numbers given in a table but also all possible numbers between the smallest and largest of these given numbers, we say that the independent variable varies continuously. In such cases we are often able to obtain by interpolation rather good approximations of the values of the function for numbers in the domain but not in the table. The most common type of interpolation is called interpolation by proportional parts and is best explained by means of an example. We shall use the table of temperatures given above and find by interpolation the approximate temperature at $t = 3.2$.

Write in tabular form the two entries, the interpolated entry, and the distances between them, as shown below:

$$
\begin{array}{cc}
t \text{ (min)} & T \text{ (deg)} \\
\hline
2\left\{1.2\left\{\begin{array}{c} 2 \\ 3.2 \end{array}\right. \atop 4\right. & \left.\left.\begin{array}{c} 74° \\ ? \\ 62° \end{array}\right\}d \right\}12
\end{array}
$$

The method assumes that the ratios of the corresponding differences on the two sides of the table are equal. Thus

$$\frac{d}{12} = \frac{1.2}{2}$$

from which it follows that $d = 7.2°$ or, rounded off to the nearest degree, $d = 7°$. The desired approximation to T is found by subtracting $7°$ from $74°$, or $T \cong 74° - 7° = 67°$.

Exercises 4.1

1. **a.** The set $A = \{(-1, 2), (0, 2), (2, 2), (3, 2), (5, 2)\}$ is a relation. Is A a function? Explain.
    b. Give an example of a relation that is not a function.
    c. Give an example of two different functions that have the same domain and the same range.

2. Determine which of the following relations are also functions. For those that are functions give the domain and range.
    a. $A = \{(-2, -9), (-1, -5), (0, 2), (1, 3), (2, 5), (3, 7)\}$
    b. $B = \{(-3, 5), (-1, 2), (-1, -3), (2, 2), (4, 7), (7, 10)\}$
    c. $C = \{(1, -1), (2, 0), (3, 3), (1, 7), (4, 10), (5, 14)\}$
    d. $D = \{(-1, -6), (0, -5), (1, -2), (2, 1), (3, 4), (4, 9)\}$

3. A function $f: (x, y)$ is completely defined by the given table. Write down the elements of the function. Find $f(2)$ and $f(20)$.

x	-1	2	5	7	10	20
y	-8	1	10	16	25	55

4. If a function f is defined by the equation $y = f(x) = \sqrt{25 - x^2}$, find $f(0), f(3), f(-3), f(4), f(5)$.

5. If $f(x) = 8x^3 - 12x^2 + 10$, find $f(\tfrac{1}{2}), f(-1), f(a), f(x + h)$.

6. If $f(x) = 3x^2 - 5x + 7$, find:

    a. $f(2)$ **b.** $f(2 + h)$ **c.** $\dfrac{f(2 + h) - f(2)}{h}$

    d. $f(x + h)$ **e.** $\dfrac{f(x + h) - f(x)}{h}$

7. If $f(x) = 4 + 5x - 3x^2$, find:

    a. $f(3)$ **b.** $f(3 + h)$ **c.** $\dfrac{f(3 + h) - f(3)}{h}$

d. $f(-3)$ **e.** $f(-3+h)$ **f.** $\dfrac{f(-3+h)-f(-3)}{h}$

8. For each of the following functions, find $\dfrac{f(x+h)-f(x)}{h}$:

   a. $f(x) = x^3$ **b.** $f(x) = x^4$ **c.** $f(x) = \dfrac{10}{x}$

   d. $f(x) = \dfrac{4}{x+1}$ **e.** $f(x) = \dfrac{x+1}{x+2}$

9. If $f(x) = \sqrt{x}$, show that
$$\dfrac{f(x+h)-f(x)}{h} = \dfrac{1}{\sqrt{x+h}+\sqrt{x}}$$
(*Hint*: Rationalize the numerator.)

10. If $F(x) = \dfrac{1-x}{1+x}$, find $F(x^2)$, $[F(x)]^2$, $F(2x+3)$, $\dfrac{1}{F(x)}$, $F\left(\dfrac{1}{x}\right)$, $F[F(x)]$.

11. If $f(t) = \dfrac{8}{t+2}$, find:

   a. $f(3)$ **b.** $f(3+h)$ **c.** $\dfrac{f(3+h)-f(3)}{h}$ **d.** $\dfrac{f(x+h)-f(x)}{h}$

12. Which of the following equations define a function $f: (x, y)$?
   a. $y = 2x + 7$ **b.** $3x - 7y = 10$ **c.** $y = x^2$ **d.** $y^2 = 4x$
   e. $y = 4$ **f.** $x = 7$ **g.** $x^2 + y^2 = 16$ **h.** $xy = 5$

13. Assuming that each of the following equations defines y as a function of x, find the domain of each.

   a. $2x - y = 5$ **b.** $y = (x-1)^2$ **c.** $y = \sqrt{16 - x^2}$

   d. $y = \sqrt{5+x}$ **e.** $y = \dfrac{4}{x-2}$ **f.** $y = \dfrac{7}{x^2 - 9}$

14. We say that the equation $A(r) = \pi r^2$ expresses the area of a circle as a function of the radius of the circle.

   a. Express the area $A(s)$ of a square as a function of s, the length of one side.
   b. Express the volume $V(x)$ of a cube as a function of x, the length of one edge.
   c. Express the area $A(s)$ of an equilateral triangle as a function of s, the length of one side.

d. A rectangle is inscribed in a circle of radius 10. Express the area $A(x)$ of this rectangle as a function of x, the length of one side of the rectangle.

e. An open cylindrical can has a volume of 100π cubic centimeters. Express the total surface area of the can as a function of r, the radius of the base.

15. The given table shows the amount (A dollars) at the end of 20 years for a principal of $100.00 at compound interest and various rates (r percent).

r	2	3	4	5	6	7
A	148.59	180.61	219.11	265.33	320.71	386.97

Assuming r to be a continuous variable, use interpolation by proportional parts to approximate:

a. The value of A when $r = 3.5\%$
b. The value of r when $A = \$250.00$
c. The value of A when $r = 4.6\%$

16. The table shows the positive square roots ($\sqrt{N}$) for several numbers N. Tell whether interpolation in this table is justified; if so, approximate:

a. $\sqrt{4.64}$ **b.** $\sqrt{4.76}$

N	4.4	4.5	4.6	4.7	4.8
$\sqrt{N}$	2.0976	2.1213	2.1448	2.1679	2.1909

17. The temperature $T°$ in a certain city at 12:00 noon each December 25 for a period of t years is shown in the following table.

t	0	1	2	3	4	5	6	7
$T°$	29	52	59	48	43	57	53	51

a. Would you consider t a continuous variable in this case?
b. If this table defines a function $F: (t, T)$, what is the domain of definition of F?
c. What is $F(0)$? $F(3)$?

18. Find the range of each of Examples 8 to 11.

Solution for Example 9: First observe that for each x in the domain of this function (the domain is $X = \{x | x \geq 3\}$), $y = \sqrt{x-3} \geq 0$. Hence the range Y is a subset of $\{y | y \geq 0\}$, that is, $Y \subseteq \{y | y \geq 0\}$. Conversely, if $y \geq 0$, then $y = \sqrt{(y^2 + 3) - 3}$ and $y^2 + 3 \in X$. Therefore, $y = f(y^2 + 3)$, and we have the reverse inclusion $\{y | y \geq 0\} \subseteq Y$. Hence the range of f is the set $Y = \{y | y \geq 0\}$. (It should be emphasized here that, ordinarily, in order to show that two sets are equal we show that each set is a subset of the other. This is exactly what was done here.)

4.2 GRAPHS OF RELATIONS AND FUNCTIONS

We described in Section 1.6 how a rectangular coordinate system may be used to establish a one-to-one correspondence between points of the plane and ordered pairs (x, y) of real numbers. Thus, if we are given a rectangular coordinate system and an ordered pair of real numbers, we can locate, in the plane, the point having this ordered pair of numbers as its coordinates. This procedure obviously can be extended to any given set of ordered pairs of real numbers, thus obtaining a corresponding set of points, called the *graph* of the given set of ordered pairs.

DEFINITION 4.5. The *graph* of a relation A is the set of all the points in the plane whose coordinates are elements of A.

Example 1 The graph of the relation $A = \{(-2, 1), (-1, 2), (1, 3), (1, 1), (+2, -1)\}$ is shown in Figure 4.1.

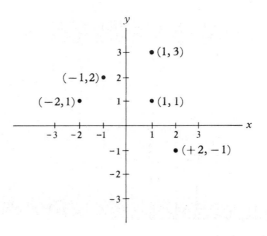

Figure 4.1

Since a function is a relation, the graph of a function f is the set of all points in the plane whose coordinates are elements of f. If the function is defined by means of a table of values, we plot the points whose coordinates are the ordered pairs of numbers given in the table. It is conventional, in plotting the graphs of functions, to associate the values of the dependent variable with points on the vertical axis. It is not necessary that the scales on the two axes be the same; hence we try to choose the scale for each axis that best fits the range of values of the variable to be represented. This is illustrated in Figures 4.2 and 4.3, which show graphs of the function defined in Example 14 in Section 4.1.

Figure 4.2 shows only the set of points whose coordinates are actually listed in the given table of values. If it is assumed that the table is the complete rule, then this set of *points* is the graph of the function. However, if it is assumed that the domain of the independent variable t includes all the numbers on the closed interval [0, 20], then we connect the plotted points by a smooth continuous curve and refer to the *curve* as an approximation to the graph of the function; see Figure 4.3.

Graphical interpolation. If the graph of a function is a smooth curve, we can read from it approximations to those values of the function corresponding to values of the independent variable that are not given in the table. Also, we can read approximations to the value or values of the independent variable for which the function has some particular value in its range and which are not listed in the table. Such readings are called graphical interpolations. How accurate these interpolated values are depends, of course, on how accurately the graph is drawn and on how accurately we can read the scales. Thus, at the very best, an interpolated result is an approximation, but if care has been exercised in getting it, the accuracy is often sufficient for most purposes.

To illustrate the use of graphical interpolation, let us find from the graph of Figure 4.3 an approximation to the temperature when $t = 1$. Reading from the

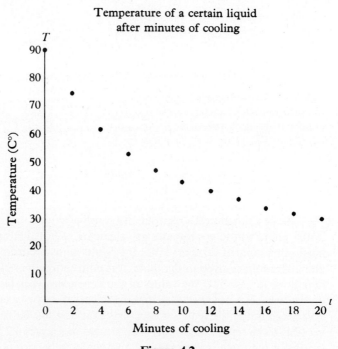

Figure 4.2

SEC. 4.2 Graphs of Relations and Functions

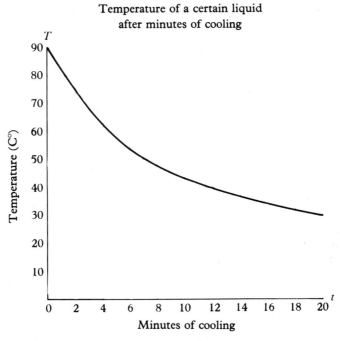

Figure 4.3

graph, we see that $T = 82°$, approximately. When was the temperature 45°? Reading again from the graph, we see that when $T = 45°$, $t \cong 9$. That is, the temperature was 45° after approximately 9 minutes of cooling.

A curve which is the graph of a function is often referred to as a *function graph*. A characteristic property of a function graph is that, if (x_1, y_1) is a point of the graph, then (x_1, y_2), $y_1 \neq y_2$ is not a point of the graph. That is, a vertical line cannot intersect a function graph in more than one point, but a horizontal line can. See Figure 4.4.

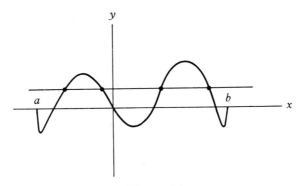

Figure 4.4

Just as not all relations are functions, not all graphs are function graphs. For example, the graph of Figure 4.5 is not a function graph.

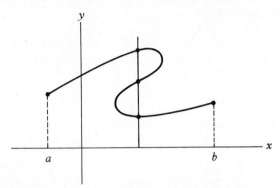

Figure 4.5

We shall end this section with some suggestions concerning the mechanics of constructing a graph. One purpose of the graph of a function is to show at a glance how one quantity varies with another. Hence it is most important that each axis be properly labeled showing what the varying quantities are and what the units on each scale represent. The first step in drawing the graph of a function is to draw and label the horizontal and vertical axes and to mark on each of them a scale suited to the table of data. Graph paper with 20 squares to the inch is probably best for most purposes. For the horizontal axis, select a scale unit of such length that the range of values of the independent variable will cover most of the page width, at least more than half. For the vertical axis, the unit should be so chosen that the total length of the axis need not be more than half the height of a sheet of regular graph paper. There are, of course, exceptions to this. The important thing is to spread the graph across the width of the paper as much as can be done conveniently and try to limit the height without distorting the shape of the graph. If the independent variable is not a continuous variable, then the graph of the function is just a set of isolated points. However, to make the graph easier to follow with the eye and to show the mode of variation a little better, the isolated points are usually joined by a series of segments of straight lines. This is often called a *broken-line* graph. It should be emphasized that on the broken-line graph only the plotted points have significance; the line segments connecting them serve only to aid the eye in following the series of points.

Graphs of Equations Involving Two Variables

To plot the graph of a function defined by means of a table of values, we simply plot the points whose coordinates are the ordered pairs of numbers listed in the

SEC. 4.2 Graphs of Relations and Functions

table. If, however, we wish to plot the graph of a function defined by means of an equation, we must first make a table of ordered pairs which are elements of the function. These ordered pairs are determined from the equation defining the function; we substitute, for x in the equation, selected values from the domain of definition and then solve the resulting equation for the corresponding value of y. The ordered pair is called a *solution* of the equation. The set consisting of all the possible solutions of the equation is called the *solution set* of the equation. In this discussion we are concerned primarily with functions of a single independent variable, which means that any equation defining a function will involve exactly two variables, whether these are actual or implied.

In general, when a function is defined by an equation, the independent variable is a continuous variable and, hence, the solution set of the defining equation is infinite. To obtain the graph of a function in such a case, we usually find a small number of solutions of the defining equation, plot the corresponding points, and then join these points by a smooth graph.

A similar discussion could be given for graphs of equations that define relations which are not functions. Examples of such equations are given in Sections 4.6 through 4.9. Our discussion suggests the following general definition.

DEFINITION 4.6. The *graph* of an equation involving two variables is the set of all points (x, y) in the plane whose coordinates satisfy the equation. Hence the graph consists of those points in the plane corresponding to the ordered pairs (x, y) in the solution set of the equation.

Example 2   Plot the graph of the equation $y = x + 120/x$, assuming $x > 0$.

Solution: We obtain the following table of ordered pairs (x, y).

x	4	6	8	10	12	15	20
y	34	26	23	22	22	23	26

An examination of the table shows that, as x increases, y decreases at first and then increases. The turning point occurs between $x = 10$ and $x = 12$. Hence it would probably be desirable to enlarge our table by finding at least the value of y for $x = 11$. With our equation we can get as many points on the graph between $x = 10$ and $x = 12$ as we need for locating the turning point fairly accurately. The graph is shown in Figure 4.6.

Example 3   Plot the graph of the equation $y = 2$.

Solution: This problem may be restated as: Plot the graph of the function $f: (x, y)$ defined by the equation $y = 2$. This is a case in which the defining equation actually exhibits only the variable y; however, the second variable, x, needed to produce ordered pairs (x, y), is implied. Freely translated, our problem says, "Given a

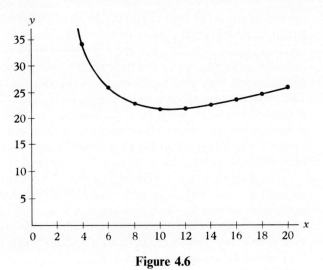

Figure 4.6

rectangular coordinate system, locate all the points whose y coordinate, referred to this system, is 2." The solution set of the equation $y = 2$ obviously consists of ordered pairs, such as $(-25, 2), (-10, 2), (-3, 2), (0, 2), (5, 2), (50, 2)$, and so on. The graph of this set is a line two units above the x axis and parallel to it, as in Figure 4.7. This is a typical example of a constant function with the domain $X = R$ and the range $Y = \{2\}$.

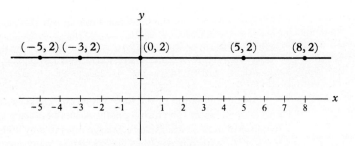

Figure 4.7

Example 4 Plot the graph of the equation $x = -2$.

Solution: Here the domain of the independent variable is $X = \{-2\}$, and the range is $Y = R$. Hence the graph is a line two units to the left of the y axis and parallel to it, as shown in Figure 4.8. It should be noted that the equation $x = -2$ defines a relation but does not define a function.

SEC. 4.2   Graphs of Relations and Functions   215

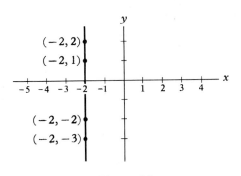

Figure 4.8

Graphs of a large number of functions and equations will be sketched in the remaining sections of this chapter. In Chapter 6 we shall use calculus to aid us in sketching rather complicated graphs of equations. For extremely accurate plotting of graphs of functions and equations, a hand calculator can be a valuable tool (see Section 3.14).

Exercises 4.2

In each of Exercises 1 through 10, draw the graph of the given relation and specify which of the graphs are function graphs.

1. $A_1 = \{(x,y) | y = 2x\}$
2. $A_2 = \{(x,y) | x = 3\}$
3. $A_3 = \{(x,y) | y = 4\}$
4. $A_4 = \{(x,y) | y = |x|\}$
5. $A_5 = \{(x,y) | y = \sqrt{25 - x^2}\}$
6. $A_6 = \{(x,y) | |x| + |y| = 2\}$
7. $A_7 = \{(x,y) | y = x + 2 \text{ and } -2 \leq x \leq 2\}$
8. $A_8 = \{(x,y) | x = \sqrt{25 - y^2}\}$
9. $A_9 = \{(x,y) | y = x - |x - 3|\}$
10. $A_{10} = \{(x,y) | y^2 = x \text{ and } 0 \leq x \leq 4\}$

In each of Exercises 11 through 26, sketch the graph of the given equation.

11. $y = 10 - 5x$
12. $y = 2x - 6$
13. $y = 4 - 3x$
14. $y = 3x + 6$
15. $y = -x$
16. $y = 3$
17. $y = x^2 - 2x - 3$
18. $x = -5$
19. $y = x^2 - 4x + 4$
20. $y = 2x - x^2$
21. $y = 1 - 8x - x^2$
22. $y = x^2 - 4$
23. $y = x^3 - 2x^2 - 5x + 6$
24. $y = x^3$
25. $y = x^3 - 3x + 1$
26. $y = 2x^3 + 3x^2 - 4x - 6$

27. Study the following four graphs and state for each whether it is or is not a function graph.

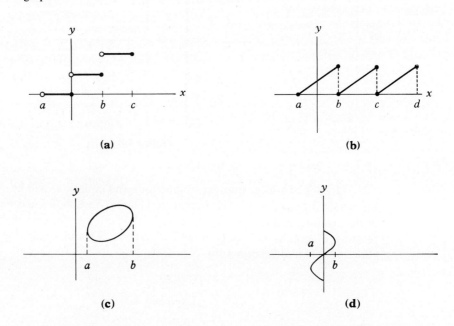

28. A manufacturing firm finds that the total cost, C, of producing x units of a certain product is that shown in the following table. Draw the graph exhibiting the relationship between C and x. The curve thus obtained is called a *cost curve*.

x	10	20	30	40	50	60
C	127	233	343	457	575	697

About how many units could be produced for $300? Approximately what would be the cost of producing 35 units?

29. The average cost, A per unit, for an output of x units of the product in Exercise 28 is given in the following table. Plot the graph.

x	10	20	30	40	50	60
A	12.70	11.60	11.40	11.40	11.50	11.60

Approximately what output will give lowest average cost?

30. Assume the postage (P cents) required to send a letter or package weighing x ounces by first-class mail is determined by the following rule. The postage is 20 cents for the first ounce or fraction thereof and 17 cents for each additional ounce or fraction thereof up to a maximum of 70 pounds. Plot the graph showing the postage for the first 6 ounces.

a. Is this a function graph?
b. What would be the first-class postage on a package weighing 4.7 ounces?
c. What would be the postage for a package weighing 50 pounds = 800 ounces? [*Hint:* The nth member of the set $\{a, a+d, a+2d, ..., a+(n-1)d\}$ is given by the formula $a_n = a + (n-1)d$.]

4.3 AN INTRODUCTION TO ANALYTIC GEOMETRY

We have seen how a rectangular coordinate system enables us to locate in the xy plane a point whose coordinates are any given ordered pair of real numbers and how it enables us to describe lines and curves by means of equations. Since plane geometry is basically a study of points, lines, and curves in a plane, it is seen that a rectangular coordinate system could be a very important tool in studying that subject. In fact, geometry studied by means of algebra is a subject of study by itself, called analytic geometry; in it, the basic connecting link between the geometry and the algebra is a rectangular coordinate system. However, to use such a system in the study of geometry, it is necessary that the unit used in marking the scale on each axis be a unit of distance. Furthermore, to use a rectangular coordinate system in calculating the lengths of line segments which are not parallel to a coordinate axis, we must use the same scale on both axes.

Analytic geometry is essentially concerned with two major problems:

Problem 1. Given an equation involving x and y, plot its graph.
Problem 2. Given a set of points in the plane, find an equation having this set and only this set of points as its graph.

In the previous section we gave, by means of examples, a procedure for handling the first of these. This procedure may be briefly summarized as follows.

Step 1. Solve the equation for y in terms of x.
Step 2. Substitute, for x, convenient positive and negative values (usually integers) from its domain and compute the corresponding values of y.
Step 3. Make a table of values by arranging these pairs in order of magnitudes of the values of x.
Step 4. Plot the points whose coordinates are these ordered pairs and connect them by a smooth curve in the order their coordinates appear in the table.

Remarks: In some cases it may be easier or more convenient to solve for x rather than y. If so, solve for x and then apply the rules given above, interchanging the roles of x and y.

If solving for y in terms of x, or x in terms of y, involves a square-root radical, the double sign $\pm$ must be used with the radical. This usually results in giving two values of y for each value of x.

Although this procedure may be quite laborious, it should always work. We shall, however, as we proceed, learn ways of obtaining the graphs of certain types of equations with a minimum amount of point-by-point plotting.

In the discussion and the problems that follow, we shall assume a knowledge of the following definitions from geometry.

1. A right triangle is a triangle one of whose angles is a right angle (a 90° angle).
2. An isosceles triangle is a triangle having two equal sides.
3. An equilateral triangle is a triangle having three equal sides.
4. A quadrilateral is a closed plane figure consisting of four points (called vertices), no three of which lie on the same straight line, and four line segments which join these vertices in such a way that each vertex is joined to exactly two other vertices and no two sides intersect except at a vertex.
5. A parallelogram is a quadrilateral whose opposite sides are parallel.
6. A rectangle is a parallelogram all of whose angles are right angles.
7. A square is a rectangle all of whose sides are equal.

Distance between Two Points

Before turning our attention to Problem 2 mentioned above, we shall derive some formulas which we shall need. The first and most important of these formulas is called the *distance formula*. It is a formula for finding the distance between two points whose coordinates are known. To derive this formula, we shall need to recall a very important theorem from plane geometry regarding right triangles, which we shall state without proof.

THEOREM 4.1 (PYTHAGOREAN THEOREM). *The sum of the squares of the lengths of the legs of a right triangle is equal to the square of the length of the hypotenuse.*

See Figure 4.9. Our theorem may be stated in the simple form

$$a^2 + b^2 = c^2$$

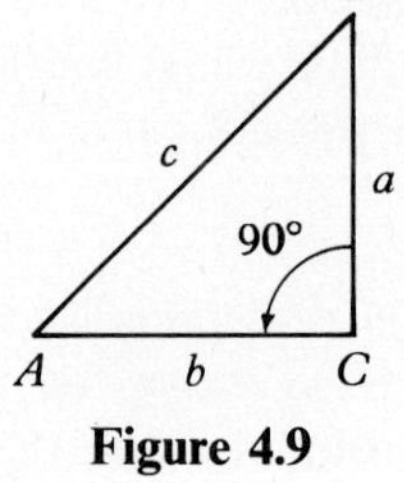

Figure 4.9

SEC. 4.3   An Introduction to Analytic Geometry

The converse of this theorem is also true: *If the sum of the squares of the lengths of two sides of a triangle is equal to the square of the length of the third side, then the triangle is a right triangle.*

We are now ready to derive the distance formula. Let the points $P_1(x_1, y_1)$ and $P_2(x_2, y_2)$ be any two points in the plane and let $d(P_1, P_2)$ be the distance between P_1 and P_2. Draw a line through P_1 parallel to the x axis and a line through P_2 parallel to the y axis. These two lines meet in some point M (see Figure 4.10) whose coordinates are easily found to be (x_2, y_1). The distance $d(P_1, M)$ between two points $P_1(x_1, y_1)$ and $M(x_2, y_1)$ on a line parallel to the x axis was defined in Section 4.1 as $d(P_1, M) = |x_2 - x_1|$. Similarly, the distance $d(P_2, M)$ between two points $P_2(x_2, y_2)$ and $M(x_2, y_1)$ on a line parallel to the y axis was defined as $d(P_2, M) = |y_2 - y_1|$. The triangle $P_1 P_2 M$ is a right triangle, the vertex of whose right angle is at M. Hence, by the Pythagorean theorem,

$$[d(P_1, P_2)]^2 = |x_2 - x_1|^2 + |y_2 - y_1|^2$$
$$= (x_2 - x_1)^2 + (y_2 - y_1)^2$$

or (1) $\qquad d(P_1, P_2) = \sqrt{(x_2 - x_1)^2 + (y_2 - y_1)^2}$

The last equation is known as the *distance formula*. It should be observed from the distance formula that, first, $d(P_1, P_2)$ is always positive or zero and that, second, since $(x_2 - x_1)^2 = (x_1 - x_2)^2$ and $(y_2 - y_1)^2 = (y_1 - y_2)^2$, the relation $d(P_1, P_2) = d(P_2, P_1)$ obtains. In other words, when the distance formula is used to find the distance between two given points, it makes no difference which of the points is denoted by P_1 and which by P_2.

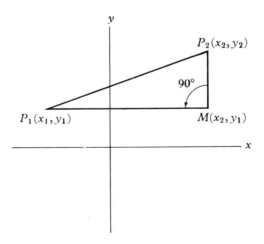

Figure 4.10

Example 1 | Use the distance formula to calculate the distance between the two points $A(-2, 1)$ and $B(4, 3)$.

Solution: Let $(x_1, y_1) = (-2, 1)$ and $(x_2, y_2) = (4, 3)$. Then $x_1 = -2, y_1 = 1, x_2 = 4$, and $y_2 = 3$. Substituting in the distance formula, we get

$$d(A, B) = \sqrt{[4-(-2)]^2 + (3-1)^2} = \sqrt{6^2 + 2^2} = \sqrt{40}$$

Remark: Since the pair of coordinates of a point completely specifies the point, it is common practice to designate the point $P_1(x_1, y_1)$ simply as the point (x_1, y_1), the point $P_2(x_2, y_2)$ as (x_2, y_2), and the distance between two points by the single letter d. Then the distance formula is written

$$d = \sqrt{(x_2 - x_1)^2 + (y_2 - y_1)^2}$$

Midpoint Formulas

A second formula, or actually a pair of formulas, which plays a very important role in the study of analytic geometry is a formula which enables us to calculate the coordinates of the midpoint of a line segment when the coordinates of the endpoints are known. It is easy to show, by equating the ratios of the lengths of corresponding sides of similar triangles, that the coordinates $(\bar{x}, \bar{y})$ of the midpoint of the line segment joining the two distinct points $P_1(x_1, y_1)$ and $P_2(x_2, y_2)$ are given by the formulas

(2) $$\boxed{\bar{x} = \frac{x_1 + x_2}{2} \quad \text{and} \quad \bar{y} = \frac{y_1 + y_2}{2}}$$

This pair of formulas is often expressed as the single formula

(3) $$\boxed{(\bar{x}, \bar{y}) = \left(\frac{x_1 + x_2}{2}, \frac{y_1 + y_2}{2}\right)}$$

Formulas (2) are called the *midpoint formulas* and their derivation is left to the student (see Exercise 10).

Example 2 | Find the coordinates of the midpoint of the line segment joining the two points $(-3, 2)$ and $(5, 5)$.

Solution: Let $(x_1, y_1) = (-3, 2)$ and $(x_2, y_2) = (5, 5)$. Applying the midpoint formulas, we have

$$\bar{x} = \frac{-3 + 5}{2} = 1 \quad \text{and} \quad \bar{y} = \frac{2 + 5}{2} = \frac{7}{2}$$

The coordinates of the midpoint of the given line segment are therefore $(1, \frac{7}{2})$.

SEC. 4.3   An Introduction to Analytic Geometry

Example 3   Using the distance formula only, show that the points $A(3, -3)$, $B(4, 0)$, $C(7, 3)$, and $D(6, 0)$ are the vertices of a parallelogram.

Solution: We need to recall a theorem from plane geometry: "If the opposite sides of a quadrilateral are equal, the quadrilateral is a parallelogram." Our problem now is simply to show that the opposite sides of our parallelogram are equal. First let us plot the given points in Figure 4.11. Now we use the distance formula to calculate the length of each side:

$$d(A, B) = \sqrt{(3-4)^2 + (-3-0)^2} = \sqrt{1+9} = \sqrt{10}$$
$$d(B, C) = \sqrt{(7-4)^2 + (3-0)^2} = \sqrt{9+9} = \sqrt{18}$$
$$d(C, D) = \sqrt{(7-6)^2 + (3-0)^2} = \sqrt{1+9} = \sqrt{10}$$
$$d(A, D) = \sqrt{(6-3)^2 + [0-(-3)]^2} = \sqrt{9+9} = \sqrt{18}$$

Our calculations show that

$$d(A, B) = d(C, D) \quad \text{and} \quad d(B, C) = d(A, D)$$

and that, hence, the quadrilateral $ABCD$ is a parallelogram.

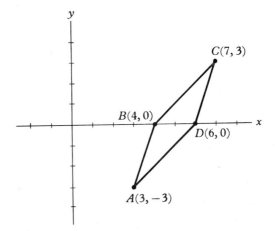

Figure 4.11

Exercises 4.3

1. Find $d(A, B)$ for each given pair of points.
   a. $A(5, 3)$, $B(-2, 1)$
   b. $A(5, 3)$, $B(-6, -3)$
   c. $A(4, -7)$, $B(-6, -3)$
   d. $A(-6, -3)$, $B(0, 4)$
   e. $A(0, 4)$, $B(-5, 0)$
   f. $A(-2, 1)$, $B(0, 4)$

2. Find the coordinates of the midpoint of the line segment joining each of the pairs of points in Exercise 1.

3. Show that the points $(-3, 0)$, $(1, -2)$, and $(5, 6)$ are the vertices of a triangle.

4. Show that the quadrilateral whose vertices are $(8, 0)$, $(6, 6)$, $(-3, 3)$, and $(-1, -3)$ is a parallelogram. Is it a rectangle? (*Hint:* Show that the diagonals are equal.)

5. If the lengths of two sides of a triangle are equal, the triangle is isosceles. Show that the points $(2, 4)$, $(5, 1)$, and $(6, 5)$ are the vertices of an isosceles triangle.

6. Plot the graph of each given equation.
   a. $y = 2x$ b. $y = x^2$ c. $y = x^3$
   d. $y = \sqrt{25 - x^2}$ e. $2x - 3y = 6$ f. $x^2 + y^2 = 4$

7. Three points $P_1(x_1, y_1)$, $P_2(x_2, y_2)$, and $P_3(x_3, y_3)$, such that $x_1 < x_2 < x_3$, lie on the same straight line if and only if $d(P_1, P_2) + d(P_2, P_3) = d(P_1, P_3)$. Show that the points $(-2, 4)$, $(1, 1)$, and $(3, -1)$ all lie on the same straight line.

8. A median of a triangle is a line segment joining a vertex to the midpoint of the side opposite that vertex. Find the lengths of the medians of the triangle whose vertices are the points $A(1, -2)$, $B(5, 4)$, and $C(9, 8)$.

9. The point P is the midpoint of the line segment joining the points P_1 and P_2. Find the coordinates of (a) P_1 if given $P(2, 3)$ and $P_2(7, 5)$ and (b) P_2 if given $P_1(-4, -7)$ and $P(2, -1)$.

10. Derive the midpoint formulas.

11. Determine the value of c so that the graph of the equation $3x + 5y + c = 0$ will pass a. through the point $(3, -2)$; b. through the point $(-4, 1)$.

12. Which of the given points are on the graph of the equation $y^2 = 12x$: $(3, 6)$, $(0, 0)$, $(-3, 6)$, $(12, -12)$, $(\frac{4}{3}, 4)$, $(48, 24)$?

13. How far is $(7, 4)$ from the midpoint of the line segment joining $(3, 2)$ and $(5, 14)$?

14. Express by an equation free from radicals the fact that the point (x, y) is equidistant from the two points $(1, 0)$ and $(6, 4)$.

15. Find the coordinates of a point 10 units distance from the point $(-3, 6)$ and having 3 as its abscissa (two solutions).

4.4 EQUATIONS OF SETS OF POINTS IN THE PLANE

In the previous section we discussed the problem of finding an equation whose graph is a given set of points in the xy plane. In this regard it is helpful to formulate the next definition.

DEFINITION 4.7. For a given set of points in the xy plane, an *equation of this set of points* is any equation whose graph is precisely the given set of points.

SEC. 4.4 Equations of Sets of Points in the Plane

Example 1   If we consider the line consisting of those points located two units above the x axis, then an equation of this set of points is $y = 2$ (see Example 3 of Section 4.2). In this case we say that $y = 2$ is "an equation of the line." Another equation of this particular line is $-3y = -6$. In fact, if c is any nonzero real number, then $cy = 2c$ is an equation of this line. We can also write an equation of this line which involves both x and y; that is $0 \cdot x + y = 2$. An even more complicated equation for this line which involves both x and y is $y(|x| + 1) = 2(|x| + 1)$.

As Example 1 illustrates, one significant difference between graphs of equations and equations of sets of points is:

Each equation has exactly one graph, whereas a set of points in the xy plane may have many equations.

Any two equations of the same set of points in the xy plane are equivalent in the sense that they have the same graph. Ordinarily, we are satisfied if we can find one equation of a given set of points. Moreover, we generally seek as simple an equation as possible. Often equations of a set of points can be written in several different ways, all equally correct. Because of this, a student might obtain a correct answer to an exercise which does not agree exactly with the answer found in the back of the book. In this case, algebraic manipulation will usually transform the student's answer to the answer in the back of the book. It should be emphasized, however, that unless the exercise specifically asked for an answer in the form given in the back of the book, the student's answer is entirely acceptable. The next example illustrates the above discussion.

Example 2   Find an equation of the set of points shown in Figure 4.12.

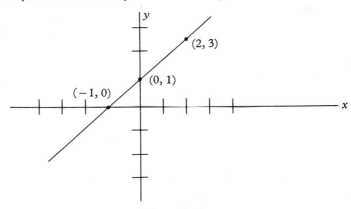

Figure 4.12

Solution: We shall find in Section 4.6 that one equation of this line is $y = x + 1$. Equally correct answers to this problem would include $x - y + 1 = 0$, $x = y - 1$, $y - x = 1$, $2y = 2x + 2$, and $-2y = -2x - 2$.

In Sections 4.5 through 4.10, our attention will be directed toward finding equations of various sets of points in the xy plane. In particular, we treat straight lines, circles, parabolas, ellipses, and hyperbolas. These are the most common sets of points studied in analytic geometry. The most important of these sections, for the purpose of later chapters in this text, are the next two, which deal with straight lines and their equations. Chapters 5 and 6 will depend heavily on Sections 4.5 and 4.6.

Exercises 4.4

1. An equation for the x axis is $y = 0$, and an equation for the y axis is $x = 0$. Find an equation for the union of the x axis with the y axis (that is, those points which lie either on the x axis or on the y axis or on both). *Hint:* $a \cdot b = 0$ if, and only if, $a = 0$ or $b = 0$.

2. **a.** Write an equation, free of radicals, which expresses the fact that the point (x, y) is five units from the origin.
   b. Draw the curve in the xy plane that passes through every point in that plane that is five units from the origin. *Note:* The curve of part **b** is a circle and the equation of part **a** is called an equation of that circle.

3. Find an equation for those points which lie (**a**) on the horizontal line through the point $(0, 5)$ and (**b**) on the vertical line through the point $(-2, 0)$.

4. Find an equation for those points which are equidistant from the x axis and the y axis. Sketch the graph of this set of points.

4.5 STRAIGHT LINES

The Slope of a Line

Let L be any line in the xy plane not parallel to the y axis, and let $P_1(x_1, y_1)$ and $P_2(x_2, y_2)$ be any two distinct points on L. We associate with the line L a number m, called the *slope* of L, defined by the equation

(1) $$m = \frac{y_2 - y_1}{x_2 - x_1}$$

It is important to observe that Formula (1) expresses the slope m of the line L as the ratio of the difference in the y coordinates to the difference in the x coordinates, taken in the same order, of the points P_1 and P_2. This means that, if we take as the numerator in Formula (1) the difference $y_2 - y_1$, then we must take for the denominator $x_2 - x_1$. It would be equally correct to use $y_1 - y_2$ as the numerator, provided we use $x_1 - x_2$ as the denominator. Thus, in using

Formula (1) to find the slope of a line when the coordinates of two points on the line are known, it does not matter which point we label (x_1, y_1) if we label the other (x_2, y_2). As was stated at the beginning of this section, Formula (1) applies only to lines that are not parallel to the y axis, that is, nonvertical lines. In fact, the slope of a vertical line is not defined, for Formula (1) is meaningless when $x_1 = x_2$.

The importance of the ratio $\dfrac{y_2 - y_1}{x_2 - x_1}$ stems from the fact that the number m is independent of the two distinct points on L that are used to calculate it. As an indication of why this statement is true, let $P_3(x_3, y_3)$ be any third point on L distinct from P_1 and P_2 (see Figure 4.13). Draw a line through P_1 parallel to the x axis, and draw lines through P_2 and P_3 parallel to the y axis and intersecting the first line in the points M_2 and M_3, respectively. The coordinates of M_2 are (x_2, y_1) and of M_3 they are (x_3, y_1). The triangles $P_1 M_2 P_2$ and $P_1 M_3 P_3$ are both right triangles that have an acute angle in common and are therefore similar. The ratios of the lengths of corresponding sides of similar triangles are equal. Hence

$$\frac{y_2 - y_1}{x_2 - x_1} = \frac{y_3 - y_1}{x_3 - x_1}$$

which suggests the validity of our statement.

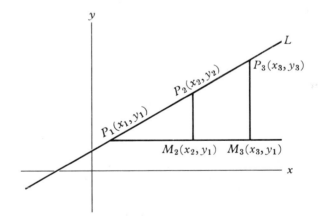

Figure 4.13

If we choose P_1 arbitrarily, then choose P_2 such that $x_2 - x_1 = 1$, our slope formula becomes $m = y_2 - y_1$. The slope of a line is therefore the algebraic change in y per unit increase in x, as the point P_1 moves along the line. Thus, if y increases as x increases, the slope is positive and the line rises to the right. On the other hand, if y decreases as x increases, the slope is negative and the line falls to the right, or rises to the left. If y remains constant as x increases, then the slope is zero and the line is parallel to the x axis.

Important

1. *The slope of a line parallel to the x axis is zero.*
2. *The slope of a line parallel to the y axis is not defined.*

Note that we deliberately avoided saying "a vertical line has no slope," because such a statement could lead some to the false conclusion that a vertical line has zero slope. Only lines parallel to the x axis have zero slope.

Since the slope m of a nonvertical line can be interpreted as the change in y per unit increase in x as the point (x, y) moves along the line, we may also say that m is the rate of change of y per unit increase in x. This leads us to the following observation regarding functions: Every function whose graph is a straight line increases (or decreases) at a constant rate. Conversely, if a function increases (or decreases) at a constant rate, its graph is a straight line.

Parallel Lines

The common concept that most of us have about parallel lines is the following: Two distinct lines in the same plane are parallel if they do not intersect. We also will agree that any line is parallel to itself. The following facts characterize parallel lines:

Two lines are parallel if and only if either:

(i) *Both lines are vertical; or*
(ii) *The slopes of the two lines are equal.*

Example 1 Find the slope of the line passing through the two points whose coordinates are $(-2, 5)$ and $(4, -1)$.

Solution: Let P_1 be the point $(-2, 5)$ and P_2 be the point $(4, -1)$; then $x_1 = -2$, $y_1 = 5$, $x_2 = 4$, and $y_2 = -1$. Hence, by definition,

$$m = \frac{y_2 - y_1}{x_2 - x_1} = \frac{(-1) - (5)}{4 - (-2)} = \frac{-1 - 5}{4 + 2} = -\frac{6}{6} = -1$$

Example 2 Show that the line L_1 determined by the two points $(3, -1)$ and $(6, 8)$ is parallel to the line L_2 determined by the two points $(5, 2)$ and $(3, -4)$.

Solution: Let m_1 and m_2 be the slopes of L_1 and L_2, respectively; then

$$m_1 = \frac{8 + 1}{6 - 3} = 3 \quad \text{and} \quad m_2 = \frac{-4 - 2}{3 - 5} = 3$$

Since $m_1 = m_2$, the lines L_1 and L_2 are parallel.

SEC. 4.5 Straight Lines

Perpendicular Lines

What is the relation between the slopes of two lines that are perpendicular? To answer this question, let us consider the two lines, L and L', which intersect at $P(h, k)$ and neither of which is parallel to either of the coordinate axes. Let $A(a, b)$ be a second point on L and $B(c, d)$ a second point on L', both distinct from P. Draw the line segment joining points A and B (see Figure 4.14). The lines L and L' are mutually perpendicular if and only if the triangle PAB is a right triangle with the segment AB as the hypotenuse. This will be true if and only if

$$[d(P, A)]^2 + [d(P, B)]^2 = [d(A, B)]^2$$

that is, if and only if

$$(h-a)^2 + (k-b)^2 + (h-c)^2 + (k-d)^2 = (a-c)^2 + (b-d)^2$$

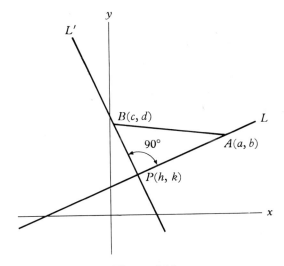

Figure 4.14

When the squared terms in the last equation are expanded, this equation becomes

$$h^2 - 2ah + a^2 + k^2 - 2bk + b^2 + h^2 - 2ch + c^2 + k^2 - 2kd + d^2$$
$$= a^2 - 2ac + c^2 + b^2 - 2bd + d^2$$

which is true if and only if

$$(h-a)(h-c) + (k-b)(k-d) = 0$$

or if and only if

$$\frac{k-b}{h-a} = -\frac{h-c}{k-d}$$

But the slope of L is
$$m = \frac{k-b}{h-a}$$

and the slope of L' is
$$m' = \frac{k-d}{h-c}$$

Hence we have shown that the triangle PAB is a right triangle with AB as hypotenuse if and only if

$$m = -\frac{1}{m'}$$

The above argument justifies part (ii) of the following characterization of perpendicular lines:

Two lines are perpendicular if and only if either:

(i) *One line is horizontal (has slope 0) and the other is vertical; or*
(ii) *The slope of one line is the negative reciprocal of the slope of the other.*

Example 3 Show that the line L_1 determined by $(3, -1)$ and $(9, 2)$ is perpendicular to the line L_2 passing through $(-2, 10)$ and $(5, -4)$.

Solution: Let m_1 be the slope of L_1 and m_2 the slope of L_2; then

$$m_1 = \frac{2-(-1)}{9-3} = \frac{3}{6} = \frac{1}{2} \quad \text{and} \quad m_2 = \frac{10-(-4)}{-2-5} = \frac{14}{-7} = -2$$

Since $m_1 = -1/m_2$, the lines L_1 and L_2 are perpendicular.

Exercises 4.5

1. In each of the following, plot the given points and find the slope of the line determined by them in each case.
   a. $(2, 1), (6, 4)$ **b.** $(8, 7), (2, -1)$ **c.** $(-2, 5), (1, 1)$
   d. $(1, 7), (-3, 3)$ **e.** $(-5, -3), (-1, 7)$
   f. $(1.4835, 2.9531), (3.5862, 6.1384)$

2. Find the length, slope, and midpoint of the line segment joining the given points
   a. $(2, 6), (8, 4)$ **b.** $(4, -3), (-10, 7)$ **c.** $(3, -5), (5, 9)$

3. Find the slope of the perpendicular bisector of the line segment joining $(2, 3)$ and $(6, 5)$.

4. The line L passes through the point $(7, 1)$ and has slope $\frac{3}{2}$.

SEC. 4.5 Straight Lines

a. Find the coordinates of a second point on L.
b. Draw the line L.

5. By calculating slopes, show that the quadrilateral whose vertices are $A(-3, 1)$, $B(5, 3)$, $C(3, 0)$ and $D(-5, -2)$ are the vertices of a parallelogram.

6. Draw a line through the given point which has the given slope.

   a. $(0, 0)$ with slope $\frac{3}{5}$ b. $(2, 1)$ with slope $\frac{1}{2}$

   c. $(-1, 2)$ with slope $-\frac{2}{3}$ d. $(3, -2)$ with slope $-\frac{1}{2}$

   e. $(0, 0)$ with slope 4 f. $(0, 3)$ with slope 2

7. Show that $A(-11, 12)$, $B(6, -5)$, $C(1, -8)$, and $D(-6, 15)$ are the vertices of a rectangle.

8. Prove by means of slopes that the points $(-5, -2)$, $(2, 1)$, and $(12, 7)$ all lie on the same straight line. Do the same for the three pairs $(1, -2)$, $(2, 0)$, and $(3, 2)$.

9. Prove by means of slopes that the points $(3, -1)$, $(-1, 2)$, and $(2, 6)$ are the vertices of a right triangle. Do the same for the three points $(4, -7)$, $(5, -4)$ and $(-8, -3)$.

10. Three vertices of a parallelogram are in the order $(2, 5)$, $(4, -2)$, and $(10, -1)$. Find the coordinates of the fourth vertex.

11. Show that the points $(2, 5)$, $(7, 1)$, $(11, 6)$, and $(6, 10)$ are the vertices of a square and find its area.

12. Given the triangle whose vertices are $A(-2, 3)$, $B(4, -5)$, and $C(8, 1)$, show that the line segment joining the midpoints of AB and BC is parallel to the side AC and is equal in length to $\frac{1}{2} d(A, C)$.

13. Write an equation which expresses the fact that the line segment joining the point (x, y) to the point $(-4, 0)$ is perpendicular to the line segment joining (x, y) to the point $(4, 0)$.

14. Write an equation which expresses the fact that the point (x, y) is twice as far from the point $(1, -2)$ as it is from the point $(6, 4)$.

15. THEOREM. *The diagonals of a rectangle are equal.* Prove this theorem by using a rectangular coordinate system and a rectangle whose length is a units and width b units. (*Hint:* Choose one vertex of your rectangle at the origin and the axes to lie along adjacent sides of the rectangle.)

16. Use a rectangular coordinate system to prove that the diagonals of a parallelogram bisect each other.

17. Given a quadrilateral $ABCD$, show that the line segments joining the midpoints of adjacent sides form a parallelogram.

4.6 EQUATIONS OF STRAIGHT LINES

We proceed now to show that every straight line has an equation, in the sense of Definition 4.7, and to indicate several procedures for finding such an equation. The method used will depend on the information given about the line.

First we consider the lines in the xy plane that are parallel to the y axis. If a line L is parallel to the y axis, then all points on L have the same x coordinate. If this coordinate is k, then the coordinates of every point on L satisfy the equation

(1) $$x = k$$

Furthermore, if a point is not on L, then its x coordinate is different from k and, hence, its coordinates do not satisfy Equation (1). Thus, in accordance with Definition 4.7, Equation (1) is an equation of L.

If L is a line parallel to the y axis and d units to the right of the y axis, then (using the notation of the previous paragraph) $k = d$ and an equation of L is $x = d$. If L is a line parallel to the y axis and d units to the left of the y axis, then $k = -d$ and an equation of L is $x = -d$. If L coincides with the y axis, then $k = 0$, and so an equation for the y axis is $x = 0$.

Example 1 An equation for the line which is parallel to the y axis and five units to the right of the y axis is $x = 5$.

Example 2 An equation for the line which is parallel to the y axis and seven units to the left of the y axis is $x = -7$.

Example 3 Write an equation of the line which is parallel to the y axis and passes through the point $(-3, 2)$.

Solution: Since the line passes through $(-3, 2)$, the x coordinate of every point on the line is -3. Hence an equation of the line is $x = -3$.

Next consider a line L not parallel to the y axis, passing through the fixed point $P_1(x_1, y_1)$ and having the slope m. Let $P(x, y)$ be another point on L distinct from P_1. It follows from Formula (1) of the previous section that

$$\frac{y - y_1}{x - x_1} = m$$

or

(2) $$\boxed{y - y_1 = m(x - x_1)}$$

It is obvious from the manner in which Equation (2) was derived that the coordinates of every point on L, with the possible exception of (x_1, y_1), satisfy Equation (2). It is easily shown by direct substitution that (x_1, y_1) also satisfies

SEC. 4.6 Equations of Straight Lines

Equation (2). Hence the coordinates of every point on L satisfy Equation (2) and, conversely, every ordered pair (x, y) of real numbers which satisfy Equation (2) represents the coordinates of a point on L. Thus L is the graph of Equation (2), and it follows from Definition 4.7 that Equation (2) is an equation of L.

Equation (2) is called the *point-slope* form of the equation of L because it exhibits the coordinates (x_1, y_1) of a point on L and the slope m of L. The importance of the point-slope form of the equation of a line lies in the fact that it furnishes us with a simple and easy form to use in writing the equation of a line when the coordinates of a point on the line and the slope of the line are known or can be found. Note that the letters x_1, y_1, and m in Equation (2) represent fixed specific real numbers, while the letters x and y represent variables. For example, to use Equation (2) to write an equation of the line containing the point $(4, 3)$ and whose slope is 7, we let $x_1 = 4$, $y_1 = 3$, and $m = 7$, giving $y - 3 = 7(x - 4)$ as an equation of the line.

Example 4 Write an equation of the line passing through the point $(5, 3)$ and having the slope -2.

Solution: Here $x_1 = 5$, $y_1 = 3$, and $m = -2$. Substituting in Equation (2), we obtain $y - 3 = -2(x - 5)$, or $2x + y = 13$.

Example 5 Write an equation of the line determined by the points $(2, 3)$ and $(-1, 5)$.

Solution: We first find the slope $m = \dfrac{3 - 5}{2 - (-1)} = \dfrac{-2}{2 + 1} = -\dfrac{2}{3}$. Now, using either of the two points, say $(2, 3)$, we substitute in Equation (2) and we get

$$y - 3 = -\frac{2}{3}(x - 2) \quad \text{or} \quad 3y - 9 = -2x + 4$$

which, when simplified, becomes $2x + 3y = 13$.

Check: We may check our solution for errors in algebra by substituting the coordinates of each given point for x and y in our final equation:

$$2(2) + 3(3) = 4 + 9 = 13 \quad \text{and} \quad 2(-1) + 3(5) = -2 + 15 = 13$$

Since the coordinates of each given point are a solution of our equation, the solution must be correct.

The y coordinate of the point of intersection of a line L and the y axis is called the y *intercept* of the line L. Suppose a line L with slope m intersects the y axis at the point $(0, b)$. The y intercept of L is b. An equation of L is

$$y - b = m(x - 0)$$

or

(3) $$y = mx + b$$

Equation (3) is called the *slope-intercept* form of the equation of a line. This form is of interest because it enables us to obtain by inspection the slope of a line from the equation of the line. To do this, we solve an equation of the line for y and then compare our equation with Equation (3). For example, suppose an equation of a given line is

$$3x + 5y = 20$$

Solving for y, we write $5y = -3x + 20$, or

$$y = -\frac{3}{5}x + 4$$

Comparing this last equation with Equation (3), we see that

$$m = -\frac{3}{5} \quad \text{and} \quad b = 4$$

Hence $3x + 5y = 20$ is an equation of the line with slope $-\frac{3}{5}$ and y intercept 4.

It should be noted that if $m = 0$ then Equation (3) becomes $y = b$. That is, if a line is parallel to the x axis, its equation can always be written in the simple form $y = b$. Then if b is positive, the line is b units above the x axis; if b is negative, the line is $|b|$ units below the x axis; if $b = 0$, the line is the x axis. Thus an equation of the x axis is

$$y = 0$$

It follows from the above discussion that an equation of any vertical line can be written in the form

$$x = k$$

and that an equation of every nonvertical line can be written in either of the equivalent forms

$$y - y_1 = m(x - x_1) \quad \text{or} \quad y = mx + b$$

Since each of these equations is an equation of the first degree in x and y, that is, it contains only first powers of x and y, we conclude that every straight line has an equation which is of the first degree in x and y. Such an equation is said to be "linear in x and y." We wish, next, to show that every equation linear in x and y has a straight line as its graph.

Consider the general linear equation in x and y

(4) $$ax + by + c = 0$$

where a, b, and c are fixed real numbers but a and b are not both zero. If $b \neq 0$, this equation can be written in the form

(5) $$y = -\frac{a}{b}x - \frac{c}{b}$$

SEC. 4.6 Equations of Straight Lines

In comparing this equation with Equation (3), we see that it is an equation of the line with slope $-\dfrac{a}{b}$ and y intercept $-\dfrac{c}{b}$. If $b = 0$ and $a \neq 0$, then Equation (4) can be written in the form

$$x = -\dfrac{c}{a}$$

which, according to Equation (1), is an equation of a line parallel to the y axis and $\left|-\dfrac{c}{a}\right|$ units from it. These results may be summarized in the following theorem.

THEOREM 4.2. *Every straight line has an equation which is linear in x and y and, conversely, every linear equation in x and y has a straight line as its graph.*

We are now able to solve the following two problems: **(a)** given a straight line, find an equation for it; **(b)** given a linear equation in x and y, draw its graph.

Example 6 Write an equation of the line passing through the point $(2, -3)$ and parallel to the line which has $5x - 2y = 7$ as an equation.

Solution: To write an equation of a line, it suffices to know the coordinates of a point on the line and the slope of the line if it has a slope. Since our line is parallel to the line with equation $5x - 2y = 7$, it has the same slope as this line. We find the slope by solving for y:

$$2y = 5x - 7 \quad \text{or} \quad y = \dfrac{5}{2}x - \dfrac{7}{2}$$

Hence the slope is $\dfrac{5}{2}$. Now, by using this slope and the given point $(2, -3)$ we can write

$$y + 3 = \dfrac{5}{2}(x - 2) \quad \text{or} \quad 2y + 6 = 5x - 10$$

which when simplified becomes $5x - 2y = 16$.

Example 7 Draw the graph of the line which has $3x - 5y = 10$ as an equation.

Solution: Since we know the graph is a straight line, two points are sufficient to determine the line. A third point is desirable, as a check. To find the coordinates of three points on our line, we solve the given equation for y, obtaining $y = \dfrac{3}{5}x - 2$, and then substitute any three convenient values for x, preferably ones that give integers as coordinates. In this case x equals −5, 0 and 5 gives the table.

x	−5	0	5
y	−5	−2	1

The graph of the line is shown in Figure 4.15.

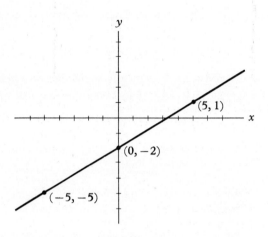

Figure 4.15

Point of Intersection of Two Lines

It was observed earlier that two lines in the xy plane having the same slope are parallel and hence do not intersect. It seems reasonable to assume that two lines in the same plane with different slopes do intersect. In fact, it will be shown in Chapter 8 that two such lines not only intersect, but they intersect in exactly one point. It is often desirable to find the coordinates of the point of intersection of two lines whose equations are known or can be found. The simplest procedure for this is a method most students learn in high school algebra called the *method of elimination by addition or subtraction*. We illustrate this method with the following example.

Example 8 Find the coordinates of the point of intersection of the lines having equations $3x - 2y = 12$ and $2x - 5y = -3$.

Solution: Multiply both sides of the first equation by 5 and both sides of the second equation by 2, and subtract the resulting second derived equation from the first derived equation as follows.

SEC. 4.6 Equations of Straight Lines

$$15x - 10y = 60$$
$$4x - 10y = -6$$

Subtracting $11x 66$
Solving for x $ x = 6$

Now substitute 6 for x in either of the two given equations. We shall use the first and solve for y. Thus

$$3(6) - 2y = 12$$
$$18 - 2y = 12$$
$$-2y = -6$$
$$y = 3$$

Check by substituting 6 for x and 3 for y in the second given equation:

$$2(6) - 5(3) = -3$$

a numerical identity. Hence the coordinates of the point of intersection of the two given lines are (6, 3). A procedure such as the one used in this example is often called solving the two given equations simultaneously. A complete discussion of solving systems of linear equations is given in Chapter 8.

Example 9 Show that the three medians of the triangle whose vertices are $A(5, 4)$, $B(1, -4)$, and $C(-9, 6)$ intersect in a common point. See Figure 4.16.

Solution: We first write equations of the medians of triangle ABC. They are

$$AF: \quad x - 3y = -7$$
$$BE: \quad 3x + y = -1$$
$$CD: \quad x + 2y = 3$$

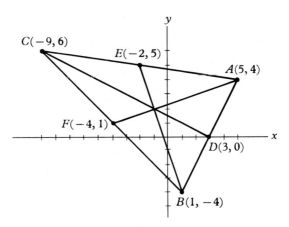

Figure 4.16

We next find the coordinates of the point of intersection of medians BE and CD. To do this, we multiply the equation of BE by 2 and subtract from this derived equation the equation of median CD; thus

$$6x + 2y = -2$$
$$x + 2y = 3$$
$$\overline{5x = -5}$$

Solving for x $ x = -1$

Substituting -1 for x in the equation of CD, we get

$$-1 + 2y = 3$$
$$2y = 4$$
$$y = 2$$

Therefore, the coordinates of the point of intersection of BE and CD are $(-1, 2)$. The three medians intersect in a common point if these coordinates satisfy the equation of AF. Checking, we have

$$-1 - 3(2) = -7$$

which is a numerical identity. Therefore, the three medians do intersect in a common point. If three lines intersect in a common point, the lines are said to be *concurrent*.

Note: For convenience we shall write statements such as "the line $y = 3x + 2$" as an abbreviation for "the line which has $y = 3x + 2$ as an equation."

Exercises 4.6

Find an equation of each of the straight lines described in Exercises 1 through 6.

1. Passing through (1, 1) with slope

    a. $m = 2$ **b.** $m = -1$ **c.** $m = \dfrac{3}{5}$ **d.** $m = -\dfrac{2}{3}$ **e.** $m = 0$

2. a. Having y intercept 3 and slope $\dfrac{1}{2}$.

   b. Having y intercept 5 and slope $-\dfrac{4}{3}$.

3. a. Having y intercept 3 and x intercept 4.

   b. Having y intercept -2 and x intercept 7.

4. a. Parallel to y axis and 4 units to the left of it.

   b. Parallel to y axis and 7 units to the right of it.

   c. Parallel to x axis and 5 units below it.

SEC. 4.6 Equations of Straight Lines

5. Passing through $(3, -2)$ and
   a. Parallel to y axis
   b. Parallel to x axis
   c. Through $(7, 4)$
   d. Parallel to the line $5x + 7y = 8$
   e. Perpendicular to the line $4x + 5y = 10$
   f. Parallel to the line $4x - y = 3$
   g. Perpendicular to the line $6x + 2y = 3$

6. a. Passing through $(2, 4)$ and parallel to the line $3x - 2y = 8$
   b. Passing through $(-5, 6)$ and perpendicular to the line $2x - y = 3$
   c. Passing through $(-3, -1)$ and perpendicular to the line $4x + 3y = 6$

7. Each of the following pairs of points determines a line. Write an equation of each.
   a. $(2, 1), (7, 4)$
   b. $(-5, -2), (-1, -4)$
   c. $(-3, 2), (4, 7)$
   d. $\left(\frac{5}{2}, \frac{1}{2}\right), (8, 5)$
   e. $(0, 0), (4, 3)$
   f. $(0, 0), (a, b)$

8. Write an equation of the perpendicular bisector of the line segment joining the pair of points.
   a. $(1, 2), (5, 6)$
   b. $(-2, 3), (4, -1)$
   c. $(-8, 3), (2, 7)$

9. Show that an equation of the line whose x intercept is $a \neq 0$ and whose y intercept is $b \neq 0$ is $\frac{x}{a} + \frac{y}{b} = 1$.

10. Express by an equation free from radicals the condition that the point (x, y) is equidistant from the two points $(3, 8)$ and $(7, 4)$.

11. Show that if $a \neq 0$ and $b \neq 0$, the lines $ax + by = c$ and $bx - ay = d$ are perpendicular. Use this fact to write an equation of the line passing through the point $(1, 1)$ and perpendicular to the line $3x - 5y = 10$.

12. Find by inspection the slope of each of the following lines.
    a. $2x - y = 5$
    b. $3x + 2y = 11$
    c. $x + 3y + y = 0$
    d. $4x + 7y = 20$
    e. $3x - 4y = 0$
    f. $5x - 7y = 18$

13. Sketch the graph of each given equation.
    a. $5x - 2y = 10$
    b. $4x + 3y = 12$
    c. $4x - 5y = 20$
    d. $y + 3 = 0$
    e. $x - 5 = 0$
    f. $3x - 2y = 0$
    g. $7x + 3y = 16$
    h. $y^2 - 4 = 0$

14. Write equations of the medians of the triangle with vertices at $A(0, 0)$, $B(2, 4)$, and $C(10, 10)$, and show they are concurrent.

15. Write equations of the perpendicular bisectors of the sides of the triangle with vertices at $A(-5, -10)$, $B(9, -8)$, and $C(9, 4)$, and show that the lines are concurrent.

16. A line segment drawn from a vertex of a triangle perpendicular to the side opposite

that vertex is called an *altitude* of the triangle. Each triangle has three altitudes. Write an equation for each altitude of the triangle whose vertices are $A(0, 1)$, $B(4, 5)$, and $C(6, -3)$. Show that these altitudes are concurrent.

17. Show that the line $2x - 3y + 4 = 0$ is:
    a. Parallel to the line $10x - 15y = 22$.
    b. Perpendicular to the line $12x + 8y = 15$.
    c. The same as the line $6x - 9y + 12 = 0$.

18. Show that the following lines form the sides of a rectangle: $x - 2y = 9$, $6x + 3y = 14$, $4x - 8y = 15$, and $2x + y + 5 = 0$.

19. Any three noncollinear points determine a circle. A line segment joining any two points on a circle is called a *chord* of the circle. The perpendicular bisector of a chord of a circle passes through the center of the circle. Write an equation for the perpendicular bisector of each of two of the chords of the circle determined by the three points $(-2, -2)$, $(10, -8)$, and $(7, 1)$, and by solving these two equations find the center of the circle.

20. a. Find an equation of the line L passing through the point $(1, 5)$ and perpendicular to the line L'. An equation of L' is $x + y = 2$.
    b. Find the coordinates of the point of intersection of L and L' in part **a**.
    c. Use the distance formula to find the perpendicular distance from the point $(1, 5)$ to the line L' in part **a**.

4.7 CIRCLES AND THEIR EQUATIONS

In the previous section we learned how to write an equation of the set of all the points lying on the same straight line in the xy plane having given enough conditions to completely determine the line. In this section, and the three sections that follow, we are going to define four more important point sets, a circle, a parabola, an ellipse, and a hyperbola, and determine an equation for each.

DEFINITION 4.8. The set of all points in the plane equidistant from a fixed point is called a *circle*. The fixed point is called the *center* of the circle, and the common distance of each point of the circle from the center is called the *radius* of the circle.

To find an equation for a given circle of radius r, let us choose a rectangular coordinate system in the plane such that the center of the circle is at the point (h, k), where h and k are measured in the same units as the given radius r. A point $P(x, y)$ is a point on the circle if and only if (see Figure 4.17)

$$\sqrt{(x-h)^2 + (y-k)^2} = r$$

SEC. 4.7 Circles and Their Equations

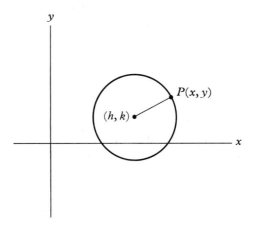

Figure 4.17

and by squaring, this equation reduces to

(1) $$(x-h)^2 + (y-k)^2 = r^2$$

We call this the *standard form* for the equation of the circle. This form exhibits the coordinates of the center and the radius of the circle and hence is most useful as a formula for writing equations of specific circles.

Example 1 Write an equation of the circle with center at $(2, -3)$ and radius equal to 4.

Solution: Here $h = 2$, $k = -3$, and $r = 4$. Substituting these values in Equation (1), we obtain

$$(x-2)^2 + (y+3)^2 = 16$$

Any equation that can, by the performing of algebraic operations, be reduced to the form of Equation (1) is the equation of a circle. For example, consider the equation

(2) $$Ax^2 + Ay^2 + Bx + Cy + D = 0, \quad A \neq 0$$

Now divide both sides by A, which gives

(3) $$x^2 + y^2 + ax + by + c = 0$$

where $a = B/A$, $b = C/A$, and $c = D/A$. Then completing the square in x and in y, Equation (3) can be reduced to

(4) $$\left(x + \frac{a}{2}\right)^2 + \left(y + \frac{b}{2}\right)^2 = \frac{a^2}{4} + \frac{b^2}{4} - c$$

Comparing this equation with Equation (1), we see that it is the equation of a circle with center at $(-a/2, -b/2)$ and radius

$$r = \sqrt{\frac{a^2}{4} + \frac{b^2}{4} - c}$$

provided, of course, that $a^2/4 + b^2/4 - c$ is positive. If $a^2/4 + b^2/4 - c$ is negative, then Equation (3), and hence Equation (2), has no graph. That is, Equation (2) is not the equation of any set of points. If $a^2/4 + b^2/4 - c = 0$, then the graph consists of the single point $(-a/2, -b/2)$. This can be considered as the circle with center at $(-a/2, -b/2)$ with radius 0. Hence, if Equation (2) has a graph, it is a circle (or a single point). To find the center and radius of a circle from its equation, we proceed as we did in obtaining Equation (4). We continue to use the convention of writing "the circle $x^2 + y^2 = 1$" in place of "the circle which has $x^2 + y^2 = 1$ as an equation."

Example 2 Find the center and radius of the circle $x^2 + y^2 - 6x + 4y - 12 = 0$.

Solution: Completing the square in x and in y, we obtain

$$x^2 - 6x + 9 + y^2 + 4y + 4 = 12 + 9 + 4$$

or

$$(x - 3)^2 + (y + 2)^2 = 25$$

and hence the center is at $(3, -2)$ and the radius is $r = 5$.

Remark: It follows from the above discussion that the distinguishing features of an equation of a circle are (1) the equation contains an x^2 term and a y^2 term, (2) the coefficients of x^2 and y^2 are identical, (3) it may or may not contain a first-degree term in x or in y, and (4) it contains no xy term.

Circles Determined by Geometric Conditions

We have learned how to write equations of a line from certain information which completely determines the position of the line in the xy plane, such as the coordinates of two points on the line, or a point and the slope. We shall now show that a similar statement is also true regarding circles. We already know how to write an equation of a circle having given the coordinates of its center and its radius. It follows, therefore, that if we have enough information about a circle to enable us to find the coordinates of its center and its radius then we can certainly write an equation for it. Those who have studied plane geometry in high school will recall that given any three points which are not on the same straight line, one and only one circle can be drawn which passes through each of these points. This suggests that if given the coordinates of three distinct points on a circle we should be able to use this information to find the coordinates of

SEC. 4.7   Circles and Their Equations

the center and the radius of the circle, and hence write an equation for it. A method for doing this is illustrated in the following example.

Example 3 Find an equation of the circle which passes through the points $P(0, -2)$, $Q(8, 2)$, and $R(-1, 5)$. See Figure 4.18.

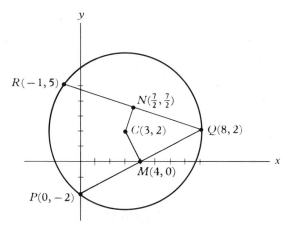

Figure 4.18

Solution: To find the coordinates of the center of the circle determined by the three given points, we make use of the fact that the perpendicular bisectors of two chords of a circle intersect at the center of the circle. We need to write equations of the perpendicular bisectors of two chords. We shall use chords PQ and QR. The slope of PQ is $\frac{1}{2}$; hence the slope of the perpendicular bisector is -2. The midpoint of PQ is $(4, 0)$. An equation of the perpendicular bisector of chord PQ is therefore $2x + y = 8$. Similarly, we find an equation of the perpendicular bisector of chord QR to be $3x - y = 7$. Solving these two equations simultaneously, we obtain $x = 3$ and $y = 2$. Thus the center of our circle is the point $C(3, 2)$. The radius of the circle is the distance from the center to any one of the three given points. We use the point $P(0, -2)$ and find

$$r = d(P, C) = \sqrt{9 + 16} = 5$$

The equation in standard form is therefore

$$(x - 3)^2 + (y - 2)^2 = 25$$

Thus, to find an equation of a circle, assuming sufficient conditions are given to completely determine the circle, the best procedure is usually the following. Use the given conditions to find the center and radius of the circle, and then write its equation in standard form.

Example 4 A circle is tangent to the line $x - 2y = 1$ at the point $P(5, 2)$, and the center of the circle is on the line $x + y = 9$. Find an equation of the circle. See Figure 4.19.

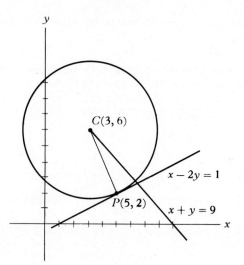

Figure 4.19

Solution: Here we need to use a theorem from plane geometry which states that, if a line L is tangent to a circle at the point P, then a line drawn through P and perpendicular to L will pass through the center of the circle. An equation of the line passing through $P(5, 2)$ and perpendicular to the line $x - 2y = 1$ is $2x + y = 12$. We now have equations of two lines which intersect at the center of the required circle,

$$2x + y = 12 \quad \text{and} \quad x + y = 9$$

Solving these two equations simultaneously, we obtain the coordinates $x = 3$, $y = 6$; that is, the center is the point $C(3, 6)$. The radius is $r = d(P, C) = \sqrt{4 + 16} = \sqrt{20}$, and the standard form for the equation of the circle is

$$(x - 3)^2 + (y - 6)^2 = 20$$

Exercises 4.7

Write an equation in standard form of the circle described in each of Exercises 1 through 8.

1. Center $(0, 0)$, radius 7
2. Center $(2, 3)$, radius 6
3. Center $(5, -3)$, radius 8
4. Center $(-4, -2)$, radius 5
5. Center $\left(-\frac{3}{2}, \frac{1}{2}\right)$, radius 4
6. Center $\left(-4, \frac{2}{3}\right)$, radius $\sqrt{8}$
7. Center at $(2, 5)$ and passing through $(5, 1)$
8. With ends of a diameter at $(-2, 4)$ and $(6, -2)$

9. Write an equation of the circle determined by the three given points:
   a. (2, 3), (4, −1), (5, 2)   b. (4, 3), (2, 7), (−3, −8)
10. Write an equation of the circle tangent to the line $3x - 4y = 20$ at the point $(4, -2)$ and having its center on the line $x + y = 3$.
11. Find an equation of the circle passing through (6, 3) and (−2, 7) and having its center on the line $3x - 5y = 16$.
12. Find an equation of the circle which passes through (7, −8) and (0, 9) and has its center on the line $x - 2y = 1$.
13. Find an equation of the line tangent to the circle $x^2 + y^2 = 25$ at the point $(-3, 4)$.
14. Find the center and radius of each of the following circles by first writing the equation in standard form $(x - h)^2 + (y - k)^2 = r^2$.
   a. $x^2 + y^2 - 4x + 6y - 3 = 0$ b. $x^2 + y^2 + 10x - 8y + 5 = 0$
   c. $4x^2 + 4y^2 - 12x - 20y + 9 = 0$ d. $x^2 + y^2 - 6x = 0$
   e. $x^2 + y^2 + 4y = 0$ f. $x^2 + y^2 - 3x - 7y = 0$

4.8 PARABOLAS AND THEIR EQUATIONS

In this section we define and discuss a set of points called a parabola.

DEFINITION 4.9. The set of all points in a plane which are equidistant from a fixed point and a fixed line is called a *parabola*. The fixed point is called the *focus* of the parabola, and the fixed line is called the *directrix* of the parabola.

Suppose that the point F is the focus and the line L is the directrix of a given parabola. Obviously, the form of an equation of this parabola will depend on the location of the x and y axes relative to the focus and directrix. The simplest form of such an equation is obtained if we choose the origin of our coordinate system at the midpoint of the perpendicular drawn from F to the line L, and choose the x axis parallel to L. See Figure 4.20. Denote the distance between F and L by $2p$, $p > 0$. A point $P(x, y)$ is a point on the parabola if and only if

$$FP = DP$$

That is,

$$\sqrt{x^2 + (y - p)^2} = |y + p|$$

Squaring both sides, we obtain

$$x^2 + y^2 - 2py + p^2 = y^2 + 2py + p^2$$

and simplifying we obtain

(1) $$x^2 = 4py$$

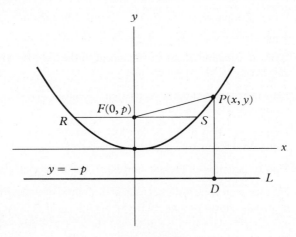

Figure 4.20

as an equation of the given parabola. The line which passes through the focus and is perpendicular to the directrix is called the *axis* of the parabola. The axis of the parabola discussed above is the y axis. An inspection of Equation (1) reveals that the origin, $(0, 0)$, is a point of the parabola. Since $p > 0$, then $y \geq 0$, and hence $(0, 0)$ is the lowest point on the parabola. This is the point on the parabola nearest the directrix. Such a point is called the *vertex* of the parabola. Also, if (x_1, y_1) is a point on the parabola, then $(-x_1, y_1)$ is also on the parabola. This shows that the axis of the parabola bisects every chord of the parabola that is perpendicular to the axis. Two points are said to be *symmetric with respect to a line* if the line is a perpendicular bisector of the segment joining them. Furthermore, a set S of points is said to be *symmetric with respect to a line L* if for each point $A \in S$ there is a point $B \in S$ such that A and B are symmetric with respect to L. It follows from this definition that a parabola is symmetric with respect to its axis.

The chord RS in Figure 4.20 through the focus and perpendicular to the axis is called the *latus rectum*. Its length is $4p$.

If the point $(0, -p)$ is chosen for the focus and the line $y = p$ for the directrix, the equation of the parabola is then

$$x^2 = -4py$$

Example 1 Discuss the parabola $x^2 = -8y$ and sketch its graph.

Solution: The axis of the parabola is the y axis, and the vertex is at the origin. Since y must be negative to give real values to x, the parabola opens downward. Also, $-4p = -8$; hence $p = 2$. Therefore, the focus is at $(0, -2)$, and the directrix is the line $y = 2$. The ends of the latus rectum are the points $(-4, -2)$ and $(4, -2)$. See Figure 4.21.

SEC. 4.8   Parabolas and Their Equations

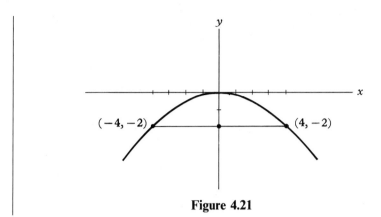

Figure 4.21

Suppose we choose a new coordinate system so that its axes are parallel to the corresponding axes in Figure 4.21 but so that the vertex of the parabola is at the point (h, k) referred to as the new axis. See Figure 4.22. The equation of the parabola now becomes

(2) $$(x - h)^2 = 4p(y - k)$$

To see that this is true, draw in the old coordinate axes using dashed lines and denote them by x' and y'. In terms of x' and y' the equation of the parabola, Equation (1), is

(3) $$x'^2 = 4py'$$

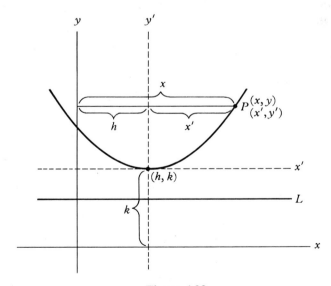

Figure 4.22

Let P be a point on the parabola whose old coordinates are (x', y') and whose new coordinates are (x, y). It is obvious from the figure that

$$x = x' + h \quad \text{and} \quad y = y' + k$$

or

$$x' = x - h \quad \text{and} \quad y' = y - k$$

Substituting these values into Equation (3), it becomes Equation (2).

Expanding Equation (2) and solving for y, we obtain

$$y = \frac{1}{4p}x^2 - \frac{h}{2p}x + \frac{h^2 + 4pk}{4p}$$

or

$$y = ax^2 + bx + c$$

where $a = 1/4p$, $b = -h/2p$, and $c = (h^2 + 4pk)/4p$; since $p > 0$, then $a > 0$. This helps to verify a statement made earlier about the graph of $y = ax^2 + bx + c$.

If we interchange the positions of the focus and directrix and keep the same coordinate system, the equation of the parabola will be

(4) $$(x - h)^2 = -4p(y - k)$$

and the parabola will open downward. If we had chosen the y axis to be parallel to the directrix and kept the vertex at (h, k), the equation of the parabola would have been

(5) $$(y - k)^2 = 4p(x - h)$$

or

(6) $$(y - k)^2 = -4p(x - h)$$

Equations (2), (4), (5), and (6) are said to be in standard form.

Example 2 Write the equation $y^2 - 8y = 8x$ in standard form and then sketch its graph.

Solution: To write the given equation in standard form, we complete the square in y by adding 16 to each side; thus

$$y^2 - 8y + 16 = 8x + 16$$

Factoring, we obtain the standard form

$$(y - 4)^2 = 8(x + 2)$$

Comparing this equation with Equation (5), we see that its graph is that of a parabola whose vertex is at $(-2, 4)$, the focus is at $(0, 4)$, and the ends of the latus rectum are at $(0, 0)$ and $(0, 8)$. The graph is shown in Figure 4.23.

SEC. 4.8   Parabolas and Their Equations

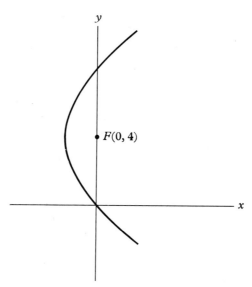

Figure 4.23

Exercises 4.8

1. Find the vertex, the focus, and the coordinates of the ends of the latus rectum of each of the following parabolas.
   a. $y = x^2$
   b. $x^2 = 8y$
   c. $y^2 = 4x$
   d. $x^2 = 4y$
   e. $(x-2)^2 = 12(y-1)$
   f. $(y-2)^2 = 8(x+3)$
   g. $y = 4 - x^2$
   h. $y^2 = 4 + x$
   i. $y = x^2 - 9$

2. Write an equation of each of the following parabolas. Find the coordinates of the ends of the latus rectum in each case, and sketch the graph of the parabola.
   a. Directrix $x + 2 = 0$, focus $(2, 0)$
   b. Directrix $y = 3$, focus $(0, -3)$
   c. Vertex at $(0, 0)$, focus at $(-3, 0)$
   d. Vertex at $(1, 1)$, focus at $(4, 1)$
   e. Vertex at $(-3, 2)$, focus at $(-3, 4)$

3. Find the coordinates of the points of intersection of each given parabola and the given line.
   a. $y = x^2$, $y = 4$
   b. $y = 9 - x^2$, $y = 0$
   c. $y = x^2 - 4x$, $y = x$
   d. $y = x^2$, $y = x + 2$

4. Find the vertex, the focus, the ends of the latus rectum, and an equation of the directrix of each of the following parabolas, and sketch it.
   a. $x^2 + 4x = 10y + 16$
   b. $y^2 - 4x + 4y = 4$
   c. $x^2 - 4x + y = 0$
   d. $y^2 + 6y + 6x = 0$

4.9 ELLIPSES AND THEIR EQUATIONS

Another interesting set of points is a set that forms an oval-shaped curve called an ellipse. It is defined as follows.

DEFINITION 4.10. If F and F' are two fixed points in the plane whose distance apart is denoted by FF' and k is a given positive number greater than the distance FF', then the set of all points P in the plane the sum of whose distances from F and F' equals k is called an *ellipse*. Stated in symbols, the set of all points P such that $PF + PF' = k$ is called an ellipse. The two fixed points F and F' are called the *foci* of the ellipse, and the midpoint of the segment joining the foci is called the *center* of the ellipse.

To write an equation of an ellipse, let F and F' be the two fixed points, and let the distance $FF' = 2c$ be greater than zero. For convenience, let the given positive number $k = 2a$. Choose the line determined by F and F' as the x axis, and choose the midpoint of the segment joining F and F' as the origin. In this coordinate system the coordinates of F and F' are those shown in Figure 4.24.

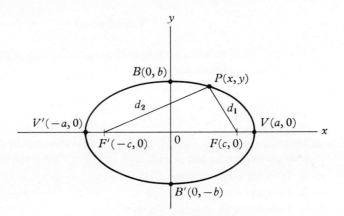

Figure 4.24

A point $P(x, y)$ will be a point on the ellipse if and only if

$$\frac{1}{2}(F'P + FP) = a$$

or

(1) $$F'P + FP = 2a$$

$$PF' = \sqrt{(x+c)^2 + y^2} \quad \text{and} \quad PF = \sqrt{(x-c)^2 + y^2}$$

Substituting these values in Equation (1), we get

$$\sqrt{(x+c)^2 + y^2} + \sqrt{(x-c)^2 + y^2} = 2a$$

To obtain an equivalent equation free from radicals, we transpose the second radical to the right side and square, which produces

$$(x+c)^2 + y^2 = 4a^2 - 4a\sqrt{(x-c)^2 + y^2} + (x-c)^2 + y^2$$

When we simplify the last equation, we have

$$a\sqrt{(x-c)^2 + y^2} = a^2 - cx$$

Squaring again gives

$$a^2[(x-c)^2 + y^2] = a^4 - 2a^2cx + c^2x^2$$

which when simplified becomes

$$(a^2 - c^2)x^2 + a^2y^2 = a^2(a^2 - c^2)$$

Since $a > c$, $a^2 - c^2 > 0$; so for convenience we let $b = \sqrt{a^2 - c^2}$, and we obtain as an equation of our ellipse

(2) $$b^2x^2 + a^2y^2 = a^2b^2$$

Equation (2) is often written in the form

(3) $$\frac{x^2}{a^2} + \frac{y^2}{b^2} = 1$$

which is obtained by dividing both sides of Equation (2) by a^2b^2.

Any point whose coordinates satisfy Equation (3) is a point on the ellipse which is the graph of this equation. It is readily seen that if the coordinates of the point (x_1, y_1) satisfy Equation (3), then so do the coordinates of the points $(-x_1, y_1)$ and $(x_1, -y_1)$; hence our ellipse is symmetric with respect to both the x axis and the y axis. When $y = 0$, then $x = \pm a$, and when $x = 0$, $y = \pm b$. Hence the ellipse crosses the x axis at the points $V(a, 0)$ and $V'(-a, 0)$, and it crosses the y axis at the points $B(0, b)$ and $B'(0, -b)$. The segment VV' of length $2a$ is called the major axis of the ellipse, and the segment BB' of length $2b$ is called the minor axis. The ends of the major axis, V and V', are called the vertices of the ellipse. If Equation (3) is solved in turn for x and then for y, we get

$$x = \pm \frac{a}{b}\sqrt{b^2 - y^2} \quad \text{and} \quad y = \pm \frac{b}{a}\sqrt{a^2 - x^2}$$

which shows that the permissible values of x and y are $-a \leq x \leq a$ and $-b \leq y \leq b$. Thus all the points of the ellipse lie inside or on the rectangle determined by the lines $x = -a$, $x = a$, $y = -b$, and $y = b$.

It should be noted that, had we taken the line determined by the foci as the

y axis instead of the x axis but retained the origin at the same place, an equation of the ellipse would be

(4) $$\frac{x^2}{b^2} + \frac{y^2}{a^2} = 1$$

where a and b have the same significance as before.

Example 1 Sketch the ellipse $\frac{x^2}{16} + \frac{y^2}{25} = 1$ and indicate the location of the foci.

Solution: Remembering that $a > b$, and comparing our equation with Equation (4), we find that $a^2 = 25$, $b^2 = 16$, and $c^2 = 25 - 16 = 9$. Therefore, $a = 5$, $b = 4$, and $c = 3$. The major axis is vertical, and hence the coordinates of the foci are $(0, -3)$ and $(0, 3)$. See Figure 4.25.

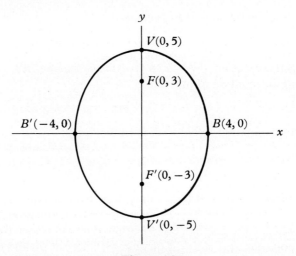

Figure 4.25

Suppose in Figure 4.24 we relabel the coordinate axes shown there as x' and y' and then draw new x and y axes parallel to the x' axis and the y' axis, respectively (see Figure 4.26), and let (h, k) be the coordinates of the center of the ellipse referred to the new axes. An equation of our ellipse in the x', y' coordinate system is

$$\frac{x'^2}{a^2} + \frac{y'^2}{b^2} = 1$$

Substituting the relations $x' = x - h$, $y' = y - k$ for x', y', we see that an equation of the ellipse referred to the x, y system is

(5) $$\frac{(x-h)^2}{a^2} + \frac{(y-k)^2}{b^2} = 1, \qquad a > b$$

SEC. 4.9 Ellipses and Their Equations

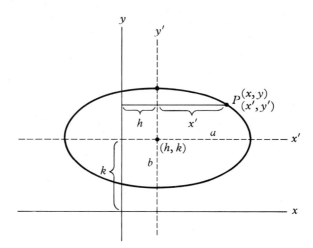

Figure 4.26

A similar generalization of Equation (4) would give

(6) $$\frac{(x-h)^2}{b^2} + \frac{(y-k)^2}{a^2} = 1, \quad a > b$$

Equation (5) is called the *standard form* for the equation of an ellipse with center at (h, k) and major axis parallel to the x axis. If the major axis is parallel to the y axis, then Equation (6) is the standard form.

In expanded form, Equation (5) is

$$b^2 x^2 + a^2 y - 2b^2 hx - 2a^2 ky + b^2 h^2 + a^2 k^2 - a^2 b^2 = 0$$

or $$Ax^2 + By^2 + Cx + Dy + E = 0$$

where $A = b^2$, $B = a^2$, $C = -2b^2 h$, $D = -2a^2 k$, and $E = b^2 h^2 + a^2 k^2 - a^2 b^2$. Note that A and B are both positive but unequal.

Exercises 4.9

1. Write equations of the following ellipses.
   a. Major axis 10, foci $(\pm 3, 0)$
   b. Minor axis 4, foci $(0, \pm 2)$
   c. Minor axis 6, vertices $(\pm 5, 0)$
   d. Center at $(0, 0)$, a vertex at $(5, 0)$, and passing through $(4, 2)$

2. Sketch each of the following ellipses.

   a. $\dfrac{x^2}{16} + \dfrac{y^2}{4} = 1$
   b. $\dfrac{x^2}{16} + \dfrac{y^2}{49} = 1$

c. $4x^2 + 9y^2 = 36$ d. $25x^2 + 9y^2 = 225$

e. $\dfrac{(x-2)^2}{25} + \dfrac{(y+1)^2}{9} = 1$ f. $\dfrac{(x+2)^2}{4} + \dfrac{(y+1)^2}{16} = 1$

3. Given the two points $F'(1, 3)$ and $F(7, 3)$, find an equation of the set of all points P such that $PF' + PF = 10$.

4. An arch is the upper half of an ellipse with major axis horizontal and 100 feet long. The arch is 30 feet high in the middle. Find the height of the arch 20 feet from either end.

5. An ellipse has its center at $(-5, -3)$, its major axis is horizontal, and it is tangent to the x and y axes. Write an equation for it.

6. Explain how to use a loop of string and two pins to draw an ellipse if its equation is known. What length loop is needed and where do we place the pins to draw the ellipse $9x^2 + 25y^2 = 225$?

7. **a.** Show how to draw the ellipse $b^2x^2 + a^2y^2 = a^2b^2$ using two pins and a piece of string $2a$ units long.

   b. Using pins 8 inches apart and a loop of string whose total length is 18 inches, draw an ellipse and find an equation for it.

8. An ellipse is drawn with string 400 centimeters long whose pins are 24 centimeters apart. Find as simple an equation as you can for this ellipse.

9. Find how long a loop of string should be used to lay out an elliptical flower bed 20 feet long and 16 feet wide; how far apart should the fixed pins or stakes be?

10. Write each of the following equations in standard form, and sketch its graph.

    a. $x^2 + 4y^2 - 2x - 24y + 21 = 0$ **b.** $9x^2 + 4y^2 + 54x - 32y + 1 = 0$
    c. $x^2 + 4y^2 - 48y - 52 = 0$ **d.** $16x^2 + 25y^2 + 50y = 311$

11. Assuming that the area of the ellipse $b^2x^2 + a^2y^2 = a^2b^2$ is πab, find the area of the flower bed in Exercise 9.

4.10 HYPERBOLAS AND THEIR EQUATIONS

We come now to the fourth and last special set of points mentioned in Section 4.4, the hyperbola. It is defined as follows.

DEFINITION 4.11. If F and F' are two fixed points in a plane and $2a$, $0 < 2a < FF'$ is a given positive number, then the set of all points P in the plane such that $|PF - PF'| = 2a$ is called a *hyperbola*. The points F and F' are called *foci* of the hyperbola, and the midpoint of the line segment joining F and F' is called the *center* of the hyperbola.

SEC. 4.10 Hyperbolas and Their Equations

Consider the hyperbola whose foci are $F(c, 0)$ and $F'(-c, 0)$. A point (x, y) is a point of the hyperbola if and only if

$$|F'P - FP| = 2a$$

or

$$|\sqrt{(x+c)^2 + y^2} - \sqrt{(x-c)^2 + y^2}| = 2a$$

If we rationalize this last equation as we did in the case of the ellipse, the result is

(1) $$\frac{x^2}{a^2} - \frac{y^2}{b^2} = 1$$

where $b = \sqrt{c^2 - a^2}$. Since the points F, F', and P are the vertices of a triangle and since the difference in the lengths of any two sides of a triangle is less than the length of the third side, we must have $a < c$. The hyperbola crosses the x axis at the points $V(a, 0)$ and $V'(-a, 0)$. It does not intersect the y axis. The segment VV' is called the *transverse axis*, and the points V and V' are called the *vertices*. If we solve Equation (1) for x and for y, we get

$$x = \pm \frac{a}{b}\sqrt{b^2 + y^2} \quad \text{and} \quad y = \pm \frac{b}{a}\sqrt{x^2 - a^2}$$

which shows that y can be any real number, but x^2 cannot be less than a^2. In other words, the graph extends infinitely far to the left and to the right, but no point of the hyperbola lies between the lines $x = a$ and $x = -a$.

If the center of the hyperbola is at (h, k), and the x axis is parallel to the transverse axis, then an equation of the hyperbola is

(2) $$\frac{(x-h)^2}{a^2} - \frac{(y-k)^2}{b^2} = 1$$

But if the transverse axis is parallel to the y axis, then an equation is

(3) $$\frac{(y-k)^2}{a^2} - \frac{(x-h)^2}{b^2} = 1$$

By expanding and renaming the coefficients, Equation (2) may be written in the form

$$Ax^2 + By^2 + Cx + Dy + E = 0$$

where $A \neq 0$, $B \neq 0$, and A and B are opposite in sign.

We end this section by calling attention to two very important lines associated with a hyperbola, the *asymptotes* of the hyperbola. These lines are the extended diagonals of the rectangle determined by the lines $x = a$, $x = -a$, $y = b$, and $y = -b$. See Figure 4.27. Equations for the asymptotes are

$$y = \frac{b}{a}x \quad \text{and} \quad y = -\frac{b}{a}x$$

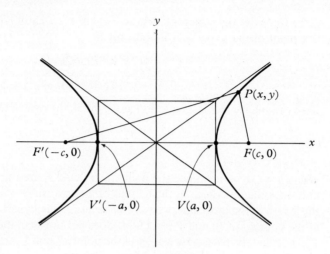

Figure 4.27

To see how the asymptotes are related to the hyperbola, consider the height of the asymptote and the height of the hyperbola in the first quadrant. For a given x, these heights are

$$y = \frac{b}{a}x \quad \text{and} \quad y = \frac{b}{a}\sqrt{x^2 - a^2}$$

These two equations show that in the first quadrant the asymptote is above the hyperbola at each point for which $x \geq a$. However, as x increases, the difference in the heights of the asymptote and the hyperbola decreases, and the difference in heights approaches zero. To see that this is true, note that

$$\frac{b}{a}x - \frac{b}{a}\sqrt{x^2 - a^2} = \frac{b}{a}(x - \sqrt{x^2 - a^2})$$

$$= \frac{b}{a}\left[\frac{x^2 - (x^2 - a^2)}{x + \sqrt{x^2 - a^2}}\right] = \frac{ab}{x + \sqrt{x^2 - a^2}}$$

Since the numerator of the fraction on the right side of the last equation is a constant, it is obvious that, as x gets larger and larger, the value of the fraction becomes smaller and smaller; that is, it approaches zero.

Example 1 Write the following equation in standard form and sketch its graph.

$$4x^2 - 9y^2 - 32x + 72y - 116 = 0$$

Solution: First, rewrite the equation in the form

$$4(x^2 - 8x) - 9(y^2 - 8y) = 116$$

SEC. 4.10 Hyperbolas and Their Equations

Next, complete the square in x in the first parentheses, and complete the square in y inside the second parentheses, and obtain

$$4(x^2 - 8x + 16) - 9(y^2 - 8y + 16) = 116 + 64 - 144$$

Simplifying,

$$4(x - 4)^2 - 9(y - 4)^2 = 36$$

or

$$\frac{(x - 4)^2}{9} - \frac{(y - 4)^2}{4} = 1$$

the required standard form. Comparing this equation with Equation (2) shows that our equation is that of a hyperbola with center at (4, 4), vertices at (1, 4) and (7, 4), $a = 3$, and $b = 2$. The graph is shown in Figure 4.28.

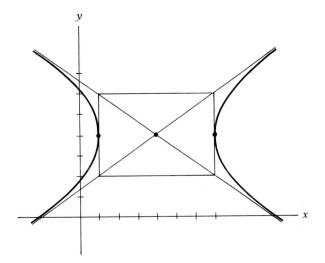

Figure 4.28

Exercises 4.10

1. Write equations of the following hyperbolas.
   a. Transverse axis 6, foci (± 5, 0)   **b.** Vertices (0, ± 2), foci (0, ± 4)
   c. Vertices (0, ± 5), foci (0, ± 6)
   d. Transverse axis 2, foci (1, -1) and (7, -1)

2. Sketch the graph of each of the following hyperbolas and show its asymptotes.
   a. $x^2 - y^2 = 4$ **b.** $y^2 - x^2 = 1$ **c.** $\dfrac{x^2}{4} - \dfrac{y^2}{9} = 1$ **d.** $\dfrac{y^2}{16} - \dfrac{x^2}{9} = 1$

4.11 GRAPHS OF LINEAR INEQUALITIES IN TWO VARIABLES

We conclude this chapter with a discussion of a method of plotting the graph of the solution set of a linear inequality in two variables. An inequality of the form

$$ax + by + c > 0 \quad \text{or} \quad ax + by + c < 0$$

where a, b, and c are arbitrary but fixed real numbers and a and b are not both zero, is called a linear inequality in x and y. Some examples of linear inequalities are

$$2x + 3y > 5$$

$$x + 2y < 3x - 4$$

$$x + 7 > 0$$

$$y - 2 < 0$$

$$2y < 5x$$

A solution of a linear inequality in x and y is an ordered pair (x_1, y_1) of real numbers such that, if x_1 is substituted for x and if y_1 is substituted for y in the inequality, the resulting numerical inequality is a true statement. For example, the pair (3, 1) is a solution of $2x + 3y > 5$, because the inequality $6 + 3 > 5$ is a true statement. The set consisting of all the solutions of a given inequality is called the *solution set* of that inequality. The graph of a linear inequality is the set of points consisting of all the points whose coordinates are elements of the solution set of the inequality. In plotting the graphs of linear inequalities, we shall find the following theorem useful.

THEOREM 4.3. *Let $f(x, y)$ denote the value of the linear form $ax + by + c$ at (x, y), that is, $f(x, y) = ax + by + c$, and let L be the graph of $ax + by + c = 0$. If P and Q are any two points, neither of which is on L, that can be joined by a straight line segment having no point in common with L, then the values of $f(x, y)$ at P and Q have the same sign.*

A rigorous proof of this theorem is beyond the scope of this text; however, the following intuitive argument may make it seem plausible. It is understood that by a straight line segment joining P and Q we mean only the points on the line determined by P and Q that lie between P and Q. Now consider the points $P(x_1, y_1)$ and $Q(x_2, y_2)$ and the straight line segment joining them. Assume that the line segment PQ has no point in common with L. Suppose that $f(x_1, y_1)$ and $f(x_2, y_2)$ are opposite in sign and, for the sake of the argument, that $f(x_1, y_1)$ is positive. If $f(x, y)$ is positive at P and negative at Q and if the point (x, y) moves from P to Q along the segment Q, then $f(x, y)$ must be zero at some point R between P and Q. But every point at which $f(x, y)$ is zero is a point on L, and hence R would be a point on L. This contradicts our assumption that the

SEC. 4.11 Graphs of Linear Inequalities in Two Variables

segment PQ has no point in common with L. We conclude therefore that the supposition that $f(x_1, x_2)$ and $f(x_2, y_2)$ are opposite in sign is false. We shall assume the validity of the theorem.

Every straight line L in the xy plane divides the plane into portions, called *half-planes*. It follows from Theorem 4.3 that if the point $P(x_1, y_1)$ is any point in one of the half-planes determined by the line L, where L is the graph of the linear equation $ax + by + c = 0$, and if $ax_1 + by_1 + c > 0$ at P, then $ax + by + c > 0$ at every point in the same half-plane with P, and $ax + by + c < 0$ at every point in the other half-plane. Thus, if we wish to plot the graph of the inequality $ax + by + c > 0$, we first draw L, the graph of $ax + by + c = 0$, and then check the value of $ax + by + c$ at any convenient point in either half-plane determined by L. This will determine the half-plane in which $ax + by + c > 0$. The graph is this half-plane, which is indicated by the shading in Figure 4.29 for $a = 1$, $b = 0$, and $c = 2$. Some examples will show how such a solution is arrived at.

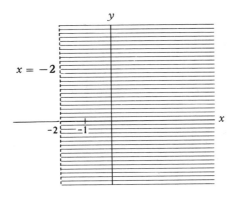

Figure 4.29

Example 1 Draw the graph of the solution set of $x > -2$.

Solution: We draw the line $x = -2$ dashed to show that it is not part of the graph. Our graph then consists of the right half-plane, which we indicate by shading in Figure 4.29.

Example 2 Draw the graph of the solution set of $2x - 3y + 6 < 0$.

Solution: We plot the graph of $2x - 3y + 6 = 0$. The point $(0, 0)$ is in the lower half-plane determined by the line $2x - 3y + 6 = 0$, and, since $0 - 0 + 6 < 0$ is false, we know that the desired graph is the upper half-plane, which we shade as shown in Figure 4.30. Again, the line $2x - 3y + 6 = 0$ is broken to show that it is not part of the graph.

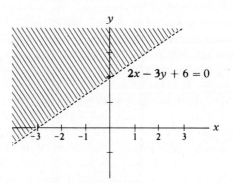

Figure 4.30

Example 3 | Draw the graph of the solution set of $3x + y + 3 \geq 0$.

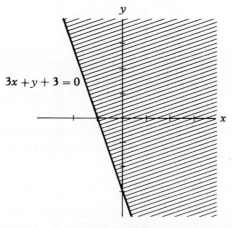

Figure 4.31

Solution: The line $3x + y + 3 = 0$ passes through the points $(-1, 0)$ and $(0, -3)$. We draw this line in heavily to indicate that it is part of the graph. The point $(0, 0)$ is in the upper half-plane, and, since $0 + 0 + 3 \geq 0$ is true, $(0, 0)$ belongs to the solution set; therefore, the rest of the graph is also in the upper half-plane, as shown in Figure 4.31.

Exercises 4.11

Draw the graph of the solution set of each of the following.

1. $y + 2 \geq 0$
2. $x - 1 < 0$
3. $2x + y - 4 > 0$
4. $x - y + 1 \leq 0$
5. $5x + 4y \leq 20$
6. $3x - y > 6$
7. $|2y + 1| < 3$
8. $|x - 1| < 3$
9. $2x - 3y < 0$
10. $5x - 3y + 10 < 0$

5

Some Basic Concepts of Elementary Calculus

5.1 AVERAGE RATES

It was pointed out in Chapter 4 that elementary calculus is concerned with two major problems, the first of which is the rate problem. The purpose of this chapter is to explain precisely what is meant by the rate problem, that is, what is meant by the rate at which one of two related physical quantities is changing per unit change in the other at some particular value of the latter. Actually, we need to consider two types of rates of change. The first type is called an *average rate of change*, and the second type is called an *instantaneous rate of change*, or, for brevity, *average rate* and *instantaneous rate*. Although in elementary calculus we are primarily concerned with instantaneous rates, we cannot avoid, or ignore, the concept of an average rate because the definition of an instantaneous rate is dependent on the definition of an average rate. Hence we first turn our attention to the meaning of an average rate. However, before giving a formal definition, let us look at some special cases of average rates which should help to make the definition more meaningful.

CASE 1: The notion of an average rate is not completely new to us for most can recall working problems in high school mathematics courses which involved the physical quantities, *distance*, *rate*, and *time*. In fact, you probably remember using the formula

$$\text{Rate} = \frac{\text{distance}}{\text{time}}$$

The rate found by using this formula is a typical example of an average rate. For instance, if a car is driven 200 miles in 5 hours, we say it averaged 40 miles per hour, or that the average velocity, or average speed, over the 5-hour interval of time was 40 miles per hour. To obtain this answer, we used the formula just mentioned above.

CASE 2: Suppose that a large water tank is equipped with a gauge from which you can read the number of gallons of water in the tank at any time and a small faucet near the bottom of the tank which can be opened to let the water run out. Suppose also that just before the faucet is opened the gauge shows that this tank contains 100 gallons of water. The faucet is then opened and 10 minutes later the gauge shows only 80 gallons of water remaining in the tank. We say that the volume of water in the tank decreased at the average rate of 2 gallons per minute during the 10-minute period.

CASE 3: If the length x of the side of a square is increased from $x = 4$ inches to $x = 6$ inches, the area A of the square will increase from $A = 16$ square inches to $A = 36$ square inches, producing a total change in A of 20 square inches. We divide this increase by the increase in length, $6 - 4 = 2$ inches, and find that the average rate of increase of A from $x = 4$ to $x = 6$ is 10 square inches per inch increase in x.

CASE 4: A company that makes and sells a certain type of small radio found that by selling the radios at a certain fixed price they could make and sell 500 radios per week at a total weekly profit of \$2,000. An experiment showed that by reducing the selling price slightly they could make and sell 600 radios per week at a total weekly profit of \$2,700. By doing this, the company's weekly profit was increased at an average rate of \$7 per radio increase in their weekly production.

A careful study of the four examples of average rates discussed above reveals certain important features shared by all four such rates. For example, each case involved basically two related physical quantities, which, named in the order discussed, were (1) distance and time, (2) volume and time, (3) area and length, and (4) weekly profit and the number of radios made and sold. If we examine each of these related pairs of physical quantities, we see that one of the quantities is obviously a function of the other, and in each case we found the average rate of change of the value of a function over an interval of values of its independent variable. This implies that whenever we speak of an average rate we always mean the average rate of change of the value of *some function* over *some interval* of values of its independent variable.

The above discussion suggests the following formal definition of an average rate of change of a function over an interval.

DEFINITION 5.1. Let the equation $y = f(x)$ define a function f over an interval (a, b) of real numbers, and let x_1 and x_2 be any two distinct numbers, that is, $x_1 \neq x_2$, belonging to the interval (a, b). We define the average rate of

SEC. 5.1 Average Rates

change of y, the value of f, over the interval from $x = x_1$ to $x = x_2$ to be the value of y at $x = x_2$ minus the value of y at $x = x_1$ divided by $x_2 - x_1$. If we denote the average rate by $\bar{R}$ and let $y_1 = f(x_1)$ and $y_2 = f(x_2)$, our definition can be expressed in the convenient form

(1) $$\bar{R} = \frac{y_2 - y_1}{x_2 - x_1} = \frac{f(x_2) - f(x_1)}{x_2 - x_1} \quad \text{units/unit}$$

The symbol "units/unit" should be read "units per unit." This is a very common and convenient form of notation and will be used hereafter for expressing the units involved in a rate problem. Of course, in a particular problem where specific units are indicated, these units will be used. In Case 1 above, we would express the average velocity in the units "mi/hr" and read "miles per hour." Similarly, in Case 2 we would use the symbol "gal/min" which is read "gallons per minute." If no specific units are indicated in a given problem, we shall use the symbol units/unit as in Equation (1) above.

Example 1

For the function defined by the equation $f(x) = x^2 + 4$, find the average rate of change, $\bar{R}$, of $f(x)$ over the interval from $x = 2$ to $x = 6$.

Solution: Here $x_1 = 2$ and $x_2 = 6$. Hence

$$f(x_2) = f(6) = (6)^2 + 4 = 36 + 4 = 40$$
$$f(x_1) = f(2) = (2)^2 + 4 = 4 + 4 = 8$$

Now, using our definition, that is, Equation (1), we have

$$\bar{R} = \frac{f(x_2) - f(x_1)}{x_2 - x_1} = \frac{40 - 8}{6 - 2} = \frac{32}{4} = 8 \text{ units/unit}$$

If the two physical quantities involved in a rate problem are position and time, it is common practice to use the letter s (space) to represent the position and the letter t to represent time. We call the average rate of change of position per unit of time *average velocity* and denote it by the symbol $\bar{v}$. Suppose an object is moving along a straight line in such a way that its position relative to a certain fixed point t units of time after starting is $s(t)$. The average velocity of the object from $t = t_1$ to $t = t_2$ is given by the formula

(2) $$\bar{v} = \frac{s(t_2) - s(t_1)}{t_2 - t_1} \quad \text{units of distance/unit of time}$$

Equation (2) could be taken as the definition of average velocity.

Example 2

A ball, after being hit sharply, rolled up an inclined plane, its position s feet from the starting point t seconds after being hit being given by $s(t) = 60t - 3t^2$. Find the average velocity of the ball from $t = 3$ to $t = 8$.

Solution: Here $t_1 = 3$ and $t_2 = 8$. Hence

$$s(t_2) = s(8) = 60(8) - 3(8)^2 = 480 - 192 = 288$$

$$s(t_1) = s(3) = 60(3) - 3(3)^2 = 180 - 27 = 153$$

Using Formula (2), we get

$$\bar{v} = \frac{s(t_2) - s(t_1)}{t_2 - t_1} = \frac{288 - 153}{8 - 3} = \frac{135}{5} = 27 \text{ ft/sec}$$

A very important special case of an average rate is the following. Let $y = f(x)$ be the equation of some curve, and let (x_1, y_1) and (x_2, y_2) be two points on the curve. This means that $y_1 = f(x_1)$ and $y_2 = f(x_2)$. The average rate of change of y per unit change in x over the interval from $x = x_1$ to $x = x_2$ is called the *average slope* of the curve over the given interval and will be denoted by the symbol $\bar{m}$. Obviously,

(3) $$\bar{m} = \frac{y_2 - y_1}{x_2 - x_1} = \frac{f(x_2) - f(x_1)}{x_2 - x_1} \text{ units/unit}$$

It is very important to note here that $\bar{m}$ is the slope of the line determined by the points (x_1, y_1) and (x_2, y_2). This line is called the *secant line* of the curve joining these two points on the curve.

Example 3 An equation of a curve is $y = x^2 - 5x + 7$. Find the average slope $\bar{m}$ of this curve over the interval from $x = 2$ to $x = 5$.

Solution: We use Equation (3) with $f(x) = x^2 - 5x + 7$, $x_1 = 2$, and $x_2 = 5$. We find

$$f(x_2) = f(5) = (5)^2 - 5(5) + 7 = 7$$

$$f(x_1) = f(2) = (2)^2 - 5(2) + 7 = 1$$

Hence $$\bar{m} = \frac{7 - 1}{5 - 2} = \frac{6}{3} = 2 \text{ units/unit}$$

The units involved here are, of course, the same as the units in which x and y are measured and usually are not specified.

An application of the concept of an average rate which is of interest to students in Economics has to do with the average rate of change of the cost of manufacturing a product when the number of units of the item produced is changed. For example, the total cost C (dollars) of manufacturing a product is often found to depend on the number of units n produced; that is, C is a function of n, $C(n)$. If the number produced is changed from $n = n_1$ to $n = n_2$, the average rate of change of the cost per unit change in the number produced is called the *average marginal cost*, denoted by $\bar{M}_C$, and is given by the equation

$$\bar{M}_C = \frac{C(n_2) - C(n_1)}{n_2 - n_1}$$

SEC. 5.1 Average Rates

Example 4 If the total cost C (dollars) of producing n units of a certain product is given by
$$C(n) = 2n^2$$
find the average marginal cost when the number of units produced is increased from $n = 50$ to $n = 80$.

Solution: Here $n_1 = 50$ and $n_2 = 80$, and
$$C(n_2) = 2(80)^2 = 12{,}800$$
$$C(n_1) = 2(50)^2 = 5{,}000$$

Therefore,
$$\bar{M}_C = \frac{C(n_2) - C(n_1)}{n_2 - n_1} = \frac{12{,}800 - 5{,}000}{80 - 50} = \$260/\text{unit}$$

Let us next consider an example of finding the average rate of change of a function over a given interval when the function is defined by means of a graph.

Example 5 The graph shown in Figure 4.3 exhibits the temperature $T°$ Celsius of a certain liquid after t minutes of cooling. Approximate the average rate at which the temperature is changing over the interval from $t = 3$ to $t = 7$.

Solution: Let T_1 be the value of T at $t_1 = 3$ and T_2 be the value of T at $t_2 = 7$. Reading the values of T_1 and T_2 from the graph, we estimate that $T_1 = 67$ and $T_2 = 51$. Hence
$$\bar{R} = \frac{T_2 - T_1}{t_2 - t_1} = \frac{51 - 67}{7 - 3} = -4 \text{ deg/min}$$

The minus sign indicates that the temperature of the liquid is decreasing over the interval from $t = 3$ to $t = 7$.

Remark: When a function is defined by a graph and graphical interpolation has to be used to find values of the function at particular values of the independent variable, it is important to understand that the accuracy of such interpolated values depends on (1) how accurately the graph is drawn and (2) how accurately we can read the graph. Thus, at the very best, an interpolated result is an approximation. However, if care has been exercised in drawing and reading the graph, the accuracy is often sufficient for most purposes.

Summary. In this section we gave a formal definition of the general concept of the average rate of change of a function over a given interval of values of the independent variable involved. We then used the definition to calculate the average rates of change of some specific functions which represented particular physical quantities. In fact, to certain of these average rates we attached a special name and a special symbol in keeping with common practice and the particular physical quantities involved. For example, if the two related physical quantities were *distance* and *time*, we called the average rate *average velocity* and denoted it by the special symbol $\bar{v}$. The average rate of change of the height y of a curve over an interval was given the special name *average slope of the curve* and denoted by the special symbol $\bar{m}$. The average rate of change of the cost of producing n units of a product due to a change in the number of units produced was given the name *average marginal cost* and denoted by the special symbol $\bar{M}_C$, the name and symbol commonly used by economists in discussing such rates. The practice of using these special names and symbols, and in general using a special name and symbol to denote the average rate of change of a particular function, will be followed hereafter whenever and wherever to do so will add meaning to such rates. The student should learn these special names and symbols when they are introduced in the text and should use them in working problems if they apply.

Exercises 5.1

1. A tourist stopped for gas at 1:00 P.M. and noted that his odometer registered 41,284.6 miles. He stopped again at 5:00 P.M. and at this time his odometer registered 41,478.2 miles. What was his average speed from 1:00 P.M. to 5:00 P.M.?

2. Farmer Jones harvested 1,875 bushels of corn from 500 acres. What was the average yield per acre?

3. Find the average rate of change, $\bar{R}$, of each of the following functions over the indicated interval.
   a. $f(x) = x^2 + 6x$ from $x = 2$ to $x = 4$
   b. $f(x) = 4x + 7$ from $x = 3$ to $x = 7$
   c. $f(x) = 4x^2$ from $x = 1$ to $x = 1.2$
   d. $f(x) = 5 - 3x$ from $x = 2$ to $x = 5$

4. Assuming that each of the following equations is an equation of some curve, find the average slope of each curve over the indicated interval.
   a. $y = 5x - x^2$ from $x = 2$ to $x = 4$
   b. $y = x^3 - 2x + 5$ from $x = -1$ to $x = 2$
   c. $y = \dfrac{12}{x}$ from $x = 2$ to $x = 6$
   d. $y = 3x^2 - 7x + 5$ from $x = x_1$ to $x = x_2$ ANSWER: $\bar{m} = 3(x_2 + x_1) - 7$
   e. $y = \dfrac{x}{x+1}$ from $x = x_1$ to $x = x_2$ ANSWER: $\bar{m} = \dfrac{1}{(x_1 + 1)(x_2 + 1)}$

5. The distance, s feet, fallen by an object in t seconds is given by $s(t) = 16t^2$. Find the average velocity $\bar{v}$:
   a. During the first 2 seconds   b. During the first 3 seconds
   c. During the third second, that is, from $t = 2$ to $t = 3$
   d. How long will it take the object to fall 256 feet?

SEC. 5.1   Average Rates   265

6. The cost C (dollars) of producing n units of a certain commodity is given by $C(n) = 10n^2 - 14n + 130$. Find the average marginal cost when n increases from $n = 2$ to $n = 5$.

7. A train traveled 100 miles in 3 hours and 20 minutes. What was its average velocity?

8. An equation of a curve is $y = \sqrt{25 - x^2}$. Find the average slope of the curve over the interval from $x = 3$ to $x = 4$.

9. On a 120-mile trip Mr. Jones averaged 30 miles per hour, but on his return trip home over the same route he averaged 40 miles per hour. What was his average velocity over the 240-mile round trip?

10. A tank initially empty is filled with water by opening a valve on a small inlet pipe. The volume V (liters) of water in the tank t minutes after the valve is opened is $V(t) = 10t^2 + 20t$. What is the average rate of increase in V from $t = 1$ to $t = 3$?

11. A straight stretch of a certain highway is 2 kilometers in length. If the first kilometer is driven at an average velocity of 30 kilometers per hour, at what average velocity must the second kilometer be driven to average, over the 2-kilometer stretch,
    a. 40 km/hr   **b.** 50 km/hr   **c.** 60 km/hr

12. If $y = 10x^2$, find the exact average rate of change of y per unit change in x over the interval from $x = 3$ to $x = 3.0002$.

13. A circular metal plate is heated uniformly, causing the radius to increase from $r = 4$ centimeters to $r = 5$ centimeters. Find the average rate at which the area of the plate increases over this interval.

14. Table 1 expresses y as a function of x. Find the average rate of change of y per unit increase in x from $x = 2$ to $x = 6$, also over the interval from $x = 4$ to $x = 8$, and from $x = 2$ to $x = 10$. What do your results suggest about the graph of the function?

15. An object travels along a straight line. Its distance (y kilometers) from a fixed point after t seconds is given in Table 2 for various values of t. Plot the graph. Find the average velocity of the object during the interval $t = 6$ to $t = 12$. Also, approximate the average velocity during the interval from $t = 3$ to $t = 7$.

16. Table 3 expresses y as a function of x. Assuming x is a continuous variable, plot the graph. Approximate the average rate of change of y per unit increase in x from $x = 2$ to $x = 5$.

TABLE 1		TABLE 2		TABLE 3	
x	y	t	y	x	y
0	2	0	0	0	0
2	8	2	72	1	2
4	14	4	128	4	4
6	20	6	168	9	6
8	26	8	192	16	8
10	32	10	200	25	10
		12	192	36	12

5.2 AVERAGE RATES OVER A FLEXIBLE INTERVAL

In Section 5.1 a formal definition was given of the average rate of change of the value of a function over an interval of values of the independent variable. To obtain this rate, we divided the difference in the values of the function at the endpoints of the interval by the length of the interval. This gives, as the name implies, the average rate for the whole interval but very little information about the behavior of the function at specific points of the interval. For instance, suppose you drive your car a distance of 80 miles in exactly 2 hours. The average velocity of the car over this 2-hour interval of time is 40 miles per hour. This does not mean that you drive the 80 miles at the constant rate of 40 miles per hour. In fact, part of the time you may have been driving much more slowly than 40 miles per hour and part of the time much faster. This brings us around to the second type of rate mentioned in the last section, an *instantaneous rate*, the rate at some instant or point. As you are driving your car along the highway you look at the speedometer and see the indicator pointing to the number 40 on the dial. You say, "I am driving 40 miles per hour at this instant." You press down on the accelerator and your car starts moving faster and faster. The indicator on the speedometer moves around to 45, then to 50, and then to 60. You observe as the indicator passes each of the numbers 45, 50, and 55 on the dial that you are driving 45, then 50, and then 55 miles per hour. Are these velocities average velocities? Hardly, for an average velocity implies that an interval of time was involved, which is certainly not the case here, for the indicator did not stop at 45 or 50 or 55 but was in continuous motion. When you say, "I am driving at 45 miles per hour at this instant," you do not mean that you have driven 45 miles per hour in the last hour or that your average velocity for the last 10 minutes was 45 miles per hour or even that this was your average velocity over the last 2 minutes or the last minute or the last 10 seconds. You mean that your velocity at the instant the indicator on the speedometer passed 45 on the dial was 45 miles per hour. That is, your velocity at the instant, your instantaneous velocity, was 45 miles per hour. This suggests that the speedometer on a car registers what we shall call *instantaneous velocities*. An instantaneous velocity is therefore a typical example of what we mean by an instantaneous rate. For the time being we shall content ourselves with this intuitive notion of an instantaneous rate and postpone a formal definition to a later section.

It seems logical that the instantaneous velocity of a moving object at the instant $t = t_1$ could be approximated by an average velocity calculated over a short interval of time containing t_1 and that the shorter the interval of time the better the approximation, as long as the interval contains t_1. One easy way to guarantee that such an interval contains t_1 is to take t_1 as one end of the interval. Stated in more general terms, assuming that $y = f(x)$ defines a function on an interval (a, b) of real numbers, the instantaneous rate of change, R, of y at $x = x_1$, $a < x_1 < b$, can be approximated by calculating the average rate of change of y

SEC. 5.2 Average Rates Over a Flexible Interval

over an interval containing x_1. Also, the shorter the interval used containing x_1, the better the approximation. Experience shows that the best procedure for obtaining such approximations is to use a flexible interval, say from $x = x_1$ to $x = x_1 + h$ and to express the average rate in terms of h. Then by assigning particular values to h we can make our interval as short as we please and hence can obtain as accurate an approximation as is desired to the instantaneous rate of change of y at $x = x_1$. More importantly, however, the use of a flexible interval in calculating average rates will lead us in a very simple and natural way to a formal definition of an instantaneous rate, as is illustrated in the following example.

Example 1   Given the function $y = f(x) = x^2 + 5x - 7$, find the average rate of change, $\bar{R}$, of y over the interval from $x = 2$ to $x = 2 + h$. Use the result to make a table showing values of $\bar{R}$ for $h = 0.1$, $h = 0.001$, $h = 0.000001$, and $h = 0.0000000001$.

Solution: The average rate of change, $\bar{R}$, of $f(x)$ over the interval from $x = 2$ to $x = 2 + h$ (where $h \neq 0$) is given by

$$\bar{R} = \frac{f(2+h) - f(2)}{h}$$

$$\begin{aligned} f(2+h) &= (2+h)^2 + 5(2+h) - 7 \\ &= 4 + 4h + h^2 + 10 + 5h - 7 \\ f(2) &= 4 \qquad\qquad\qquad\quad + 10 \qquad -7 \end{aligned}$$

$$\begin{aligned} f(2+h) - f(2) &= 4h + h^2 + 5h \\ &= 9h + h^2 \end{aligned}$$

and hence

$$\bar{R} = \frac{9h + h^2}{h} = 9 + h, \qquad h \neq 0$$

The required table is

h	0.1	0.001	0.000001	0.0000000001
$\bar{R}$	9.1	9.001	9.000001	9.0000000001

In the above example we found the value of $\bar{R}$, the average rate of change of $f(x)$ over the interval from $x = 2$ to $x = 2 + h$, to be $\bar{R} = 9 + h$. Note that the equation $\bar{R} = 9 + h$ defines $\bar{R}$ as a function of h. We observe that by taking values of h nearer and nearer to zero the value of $\bar{R}$ comes closer and closer to 9. In fact, by taking a value of h sufficiently near zero we can obtain a value of $\bar{R}$ as close to 9 as we please. We say that $\bar{R}$ approaches 9 as h approaches zero, and that 9 is the limit approached by $\bar{R}$, or that 9 is the limit of $\bar{R}$, as h approaches zero. A convenient notation for expressing this is

(1) $$\lim_{h \to 0} \bar{R} = 9$$

which is read "the limit of $\bar{R}$ as h approaches zero is 9." We call this limit, since $\bar{R}$ was calculated over the interval from $x = 2$ to $x = 2 + h$, the instantaneous rate of change of $f(x)$ at $x = 2$ or at the point $x = 2$. If we use the letter R to represent this instantaneous rate, and the notation of Equation (1), we can summarize the above discussion by writing

(2) $$R = \lim_{h \to 0} \bar{R} = \lim_{h \to 0} (9 + h) = 9$$

The discussion leading to Equation (2) was concerned with the particular function $f(x) = x^2 + 5x - 7$ and the specific value of x, $x = 2$. We can generalize this discussion in the following way. Let $y = f(x)$ define a function on the interval (a, b), and let $\bar{R}$ denote the average rate of change of y over the flexible interval from $x = x_1$ to $x = x_1 + h$, where x_1 and $x_1 + h$ belong to the interval (a, b). The instantaneous rate of change, R, of y at $x = x_1$ is given by

$$R = \lim_{h \to 0} \bar{R}$$

if $\bar{R}$ approaches a limit as h approaches zero. It should be pointed out that, strictly speaking, $\bar{R}$ will, in general, be a function of x_1 and h, but in the process of evaluating the $\lim_{h \to 0} \bar{R}$, x_1 is considered a constant as h approaches zero.

Example 2 If $y = f(x) = 3x^2 - 6x + 4$, find the instantaneous rate of change, R, of y at $x = 2$.

Solution: First we find $\bar{R}$, the average rate of change of y, over the interval from $x = x_1$ to $x = x_1 + h$. We use the formula

$$\bar{R} = \frac{f(x_1 + h) - f(x_1)}{h}$$

$$f(x_1 + h) = 3(x_1 + h)^2 - 6(x_1 + h) + 4$$
$$= 3x_1^2 + 6x_1 h + 3h^2 - 6x_1 - 6h + 4$$
$$f(x_1) = 3x_1^2 \qquad\qquad\qquad - 6x_1 \qquad + 4$$

Subtracting, we get

$$f(x_1 + h) - f(x_1) = 6x_1 h + 3h^2 - 6h$$
$$= 6x_1 h - 6h + 3h^2$$

Hence

$$\bar{R} = \frac{f(x_1 + h) - f(x_1)}{h} = \frac{6x_1 h - 6h + 3h^2}{h} = 6x_1 - 6 + 3h, \quad \text{if } h \neq 0$$

It follows from Equation (2) that

$$R = \lim_{h \to 0} \bar{R} = \lim_{h \to 0} (6x_1 - 6 + 3h) = 6x_1 - 6$$

This result can now be used to find the instantaneous rate of change of y at any permissible value of x_1. In particular, the instantaneous rate of change of y at $x = 2$ is $(6)(2) - 6 = 6$ units/unit.

To understand the fundamental ideas of calculus, we must understand the limit concept. In the next section we shall study this concept in some detail.

Exercises 5.2

1. If $f(x) = 7 - 5x - 3x^2$, find the average rate of change of $f(x)$ over the interval from $x = 1$ to $x = 1 + h$. Use your result to find $R = \lim_{h \to 0} \bar{R}$, the instantaneous rate of change of $f(x)$ at $x = 1$.

2. If $f(x) = x^3 - 4x + 6$, find
   a. $\bar{R}$ over the interval from $x = 2$ to $x = 2 + h$.   *book is wrong. answer = 8*
   b. $R = \lim_{h \to 0} \bar{R}$ at $x = 2$.

3. A function is defined by the equation $y = f(x) = 4x^2 + 5x - 7$. Find $\bar{R}$ over the interval from $x = x_1$ to $x = x_1 + h$. Use your result to find $R = \lim_{h \to 0} \bar{R}$ at
   a. $x_1 = -1$ b. $x_1 = 0$ c. $x_1 = 2$ d. $x_1 = 5$

5.3 THE LIMIT CONCEPT AND SOME LIMIT THEOREMS

In Section 5.2 we received an intuitive notion of the meaning of such statements as

(a) $\lim_{h \to 0} (9 + h) = 9$

(b) $\lim_{h \to 0} (6x_1 - 6 + 3h) = 6x_1 - 6$

(c) $\lim_{h \to 0} (2h - 3) = -3$

No formal explanation is needed to understand the meaning in such simple cases as those above. But, since this limit process, or limit concept, is such an essential basic tool in elementary calculus, it is necessary to consider it in more detail and in a more general sense. Looking again at Examples 1 and 2 of the last section, we are reminded that in each case, the average rate, $\bar{R}$, is a function of h, provided in Example 2 we accept x_1 as representing a specific value of x. This suggests that the limit process as used in these examples might be applied to functions in general. To do this, we must have a clear and explicit definition of what is meant by a statement such as "$f(x)$ approaches the real number L as x approaches the real number c," or, in limit notation, what is meant by

$$\lim_{x \to c} f(x) = L$$

One form of such a definition is the following.

DEFINITION 5.2. Let f be a function of an independent variable x defined for all values of x in the open interval (a, b) of real numbers, except perhaps for the real number c, which is in (a, b). Recall that the open interval (a, b) of real numbers consists of all the real numbers lying between a and b but including neither a nor b. Whether f has a value at $x = c$ is immaterial. If there exists a real number L such that the value of $f(x)$ can be made as close to L as we please by choosing values of x sufficiently close to c, but different from c, then we say that $f(x)$ approaches L as a limit as x approaches c or, stated symbolically,

$$\lim_{x \to c} f(x) = L$$

It is important to emphasize the following points in regard to the definition just given.

1. It makes no difference whether f has or does not have a value at $x = c$. That is, the definition is not concerned with the value of $f(x)$ at $x = c$ but is very much concerned with the values of $f(x)$ for all other values sufficiently near c. For example, suppose that the function f is defined by the equation

$$f(x) = \frac{x^2 - 1}{x - 1}$$

Here $f(x)$ has a value for all values of x except $x = 1$. According to Definition 5.2, does $f(x)$ approach a limit as x approaches 1, and if so, what limit? That $f(x)$ does have a limit can be seen as follows.

$$f(x) = \frac{x^2 - 1}{x - 1} = \frac{(x + 1)(x - 1)}{x - 1} = x + 1 \quad \text{if } x \neq 1$$

This tells us $(x^2 - 1)/(x - 1)$ and $(x + 1)$ have the same value for all values of x except $x = 1$. When $x = 1$ the fraction $(x^2 - 1)/(x - 1)$ has no value, but $(x + 1)$ has the value 2. Now since the existence of the limit of $f(x)$ as x approaches 1 does not depend upon the value of $f(x)$ at $x = 1$, it follows that $(x^2 - 1)/(x - 1)$ will have a limit as x approaches 1 if $(x + 1)$ has such a limit, and furthermore these limits will be the same. We observe that by taking values of x closer and closer to 1 we can make the value of $(x + 1)$ come as close to 2 as we please, that is, $\lim_{x \to 1} (x + 1) = 2$. Hence

$$\lim_{x \to 1} \frac{x^2 - 1}{x - 1} = \lim_{x \to 1} (x + 1) = 2$$

We conclude, therefore, that $f(x)$ does approach a limit as x approaches 1 and that the limit is 2.

2. In order to be entirely precise, we need to describe more clearly what is meant by the statement "the value of $f(x)$ may be made as close to L as we

please by taking values of x sufficiently close to c." This can be done in the following manner. Let d be any positive real number and let I represent the open interval of real numbers with L as its midpoint and having length $2d$. Hence, $I = (L - d, L + d)$, the set of all real numbers lying between $L - d$ and $L + d$. We can express the fact that $f(x)$ belongs to I by writing $L - d < f(x) < L + d$ or, more compactly, $|f(x) - L| < d$. Let k be a positive real number, and let J represent the open interval $(c - k, c + k)$, the set of all the real numbers lying between $c - k$ and $c + k$. Note that c is the midpoint of $J = (c - k, c + k)$. We express the fact that x belongs to J by writing $c - k < x < c + k$ or $|x - c| < k$. To exclude c as a possible value of x, we write $0 < |x - c| < k$. Thus to say that "the value of $f(x)$ can be made as close to L as we please by taking values of x sufficiently close to c" means that for each positive real number d there corresponds a positive real number k such that $|f(x) - L| < d$ if $0 < |x - c| < k$. We may now rewrite Definition 5.2 in precise notation as follows.

DEFINITION 5.3. If f is a function defined for all values of x in an open interval (a, b), except perhaps at $x = c$, then $f(x)$ approaches the real number L as a limit as x approaches c if and only if for each positive real number d there corresponds some positive real number k such that

$$|f(x) - L| < d \quad \text{if } 0 < |x - c| < k$$

Example 1

Suppose that $f(x) = 10 + 2x$. Let us show that $\lim_{x \to 0} f(x) = 10$. To do this, we need to show that for any given positive number d we can find a positive number k such that

$$|(10 + 2x) - 10| < d \quad \text{when } 0 < |x - 0| < k$$

Since $|(10 + 2x) - 10| = |2x| = 2|x|$ and $|x - 0| = |x|$, our problem is to find a number k such that

$$2|x| < d \quad \text{when } |x| < k$$

It follows at once that

$$2|x| < d \quad \text{if } |x| < \frac{d}{2}$$

Hence, no matter what value d may have, if we let $k = \frac{d}{2}$, we have

$$2|x| < d \quad \text{when } |x| < k$$

or

$$|f(x) - 10| < d \quad \text{when } 0 < |x - 0| < k$$

This by definition proves that $\lim_{x \to 0} f(x) = 10$.

Some Theorems Concerning Limits

It should be noted that our definition of a limit provides no method for finding the limit of a given function. It only affords a method for determining whether or not a number L is the limit of a function f. How does one find the limit of a given function? One possibility is to make an educated guess and then use Definition 5.3 to verify that your guess is correct. Such a procedure is rather laborious even where it is possible. To expedite the process of finding limits we find it convenient to make use of the following theorems, which can be proved from the definition.

THEOREM 5.1. *Limit of the constant function.* Let the function f be defined by the equation $f(x) = K$ for all x such that $-\infty < x < \infty$, where K is any real number. If c is a real number, then

$$\lim_{x \to c} f(x) = \lim_{x \to c} K = K$$

Example 2 Find $\lim_{x \to 3} f(x)$ if $f(x) = 17$.

Solution: Using Theorem 5.1, we have

$$\lim_{x \to 3} f(x) = \lim_{x \to 3} 17 = 17$$

THEOREM 5.2. *Limit of the identity function.* Let the function f be defined by the equation $f(x) = x$ for $-\infty < x < \infty$. If c is a real number, then

$$\lim_{x \to c} f(x) = \lim_{x \to c} x = c$$

Example 3 Find $\lim_{x \to 5} f(x)$ if $f(x) = x$.

Solution: $\lim_{x \to 5} f(x) = \lim_{x \to 5} x = 5$

In the following theorems we shall assume that $\lim_{x \to c} f(x) = L$, that $\lim_{x \to c} g(x) = M$, and that n is a positive integer whenever it occurs.

THEOREM 5.3. *If K is any constant, then*

$$\lim_{x \to c} Kf(x) = K \lim_{x \to c} f(x) = KL$$

COROLLARY. $\lim_{x \to c} [-f(x)] = -\lim_{x \to c} f(x) = -L$

Example 4 Find $\lim_{x \to 2} 7x$.

Solution: $\lim_{x \to 2} 7x = 7 \lim_{x \to 2} x = 7 \cdot 2 = 14$

SEC. 5.3 The Limit Concept and Some Limit Theorems

THEOREM 5.4. $\lim_{x \to c} [f(x) + g(x)] = \lim_{x \to c} f(x) + \lim_{x \to c} g(x) = L + M$

In words: The limit of the sum of two functions is the sum of the limits of the functions.

COROLLARY. $\lim_{x \to c} [f_1(x) + f_2(x) + \cdots + f_n(x)]$
$$= \lim_{x \to c} f_1(x) + \lim_{x \to c} f_2(x) + \cdots + \lim_{x \to c} f_n(x)$$

Example 5 Find $\lim_{x \to 3} (4x + 7)$.

Solution: Using Theorems 5.1 through 5.4, we get
$$\lim_{x \to 3} (4x + 7) = \lim_{x \to 3} 4x + \lim_{x \to 3} 7$$
$$= 4 \lim_{x \to 3} x + \lim_{x \to 3} 7$$
$$= 4 \cdot 3 + 7 = 12 + 7 = 19$$

THEOREM 5.5. $\lim_{x \to c} [f(x) \cdot g(x)] = \lim_{x \to c} f(x) \cdot \lim_{x \to c} g(x) = LM$

In words: The limit of the product of two functions equals the product of the limits of the functions.

Example 6 Find $\lim_{x \to c} x^2$.

Solution: Let $f(x) = x$; then $f(x) \cdot f(x) = x \cdot x = x^2$. Hence, by Theorem 5.5,
$$\lim_{x \to c} x^2 = \lim_{x \to c} [f(x) \cdot f(x)] = \lim_{x \to c} f(x) \cdot \lim_{x \to c} f(x)$$
$$= \lim_{x \to c} x \cdot \lim_{x \to c} x = c \cdot c = c^2$$

THEOREM 5.6. $\lim_{x \to c} \dfrac{f(x)}{g(x)} = \dfrac{\lim_{x \to c} f(x)}{\lim_{x \to c} g(x)} = \dfrac{L}{M}$ if $M \neq 0$

In words: The limit of a fraction is the limit of the numerator divided by the limit of the denominator, provided the limit of the denominator is not zero.

Remark: Theorem 5.6 is concerned only with the case $M \neq 0$. It can be shown that if $M = 0$ and $L \neq 0$ then $\lim_{x \to c} \dfrac{f(x)}{g(x)}$ does not exist. That is, $\dfrac{f(x)}{g(x)}$ does not approach a definite finite limit as x approaches c.

Example 7 Find $\lim_{x \to 3} \dfrac{4x + 7}{x^2}$.

Solution: Let $f(x) = 4x + 7$ and $g(x) = x^2$. Then

$$\lim_{x \to 3} \frac{4x + 7}{x^2} = \lim_{x \to 3} \frac{f(x)}{g(x)} = \frac{\lim_{x \to 3} f(x)}{\lim_{x \to 3} g(x)}$$

$$= \frac{\lim_{x \to 3}(4x + 7)}{\lim_{x \to 3} x^2} = \frac{4 \cdot 3 + 7}{3 \cdot 3} = \frac{19}{9}$$

THEOREM 5.7. $\lim_{x \to c}[f(x)]^n = [\lim_{x \to c} f(x)]^n = L^n$

Example 8 Find $\lim_{x \to c} x^7$.

Solution: Let $f(x) = x$. By Theorem 5.7,

$$\lim_{x \to c}[f(x)]^7 = \lim_{x \to c} x^7 = (\lim_{x \to c} x)^7 = c^7$$

THEOREM 5.8. $\lim_{x \to c} \sqrt[n]{f(x)} = \sqrt[n]{\lim_{x \to c} f(x)} = \sqrt[n]{L}$ if $L > 0$

Example 9 Find $\lim_{x \to 5} \sqrt[3]{4x + 7}$.

Solution: We use Theorem 5.8 with $f(x) = 4x + 7$ and we get

$$\lim_{x \to 5} \sqrt[3]{4x + 7} = \sqrt[3]{\lim_{x \to 5}(4x + 7)} = \sqrt[3]{27} = 3$$

THEOREM 5.9. *If $F(x) = f(x)$ for every x, except possibly $x = c$, in some interval containing c, and $\lim_{x \to c} f(x) = L$, then*

$$\lim_{x \to c} F(x) = \lim_{x \to c} f(x) = L$$

Example 10 Find $\lim_{x \to 0} \frac{x^2 + 3x}{x}$.

Solution: Let $F(x) = \frac{x^2 + 3x}{x}$ and let $f(x) = x + 3$. Since $\frac{x^2 + 3x}{x} = x + 3$ for all values of x except $x = 0$, then, by Theorem 5.9,

$$\lim_{x \to 0} \frac{x^2 + 3x}{x} = \lim_{x \to 0}(x + 3) = 0 + 3 = 3$$

To indicate how the definition is used in proving these theorems, we shall prove Theorem 5.4. We shall need the following rather obvious property of the absolute-value symbol. If two numbers a and b have the same sign, then

SEC. 5.3   The Limit Concept and Some Limit Theorems

$|a + b| = |a| + |b|$, but if they have opposite signs, then $|a + b| < |a| + |b|$. Both of these cases are expressed by the inequality

$$|a + b| \leq |a| + |b|$$

Proof of Theorem 5.4. It is assumed that

$$\lim_{x \to c} f(x) = L \quad \text{and} \quad \lim_{x \to c} g(x) = M$$

If d_1 and d_2 are any two positive numbers, there must exist corresponding positive numbers k_1 and k_2 such that

$$|f(x) - L| < d_1 \quad \text{when } 0 < |x - c| < k_1$$

$$|g(x) - M| < d_2 \quad \text{when } 0 < |x - c| < k_2$$

We wish to show that, if d is any positive number, there exists a positive number k such that

$$|[f(x) + g(x)] - (L + M)| < d \quad \text{when } 0 < |x - c| < k$$

Let $d_1 = d_2 = \dfrac{d}{2}$. Now, if k is the smaller of k_1 and k_2, then

$$|f(x) - L| < \frac{d}{2} \quad \text{and} \quad |g(x) - M| < \frac{d}{2} \quad \text{when } 0 < |x - c| < k$$

By using the property of the absolute-value symbol, mentioned above, we have

$$\begin{aligned}|[f(x) + g(x)] - (L + M)| &= |(f(x) - L) + [g(x) - M]| \\ &\leq |f(x) - L| + |g(x) - M| \\ &= \frac{d}{2} + \frac{d}{2} = d \quad \text{when } 0 < |x - c| < k\end{aligned}$$

Hence, by definition, $\lim_{x \to c} [f(x) + g(x)] = L + M$.

We shall assume the validity of the other theorems given. For convenience, we collect the limit theorems in the following table:

The Limit Theorems

THEOREM 5.1.	$\lim_{x \to c} K = K$	where K is a constant
THEOREM 5.2.	$\lim_{x \to c} x = c$	
THEOREM 5.3.	$\lim_{x \to c} Kf(x) = K \lim_{x \to c} f(x)$	where K is a constant

THEOREM 5.4. $\lim_{x \to c} [f(x) + g(x)] = \lim_{x \to c} f(x) + \lim_{x \to c} g(x)$

THEOREM 5.5. $\lim_{x \to c} [f(x) \cdot g(x)] = \lim_{x \to c} f(x) \cdot \lim_{x \to c} g(x)$

THEOREM 5.6. $\lim_{x \to c} \dfrac{f(x)}{g(x)} = \dfrac{\lim_{x \to c} f(x)}{\lim_{x \to c} g(x)}$ if $\lim_{x \to c} g(x) \neq 0$

THEOREM 5.7. $\lim_{x \to c} [f(x)]^n = \left[\lim_{x \to c} f(x) \right]^n$

THEOREM 5.8. $\lim_{x \to c} \sqrt[n]{f(x)} = \sqrt[n]{\lim_{x \to c} f(x)}$ if $\lim_{x \to c} f(x) > 0$

THEOREM 5.9. If $F(x) = f(x)$ for every x, except possibly $x = c$, in some interval containing c, then

$$\lim_{x \to c} F(x) = \lim_{x \to c} f(x)$$

Exercises 5.3

Using any of the theorems of this section which are applicable, find each of the following limits.

1. $\lim_{x \to 5} 17$
2. $\lim_{x \to 3} -7x$
3. $\lim_{x \to 1} (10x - 6)$
4. $\lim_{x \to -2} x^3$
5. $\lim_{x \to 1} [(3x - 2)(2x + 8)]$
6. $\lim_{x \to -1} (x^4 - 3x)$
7. $\lim_{x \to 2} \dfrac{4x - 1}{x + 2}$
8. $\lim_{h \to 0} (3x_1 + 2h)$
9. $\lim_{x \to 0} \dfrac{3x^2 + 4x}{2x}$
10. $\lim_{x \to 3} (x^2 - 4x)$
11. $\lim_{x \to -1} \dfrac{x^2 + 5}{x + 4}$
12. $\lim_{x \to 1} [(3x + 8)(5x + 1)]$
13. $\lim_{x \to 2} \sqrt{5x + 6}$
14. $\lim_{x \to 3} \sqrt[3]{8x + 3}$
15. $\lim_{h \to 0} \dfrac{5h^2 + 6h}{h}$
16. $\lim_{h \to 0} (4x_1 - 5 + 3h)$

5.4 CONTINUOUS FUNCTIONS

In this section we wish to define a concept, called the *continuity* of a function, which is very basic in mathematics and one which we shall find most useful in our further discussion of limits. We shall then show that this concept applies to a large class of functions called polynomial functions.

SEC. 5.4 Continuous Functions

DEFINITION 5.4. A function f which is defined for all values of x in some interval containing the number c is said to be *continuous at* c provided

(1) f has a definite value $f(c)$ at c, and
(2) $\lim_{x \to c} f(x) = f(c)$

If f is continuous at each point of a closed interval $[a, b]$ or at each point of the open interval (a, b), then it is said to be continuous on the closed interval or continuous on the open interval.

Example 1   The function f is defined by

$$f(x) = \begin{cases} \dfrac{x^2 - 4}{x - 2} & \text{if } x \neq 2 \\ 4 & \text{if } x = 2 \end{cases}$$

Is f continuous at $x = 2$?

Solution: According to Definition 5.4, f is continuous at $x = 2$ if and only if f has a value $f(2)$ at $x = 2$, $f(x)$ approaches a limit as x approaches 2, and that limit is $f(2)$. We are given that $f(2) = 4$. Also

$$\lim_{x \to 2} f(x) = \lim_{x \to 2} \frac{x^2 - 4}{x - 2} = \lim_{x \to 2} (x + 2) = 4 = f(2)$$

Therefore, f is continuous by definition at $x = 2$.

Example 2   Suppose a function f is defined by

$$f(x) = \begin{cases} \dfrac{x^2 - 4}{x - 2} & \text{if } x \neq 2 \\ 10 & \text{if } x = 2 \end{cases}$$

Is f continuous at $x = 2$?

Solution: Here we see that f has a value at $x = 2$, that is, $f(2) = 10$. We found in Example 1 that $\lim_{x \to 2} f(x) = 4$; since $\lim_{x \to 2} f(x) \neq f(2)$ we conclude that f is not continuous at $x = 2$. In such a case we say that f is *discontinuous* at $x = 2$.

Example 3   The function defined by the equation $f(x) = \dfrac{x^2 - 4}{x - 2}$ is discontinuous at $x = 2$ because $f(2)$ is not defined.

Example 4   Test for continuity at $x = 2$ the function h defined by

$$h(x) = \begin{cases} \dfrac{x^2 - 4}{(x - 2)^2} & \text{if } x \neq 2 \\ 20 & \text{if } x = 2 \end{cases}$$

Solution: Applying Definition 5.4, we see that h has a value at $x = 2$, that is, $h(2) = 20$. Next we attempt to evaluate the $\lim_{x \to 2} h(x)$ as follows:

$$\lim_{x \to 2} h(x) = \lim_{x \to 2} \frac{x^2 - 4}{(x - 2)^2} = \lim_{x \to 2} \frac{x + 2}{x - 2}$$

It follows from the Remark following Theorem 5.6, since $\lim_{x \to 2} x + 2 = 4$ and $\lim_{x \to 2} x - 2 = 0$, that $h(x)$ does not approach a limit as x approaches 2; hence h is discontinuous at $x = 2$.

The very definition of a continuous function exhibits a property of such functions that makes them important: the $\lim_{x \to c} f(x)$ is the value of the function at c, which is $f(c)$. Thus, if we know that a particular function is continuous and need its limit as $x \to c$, we simply find $f(c)$. In the next subsection we shall discover a very important class of continuous functions.

Continuity of Polynomial Functions

Denote by $P(x)$ a general polynomial thus,

(1) $$P(x) = A_n x^n + A_{n-1} x^{n-1} + \cdots + A_1 x + A_0$$

where n is a positive integer or zero, and $A_n, A_{n-1}, \ldots, A_1, A_0$ are arbitrary real numbers. For each positive integer n and for each set of real numbers, $A_n, A_{n-1}, \ldots, A_1, A_0$, Equation (1) defines a function, called a polynomial function (see Section 2.4). Thus, by letting n vary through the set of positive integers by considering all the possible choices for the coefficients $A_n, A_{n-1}, \ldots, A_1, A_0$, we see that Equation (1) defines a whole class of polynomial functions and that each such function is defined for all real numbers x. Furthermore, if $f(x) = x$ and n is a positive integer, then

$$\lim_{x \to c} f(x) = \lim_{x \to c} x = c \qquad (Theorem\ 5.2)$$

Therefore, $\quad \lim_{x \to c} [f(x)]^n = \lim_{x \to c} x^n = c^n \qquad (Theorem\ 5.7)$

Hence $\quad \lim_{x \to c} k x^n = k \lim_{x \to c} x^n = kc^n \qquad (Theorem\ 5.3)$

So if P is a polynomial defined by

$$P(x) = A_n x^n + A_{n-1} x^{n-1} + \cdots + A_1 x + A_0$$

then
$$\lim_{x \to c} P(x) = \lim (A_n x^n + A_{n-1} x^{n-1} + \cdots + A_1 x + A_0)$$
$$= \lim_{x \to c} A_n x^n + \lim_{x \to c} A_{n-1} x^{n-1} + \cdots + \lim_{x \to c} A_1 x + \lim_{x \to c} A_0$$
$$= A_n c^n + A_{n-1} c^{n-1} + \cdots + A_1 c + A_0$$
$$= P(c)$$

This proves the following theorem.

SEC. 5.4 Continuous Functions

THEOREM 5.10. *If P is a polynomial function in one variable and c is any real number, then P is continuous at c. That is, P is continuous in the interval $(-\infty, \infty)$.*

COROLLARY. If P and Q are polynomial functions in one variable, then the function f defined by the equation

$$f(x) = \frac{P(x)}{Q(x)}$$

is continuous at c, provided $Q(c) \neq 0$.

Also

$$\lim_{x \to c} f(x) = \lim_{x \to c} \frac{P(x)}{Q(x)}$$
$$= \frac{P(c)}{Q(c)} \quad \text{if } Q(c) \neq 0 \quad (\textit{Theorem 5.6})$$

Example 5

$$\lim_{x \to -2} (3x^2 - 5x + 1) = 3(-2)^2 - 5(-2) + 1$$
$$= 12 + 10 + 1 = 23$$

Example 6

$$\lim_{x \to 1} \frac{5x^2 + 8x - 3}{x^3 + x} = \frac{5 + 8 - 3}{1 + 1} = \frac{10}{2} = 5$$

Example 7 Evaluate the limit $\lim_{x \to 2} (x^2 + 1)^2$.

Solution: By Theorem 5.7,

$$\lim_{x \to 2} (x^2 + 1)^2 = \left[\lim_{x \to 2} (x^2 + 1) \right]^2 = (5)^2 = 25$$

Example 8 Evaluate the limit $\lim_{x \to -1} \sqrt{3x + 19}$.

Solution: By Theorem 5.8,

$$\lim_{x \to -1} \sqrt{3x + 19} = \sqrt{\lim_{x \to -1} (3x + 19)} = \sqrt{16} = 4$$

Example 9 Evaluate the limit $\lim_{x \to 2} \frac{5x - 2}{x + 2}$.

Solution: By Theorem 5.6,

$$\lim_{x \to 2} \frac{5x - 2}{x + 2} = \frac{\lim_{x \to 2} (5x - 2)}{\lim_{x \to 2} (x + 2)} = \frac{10 - 2}{2 + 2} = \frac{8}{4} = 2$$

Example 10 | Evaluate the limit $\lim_{x \to 0} (x^2 + 5)\sqrt{2x+9}$.

Solution: By Theorem 5.5,

$$\lim_{x \to 0} (x^2 + 5)\sqrt{2x+9} = \lim_{x \to 0} (x^2 + 5) \lim_{x \to 0} \sqrt{2x+9}$$
$$= \lim_{x \to 0} (x^2 + 5)\sqrt{\lim_{x \to 0} (2x+9)}$$
$$= 5\sqrt{9} = 15$$

Example 11 | Evaluate the limit $\lim_{x \to 2} \dfrac{x^2 + 5x - 14}{x^2 - x - 2}$.

Solution: We have

$$\frac{x^2 + 5x - 14}{x^2 - x - 2} = \frac{(x-2)(x+7)}{(x-2)(x+1)}$$

Now let

$$f(x) = \frac{x+7}{x+1} \quad \text{and} \quad F(x) = \frac{x^2 + 5x - 14}{x^2 - x - 2}$$

Then $F(x) = f(x)$, except when $x = 2$. By Theorem 5.9,

$$\lim_{x \to 2} F(x) = \lim_{x \to 2} f(x) = \lim_{x \to 2} \frac{x+7}{x+1} = \frac{\lim_{x \to 2}(x+7)}{\lim_{x \to 2}(x+1)} = \frac{9}{3} = 3$$

Example 12 | Evaluate $\lim_{h \to 0} \dfrac{\sqrt{x+h} - \sqrt{x}}{h}$

Solution: We first rationalize the numerator by multiplying both the numerator and denominator by $\sqrt{x+h} + \sqrt{x}$

$$\lim_{h \to 0} \frac{\sqrt{x+h} - \sqrt{x}}{x} \cdot \frac{\sqrt{x+h} + \sqrt{x}}{\sqrt{x+h} + \sqrt{x}} = \lim_{h \to 0} \frac{x+h-x}{h[\sqrt{x+h} + \sqrt{x}]}$$
$$= \lim_{h \to 0} \frac{h}{h[\sqrt{x+h} + \sqrt{x}]}$$
$$= \lim_{h \to 0} \frac{1}{[\sqrt{x+h} + \sqrt{x}]} = \frac{1}{2\sqrt{x}}$$

Exercises 5.4

By using the theorems on limits and the properties of polynomials, evaluate the following.

1. $\lim_{x \to 2} (3x^2 - 7x)$
2. $\lim_{x \to 1} \sqrt{x^2 - 5x + 13}$
3. $\lim_{x \to 5} (x^3 - 100)(3x - 12)$

SEC. 5.5   Instantaneous Rates

4. $\lim\limits_{x \to 0} \dfrac{2x^3 + 7x - 4}{x^2 + 8x - 2}$

5. $\lim\limits_{x \to 2} \dfrac{x^2 - 4}{x - 2}$

6. $\lim\limits_{x \to 0} \dfrac{2x^2 + 3x}{x}$

7. $\lim\limits_{x \to 0} \dfrac{4}{x + 2}$

8. $\lim\limits_{x \to 1} \dfrac{x^2 - x}{x^2 - x - 2}$

9. $\lim\limits_{h \to 0} \dfrac{5h - 2h^2}{h}$

10. $\lim\limits_{x \to 1/2} \dfrac{3}{x^2}$

11. $\lim\limits_{x \to 3} \dfrac{x^2 - 2x - 3}{x^2 - 9}$

12. $\lim\limits_{h \to 0} \dfrac{4xh - 7h - 2h^2}{h}$

13. $\lim\limits_{x \to 3} \dfrac{x + 2}{x^2 + 2x - 3}$

14. $\lim\limits_{x \to 2} \dfrac{x^3 - 8}{x - 2}$

15. $\lim\limits_{x \to 1} \dfrac{x^3 + 1}{x + 1}$

16. $\lim\limits_{x \to 2} \dfrac{x^2 - 5x + 6}{x^2 - 2x - 2}$

17. $\lim\limits_{x \to -1} \dfrac{x^3 + 1}{x^2 - 1}$

18. $\lim\limits_{x \to 0} \dfrac{4x^3 - 6x^2}{2x^3 + 3x^2}$

19. $\lim\limits_{h \to 0} \dfrac{(x + h)^3 - x^3}{h}$

20. $\lim\limits_{h \to 0} \dfrac{\sqrt{5 + h} - \sqrt{5}}{h}$

5.5 INSTANTANEOUS RATES

The limit idea is essential in defining many mathematical concepts, one of the most important of which is that of an instantaneous rate. In Section 5.2 we gave an intuitive notion of what is meant by the instantaneous velocity of a moving object, the slope of a curve at a point, and the instantaneous rate of change of a function at some particular value of its independent variable. In Section 5.3 we gave a precise definition of what is meant by a statement such as "$\lim_{x \to c} f(x) = L$" and listed a set of rules in the form of theorems for evaluating limits. We now have the proper background and a convenient notation for stating a formal definition for each of the instantaneous rates mentioned above. The purpose of this section is to state these definitions.

Instantaneous Velocity

In this text we shall be concerned only with moving objects that move along a straight line. It is customary in describing such motion to represent the line of motion, that is, the path of the object, by a coordinate axis which we shall call the s axis. In most of our considerations the s axis will be either vertical or horizontal. It will be helpful to assume that as a body moves along the s axis the entire body at any given instant is located at a point. We denote the position on the s axis of a moving object at time t by $s(t)$. This, in effect, expresses the position on the s axis of the object as a function of time. We select a reference point 0 on the s axis as the origin, and then adopt a positive direction and a unit of distance. If the s axis is vertical, we choose the positive direction to be upward, and if the s axis is horizontal, we choose the positive direction to be toward the right. For example, if the s axis is vertical and at time $t = 3$ seconds a moving object is at a point 5 units above the origin, we write $s(3) = 5$. But if at time $t = 3$ the object is 7 units below the origin, we write $s(3) = -7$. Similarly, if an object is moving along a horizontal s axis and at time $t = 4$ it is 6 units to the right of the origin, we write $s(4) = 6$. If at time $t = 1$ the object is at a point 3 units to the left of the origin, we write $s(1) = -3$.

The motion of a moving object is usually described by means of an equation which expresses the position $s(t)$, on the s axis, of the object as a function of the time t that has elapsed since the start of the motion. In keeping with this idea, we shall assume that the motion started at time $t = 0$. Let us suppose that we have an object, or particle, moving along the s axis and that at time $t = t_1$ the position of the object is given by $s(t_1)$, and at time $t = t_1 + h$ the object has moved to a new position given by $s(t_1 + h)$ (see Figure 5.1). During the interval of time from $t = t_1$ to $t = t_1 + h$, the object has traveled a distance $s(t_1 + h) - s(t_1)$. Its average velocity $\bar{v}$ over this interval of time is

$$\bar{v} = \frac{s(t_1 + h) - s(t_1)}{h}$$

If this average velocity $\bar{v}$ approaches a limit as h approaches zero, we define the velocity $v(t_1)$ at time $t = t_1$ to be that limit. We now write our definition in limit notation.

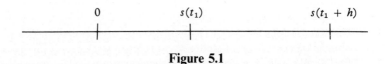

Figure 5.1

DEFINITION 5.5. $\quad v(t_1) = \lim\limits_{h \to 0} \bar{v} = \lim\limits_{h \to 0} \dfrac{s(t_1 + h) - s(t_1)}{h}$

Example 1 A ball is thrown vertically upward from the ground in such a way that its height, $s(t)$ feet, above the ground t seconds after being thrown is given by the equation $s(t) = 64t - 16t^2$. Find

  a. the velocity of the ball at time $t = t_1$.
  b. By dropping the subscript 1 from t in the result of part **a**, obtain a formula for the velocity $v(t)$ in terms of t.
  c. Use the velocity formula found in part **b** to find the velocity of the ball at time $t = 1$ second and at time $t = 3$ seconds.

Solution: To obtain the equation $s(t) = 64t - 16t^2$, the s axis was taken as vertical and the origin 0 was chosen at the ground level. We first find $\bar{v}$ over the interval from $t = t_1$ to $t = t_1 + h$. To do this, we need

$$s(t_1 + h) = 64(t_1 + h) - 16(t_1 + h)^2$$
$$= 64t_1 + 64h - 16t_1^2 - 32t_1 h - 16h^2$$
$$s(t_1) = 64t_1 \quad\quad\quad\quad - 16t_1^2$$

Subtracting, we obtain

$$s(t_1 + h) - s(t_1) = 64h - 32t_1 h - 16h^2$$

Hence

$$\bar{v} = \frac{64h - 32t_1 h - 16h^2}{h} = 64 - 32t_1 - 16h \quad \text{if } h \neq 0$$

It follows from Definition 5.5 that

$$v(t_1) = \lim_{h \to 0} \bar{v} = \lim_{h \to 0} (64 - 32t_1 - 16h) = 64 - 32t_1$$

Thus the answer to part **a** is

$$v(t_1) = (64 - 32t_1) \text{ ft/sec}$$

Following the instructions in part **b**, we get

$$v(t) = (64 - 32t) \text{ ft/sec}$$

as our formula. Using the formula just found, we see that

$$v(1) = 32 \text{ ft/sec}$$

and

$$v(3) = -32 \text{ ft/sec}$$

The sign in each of the last two equations is very important for it tells us the direction the ball is going at the given instant. When $v(t)$ is positive, the ball is moving in the positive direction along the s axis, which in this case is upward. When $v(t)$ is negative, the ball is moving in the negative direction along the s axis. In this problem the negative direction is downward. Hence at time $t = 1$ the ball is going up since $v(1) = +32$ ft/sec. But at time $t = 3$ the ball is falling, since $v(3) = -32$ ft/sec. Notice that we can find when the ball is highest; that is, we can find the value of t when the ball reaches its highest position. We know that at the instant the ball reaches its highest point it stops; that is, its velocity is $v(t) = 0$ at that instant. To find the value of t when $v(t) = 0$, we set $v(t) = 0$ in our velocity formula, thus $0 = 64 - 32t$, and solve for t. This gives $t = 2$. Hence at $t = 2$ the ball is at its highest point, and its height above the ground at that instant is $s(2) = 128 - 64 = 64$ feet. The ball then starts falling, and at the instant it reaches the ground $s(t) = 0$. By setting $s(t) = 0$ in the equation $s(t) = 64t - 16t^2$, we obtain

$$0 = 64t - 16t^2 = t(64 - 16t)$$

Solving for t we see that $s(t) = 0$ when $t = 0$, the instant it was thrown, and when $t = 4$, the instant it reaches the ground on its way down. So, strictly speaking, the equation $s(t) = 64t - 16t^2$ is only valid for $0 \leq t \leq 4$.

The discussion following Example 1 suggests a fact that can be shown to be true in general regarding the velocity $v(t)$ of an object moving along an s axis. This fact is that the velocity $v(t)$, calculated at a given instant during the motion, gives us not only the instantaneous rate at which the object is moving at that instant, but it also tells us the direction in which the object is moving. For example, if at time $t = t_1$, the velocity of an object is $v(t_1)$, the magnitude of $v(t_1)$, that is, $|v(t_1)|$ tells us how fast the object is moving and the algebraic sign tells us in which direction along the s axis it is going. If an object is moving in the positive direction of the s axis, the sign of $v(t_1)$ will be positive, but if it is moving in the negative direction, the sign of $v(t_1)$ will be negative. If a physical quantity has both magnitude and direction, it is called a *vector quantity*. Hence velocity is a vector quantity. The reader has probably noticed that thus far we have carefully avoided using the word speed. This was deliberate for we feel that the definition of speed of a moving object will be more meaningful if the reader first has a clear understanding of the full and complete meaning of the concept of velocity. We define the speed of a moving object at time t to be the magnitude of its velocity at that time, that is, $|v(t)|$. The only symbol we shall use in this text to represent speed will be $|v(t)|$.

Caution: The letter s is almost *never* used in mathematics to represent speed. To find the speed of an object at a given instant, we must first find its velocity at that instant and then find the absolute value of the velocity. Note that the speed of a moving object is always positive or zero. It might be pointed out that the speedometer on a car does not register the velocity of the car but only the magnitude of the velocity, that is, its speed. Hence, the name speedometer.

Instantaneous Acceleration

As was seen in Example 1, the velocity $v(t)$ of an object moving along the s axis may be a variable quantity; that is, it may be a function of time t. The rate at which the velocity varies with time is called the *acceleration* of the object. To be more specific, let $v(t_1)$ be the velocity of the object at time $t = t_1$, and let $v(t_1 + h)$ be the velocity of the object at time $t = t_1 + h$. The average acceleration, $\bar{a}$, over the interval of time from $t = t_1$ to $t = t_1 + h$ is defined by

$$\bar{a} = \frac{v(t_1 + h) - v(t_1)}{h}$$

The instantaneous acceleration $a(t_1)$ of the object at time $t = t_1$ is defined as the limit, if it exists, of $\bar{a}$ as h approaches zero; that is

DEFINITION 5.6. $\quad a(t_1) = \lim\limits_{h \to 0} \bar{a} = \lim\limits_{h \to 0} \dfrac{v(t_1 + h) - v(t_1)}{h}$

SEC. 5.5 Instantaneous Rates

Elementary calculus is concerned primarily with instantaneous rates of change. Hence, hereafter in this text when we use the terms velocity and acceleration, we mean instantaneous velocity and instantaneous acceleration.

Example 2 An object moves along a horizontal s axis in such a way that its velocity $v(t)$ ft/sec t seconds after starting is given by $v(t) = 10t - 2t^2$.

a. Find the acceleration of the object at the time $t = t_1$.
b. By dropping the subscript from t_1 in the result of part **a**, obtain a formula for the acceleration in terms of t.
c. Use the formula found in part **b** to find the acceleration at $t = 2$ and at $t = 4$.

Solution: We start by finding $\bar{a}$ over the interval from $t = t_1$ to $t = t_1 + h$. Since $v(t) = 10t - 2t^2$, then

$$v(t_1 + h) = 10(t_1 + h) - 2(t_1 + h)^2$$
$$= 10t_1 + 10h - 2t_1^2 - 4t_1 h - 2h^2$$

and $v(t_1) = 10t_1 - 2t_1^2$

Subtracting and dividing the result by h, we get

$$\bar{a} = \frac{v(t_1 + h) - v(t_1)}{h} = \frac{10h - 4t_1 h - 2h^2}{h} = 10 - 4t_1 - 2h \quad \text{if } h \neq 0$$

Therefore, by Definition 5.5, we have

$$a(t_1) = \lim_{h \to 0} \bar{a} = \lim_{h \to 0} \frac{v(t_1 + h) - v(t_1)}{h} = \lim_{h \to 0} (10 - 4t_1 - 2h)$$
$$= 10 - 4t_1$$

So the answer to part **a** is

$$a(t_1) = 10 - 4t_1$$

Dropping the subscript from t_1, we obtain

$$a(t) = 10 - 4t$$

the formula requested in part **b**. It follows from this formula that

$$a(2) = 10 - 8 = 2 (\text{ft/sec})/\text{sec} = 2 \text{ ft/sec}^2$$

read "$a(2) = 2$ feet per second per second," and

$$a(4) = 10 - 16 = -6 \text{ ft/sec}^2$$

It should be pointed out that the acceleration of a moving object at a given instant is, strictly speaking, the rate at which the velocity of the object is changing at that instant. It can be shown that if, at a given instant $t = t_1$, the acceleration of a moving object $a(t_1)$ is positive, then the velocity of the object is increasing; that if $a(t_1)$ is negative, then the velocity of the object is decreasing;

and that if $a(t_1) = 0$, then the velocity is neither increasing nor decreasing. Thus in the above example the velocity is increasing at $t = 2$ since $a(2) = +2$ ft/sec^2, and the velocity is decreasing at $t = 4$ since $a(4) = -6$ ft/sec^2.

5.6 THE DERIVATIVE OF A FUNCTION

In Section 5.5 it was pointed out that elementary calculus is primarily concerned with instantaneous rates of change. It seems logical, therefore, to define instantaneous rates in general and to have a common name for all such instantaneous rates. In calculus the instantaneous rate of change of any function at a fixed value of its independent variable is called the *derivative* of the function with respect to that variable at that value. To be more specific, we state the following definition.

DEFINITION 5.7. Let f be a function defined by the equation $y = f(x)$ for all values of x in the open interval (a, b), and let x_1 be a fixed number in that interval. The *derivative* of y, or of $f(x)$, with respect to x at $x = x_1$, denoted by $f'(x_1)$, read "f prime of x_1," is defined by

$$f'(x_1) = \lim_{h \to 0} \frac{f(x_1 + h) - f(x_1)}{h}$$

provided this limit exists. If $f'(x_1)$ exists, we say that f is *differentiable* at x_1, and if $f'(x)$ exists for each x in (a, b), we say that f is a *differentiable function* in (a, b).

Various other symbols are used to denote this limit; the more common ones are $y']_{x=x_1}$ (read y prime at $x = x_1$), $D_x y]_{x=x_1}$ and $\dfrac{dy}{dx}\bigg]_{x=x_1}$ (the last two are both read "the derivative of y with respect to x at $x = x_1$"). The symbols $D_x y$ and $\dfrac{dy}{dx}$ are interpreted as $D_x y = D_x(y)$ and $\dfrac{d}{dx}(y)$, where $D_x(\ldots)$ and $\dfrac{d}{dx}(\ldots)$ each denotes the operation "derivative with respect to x" of the expression following the symbols D_x and $\dfrac{d}{dx}$. Note that these symbols do not denote multiplication. Although we shall for the most part use symbols of the form $D_x y$ and $f'(x)$, the student should learn to recognize each of the other symbols mentioned as representing a derivative. As a matter of fact, most calculus texts use symbols of the form $\dfrac{dy}{dx}$ rather than $D_x y$. We prefer the latter. Also, since x_1 is an arbitrary number in (a, b), it is common practice to omit the subscript on x_1 and write

SEC. 5.6   The Derivative of a Function

DEFINITION 5.8.

$$D_x y = \lim_{h \to 0} \frac{f(x+h)-f(x)}{h} \quad \text{or} \quad f'(x) = \lim_{h \to 0} \frac{f(x+h)-f(x)}{h}$$

We shall follow this practice and use Definition 5.8 in this text. However, it must be kept in mind when we use this form of the definition of a derivative that, in performing the limit operation, h is the variable and x must be treated as a fixed number. The value of the limit will in general be expressed in terms of x. The process of finding the derivative of a function is called *differentiation*.

Δ Notation

Many texts, instead of using the single variable h to denote the length of an interval or the change in a variable, use for this purpose the Greek capital letter *delta*, Δ, along with the variable and write Δx if the variable is x. The symbol Δx, read "delta x," has precisely the same meaning as the letter h in Definition 5.7; the two may be used interchangeably by means of the simple relation $h = \Delta x$. To be more specific, suppose $P_1(x_1, y_1)$ and $P_2(x_2, y_2)$ are two distinct points in the xy plane (see Figure 5.2). If a moving particle starts at P_1 and travels to P_2, its abscissa changes from x_1 to x_2; it is common practice to use the notation Δx to represent this difference $x_2 - x_1$:

$$\Delta x = x_2 - x_1$$

Similarly, the ordinate of the moving particle as it travels from P_1 to P_2 changes

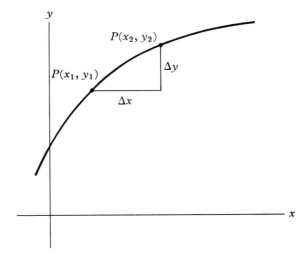

Figure 5.2

from y_1 to y_2, and we denote by Δy (read "delta y") this change $y_2 - y_1$:

$$\Delta y = y_2 - y_1$$

Suppose that points $P_1(x_1, y_1)$ and $P_2(x_2, y_2)$ are two distinct points on the graph of a function f defined by the equation $y = f(x)$. Then

$$y_2 = f(x_2)$$
$$y_1 = f(x_1)$$

and
$$y_2 - y_1 = f(x_2) - f(x_1)$$

or
$$\Delta y = f(x_2) - f(x_1)$$

Since $\Delta x = x_2 - x_1$, it follows that $x_2 = x_1 + \Delta x$, and hence

$$\Delta y = f(x_1 + \Delta x) - f(x_1)$$

The average rate of change, $\bar{R}$, of the function f over the interval $\Delta x = x_2 - x_1$ is therefore

$$\bar{R} = \frac{\Delta y}{\Delta x} = \frac{f(x_1 + \Delta x) - f(x_1)}{\Delta x}$$

If we use the Δ notation, Definition 5.8 becomes:

DEFINITION 5.9. $f'(x) = D_x y = \lim\limits_{\Delta x \to 0} \dfrac{f(x + \Delta x) - f(x)}{\Delta x}$, provided this limit exists.

If in Definition 5.9 we replace Δx by h, we obtain Definition 5.8, which shows that the two definitions are equivalent.

Using Definition 5.9 to find the derivative of a function is sometimes called the *delta process* for finding the derivative. It should be emphasized here that it makes no difference whether one uses Definition 5.8 or 5.9 to find the derivative of a function; the result is the same in either case. Actually, it is a matter of personal preference. Try both definitions and take your choice. We shall ordinarily use Definition 5.8.

Example 1   Use the delta process to find $D_x y$ if $y = x^2 + 3x - 2$.

Solution: We may write $y = f(x) = x^2 + 3x - 2$. Then

$$y_2 = f(x_2) = f(x_1 + \Delta x)$$

But
$$y_2 = y_1 + \Delta y$$

So
$$y_1 + \Delta y = f(x_1 + \Delta x)$$

SEC. 5.7 Rules for Differentiation

or, dropping subscripts, we have

$$y + \Delta y = f(x + \Delta x) = (x + \Delta x)^2 + 3(x + \Delta x) - 2$$

or
$$y + \Delta y = x^2 + 2x\,\Delta x + (\Delta x)^2 + 3x + 3\Delta x - 2$$

and
$$y \quad\quad = x^2 \quad\quad\quad\quad\quad + 3x \quad\quad - 2$$

Subtracting
$$\Delta y = 2x\,\Delta x + (\Delta x)^2 + 3\Delta x$$

Hence
$$\frac{\Delta y}{\Delta x} = 2x + \Delta x + 3 \quad\quad \Delta x \neq 0$$

and
$$D_x y = \lim_{\Delta x \to 0} \frac{\Delta y}{\Delta x} = \lim_{\Delta x \to 0} (2x + \Delta x + 3) = 2x + 3$$

Exercises 5.6

In each of the following exercises, use either Definition 5.8 or Definition 5.9 to find the derivative of the given function and denote it by either $D_x y$ or $f'(x)$.

1. $y = f(x) = x^2$
2. $y = f(x) = 3x^2 - 8x + 9$
3. $y = f(x) = 8x^2 - 7x + 3$
4. $y = f(x) = 5 + 6x - 2x^2$
5. $y = f(x) = x^3$
6. $y = f(x) = 7x^3 - 5x + 10$
7. $y = f(x) = x^4$
8. $y = f(x) = \dfrac{4}{x}$
9. $y = f(x) = \dfrac{10}{x^2}$
10. $y = f(x) = (2x + 3)(5x - 1)$ (First multiply out.)
11. $y = f(x) = (7x + 2)(3x - 4)$
12. $y = f(x) = (1 - 2x)^2$
13. $y = f(x) = \sqrt{x}$
14. $y = f(x) = \dfrac{x + 3}{x + 2}$

15. If $y = 8 + x - 5x^2$ is an equation of a curve, find the slope of the curve, that is, $D_x y$ at the point $(1, 4)$.

5.7 RULES FOR DIFFERENTIATION

In each of the exercises in Section 5.6 the student was expected to find the derivative by each time applying the definition directly to the specific function involved. The procedure is laborious and rather time consuming. We shall derive some simple rules that will reduce the labor involved in finding a derivative and at the same time speed up the operation.

Derivative of a Constant

Let $y = f(x) = C$ for $a < x < b$;

then
$$f(x+h) = C, \quad f(x) = C$$
and
$$f(x+h) - f(x) = 0$$

Hence
$$D_x y = \lim_{h \to 0} \frac{f(x+h) - f(x)}{h} = \lim_{h \to 0} \frac{0}{h} = \lim_{h \to 0} 0 = 0$$

Hence the derivative of any constant with respect to a variable is zero. In symbols, this is written as follows.

RULE 5.1. $D_x C = 0$, where C is a constant.

Derivative of $y = f(x) = x$ with Respect to x

We have the following equations, which express the fact that the derivative of any variable with respect to itself is equal to 1 (Rule 5.2 below).

$$f(x+h) = x+h, \quad f(x) = x$$
and
$$f(x+h) - f(x) = h$$

Hence
$$D_x y = \lim_{h \to 0} \frac{f(x+h) - f(x)}{h} = \lim_{h \to 0} \frac{h}{h} = 1$$

RULE 5.2. $D_x x = 1$.

Derivative of x^n Where n Is a Positive Integer

From Rule 5.2 we have that $D_x x = 1$. In Exercises 1, 5, and 7 of Section 5.6, we found that

$$D_x [x^2] = 2x$$
$$D_x [x^3] = 3x^2$$
$$D_x [x^4] = 4x^3$$

These equations suggest a third rule, which we shall prove.

RULE 5.3. $D_x(x^n) = n x^{n-1}$, where n is a positive integer.

Proof: To prove this rule, we shall apply Definition 5.8 to the function $f(x) = x^n$. First, however, we need to recall a special product discussed in

SEC. 5.7 Rules for Differentiation

Chapter 2: if n is a positive integer, then

$$A^n - B^n = (A - B)(A^{n-1} + A^{n-2}B + A^{n-3}B^2 + \cdots + AB^{n-2} + B^{n-1})$$

where the second factor on the right consists of exactly n terms. Now if $f(x) = x^n$, then $f(x + h) = (x + h)^n$ and

$$\frac{f(x+h) - f(x)}{h}$$

$$= \frac{(x+h)^n - x^n}{h}$$

$$= \frac{[(x+h) - x][(x+h)^{n-1} + (x+h)^{n-2}x + (x+h)^{n-3}x^2 + \cdots + (x+h)x^{n-2} + x^{n-1}]}{h}$$

$$= (x+h)^{n-1} + (x+h)^{n-2}x + (x+h)^{n-3}x^2 + \cdots + (x+h)x^{n-2} + x^{n-1}$$

using the formula above with $A = x + h$ and $B = x$ to obtain the second equality. Applying Definition 5.8, we get

$$D_x(x^n) = \lim_{h \to 0} \frac{f(x+h) - f(x)}{h}$$
$$= \lim_{h \to 0} [(x+h)^{n-1} + (x+h)^{n-2}x + (x+h)^{n-3}x^2 + \cdots + (x+h)x^{n-2} + x^{n-1}]$$
$$= x^{n-1} + x^{n-2} \cdot x + x^{n-3} \cdot x^2 + \cdots + x \cdot x^{n-2} + x^{n-1}$$
$$= nx^{n-1}$$

since there are exactly n terms and each is equal to x^{n-1}.

Example 1 Find the derivative of $y = x^{20}$.

Solution: By Rule 5.3, $D_x y = 20x^{19}$.

Derivative of $kf(x)$ Where k Is a Constant

We wish to show that the derivative of a constant times a function is equal to the constant times the derivative of the function. This is expressed symbolically in the following rule.

RULE 5.4. $D_x[kf(x)] = kD_x[f(x)]$

Proof: $kf(x + h) - kf(x) = k[f(x + h) - f(x)]$

$$D_x[kf(x)] = \lim_{h \to 0} \frac{k[f(x+h) - f(x)]}{h}$$
$$= k \lim_{h \to 0} \frac{f(x+h) - f(x)}{h} \quad \text{(Theorem 5.3)}$$
$$= kD_x[f(x)]$$

Example 2 | Find the derivative of $7x^{10}$.

Solution: By Rule 5.4, $D_x[7x^{10}] = 7D_x(x^{10})$
By Rule 5.3 $= 7(10x^9)$
               $= 70x^9$

Derivative of $f(x) + g(x)$

RULE 5.5. $D_x[f(x) + g(x)] = D_x[f(x)] + D_x[g(x)]$

Proof: $[f(x+h) + g(x+h)] - [f(x) + g(x)]$
$$= [f(x+h) - f(x)] + [g(x+h) - g(x)]$$

Applying the definition of a derivative and assuming each limit exists, we may write

$$D_x[f(x) + g(x)] = \lim_{h \to 0} \left\{ \left[\frac{f(x+h) - f(x)}{h} \right] + \left[\frac{g(x+h) - g(x)}{h} \right] \right\}$$

$$= \lim_{h \to 0} \frac{f(x+h) - f(x)}{h} + \lim_{h \to 0} \frac{g(x+h) - g(x)}{h}$$

(*Theorem* 5.4)

$$= D_x[f(x)] + D_x[g(x)]$$

Stated in words, Rule 5.5 says that the derivative of the sum of two differentiable functions is equal to the sum of the derivatives of the functions. The following generalization of Rule 5.5 can be shown to be true.

RULE 5.6. $D_x[f_1(x) + f_2(x) + \cdots + f_n(x)]$
$$= D_x[f_1(x)] + D_x[f_2(x)] + \cdots + D_x[f_n(x)]$$

or, in words, the derivative of the algebraic sum of any finite number of differentiable functions is equal to the algebraic sum of the derivatives of the individual functions.

Example 3 | Find the derivative of $y = 2x^3 - 7x^2 + 3x - 10$.

Solution: $D_x y = D_x[2x^3 - 7x^2 + 3x - 10]$
By Rule 5.6 $= D_x[2x^3] + D_x[-7x^2] + D_x[3x]$
                       $+ D_x[-10]$
By Rules 5.1 through 5.4 $= 6x^2 - 14x + 3 + 0$
                           $= 6x^2 - 14x + 3$

In actual practice, all intermediate steps in the above solution would be omitted and the final result written by inspection wherever possible.

SEC. 5.7 Rules for Differentiation

Differentiating Any Polynomial

It should be observed that the six special rules listed above enable us to differentiate at sight any polynomial

$$y = A_n x^n + A_{n-1} x^{n-1} + \cdots + A_1 x + A_0$$

and, consequently, to differentiate any expression that can be reduced to a polynomial. Some examples of how this is done follow:

Example 4 Find $D_x y$, where $y = x^2(3x - 2)$.

Solution: Multiplying, $y = 3x^3 - 2x^2$. Differentiating, $D_x y = 9x^2 - 4x$.

Example 5 If $y = \dfrac{x^4 - 2x^3 + 7x^2}{3x}$, $x \neq 0$, find $D_x y$.

Solution: Rewriting, we have $y = \dfrac{1}{3} x^3 - \dfrac{2}{3} x^2 + \dfrac{7}{3} x$; hence

$$D_x y = x^2 - \dfrac{4}{3} x + \dfrac{7}{3}$$

Although the rules for differentiation listed in this section are expressed in terms of the variable x, they are valid for any independent variable.

Example 6 If $s = 3t^4 - 5t^2$, then $D_t s = 12t^3 - 10t$.

Example 7 If $A = 5r^2$, then $D_r A = 10r$.

Example 8 If $u = v^3 - 4v^2 + 5v + 7$, then $D_v u = 3v^2 - 8v + 5$.

Rules for Differentiation

> RULE 5.1. $D_x C = 0$, where C is a constant
> RULE 5.2. $D_x x = 1$
> RULE 5.3. $D_x(x^n) = nx^{n-1}$, where n is a positive integer
> RULE 5.4. $D_x[kf(x)] = k D_x[f(x)]$, where k is a constant
> RULE 5.5. $D_x[f(x) + g(x)] = D_x[f(x)] + D_x[g(x)]$
> RULE 5.6. $D_x[f_1(x) + f_2(x) + \cdots + f_n(x)]$
> $\qquad = D_x[f_1(x)] + D_x[f_2(x)] + \cdots + D_x[f_n(x)]$

Exercises 5.7

In Exercises 1 through 20, find by inspection, using the above rules for differentiation, the derivative of each given function with respect to its independent variable. The letters a, b, and c represent constants wherever they occur.

1. $y = 4x - 3$
2. $y = 5x^4 + 7x^2 - 8$
3. $y = x^2(7 + 2x^3)$
4. $v = 10t^3 - t^4$
5. $s = 4\pi r^2$
6. $Q = (u^2 - 1)(2u + 3)$
7. $y = ax^2 + bx + c$
8. $y = ax^2 + bz$
9. $w = 4y^3 + 2y^2 - 6y - 5$
10. $y = \frac{3}{5}v^5 + \frac{1}{3}v^3 + 6^2$
11. $y = (3x - 2)^2$
12. $x = 2t^4 + 5t^3 - 3t^2 + 7x - 5$
13. $z = y^3 - a^3$
14. $S = \frac{at^2 + bt}{c}, \quad c \neq 0$
15. $y = \frac{x^5 - 5x^3 + 3x^2}{x^2}, \quad x \neq 0$
16. $y = x + 0.3x^2 - 0.5x^3 - 0.4x^4$
17. $y = (5x - 2)^2$
18. $y = (3x + 2)(5x - 7)$
19. $y = (x + 1)^3$
20. $y = x^4\left(3 - \frac{2}{x^2}\right)$

21. Use Definition 5.8 to prove that if $y = kx^{-n}$, where k is a constant and n is a positive integer, then $D_x y = nkx^{-n-1}$. $\left(\text{Suggestion: First write } y = \frac{k}{n} \text{ and use the method used in proving Rule 5.3.}\right)$

22. Use the rule established in Exercise 21, $D_x(kx^{-n}) = -nkx^{-n-1}$, n a positive integer, to find $D_x y$ if

    a. $y = x^{-8} + 6x^{-4}$
    b. $y = 3x^{-3} + 8x^{-2} - 2x^{-1}$
    c. $y = 4x^{-2} - 2x^{-5}$
    d. $y = \frac{1}{2}x^{-4} - \frac{2}{3}x^{-3}$

23. Use the rule used in Exercise 22 to find $D_x y$ if

    a. $y = \frac{7}{x^2}$
    b. $y = \frac{2}{3x}$
    c. $y = -\frac{4}{x^5}$
    d. $y = \frac{6}{x^3}$

5.8 DERIVATIVES OF PRODUCTS, QUOTIENTS, AND POWERS

In Section 5.7 we produced a list of six rules for differentiation that enables us to write by inspection the derivative of any polynomial function in one variable with respect to that variable. If we add to that list the rule suggested in Exercise 21 of Section 5.7, we can now write by inspection the derivative of any algebraic expression in one variable, if that expression can be written as a sum of terms of the form kx^n, where k is a constant and n is an integer. In this section we shall state and prove three more very important rules for differentiating (1) the product of two functions, (2) the quotient of two functions, and (3) a function raised to a power.

For the proof in each case, we need the following important result. It can be shown that Theorem 5.11 is equivalent to "differentiable functions are continuous."

THEOREM 5.11. *If f is differentiable at x, then $\lim_{h \to 0} f(x+h) = f(x)$.*

Proof: We are given that f is differentiable at x, and so

$$f'(x) = \lim_{h \to 0} \frac{f(x+h) - f(x)}{h}$$

exists. Also, $\lim_{h \to 0} h = 0$ by Theorem 5.2.

Therefore,

$$\lim_{h \to 0} [f(x+h) - f(x)] = \lim_{h \to 0} \left[\frac{f(x+h) - f(x)}{h} \cdot h \right]$$

$$= \left[\lim_{h \to 0} \frac{f(x+h) - f(x)}{h} \right] \left[\lim_{h \to 0} h \right]$$

$$= f'(x) \cdot 0$$

$$= 0$$

Now

$$\lim_{h \to 0} [f(x+h) - f(x)] = \left[\lim_{h \to 0} f(x+h) \right] - \left[\lim_{h \to 0} f(x) \right]$$

$$= \left[\lim_{h \to 0} f(x+h) \right] - f(x)$$

[$f(x)$ is a constant relative to h]. Thus $\left[\lim_{h \to 0} f(x+h) \right] - f(x) = 0$ and, upon adding $f(x)$ to both sides, we obtain the desired result:

$$\lim_{h \to 0} f(x+h) = f(x)$$

Derivative of a Product

RULE 5.7. THE PRODUCT RULE. If f and g are any two differentiable functions of x, then
$$D_x[f(x)g(x)] = f(x)D_x[g(x)] + g(x)D_x[f(x)]$$

Proof: Let $y = f(x)g(x)$. Then, applying Definition 5.8,
$$D_x y = D_x[f(x)g(x)] = \lim_{h \to 0} \frac{f(x+h)g(x+h) - f(x)g(x)}{h}$$

provided this limit exists. To evaluate this limit, we would like, if possible, to separate the functions f and g as was done in the proof of Rule 5.5 in Section 5.2. The following rather tricky, but legal, device of adding and subtracting the term $f(x+h)g(x)$ in the numerator of the above fraction allows us to make such a separation, for

$$\frac{f(x+h)g(x+h) - f(x)g(x)}{h}$$
$$= \frac{f(x+h)g(x+h) - f(x+h)g(x) + f(x+h)g(x) - f(x)g(x)}{h}$$
$$= f(x+h)\left[\frac{g(x+h) - g(x)}{h}\right] + g(x)\left[\frac{f(x+h) - f(x)}{h}\right]$$

Hence, by the limit theorems, and remembering that $g(x)$ is treated as a constant in the above limit operation,

$$D_x y = D_x[f(x)g(x)]$$
$$= \lim_{h \to 0} f(x+h) \lim_{h \to 0} \frac{g(x+h) - g(x)}{h} + g(x) \lim_{h \to 0} \frac{f(x+h) - f(x)}{h}$$

Since f and g are differentiable,
$$\lim_{h \to 0} \frac{f(x+h) - f(x)}{h} = D_x[f(x)] \quad \text{and} \quad \lim_{h \to 0} \frac{g(x+h) - g(x)}{h} = D_x[g(x)]$$

Also, by Theorem 5.4, $\lim_{h \to 0} f(x+h) = f(x)$. Therefore,
$$D_x y = D_x[f(x)g(x)] = f(x)D_x[g(x)] + g(x)D_x[f(x)]$$

and our Product Rule is proved.

Rule 5.7 states that *the derivative of the product of two functions is equal to the first function times the derivative of the second function, plus the second function times the derivative of the first function.* (Note that the derivative of a product is not the product of the derivatives.)

For ease in applying Rule 5.7 to specific products, it should be memorized in words. This rule can be generalized into a rule for finding the derivative of any finite number of functions, but such a rule will not be needed in this text.

SEC. 5.8 Derivatives of Products, Quotients, and Powers

Example 1 If $y = (3x + 7)(5 - 2x)$, find $D_x y$.

Solution: We could, of course, easily multiply the product and then differentiate, but it is also easy to use Rule 5.7. Thus

$$D_x y = (3x + 7)D_x(5 - 2x) + (5 - 2x)D_x(3x + 7)$$
$$= (3x + 7)(-2) + (5 - 2x)(3)$$
$$= -6x - 14 + 15 - 6x = 1 - 12x$$

Example 2 If $y = (x^2 + 4)(2x^3 - 3x + 7)$, find $D_x y$.

Solution: Using the Product Rule, we have

$$D_x y = (x^2 + 4)D_x(2x^3 - 3x + 7) + (2x^3 - 3x + 7)D_x(x^2 + 4)$$
$$= (x^2 + 4)(6x^2 - 3) + (2x^3 - 3x + 7)(2x)$$
$$= 6x^4 + 21x^2 - 12 + 4x^4 - 6x^2 + 14x$$
$$= 10x^4 + 15x^2 + 14x - 12$$

Derivative of a Quotient

RULE 5.8. QUOTIENT RULE. If f and g are differentiable functions of x, and $g(x) \neq 0$, then

$$D_x \left[\frac{f(x)}{g(x)} \right] = \frac{g(x)D_x[f(x)] - f(x)D_x[g(x)]}{[g(x)]^2}$$

Rule 5.8 states that *the derivative of a quotient is equal to the denominator times the derivative of the numerator, minus the numerator times the derivative of the denominator, all divided by the square of the denominator.* $\Big($Note that $D_x \left[\dfrac{f(x)}{g(x)} \right]$ is not $\dfrac{D_x[f(x)]}{D_x[g(x)]}.\Big)$

Proof: Let $y = \dfrac{f(x)}{g(x)}$. Then, applying Definition 5.8,

$$D_x y = \lim_{h \to 0} \frac{\frac{f(x+h)}{g(x+h)} - \frac{f(x)}{g(x)}}{h} = \lim_{h \to 0} \frac{g(x)f(x+h) - f(x)g(x+h)}{hg(x+h)g(x)}$$
$$= \lim_{h \to 0} \frac{1}{g(x+h)g(x)} \lim_{h \to 0} \frac{g(x)f(x+h) - f(x)g(x+h)}{h}$$

To separate the f and g functions in the numerator of the second limit of the last equation above, we add and subtract the term $f(x)g(x)$ to that numerator.

This allows us to write

$$D_x y = D_x \left[\frac{f(x)}{g(x)} \right]$$

$$= \lim_{h \to 0} \frac{1}{g(x+h)g(x)}$$

$$\times \left[g(x) \lim_{h \to 0} \frac{f(x+h) - f(x)}{h} - f(x) \lim_{h \to 0} \frac{g(x+h) - g(x)}{h} \right]$$

Since f and g are differentiable,

$$\lim_{h \to 0} \frac{f(x+h) - f(x)}{h} = D_x[f(x)] \quad \text{and} \quad \lim_{h \to 0} \frac{g(x+h) - g(x)}{h} = D_x[g(x)]$$

Also, by Theorem 5.7, $\lim_{h \to 0} g(x+h) = g(x)$. Therefore,

$$D_x y = D_x \left[\frac{f(x)}{g(x)} \right] = \frac{g(x) D_x[f(x)] - f(x) D_x[g(x)]}{[g(x)]^2}$$

which is Rule 5.8. This rule is most easily applied when memorized in words.

Example 3 If $y = \frac{3x + 7}{5 - 2x}$, find $D_x y$.

Solution: Applying Rule 5.8, we have

$$D_x y = \frac{(5 - 2x) D_x[3x + 7] - (3x + 7) D_x[5 - 2x]}{(5 - 2x)^2}$$

$$= \frac{(5 - 2x)(3) - (3x + 7)(-2)}{(5 - 2x)^2}$$

$$= \frac{(15 - 6x + 6x + 14)}{(5 - 2x)^2} = \frac{29}{(5 - 2x)^2}$$

Example 4 If $y = \frac{x^2 - 5x}{x^2 + 3x}$, find $D_x y$.

Solution: Applying Rule 5.8, we get

$$D_x y = \frac{(x^2 + 3x) D_x[x^2 - 5x] - (x^2 - 5x) D_x[x^2 + 3x]}{(x^2 + 3x)^2}$$

$$= \frac{(x^2 + 3x)(2x - 5) - (x^2 - 5x)(2x + 3)}{(x^2 + 3x)^2}$$

$$= \frac{2x^3 + x^2 - 15x - 2x^3 + 7x^2 + 15x}{(x^2 + 3x)^2} = \frac{8x^2}{(x^2 + 3x)^2}$$

SEC. 5.8 Derivatives of Products, Quotients, and Powers

Derivative of a Function Raised to a Power

RULE 5.9. If f is a differentiable function and n is a positive integer or zero, then

$$D_x[f(x)^n] = n[f(x)]^{n-1}D_x[f(x)]$$

Proof: Let $y = f(x)^n$. It follows from Definition 5.7 that

$$D_x y = D_x[f(x)^n] = \lim_{h \to 0} \frac{[f(x+h)]^n - [f(x)]^n}{h}$$

Since n is a positive integer or zero, we can use the same special product used in the proof of Rule 5.3, in Section 5.2, to factor $[f(x+h)]^n - [f(x)]^n$, as follows:

$$[f(x+h)]^n - [f(x)]^n$$
$$= [f(x+h) - f(x)] \cdot \{[f(x+h)]^{n-1} + [f(x+h)]^{n-2}f(x)$$
$$+ \cdots + f(x+h)[f(x)]^{n-2} + [f(x)]^{n-1}\}$$

Hence

$$D_x y = \left[\lim_{h \to 0} \frac{f(x+h) - f(x)}{h}\right] \cdot \left[\lim_{h \to 0} \{[f(x+h)]^{n-1} + [f(x+h)]^{n-2}f(x) \right.$$
$$\left. + \cdots + f(x+h)[f(x)]^{n-2} + [f(x)]^{n-1}\}\right]$$

Since f is a differentiable function,

$$\lim_{h \to 0} \frac{f(x+h) - f(x)}{h} = D_x[f(x)]$$

Also, by Theorem 5.4, $\lim_{h \to 0} f(x+h) = f(x)$. Hence, by the limit theorems, we have

$$D_x y = D_x[f(x)]^n = n[f(x)]^{n-1}D_x[f(x)]$$

It follows from Rule 5.4 that:

RULE 5.9'. $D_x[kf(x)^n] = nk[f(x)]^{n-1}D_x[f(x)]$.

Example 5 If $y = (x^3 + 1)^{10}$, find $D_x y$.

Solution: Here $f(x) = x^3 + 1$ and $D_x[f(x)] = 3x^2$, hence applying Rule 6.9, we have

$$D_x y = 10(f(x))^9 D_x[f(x)] = 10(x^3 + 1)^9(3x^2) = 30x^2(x^3 + 1)^9$$

Example 6 If $y = 7(x^2 - 1)^{40}$, find $D_x y$.

Solution: Applying Rule 6.9' with $k = 7$, $f(x) = x^2 - 1$, and $n = 40$, we find

$$D_x y = (40)(7)f(x)^{39} D_x[f(x)] = (40)(7)(x^2 - 1)^{39}(2x) = 560x(x^2 - 1)^{39}$$

In this example it would have been most laborious to raise $(x^2 - 1)$ to the 40th power before differentiating.

Example 7 If $y = (2x + 1)^5(x^2 - 3)^4$, find $D_x y$.

Solution: Using the Product Rule, we get

$$\begin{aligned}D_x y &= (2x + 1)^5 D_x[(x^2 - 3)^4] + (x^2 - 3)^4 D_x[(2x + 1)^5] \\ &= (2x + 1)^5[4(x^2 - 3)^3(2x)] + (x^2 - 3)^4[5(2x + 1)^4(2)]\end{aligned}$$

Simplifying, by factoring, we obtain

$$\begin{aligned}D_x y &= (2x + 1)^4(x^2 - 3)^3[8x(2x + 1) + 10(x^2 - 3)] \\ &= (2x + 1)^4(x^2 - 3)^3(26x^2 + 8x - 30)\end{aligned}$$

Let us now show that Rule 6.9 is valid when n is a negative integer. To do this we let $y = [f(x)]^n$, where f is a differentiable function of x, $f(x) \neq 0$, and n is a negative integer. Since n is a negative integer, we can write $n = -m$, where m is a positive integer, then

$$y = [f(x)]^n = [f(x)]^{-m} = \frac{1}{[f(x)]^m}$$

We now apply Rule 6.8 to this quotient as follows,

$$D_x y = \frac{[f(x)]^m D_x(1) - (1)D_x[f(x)]^m}{[f(x)]^{2m}}$$

Since $D_x(1) = 0$ and $D_x[f(x)]^m = m[f(x)]^{m-1} D_x[f(x)]$ we can now write

$$D_x y = \frac{-m[f(x)]^{m-1} D_x[f(x)]}{[f(x)]^{2m}} = -m[f(x)]^{m-1-2m} \cdot D_x[f(x)]$$

Simplifying we get

$$D_x y = -m[f(x)]^{-m-1} \cdot D_x[f(x)]$$

But $-m = n$; hence

$$D_x y = n[f(x)]^{n-1} D_x[f(x)]$$

which shows that Rule 5.9 is valid for any integer n, positive, negative, or zero. It follows that Rule 5.9' is also valid for any integer n.

SEC. 5.8 Derivatives of Products, Quotients, and Powers

In the special case $f(x) = x$, $[f(x)]^n = x^n$ and $D_x[f(x)] = 1$; then

$$y = x^n$$

and

$$D_x y = n x^{n-1}$$

where n is any integer. The reader will probably recognize the last equation above as Rule 5.3 of Section 5.7 which was shown there to be valid for n a positive integer. We have now shown that Rule 5.3 is valid for any integer n.

Example 8 Given $y = 3x^{-7}$, find $D_x y$.

Solution: $D_x y = (-7)(3)x^{-7-1} = -21x^{-8}$

Example 9 If $y = \dfrac{10}{(x^2+1)^3}$, find $D_x y$.

Solution: $y = \dfrac{10}{(x^2+1)^3} = 10(x^2+1)^{-3}$

Applying Rule 5.9′, we obtain

$$D_x y = -30(x^2+1)^{-4} D_x[x^2+1]$$
$$= -30(x^2+1)^{-4}(2x) = -\dfrac{60x}{(x^2+1)^4}$$

Example 10 If $y = 4x^3 + \dfrac{3}{x^2} - \dfrac{7}{x} + 10$, find $D_x y$.

Solution: Rewriting the given equation in the form

$$y = 4x^3 + 3x^{-2} - 7x^{-1} + 10$$

and differentiating term by term, we get

$$D_x y = 12x^2 - 6x^{-3} + 7x^{-2}$$
$$= 12x^2 - \dfrac{6}{x^3} + \dfrac{7}{x^2}$$

Rules of Differentiation

RULE 5.1. $D_x C = 0$, where C is a constant
RULE 5.2. $D_x x = 1$
RULE 5.3. $D_x[x^n] = nx^{n-1}$, where n is a positive integer
RULE 5.4. $D_x[kf(x)] = k D_x[f(x)]$, where k is a constant
RULE 5.5. $D_x[f(x) + g(x)] = D_x[f(x)] + D_x[g(x)]$

RULE 5.6. $D_x[f_1(x) + f_2(x) + \cdots + f_n(x)]$
$= D_x[f_1(x)] + D_x[f_2(x)] + \cdots + D_x[f_n(x)]$

RULE 5.7. $D_x[f(x)g(x)] = f(x)D_x[g(x)] + g(x)D_x[f(x)]$

RULE 5.8. $D_x\left[\dfrac{f(x)}{g(x)}\right] = \dfrac{g(x)D_x[f(x)] - f(x)D_x[g(x)]}{[g(x)]^2}$, provided $g(x) \neq 0$

RULE 5.9. $D_x[f(x)^n] = n[f(x)]^{n-1} D_x[f(x)]$, where n is any integer

RULE 5.9'. $D_x[kf(x)^n] = nk[f(x)]^{n-1} D_x[f(x)]$, where k is a constant and integer

Exercises 5.8

Find $D_x y$ in Exercises 1 through 25.

1. $y = 4x^{-3}$

2. $y = -7x^{-2}$

3. $y = \dfrac{6}{x^2}$

4. $y = (2 + x^3)(x^2 - 4)$

5. $y = \dfrac{x-1}{x+1}$

6. $y = \dfrac{3x+2}{2x-5}$

7. $y = (2x+1)^3(x^2 - 4)$

8. $y = (x^3 - 1)(2x + 3)$

9. $y = (2x^2 + 3)^5$

10. $y = \dfrac{10}{(1-x)^2}$

11. $y = 2(x^3 + 3)^{20}$

12. $y = 8(1 - x^2)^5$

13. $y = \dfrac{x^2}{2x+3}$

14. $y = \dfrac{5x - 2}{4 - 3x}$

15. $y = x^3 + \dfrac{4}{x^2}$

16. $y = \dfrac{8}{4 + x^2}$

17. $y = \dfrac{8}{(4 + x^2)^2}$

18. $y = \dfrac{10}{(1-x)^2}$

19. $y = x^3(1 + 2x)^{20}$

20. $y = \dfrac{3x^4 - 2x^3 + 5x^2 - 4}{x^2}$

21. $y = \left(\dfrac{x-2}{x+3}\right)^2$

22. $y = \dfrac{10}{x}$

SEC. 5.9   Implicit Differentiation

23. $y = \dfrac{4}{3x}$ 24. $y = \dfrac{3x+2}{2-5x}$

25. $y = x^{10} + \dfrac{6}{x^3} + \dfrac{4}{x^2} - \dfrac{7}{x} - 3$

26. Find $D_t S$ if

   a. $S = (5t^2 + 3t)^2$ b. $S = \dfrac{6t}{4t^2 - 3}$ c. $S = t^3(t^2 - 1)^1$

   d. $S = (t^2 - t^{-2})^2$

5.9 IMPLICIT DIFFERENTIATION

Most of the functions we have encountered thus far have been of the form $y = f(x)$. We say that in this form y is expressed explicitly as a function of x, that is, that our equation is solved for y in terms of x. In contrast to this we often need to work with equations such as the following.

$$x^2 + y^2 = 25$$
$$x^2 - 2xy + y^2 = 1$$
$$3x^2 - 5xy + 9y^2 = 40$$
$$x^3 + x^2 y^2 + 3xy^4 - 8y^5 + 6 = 0$$

Each of these equations defines y as a function of x in the sense that if x be given a value the resulting equation determines one or more corresponding values of y. In such a case we say that the equation expresses y implicitly as a function of x or that y is an implicit function of x.

If often happens that an equation which determines y as an implicit function of x can be solved for y in terms of x. For example, the first three equations above are of this type. The last equation cannot be solved for y. However, it is possible to find $D_x y$ when it exists without solving for y.

Suppose we have the equation

$$y^2 = x^3$$

and we wish to find $D_x y$. Since y^2 is constantly equal to x^3, each of these quantities must change at the same rate with respect to x; that is,

$$D_x[y^2] = D_x[x^3]$$

Now, if we apply the rule for finding the derivative of a power function

(Section 5.7), we have

$$D_x[y^2] = 2yD_xy$$

and

$$D_x[x^3] = 3x^2$$

Hence

$$2yD_xy = 3x^2$$

Dividing by $2y$, we obtain

$$D_xy = \frac{3x^2}{2y} \qquad y \neq 0$$

This method of differentiation is called *implicit differentiation*.

Example 1 Find D_xy at the point $(1, 2)$, given $x^2 - 3xy + 2y^2 = 3$.

Solution: $D_x(x^2 - 3xy + 2y^2) = D_x(3)$. Now, by applying the rules for differentiating a sum, a power, $[f(x)]^n$, and a product $f(x)g(x)$, we can find D_xy thus:

$$D_x[x^2] + D_x[-3xy] + D_x[2y^2] = D_x[3]$$

$$2x - 3xD_xy + yD_x[-3x] + 4yD_xy = 0$$

$$2x - 3xD_xy - 3y + 4yD_xy = 0$$

or, combining like terms, we get

$$2x - 3y - (3x - 4y)D_xy = 0$$

Solving for D_xy, we have

$$D_xy = \frac{2x - 3y}{3x - 4y}$$

At the point $(1, 2)$,

$$D_xy = \frac{2 - 6}{3 - 8} = \frac{-4}{-5} = \frac{4}{5}$$

We have shown that the equation $D_x(x^n) = nx^{n-1}$ is valid when n is either a positive or a negative integer. We shall now use the method of implicit differentiation to show that this rule is valid for all rational values of n, provided we restrict our attention to positive values of x.

To do this, let $n = p/q$, where $q > 0$, and consider

$$y = x^{p/q}$$

where p and q are integers. This, of course, means that

$$y^q = x^p$$

Differentiating implicitly with respect to x, we obtain

$$qy^{q-1}D_xy = px^{p-1}$$

SEC. 5.9 Implicit Differentiation

If $y \neq 0$, then

$$D_x y = \frac{px^{p-1}}{qy^{q-1}}$$

But

$$y^{q-1} = (x^{p/q})^{q-1} = x^{p-(p/q)}$$

Hence

$$D_x y = \frac{p}{q} \cdot \frac{x^{p-1}}{x^{p-(p/q)}} = \frac{p}{q} x^{(p/q)-1}$$

which is equivalent to $D_x(x^n) = nx^{n-1}$ with n replaced by p/q. By a similar argument, it is easy to generalize this result and thus show that

$$D_x[f(x)]^n = n[f(x)]^{n-1} D_x[f(x)]$$

if f is any positive differentiable function of x and n is a rational number. Finally, it can be shown that the same equation holds for n any real number whatever.

Example 2

Given $y = x^{5/3}$, find $D_x y$.

Solution: $D_x y = \frac{5}{3} x^{5/3-1} = \frac{5}{3} x^{2/3}$

Example 3

Given $y = 7\sqrt{x^3}$, find $D_x y$.

Solution: In exponential form, $y = 7x^{3/2}$; hence $D_x y = \frac{21}{2} x^{1/2} = \frac{21}{2} \sqrt{x}$.

Example 4

Given $y = \frac{10}{\sqrt{x}}$, find $D_x y$.

Solution: $y = \frac{10}{\sqrt{x}} = \frac{10}{x^{1/2}} = 10x^{-1/2}$. Differentiating, we get

$$D_x y = -5x^{-3/2} = -\frac{5}{x^{3/2}} = -\frac{5}{\sqrt{x^3}}$$

Example 5

Given $y = \sqrt[3]{7 + 2x^4}$, find $D_x y$.

Solution: $y = \sqrt[3]{7 + 2x^4} = (7 + 2x^4)^{1/3}$. Using the rule for differentiating $[f(x)]^n$, we obtain

$$D_x y = \frac{1}{3}(7 + 2x^4)^{-2/3} D_x[7 + 2x^4]$$

$$= \frac{1}{3}(7 + 2x^4)^{-2/3}(8x^3)$$

$$= \frac{8x^3}{3(7 + 2x^4)^{2/3}} = \frac{8x^3}{3\sqrt[3]{(7 + 2x^4)^2}}$$

We have now proved that the rules $D_x(x^n) = nx^{n-1}$ and $D_x[f(x)]^n = n[f(x)]^{n-1} D_x[f(x)]$ are valid when n is a rational number and f is a positive differentiable function of x. An important special case of the second of these two rules is obtained by taking $n = \frac{1}{2}$. In this case, Rule 5.9 reduces to

$$D_x \sqrt{f(x)} = \frac{D_x[f(x)]}{2\sqrt{f(x)}}$$

Example 6 Find $D_x y$ if $y = \sqrt{7 + x^2}$.

Solution: Applying the rule just found, noting that $f(x) = 7 + x^2$ and $D_x[f(x)] = 2x$, we obtain

$$D_x y = \frac{2x}{2\sqrt{7 + x^2}} = \frac{x}{\sqrt{7 + x^2}}$$

Exercises 5.9

Find $D_x y$ in each of the following problems.

1. $y = 10x^{3/2}$
2. $y = 6x^{-1/2}$
3. $y = \dfrac{4}{\sqrt{x}}$
4. $y = \sqrt[8]{x^5}$
5. $y = 6\sqrt[3]{x^2}$
6. $y = \sqrt{9 - x^2}$
7. $y = \dfrac{4}{\sqrt{4 - x^2}}$
8. $y = \sqrt[3]{2x^3 + 3x^2}$
9. $y = \sqrt{x}\left(2x + \dfrac{5}{\sqrt{x}}\right)$
10. $y = 2\sqrt{x} + 3\sqrt[3]{x} + 4\sqrt[4]{x}$
11. $y = (2x^2 + 5)(x + \sqrt{x})$
12. $y = \sqrt[3]{x^3 + 2}$
13. $y = \sqrt{4x^2 - 6x + 7}$
14. $y = \dfrac{12}{\sqrt[3]{x^2}}$
15. $y = 2\sqrt{x} + \dfrac{4}{\sqrt{x}}$
16. $y = \sqrt{4 - x^4}$
17. $y = \dfrac{7}{\sqrt{2x - 3}}$
18. $x^2 + y^2 = 25$

SEC. 5.9　Implicit Differentiation

19. $x^2 - y^2 = 10$
20. $3x^2 + 2xy = 5$
21. $2xy - 2x + 3y = 15$
22. $x^2 + y^2 - 6x - 4y - 7 = 0$
23. $3x^2 + 2y^2 = 7$
24. $x^{2/3} + y^{2/3} = 8^{2/3}$
25. $y = (3x^2 + 4x)^{20}$
26. $y = x^2\sqrt{1 - x^2}$
27. $y = (x^3 + 5)^{4/3}$
28. $y = \dfrac{2x^3 - x^2 + 4}{\sqrt{x}}$
29. $y = x^3\left(3x + \dfrac{4}{\sqrt{x}}\right)$
30. $y = x\sqrt{4 - x^2}$
31. $y = \dfrac{x}{\sqrt{4 - x^2}}$
32. $xy = 8$
33. $x^2 - 3xy - 2y^2 = 10$

6
Applications of the Derivative

6.1 INTERPRETATION OF A DERIVATIVE

In Section 5.6 it was pointed out that the *derivative* was a general name given to all instantaneous rates. Hence, if we wish to find the instantaneous rate of change of a function, we find its derivative. The interpretation given to such a derivative depends on the kind of quantity represented by the function. We have previously introduced the word *velocity* to denote the instantaneous rate of change of position with respect to time. Thus, if the formula $s(t) = f(t)$ gives the position s in feet relative to the origin at time t in seconds of an object moving along the s axis, then $D_t s$ is the velocity of the object at time t, and we write $v = D_t s$. Similarly, the instantaneous rate of change of velocity is called *acceleration* and is written $a = D_t v$.

Example 1 A particle moves along an s axis in such a way that its position s feet relative to the origin after t seconds is given by the formula $s(t) = t^4 - 12t^3 + 40t^2$. Find

a. Velocity at any time t
b. Acceleration at any time t
c. Velocity and acceleration at $t = 2$

Solution: $s(t) = t^4 - 12t^3 + 40t^2$

a. Velocity: $v(t) = D_t[s(t)] = 4t^3 - 36t^2 + 80t$
b. Acceleration: $a(t) = D_t[v(t)] = 12t^2 - 72t + 80$

SEC. 6.1 Interpretation of the Derivative

c. When $t = 2$: $v(2) = 4(2)^3 - 36(2)^2 + 80(2) = 32 - 144 + 160$
$= 48$ ft/sec
$a(2) = 12(2)^2 - 72(2) + 80 = 48 - 144 + 80$
$= -16$ ft/sec^2

The negative sign here means that the velocity was decreasing at $t = 2$; that is, the object was slowing down.

Example 2 An object moves along an s axis in such a way that its position s meters relative to the origin at time t minutes is given by $s(t) = 40t - 5t^2$. Find

a. Its velocity
b. Its speed at $t = 5$

Solution: $s(t) = 40t - 5t^2$

a. Velocity: $v(t) = D_t[s(t)] = 40 - 10t$
b. First $v(5) = 40 - 50 = -10$ m/min. We recall that speed $= |v(t)|$; hence speed at $t = 5$ is $|-10| = 10$ m/min.

Slope of a Curve at a Point

The average slope of a curve over an interval along the x axis was defined in Section 5.2 as the difference in the heights of the curve at the endpoints of the interval divided by the length of the interval. Let $y = f(x)$ be an equation of the curve C in Figure 6.1. By definition, the average slope $\bar{m}$ of the curve C over the

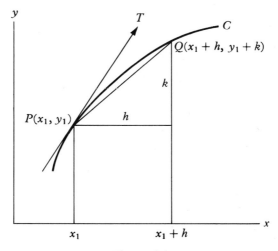

Figure 6.1

interval from $x = x_1$ to $x = x_1 + h$ is

(1) $$\bar{m} = \frac{f(x_1 + h) - f(x_1)}{h}, \qquad h \neq 0$$

We define the slope m of the curve C at the point $P(x_1, y_1)$ as follows:

DEFINITION 6.1.

(2) $$m = \lim_{h \to 0} \bar{m} = \lim_{h \to 0} \frac{f(x_1 + h) - f(x_1)}{h}$$

provided this limit exists. Thus the slope of the curve, m, at the point $p(x_1, y_1)$ is the value of $f'(x)$ at the point (x_1, y_1) and can be written $m = f'(x_1) = D_x y]_{x = x_1}$.

It should be noted that $\bar{m}$, defined by Equation (1), is precisely the slope of the secant PQ; see Figure 6.1. If the limit in Definition 6.1 exists, the secant PQ approaches a limiting position PT as h approaches zero. Such a line PT is called the *tangent line* to the curve C at the point P. The slope of PT is obviously m, the slope of the curve at P. This suggests the following definition.

DEFINITION 6.2.   A tangent line to a curve at a point P on the curve is the line passing through P and having the same slope as the slope of the curve at that point, assuming the curve has a slope at the point.

It should be pointed out that the above definition applies only to nonvertical tangent lines. Also, it is important to note that Definition 6.2 does not preclude the possibility of a line tangent to a curve at a point crossing the curve at that point or of a curve having two tangent lines at the same point; see Figure 6.2.

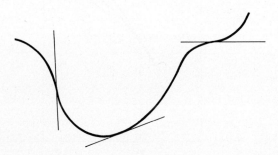

Figure 6.2

SEC. 6.1 Interpretation of the Derivative

Example 3 Find the slope of the curve $y = x^2$ at the point $(1, 1)$ and write an equation of the tangent line to the curve at the point $(1, 1)$.

Solution: Here $y = f(x) = x^2$ and $(x_1, y_1) = (1, 1)$. Let m be the slope of the given curve at the given point. It follows from Definition 6.1 that

$$m = D_x y = 2x]_{(1,1)} = 2$$

Hence the slope of the tangent line to the curve $y = x^2$ at the point $(1, 1)$ is also $m = 2$ by Definition 6.2. Using the point slope formula, $y - y_1 = m(x - x_1)$, we obtain $y - 1 = 2(x - 1) \to 2x - y = 1$ as an equation of the tangent line of the given curve at the given point.

The instantaneous rate of change of slope is called the *flexion* of the curve. Thus, flexion = $D_x m$.

Example 4 Find the slope of the curve $y = x^3 - 2x^2 + 3x - 5$ at the point for which $x = 1$.

Solution:
$y = x^3 - 2x^2 + 3x - 5$
$m = D_x y = 3x^2 - 4x + 3$
$m = 3 - 4 + 3 = 2$ when $x = 1$

Example 5 As water leaks out of a tank, the quantity, Q gallons, remaining in the tank after t minutes is given by the formula $Q = 500 - 40t + 0.6t^2$. How fast was Q changing at the end of 10 minutes?

Solution:
$Q = 500 - 40t + 0.6t^2$
$D_t Q = -40 + 1.2t$ (rate of change for any t)
$D_t Q = -40 + 12 = -28$ gal/min when $t = 10$

In economics it is frequently necessary to make use of the rate of change of one variable with respect to another. For instance, the *marginal cost* M_C is defined as the rate of change of the total cost of production with respect to the number of units produced. Thus, if n is the number of units produced and C is the total cost of producing these n units, the marginal cost indicates the increase in total cost of production per additional unit produced. If we treat n as a continuous variable, we see that the marginal cost is just the derivative of the cost C with respect to n or, in symbols,

$$M_C = D_n C$$

If M_C is less than the selling price, it pays to increase production, whereas if M_C is greater than the selling price, there is a loss on each additional item produced.

Similarly, *marginal revenue* M_R is defined as the rate of increase of total revenue R with respect to the increase in output x. Hence

$$M_R = D_x R$$

where, again, we have a particular interpretation of a derivative.

Example 6 | Find the marginal cost of producing n thousand units when the cost is given by the relation

$$C = 15 + 8n - 4n^2 + n^3$$

Solution: By definition, $M_C = D_n C$
$$= 8 - 8n + 3n^2$$

Figure 6.3 shows the graphs of the cost and the marginal cost plotted on the same coordinate axes. We see from these graphs that, while the cost continues to rise, the marginal cost decreases at first and then starts to rise again between $n = 1$ and $n = 2$. Thus after a certain point the marginal cost increases as more units are produced.

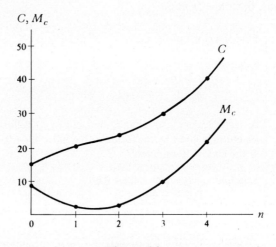

Figure 6.3

Exercises 6.1

1. An object is thrown vertically upward from the top of a building in such a way that its height, s feet, above the ground t seconds after it is thrown is given by the formula $s(t) = 100 + 80t - 16t^2$. Find
    a. A formula for its velocity v ft/sec at any time t
    b. A formula for the acceleration, a ft/sec^2, at any time t
    c. The initial velocity (velocity at time $t = 0$)
    d. The height above the ground, the velocity, and the acceleration of the object at the end of 2 seconds and at the end of 6 seconds.

2. The distance, x kilometers, of a car from a certain point t hours after starting was $x(t) = 60t - 5t^2$. Find formulas for its velocity and acceleration at any time t.

SEC. 6.1 Interpretation of the Derivative

Suppose that for a second car, starting at the same time as the first car, the distance $y(t)$ kilometers from the same point and at the same time t is $y(t) = 12 + 60t - 5t^2$. Find formulas for the velocity and acceleration of the second car. Compare results for the two cars. What is the significance of the constant term, 12, in the distance formula for the second car?

3. In Exercise 2, when did the first car stop and reverse its direction? How far was it from the starting point when this happened?

4. A car is traveling along a straight road, and its distance, s miles, from the starting point after t hours is given by the formula $s(t) = t^3 - 0.25t^4$. Find the velocity v and the acceleration a at the instants $t = 0$, $t = 1$, $t = 2$, and $t = 3$. On the same axes, sketch a graph representing s, v, and a, each as a function of t. Could you have anticipated the graph of v from the graph of s?

5. A ball was made to roll up an inclined plane and its distance, s feet, from its starting point after t seconds was $s(t) = 20t - 2t^2$. What was its velocity when $t = 6$? What was its speed when $t = 6$? How far up the incline did it go, measured from the starting point?

6. The cost of producing N radios in a certain factory is given by the relation $C(N) = 2,000 + 25N + 2N^2$, where C is in dollars. What is the marginal cost when the factory is producing 100 radios?

7. The revenue is defined as the product of the price of a commodity and the number of items of that commodity sold. If the price, P dollars, of a certain commodity is $P = 45 - 0.01x$, where x is the output per week, what are the weekly revenue R and the marginal revenue M_R when $x = 500$?

In each of the following exercises, find the slope of the given curve at the point indicated.

8. $y = \frac{1}{4}x^2$ at (2, 1)
9. $y = x^3$ at (0, 0)
10. $y = x^2 - 3x + 2$ at (2, 0)
11. $y = 3x - 5$ at (4, 7)
12. $y = x^3 - x^2 + 6x - 8$ at (−1, −16)
13. $y = x^2 - 6x$ at (3, −9)
14. $y = 1/x$ at (1, 1)
15. $y = 2x^3 - x^4$ at (2, 0)
16. $x^2 + y^2 = 25$ at (3, −4)
17. $x^2 - y^2 = 7$ at (4, 3)
18. $x^2 + y^2 - 2x + 4y - 4 = 0$ at (1, 1)
19. $y^2 = 4x$ at (4, 4)
20. $x^2 - 2xy + 3y^2 = 11$ at (−2, 1)
21. $3x^2 + 4y^2 = 7$ at (1, −1)

22. Write an equation of the tangent line to the given curve at the given point in each of Exercises 20 and 21.

23. A formula for the volume of a sphere is $V = \frac{4}{3}\pi r^3$. Find the rate of change of the volume with respect to its radius.

24. At what points on the curve $y = x^3 - 3x^2 - 12x + 10$ is the slope -3?

25. At what point on the curve $y = 4x - x^2$ is the tangent line parallel to the line $2x + y = 7$?

6.2 SUCCESSIVE DIFFERENTIATION

The derivative $D_x y$ of a function $y = f(x)$ is, in general, another function of x, say $f'(x)$. This function $f'(x)$ may have a derivative $f''(x)$, which in turn may have a derivative $f'''(x)$, and so on. In this connection we define the following symbols:

$$f''(x) = D_x[f'(x)] = D_x[D_x y] = D_x^2 y = \frac{d^2 y}{dx^2}$$

where the symbols $D_x^2 y$ and $\dfrac{d^2 y}{dx^2}$ are each read "the second derivative of y with respect to x." (The superscript 2's appearing in the symbols $D_x^2 y$ and $\dfrac{d^2 y}{dx^2}$ are not exponents and do not indicate a power of D_x or d or x. Note also the relative position of these superscripts to D_x and y and to d and y.) The *second derivative* is the derivative of the first derivative, $D_x y$. In the same manner,

$$f'''(x) = D_x[f''(x)] = D_x[D_x^2 y] = D_x^3 y = \frac{d^3 y}{dx^3}$$

where $D_x^3 y$ and $d^3 y/dx^3$ are read "the third derivative of y with respect to x," the third derivative being the derivative of the second derivative. It is possible for one to differentiate a function successively any number of times, say n times, so we make the general definition

$$D_x^n y = D_x[D_x^{n-1} y] = \frac{d^n y}{dx^n}$$

where n is a positive integer.

We have already met situations in which it was necessary to find the second derivative of a given function. For example, to find the acceleration of any object when its position relative to some fixed point is given as a function of the elapsed time, we first differentiated the position function to find the velocity v and then differentiated the velocity to obtain the acceleration. Thus, if the position s be given by the relation $s = f(t)$, the velocity v is

$$v = D_t s$$

and the acceleration a is

$$a = D_t v = D_t[D_t s] = D_t^2 s$$

Similarly, if $y = f(x)$ is the equation of a given curve, the slope m is

$$m = D_x y$$

and the flexion m' is

$$m' = D_x m = D_x[D_x y] = D_x^2 y$$

SEC. 6.3 Some Important Theorems

To find the rate at which the acceleration is changing, we take the derivative of the acceleration, which gives

$$D_t a = D_t[D_t^2 s] = D_t^3 s$$

a third derivative. So derivatives of higher order than the first are not at all unusual in mathematics.

Example 1 A particle traveled in such a way that its position relative to s feet from a given point after t seconds was $s = 20t^3 - t^4$. Find how fast the acceleration was changing at $t = 2$.

Solution:
$$s = 20t^3 - t^4$$
$$v = D_t s = 60t^2 - 4t^3$$
$$a = D_t v = D_t^2 s = 120t - 12t^2$$
$$r = D_t a = D_t^3 s = 120 - 24t$$

Therefore, when $t = 2$, $r = 120 - 48 = 72$ ft/sec^3.

Exercises 6.2

1. Find $D_x^4 y$ if:
    a. $y = 2x^5 - 5x^3$ b. $y = 6x^2 - x^5$ c. $y = x^5 - 2x^4 - x^3 + 5x^2 + 10x - 12$
    d. $y = 3x^4$

2. Find $D_x^2 y$ if:
    a. $y = \dfrac{4}{x}$ b. $y = \dfrac{x+2}{x-3}$ c. $y = 3x^2 - (2/x)$
    d. $y = 5x + 7 - \dfrac{100}{x^2}$ e. $x^2 - 3x + y^2 = 20$

3. The distance (S feet) traveled by an object in t minutes is given by the formula $S(t) = 10t^2 - 2t^4$. Find the acceleration of the object at the instant $t = 2$.

4. Find the instantaneous rate of change of the slope of the curve $y = 2x^3 + 3x - 5$ with respect to x when $x = 1$.

5. How fast is the acceleration of the object in Exercise 3 changing at the instant $t = 2$?

6.3 SOME IMPORTANT THEOREMS

In this section we shall have a look at some theorems which are basic to the discussions to follow in the remainder of this chapter. The first two of these theorems we shall state without proof. However, both seem true intuitively.

THEOREM 6.1. *If f is a function continuous on the closed interval $[a, b]$, then f has a maximum value, M, and a minimum value, m, on $[a, b]$.*

THEOREM 6.2 (INTERMEDIATE VALUE THEOREM). *If f is a continuous function on the closed interval $[a, b]$, with maximum value M and minimum value m, and if k is a number between m and M, then there is at least one number c on $[a, b]$ such that $f(c) = k$.*

THEOREM 6.3. *Suppose f is a function continuous on the closed interval $[a, b]$ with maximum value M and minimum value m on $[a, b]$. If $f(x_1) = M$ (or $f(x_1) = m$), where $a < x_1 < b$, and if f is differentiable at x_1, then $f'(x_1) = 0$.*

We shall prove the theorem for the case $f(x_1) = M$. The proof for the case $f(x_1) = m$ is obtained with precisely the same type of argument.

Proof: Since $f(x_1) = M$, the maximum value of f on $[a, b]$, then $f(x_1 + h) - f(x_1) \leq 0$ for all values of $h \neq 0$, and such that $a < x_1 + h < b$. Hence

$$\frac{f(x_1 + h) - f(x_1)}{h} \leq 0, \quad \text{if } h > 0$$

and

$$\frac{f(x_1 + h) - f(x_1)}{h} \geq 0, \quad \text{if } h < 0$$

It follows that

$$\lim_{h \to 0} \frac{f(x_1 + h) - f(x_1)}{h} \leq 0, \quad \text{if } h > 0$$

and

$$\lim_{h \to 0} \frac{f(x_1 + h) - f(x_1)}{h} \geq 0, \quad \text{if } h < 0$$

Now if f is differentiable at x_1, these two limits must have the same value, $f'(x_1)$. Therefore, $f'(x_1) \leq 0$ and $f'(x_1) \geq 0$; that is, $f'(x_1) = 0$, which proves our theorem.

THEOREM 6.4 (ROLLE'S THEOREM). *Let f be a function continuous on $[a, b]$ and differentiable in (a, b), and suppose $f(a) = f(b)$. Then there is at least one number $c \in (a, b)$ such that $f'(c) = 0$.*

Proof: Let M and m be the maximum and minimum values, respectively, of f on $[a, b]$. We need to consider two cases.

SEC. 6.3 Some Important Theorems

Case 1. If $M = m$, then $f(x) = M = m$ for all $x \in [a, b]$ and $f'(x) = 0$ at each point of $[a, b]$; the theorem is certainly true in this case.

Case 2. Suppose $M \neq m$; that is, $M > m$. Since $f(a) = f(b)$, f cannot have both its maximum value and its minimum value at an endpoint. Suppose $f(a) = f(b) \neq M$. Hence there is at least one point $x_1 \in (a, b)$ such that $f(x_1) = M$. By Theorem 6.2, $f'(x_1) = 0$ and our theorem is proved.

THEOREM 6.5 (THE MEAN VALUE THEOREM). *Let f be a function continuous on $[a, b]$ and differentiable in (a, b). Then there is at least one number $c \in (a, b)$ such that*

$$\frac{f(b) - f(a)}{b - a} = f'(c)$$

Proof: The endpoints of the graph of f are $A(a, f(a))$ and $B(b, f(b))$. See Figure 6.4. The slope of the segment AB is

$$m = \frac{f(b) - f(a)}{b - a}$$

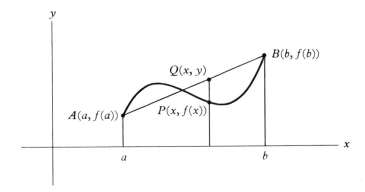

Figure 6.4.

Let $Q(x, y)$ and $P(x, f(x))$ be points on line AB and the graph of f, respectively, for each $x \in (a, b)$. Then

$$y = \frac{f(b) - f(a)}{b - a}(x - a) + f(a) \qquad \text{(equation of line } AB\text{)}$$

Now let $F(x) = y - f(x)$:

$$F(x) = \frac{f(b) - f(a)}{b - a}(x - a) + f(a) - f(x)$$

It can be shown that F is continuous on $[a, b]$ and differentiable in (a, b). Also it is easily seen that $F(a) = 0 = F(b)$. Thus F satisfies the hypothesis of Rolle's Theorem and hence, for some $c \in (a, b)$, $F'(c) = 0$. But if

$$F'(x) = \frac{f(b) - f(a)}{b - a} - f'(x) \quad \text{and} \quad F'(c) = 0$$

then

$$\frac{f(b) - f(a)}{b - a} = f'(c)$$

This proves our theorem.

A simple geometrical interpretation of the Mean Value Theorem is that there is at least one point of the graph of f between $A(a, f(a))$ and $B(b, f(b))$ at which the tangent to the graph is parallel to the chord AB.

THEOREM 6.6. *If the functions f and g are differentiable on $[a, b]$ and $f'(x) = g'(x)$ for all $x \in [a, b]$, then $g(x) = f(x) + C$, where C is a constant.*

Proof: Let $F(x) = g(x) - f(x)$; then F is continuous and differentiable on $[a, b]$ and the Mean Value Theorem applies. Hence, for each x on $[a, b]$, there is a $C \in (a, b)$ such that

$$\frac{F(x) - F(a)}{x - a} = F'(C)$$

But since $f'(x) = g'(x)$ for all $x \in [a, b]$, $F'(C) = 0$, and so

$$F(x) = F(a)$$

for all $x \in [a, b]$; that is,

$$g(x) - f(x) = F(a)$$

or

$$g(x) = f(x) + C$$

where $C = F(a)$.

6.4 INCREASING AND DECREASING FUNCTIONS

A function $f(x)$ is said to be an *increasing function* on the interval (a, b) if $f(x_1) < f(x_2)$ for any two numbers x_1 and x_2 in the interval such that $x_1 < x_2$. Similarly, $f(x)$ is said to be a *decreasing function* on the interval (a, b) if $f(x_1) > f(x_2)$ for any two numbers x_1 and x_2 in the interval such that $x_1 < x_2$.

It follows from these definitions that the graph of a function $y = f(x)$ rises

SEC. 6.4 Increasing and Decreasing Functions

as x increases over any interval where $f(x)$ is an increasing function, and that it falls as x increases over any interval where $f(x)$ is a decreasing function. In Figure 6.5 the curve rises as x increases from $x = b$ to $x = c$, falls from $x = c$ to $x = d$, and then rises again for values of x greater than $x = d$. A study of this graph suggests that the value of the derivative of a function for a particular value of x should determine whether the function is increasing or decreasing as x increases through that value. Some of these facts are expressed by the following theorems.

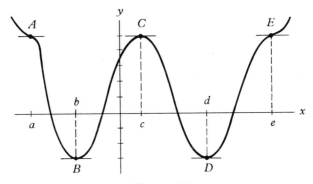

Figure 6.5

THEOREM 6.7. *If $f'(x) > 0$ for all x in some open interval (a, b), then $f(x)$ is an increasing function on the interval (a, b).*

THEOREM 6.8. *If $f'(x) < 0$ for all x in the open interval (a, b), then $f(x)$ is a decreasing function on the interval (a, b).*

We shall prove Theorem 6.7. The proof of Theorem 6.8 is obtained in a similar manner.

Proof: Let x_1 and x_2 be any two numbers such that $a < x_1 < x_2 < b$. Applying the Mean Value Theorem to the interval $[x_1, x_2]$, there is a number $c \in (x_1, x_2)$ such that

$$f(x_2) - f(x_1) = (x_2 - x_1)f'(c)$$

We have given that $f'(x) > 0$ for all $x \in (a, b)$, so $f'(c) > 0$. Also, since $x_1 < x_2$, $x_2 - x_1 > 0$, we have

$$f(x_2) - f(x_1) > 0 \quad \text{or} \quad f(x_1) < f(x_2)$$

Thus, by definition, f is an increasing function in (a, b).

It follows from Theorem 6.7 that if $f'(x_1) > 0$ and $f'(x)$ is continuous, there must exist some open interval including x_1 on which $f(x)$ is increasing.

Hereafter we shall imply this by the briefer statement "if $f'(x_1) > 0$, $f(x)$ is increasing at $x = x_1$." Similarly, the statement "$f(x)$ is decreasing at $x = x_1$" will mean that there exists an open interval including x_1 on which $f(x)$ is a decreasing function.

If $f'(x_1) = 0$, the function is said to be stationary at $x = x_1$. At such a point (see points A, B, C, D, and E, Figure 6.5) the curve has zero slope, and hence the line tangent to the curve at that point is parallel to the x axis. We shall refer to such a tangent line as a horizontal tangent.

Furthermore, if $f'(x_1) = 0$, the function could be increasing at $x = x_1$ (see point E, Figure 6.5). It could also be decreasing at $x = x_1$ (see point A, Figure 6.5). On the other hand, the function may be neither increasing nor decreasing at $x = x_1$ (see points B, C, and D, Figure 6.5).

Example 1

Determine whether the function $y = f(x) = x^3 - 4x^2 + 3x - 5$ is increasing or decreasing at $x = 1$.

Solution: $D_x y = f'(x) = 3x^2 - 8x + 3$
$f'(1) = 3 - 8 + 3 = -2$

Hence $f(x)$ is decreasing at $x = 1$.

Example 2

A stone is thrown vertically upward and its height, y feet, above the ground t seconds later is given by the formula $y = 80t - 16t^2$. Was the stone rising or falling when $t = 2$, when $t = 3$, and when $t = 2.5$?

Solution: $y = f(t) = 80t - 16t^2$
$D_t y = 80 - 32t$
$f'(2) = 80 - 64 = 16$ ft/sec

Since $f'(2) > 0$, the function $f(t)$ was increasing at $t = 2$; that is, the stone was rising at $t = 2$.

For $t = 3$ we have $f'(3) = 80 - 96 = -16$ ft/sec. Since $f'(3) < 0$, the stone was falling at $t = 3$.

For $t = 2.5$ we have $f'(2.5) = 80 - 80 = 0$. Thus at $t = 2.5$ the velocity $D_t y = 0$, so the stone was neither rising nor falling. Here it reached its highest point and was instantaneously at rest.

Example 3

For the curve $y = x^2 - 4x - 2$, find the range of values of x where the derivative is positive and the range where the derivative is negative. With the aid of this information, graph the function.

Solution: $D_x y = 2x - 4 = 2(x - 2)$

Therefore, $\qquad\qquad D_x y > 0 \qquad$ where $x > 2$

and $\qquad\qquad\qquad D_x y < 0 \qquad$ where $x < 2$

Also, $\qquad\qquad\qquad D_x y = 0 \qquad$ where $x = 2$

This analysis shows that the curve has a horizontal tangent at the point $(2, -6)$, that for all points to the left of $(2, -6)$ y decreases as x increases, and that for all points to the right of $(2, -6)$ y increases as x increases (see Figure 6.6).

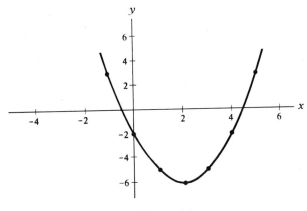

Figure 6.6

Concavity

It follows from Theorem 6.7 that, if $f''(x) > 0$ for all values of x in an open interval, then $f'(x)$ increases with x on that interval. Likewise it follows from Theorem 6.8 that, if $f''(x) < 0$ for all values of x on an open interval, $f'(x)$ is a decreasing function on that interval. Also, if $f'(x)$ exists for each x in an interval, the graph of the function f is a continuous curve with a tangent line at each point. If we think of a point P as moving along the curve, the tangent line at P will also move along the curve having at each point the direction of the curve at that point. We say the curve has a continuously turning tangent. If $f'(x)$ increases with x, the tangent line turns in a counterclockwise manner and hence always lies below the curve, whereas if $f'(x)$ decreases with increasing x, the tangent line turns in a clockwise manner and lies above the curve. A curve is said to be *concave upward* at a point if the curve lies above its tangent line at the point and is said to be *concave downward* if the curve lies below its tangent line at the point. For example, the curve in Figure 6.7 is concave downward at A and at all the points between A and C. It is concave upward at all the points between C and E. However, at C the curve is changing from concave downward to concave upward and the tangent line to the curve at C crosses the curve there.

The curve is said to have a *point of inflection* at C. In general, a point at which a curve changes from concave downward to being concave upward, or vice versa, is called a point of inflection. More precisely, a curve $y = f(x)$, with a continuously turning tangent, has a point of inflection at $(x_0, f(x_0))$ if for some interval containing x_0 the curve has one direction of concavity at all the points

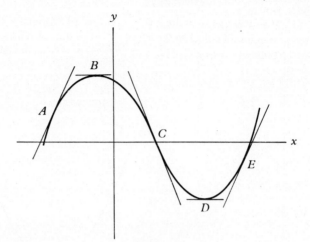

Figure 6.7

$(x, f(x))$ in the interval such that $x < x_0$ and has the opposite direction of concavity at all points $(x, f(x))$ in the interval such that $x > x_0$.

The results of the above discussion may be summarized in the following two theorems.

THEOREM 6.9. *The graph of the function $y = f(x)$ is concave downward at $x = x_1$ if $f''(x_1) < 0$ and is concave upward at $x = x_1$ if $f''(x_1) > 0$.*

This theorem follows from the fact that, if $f''(x_1) < 0$, then $f'(x)$ is a decreasing function in some interval containing x_1, and if $f''(x_1) > 0$, then $f'(x)$ increases with increasing x in some interval containing x_1.

THEOREM 6.10. *If (x_1, y_1) is a point of inflection of the curve $y = f(x)$ and if $f''(x_1)$ exists, then $f''(x_1) = 0$.*

Proof: If (x_1, y_1) is a point of inflection then there exists an open interval containing x_1 and such that $f'(x_1)$ is either a maximum or minimum value of the function $f'(x)$ in that interval, and hence it follows from Theorem 6.9 that $f''(x_1) = 0$, if it exists.

Theorems 6.9 and 6.10 suggest the following test for a point of inflection. The point (x_1, y_1) is a point of inflection of the curve $y = f(x)$ if both

1. $f''(x_1) = 0$
2. $f''(x)$ changes sign as x increases through x_1

Furthermore, if $f''(x_1) = 0$ but $f''(x)$ does not change sign as x increases through x_1, then (x_1, y_1) is not a point of inflection.

SEC. 6.4 Increasing and Decreasing Functions

Example 4

Test the curve $y = f(x) = x^3 - 3x^2 - 9x + 7$ for points of inflection.

Solution: We apply the test suggested above as follows.
$$y = f(x) = x^3 - 3x^2 - 9x + 7$$
$$f'(x) = 3x^2 - 6x - 9$$
$$f''(x) = 6x - 6 = 6(x - 1)$$
So $f''(x) = 0$ when $x = 1$

If $x < 1$, $f''(x) < 0$, and if $x > 1$, $f''(x) > 0$; hence $f''(x)$ changes sign as x increases through $x = 1$. Therefore, $(1, -4)$ is a point of inflection.

Exercises 6.4

1. The position relative to the origin of a particle moving along the s axis in t seconds was $s = 10t^3 - t^4$. Was the velocity increasing or decreasing at $t = 3$? Was the acceleration increasing or decreasing at $t = 3$?

2. Find the slope and the flexion of the curve $y = x^3 - 4x + 10$ at the point $(1, 7)$. Was the slope increasing or decreasing at this point?

3. In t seconds after the brakes were applied, a train traveled a distance, s feet, given by the formula $s = 35t - 2t^2$. How fast was the velocity decreasing at $t = 5$?

4. A particle moves along the x axis according to the law $x = 7 - 15t + 9t^2 - t^3$. Find the velocity at the instant the acceleration is zero. Also find the acceleration at each instant the velocity is zero.

5. A ball was thrown vertically upward from a rooftop 112 feet high. Its height, s feet, above the ground after t seconds was $s = 112 + 96t - 16t^2$. Find the interval of time during which s was increasing. When was the ball highest; that is, when was the velocity zero? How high above the ground did the ball rise? When did the ball reach the ground and with what velocity?

6. A particle moves along the x axis in such a way that its position, x, relative to the origin t seconds after starting is given by the formula $x = t^3 - 6t^2 + 9t + 4$. Find the position, velocity, and direction of motion of the particle at the start. Also find when and where the particle stopped and reversed its direction.

7. Same as Exercise 6, for $x = 2t^3 - 21t^2 + 60t - 10$.

For each of the following functions, find the intervals in which f is increasing, f is decreasing, the graph is concave upward, concave downward, and the values of x where $f'(x) = 0$. Sketch the graph of each function.

8. $f(x) = 3 + 4x - x^2$

9. $f(x) = x^2 - 10x$

10. $f(x) = x^4$

11. $f(x) = x^3$

12. $f(x) = x^3 - 3x^2 + 3x + 5$

13. $f(x) = 2x^3 - 3x^2 - 12x + 5$

14. $f(x) = 3x^4 - 4x^3 - 5$ 15. $f(x) = x^3 - 12x - 10$

16. $f(x) = \sqrt{4 - x^2}$ 17. $f(x) = x + \dfrac{4}{x}$

18. For the graph of each function listed in Exercises 10 through 15, find the point(s) of inflection if such exist.

6.5 MAXIMA AND MINIMA

Let f be a function continuous in a closed interval $[a, b]$. The function f is said to have a *relative maximum* value at $x = x_1$ if any one of the following three sets of conditions hold.

1. x_1 is not an endpoint of the interval $[a, b]$ and for some $h > 0, f(x_1) \geq f(x)$ when $x_1 - h < x < x_1 + h$.
2. x_1 is the left-hand endpoint of the interval $[a, b]$, that is, $x_1 = a$, and for some $h > 0, f(x_1) \geq f(x)$ when $x_1 < x < x_1 + h$.
3. x_1 is the right-hand endpoint of $[a, b]$, that is, $x_1 = b$, and for some $h > 0$, $f(x_1) \geq f(x)$ when $x_1 - h < x < x_1$.

Similarly, the function f is said to have a *relative minimum* value at $x = x_2$ if one of the following three sets of conditions holds.

1. x_2 is not an endpoint of the interval $[a, b]$ and for some $h > 0, f(x_2) \leq f(x)$ when $x_2 - h < x < x_2 + h$.
2. x_2 is the left-hand endpoint of the interval $[a, b]$, that is, $x_2 = a$, and for some $h > 0, f(x_2) \leq f(x)$ when $x_2 < x < x_2 + h$.
3. x_2 is the right-hand endpoint of the interval $[a, b]$, that is, $x_2 = b$, and for some $h > 0$, $f(x_2) \leq f(x)$ when $x_2 - h < x < x_2$.

It is important to understand that the word *relative* as used in the above definitions means that $f(x_1)$ is a maximum, or minimum, when compared with $f(x)$ at values of x nearby x_1. In such a case, $f(x_1)$ may or may not be the largest, or smallest, value of f on $[a, b]$. If it happens that $f(x_1) \geq f(x)$ for all values of x on $[a, b]$, that is, $f(x_1) = M$ (see Theorem 6.1), then $f(x_1)$ is called the *absolute maximum* value of f on $[a, b]$. If $f(x_2) \leq f(x)$ for $x \in [a, b]$, that is, if $f(x_2) = m$, then $f(x_2)$ is called the *absolute* minimum value of f on $[a, b]$. It should be noted that every absolute maximum is automatically a relative maximum, and every absolute minimum is at the same time a relative minimum.

It can be seen from Figure 6.8 that, for a function f defined on a closed interval, the values of f at the endpoints can be among the relative maxima and relative minima of the function, and in many instances they are. Thus, if we are investigating a function defined on a closed interval to find all its relative

SEC. 6.5 Maxima and Minima

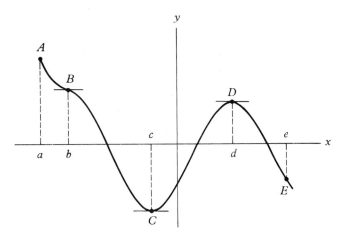

Figure 6.8

maxima and relative minima, *we must always check the values of the endpoints.* Checking the endpoints is routine and easily done. However, relative maxima and relative minima may occur at values of *x* within the interval, and these are not so easily found.

To aid us in finding relative maxima and relative minima which occur at interior points of the interval, we prove the following theorem.

THEOREM 6.11. *Suppose f is a function continuous on a closed interval $[a, b]$, and suppose $a < x_1 < b$. If $f(x_1)$ is a relative maximum (minimum) value of f, and $f'(x_1)$ exists, then $f'(x_1) = 0$.*

Proof: This theorem follows as a direct consequence of Theorem 6.3 and the definitions of maximum and minimum values at interior points of an interval. For if $f(x_1)$ is a relative maximum value of f and $a < x_1 < b$, there exists an interval $(x_1 - h, x_1 + h)$, contained in (a, b), such that $f(x_1)$ is the *maximum* value of f in $(x_1 - h, x_1 + h)$. Hence, by Theorem 6.3, if $f'(x_1)$ exists, then $f'(x_1) = 0$. A similar statement holds if $f(x_1)$ is a relative minimum.

Unfortunately, the converse of Theorem 6.11 is not true. Referring again to Figure 6.8, we see that the curve has a horizontal tangent at the point *B*. Hence $f'(b) = 0$, but $f(b)$ is neither a relative maximum nor a relative minimum value. Thus $f'(x_1) = 0$ does not imply that $f(x_1)$ is a relative maximum or a relative minimum; it may be neither. For example, let f be defined by the equation $f(x) = x^3$. Then $f'(x) = 3x^2$ and $f'(0) = 0$. However, $f(0)$ is neither a relative maximum nor a relative minimum because $f'(x) > 0$ if $x \neq 0$; hence f is an increasing function on both sides of $x = 0$. See Figure 6.9.

The above discussion and example show that $f'(x_1) = 0$ is no guarantee

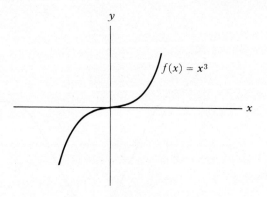

Figure 6.9

that $f(x_1)$ is a relative maximum value or a relative minimum value of f, even though $x = x_1$ is not an endpoint of the domain of definition of f. Conversely, $f(x_1)$, being a relative maximum or relative minimum, is no guarantee that $f'(x_1) = 0$. In fact, it is possible for $f(x_1)$ to be a relative maximum (or relative minimum) and for $f'(x_1)$ not to exist. To see that this is true, consider the function $f(x) = |x|$ defined on the closed interval $[-2, 2]$. Since $|x| = x$ if $x \geq 0$ and $|x| = -x$ if $x < 0$, we see that the graph of f, shown in Figure 6.10 consists of two straight line segments. The portion of the graph that lies in the first quadrant is a segment of the line $y = x$, and the portion of the graph that lies in the second quadrant is a segment of the line $y = -x$. It follows that $f'(x) = D_x(x) = 1$ for each $x > 0$, and $f'(x) = D_x(-x) = -1$ for each $x < 0$. Note also that $f(0) = 0$, and hence the origin is a point on the graph of f. However, it can be shown that f is not differentiable at $x = 0$; that is, $f'(0)$ does not exist. And yet $f(0) = 0$ is a relative minimum value of f. In fact, $f(0) = 0$ is an absolute minimum value of f in the closed interval $[-2, 2]$.

We shall not stress nondifferentiability in this text. We point out, however,

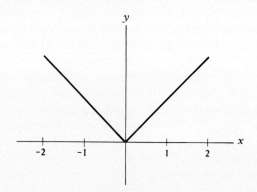

Figure 6.10

that, geometrically, a function f is not differentiable at $x = x_1$ if the graph of f has a sharp corner at the point $(x_1, f(x_1))$. The point $(0, 0)$ is such a point on the graph in Figure 6.10.

Summary. For a function f defined on a closed interval $[a, b]$, in order to find all the relative maximum values and all the relative minimum values of f on $[a, b]$, we must examine

a. all points in (a, b) at which $f'(x) = 0$
b. all points in (a, b) at which $f'(x)$ fails to exist
c. the two endpoints $x = a$ and $x = b$

A number $x = x_1$ in (a, b) at which $f'(x_1) = 0$, or any number x on $[a, b]$ at which $f'(x)$ does not exist is called a *critical value* of x for the function f. Note that this includes the endpoints, since for any function f defined on a closed interval $[a, b]$, $f'(a)$ and $f'(b)$ do not exist. The corresponding points $(a, f(a))$, $(b, f(b))$, and all other points $(x, f(x))$, where x is a critical value, are called *critical points* on the graph of f. Hence all the relative maximum points and all the relative minimum points which lie on the graph of f are critical points, but the converse is not true. Some critical points are neither relative maximum points nor relative minimum points.

For the important applied problems of the next section, we shall be interested only in absolute maxima and absolute minima. Moreover, each function will be defined on a closed interval which makes the determination of the absolute maximum and the absolute minimum quite easy. One simply locates all critical values (including the endpoints), evaluates the function at each critical value, and selects the largest such functional value as the absolute maximum and the smallest such functional value as the absolute minimum. The question of whether or not a particular critical value yields a relative maximum or a relative minimum is a bit more complicated. Such information is particularly useful in sketching graphs of functions.

In each problem we deal with, every point x at which $f'(x)$ does not exist (including the endpoints) will yield either a relative maximum or a relative minimum point. We hasten to add that this need not be the case in general. If x is a critical value of f and $f'(x) = 0$, it is desirable to have simple tests we can apply and thereby determine whether $f(x)$ is a relative maximum value of f, a relative minimum value of f, or neither. The following two tests are designed for that purpose.

The First Derivative Test

Suppose f is a function continuous on some interval $[a, b]$ and differentiable in (a, b), except perhaps at finitely many points at each of which the derivative fails to exist. Let x_1 be a critical value of x in (a, b). Select two numbers, t_1 and t_2, in (a, b) such that $t_1 < x_1 < t_2$ and x_1 is the only critical value of x on $[t_1, t_2]$. Then

a. $f(x_1)$ is a relative maximum value of f if
   1. $f'(t_1) > 0$
   2. $f'(t_2) < 0$
b. $f(x_1)$ is a relative minimum value of f if
   1. $f'(t_1) < 0$
   2. $f'(t_2) > 0$
c. $f(x_1)$ is neither a relative maximum value of f nor a relative minimum value of f if $f'(t_1)$ and $f'(t_2)$ have the same sign.

Important: Note, in the above test, that the requirement that x_1 be the only critical value of x on the closed interval $[t_1, t_2]$ means that neither t_1 nor t_2 can be a critical value of x. This restriction on t_1 and t_2 is in most cases no handicap in the selection of these numbers, for usually a wide variety of choices is readily available. However, in certain special cases such a selection is impossible. For example, if f is a constant function defined by $f(x) = k$ and x_1 is a critical value of x, then numbers t_1 and t_2 do not exist such that x_1 is the only critical value on the closed interval $[t_1, t_2]$. This is because $f'(x) = 0$ for all values of x; hence every value of x is a critical value. We shall illustrate, with the following example, a typical procedure for applying the first derivative test.

Example 1

For the function f defined for all values of x by the equation $f(x) = x^4 - 8x^2 + 10$, find all the critical values of x and determine for each whether the corresponding value of f is a relative maximum, a relative minimum, or neither.

Solution: $f(x) = x^4 - 8x^2 + 10 \quad (-\infty, \infty)$
$f'(x) = 4x^3 - 16x$
$\quad\quad = 4x(x+2)(x-2)$

The critical values in this case are the solutions of $f'(x) = 0$, that is, of

$$x(x+2)(x-2) = 0$$

that is, -2, 0, and 2. We now apply the First Derivative Test to each of these critical values separately and then indicate a procedure for testing all at the same time. We start by letting $x_1 = -2$. We can choose for t_1 any number less than -2 and for t_2 any number between -2 and 0. For convenience, we take $t_1 = -3$ and $t_2 = -1$. Now we have $f'(-2) = 0$, $f'(-3) = -12(-1)(-5) = -60 < 0$, and $f'(-1) = -4(1)(-3) = 12 > 0$. Thus, by the First Derivative Test, $f(-2) = -6$ is a relative minimum. Similarly, if we let $x_1 = 0$, $t_1 = -1$, and $t_2 = 1$, then $f'(0) = 0$, $f'(-1) = 12$, and $f'(1) = -12$; hence $f(0) = 10$ is a relative maximum. Likewise, if we let $x_1 = 2$, $t_1 = 1$, and $t_2 = 3$, then $f'(2) = 0$, $f'(1) = -12$, and $f'(3) = 60$. Therefore, $f(2) = -6$ is also a relative minimum. We could have tested all three

SEC. 6.5 Maxima and Minima

critical values at once by recording the desired information in the following tabular form:

x	-3	-2	-1	0	1	2	3
$f'(x)$	-60	0	12	0	-12	0	60

This table shows at a glance the sign of $f'(x)$ for values of x on both sides of each critical value. Now applying the First Derivative Test we see immediately that $f(-2) = -6$ is a relative minimum value of f, $f(0) = 10$ is a relative maximum value of f, and $f(2) = -6$ is a relative minimum value of f. The graph of the given function is shown in Figure 6.11.

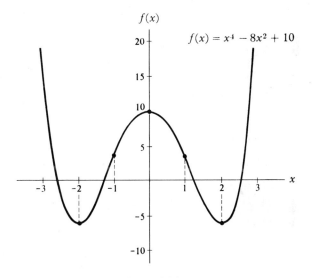

Figure 6.11

Second Derivative Test

Suppose f is a function continuous on $[a, b]$ and differentiable in (a, b). Let x_1 be a point in (a, b) such that $f'(x_1) = 0$. Then

(a) $f(x_1)$ is a relative maximum if $f''(x_1) < 0$
(b) $f(x_1)$ is a relative minimum if $f''(x_1) > 0$
(c) the test cannot be applied if $f''(x_1) = 0$

This is the simpler test to use if the second derivative is easy to calculate. The last example above is a case in point. We repeat the solution of that example here, but using the Second Derivative Test.

Solution: $f(x) = x^4 - 8x^2 + 10$
$f'(x) = 4x^3 - 16x = 4x(x+2)(x-2)$
$f''(x) = 12x^2 - 16$

The critical values of x are $-2, 0$, and 2. $f''(-2) = 32 > 0$, $f''(0) = -16 < 0$, and $f''(2) = 32 > 0$, showing that $f(-2) = -6$ and $f(2) = -6$ are relative minimum values of f and that $f(0) = 10$ is a relative maximum value of f. This agrees with the results found above by the First Derivative Test.

We point out that the First Derivative Test will enable one to answer the question, "Is $f(x)$ an absolute maximum or an absolute minimum of f?" for each function f and each critical value x of f in this text. The second derivative test is ordinarily easier to check (an advantage), but it does not apply in the case $f''(x) = 0$ (a disadvantage). So a recommended procedure is to apply the Second Derivative Test first if f'' can be easily calculated and, if the Second Derivative Test does not apply, that is, if $f''(x) = 0$, then apply the First Derivative Test.

Finding Absolute Maximum and Absolute Minimum Values of a Function f Continuous on a Closed Interval $[a, b]$

Procedure:
1. Find $f(a)$ and $f(b)$.
2. Find the value of f at each critical value of x in (a, b).
3. The largest value of f found in Steps 1 and 2 is the absolute maximum value of f on $[a, b]$.
4. The smallest of the values of f found in steps 1 and 2 is the absolute minimum value of f on $[a, b]$.

Example 2 | Find the absolute maximum value of f and the absolute minimum value of f on the closed interval $[0, 3]$ if $f(x) = 2 + 6x - 3x^2$.

Solution: The endpoints of the given closed interval are $a = 0$ and $b = 3$. Since $f'(x) = 6 - 6x = 6(1 - x)$, the only critical value of x is $x = 1$. We now calculate $f(0) = 2$, $f(3) = -7$, and $f(1) = 5$. Comparing these numbers, we see that the absolute maximum value of f is $f(1) = 5$ and the absolute minimum of f is $f(3) = -7$.

Exercises 6.5

For each of Exercises 1 through 16, find all the critical values of x and determine for each whether the corresponding value of f is a relative maximum, a relative minimum, or neither. Use this information to sketch the graph in each case.

1. $f(x) = 4x - x^2$ **2.** $f(x) = 3x^2 - 12x + 7$
3. $f(x) = x^3 - 12x + 2$ **4.** $f(x) = x^3 - 9x^2 + 24x - 7$

SEC. 6.6   Applied Problems in Maxima and Minima   331

5. $f(x) = x^4 - 4x^3 + 5$ 6. $f(x) = 2x^3 - 12x^2 - 15$

7. $f(x) = 3x^4 + 4x^3 - 12x^2 + 10$ 8. $f(x) = |x - 2|$

9. $f(x) = x^3 + 3x^2 - 9x + 4$ 10. $f(x) = 5x^3 - 3x^5 + 1$

11. $f(x) = x^3 - 3x^2 + 3x + 2$ 12. $f(x) = 12x^5 - 15x^4 - 40x^3 + 5$

13. $f(x) = x + \dfrac{9}{x}, \; x > 0$ 14. $f(x) = x + \dfrac{4}{x - 2}, \; x \neq 2$

15. $f(x) = 4\sqrt{x} - x$ 16. $f(x) = |x^2 - 1|$

17. Find the absolute maximum value and the absolute minimum value of the function $f(x) = x^2 - 4x + 5$ on the closed interval $[-1, 3]$.

18. Find the absolute maximum value and the absolute minimum value of the function $f(x) = x^3 - 3x^2 - 9x + 10$ on the closed interval $[-2, 4]$.

19. Find the absolute minimum value of f on the closed interval $[2, 8]$ if $f(x) = 2x + \dfrac{32}{x}$.

20. A certain rod is 20 inches long. Its temperature, T degrees, x inches from one end is given by the formula $T = 20x - x^2$. Find where the rod is hottest and how hot it is at that point.

6.6 APPLIED PROBLEMS IN MAXIMA AND MINIMA

The theory of relative maximum and relative minimum values of functions discussed in the last section is of great use in the solution of applied problems that require the determination of maximum (or minimum) values of a quantity which is expressible as a function of a single independent variable defined on a closed interval. For convenience in our discussion here, let us represent by the letter Q the quantity that is to be made a maximum (or minimum). Strictly speaking, the letter used to represent such a quantity is a matter of personal preference. It is conventional, however, to use a letter that seems appropriate in each case. For example, if the quantity to be made a maximum (or minimum) is an area, we would probably use the letter A to represent it; if the quantity is a volume, we would perhaps represent it by the letter V; or if it was a cost, we might use the letter C; and so on.

In this text we shall consider two cases.

CASE 1:   The quantity Q involves only one variable. We illustrate this case with the following example.

Example 1   A rectangular box with no top is to be made from a piece of cardboard 8 inches long and 5 inches wide by cutting from each corner the same-sized square and turning up the resulting edges to form the sides and ends of the box. Find the dimensions of the square that should be cut from each corner so that the resulting box will have the maximum volume of any such box that can be made from the given piece of cardboard.

Solution:
Step 1. We read the problem carefully to determine what quantity is to be made a maximum (or minimum). In this problem we see that the quantity under consideration is the volume of a rectangular box and that this volume is to be made a maximum. We shall use the letter V to represent this volume.

Step 2. Next we draw a figure to illustrate the problem and label the important parts, letting x (inches) represent the length of a side of the square to be cut from each corner of the piece of cardboard. See Figure 6.12

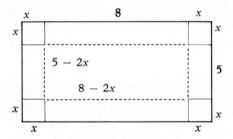

Figure 6.12

Step 3. We observe from Figure 6.12 that the volume V involves only one variable, x, and that, when we remove a square from each corner as indicated and fold up the edges along the dashed lines, the bottom of our box will be a rectangle $(8 - 2x)$ inches long and $(5 - 2x)$ inches wide. Also, the box will be x inches deep. Hence the volume of the box is $V(x) = x(8 - 2x)(5 - 2x)$ cubic inches.

Step 4. Since each square cut from a corner is to have the same size, it follows that x cannot have a value greater than 2.5 inches. Also, x cannot be negative. Thus the domain of definition of $V(x)$ is the closed interval $[0, 2.5]$.

Step 5. We next find the critical values of x for the function $V(x)$. Two critical values of x are the endpoints $x = 0$ and $x = 2.5$. We must now find all the values of x, if any, for which $V'(x) = 0$. We have

$$V(x) = x(8 - 2x)(5 - 2x) = 40x - 26x^2 + 4x^3, \quad 0 \leq x \leq 2.5$$

Hence

$$V'(x) = 40 - 52x + 12x^2$$

Factoring, we get

$$V'(x) = 4(10 - 3x)(1 - x)$$

Setting $V'(x) = 0$ and solving for x, we see that $V'(x) = 0$ if $x = \dfrac{10}{3}$ or $x = 1$.

But $x = \dfrac{10}{3}$ is outside the domain of definition. Hence the only critical value of x in $[0, 2.5]$ is $x = 1$. Therefore, the critical values of x for $V(x)$ are $x = 0$, $x = 2.5$, and $x = 1$.

Step 6. We find that $V(0) = V(2.5) = 0$ and $V(1) = 18$.

Step 7. Since $V(1) = 18$ is the largest of the values found in step 6, we have now shown that the maximum value of V for any x in the domain of definition is $V = 18$ cubic inches.

Step 8. Rereading our problem, we are reminded that we were asked to find the dimensions of the square to be cut from each corner of the given piece of cardboard so that the volume of the resulting box would be a maximum. Since $V(1) = 18$ is the maximum volume, the answer to our problem is: To obtain a box of maximum volume, we should cut from each corner a square of 1 inch on each side.

CASE 2: The quantity Q involves two variables, and the statement of the problem contains a restraining condition which makes it possible to express one of these variables in terms of the other. This case is illustrated in Example 2.

Example 2

A farmer has 100 yards of fencing with which he wishes to enclose a rectangular plot. What dimensions should be used if the area of the enclosed plot is to be a maximum?

Solution:

Step 1. From reading the problem, we learn that the quantity to be made a maximum is the area of rectangular plot. Let us represent this area by the letter A.

Step 2. We draw a figure to represent the problem, using x and y to represent the dimensions of the rectangle. See Figure 6.13.

Figure 6.13

Step 3. Here we observe from Figure 6.13 that there are two variables involved, x and y. Since the plot is to be rectangular, it follows that the area is given by the formula $A = xy$. We now have A expressed in terms of two variables. But the problem states that the farmer has only 100 yards of fencing to enclose this rectangular plot. Hence $2x + 2y = 100$ or solving for y ($y = 50 - x$), which is the restraining condition which enables us to express A in terms of a single variable, say x. Thus $A(x) = x(50-x) = 50x - x^2$.

Step 4. It follows from the equation $y = 50 - x$ and the physical fact that neither x nor y can be negative, that the domain of definition of $A(x)$ is $0 \leq x \leq 50$.

Step 5. Next we find the critical values of x for the function $A(x)$. Again we have the endpoints $x = 0$ and $x = 50$ as critical values. Let us now find all the values of x in $[0, 50]$ for which $A'(x) = 0$. Starting with

$$A(x) = 50x - x^2, \qquad 0 \leq x \leq 50$$

and differentiating we find

$$A'(x) = 50 - 2x$$

and hence $A'(x) = 0$ if $x = 25$. So the critical values of x are $x = 0$, $x = 50$, and $x = 25$.

Step 6. Substituting the critical values of x into our formula for $A(x)$ we obtain

$$A(0) = A(50) = 0 \quad \text{and} \quad A(25) = 625 \text{ square yards}$$

Step 7. The largest value of A found in step 6 was $A(25) = 625$. Hence the maximum area of a rectangular plot that the farmer can enclose with 100 yards of fencing is 625 square yards.

Step 8. A rereading of the problem shows that we are asked to find the dimensions of the largest rectangular plot that can be enclosed with 100 yards of fencing. We found one of the dimensions in step 5, $x = 25$ yards. To find the other dimension, we use the equation $y = 50 - x$ and substitute $x = 25$, getting $y = 25$. Hence the dimensions requested are $x = 25$ and $y = 25$; that is, the plot must be a square.

Suggested Steps for Solving Applied Maxima and Minima Problems

Step 1. Use an appropriate letter to represent the quantity to be made a maximum (or minimum).

Step 2. If possible, draw a figure to illustrate the problem and label the variable parts. If only one variable is involved, label it x. If two variables are involved, label one x and the other y.

Step 3. CASE 1. Only one variable is involved. Write an equation expressing the quantity to be made a maximum (or minimum) in terms of x.

CASE 2. Two variables are involved. First write an equation expressing the quantity to be made a maximum (or minimum) in terms of x and y. Next, use the condition stated in the problem relating x and y to express y in terms of x. Now eliminate y from the first equation and obtain an equation in the single independent variable x.

Step 4. Determine the closed interval $[a, b]$ on which the equation in step 3 is valid.

SEC. 6.6 Applied Problems in Maxima and Minima

Step 5. Find all the critical values of x on $[a, b]$ for the given quantity.

Step 6. Evaluate the given quantity at each critical value of x on $[a, b]$ including the endpoints $x = a$ and $x = b$.

Step 7. Select the largest (or the smallest, depending on the problem) of the values found in Step 6. This is the maximum (or minimum) value of the given quantity on $[a, b]$.

Step 8. Reread the problem and be sure to answer the question that is asked there.

Exercises 6.6

1. Find the dimensions of a rectangle with maximum area if its perimeter is 60 inches.

2. Find the dimensions of a rectangle of maximum area if the sum of three of its sides is 40 inches.

3. Find two nonnegative numbers whose sum is 15, if the product of one of the numbers and the square of the other is a maximum.

4. A rectangular field of area 800 square yards is to be fenced off adjacent to a straight expressway whose right-of-way is already fenced. If the right-of-way fence can be used for one side of the field, find the dimensions of the field so that the amount of fence required for the remaining three sides will be a minimum. No side is to be less than 10 yards or greater than 50 yards.

5. Find the highest point and the lowest point on the curve $y = x/25 + 1/x$ on the interval $[1, 10]$.

6. A farmer wishes to lay out a rectangular lot whose area is to be 400 square yards. If no dimension is to be less than 10 yards or greater than 40 yards, what dimensions should he use so that the total length of the fence needed to enclose the lot will be a minimum?

7. A long rectangular piece of tin 20 inches wide is to be made into a gutter by turning up vertically the two long sides to the same depth, thus forming the gutter. How many inches should be turned up on each side to give the gutter the greatest vertical cross-sectional area?

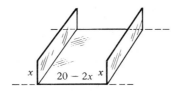

8. A rectangular box with open top is to be made from a square piece of cardboard 12 inches on a side by cutting equal squares from each corner and turning up the

resulting edges vertically. What should be the length of a side of the square cut out so that the resulting box has a maximum volume?

9. Find two nonnegative numbers whose sum is 12 and whose product is a maximum.

10. A rectangular box with a square base and no top is to be made. Find the dimensions of the box with maximum volume that can be made from 300 square feet of lumber.

11. The distance, s feet, traveled by an object in t minutes is given by the formula $s(t) = 30t^4 - t^5$, $0 \leq t \leq 30$.
    a. Find formulas for the velocity and acceleration in terms of t.
    b. Find when the velocity was a maximum.
    c. Find when the acceleration was a maximum.

12. A rectangular box with a square base and no top is to be built to contain a volume of 32 cubic meters. If no dimension is to be less than 1 meter or more than 10 meters and neglecting thickness of material and waste in cutting, what dimensions should be used to require the least amount of material?

13. A rectangular garden is to have an area of 800 square yards. If no dimension is to be less than 10 yards or more than 50 yards, what dimensions should be used so that the total cost of fence needed to enclose the garden will be a minimum, assuming that the cost for fencing the two sides will be $1 per yard and the cost of fencing the two ends will be $2 per yard?

14. The cost, C dollars, of producing x units of a certain item per day is $C(x) = 400 + 12x - (4-x)^3$. The capacity of the plant is 10 items per day.
    a. Find a formula for the marginal cost.
    b. For what value of x will the marginal cost be a minimum?

15. A particle moves along the x axis in such a way that its distance, x (in centimeters), from the origin after t seconds is given by the formula $x(t) = 10t^3 - t^4$, $0 \leq t \leq 10$. Find the velocity at the instant the acceleration is a maximum.

16. Postal regulations specify that for a package to be sent by parcel post the sum of its length and girth (distance around) must not exceed 100 inches. Find the volume of the largest rectangular package with square ends that can be sent by parcel post.

17. A printed page is to allow 108 square inches for printed matter and have a margin 1.5 inches on each side and 2 inches at the top and bottom. If the page must be at least 6 inches wide but not more than 21 inches wide, what dimensions of paper are required if the area is to be a minimum?

18. An open rectangular box with a square base is to contain a volume of 18 cubic feet. The material for the base is to cost 8 cents per square foot and for the sides 6 cents per square foot. Find the dimensions of such a box that will make the cost of materials a minimum.

19. A cylindrical can with a top and bottom is to be made from tin and is to have a volume of 250π cubic inches. What dimensions should be used if the amount of material needed for the can is to be a minimum? The diameter of the top must be at least 2 inches and not more than 20 in.

SEC. 6.7 The Chain Rule for Differentiation and Related Rates

20. Find the dimensions of the largest rectangular flower bed that can be made in a right-triangular lot whose hypotenuse is 50 feet and whose legs are 30 feet and 40 feet, respectively, if two sides of the bed lie along the two sides of the triangular lot.

21. At a price of $x each, the manufacturer of a certain type of transistor radio can sell weekly a number $n = 180 - 5x$, the cost, $C, of which to him is $C = 600 + 4n$. What price x will maximize his profit? The selling price may not exceed $50.

22. A man is stranded on a level desert 5 kilometers from A, the nearest point on a straight highway. A town B on this highway is 12 kilometers from A. If the man can walk 3 kilometers per hour through the sand and 5 kilometers per hour along the highway, what is the shortest possible time for him to get to point B, assuming he walks all the way?

23. The manager of a theater found that with an admission price of $2.50 per person the daily attendance was 1,200, while with every increase of 25 cents in admission the attendance dropped 60 persons. Assuming the manager wishes to keep the admission at not more that $5.00 per person, what should be the admission price for the daily receipts to be a maximum?

24. Farmer Jones estimates that if he digs his potatoes now he will have 150 bushels worth $2.00 per bushel, but if he waits the crop will grow 25 bushels per week for the next four weeks, while the price will drop 20 cents per bushel per week. When should he dig the potatoes to get the largest cash return?

6.7 THE CHAIN RULE FOR DIFFERENTIATION AND RELATED RATES

Let us suppose that $y = f(u)$ is a differentiable function of u and $u = g(x)$ is a differentiable function of x. We wish to show that $y = f[g(x)]$ is a differentiable function of x and that

$$D_x y = D_u y D_x u$$

This equation is called the *chain rule for differentiation*.

To establish this result, give x the increment $\Delta x \neq 0$, and denote by Δy and Δu the corresponding changes in y and u, respectively. Since y is a differentiable function of u and u is a differentiable function of x, we have

$$\lim_{\Delta u \to 0} \frac{\Delta y}{\Delta u} = D_u y$$

$$\lim_{\Delta x \to 0} \frac{\Delta u}{\Delta x} = D_x u$$

We now define the number d as follows:

$$d = \frac{\Delta y}{\Delta u} - D_u y, \quad \Delta u \neq 0$$
$$= 0 \quad \quad\quad\quad\quad \Delta u = 0$$

Hence

$$\Delta y = D_u y \, \Delta u + d \Delta u$$

where d approaches zero as Δu approaches zero. Now divide this by Δx:

$$\frac{\Delta y}{\Delta x} = D_u y \frac{\Delta u}{\Delta x} + d \frac{\Delta u}{\Delta x}$$

and take the limit as Δx approaches zero. Since u is a differentiable function of x, it follows that Δu approaches zero as Δx does, and hence that

$$\lim_{\Delta x \to 0} d = 0$$

Therefore,

$$\lim_{\Delta x \to 0} \frac{\Delta y}{\Delta x} = \lim_{\Delta x \to 0} D_u y \frac{\Delta u}{\Delta x} + \lim_{\Delta x \to 0} d \frac{\Delta u}{\Delta x}$$

and we have $D_x y = D_u y D_x u$, the desired result. Observe that the rule we established for differentiating a power u^n is a special case of the chain rule with $f(u) = u^n$.

Example 1 Given $y = u^3$ and $u = x^2 + x$, find $D_x y$ at $x = 1$.

Solution: Using the chain rule, $D_x y = D_u y D_x u$, we find $D_x y = 3u^2(2x + 1)$. But $u = 2$ when $x = 1$; therefore, $D_x y = (12)(3) = 36$.

Related Rates

If two or more related variables are functions of time t, and if we know rates at which all of these variables, except one, are changing, then the rate at which the one variable is changing with respect to time can often be found by applying the chain rule.

Example 2 The area, A square inches, of a square is related to the length of one of its sides, x inches, according to the relation $A = x^2$. If the length of each side is increased at the constant rate of 2 inches per minute, how fast is the area increasing at the instant $x = 10$?

Solution: Given $A = x^2$ and $D_t x = 2$, we are to find $D_t A$ when $x = 10$.
Using the chain rule, $D_t A = D_x A D_t x$, we have

$$D_t A = D_x(x^2) D_t x = 2x D_t x = 20(2) = 40 \text{ in.}^2/\text{min}$$

SEC. 6.7 The Chain Rule for Differentiation and Related Rates

Example 3 A north–south highway and an east–west highway intersect at A. Two trucks leave A at the same time, one traveling north at 30 miles per hour and the other traveling east at 40 miles per hour. How fast was the distance between them increasing at the end of 2 hours?

Solution: Let B and C be the respective positions of the two trucks at the end of t hours. Denote the distance AC by x, the distance AB by y, and the distance BC by z (see Figure 6.14). Then

$$z^2 = x^2 + y^2, \qquad y = 30t, \qquad x = 40t$$

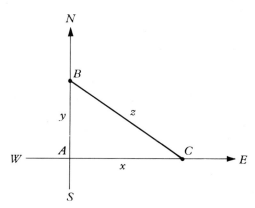

Figure 6.14

Differentiating implicitly with respect to t, we get

$$2z\frac{dz}{dt} = 2x\frac{dx}{dt} + 2y\frac{dy}{dt}, \qquad \frac{dy}{dt} = 30, \qquad \frac{dx}{dt} = 40$$

When $t = 2$, then $x = 80$, $y = 60$, and $z = 100$; therefore,

$$200\frac{dz}{dt} = 160(40) + 120(30)$$

$$\frac{dz}{dt} = \frac{6{,}400 + 3{,}600}{200} = 50 \text{ mph}$$

Example 4 A north–south highway and an east–west highway intersect at A; a truck leaves A, traveling east at 40 miles per hour, 2 hours before a second truck traveling at 30 miles per hour and coming from the south arrives at A. How fast is the distance between the two trucks changing 1 hour after the eastbound truck leaves A?

Solution: Let B and C be the respective positions of the two trucks t hours after the eastbound truck leaves A. Denote the distance AB by x, the distance AC by y, and the distance BC by z (see Figure 6.15). Then

$$z^2 = x^2 + y^2, \qquad x = 40t, \qquad y = 60 - 30t$$

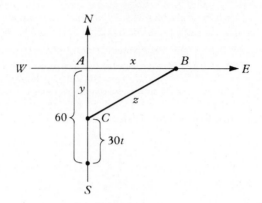

Figure 6.15

Differentiating with respect to t, we have

$$2z\frac{dz}{dt} = 2x\frac{dx}{dt} + 2y\frac{dy}{dt} \quad \text{or} \quad z\frac{dz}{dt} = x\frac{dx}{dt} + y\frac{dy}{dt}$$

$$\frac{dx}{dt} = 40, \quad \frac{dy}{dt} = -30$$

When $t = 1$, then $x = 40$, $y = 30$, and $z = 50$; therefore,

$$50\frac{dz}{dt} = 40(40) + 30(-30) \quad \text{and} \quad \frac{dz}{dt} = \frac{1{,}600 - 900}{50} = \frac{700}{50} = 14 \text{ mph}$$

Exercises 6.7

1. A ladder 30 feet long has one end resting on the level ground and the top end resting against a vertical wall. If the lower end is pulled away from the wall at the rate of 0.2 foot per minute in a direction perpendicular to the wall, how fast is the top descending when 18 feet high?

2. A baseball diamond is a square 90 feet on each side. If the batter bats a ball down the third-base line at the rate of 120 feet per second, how fast is its distance from first base changing at the end of 1 second? When it passes third base?

3. A cube was expanding at the rate of 300 cubic centimeters per minute at the instant the edge was 20 centimeters. How fast was the edge then changing?

4. The radius of a circle is being increased at the constant rate of 0.4 feet per second. How fast is the area increasing at the instant the radius is 2 feet?

5. Two cars leave the same intersection at 10 o'clock. One is traveling north at 40 miles per hour and the other is traveling east at 30 miles per hour. How fast is the distance between them changing 1 hour later, assuming they travel at a constant speed?

SEC. 6.7 The Chain Rule for Differentiation and Related Rates

6. A north–south and an east–west highway intersect at point A. At 9 o'clock one car is 90 kilometers west of A and traveling toward A at the constant speed of 50 kilometers per hour. At this same time a second car is leaving A and traveling north at a constant speed of 30 kilometers per hour. How fast is the distance between them changing at 10 o'clock? Is this distance increasing or decreasing at that time?

7. A rectangular tank has a square base 10 by 10 feet, and water is running into the tank at the uniform rate of 40 cubic feet per minute. Show that the surface is rising at a constant rate and find that rate.

8. A man 6 feet tall is walking directly away from a street light at the constant rate of 3 feet per second. If the street light is 15 feet above the street, how fast is the tip of his shadow moving?

9. Gas is being forced into a spherical balloon in such a way that the radius is increasing at the constant rate of 0.2 meter per minute. How fast is the volume of the balloon increasing at the instant the radius is 10 meters?

10. A man is standing on a river bridge 30 feet above the water in the river below just as a boat passes beneath him. If the boat travels in a straight line and at a rate of 8 feet per second, how fast is the distance between the boat and the man on the bridge changing 5 seconds later?

11. Sand falling at the rate 100 cubic centimeters per minute forms a conical pile whose radius always equals twice the height. Find the rate at which the height is changing at the instant the height is 10 centimeters. (*Hint:* $V = \frac{1}{3}\pi r^2 h$.)

7

Integration

Introduction

In Chapter 4 it was stated that one of the objectives in this course was to investigate two major problems concerning the mode of variation of two related quantities. The first problem, that of finding the instantaneous rate at which one variable is changing with respect to another when a relation between the two variables is known, was studied in some detail in Chapters 5 and 6. We learned to find various instantaneous rates and observed that such rates were called derivatives. The process of finding a derivative we called *differentiation*. Thus the first of the two problems can now be stated briefly as, "given a function of one independent variable, find its derivative with respect to that variable." We were able to derive several simple rules for differentiating various types of algebraic functions, and we became acquainted with many applications of derivatives. In this chapter we wish to study the second major problem, the inverse, or reverse, of the problem of differentiation. This second problem is that of finding the functional relation between two variables when the derivative of one of the variables with respect to the other is known. For example, suppose we have a formula which expresses the velocity of a certain object as a function of t, the number of seconds the object has been traveling since it started, and suppose we need to find the position of the object at a given time. Since velocity is the instantaneous rate of change of position with respect to time, our problem may be considered as one of finding a function of t whose derivative is the given formula.

7.1 ANTIDERIVATIVES AND INTEGRATION FORMULAS

Let us suppose that it is desired to express y as a function of x when the only information we have concerning y and x is that $D_x y = 6x^2$. Our problem then is to find a function of x whose derivative is $6x^2$. Recalling the rule for differentiating kx^n and reversing the operation, it is easy to see that $y = 2x^3$ defines a function whose derivative is $6x^2$. Could there be other functions having this same derivative? The answer to this question is given in Theorem 6.6, which says that if two functions, $f(x)$ and $g(x)$, are so related that $f'(x) = g'(x)$ for all x in a closed interval $[a, b]$, then $g(x) = f(x) + C$. Therefore, $y = 2x^3 + C$, where C is an arbitrary constant, represents every possible function whose derivative is $6x^2$. This shows that the derivative alone is not sufficient to completely determine a function. If, however, in addition to knowing the derivative, we also know the value of the function for at least one value of the independent variable, then the function is completely determined. For example, suppose we have given

$$D_x y = 6x^2 \quad \text{and} \quad y = 3 \quad \text{when } x = 1$$

We first write

$$y = 2x^3 + C$$

Substituting $y = 3$ and $x = 1$, we get

$$3 = 2(1^3) + C$$

or

$$C = 1$$

and hence $y = 2x^3 + 1$ is the desired function.

A function F is called an *antiderivative* of the function f if $F'(x) = f(x)$. As was pointed out above, if a given function has one antiderivative it has infinitely many, any two of which can differ at most by a constant. In other words, if the functions F and G are antiderivatives of the same function f, then $G(x) = F(x) + C$, where C is a constant. We call $F(x) + C$ the *general* antiderivative of f. Hence, if $D_x y = f(x)$ and $F'(x) = f(x)$, then $y = F(x) + C$. The process of finding an antiderivative of a given function is called *integration*. The constant C in the general antiderivative $F(x) + C$ is called a *constant* of integration. Usually, in solving problems involving integration, the constant of integration plays a very important role; to omit it gives us a particular antiderivative, the one for which $C = 0$.

Some Integration Formulas

Based on the definition of integration given in the previous subsection, the process of integration is in a sense a trial-and-error type of operation. In this operation we are given a function f and are required to find a function F such that $F'(x) = f(x)$. It is often helpful to think of integration as the reverse of the operation of differentiation. For example, to differentiate x^3 with respect to x, we multiply by the exponent 3 and decrease the exponent of x by 1, getting as the derivative $3x^2$. Now suppose we reverse this operation. We start with $3x^2$, increase the exponent of x by 1, and divide by the new exponent, giving $3x^3/3 = x^3$, the original function. Using this general idea and reversing the rules for differentiation from Chapter 5, we obtain the following formulas for finding the general antiderivative in some important special cases.

I. If $D_x y = f(x) = 0$, then $y = 0 + C = 0$ (C a constant) for $D_x 0 = 0$
II. If $D_x y = f(x) = k$, $k \neq 0$, then $y = kx + C$ for $D_x(kx) = k$
III. If $D_x y = f(x) = 1$, then $y = x + C$ for $D_x(x) = 1$
IV. If $D_x y = kx^n$, then $y = \dfrac{kx^{n+1}}{n+1} + C$ for $D_x\left(\dfrac{kx^{n+1}}{n+1}\right) = kx^n$ provided $n \neq -1$
V. If $D_x y = kf(x)$ and $F'(x) = f(x)$, then $y = kF(x) + C$, for $D_x(kF(x)) = kF'(x) = kf(x)$
VI. If $D_x y = f(x) + g(x)$ and $F'(x) = f(x)$ and $G'(x) = g(x)$, then $y = F(x) + G(x) + C$ for $D_x(F(x) + G(x)) = F'(x) + G'(x) = f(x) + g(x)$

In words, Formula V states, "An antiderivative of the product of a constant and a function is the product of that same constant and the antiderivative of the function." Similarly, Formula VI states, "An antiderivative of the sum of two functions is the sum of the antiderivatives of the functions." In fact, this rule can be generalized to state, "An antiderivative of the algebraic sum of any finite number of functions is the algebraic sum of the antiderivatives of these functions."

It should be emphasized that the acid test of whether or not a function F is an antiderivative of a function f is the following: The function F is an antiderivative of the function f if and only if $F'(x) = f(x)$. The student is encouraged to guess an antiderivative when not sure, but to always check the guess by differentiation. The following examples may help the student to understand how the above formulas are used. In the examples and problems which follow in this and other sections, a statement of the form, "Find y if $D_x y = f(x)$," will mean, unless otherwise specified, to find the general antiderivative of f expressed in the form $y = F(x) + C$.

Example 1

Find y if $D_x y = 5$.

Solution: By Formula II, $y = 5x + C$.

Check: $D_x y = D_x(5x + C) = 5$

SEC. 7.1   Antiderivatives and Integration Formulas

Example 2   Find y if $D_x y = 7x^4$.

Solution: By Formula IV, $y = \dfrac{7x^{4+1}}{4+1} + C = \dfrac{7}{5}x^5 + C$.

Check: $D_x y = D_x\left(\dfrac{7}{5}x^5 + C\right) = \dfrac{35}{5}x^4 = 7x^4$

Example 3   Find s if $D_t s = 20 - 32t$.

Solution: By Formulas II, IV, and VI, $s = 20t - 16t^2 + C$.

Check: $D_t s = D_t(20t - 16t^2 + C) = 20 - 32t$

Example 4   Find y if $D_x y = \dfrac{3}{\sqrt{x}}$.

Solution: Rewriting gives $D_x y = 3x^{-1/2}$. By Formula IV,

$$y = \dfrac{3x^{-1/2+1}}{-(1/2)+1} + C = \dfrac{3x^{1/2}}{1/2} + C = 6x^{1/2} + C = 6\sqrt{x} + C$$

Check: $D_x y = D_x(6\sqrt{x} + C) = D_x(6x^{1/2} + C) = 3x^{-1/2} = \dfrac{3}{\sqrt{x}}$

Example 5   Find Z if $D_y Z = 8y^3 - 3y^2 + 6y + 5$.

Solution: By Formulas IV and VI,

$$Z = \dfrac{8y^4}{4} - \dfrac{3y^3}{3} + \dfrac{6y^2}{2} + 5y + C$$
$$= 2y^4 - y^3 + 3y^2 + 5y + C$$

Check: $D_y Z = D_y(2y^4 - y^3 + 3y^2 + 5y + C)$
$= 8y^3 - 3y^2 + 6y + 5$

The student should find that with practice he or she will be able to write the antiderivatives of polynomial functions by inspection in simplified form in one step. See the next example.

Example 6   Find y if $D_x y = 3x^2 - 4x + 5$ and $y = 3$ when $x = 1$.

Solution: $y = x^3 - 2x^2 + 5x + C$. Substituting $y = 3$ and $x = 1$ gives

$$3 = 1 - 2 + 5 + C$$

Solving for C we obtain

$$C = -1$$

Hence

$$y = x^3 - 2x^2 + 5x - 1$$

Exercises 7.1

In Exercises 1 through 12, find y and check your answer in each case.

1. $D_x y = 4x^5$
2. $D_x y = 5x^{3/2}$
3. $D_x y = 2x^3 - 4x^2 + 5x - 7$
4. $D_x y = \sqrt[3]{x^2}$
5. $D_x y = 4x^{-3}$
6. $D_x y = 4x^3 - \dfrac{x}{x^2}$
7. $D_x y = x^2(4x - 3)$
8. $D_x y = (3x - 5)(2x + 4)$

(*Hint:* In Exercises 7 and 8, multiply before integrating.)

9. $D_x y = 12x^{-3} - 10x^{-5/3}$
10. $D_x y = 6x + 7\sqrt[3]{x^4}$
11. $D_x y = 8x^{-2/3} + 2x^{-1/2}$
12. $D_x y = \dfrac{3}{2}x^{1/2} + \dfrac{5}{4}x^{2/3}$

13. Find y if $D_x y = 10x^4 - 3x^2 + 4$ and $y = 6$ when $x = -1$.
14. Find y if $D_x y = x^3 + 4x^2 - 2x + 3$ and $y = 4$ when $x = 2$.
15. Find S if $D_x s = 40 - 32t$.
16. Find U if $D_v U = 4v^3 + 3v^2 + 2v - 8$.
17. Find v if $D_t v = -32$ and $v = 40$ when $t = 0$.
18. Find z if $D_y z = 5y^4 + 2y^3 - 4y$.
19. Find v if $D_r v = 4\pi v^2$ and $v = 0$ when $r = 0$.
20. Find s if $D_t s = 96 - 32t$ and $s = 256$ when $t = 0$.

7.2 APPLICATIONS OF INTEGRATION

Velocity and Distance

Let us now return to the problem mentioned at the end of the introduction. In that problem it is supposed that we are given a formula which expresses the velocity $v(t)$ of a moving object in terms of t (the number of units of time the object has been traveling since starting), and we are asked to find a formula that will give the position $s(t)$ of the object at time t units. The concept of integration and the integration formulas discussed in Section 7.1 provide a method for solving such a problem. Since, as we learned in Chapter 6, $v(t) = D_t[s(t)]$, we can find $s(t)$ in terms of t by integrating $v(t)$, that is, finding the antiderivative of $v(t)$. This procedure is illustrated in the following example.

SEC. 7.2   Applications of Integration

Example 1   A car is traveling along a highway and its velocity at any time t hours after starting is given by the formula $v(t) = 120t - 60t^2$ miles/hour.

(a) Find a formula for the position $s(t)$ of the car t hours after starting.
(b) Find how far the car traveled during the first 2 hours.
(c) Find the total distance the car traveled during the first 3 hours.

Solution
(a) We are given that $v(t) = 120t - 60t^2$ and since

$$D_t s(t) = v(t) = 120t - 60t^2$$

we find by integration that

$$s(t) = 60t^2 - 20t^3 + C$$

We know that $s(0) = 0$, measured from the starting point, hence

$$0 = 0 - 0 + C$$

or

$$C = 0$$

Therefore, the desired formula is

$$s(t) = 60t^2 - 20t^3$$

(b) We have found that $s(t) = 60t^2 - 20t^3$ and so

$$s(2) = 240 - 160 = 80 \text{ miles}$$

Also, since $v(t) = 60t(2-t) \geq 0$ for all t between 0 and 2 (see Section 3.18), the car was moving away from the starting point during this entire period of time. It follows that the car traveled 80 miles during the first 2 hours.

(c) Here one might be tempted to simply evaluate

$$s(3) = 60(3)^2 - 20(3)^3 = 540 - 540 = 0$$

and say that the car traveled 0 miles during the first three hours. This is not the case, however, since the actual situation is that the car traveled in one direction during the first two hours and in the opposite direction during the third hour. This follows from the fact that $v(t) = 60t(2-t)$ is positive if $0 < t < 2$, and negative if $2 < t < 3$. Hence, to find the total distance traveled during the first 3 hours, we must find the distance traveled during the third hour and add this to the distance traveled during the first 2 hours. We already know from part **b** that 80 miles were traveled during the first two hours. Now, we saw above that at $t = 3$ the car had returned to its starting point, since $s(3) = 0$. Hence an additional 80 miles were traveled (unfortunately the same 80 miles; perhaps the travelers forgot something or realized they were going in the wrong direction) and a total of 160 miles was traveled during the first 3 hours.

Example 2   Suppose that in Example 1 the highway passes through Knoxville, Tennessee, and that the car starts at 12:00 noon at a point 25 miles from Knoxville and is headed away from Knoxville.

(a) Find a formula that will give the position of the car relative to Knoxville t hours after starting.
(b) Find the position of the car relative to Knoxville at one o'clock and at two o'clock the same afternoon.

Solution: In Example 1 we found that $s(t) = 60t^2 - 20t^3 + C$. The car started at a point 25 miles from Knoxville. Thus, when $t = 0$, $s = 25$, and hence

$$25 = 0 - 0 + C$$

or
$$C = 25$$

Therefore,
$$s(t) = 60t^2 - 20t^3 + 25$$

is a formula which gives the position of the car relative to Knoxville t hours after starting. At one o'clock $t = 1$ and at two o'clock $t = 2$.

$$s(1) = 60 - 20 + 25 = 65 \text{ miles from Knoxville}$$

$$s(2) = 240 - 160 + 25 = 105 \text{ miles from Knoxville}$$

In this example the condition that $s = 25$ when $t = 0$ is called an *initial condition*. Thus our example suggests that, in general, if a formula for the velocity $v(t)$ is given and an initial condition is known, a formula for the position $s(t)$ can be found by integration.

Similarly, if a formula for the acceleration, $a(t)$, of a moving object is given and initial conditions are known, formulas for the velocity $v(t)$ and the position $s(t)$ can be found by integration. It should be noted that in such a case the initial conditions are usually the initial velocity and the initial position of the moving object, that is, the value of v and the value of s when $t = 0$. However, it is also possible to find a formula for v if the velocity is given at a time $t \neq 0$. Similarly, it is possible to find a formula for s if the position is given at a time $t \neq 0$.

Example 3 At the instant the brakes were applied, a train was moving at the rate of 44 feet per second, and thereafter, until it stopped, the velocity decreased at the constant rate of 4 feet per second per second. Find (a) formulas for the position and the velocity t seconds after the brakes were applied, (b) the number of seconds required for the train to stop, and (c) how far the train traveled after the brakes were applied.

Solution
(a) The velocity was decreasing at the constant rate of 4 feet per second per second. This is equivalent to the statement

(1) $$D_t v = -4$$

The initial conditions are $v(t) = 44$ feet per second and $s(t) = 0$ when $t = 0$. Integrating the expression in Equation (1), we get

$$v(t) = -4t + C_1$$

But $v(t) = 44$ when $t = 0$; therefore,

$$44 = 0 + C_1$$

or $\qquad C_1 = 44$

and (2) $\qquad v(t) = 44 - 4t$

This is our formula for the velocity at any time t seconds after the brakes were applied. Since $v(t) = D_t s$, it follows from Equation (2) that

$$D_t s = 44 - 4t$$

By integration, $\qquad s(t) = 44t - 2t^2 + C_2$

Notice here that subscripts are used to indicate the fact that the two constants of integration, C_1 and C_2, are in general different. Since $s(t) = 0$ when $t = 0$, we have the following formula, which gives the position $s(t)$ in feet, t seconds after the brakes were applied:

(3) $\qquad s(t) = 44t - 2t^2$

(b) The train stopped when $v(t) = 0$. So in Equation (2) we set $v(t)$ equal to 0 and solve for t, which will be the time the train required before coming to a stop after the brakes were applied.

$$44 - 4t = 0$$
$$4t = 44$$
$$t = 11 \text{ sec}$$

(c) To find how far the train traveled after the brakes were applied, we substitute $t = 11$ in Equation (3) and set

$$s(t) = 44(11) - 2(11)^2 = 484 - 242 = 242 \text{ ft}$$

Hence the train moved 242 feet after the brakes were applied before it stopped.

Freely Falling Bodies

Another application of integration, which is really just a special case of the one discussed in the previous subsection, is that of finding the height, s feet, of a body t seconds after it has been released from rest at a certain height or has been projected vertically upward from a fixed point. We shall assume that the only force acting on the body after its release or projection is the pull of gravity. A body in flight under such conditions is called a *freely falling body*. The acceleration of a freely falling body is known to be approximately constant, and we shall take it to be numerically equal to 32 feet per second per second.

To express this fact mathematically, let s feet represent the height of the body above the ground at any time t seconds after release or projection. Recalling that acceleration is the instantaneous rate of change of velocity, that is,

$a(t) = D_t[v(t)]$, and that $v(t) = D_t[s(t)]$, it follows that $a(t) = D_t^2[s(t)]$. Hence we can write

(1) $$D_t^2[s(t)] = -32$$

in the case of a freely falling body. The negative sign is due to the fact that s is positive upward and the pull of gravity is toward the center of the earth.

Example 4 An object is thrown vertically downward from an airplane, at an altitude of 7,200 feet, with an initial velocity of $v(0) = -40$ feet per second. The object is moving in the negative direction along the s axis; hence v is negative. What will be the velocity of the object after 8 seconds, and how far will it have fallen during these 8 seconds?

Solution: We start with Equation (1), after substituting $D_t[v(t)]$ for $D_t^2[s(t)]$, and write

(2) $$D_t^2[s(t)] = D_t[v(t)] = -32$$

Integrating, we have

$$v(t) = -32t + C_1$$

But $v = -40$ when $t = 0$; that is, $v(0) = -40$. Hence

$$-40 = -0 + C_1$$

or

$$C_1 = -40$$

giving the formula

(3) $$v(t) = -32t - 40$$

for the velocity of the object, t seconds after being thrown from the plane. After 8 seconds, when $t = 8$, we obtain

$$v(8) = -32(8) - 40 = -296 \text{ ft/sec}$$

Hence, the velocity of the object after 8 seconds is -296 ft/sec and the speed of the object is 296 ft/sec. Now since $v(t) = D_t[s(t)]$, Equation (3) becomes

$$D_t[s(t)] = -32t - 40$$

Integrating again, we have

$$s(t) = -16t^2 - 40t + C_2$$

Here $s(t)$ represents the height of the object above the ground t seconds after being thrown from the plane. It follows from the statement of the problem that $s = 7{,}200$ when $t = 0$; thus

$$C_2 = 7{,}200$$

and

(4) $$s(t) = -16t^2 - 40t + 7{,}200$$

SEC. 7.2 Applications of Integration

Substituting $t = 8$ in Equation (4), we get

$$s(8) = -16(8)^2 - 40(8) + 7{,}200$$
$$= -1{,}024 - 320 + 7{,}200 = 5{,}856 \text{ ft}$$

which shows that at the end of 8 seconds the object was 5,856 feet above the ground.

Suppose in Example 1 that we wished to find when the object reached the ground. To do this, we would replace $s(t)$ by 0 in Equation (4) and obtain the quadratic equation

$$0 = -16t^2 - 40t + 7{,}200$$

or, upon simplifying,

$$2t^2 + 5t - 900 = 0$$

Factoring,

$$(2t + 45)(t - 20) = 0$$

Hence $t = 20$ or $t = -22.5$. Since t must be positive to satisfy the physical conditions of the problem, $t = 20$ is the required answer.

Example 5 A ball is thrown vertically upward from a tower 100 feet high with an initial velocity of 64 feet per second. Find when the ball was highest and its maximum height.

Solution: Again we start with the basic Equation (1), $D_t^2[s(t)] = D_t[v(t)] = -32$, and integrate to obtain

$$v(t) = -32t + C_1$$

But $v = 64$ when $t = 0$; therefore, $C_1 = 64$ and

(5) $$v(t) = -32t + 64$$

Integrating $v(t) = D_t[s(t)] = -32t + 64$ gives

$$s(t) = -16t^2 + 64t + C_2$$

From the statement of the problem we know that $s = 100$ when $t = 0$; thus $C_2 = 100$ and

(6) $$s(t) = -16t^2 + 64t + 100$$

When the ball reached its highest point, $v = 0$; hence from Equation (5) we have

$$0 = -32t + 64 \quad \text{or} \quad t = 2$$

The ball reached its highest point above the ground at the end of 2 secs. To find the height at that time, we substitute $t = 2$ in Equation (6) and obtain

$$s(2) = -16(2)^2 + 64(2) + 100$$
$$= -64 + 128 + 100 = 164 \text{ ft}$$

The maximum height of the ball above the ground was 164 ft.

Exercises 7.2

1. A certain object moves along the s axis, and its velocity t seconds after starting is given by the formula $v(t) = 30t^2 - 4t^3$.
   a. Find the distance, s feet, traveled by the object from $t = 0$ to $t = 4$, if $s = 0$ when $t = 0$.
   b. Find this distance if $s = 25$ when $t = 0$, and again if $s = -35$ when $t = 0$.
   c. Does the value of s when $t = 0$ affect the distance traveled between $t = 0$ and $t = 4$?
   d. Find the total distance traveled by the object from $t = 0$ to $t = 10$. (*Warning:* The answer is not 0 ft. See Example 1, part **c**.)

2. An object initially at the point (18, 0) on the x axis is projected toward the left along the x axis with an initial speed of 20 feet per second ($v = -20$, $t = 0$). If the acceleration is thereafter constantly equal to 4 feet per second per second, what is a formula for the distance (x feet) of the object from the origin after t seconds? When does the object pass through the origin? When and where does it stop and reverse its direction?

3. An object moves along the s axis in such a way that its acceleration is given by the formula $a(t) = 6t + 4$. If $v = 4$ and $s = 5$ when $t = 1$, find the value of v and the value of s when $t = 5$.

4. A ball is thrown vertically upward from the edge of a roof 128 feet high with an initial velocity of 32 feet per second. With what velocity does the ball hit the ground? With what speed?

5. If an object travels along the s axis with the velocity $v(t) = 100 - 20t$, and if $s = 40$ when $t = 1$, find s and a when $t = 9$.

6. An antiaircraft shell is fired vertically upward at an initial velocity of 4,096 feet per second. Neglecting air resistance, how long will it take to reach the highest point of its path? How high will this be?

7. If the velocity of an object moving along the s axis is $v(t) = 12t - 3t^2$, and $s = 5$ when $t = 0$, where is the object when the velocity is a maximum?

8. What is the distance traveled by an object in the first 10 seconds if its acceleration t seconds after starting is given by the formula $a(t) = (12t - 4)$ feet per second squared and the initial velocity is 2 feet per second?

9. An object falls from rest from a height of 256 feet. With what velocity does it hit the ground?

10. A train was traveling at the rate of 60 miles per hour at the instant the brakes were applied. The velocity thereafter decreased at the rate of 1,200 miles per hour per hour. Find when the train stopped and how far it moved in stopping.

11. Babe Ruth is supposed to have caught a baseball dropped from the top of the Washington Monument, 576 feet high. Assuming no air resistance and that the ball was caught at ground level, find the velocity of the ball when caught.

12. What should be the initial upward velocity of a body if it goes to a maximum height of 4,900 feet?

SEC. 7.3 Equations of Curves

7.3 EQUATIONS OF CURVES

We have made frequent use, heretofore, of the fact that the slope of the line tangent to a curve at a point (or the slope of the curve at the point) is the value of the derivative $D_x y$ at the point. Thus, if m is the slope of the curve $y = f(x)$ at a general point (x, y) on the curve, then

$$D_x y = m$$

If m is a function of x, we can frequently find y by integration and so find an equation of the curve, if we have sufficient information to determine the constant of integration. For the special case $m = $ constant, the curve represented is always a straight line, as was pointed out in Chapter 6. Hence, if a formula for the slope m of a curve is known, an equation of the curve can often be found by integration.

Example 1 Find an equation of the curve whose slope is given by the equation $m = 2x - 3$ and whose y intercept is -2.

Solution: Here $D_x y = 2x - 3$ and, by integration, we get $y = x^2 - 3x + C$. This last equation does not represent one curve, but a family of curves: one curve for each different value of C. Figure 7.1 shows the graph of this equation for several

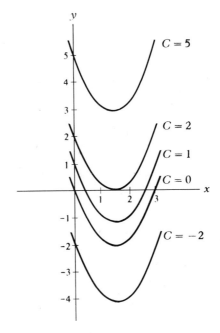

Figure 7.1

values of C. Note that the curves all have the same shape. Also, for the same value of x each curve has the same slope. So far we have not made use of the second condition stated in our problem, that the y intercept is -2. This condition, when applied, has the effect of singling out one particular curve of the family in question. To completely determine a curve, we need to be given a formula for its slope at any point and the coordinates of one point on the curve. In this case, knowing that the y intercept is -2 is equivalent to knowing that the curve passes through the point $(0, -2)$. Thus our second condition is $y = -2$ when $x = 0$. From this we find that $C = -2$; hence

$$y = x^2 - 3x - 2$$

is an equation of the curve. Notice that its graph is the bottom curve in the figure.

From the discussion in this example we can conclude that, if a formula for the slope of a curve is given, integration will provide an equation of a whole family of curves having the same slope formula. An equation for a single curve is obtained only when sufficient information is available to determine the constant of integration.

Exercises 7.3

In each of the following exercises, write an equation of the curve, or straight line, having the given slope formula, $m = D_x y$, and passing through the given point.

1. $m = 4x; (1, 1)$
2. $m = \frac{3}{2}; (2, -3)$
3. $m = -3; (-2, 1)$
4. $m = -\frac{2}{3}; (2, -1)$
5. $m = 2; (0, 4)$
6. $m = -1; (0, -2)$
7. $m = 2x + 5; (1, 4)$
8. $m = 4x - 2; (1, -1)$
9. $m = 9x^2 - 12x; (0, 4)$
10. $m = 2x^2 - 3x + 1; (-3, -2)$
11. $m = 5x^4 - 4x^3 + 2x - 3; (2, -1)$
12. $m = -\frac{2}{x^2}; (1, 1)$

7.4 INTEGRALS AND DIFFERENTIALS

We are already familiar with the use of symbols in mathematics to indicate operations that are to be performed. For example, the symbol $+$ indicates the operation of addition, the radical, $\sqrt{}$, indicates the operation of extracting a

SEC. 7.4 Integrals and Differentials

square root, and in Chapter 6 we used a capital letter D with a subscript, say D_x, to indicate the operation of taking a derivative. We now find it convenient to introduce a symbol to indicate the operation of integration, that is, the operation of finding an antiderivative.

The general antiderivative of a function f with respect to x will be denoted by the symbol

$$\int f(x)\,dx$$

which is read "the integral of $f(x)$ with respect to x." To be more specific, we define this new symbol in the following way:

$$\int f(x)\,dx = F(x) + C$$

where F is a function such that $F'(x) = f(x)$. The symbol $\int$ is called an *integral sign*, and the equation is read "the integral, with respect to x, of $f(x)$ is $F(x)$ plus C." We are thus interpreting the symbol $\int (\ldots)\,dx$ as the inverse of the symbol $D_x(\ldots)$. The function between the integral sign and dx, such as $f(x)$ in the integral $\int f(x)\,dx$, is called the *integrand*. We have used x as the independent variable in stating our definition, but any other variable would have been equally good, for instance,

$$\int f(t)\,dt = F(t) + C \qquad \text{if } F'(t) = f(t)$$

$$\int g(y)\,dy = G(y) + C \qquad \text{if } G'(y) = g(y)$$

Using this new notation, we may write

$$\int 3x^2\,dx = x^3 + C$$

$$\int 5t^4\,dt = t^5 + C$$

$$\int (6y^2 - 4y + 7)\,dy = 2y^3 - 2y^2 + 7y + C$$

The integration formulas of Section 7.1, when written in integral notation, become

I. $$\int 0\,dx = C$$

II. $$\int dx = x + C$$

III. $$\int k\,dx = kx + C$$

IV. $$\int kx^n\,dx = \frac{kx^{n+1}}{n+1} + C, \quad \text{if } n \neq -1$$

V. $$\int kf(x)\,dx = k\int f(x)\,dx, \quad \text{if } k \neq 0$$

VI. If $F'(x) = f(x)$ and $G'(x) = g(x)$, then
$$\int [f(x) + g(x)]\,dx = F(x) + G(x) + C$$

Let us add to our list a new formula which is a generalization of Formula IV, Section 7.1. If $f(x)$ is a differentiable function of x, then

$$D_x\left[\frac{(f(x))^{n+1}}{n+1}\right] = [f(x)]^n \cdot f'(x) \quad \text{(Rule 5.9, Section 5.8)}$$

Therefore, by the definition of an integral,

VII. $$\int [f(x)]^n \cdot f'(x)\,dx = \frac{[f(x)]^{n+1}}{n+1} + C, \quad n \neq -1$$

For the special case $n = -\frac{1}{2}$, we have the following useful formula.

VIII. $$\int \frac{f'(x)\,dx}{\sqrt{f(x)}} = 2\sqrt{f(x)} + C$$

For the sake of brevity, we shall write "Integrate $\int f(x)\,dx$" to mean "find the expression $F(x) + C$ represented by the symbol $\int f(x)\,dx$."

Example 1 Integrate $\int (x^4 + 1)^{19} 4x^3\,dx$.

Solution: If we let $f(x) = x^4 + 1$, then $f'(x) = 4x^3$, and hence

$$\int (x^4 + 1)^{19} 4x^3\,dx = \int [f(x)]^{19} f'(x)\,dx$$

$$= \frac{[f(x)]^{20}}{20} + C \quad \text{(Formula VII)}$$

$$= \frac{(x^4 + 1)^{20}}{20} + C$$

Check: $D_x\left[\frac{1}{20}(x^4 + 1)^{20} + C\right] = (x^4 + 1)^{19} 4x^3$

SEC. 7.4 Integrals and Differentials

Example 2

Integrate $\int x\sqrt{x^2+4}\,dx$.

Solution: $\int x\sqrt{x^2+4}\,dx = \int (x^2+4)^{1/2}x\,dx$. Let $f(x)=x^2+4$, then $f'(x)=2x$ or $x = \frac{1}{2}f'(x)$. Substituting in the last integral above, we get

$$\int (x^2+4)^{1/2}x\,dx = \int [f(x)]^{1/2}\frac{1}{2}f'(x)\,dx = \frac{1}{2}\int [f(x)]^{1/2}f'(x)\,dx$$

$$= \frac{1}{2}\frac{[f(x)]^{3/2}}{3/2} + C \quad \text{(Formula VII)}$$

$$= \frac{1}{3}[f(x)]^{3/2} + C$$

$$= \frac{1}{3}(x^2+4)^{3/2} + C$$

Check: $D_x\left[\frac{1}{3}(x^2+4)^{3/2}\right] = \frac{1}{3}\cdot\frac{3}{2}(x^2+4)^{1/2}(2x) = x\sqrt{x^2+4}$

Example 3

Integrate $\int \dfrac{x^2\,dx}{\sqrt{x^3-7}}$

Solution: Let $f(x)=x^3-7$; then $f'(x)=3x^2$ or $x^2 = \frac{1}{3}f'(x)$. Substituting in the given integral, we obtain

$$\int \frac{x^2\,dx}{\sqrt{x^3-7}} = \int \frac{\frac{1}{3}f'(x)\,dx}{\sqrt{x^3-7}} = \frac{1}{3}\int \frac{f'(x)\,dx}{\sqrt{f(x)}}$$

$$= \frac{1}{3}\cdot 2\sqrt{f(x)} + C \quad \text{(Formula VIII)}$$

$$= \frac{1}{3}\cdot 2\sqrt{x^3-7} + C$$

$$= \frac{2}{3}\sqrt{x^3-7} + C$$

Check: $D_x\left[\frac{2}{3}\sqrt{x^3-7}\right] = D_x\left[\frac{2}{3}(x^3-7)^{1/2}\right] = \frac{2}{3}\cdot\frac{1}{2}(x^3-7)^{-1/2}(3x^2)$

$$= \frac{x^2}{\sqrt{x^3-7}}$$

The Differential of a Function

Let $y=f(x)$ be the value of a function of x defined on the interval $a \leq x \leq b$ and having a derivative at each point of this interval. We wish now to define a quantity which we shall represent by dy and call the differential of $f(x)$ or, more simply, the differential of y. Before giving our definition, let us first introduce the new symbol dx for an arbitrary increment of the independent variable x. Note that since dx is arbitrary, that is, can be assigned any value whatever, it is in reality an independent variable. In fact, dx is the same type of variable as was h or Δx in Chapters 5 and 6, except that $x + h$ and $x + \Delta x$ must belong to the domain of f, while dx is unrestricted. This symbol dx is called the *differential of x* and its range is $-\infty < dx < \infty$.

The *differential* dy of the function $y = f(x)$ is defined by the relation

$$dy = D_x y \, dx$$

An equivalent notation is $df(x) = D_x y \, dx$.

Again, it should be emphasized that dx, as also dy, represents one number only, and is not d times x and should never be thought of as such.

It follows from our definition that if $dx \neq 0$, then

$$\frac{dy}{dx} = D_x y$$

This is the source of the notation $\dfrac{dy}{dx}$ for the derivative of y with respect to x mentioned in the first section of Chapter 6. The notation has the advantage that with it the derivative can always be regarded as a quotient of differentials.

Example 4   If $y = 3x^2$, then $D_x y = 6x$ and $dy = 6x \, dx$. We now list a few numerical values of dy for various values of x and dx.

When $x = 1$	and $dx = 0.2$	then $dy = 1.2$
$x = 3$	$dx = 5$	$dy = 90$
$x = 2$	$dx = -1$	$dy = -12$
$x = 2$	$dx = 0.01$	$dy = 0.12$
$x = -1$	$dx = 100$	$dy = -600$

Thus dy is, in fact, a function of two independent variables, x and dx.

Example 5   If $y = 2x^2 - 7x + 8$, what is dy?

Solution: By definition, $dy = D_x y \, dx = (4x - 7) \, dx$.

A geometrical interpretation of the differential may be given as follows. Consider the graph of the curve $y = f(x)$; see Figure 7.2. Let $P_1(x_1, y_1)$ be a point on the curve. For $x = x_1$ the ratio of dy to dx is the slope of the curve $y = f(x)$ at the point P_1. Hence the point R with coordinates $(x_1 + dx, y_1 + dy)$ lies on the tangent to the curve at the point P_1, no matter what value may be assigned dx.

SEC. 7.4 Integrals and Differentials

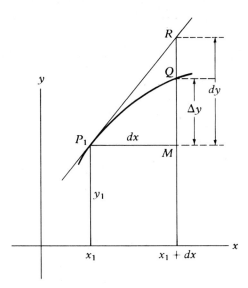

Figure 7.2

So in Figure 7.2 the differentials dx and dy are represented by the directed segments

$$dx = P_1 M, \qquad dy = MR$$

In the figure, Q is the point where the segment MR intersects the curve $y = f(x)$. Denote MQ by Δy. It follows that

$$\Delta y = f(x_1 + dx) - f(x_1)$$

As dx approaches zero, so does Δy. Similarly, as dx approaches zero, so does dy. This suggests the application of the differential dy as an approximation for Δy when dx is sufficiently small but not zero.

The application we are particularly interested in at the moment, however, is that of representing any derivative as the ratio of two differentials. This trick is a boon to integration. For example, if we have been given

$$D_x y = f(x)$$

we can replace this with the equivalent equation

$$\frac{dy}{dx} = f(x)$$

Since the left side of this equation is a fraction, dy divided by dx, we can multiply both sides of the equation by dx and get

$$dy = f(x)\, dx$$

Integrating both sides,

$$\int dy = \int f(x)\,dx$$

or
$$y + C_1 = F(x) + C_2$$

and finally
$$y = F(x) + C$$

where $F'(x) = f(x)$ and $C_2 - C_1 = C$.

The advantage of using differentials is much more obvious in cases like the following. Suppose we have been given

$$D_x y = \frac{f(x)}{g(y)}$$

To integrate such a problem seems almost hopeless. But let us rewrite this equation in the form

$$\frac{dy}{dx} = \frac{f(x)}{g(y)}$$

Clearing fractions, we have

$$g(y)\,dy = f(x)\,dx$$

Integration can now be indicated thus:

$$\int g(y)\,dy = \int f(x)\,dx$$

Example 6   Find the equation of the curve passing through the point (2, 1), whose slope is given by the equation

$$m = \frac{2x+1}{2y-1}$$

Solution:
$$m = \frac{dy}{dx}$$

Therefore,
$$\frac{dy}{dx} = \frac{2x+1}{2y-1}$$

Clearing fractions, $(2y-1)\,dy = (2x+1)\,dx$

Integrating, $\int (2y-1)\,dy = \int (2x+1)\,dx$

or
$$y^2 - y = x^2 + x + C$$

Since $y = 1$ when $x = 2$, we find, by substitution, that $C = -6$. The required equation can now be written: $x^2 - y^2 + x + y = 6$.

SEC. 7.4 Integrals and Differentials

An equation involving one or more derivatives is called a *differential equation*. $D_x y = 6x$, $D_t v = 4 - 2t$, $D_t^2 y = -32$, and $\dfrac{dy}{dx} = \dfrac{2x+1}{2y-1}$ are examples of differential equations. These are, of course, of the simplest possible type, but they are sufficient for our present purposes. The equation free of derivatives, obtained by integrating a differential equation, is called a *solution of the differential equation*. Thus $x^2 - y^2 + x + y = 6$ is a solution of the differential equation $\dfrac{dy}{dx} = \dfrac{2x+1}{2y-1}$.

Example 7 If $y = x^3$, approximately how much does y increase when x increases from $x = 10$ to $x = 10.02$? Exactly how much?

Solution: Since the increase in x is small, dy is probably a good approximation of the change in y. Let us calculate dy by letting $dx = \Delta x = 0.02$:

$$dy = 3x^2\, dx = 3(10)^2(0.02) = 6$$

This says that y increases about 6 units when x increases from $x = 10$ to $x = 10.02$. The exact increase in y is, of course,

$$\Delta y = (10.02)^3 - 10^3$$

This difference can be found by direct calculation, but the following procedure may be easier in many cases.

$$y = x^3$$
$$y + \Delta y = (x + dx)^3$$
$$= x^3 + 3x^2(dx) + 3x(dx)^2 + (dx)^3$$

whence
$$\Delta y = 3x^2(dx) + 3x(dx)^2 + (dx)^3$$

Now let $x = 10$ and $dx = 0.02$:

$$\Delta y = 3(10^2)(0.02) + 3(10)(0.02)^2 + (0.02)^3 = 6 + 0.012 + 0.000008$$
$$= 6.012008$$

from which it is seen that the error made by taking dy as an approximation of Δy is $\Delta y - dy = 0.012008$.

Exercises 7.4

Perform each indicated integration.

1. $\displaystyle\int 4x^3\, dx$

2. $\displaystyle\int 5\, dx$

3. $\displaystyle\int 6t^2\, dt$

4. $\displaystyle\int (7 - 4t)\, dt$

5. $\int dx$

6. $\int 4x^{-2} dx$

7. $\int \frac{20}{x^2} dx$

8. $\int 6x^{1/2} dx$

9. $\int x^3 dx$

10. $\int x\sqrt{x}\, dx$

11. $\int \sqrt[3]{y^2}\, dy$

12. $\int \frac{dx}{\sqrt{x}}$

13. $\int 4x\, dx$

14. $\int (ax^2 + bx)\, dx$

15. $\int (5 + 2t - 6t^2)\, dt$

16. $\int x(6x^2 - 4)\, dx$

17. $\int (60 - 18t + 12t^3)\, dt$

18. $\int \left(\frac{2}{3}x^2 - \frac{5}{7}x + 4^2\right) dx$

19. $\int (x+2)^4\, dx$

20. $\int \sqrt{2x-5}\, dx$

21. $\int x^2(6 + 10x^3)\, dx$

22. $\int x^2(4 + x^3)^4\, dx$

23. $\int \left(\frac{x^2}{2} - \frac{2}{x^2}\right) dx$

24. $\int \frac{(6x+5)\, dx}{\sqrt{3x^2 + 5x - 7}}$

25. $\int \frac{3x^4 - 5x^2 + 4}{x^2}\, dx$

26. $\int \frac{1}{\sqrt{2x+3}}\, dx$

27. $\int \frac{x\, dx}{\sqrt{1-x^2}}$

28. $\int \frac{x\, dx}{(4+x^2)^3}$

29. $\int (t^{15} + 10t^6 + 8)\, dt$

30. $\int (2x+1)(3x-4)\, dx$

31. $\int (10x^4 - 4x^3 - 6x^2 + 8x - 5)\, dx$

32. $\int x\sqrt{4-x^2}\, dx$

33. $\int x(3x^2 - 1)^9\, dx$

34. $\int \frac{5t}{\sqrt{2-t^2}}\, dt$

35. $\displaystyle\int 7x(5+3x^2)^6\, dx$

36. $\displaystyle\int \frac{(x+2)\, dx}{\sqrt{x^2+4x+7}}$

37. $\displaystyle\int \frac{dx}{(3+x)^4}$

38. $\displaystyle\int \sqrt{x}\left(2x+\frac{3}{\sqrt{x}}\right) dx$

39. $\displaystyle\int t^2(5-t^3)^4\, dt$

40. $\displaystyle\int \frac{4x\, dx}{\sqrt[3]{6-x^2}}$

41. Find an equation of the curve which passes through (2, 1) and whose slope at any point (x, y) on the curve is $m = \dfrac{x+5}{y}$.

42. Find an equation of the curve which passes through (0, 5) and whose slope at any point on the curve is $m = \dfrac{2}{y}$.

43. Find a solution of the differential equation $\dfrac{dy}{dx} = \dfrac{4+x}{2-y}$ if $y = 3$ when $x = 0$.

44. About how much would y change if x increased from $x = 1$ to $x = 1.01$ and $y = x^3 + 3x^2 + 10x - 7$?

45. Find a solution of the differential equation $\dfrac{dy}{dx} = x^2 y^2$ if $y = 1$ when $x = 1$.

7.5 AREAS FOUND BY INTEGRATION

Almost everyone has some familiarity with the concepts of length and area and the methods that have been developed to measure these quantities. As a review we recall that the measure of a line segment is called *length* and the measure of a line segment is relative to some established unit. Typical units are one centimeter, one inch, one foot, and one yard, etc. As an example of how such units are established, consider the definition of the unit one yard. One yard is the distance between two lines on a specially prepared and carefully preserved metal bar kept at the constant temperature of 62°F in a vault at the Bureau of Standards in Washington, D.C. The unit one foot is one-third of the unit one yard; that is, if a line segment 3 feet in length were placed on this bar with one end on one of the lines, then the other end would coincide with the other line on the bar. The unit one inch is $\frac{1}{12}$ of the unit one foot or $\frac{1}{36}$ the unit one yard. To express the length of a line segment requires two things: a number and the type of unit being used. It would be meaningless to say that the length of a line segment is 20, but quite meaningful to say its length is 20 inches or 20 feet. Similarly, the area of a region is relative to an established unit. A typical unit of area is one square foot. It is defined as the area of a square the length of each of whose sides is one foot. Other such units in abbreviated form are 1 cm^2, 1 in.^{2}, 1 yd^2, etc. We define the area, A, of a rectangle l units long and w units wide as

$$A = lw \text{ (units)}^2$$

Every triangle is "half" a rectangle [Figure 7.3(a)] and hence its area is defined as $\frac{1}{2}bh$, where b is its base and h is its altitude. Since every polygon can be subdivided into nonoverlapping triangles [Figure 7.3(b)], the area of the polygon can be defined as the sum of the areas of the triangles into which it has been subdivided. It can be proved that the area so obtained is the same regardless of how the polygon has been subdivided. Suppose in a circle of radius r we inscribe a regular polygon of n sides. It can be shown that the area of the inscribed polygon approaches the number πr^2 as a limit as n becomes infinite. We define the area, A, of the circle to be this limit; that is, $A = \pi r^2$. What about areas of regions bounded by portions of lines or curves or combinations of both? Does each such region always have an area? It can be shown by a rather involved argument, making use of inscribed or circumscribed rectangles and a limiting process, that such regions do have areas. In what follows we shall assume this to be true. We now shall show how the process of integration may often be used to find the area of a region when the equations of the curves which enclose it are known.

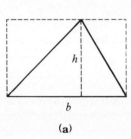
(a)

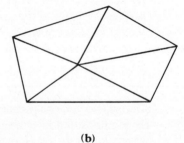
(b)

Figure 7.3

Area under a Curve

The curve C in Figure 7.4 is the graph of a continuous function f which is an increasing function on the closed interval $[a, b]$. This means that if x_1 and x_2 are any two points on $[a, b]$ such that $a \le x_1 < x_2 < b$, then $f(x_1) < f(x_2)$. The equation of the curve C is $y = f(x)$.

Let us now consider the area of the region in Figure 7.4, that is bounded above by the curve C, below by the x axis, on the left by the line $x = a$, and on the right by the line $x = b$. We shall denote the area of that portion of the region just described which lies between the fixed ordinate at $x = a$ and a variable ordinate

SEC. 7.5 Areas Found by Integration

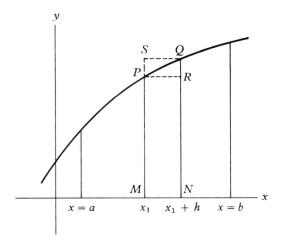

Figure 7.4

at x, $a \le x \le b$, by $A(x)$, because this area is certainly a function of x and hence changes in value as x increases. We next proceed to find the instantaneous rate of change of $A(x)$ per unit change in x at some arbitrary point $x = x_1$, $a \le x_1 \le b$. Note that $A(x_1)$ represents the area of the region under consideration lying between $x = a$ and $x = x_1$. Suppose we increase x_1 by an amount h, where $h > 0$ and $a \le x_1 + h \le b$. Then the area of the region between $x = x_1$ and $x = x_1 + h$ is $A(x_1 + h) - A(x_1)$. We see from the figure that this area is not less than the area of the rectangle $MPRN$ and not greater than the area of the rectangle $MSQN$; that is,

(1) $$hf(x_1) \le A(x_1 + h) - A(x_1) \le hf(x_1 + h)$$

or $$f(x_1) \le \frac{A(x_1 + h) - A(x_1)}{h} \le f(x_1 + h)$$

It follows that

$$\lim_{h \to 0} f(x_1) \le \lim_{h \to 0} \frac{A(x_1 + h) - A(x_1)}{h}$$

and

(2) $$\lim_{h \to 0} \frac{A(x_1 + h) - A(x_1)}{h} \le \lim_{h \to 0} f(x_1 + h)$$

Since $f(x_1)$ does not involve h, $\lim_{h \to 0} f(x_1) = f(x_1)$ and since f is a continuous function at $x = x_1$, $\lim_{h \to 0} f(x_1 + h) = f(x_1)$. (See Theorem 5.11.) Hence

$$f(x_1) \le \lim_{h \to 0} \frac{A(x_1 + h) - A(x_1)}{h} \le f(x_1)$$

and the equality sign must hold.

$$\lim_{h \to 0} \frac{A(x_1+h)-A(x_1)}{h} = f(x_1)$$

But $\dfrac{A(x_1+h)-A(x_1)}{h}$ is the average rate of change of $A(x)$ over the interval from $x=x_1$ to $x=x_1+h$, and since its limit does exist as shown above, this limit is the derivative of $A(x)$ at $x=x_1$; that is

$$D_x[A(x_1)] = f(x_1)$$

Since x_1 is an arbitrary value of x on $[a, b]$, we can drop the subscript and write

(3) $$D_x[A(x)] = f(x)$$

In words, Equation (3) says that the instantaneous rate at which the area A of the region between $x=a$ and any x in $[a, b]$ is changing is equal to the height of the curve at x. This result can be shown to be valid if in Figure 7.4 we replace the curve c by the graph of any function f continuous on the closed interval $[a, b]$ such that $f(x) > 0$ for each x on $[a, b]$.

To complete our discussion of finding the area of the region under a curve, let us return to Equation (3). This equation yields that the derivative of $A(x)$ with respect to x is $f(x)$. Suppose that we can find some function $F(x)$ whose derivative is $f(x)$; that is, suppose that we can integrate the function f. Then, using Theorem 6.6, we find that

(4) $$A(x) = F(x) + C$$

Now, since $A(x)$ represents the area of the region from the vertical line through $(a, 0)$ to the vertical line through $(x, 0)$, we see that

(5) $$A(a) = 0 = F(a) + C \quad \text{or} \quad C = -F(a)$$

Therefore,

(6) $$A(x) = F(x) - F(a)$$

A more convenient and also a more descriptive notation often used for the formula in Equation (6) is to replace $A(x)$ by the symbol A_a^x, read "the area from a to x," giving

$$A_a^x = F(x) - F(a)$$

and also to represent this same area by the symbol $\int_a^x f(x)\, dx$, read "the integral from a to x of $f(x)\, dx$" (since the function F is obtained by integrating f). Thus in our new notation we have

(7) $$A_a^x = \int_a^x f(x)\, dx = F(x) - F(a), \quad \text{where } F'(x) = f(x)$$

SEC. 7.5 Areas Found by Integration

To find the area of the region in Figure 7.4 bounded by $x = a$ on the left and by $x = b$ on the right, we substitute $x = b$ in Equation (7) and get

(8) $$A_a^b = \int_a^b f(x)\,dx = F(b) - F(a), \quad \text{where } F'(x) = f(x)$$

An integral of the form $\int_a^b f(x)\,dx$ is called a *definite integral* to distinguish it from the integral $\int f(x)\,dx$, which is often called an *indefinite integral*. The *a* at the bottom of the definite integral sign is the initial value of x, or the value of x at which the area in question begins, and is called the *lower limit of integration*. The *b* at the top of the integral sign is the terminal value of x, or the value of x at which the area stops as we move from left to right along the x axis, and is called the *upper limit of integration*. The real distinguishing feature between a definite integral and an indefinite integral is that a definite integral represents a single specific number, while the indefinite integral represents a whole class of functions, one function for each different value of the constant of integration.

Equation (8), in practice, is usually written in slightly more detail as

(9) $$A_a^b = \int_a^b f(x)\,dx = F(x)\Big]_a^b = F(b) - F(a)$$

The added detail, $F(x)\big]_a^b$, is important in that it exhibits the actual function $F(x)$ being employed.

The definite integral plays an important role in solving many problems other than those involving area.

Example 1 Find the value of the definite integral $\int_1^3 (4x - 1)\,dx$.

Solution: The rules of integration previously stated will apply in finding $F(x)$:

$$\int_1^3 (4x - 1)\,dx = 2x^2 - x\Big]_1^3 = [2(3)^2 - 3] - [2(1)^2 - 1]$$
$$= (18 - 3) - (2 - 1) = 14$$

Example 2 Find the area under the curve $y = x^3 - x^2 + 3$, above the x axis, and between the lines $x = -1$ and $x = 2$ (the shaded area in Figure 7.5).

Solution: Here $A_{-1}^2 = \int_{-1}^2 (x^3 - x^2 + 3)\,dx$

$$= \frac{x^4}{4} - \frac{x^3}{3} + 3x\Big]_{-1}^2$$
$$= \left(4 - \frac{8}{3} + 6\right) - \left(\frac{1}{4} + \frac{1}{3} - 3\right)$$
$$= \frac{39}{4}\ (\text{units})^2$$

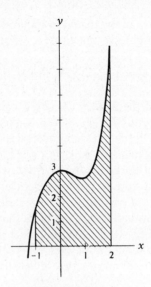

Figure 7.5

In our discussion above we restricted ourselves to areas of regions that lie wholly above the x axis. It can be shown, by arguments similar to those used above, that if the continuous curve, $y = f(x)$, lies wholly below the x axis for each x on $[a, b]$, then the area of the region below the x axis, above the curve $y = f(x)$, bounded on the left by $x = a$ and on the right by $x = b$ is given by the equation

$$A_a^b = -\int_a^b f(x)\,dx$$

The negative sign is due to the fact that the ordinate y of the curve is negative for each value of x on $[a, b]$.

If the curve $y = f(x)$ crosses the x axis at $x = a$, at $x = b$, and at one or more points between $x = a$ and $x = b$, then a part of the region bounded by this curve and the x axis lies above the x axis and a part lies below. If we wish to find the total area of the region bounded by the curve and the x axis, we must calculate separately the area of each part of the region that lies above the x axis and each part that lies below the x axis and add the results. In other words, one could not find the total area of the given region by evaluating the integral $\int_a^b f(x)\,dx$. Example 3 illustrates the procedure for handling a problem of this type.

Example 3 | Sketch the graph of the curve $y = x(x - 2)(x - 4)$, and find the area of the region bounded by this curve and the x axis.

Solution: An inspection of the given equation shows that the curve crosses the x axis at the points $(0, 0)$, $(2, 0)$, and $(4, 0)$, and only at these points. The graph is shown in Figure 7.6. The portion of the region between $x=0$ and $x=2$ lies above the x axis, and its area is given by

$$A_0^2 = \int_0^2 x(x-2)(x-4)\,dx = \int_0^2 (x^3 - 6x^2 + 8x)\,dx$$
$$= \frac{x^4}{4} - 2x^3 + 4x^2 \bigg]_0^2 = \frac{16}{4} - 16 + 16 = 4 \text{ (units)}^2$$

The portion of the region extending from $x=2$ to $x=4$ lies below the x axis, and its area is given by

$$A_2^4 = -\int_2^4 (x^3 - 6x^2 + 8x)\,dx = -\left[\frac{x^4}{4} - 2x^3 + 4x^2\right]_2^4$$
$$= -[(64 - 128 + 64) - (4 - 16 + 16)]$$
$$= -(0 - 4) = 4 \text{ (units)}^2$$

Therefore, the total area A is

$$A = A_0^2 + A_2^4 = 4 + 4 = 8 \text{ (units)}^2$$

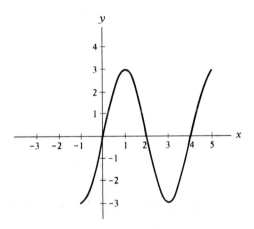

Figure 7.6

Area between Two Curves

Suppose that the graphs of the equations $y = f(x)$ and $y = g(x)$ are the curves shown in Figure 7.7. Let A_1 represent the area of the region bounded above by the curve $y = f(x)$, below by the x axis, on the left by the line $x = a$ and on the right by the line $x = b$; that is, A_1 represents the area of the region $MCDN$. Let A_2 represent the area of the region $MBEN$, and A represent the area of the region $BCDE$. It follows from our discussion regarding areas under a curve that

$$A_1 = \int_a^b f(x)\,dx \quad \text{and} \quad A_2 = \int_a^b g(x)\,dx$$

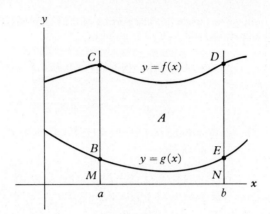

Figure 7.7

We observe that the area of the region between the two curves, A, is $A = A_1 - A_2$; that is,

$$A = \int_a^b f(x)\,dx - \int_a^b g(x)\,dx$$

Since the two definite integrals have the same limits of integration, we may write

(10) $$A = \int_a^b [f(x) - g(x)]\,dx$$

It will be noted that Equation (10) reduces to Equation (9) in the special case where the curve $y = g(x)$ is the x axis, that is, where $g(x) = 0$.

Example 4   Find the area of the region bounded above by the curve $y = x^2 + 2$ and below by the curve $y = x^2 + 1$ and lying between the line $x = 1$ and $x = 3$ (Figure 7.8).

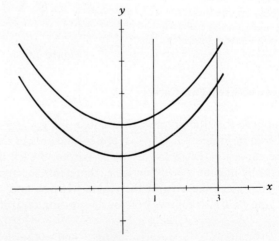

Figure 7.8

SEC. 7.5 Areas Found by Integration

Solution: Here $f(x) = x^2 + 2$, $g(x) = x^2 + 1$, $a = 1$, and $b = 3$. Substituting into Equation (10), we get

$$A = \int_1^3 [(x^2 + 2) - (x^2 + 1)]\, dx$$
$$= \int_1^3 1\, dx = x \Big]_1^3 = 3 - 1 = 2 \text{ (units)}^2$$

It can be shown that Equation (10) is valid in the following more general case. Let the two equations define two functions of x, both of which are continuous on the closed interval $[a, b]$. Suppose also that $g(x) \leq f(x)$ on $[a, b]$. Then the area of the region bounded above by the curve $y = f(x)$, below by the curve $y = g(x)$, and on the sides by the lines $x = a$ and $x = b$ can be calculated by using Equation (10). It should be observed that one or both of the given curves may lie wholly or partially below the x axis, and the curves may intersect at $x = a$ or at $x = b$ or both. If, for example, the curves intersect at $x = a$, then only one point of the boundary would lie on the line $x = a$. This is demonstrated in the following example.

Example 5 Find the area of the region bounded by the curve $y = x^2 - 2x$ and the line $y = 3$ (see Figure 7.9).

Solution: We see from Figure 7.9 that the curves intersect, and by solving the two equations simultaneously, we find they intersect at the points $(-1, 3)$ and $(3, 3)$. So, referring to Equation (10), we have $f(x) = 3$, $g(x) = x^2 - 2x$, $a = -1$, and $b = 3$. Hence

$$A = \int_{-1}^3 [3 - (x^2 - 2x)]\, dx$$
$$= \int_{-1}^3 (3 + 2x - x^2)\, dx$$
$$= 3x + x^2 - \frac{x^3}{3} \Big]_{-1}^3 = (9 + 9 - 9) - \left(-3 + 1 + \frac{1}{3}\right)$$
$$= 9 + 2 - \frac{1}{3} = \frac{32}{3} \text{ (units)}^2$$

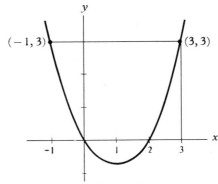

Figure 7.9

Exercises 7.5

Find the value of each of the definite integrals in Exercises 1 through 10.

1. $\int_{2}^{5} 4x \, dx$

2. $\int_{-3}^{7} 5 \, dx$

3. $\int_{-2}^{3} (6x^2 - 5) \, dx$

4. $\int_{0}^{4} (2t^3 + 4t - 7) \, dt$

5. $\int_{2}^{4} (3x^2 - 6x + 2) \, dx$

6. $\int_{-2}^{0} (5x^4 - 3x^2 + 2x) \, dx$

7. $\int_{1}^{9} \sqrt{x} \, dx$

8. $\int_{1}^{4} \frac{12}{x^2} \, dx$

9. $\int_{1}^{16} \frac{dx}{\sqrt{x}}$

10. $\int_{0}^{3} \frac{x}{\sqrt{25 - x^2}} \, dx$

In each of the following exercises, sketch the region bounded by the graphs of the given equations and find the area of the region.

11. $x + y = 10$, $x = 2$, $x = 6$ and $y = 0$

12. $y = 2x + 1$, $x = 0$, $x = 4$, $y = 0$ (*Hint:* $x = 0$ is the y axis and $y = 0$ is the x axis.)

13. $y = 4x - x^2$, $y = 0$

14. $y = 4 - x^2$, $y = 0$

15. $y = x^2$, $y = 4$

16. $y = x^2 - 2x$, $y = x$

17. $y = x^3 - 3x + 2$, $x = -2$, $x = 1$, $y = 0$

18. $y = 9 - x^2$, $y = 5$

19. $y = \dfrac{40}{x^2}$, $x = 2$, $x = 10$, $y = 0$

20. $y = \sqrt[3]{x}$, $x = 1$, $x = 8$, $y = 0$

21. $x = y^2$, $x = 9$

22. $y = x^2$, $y = x + 2$

23. $y = x\sqrt{3x^2 + 4}$, $x = 0$, $x = 2$, $y = 0$

24. $y = 2 - x^2$, $y = -x$

7.6* VOLUMES OF SOLIDS

In this section we wish to indicate how the volume of certain solids may be found by the method of integration. Let us consider the solid shown in Figure 7.10. This solid has the positive x axis as an axis of symmetry. Suppose we wish to find the volume of that portion of this solid lying between two parallel planes AB and CD. The plane AB is perpendicular to the x axis at $x = a$, and the plane CD is perpendicular to the x axis at $x = b$. Planes PQ and $P'Q'$ are two variable planes both parallel to AB, with $x = b$ at a distance x from the origin O and plane $P'Q'$ at a distance $x + h$ from O, where $h > 0$ and $a \le x + h \le b$.

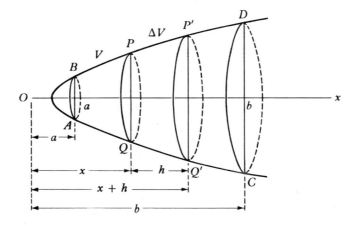

Figure 7.10

Let $V(x)$ represent the volume of that portion of the solid lying between the planes AB and PQ. Then $V(x + h)$ is the volume of the solid lying between AB and $P'Q'$. Hence the volume of the solid lying between PQ and $P'Q'$ is $V(x + h) - V(x)$. A formula from geometry states that the volume V of a horizontal right cylinder whose ends are in parallel planes perpendicular to the elements of the cylinder is $V = AL$, where A is the area of one end and L is the length of the cylinder. It follows that $V(x + h) - V(x)$ is not less than $A(x)h$ and not greater than $A(x + h)h$; that is

$$A(x)h \le V(x + h) - V(x) \le A(x + h)h$$

or, dividing by h,

$$A(x) \le \frac{V(x + h) - V(x)}{h} \le A(x + h)$$

Applying the limiting process, we have

$$\lim_{h\to 0} A(x) \le \lim_{h\to 0} \frac{V(x+h) - V(x)}{h} \le \lim_{h\to 0} A(x+h)$$

But $\lim_{h\to 0} A(x) = A(x)$, and since $A(x)$ is continuous on $[a, b]$, $\lim_{h\to 0} A(x+h) = A(x)$ (Theorem 5.11.) Therefore,

(1) $$D_x V = \lim_{h\to 0} \frac{V(x+h) - V(x)}{h} = A(x)$$

This last equation states that the instantaneous rate at which the volume, $V(x)$, is changing at any x on $[a, b]$ is equal to the area, $A(x)$, of the cross section of the solid made by a plane perpendicular to the x axis at x. By integrating both sides of Equation (1) and using the definite integral notation, we obtain

(2) $$V = \int_a^b A(x)\, dx = F(b) - F(a), \quad \text{where } F'(x) = A(x)$$

as the volume of the given solid lying between the parallel planes at $x = a$ and $x = b$. It can be shown that for Equation (2) to be valid it is not necessary that the line chosen for the x axis be an axis of symmetry or that $A(x)$ be an increasing function. It is necessary that $A(x)$ be a continuous function on some closed interval $[a, b]$. In fact, we can use Equation (2) to find the volume of any solid if, after having chosen a suitable x axis, we can do two things: (1) Express the area of a typical variable cross section of the solid made by a plane perpendicular to the x axis in terms of x and (2) integrate the expression thus obtained.

Example 1   In a certain cone, each cross section made by a plane perpendicular to the axis of the cone is a circle whose radius r varies with its distance, x inches, from the vertex according to the equation $r = 0.3x$. Find the volume of that portion of the cone extending from $x = 2$ to $x = 10$. The origin is at the vertex.

Solution: The formula for the area of a circle of radius r is $A = \pi r^2$. Since the cross section of the cone made by a plane perpendicular to the axis at a distance, x inches, from the vertex is a circle of radius $r = 0.3x$, then the area of this cross section $A(x)$ is $A(x) = \pi(0.3x)^2 = 0.09\pi x^2$. Now using Equation (2) with $A(x) = 0.09\pi x^2$, $a = 2$, and $b = 10$, we obtain

$$V = \pi \int_2^{10} 0.09 x^2 \, dx = \pi(0.03 x^3)\Big]_2^{10} = \pi[(30) - (0.24)]$$
$$= 29.76\pi \text{ in.}^3$$

Example 2   Find by integration the volume of a solid hemisphere of radius 10 inches.

Solution: First we draw a figure (see Figure 7.11) and choose the x axis as shown with the origin at O and positive x axis pointing downward. Any cross section of this hemisphere, perpendicular to the x axis, is a circle. For the circular cross

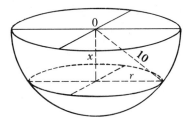

Figure 7.11

section at a distance, x inches, below the origin, let r be its radius. From the right triangle shown in Figure 7.11, we see that $r^2 = 100 - x^2$. Hence the area, $A(x)$, of this cross section is $A(x) = \pi r^2 = \pi(100 - x^2)$. The hemisphere extends from $x = 0$ to $x = 10$. Therefore,

$$V = \pi \int_0^{10} (100 - x^2)\, dx = \pi \left(100x - \frac{x^3}{3} \right)\Big]_0^{10}$$
$$= \pi \left[\left(1{,}000 - \frac{1{,}000}{3} \right) - (0 - 0) \right] = \frac{2{,}000}{3} \pi \text{ in.}^3$$

Exercises 7.6

1. Use integration to find the volume of a right circular cylinder of radius 6 inches and altitude 10 inches.

2. Use integration to find the volume of a solid hemisphere of radius 20 meters.

3. A right circular cone has an altitude of 10 inches and the radius of its base is 5 inches. Use integration to find the volume of the cone. (*Hint:* Show that a circular cross section x inches from the vertex has a radius $r = 0.5x$.)

4. The area bounded by the lines $x = y$, $x = 1$, $x = 6$, and $y = 0$ is revolved about the x axis. Find by integration the volume of the solid thus generated. [*Hint:* Show that a cross section of this solid made by a plane perpendicular to the x axis is a circle of radius y, where (x, y) is a point on the line $y = x$.]

5. A certain horn has an axis of symmetry. Each section perpendicular to this axis is a circle whose radius r varies thus with its distance x units from one end: $r = 0.04x^2$. Find the volume of the space within the horn from $x = 1$ to $x = 8$.

6. Use integration to find a formula for the volume of a right circular cone whose altitude is h and the radius of whose base is r.

7.7 INTEGRATION: A PROCESS OF SUMMATION

In Section 7.4, we defined integration as the inverse of differentiation; that is, we defined integration by the relation $\int f(x)\, dx = F(x) + C$ if and only if $F'(x) = f(x)$. In fact, we called $F(x)$ an antiderivative of $f(x)$ and showed that

$F(x)+C$ was the most general antiderivative of $f(x)$ if $F'(x)=f(x)$. In this section we shall indicate how integration may be defined as a *process of summation*. As a matter of fact, in a great many applications of the integral calculus, it is preferable to define integration as a process of summation. From an historical standpoint, the integral calculus was invented in an attempt to define and calculate the area of a region bounded by curves by dividing the given region up into a set of approximating rectangles, the sum of whose areas was the required area of the region. In fact, the integral sign, $\int$, is an elongated S, used by early mathematicians to indicate "sum."

This new definition of integration as explained in the remainder of this section is of fundamental importance, and its meaning should be thoroughly understood by all students studying calculus in order that they be able to apply integral calculus to practical problems.

In Section 7.5 we introduced the symbol $\int_a^b f(x)\,dx$ and mentioned that it was called a *definite integral* to distinguish it from the symbol $\int f(x)\,dx$, which is often referred to as an indefinite integral. This is due to the fact that the integral $\int_a^b f(x)\,dx$ always represents a definite real number, whereas the symbol $\int f(x)\,dx$ represents an indefinite set of functions, a different function for each different value of the constant of integration C. We also learned in Section 7.5 that if

(a) f is a function continuous on the closed interval $[a,b]$,
(b) $f(x) > 0$ for each x on $[a,b]$, and
(c) F is a function such that $F'(x)=f(x)$ for each x in $[a,b]$,

then $\int_a^b f(x)\,dx = F(b) - F(a)$ is the area of the region bounded by the graphs of the equations $y=f(x)$, $x=a$, $x=b$, and the x axis. Let us now reconstruct Figure 7.4 with the following modification. We divide the closed interval $[a,b]$ into any number n of subintervals, each of equal length denoted by Δx, by choosing numbers $x_0, x_1, x_2, \ldots, x_n$, where $a=x_0$, $b=x_n$, and $\Delta x = x_i - x_{i-1} = \dfrac{b-a}{n}$, for $i=1, 2, 3, \ldots, n$, as illustrated in Figure 7.12.

We next erect ordinates at each of these points of division and complete rectangles by drawing horizontal lines through the ends of the ordinates as shown in the figure. It is clear that the sum of the areas of these rectangles is an approximation for the area of the region in question, and the larger the number of rectangles, that is, the larger the value of n, the better the approximation. Suppose we denote by A_n the area of these rectangles in Figure 7.12; that is,

(1) $\qquad A_n = [f(x_0) + f(x_1) + f(x_2) + \cdots + f(x_{n-1})]\,\Delta x$

It seems intuitively obvious that by taking n sufficiently large we can make the difference between the area A and the value of A_n as small as we please. This means that

(2) $\qquad \lim_{n \to \infty} A_n = \lim_{n \to 0} [f(x_0) + f(x_1) + f(x_2) + \cdots + f(x_{n-1})]\,\Delta x = A$

SEC. 7.7   Integration: A Process of Summation

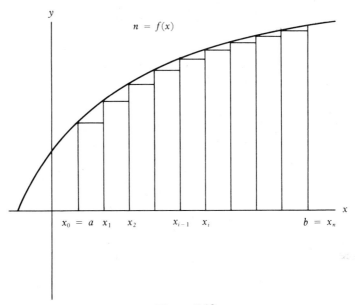

Figure 7.12

But $A = \int_a^b f(x)\,dx = F(b) - F(a)$, where $F'(x) = f(x)$. Hence $\lim_{n \to \infty} A_n = \int_a^b f(x)\,dx = F(b) - F(a)$; that is,

$$(3) \quad \int_a^b f(x)\,dx = \lim_{n \to \infty} [f(x_0)\,\Delta x + f(x_1)\,\Delta x + f(x_2)\,\Delta x + \cdots + f(x_{n-1})\,\Delta x]$$

In arriving at Equation (3), we made use of the notion of area, and for simplicity purposes we not only assumed that $f(x) > 0$ for all x in $[a, b]$, but we also assumed that $f(x)$ was an increasing function on the closed interval $[a, b]$.

In more advanced texts the following theorem, which is a generalization of the results in Equation (3) and which is known as the *Fundamental Theorem of Integral Calculus,* is shown to be true.

Fundamental Theorem of Integral Calculus

Let $f(x)$ be a function of x continuous on the closed interval $[a, b]$. Let the interval $[a, b]$ be divided into n subintervals whose lengths are $\Delta x_1, \Delta x_2, \ldots, \Delta x_n$, and points be chosen, one in each subinterval, their abscissas as being $x_1, x_2, \ldots, x_n$, respectively. Consider the sum

$$f(x_1)\,\Delta x_1 + f(x_2)\,\Delta x_2 + f(x_3)\,\Delta x_3 + \cdots + f(x_n)\,\Delta x_n = \sum_{i=1}^{n} f(x_i)\,\Delta x_i$$

Then as n increases without limit and each Δx_i approaches zero as a limit, the limit of this sum has the value of $\int_a^b f(x)\,\Delta dx$. Stated briefly,

(4) $$\int_a^b f(x)\,dx = \lim_{n\to\infty} \sum_{i=1}^n f(x_i)\,\Delta x_i$$

Note that for each i, $i = 1, 2, 3, \ldots, n$, $f(x_i)\,\Delta x_i$ is the area of a rectangle of height $f(x_i)$ and width Δx_i. In actual practice using this theorem, it is usually convenient to divide the interval $[a, b]$ into n subintervals of equal length $\Delta x = \dfrac{b-a}{n}$ and to choose x_i in each subinterval either as the left end or the right end of the subinterval. Also it is not necessary to actually carry out the subdivision but simply to draw a typical approximating rectangle to use in setting up the integral involved.

Example 1 Find the area bounded by the graphs of the equations $f(x) = 6x - x^2$, $x = 1$, $x = 4$, and the x axis.

Solution: The given region and a typical rectangle are sketched in Figure 7.13. We shall assume the interval $[1, 4]$ has been subdivided into n equal subintervals, each

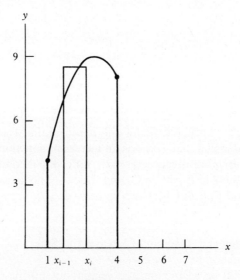

Figure 7.13

of length Δx. Let ΔA represent the area of the typical approximating rectangle shown in the figure. Then

$$\Delta A = f(x_i)\,\Delta x, \qquad f(x_i) = 6x_i - x_i^2$$

SEC. 7.7 Integration: A Process of Summation

$$A \approx \sum_{i=1}^{n} f(x_i) \Delta x = \sum_{i=1}^{n} (6x_i - x_i^2) \Delta x$$

$$A = \lim_{n \to \infty} \sum_{i=1}^{n} (5x_i - x_i^2) \Delta x$$

Apply the Fundamental Theorem of Integral Calculus:

$$A = \lim_{n \to \infty} \sum_{i=1}^{n} (6x_i - x_i^2) \Delta x = \int_{1}^{4} (6x - x^2) \, dx$$

$$= 3x^2 - \frac{x^3}{3} \Big]_1^4 = \left(48 - \frac{64}{3}\right) - \left(3 - \frac{1}{3}\right)$$

$$= 48 - \frac{64}{3} - 3 + \frac{1}{3}$$

$$= 45 - \frac{63}{3} = 45 - 21 = 24$$

Example 2 Find the area of the region bounded by the graphs of the equation $f(x) = 4x - x^2$ and $f(x) = -x$.

Solution: The given region and a typical rectangle are sketched in Figure 7.14. We think of the closed interval $[0, 5]$ as being subdivided into n equal subintervals

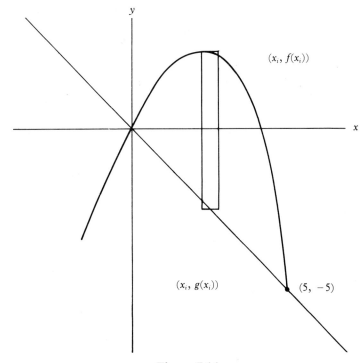

Figure 7.14

each of length Δx. The area ΔA of our typical rectangle is

$$\Delta A = [f(x_i) - g(x_i)]\, \Delta x$$
$$= [(4xi - x_i^2) - (-x_i)]\, \Delta x$$
$$= [5x_i - x_i^2]\, \Delta x$$

So
$$A = \lim_{n \to \infty} \sum_{i=1}^{n} (5x_i - x_i^2)\, \Delta x$$
$$= \int_0^5 (5x - x^2)\, dx$$
$$= \frac{5x^2}{2} - \frac{x^3}{3} \Big]_0^5 = \frac{125}{2} - \frac{125}{3} = \frac{125}{6}$$

Exercises 7.7

In each of the following exercises, sketch the region bounded by the graphs of the given equations, show a typical approximating rectangle as was done in Examples 1 and 2, and find the area of the region.

1. $y = x^3$, $y = 9$, and the y axis
2. $y = x^2$ and $y = 9$
3. $4y = x^2$ and $x - 2y + 4 = 0$
4. $y^2 = 6x$ and $x^2 = 6y$
5. $y^2 = x$ and $x = 4$
6. $y = 9 - x^2$ and $y = x + 7$
7. $y = x^2 - 3x$ and $y = x$
8. $y = x^2$ and $2x - y + 3 = 0$

8

An Introduction to Linear Algebra

Introduction

In Chapter 4 we discussed a procedure for finding the coordinates of the point of intersection of two lines in the plane when equations of the two lines are known. That is, we discussed a particular method of solving two linear equations in two variables. In this chapter we shall discuss, in some detail, methods for solving systems of two linear equations in two variables and systems of three linear equations in three variables. These methods will be used to motivate the concept of a matrix. A brief treatment of matrix algebra will be given in Sections 8.4 through 8.6. The last two sections of this chapter will be devoted to linear inequalities and linear programming.

8.1 SYSTEMS OF TWO LINEAR EQUATIONS IN TWO VARIABLES

We define the solution set of the linear equation $ax + by = r$ as the set consisting of all the ordered pairs (x, y) such that $ax + by = r$ or, stated symbolically, as the set $S = \{(x, y) | ax + by = r\}$. For example, the solution set S of the linear equation $2x - 3y = 12$ is the set $S = \{(x, y) | 2x - 3y = 12\}$. In this section we are interested in the solution set, not of one equation, but of a pair of linear equations in x and y. For example, consider the pair of equations

$$\begin{cases} 2x - 3y = 12 \\ x - y = 5 \end{cases}$$

A pair such as this is commonly referred to as a system of two simultaneous linear equations in two variables, or in two unknowns. Let S_1 be the solution set of $2x - 3y = 12$ and S_2 be the solution set of $x - y = 5$. The solution set of the system is the set of ordered pairs that are elements of both S_1 and S_2. Symbolically, the solution set of the system is the set $S = S_1 \cap S_2$. It is easily verified that $(3, -2)$ is a solution of the system and, as we shall show in the next section, it is the only solution. Hence $S = \{(3, -2)\}$. We shall now describe some simple methods of finding solution sets of systems of linear equations and systems of linear inequalities.

Graphical Method

It was shown in Chapter 4 that every linear equation in x and y has a straight line for its graph. One method of solving a system of two linear equations in x and y is to draw the two lines that are the graphs of the two equations involved, and to read from the graph the coordinates of the points the two lines have in common. Figure 8.1 shows the graphical solution of the system mentioned above. Since the two lines are distinct and not parallel, they intersect in one and only one point, and this shows that the solution set of the system consists of a single ordered pair, $(3, -2)$.

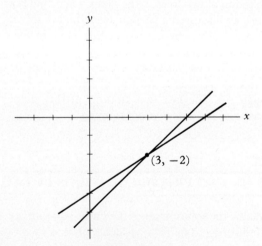

Figure 8.1

Now consider the system

$$\begin{cases} a_1 x + b_1 y = r_1 \\ a_2 x + b_2 y = r_2 \end{cases}$$

SEC. 8.1 Systems of Two Linear Equations in Two Variables

where $a_1, b_1, r_1, a_2, b_2,$ and r_2 are arbitrary real numbers, it being understood that a_1 and b_1 are not both zero and the same for a_2 and b_2. Since the coefficients $a_1, b_1,$ etc., are arbitrary, they may be assigned any real values whatever with the exceptions just mentioned. This system represents *all* systems of two linear equations in two unknowns. Let S_1 be the solution set of $a_1 x + b_1 y = r_1$, and let S_2 be the solution set of $a_2 x + b_2 y = r_2$; that is,

$$S_1 = \{(x, y) | a_1 x + b_1 y = r_1\}$$

and

$$S_2 = \{(x, y) | a_2 x + b_2 y = r_2\}$$

The solution set S of the system is defined as $S = S_1 \cap S_2$.

The graphical method of solving systems of two linear equations has the disadvantage that it is not always possible to read accurately from the graph the coordinates of the point of intersection of the two lines involved. It does, however, suggest the fact that the solution set S of the system is one of the following:

1. A single point if the two lines are not parallel
2. The empty set if the two lines are parallel and different
3. The infinite set consisting of all points (x, y) satisfying either equation if the two lines coincide.

The Method of Elimination by Addition and Subtraction

As was pointed out in the previous subsection, the graphical method for solving systems of linear equations is limited by the accuracy with which one is able to read from the graph the coordinates of the point of intersection of two lines. The graphical method is also limited for all practical purposes to systems involving two equations and two unknowns. We wish to review a method which will obviate both of these objections and also with which you are undoubtedly familiar. This method is called the method of *elimination* by addition and subtraction (see Example 8 of Section 4.6), and consists of transforming a given system into an equivalent system whose solution set can be found by inspection. Two systems of equations are said to be equivalent if they have the same solution set.

We first apply the method to a system of two linear equations in two variables. Consider the system

(System 1) $\quad \begin{cases} a_1 x + b_1 y = r_1 \\ a_2 x + b_2 y = r_2 \end{cases} \quad \begin{array}{l} a_1 \text{ and } b_1 \text{ not both zero} \\ a_2 \text{ and } b_2 \text{ not both zero} \end{array}$

It is obvious that, if either or both of these equations are replaced by a nonzero multiple of themselves, the new system thus obtained is equivalent to the original, for the solution set of each equation remains unchanged. That is, the system

(System 2) $\quad \begin{cases} k_1(a_1 x + b_1 y) = k_1 r_1 & k_1 \neq 0 \\ k_2(a_2 x + b_2 y) = k_2 r_2 & k_2 \neq 0 \end{cases}$

is equivalent to System 1. Furthermore, we can prove that the following system is also equivalent to System 1.

THEOREM 8.1. *The system*

(System 3) $\quad \begin{cases} a_1 x + b_1 y = r_1 \\ k_1(a_1 x + b_1 y) + k_2(a_2 x + b_2 y) = k_1 r_1 + k_2 r_2 & k_2 \neq 0 \end{cases}$

is equivalent to System 1.

Proof: If (x_1, y_1) is a solution of System 1, then

$$\begin{cases} a_1 x_1 + b_1 y_1 = r_1 \\ a_2 x_1 + b_2 y_1 = r_2 \end{cases}$$

obtains; hence (x_1, y_1) is a solution of System 3. To verify this, note that if x_1 and y_1 are substituted for x and y respectively in the second equation of System 3, we get

$$k_1(a_1 x_1 + b_1 y_1) + k_2(a_2 x_1 + b_2 y_1) = k_1(r_1) + k_2(r_2)$$

which shows that (x_1, y_1) is a solution of this equation. Thus (x_1, y_1) is a solution of the system. Hence every solution of System 1 is also a solution of System 3. Next, suppose that (x_2, y_2) is a solution of System 3. Then

$$\begin{cases} a_1 x_2 + b_1 y_2 = r_1 \\ k_1(a_1 x_2 + b_1 y_2) + k_2(a_2 x_2 + b_2 y_2) = k_1 r_1 + k_2 r_2 \end{cases}$$

Multiplying the first equation by k_1 and subtracting the result from the second equation, we obtain

$$k_2(a_2 x_2 + b_2 y_2) = k_2 r_2$$

and, since $k_2 \neq 0$,

$$a_2 x_2 + b_2 y_2 = r_2$$

which shows that (x_2, y_2) is also a solution of System 1.

Let us now use Theorem 8.1 to solve System 1:

$$\begin{cases} a_1 x + b_1 y = r_1 & a_1 \text{ and } b_1 \text{ not both zero} \\ a_2 x + b_2 y = r_2 & a_2 \text{ and } b_2 \text{ not both zero} \end{cases}$$

SEC. 8.1 Systems of Two Linear Equations in Two Variables

Assume that $a_1 \neq 0$ and let $k_1 = -a_2$ and $k_2 = a_1$; then System 3 becomes

$$\begin{cases} a_1 x + b_1 y = r_1 \\ -a_2(a_1 x + b_1 y) + a_1(a_2 x + b_2 y) = -a_2 r_1 + a_1 r_2 \end{cases}$$

or when simplified

(System 4) $\quad \begin{cases} a_1 x + b_1 y = r_1 \\ (a_1 b_2 - a_2 b_1) y = a_1 r_2 - a_2 r_1 \end{cases}$

which is equivalent to System 1 (Theorem 8.1). If $a_1 b_2 - a_2 b_1 \neq 0$, we can solve the second equation of System 4 for y and get

$$y = \frac{a_1 r_2 - a_2 r_1}{a_1 b_2 - a_2 b_1}$$

which, when substituted for y in the first equation of System 4, gives

$$x = \frac{r_1 b_2 - b_2 r_1}{a_1 b_2 - a_2 b_1}$$

Hence the solution of System 1 in this case is the set

$$\left\{ \left(\frac{r_1 b_2 - r_2 b_1}{a_1 b_2 - a_2 b_1}, \frac{a_1 r_2 - a_2 r_1}{a_1 b_2 - a_2 b_1} \right) \right\}$$

If $a_1 b_2 - a_2 b_1 = 0$ and $a_1 r_2 - a_2 r_1 \neq 0$, the second equation of System 4 has no solution, System 4 has no solution, and hence System 1 has no solution. This is, of course, the case in which the graphs of the two equations defining the system are parallel lines and consequently have no points in common.

Finally, if $a_1 b_2 - a_2 b_1 = 0$ and if $a_1 r_2 - a_2 r_1 = 0$, the second equation of System 4 has infinitely many solutions, System 4 therefore has infinitely many solutions, and System 1, being equivalent to System 4, likewise has a solution set containing infinitely many elements. In this case each of the two equations of System 1 has the same line as its graph.

At the outset of our solution we assumed that $a_1 \neq 0$. We could just as well have assumed that $b_1 \neq 0$; then, had we chosen $k_1 = b_2$ and $k_2 = -b_1$, we would have obtained the equivalent system

(System 5) $\quad \begin{cases} a_1 x + b_1 y = r_1 \\ (a_1 b_2 - a_2 b_1) x = (b_2 r_1 - b_1 r_2) \end{cases}$

and could then have used System 5 to arrive at the same conclusions as before regarding the solution set of System 1.

Example 1 $\quad$ Solve the system $\begin{cases} 2x - 3y = 12 \\ x - y = 5 \end{cases}$

Solution: Let $k_1 = -1$ and $k_2 = 2$. A system equivalent to the given system is

$$\begin{cases} 2x - 3y = 12 \\ -(2x - 3y) + 2(x - y) = -12 + 2(5) \end{cases}$$

or

$$\begin{cases} 2x - 3y = 12 \\ y = -2 \end{cases}$$

To complete the solution, the second equation yields $y = -2$ immediately. Then substitute $y = -2$ in the first equation and solve for x as follows:

$$2x - (3)(-2) = 12 \leftrightarrow 2x + 6 = 12$$
$$\leftrightarrow 2x = 6$$
$$\leftrightarrow x = 3$$

Now the solution set of the last system is $\{(3, -2)\}$ and, since the last system is equivalent to the first, the solution set of the original system is also $\{(3, -2)\}$.

It should be observed that the important step in solving a system of two linear equations in two variables is in choosing k_1 and k_2 such that

$$k_1(a_1 x + b_1 y) + k_2(a_2 x + b_2 y) = k_1 r_1 + k_2 r_2$$

reduces to a form that is either free of x or free of y.

Example 2 Solve the system $\begin{cases} 3x + 2y = -8 \\ 5x - y = -9 \end{cases}$

Solution: Multiply the second equation by 2, add the result to the first equation, and then replace the second equation by this sum to obtain the equivalent system

$$\begin{cases} 3x + 2y = -8 \\ 13x = -26 \end{cases}$$

Solve the second equation for x and get $x = -2$. Substitute this value of x in the first equation and solve for y, obtaining $y = -1$. The solution of the given system is the set $\{(-2, -1)\}$.

The elimination method discussed above obviously can be used to solve systems of linear equations involving more than two variables. We illustrate this fact by solving the following system:

$$\begin{cases} 3x + 4y - 2z = 17 \\ x - y + z = -4 \\ 2x - 3y + z = -7 \end{cases}$$

Solution: We obtain an equivalent system by replacing the second equation by the sum of the first equation and two times the second equation and

SEC. 8.1 Systems of Two Linear Equations in Two Variables

then replacing the third equation by the third equation minus the second equation:

$$\begin{cases} 3x + 4y - 2z = 17 \\ 5x + 2y = 9 \\ x - 2y = -3 \end{cases}$$

In this new system, replace the third equation by the sum of the second equation and third equation to get the equivalent system

$$\begin{cases} 3x + 4y - 2z = 17 \\ 5x + 2y = 9 \\ 6x = 6 \end{cases}$$

from which by inspection we see that $x = 1$, $y = 2$, and $z = -3$. The solution set of the given system is therefore the set $\{(1, 2, -3)\}$. Note that a solution of a single linear equation in three variables is an ordered triple rather than an ordered pair.

Exercises 8.1

In Exercises 1 through 6, solve graphically each given system of linear equations.

1. $\begin{cases} 2x + 3y = 12 \\ 4x - 5y = 2 \end{cases}$ 2. $\begin{cases} 2x + 3y = 5 \\ 3x - 2y = -12 \end{cases}$ 3. $\begin{cases} 4x - y = 7 \\ 8x - 2y = 3 \end{cases}$

4. $\begin{cases} 4x + 5y = 23 \\ 2x - 3y = -5 \end{cases}$ 5. $\begin{cases} x - 3y = 3 \\ 4x - 12y = 12 \end{cases}$ 6. $\begin{cases} 7x - 4y = 25 \\ x + y = 2 \end{cases}$

In Exercises 7 through 22, solve each given system of linear equations by the elimination process.

7. $\begin{cases} x + 4y = 1 \\ 2x - y = -7 \end{cases}$ 8. $\begin{cases} 5x + 6y = 17 \\ 6x + 5y = 16 \end{cases}$

9. $\begin{cases} 4x + 3y = 10 \\ 3x + 2y = 5 \end{cases}$ 10. $\begin{cases} 5x - 7y = 13 \\ 3x - 8y = 23 \end{cases}$

11. $\begin{cases} 8x - 4y + 17 = 0 \\ 16x + 12y + 9 = 0 \end{cases}$ 12. $\begin{cases} 15x - 8y = 12 \\ 11x + 5y = -9 \end{cases}$

13. $\begin{cases} 8x - 2y = -7 \\ 6x + 3y = 12 \end{cases}$ 14. $\begin{cases} 8x + 7y = -54 \\ 9x - 5y = -35 \end{cases}$

15. $\begin{cases} \dfrac{3}{x} - \dfrac{1}{y} = 5 \\ \dfrac{3}{x} + \dfrac{2}{y} = 2 \end{cases}$
16. $\begin{cases} \dfrac{2}{x} - \dfrac{1}{y} - 2 = 0 \\ \dfrac{4}{x} + \dfrac{5}{y} - \dfrac{5}{3} = 0 \end{cases}$

17. $\begin{cases} 3x - y + 2z = 9 \\ 2x + y - z = 7 \\ x + 2y - 3z = 4 \end{cases}$
18. $\begin{cases} 2x - 3y + 4z = 8 \\ 3x + 4y - 5z = -4 \\ 4x - 5y + 6z = 12 \end{cases}$

19. $\begin{cases} 4x - y + 5z = 21 \\ 3x + 7y - 2z = -17 \\ 2x + 6y + 9z = 17 \end{cases}$
20. $\begin{cases} 3x - y + z = 7 \\ 5x + 3y - 2z = 11 \\ 7x - 5z = -6 \end{cases}$

21. $\begin{cases} 2x - 4y + 5z = -7 \\ 3x + 5y - 2z = 21 \\ 8x - 3y + 7z = 11 \end{cases}$
22. $\begin{cases} 3x + 5y + 2z = 4 \\ -2x + 4y + 7z = 37 \\ 5x + 8y + 4z = 8 \end{cases}$

23. Two numbers differ by 7; the smaller number exceeds one-half the larger number by 1. Find the two numbers.

24. A person invests part of $6,000 at 4% and the rest at 6%. Find each part, given that the annual income from the first investment equals the income from the second investment.

25. An airplane travels 1,450 miles in the time required for a car to travel 300 miles. If the rates of each are constant and the plane travels 230 miles per hour faster than the car, what is the rate of each?

26. Ten horses and seven cows can be bought for $1,985 and seven horses and ten cows can be bought for $1,925. If each horse has the same value and each cow costs the same, what is the price of each?

27. A man has several coins consisting of quarters, dimes, and nickels, the total worth $3.30. He observes that if each nickel were exchanged for a quarter, each quarter for a dime, and each dime for a half-dollar, he would have $7.30, but that, if he had three times as many nickels, twice as many dimes, and the same number of quarters that he has, he would have $5.30. How many coins of each kind does he have?

28. A candy store owner wishes to mix 175 pounds of candy, to be sold at 85 cents per pound, by mixing candy worth 79 cents per pound and candy worth $1.00 per pound. How many pounds of each kind should he use?

In Exercises 29 through 32, use a hand calculator to approximate the solutions to the given system of equations. Round each coordinate to five decimal places.

29. $\begin{cases} 21.278x - 7.913y = 296 \\ 498.26x + 16.41y = 99.76 \end{cases}$

SEC. 8.1 Systems of Two Linear Equations in Two Variables

Solution: Replacing the second equation by 16.41 times the first equation plus 7.913 times the second equation (we are eliminating y from the second equation), we obtain the equivalent system:

$$\begin{cases} 21.278x - 7.913y = 296 \\ [(16.41)(21.278) + (7.913)(498.26)]x + [(16.41)(-7.913) + (7.913)(16.41)]y \\ \qquad\qquad = (16.41)(296) + (7.913)(99.76) \end{cases}$$

which, noticing that the coefficient of y in the second equation is zero, becomes

$$\begin{cases} 21.278x - 7.913y = 296 \\ [(16.41)(21.278) + (7.913)(498.26)]x = (16.41)(296) + (7.913)(99.76) \end{cases}$$

Solving for x in the second equation, we obtain

$$x = \frac{(16.41)(296) + (7.913)(99.76)}{(16.41)(21.278) + (7.913)(498.26)}$$

Using a hand calculator and rounding to five decimal places, we obtain

$$x \cong 1.31568$$

Substituting this value for x in the first equation, we obtain

$$(21.278)(1.31568) - (7.913)y = 296$$
$$\leftrightarrow -7.913y = 296 - (21.278)(1.31568)$$
$$\leftrightarrow \quad y = \frac{296 - (21.278)(1.31568)}{-7.913}$$

Again, using a hand calculator and rounding to five decimal places,

$$y \cong -33.86894$$

If we leave all the significant digits displayed for x (1.315677546) and approximate y, we obtain

$$y \cong -33.86895$$

when we round to five decimal places. This is actually the easier procedure. Hence our solution set is $S = \{(x, y)\}$, where $x \cong 1.31568$ and $y \cong -33.86895$. This can be written by using the shorthand

$$S \cong \{(1.31568, -33.86895)\}$$

30. $\begin{cases} 3.21x - 11.3y = 7.5 \\ -92x + 2.75y = 62 \end{cases}$

31. $\begin{cases} 52.98x - 41.14y = -20.16 \\ 2.3x + 8.9y = -1.7 \end{cases}$

32. $\begin{cases} 6.5x + 3.1y - 9.4z = 6.9 \\ 3.7x - 5.3y + 2.8z = -4.2 \\ -2.3x - 8.9y \quad\;\; = 1.7 \end{cases}$

8.2 THE METHOD OF DETERMINANTS

In this section we introduce the concept of a determinant, which is useful in the study of the solutions of systems of linear equations. Let us consider again a typical system of two linear equations in two variables.

(1)
$$\begin{cases} a_1 x + b_1 y = r_1 \\ a_2 x + b_2 y = r_2 \end{cases}$$

This system can be solved by the elimination method discussed in Section 8.1 and, if $a_1 b_2 - a_2 b_1 \neq 0$, the solution is

(2)
$$x = \frac{r_1 b_2 - r_2 b_1}{a_1 b_2 - a_2 b_1}, \quad y = \frac{a_1 r_2 - a_2 r_1}{a_1 b_2 - a_2 b_1}$$

This is exactly how we solved for x in the solution of Exercise 29 in Section 8.1. We (essentially) eliminated y by multiplying the first equation by the coefficient of y in the second equation, multiplying the second equation by the coefficient of y in the first equation, and subtracting. This is a very natural approach when the coefficients have no obvious relationship to each other.

The method of determinants consists of using Equations (2) as formulas, after first expressing them in terms of symbols called *determinants*. The symbol

$$\begin{vmatrix} a & c \\ b & d \end{vmatrix}$$

where a, b, c, and d are any four real numbers, is called a *determinant of the second order* and is defined by the equation

$$\begin{vmatrix} a & c \\ b & d \end{vmatrix} = ad - bc$$

Some examples are

$$\begin{vmatrix} 7 & 3 \\ 5 & 4 \end{vmatrix} = 28 - 15 = 13, \qquad \begin{vmatrix} 2 & 6 \\ 3 & 4 \end{vmatrix} = 8 - 18 = -10$$

$$\begin{vmatrix} 5 & -1 \\ 2 & 3 \end{vmatrix} = 15 - (-2) = 17, \qquad \begin{vmatrix} -6 & -4 \\ 3 & 1 \end{vmatrix} = (-6) - (-12) = 6$$

Using this notation, Equations (2) can now be written

(3)
$$x = \frac{\begin{vmatrix} r_1 & b_1 \\ r_2 & b_2 \end{vmatrix}}{\begin{vmatrix} a_1 & b_1 \\ a_2 & b_2 \end{vmatrix}}, \quad y = \frac{\begin{vmatrix} a_1 & r_1 \\ a_2 & r_2 \end{vmatrix}}{\begin{vmatrix} a_1 & b_1 \\ a_2 & b_2 \end{vmatrix}}$$

It should be noted that the denominators of the fractions in the right sides of Equations (3) are identical, and also that the numbers a_1, b_1, a_2, b_2, which

SEC. 8.2 The Method of Determinants

form the determinant appearing in the denominators of the right sides of Equations (3), are the coefficients of x and y in the two equations of the given system and occur in exactly the same order as they occur in the system. Furthermore, the determinant in the numerator of the fraction in the right side of the first of Equations (3) is the same as the determinant in the denominator of that same fraction, except that the coefficients of x, that is, a_1 and a_2, have been respectively replaced by the constant terms r_1 and r_2. Similarly, the determinant in the numerator of the fraction in the right side of the second equation of Equations (3) is obtained from the determinant in the denominator by replacing b_1 and b_2, the coefficients of y, by r_1 and r_2, respectively.

Example 1 Use determinants to solve the system $\begin{cases} 3x + 7y = 10 \\ 2x + 5y = 6 \end{cases}$

Solution

$$x = \frac{\begin{vmatrix} 10 & 7 \\ 6 & 5 \end{vmatrix}}{\begin{vmatrix} 3 & 7 \\ 2 & 5 \end{vmatrix}} = \frac{50 - 42}{15 - 14} = \frac{8}{1} = 8$$

$$y = \frac{\begin{vmatrix} 3 & 10 \\ 2 & 6 \end{vmatrix}}{\begin{vmatrix} 3 & 7 \\ 2 & 5 \end{vmatrix}} = \frac{18 - 20}{15 - 14} = \frac{-2}{1} = -2$$

Check: $3(8) + 7(-2) = 24 - 14 = 10$

$2(8) + 5(-2) = 16 - 10 = 6$

Hence the solution set consists of the single element $(8, -2)$.

Let us now extend the method of determinants to systems of three linear equations in three variables. We shall start by applying the method of elimination to a typical system of three equations which, for convenience, we write in the form

(4) $\begin{cases} a_1 x + b_1 y + c_1 z = r_1 \\ a_2 x + b_2 y + c_2 z = r_2 \\ a_3 x + b_3 y + c_3 z = r_3 \end{cases}$

where $a_1, b_1, c_1, r_1, a_2, b_2, c_2, r_2, a_3, b_3, c_3, r_3$ are arbitrary but fixed real numbers. If we replace the second equation by c_2 times the first equation, minus c_1 times the second equation, and replace the third equation by c_3 times the first equation, minus c_1 times the third equation, we obtain the equivalent system

(5) $$\begin{cases} a_1x + b_1y + c_1z = r_1 \\ (a_1c_2 - a_2c_1)x + (b_1c_2 - b_2c_1)y = r_1c_2 - r_2c_1 \\ (a_1c_3 - a_3c_1)x + (b_1c_3 - b_3c_1)y = r_1c_3 - r_3c_1 \end{cases}$$

Now, if we replace the third equation here by $b_1c_3 - b_3c_1$ times the second equation, minus $(b_1c_2 - b_2c_1)$ times the third equation, we obtain the system

(6) $$\begin{cases} a_1x + b_1y + c_1z = r_1 \\ (a_1c_2 - a_2c_1)x + (b_1c_2 - b_2c_1)y = r_1c_2 - c_1r_2 \\ [(a_1c_2 - a_2c_1)(b_1c_3 - b_3c_1) - (a_1c_3 - a_3c_1)(b_1c_2 - b_2c_1)]x \\ \quad = (r_1c_2 - c_1r_2)(b_1c_3 - b_3c_1) - (r_1c_3 - c_1r_3)(b_1c_2 - b_2c_1) \end{cases}$$

It follows from the manner in which Equations (6) were obtained that they are equivalent to Equations (5) and, hence, to Equations (4). By expanding the products in the third equation of Equations (6), combining terms where possible, and factoring out c_1 in each side, one can write it in the form

(7) $$\begin{aligned} c_1(a_1b_2c_3 + a_2b_3c_1 + a_3b_1c_2 - a_1b_3c_2 - a_2b_1c_3 - a_3b_2c_1)x \\ = c_1(r_1b_2c_3 + r_2b_3c_1 + r_3b_1c_2 - r_1b_3c_2 - r_2b_1c_3 - r_3b_2c_1) \end{aligned}$$

or, if $c_1 \neq 0$, in the form

(8) $$\begin{aligned} (a_1b_2c_3 + a_2b_3c_1 + a_3b_1c_2 - a_1b_3c_2 - a_2b_1c_3 - a_3b_2c_1)x \\ = r_1b_2c_3 + r_2b_3c_1 + r_3b_1c_2 - r_1b_3c_2 - r_2b_1c_3 - r_3b_2c_1 \end{aligned}$$

If $c_1 = 0$, then $c_2 \neq 0$ or $c_3 \neq 0$, and we can arrive at the same result by a different order of steps.

Let us examine the coefficient of x in Equation (8) a bit more closely. For convenience we denote this coefficient by D; then

(9) $$D = a_1b_2c_3 + a_2b_3c_1 + a_3b_1c_2 - a_1b_3c_2 - a_2b_1c_3 - a_3b_2c_1$$

where the right side consists of six terms, each of which is a product of the three letters a, b, and c and where the subscripts of these letters are the numbers 1, 2, and 3, arranged differently in each of the six terms. We observe also that a plus sign is attached to exactly three of the six products and a minus sign to the other three (a rule for determining which sign to attach to which product will be given later). Although these facts are helpful in theoretical considerations, they are of little use for finding the value of D in cases in which the literal coefficients of x, y, and z have been replaced by numerical coefficients. We now look for simpler ways of expressing D and of arriving at its value. We introduce a new symbol called a *determinant of the third order*. The symbol

$$\begin{vmatrix} a_1 & b_1 & c_1 \\ a_2 & b_2 & c_2 \\ a_3 & b_3 & c_3 \end{vmatrix}$$

SEC. 8.2 The Method of Determinants

where $a_1, a_2, a_3, b_1, b_2, b_3, c_1, c_2, c_3$ are any nine real numbers, is defined by the equation

$$\text{(10)} \quad \begin{vmatrix} a_1 & b_1 & c_1 \\ a_2 & b_2 & c_2 \\ a_3 & b_3 & c_3 \end{vmatrix} = a_1 b_2 c_3 + a_2 b_3 c_1 + a_3 b_1 c_2 - a_1 b_3 c_2 - a_2 b_1 c_3 - a_3 b_2 c_1$$

$$= a_1(b_2 c_3 - b_3 c_2) - a_2(b_1 c_3 - b_3 c_1) + a_3(b_1 c_2 - b_2 c_1)$$

By making use of second-order determinants, Equation (10) can be written

$$\text{(11)} \quad \begin{vmatrix} a_1 & b_1 & c_1 \\ a_2 & b_2 & c_2 \\ a_3 & b_3 & c_3 \end{vmatrix} = a_1 \begin{vmatrix} b_2 & c_2 \\ b_3 & c_3 \end{vmatrix} - a_2 \begin{vmatrix} b_1 & c_1 \\ b_3 & c_3 \end{vmatrix} + a_3 \begin{vmatrix} b_1 & c_1 \\ b_2 & c_2 \end{vmatrix}$$

The numbers $a_1, b_1, \ldots, c_3$ which form the determinant are called *elements* or *entries*. The rows are numbered from top to bottom, and the columns are numbered from left to right. This helps to describe the position of any element of the determinant. For example, the element c_2 lies in the second row and third column, the element a_3 lies in the third row and first column, and so on. It is also helpful, purely as a memory device, to assign plus and minus signs alternately, beginning in the upper left-hand corner, to each of the nine positions of the determinant, as indicated in the schematic diagram of Figure 8.2. It must be clearly understood that the signs shown in this figure have absolutely nothing to do with the algebraic signs of the elements standing in those positions. To understand the usefulness of such a device, we need to define what is called the *minor* of an element of a determinant.

Each element of a determinant can be located by giving the number of the row and the number of the column in which it lies. Now if we cross out the row and column in which a particular element is located, just four elements remain. The second-order determinant formed by these four elements in their natural positions is called the minor of that element. For example, the element c_2 lies in the second row and third column; to find its minor, cross out the second row and

$$\begin{vmatrix} + & - & + \\ - & + & - \\ + & - & + \end{vmatrix} \qquad \begin{vmatrix} a_1 & b_1 & \cancel{c_1} \\ \cancel{a_2} & \cancel{b_2} & \cancel{c_2} \\ a_3 & b_3 & \cancel{c_3} \end{vmatrix}$$

Figure 8.2   Figure 8.3

third column, as in Figure 8.3. The minor of c_2, denoted by the capital letter C_2, is

$$C_2 = \begin{vmatrix} a_1 & b_1 \\ a_3 & b_3 \end{vmatrix}$$

and the minors A_1 and B_3 of a_1 and b_3, respectively, and of the other elements, are found in the same way:

$$A_1 = \begin{vmatrix} b_2 & c_2 \\ b_3 & c_3 \end{vmatrix} \quad \text{and} \quad B_3 = \begin{vmatrix} a_1 & c_1 \\ a_2 & c_2 \end{vmatrix}$$

Equation (11) may now be written in the form

(12) $$\begin{vmatrix} a_1 & b_1 & c_1 \\ a_2 & b_2 & c_2 \\ a_3 & b_3 & c_3 \end{vmatrix} = a_1 A_1 - a_2 A_2 + a_3 A_3$$

Observe that the right side of this equation consists of three terms, each of which is the product of an element and its minor, and that the three elements a_1, a_2, and a_3 exhibited in the products are all from the same column. Note further that the products $a_1 A_1$ and $a_3 A_3$ carry plus signs, and the elements a_1 and a_3 each stand in a positive position, while the product $a_2 A_2$ carries a negative sign and the element a_2 stands in a negative position. Equation (12) is called the *evaluation of the determinant by minors of the first column*. Could we have expanded by minors of some other column or by some row? The answer is, Yes, if we use the proper sign with each product involved.

To simplify our discussion, let us define the *cofactor* of an element of a determinant.

COFACTOR.   *The cofactor of an element of a determinant is either the minor of that element or the negative of the minor of that element, depending on whether the element is in a positive position in the determinant or in a negative position. If the element is in a positive position, its cofactor is the same as its minor, and if in a negative position, its cofactor is the negative of its minor.*

Let us use primed capitals to denote cofactors. For example, the cofactor of a_1 is A'_1, of a_2 it is A'_2, of a_3 it is A'_3, and so on. Thus, expressing the cofactors in terms of corresponding minors, we can write $A'_1 = A_1$, $A'_2 = -A_2$, ..., $C'_2 = -C_2$, and $C'_3 = C_3$. In terms of the cofactor notation, the rule for expanding a determinant by minors may be summarized in the following theorem.

THEOREM 8.2.   *If each element of a row (or column) of a determinant is multiplied by its own cofactor, the sum of the resulting products is equal to the value of the determinant.*

Proof: That this theorem is true is easily verified by actual calculation with each of the six possibilities. We have already seen [in Equation (11)] that the theorem is true for elements of the first column. We shall verify the theorem

SEC. 8.2   The Method of Determinants

for elements of the second row. We wish to verify that

$$\begin{vmatrix} a_1 & b_1 & c_1 \\ a_2 & b_2 & c_2 \\ a_3 & b_3 & c_3 \end{vmatrix} = a_2 A'_2 + b_2 B'_2 + c_2 C'_2$$

By definition,

$$A'_2 = -\begin{vmatrix} b_1 & c_1 \\ b_3 & c_3 \end{vmatrix} = -(b_1 c_3 - b_3 c_1)$$

$$B'_2 = \begin{vmatrix} a_1 & c_1 \\ a_3 & c_3 \end{vmatrix} = a_1 c_3 - a_3 c_1$$

$$C'_2 = -\begin{vmatrix} a_1 & b_1 \\ a_3 & b_3 \end{vmatrix} = -(a_1 b_3 - a_3 b_1)$$

Substituting these values for A'_2, B'_2, and C'_2, we have

$$\begin{aligned} a_2 A'_2 + b_2 B'_2 + c_2 C'_2 \\ = -a_2(b_1 c_3 - b_3 c_1) + b_2(a_1 c_3 - a_3 c_1) - c_2(a_1 b_3 - a_3 b_1) \\ = -a_2 b_1 c_3 + a_2 b_3 c_1 + a_1 b_2 c_3 - a_3 b_2 c_1 - a_1 b_3 c_2 + a_3 b_1 c_2 \end{aligned}$$

Comparing this with the definition given by Equation (10), we see that this is the value of the determinant. The verifications for the other two rows and two columns are left for the student.

Another important theorem in this connection is the following.

THEOREM 8.3. *If each element of a row (or column) of a determinant is multiplied by the cofactor of the corresponding element of another row (or column) of that determinant, the sum of the resulting products is zero.*

The proof here, as in Theorem 8.2, consists of verifying the theorem for each of the six possible cases. We shall leave such verification to be made by the student.

Example 2   Find the value of the determinant $\begin{vmatrix} 3 & 5 & -2 \\ -1 & -2 & 12 \\ -4 & 1 & 7 \end{vmatrix}$.

Solution: We shall expand by minors of the elements of the first column:

$$\begin{vmatrix} 3 & 5 & -2 \\ -1 & -2 & 12 \\ -4 & 1 & 7 \end{vmatrix} = 3\begin{vmatrix} -2 & 12 \\ 1 & 7 \end{vmatrix} - (-1)\begin{vmatrix} 5 & -2 \\ 1 & 7 \end{vmatrix} + (-4)\begin{vmatrix} 5 & -2 \\ -2 & 12 \end{vmatrix}$$

$$= 3(-14 - 12) + (35 + 2) - 4(60 - 4)$$
$$= -78 + 37 - 224 = -265$$

Example 3 | Form the sum of the products of the elements of the first column and the cofactors of the corresponding elements of the second column for the determinant in the example above.

Solution: Forming the required sum, we get

$$-3\begin{vmatrix} -1 & 12 \\ -4 & 7 \end{vmatrix} - 1\begin{vmatrix} 3 & -2 \\ -4 & 7 \end{vmatrix} + 4\begin{vmatrix} 3 & -2 \\ -1 & 12 \end{vmatrix}$$
$$= -3(-7+48) - 1(21-8) + 4(36-2)$$
$$= -123 - 13 + 136 = 0$$

This result is in agreement with the conclusion of Theorem 8.3.

We are now ready to return to our original problem, that of solving a system of three linear equations in three variables by the method of determinants. Comparing Equations (9) and (10), we see that

$$(13) \qquad D = \begin{vmatrix} a_1 & b_1 & c_1 \\ a_2 & b_2 & c_2 \\ a_3 & b_3 & c_3 \end{vmatrix}$$

Now denote the right side of Equation (8) by N_x; then

$$N_x = r_1 b_2 c_3 + r_2 b_3 c_1 + r_3 b_1 c_2 - r_1 b_3 c_2 - r_2 b_1 c_3 - r_3 b_2 c_1$$

Comparing this with the right side of Equation (9), we see that we can get N_x from Equation (9) by replacing a_1, a_2, a_3 with r_1, r_2, r_3, respectively. Hence

$$(14) \qquad N_x = \begin{vmatrix} r_1 & b_1 & c_1 \\ r_2 & b_2 & c_2 \\ r_3 & b_3 & c_3 \end{vmatrix}$$

Equation (8) now becomes

$$Dx = N_x$$

and, if $D \neq 0$, we have

$$(15) \qquad x = \frac{N_x}{D}$$

We could complete our solution of Equations (4) by substituting from Equation (15) into Equations (6), but a more direct method is to use a rule called *Cramer's Rule*.

CRAMER'S RULE: If the determinant D, of the coefficients of x, y, z in a system of three linear equations in x, y, z, is not zero, then the equations have a solution set consisting of a single element. In the solution, the value of each variable

SEC. 8.2 The Method of Determinants

may be expressed as a fraction which is the ratio of two determinants. The denominator of the fraction is D and the numerator is a determinant obtained from D by replacing the column of coefficients of the variable in question by the constant terms r_1, r_2, and r_3.

Proof: Suppose (x, y, z) is an ordered triple of real numbers that satisfy the system Equations (4). If we multiply each by nonzero constants and add, we obtain a new equation. By repeating this process twice more, but using different sets of constants each time, we get two more new equations each of which has (x, y, z) as a solution. Thus the system formed by these three new equations also has the triple (x, y, z) as a solution. It can be shown that the converse is also true, and hence this new system is equivalent to Equations (4).

The determinant D of the coefficients of x, y, and z in Equations (4) is given by Equation (13). We calculate the cofactors A'_1, A'_2, and A'_3 of the elements of the first column of D. Next we multiply the first equation by A'_1, the second by A'_2, the third by A'_3, and add. The new equation is

(16) $(a_1 A'_1 + a_2 A'_2 + a_3 A'_3)x + (b_1 A'_1 + b_2 A'_2 + b_3 A'_3)y$
$+ (c_1 A'_1 + c_2 A'_2 + c_3 A'_3)z = r_1 A'_1 + r_2 A'_2 + r_3 A'_3$

It follows from Theorem 8.2 that the coefficient of x is D and from Theorem 8.3 that the coefficients of y and z are both zero. The right side, by Theorem 8.2, is N_x. Equation (16) therefore reduces to

(17) $$x \begin{vmatrix} a_1 & b_1 & c_1 \\ a_2 & b_2 & c_2 \\ a_3 & b_3 & c_3 \end{vmatrix} = \begin{vmatrix} r_1 & b_1 & c_1 \\ r_2 & b_2 & c_2 \\ r_3 & b_3 & c_3 \end{vmatrix}$$

or, if $D \neq 0$, to $x = N_x/D$, which agrees with the value found in Equation (15) and also with the statement in Cramer's Rule. Similarly, to obtain an equation involving only the value of y, we calculate B'_1, B'_2, and B'_3, the cofactors of the elements in the second column of D. Then, multiplying the first equation of Equations (4) by B'_1, the second by B'_2, and the third by B'_3, we have, after adding,

(18) $(a_1 B'_1 + a_2 B'_2 + a_3 B'_3)x + (b_1 B'_1 + b_2 B'_2 + b_3 B'_3)y$
$+ (c_1 B'_1 + c_2 B'_2 + c_3 B'_3)z = r_1 B'_1 + r_2 B'_2 + r_3 B'_3$

This time the coefficients of x and z are zero, the coefficient of y is D, and the right side is D, whose b's are replaced by r's. Let us denote the right side by N_y; then Equation (18) becomes

$$Dy = N_y$$

(19) $$y = \frac{N_y}{D}$$

Using the same type of argument and the cofactors of the elements in the third column of D, we obtain a third new equation,

(20) $$z = \frac{N_z}{D}$$

where $N_z = r_1 C'_1 + r_2 C'_2 + r_3 C'_3$. If we replace the first equation of Equations (4) by Equation (17), the second by Equation (19), and the third by Equation (20), we have a system equivalent to Equations (4). Since the solution set of this new system, if $D \neq 0$, is obviously the single triple of numbers N_x/D, N_y/D, N_z/D, it follows that Equations (4) have the same solution set, and our proof of Cramer's Rule is complete.

It should be emphasized that, if $D = 0$, Cramer's Rule does not apply. In such case the solution set would either contain infinitely many elements or be the empty set; the best procedure would be the method of elimination.

Example 4 Use Cramer's Rule to solve the system

$$\begin{cases} 2x - 3y + 4z = -1 \\ 3x + y = 5 \\ x + y - z = 3 \end{cases}$$

Solution: We first calculate D, N_x, N_y, and N_z:

$$D = \begin{vmatrix} 2 & -3 & 4 \\ 3 & 1 & 0 \\ 1 & 1 & -1 \end{vmatrix} = 2 \begin{vmatrix} 1 & 0 \\ 1 & -1 \end{vmatrix} - 3 \begin{vmatrix} -3 & 4 \\ 1 & -1 \end{vmatrix} + \begin{vmatrix} -3 & 4 \\ 1 & 0 \end{vmatrix}$$
$$= 2(-1-0) - 3(3-4) + (0-4)$$
$$= -2 + 3 - 4 = -3$$

$$N_x = \begin{vmatrix} -1 & -3 & 4 \\ 5 & 1 & 0 \\ 3 & 1 & -1 \end{vmatrix} = -1 \begin{vmatrix} 1 & 0 \\ 1 & -1 \end{vmatrix} - 5 \begin{vmatrix} -3 & 4 \\ 1 & -1 \end{vmatrix} + 3 \begin{vmatrix} -3 & 4 \\ 1 & 0 \end{vmatrix}$$
$$= -(-1-0) - 5(3-4) + 3(0-4)$$
$$= 1 + 5 - 12 = -6$$

$$N_y = \begin{vmatrix} 2 & -1 & 4 \\ 3 & 5 & 0 \\ 1 & 3 & -1 \end{vmatrix} = 2 \begin{vmatrix} 5 & 0 \\ 3 & -1 \end{vmatrix} - 3 \begin{vmatrix} -1 & 4 \\ 3 & -1 \end{vmatrix} + \begin{vmatrix} -1 & 4 \\ 5 & 0 \end{vmatrix}$$
$$= 2(-5-0) - 3(1-12) + (0-20)$$
$$= -10 + 33 - 20 = 3$$

SEC. 8.2 The Method of Determinants

$$N_z = \begin{vmatrix} 2 & -3 & -1 \\ 3 & 1 & 5 \\ 1 & 1 & 3 \end{vmatrix} = 2\begin{vmatrix} 1 & 5 \\ 1 & 3 \end{vmatrix} - 3\begin{vmatrix} -3 & -1 \\ 1 & 3 \end{vmatrix} + \begin{vmatrix} -3 & -1 \\ 1 & 5 \end{vmatrix}$$
$$= 2(3-5) - 3(-9+1) + (-15+1)$$
$$= -4 + 24 - 14 = 6$$

Hence

$$x = \frac{N_x}{D} = \frac{-6}{-3} = 2, \quad y = \frac{N_y}{D} = \frac{3}{-3} = -1, \quad z = \frac{N_z}{D} = \frac{6}{-3} = -2$$

Check: $2(2) - 3(-1) + 4(-2) = 4 + 3 - 8 = -1$
$3(2) + 1(-1) + 0(-2) = 6 - 1 = 5$
$1(2) + 1(-1) - 1(-2) = 2 - 1 + 2 = 3$

Although Cramer's Rule was stated for a system of three linear equations in three unknowns, it can be generalized to apply to a system of n linear equations in n unknowns, provided the determinant of the coefficients of the system is not zero. In fact, Theorems 8.2 and 8.3 are both valid for a determinant of any order. However, Cramer's Rule has little practical value if n is large, and even for small values of n the elimination method is often much faster. The real value of Cramer's Rule, therefore, lies in its theoretical importance.

Exercises 8.2

Evaluate each of the following determinants.

1. $\begin{vmatrix} 3 & 1 \\ 4 & 2 \end{vmatrix}$

2. $\begin{vmatrix} 0 & 3 \\ -1 & 2 \end{vmatrix}$

3. $\begin{vmatrix} ab & ac \\ bd & cd \end{vmatrix}$

4. $\begin{vmatrix} x & x+1 \\ x-1 & x \end{vmatrix}$

5. $\begin{vmatrix} 1 & 3 & 0 \\ 2 & 1 & 4 \\ -1 & 2 & -3 \end{vmatrix}$

6. $\begin{vmatrix} 2 & 1 & 5 \\ 0 & 3 & 1 \\ 1 & 1 & 2 \end{vmatrix}$

7. $\begin{vmatrix} 1 & 1 & -1 \\ 2 & 0 & 3 \\ 1 & 2 & 2 \end{vmatrix}$

8. $\begin{vmatrix} 1 & 2 & 3 \\ 4 & 5 & 6 \\ 7 & 8 & 9 \end{vmatrix}$

9. $\begin{vmatrix} 3 & -2 & 1 \\ 7 & 1 & -1 \\ 6 & -1 & 2 \end{vmatrix}$

10. $\begin{vmatrix} 1 & -2 & 3 \\ 2 & -3 & 1 \\ 4 & -8 & 12 \end{vmatrix}$

11. $\begin{vmatrix} 2 & -3 & 4 \\ 3 & 4 & -5 \\ 4 & -5 & 6 \end{vmatrix}$

12. $\begin{vmatrix} 1 & 1 & 1 & 1 \\ 1 & 2 & 3 & 4 \\ 1 & 3 & 6 & 10 \\ 1 & 4 & 10 & 20 \end{vmatrix}$

Solve systems 13 through 22 by the method of determinants (Cramer's Rule).

13. $\begin{cases} 3x - 8y = 6 \\ 6x + 2y = 3 \end{cases}$

14. $\begin{cases} x + y = 2 \\ 3x - 2y = -4 \end{cases}$

15. $\begin{cases} 9x + 5y = -6 \\ 7x + 6y = 27 \end{cases}$

16. $\begin{cases} 13x - 8y = 29 \\ 17x - 7y = 4 \end{cases}$

17. $\begin{cases} x + 3y - z = 7 \\ 5x - 7y + z = 3 \\ 2x - y - 2z = 0 \end{cases}$

18. $\begin{cases} x + y - z = 0 \\ 3x + 6y - 4z = 0 \\ x + y + z = 1 \end{cases}$

19. $\begin{cases} x - 2y + z = 12 \\ x + 2y + 3z = 48 \\ 6x + 4y + 3z = 84 \end{cases}$

20. $\begin{cases} 2x - y + 3z = 4 \\ x + 3y + 3z = -2 \\ 3x + 2y - 6z = 6 \end{cases}$

21. $\begin{cases} 4x - y - 2z = -1 \\ 3x - 5y - 4z = -1 \\ x + 6y + 3z = 1 \end{cases}$

22. $\begin{cases} 2x - 3y - z = 5 \\ 4x + 2y - 7z = 1 \\ -6x - 5y + 8z = -3 \end{cases}$

23. Prove: $\begin{vmatrix} a_1 & b_1 & c_1 \\ a_2 & b_2 & c_2 \\ a_3 & b_3 & c_3 \end{vmatrix} = \begin{vmatrix} a_1 & a_2 & a_3 \\ b_1 & b_2 & b_3 \\ c_1 & c_2 & c_3 \end{vmatrix}$

24. Prove: $\begin{vmatrix} a_1 & ka_1 & c_1 \\ a_2 & ka_2 & c_2 \\ a_3 & ka_3 & c_3 \end{vmatrix} = 0$

25. Prove: $\begin{vmatrix} a_1 + kb_1 & b_1 & c_1 \\ a_2 + kb_2 & b_2 & c_2 \\ a_3 + kb_3 & b_3 & c_3 \end{vmatrix} = \begin{vmatrix} a_1 & b_1 & c_1 \\ a_2 & b_2 & c_2 \\ a_3 & b_3 & c_3 \end{vmatrix}$

26. Show that the following is the equation of the line determined by the points (2, 3) and (−1, 5):

$$\begin{vmatrix} x & y & 1 \\ 2 & 3 & 1 \\ -1 & 5 & 1 \end{vmatrix} = 0$$

27. Show that the following is the equation of the line through (2, 3) and (5, −4):

$$\begin{vmatrix} x & 2 & 5 \\ y & 3 & -4 \\ 1 & 1 & 1 \end{vmatrix} = 0$$

28. Show that $\begin{vmatrix} a & d & g \\ b & e & h \\ c & f & j \end{vmatrix} = -\begin{vmatrix} a & d & g \\ c & f & j \\ b & e & h \end{vmatrix}$

29. Use a hand calculator to evaluate the following determinants:

a. $\begin{vmatrix} 3.21 & -11.3 \\ -92 & 2.75 \end{vmatrix}$ b. $\begin{vmatrix} 6.5 & 3.1 & -9.4 \\ 3.7 & -5.3 & 2.8 \\ -2.3 & -8.9 & 0 \end{vmatrix}$

30. Use a hand calculator and Cramer's Rule to find approximate solutions to each of the following systems of equations (compare with Exercise 30 and 32 of Section 8.1).

a. $\begin{cases} 3.21x - 11.3y = 7.5 \\ -92x + 2.75y = 62 \end{cases}$ b. $\begin{cases} 6.5x + 3.1y - 9.4z = 6.9 \\ 3.7x - 5.3y + 2.8z = -4.2 \\ -2.3x - 8.9y = 1.7 \end{cases}$

8.3 THE CONCEPT OF A MATRIX

A concept in mathematics that has become exceedingly useful in recent years, due mainly to the ever increasing use of computers, is that of a mathematical object called a matrix. It will be our purpose here to consider this basic concept. We shall define the operations with matrices, note certain of their properties, and indicate a few of their many applications. We start by giving a definition of a matrix.

DEFINITION 8.1.   A *matrix* is a rectangular array of elements from some mathematical system.

In this discussion we shall consider only matrices whose elements are real numbers. It is conventional to enclose the elements of a matrix by parentheses or by brackets. We shall use the latter. The following are examples of matrices.

(a) $\begin{bmatrix} 1 & -2 & 5 \\ 3 & 1 & 4 \end{bmatrix}$ (b) $\begin{bmatrix} 2 & 1 \\ 0 & -3 \end{bmatrix}$ (c) $[5 \ 3 \ 0]$

(d) $\begin{bmatrix} 3 \\ 2 \\ -1 \end{bmatrix}$ (e) $\begin{bmatrix} 1 & 5 & -1 \\ 3 & 4 & -2 \\ 2 & -3 & 5 \end{bmatrix}$ (f) $\begin{bmatrix} -1 & 3 \\ 0 & 7 \\ 2 & 1 \\ 5 & 4 \end{bmatrix}$

Note that the elements of a matrix are arranged in rows with the same number of elements in each row. One could also think of the elements of a matrix as being arranged in columns with the same number of elements in each column. The rows are numbered from the top down, the columns are numbered from left to right. The top row is row number 1, or simply row 1, the next row down is row 2, etc. The first column on the left is column 1, the next column is column 2, and so on. Although a matrix is made up of individual elements, it is desirable to consider the entire matrix as a single entity. In keeping with this point of view and for ease in discussion, matrices are often indicated by capital letters. The method mentioned above for numbering the rows and columns of a matrix makes it possible to give the location of a particular element in the matrix by giving its row number, called the *row index*, and its column number, called the *column index*. In working with general matrices it is convenient to use a lowercase letter with a double subscript to denote an element of a matrix. The double subscript plays a dual role. First it is used to differentiate between pairs of elements of the matrix; second, it gives the location of the element in the matrix. The first subscript is the row index of the element and the second subscript is the column index. This procedure is illustrated in matrix A shown below.

$$A = \begin{bmatrix} a_{11} & a_{12} & a_{13} & a_{14} \\ a_{21} & a_{22} & a_{23} & a_{24} \\ a_{31} & a_{32} & a_{33} & a_{34} \end{bmatrix}$$

Note that the first subscript of each element in the first row is 1, the first subscript of each element in the second row is 2, and the first subscript of each element in the third row is 3. Similarly, each element in the first column has the second subscript 1, in the second column each element has 2 as its second subscript, and so on for elements in the third and fourth columns. For example, the element denoted by a_{24} is the element in the second row and fourth column. Likewise, the element denoted by a_{31} is the element in the third row and first column. To make a general statement, the element denoted by a_{ij} is the element in the *i*th row and *j*th column. The matrix A shown above is often written in the abbreviated form $A = [a_{ij}]$. When such an abbreviated notation is used, it is necessary to know the number of rows and columns in the matrix. This brings us to the following definition.

SEC. 8.3 The Concept of a Matrix

DEFINITION 8.2. A matrix with m rows and n columns is said to have dimension, or size, $m \times n$, read "m by n." If $m = n$, the matrix is said to be a square matrix.

The dimensions of the six matrices used as examples at the beginning of this section are (a) 2×3, (b) 2×2, (c) 1×3, (d) 3×1, (e) 3×3, and (f) 4×2.

DEFINITION 8.3. A matrix with only one row is called a *row matrix*. A matrix with only one column is called a *column matrix*.

A row matrix is also called a row vector, and a column matrix is often called a column vector. The matrix (c) listed at the beginning of this section is a row matrix, and matrix (d) is a column matrix.

Matrices and the Elimination Method for Solving Systems of Linear Equations

Matrices have many applications; of particular interest are their relations to systems of linear equations. In this connection we shall now demonstrate how matrices may be used to shorten the labor normally involved in solving a system of n linear equations in n unknowns by the method of elimination discussed in Section 8.1. First, we give the following definition.

DEFINITION 8.4. A system of n linear equations in n unknowns is said to be in *echelon form* if: The nth, or last, equation of the system contains only one unknown; the $(n-1)$st equation contains two of the unknowns, including the unknown in the last equation; the $(n-2)$nd equation contains three unknowns, including the two unknowns in the $(n-1)$st equation; and so on, with the first equation of the system containing all n unknowns.

Systems (I) and (II), below, are in echelon form.

$$\text{(I)} \begin{cases} a_1 x + b_1 y = r_1 \\ b_2 y = r_2 \end{cases} \qquad \text{(II)} \begin{cases} a_1 x + b_1 y + c_1 z = r_1 \\ b_2 y + c_2 z = r_2 \\ c_3 z = r_3 \end{cases}$$

It will be recalled that two or more systems of linear equations are said to be equivalent if they have the same solution set. It follows from the discussion in Section 8.1 that a system of linear equations can be reduced to an equivalent system in echelon form by performing the following types of operations.

(a) Interchange two equations of the system.
(b) Replace an equation by a nonzero multiple of itself.
(c) Replace an equation by the sum of a nonzero multiple of that equation and some multiple of another equation of the system (see Theorem 8.1).

For example, consider the system of equations

(III)
$$\begin{cases} 3x + 4y - 2z = 17 \\ x - y + z = -4 \\ 2x - 3y + z = -7 \end{cases}$$

We shall, for convenience, first interchange rows and obtain the following equivalent system.

$$\begin{cases} x - y + z = -4 \\ 2x - 3y + z = -7 \\ 3x + 4y - 2z = 17 \end{cases}$$

Next we add to the second equation -2 times the first equation and then add to the third equation -3 times the first equation, giving the equivalent system

$$\begin{cases} x - y + z = -4 \\ -y - z = 1 \\ 7y - 5z = 29 \end{cases}$$

Now multiply the second equation by 7 and add the result to the third equation, giving the system

$$\begin{cases} x - y + z = -4 \\ -y - z = 1 \\ -12z = 36 \end{cases}$$

This system is in echelon form, and we could solve it easily by first solving for z (from the third equation), substituting this value for z in the second equation and solving for y, and then substituting the values of y and z in the first equation and solving for x. However, the common practice is to make the "first coefficient" of each equation equal to 1, which we proceed to do. Multiplying the second equation by -1 and dividing the third equation by -12, we obtain the following equivalent system in echelon form.

(IV)
$$\begin{cases} x - y + z = -4 \\ y + z = -1 \\ z = -3 \end{cases}$$

In this form the solution of the system is very easy to obtain. First, we observe that the third equation yields $z = -3$. Substituting $z = -3$ in the second equation, we obtain

$$y + (-3) = -1$$
$$y = 3 - 1$$
$$y = 2$$

SEC. 8.3   The Concept of a Matrix

Then substituting $y = 2$ and $z = -3$ in the first equation, we obtain

$$x - 2 - 3 = -4$$
$$x = 5 - 4$$
$$x = 1$$

Hence the solution set is $S = \{(1, 2, -3)\}$.

To solve the above system of equations using matrices, replace the system of equations by a matrix whose elements are the coefficients and constants occurring in the equations and in the same order as they occur in the equation. For example, system (III) of linear equations can be represented by the matrix

(III′)
$$\begin{bmatrix} 3 & 4 & -2 & 17 \\ 1 & -1 & 1 & -4 \\ 2 & -3 & 1 & -7 \end{bmatrix}$$

The operations **(a)**, **(b)**, and **(c)** listed above, which always lead to an equivalent system of equations, can be carried out by performing the following types of operations on the matrix representation of the system of equations. Let a denote the matrix representation of the system of equations.

(a′) Interchange two rows of the matrix.
(b′) Multiply the elements of a row of the matrix by a nonzero constant.
(c′) Add a nonzero constant multiple of the elements of a row of the matrix to a constant multiple of the corresponding elements of another row.

Operation of types **(a′)**, **(b′)**, and **(c′)** on a matrix are called *elementary row operations*.

To see how this works in a particular case, let us start with matrix (III′) and interchange rows to obtain

$$\begin{bmatrix} 1 & -1 & 1 & -4 \\ 2 & -3 & 1 & -7 \\ 3 & 4 & -2 & 17 \end{bmatrix}$$

Now let us add to the elements of the second row -2 times the corresponding elements of the first row and add to the elements of the third row -3 times the corresponding elements of the first row. The result is the matrix

$$\begin{bmatrix} 1 & -1 & 1 & -4 \\ 0 & -1 & -1 & 1 \\ 0 & 7 & -5 & 29 \end{bmatrix}$$

Next multiply the elements of the second row by -1 in order to make the *first nonzero element* in the second row equal to 1. The result is the matrix

$$\begin{bmatrix} 1 & -1 & 1 & -4 \\ 0 & 1 & 1 & -1 \\ 0 & 7 & -5 & 29 \end{bmatrix}$$

Now multiply the elements of the second row by -7 and add to the corresponding elements of the third row. This produces the matrix

$$\begin{bmatrix} 1 & -1 & 1 & -4 \\ 0 & 1 & 1 & -1 \\ 0 & 0 & -12 & 36 \end{bmatrix}$$

Finally, by multiplying the elements of the third row by $-\frac{1}{12}$ (again, to make the *first nonzero element* of the third row equal to 1), we obtain the matrix

$$\begin{bmatrix} 1 & -1 & 1 & -4 \\ 0 & 1 & 1 & -1 \\ 0 & 0 & 1 & -3 \end{bmatrix}$$

This last matrix is the matrix representation of the system of linear equations,

$$\begin{cases} x - y + z = -4 \\ y + z = -1 \\ z = -3 \end{cases}$$

which is precisely system (IV) arrived at above by working with the equations of system (III). One obvious advantage of working with the matrix rather than the equations of a system of linear equations is that the unknowns, the plus signs, and the equal signs are deleted, and we work only with coefficients and constants. This is somewhat analogous to using synthetic substitution as an aid in solving polynomial equations.

Exercises 8.3

Solve each of the following systems of equations by replacing each system by its matrix representation and then working with the matrix, as was done in the example worked out above.

1. $\begin{cases} x - 2y = 4 \\ 3x + 5y = 1 \end{cases}$

2. $\begin{cases} 3x + 2y = 7 \\ 2x - 7y = -12 \end{cases}$

SEC. 8.3   The Concept of a Matrix

3. $\begin{cases} 5x + 3y = -3 \\ 2x + 5y = 14 \end{cases}$
4. $\begin{cases} 5x - 7y = 13 \\ 3x - 8y = 23 \end{cases}$

5. $\begin{cases} 4x + 3y = 10 \\ 3x + 2y = 5 \end{cases}$
6. $\begin{cases} 2x + 3y = 12 \\ 4x - 5y = 2 \end{cases}$

7. $\begin{cases} 3x - 5y = 9 \\ 4x + 3y = -17 \end{cases}$
8. $\begin{cases} 9x + 4y = 7 \\ 7x - 6y = 10 \end{cases}$

9. $\begin{cases} 11x - 7y = 17 \\ 5x - 13y = -7 \end{cases}$
10. $\begin{cases} 3x - 4y + 2z = 1 \\ 2x - 3y + z = -1 \\ x + y + z = 6 \end{cases}$

11. $\begin{cases} x - y + 2z = 4 \\ 3x + y - z = 1 \\ x - 3y + z = 5 \end{cases}$
12. $\begin{cases} 2x - y - 2z = 3 \\ x + y + z = 4 \\ 4x - 3y - 3z = 2 \end{cases}$

13. $\begin{cases} 3x - y + 2z = 9 \\ 2x + y - z = 7 \\ x + 2y - 3z = 4 \end{cases}$
14. $\begin{cases} 2x - 3y + 4z = 8 \\ 3x + 4y - 5z = -4 \\ 4x - 5y + 6z = 12 \end{cases}$

15. $\begin{cases} 2x + 5y - 4z = -7 \\ 3x - 2y + 5z = 21 \\ 8x + 7y - 3z = 11 \end{cases}$
16. $\begin{cases} x + y + z + w = 5 \\ x - 2y + 2z + w = 4 \\ 2x - y + z - 2w = 9 \\ 3x + y - 2z + 3w = 3 \end{cases}$

17. $\begin{cases} x + y + z + w = 8 \\ 2x + y - 6z + 6w = 1 \\ x + 2y - 5z + 5w = 1 \\ x + 2y + 3z + w = 1 \end{cases}$

In Exercises 18 through 20, replace each system by its matrix representation and then, using a hand calculator to approximate the elements in the various steps (round each approximation to five decimal places), approximate the solutions to each system. Compare the solution for Exercises 18 and 20 with the solutions to the same exercises in Sections 8.1 and 8.2.

18. $\begin{cases} 3.21x - 11.3y = 7.5 \\ -92x + 2.75y = 62 \end{cases}$
19. $\begin{cases} 21.278x - 7.913y = 296 \\ 498.26x + 16.41y = 99.76 \end{cases}$

20. $\begin{cases} 6.5x + 3.1y - 9.4z = 6.9 \\ 3.7x - 5.3y + 2.8z = -4.2 \\ -2.3x - 8.9y = 1.7 \end{cases}$

8.4 MATRIX ALGEBRA

To gain skill in the use of matrices, a knowledge of certain elementary operations is essential. In this section we shall define these operations and illustrate by using systems of linear equations. We start by defining the equality of two matrices.

DEFINITION 8.5. Two matrices A and B are said to be *equal* if and only if the following two conditions are met:

1. A and B have the same number of rows and the same number of columns (i.e., A and B have the same dimensions).
2. Each element of A equals the corresponding element of B, that is, if $A = [a_{ij}]$ and $B = [b_{ij}]$ are both of dimensions $m \times n$ and $a_{ij} = b_{ij}$ for each i and j involved.

This definition may be made more clear by the following examples.

(a) $\begin{bmatrix} 5 & 7 & 3 \\ 2 & 1 & 4 \end{bmatrix} = \begin{bmatrix} 5 & 7 & 3 \\ 2 & 1 & 4 \end{bmatrix}$

since the matrices have the same dimension and the corresponding elements are equal.

(b) $\begin{bmatrix} 5 & 7 & 3 \\ 2 & 1 & 4 \end{bmatrix} \neq \begin{bmatrix} 5 & 7 & 3 \\ 2 & 8 & 4 \end{bmatrix}$

because the corresponding elements in the second row and second column are not equal.

(c) $\begin{bmatrix} 5 & 7 & 3 \\ 2 & 1 & 4 \end{bmatrix} \neq \begin{bmatrix} 7 & 3 \\ 1 & 4 \end{bmatrix}$

because the dimensions of the two matrices are not the same.

DEFINITION 8.6. Let $A = [a_{ij}]$ and $B = [b_{ij}]$ be matrices, both with dimensions $m \times n$. The *sum* of A and B is defined to be a matrix $C = [c_{ij}]$ with dimensions $m \times n$ such that each element of C is the sum of the corresponding elements of A and B. Symbolically, $A + B = C$, where $c_{ij} = a_{ij} + b_{ij}$ for each i and j involved.

SEC. 8.4 Matrix Algebra

It must be emphasized that the sum of two matrices with different dimensions is not defined.

Example 1 Find $A + B$ if $A = \begin{bmatrix} 1 & 3 & 2 \\ 6 & 2 & 4 \end{bmatrix}$ and $B = \begin{bmatrix} 3 & 4 & 1 \\ 2 & 7 & 2 \end{bmatrix}$.

Solution: $A + B = \begin{bmatrix} 1 & 3 & 2 \\ 6 & 2 & 4 \end{bmatrix} + \begin{bmatrix} 3 & 4 & 1 \\ 2 & 7 & 2 \end{bmatrix}$

$= \begin{bmatrix} 1+3 & 3+4 & 2+1 \\ 6+2 & 2+7 & 4+2 \end{bmatrix} = \begin{bmatrix} 4 & 7 & 3 \\ 8 & 9 & 6 \end{bmatrix}$

Example 2 Add the matrices $\begin{bmatrix} 5 & 2 \\ 1 & 3 \end{bmatrix}$ and $\begin{bmatrix} 3 & -1 & 4 \\ 2 & 7 & 5 \end{bmatrix}$.

Solution: The sum is not defined since the dimensions of the two matrices are not the same.

The definition of the sum of two matrices of like dimensions given above is easily extended to the sum of three or more matrices. For example, if $A = [a_{ij}]$, $B = [b_{ij}]$, and $C = [c_{ij}]$ are matrices each of dimensions $m \times n$, we define their sum by the equation

$$A + B + C = (A + B) + C$$

Similarly, we define the sum of four matrices A, B, C, and D by the equation

$$A + B + C + D = (A + B + C) + D$$

assuming of course that each of the four matrices has dimensions $m \times n$.

Matrix addition is commutative and associative. The truth of this statement follows from the definition of matrix addition and from the fact that addition of real numbers is commutative and associative.

To distinguish them from vectors or matrices, real numbers are usually referred to as scalars when working with real numbers and vectors or with real numbers and matrices. We now define multiplication of a matrix A of any dimensions by a scalar k.

DEFINITION 8.7. If A is a matrix and k is a scalar, the *scalar product* kA is defined as a matrix each element of which is the corresponding element of A multiplied by k. Symbolically, if $A = [a_{ij}]$ with dimensions $m \times n$, then $kA = [ka_{ij}]$.

Example 3 If $A = \begin{bmatrix} 2 & -1 & 3 \\ 4 & 2 & 1 \end{bmatrix}$, then $5A = \begin{bmatrix} 10 & -5 & 15 \\ 20 & 10 & 5 \end{bmatrix}$.

A matrix, regardless of its dimensions, each of whose elements is zero is called a *zero matrix*. Thus, for each positive integer m and each positive integer n, there is a zero matrix with dimensions $m \times n$. We shall use the capital letter Z to denote a zero matrix. For every $m \times n$ matrix A, the $m \times n$ zero matrix Z satisfies $A + Z = A$. Therefore, the $m \times n$ zero matrix Z is also called the *additive identity* for $m \times n$ matrices.

It follows from Definition 8.7 with $k = -1$ that for every matrix $A = [a_{ij}]$ there exists a matrix $(-A) = [-a_{ij}]$. Since $a_{ij} + (-a_{ij}) = 0$, $A + (-A) = Z$. The matrix $-A$ is the *additive inverse* of A. Since every matrix has an additive inverse, we can use that fact and the definition of addition to define subtraction for matrices.

DEFINITION 8.8. If A and B are matrices, each of dimensions $m \times n$, the *difference* $A - B$ is defined by the equation

$$A - B = A + (-B)$$

We have now defined the operations of addition and subtraction for matrices. The next logical step is to define the product of two matrices.

The multiplication of a matrix by a matrix is much more complicated than the operation of adding two matrices having the same dimensions or the operation of multiplying a matrix by a scalar. In fact, the rules for finding the product of two matrices may seem strange at first, but rest assured that the multiplication of one matrix by another is defined so as to produce a product matrix with the greatest possible usefulness. In the first place, the multiplication of two matrices is not always possible. To be more specific, the product AB, in that order, of matrices A and B is defined if and only if the number of columns of A equals the number of rows of B. In the special case where A is a row matrix and B is a column matrix, the product AB is a matrix which has only one row and one column. That is, the product has exactly one element. Such matrices are called *scalar matrices*.

DEFINITION 8.9. If $A = [a_{11} \quad a_{12} \quad \cdots \quad a_{1m}]$ and

$$B = \begin{bmatrix} b_{11} \\ b_{21} \\ \vdots \\ b_{m1} \end{bmatrix}$$

the *product AB* is defined by the equation

$$AB = [a_{11} \quad a_{12} \quad \cdots \quad a_{1m}] \begin{bmatrix} b_{11} \\ b_{21} \\ \vdots \\ b_{m1} \end{bmatrix} = [a_{11}b_{11} + a_{12}b_{21} + \cdots + a_{1m}b_{m1}]$$

SEC. 8.4 Matrix Algebra

A convenient way to describe this operation is to say that we have "poured the row matrix A down the column matrix B," multiplying elements consecutively and adding the products. Notice that the product matrix has *only one* element represented as a sum of the m numbers $a_{11}b_{11}, a_{12}b_{21}, \ldots, a_{1m}b_{m1}$.

Definition 8.9 is illustrated in the following examples.

Example 4
$$[a_{11} \quad a_{12}] \begin{bmatrix} b_{11} \\ b_{21} \end{bmatrix} = [a_{11}b_{11} + a_{12}b_{21}]$$

Example 5
$$[2 \quad 5] \begin{bmatrix} 3 \\ 7 \end{bmatrix} = [6 + 35] = [41]$$

Example 6
$$[a_{11} \quad a_{12} \quad a_{13}] \begin{bmatrix} b_{11} \\ b_{21} \\ b_{31} \end{bmatrix} = [a_{11}b_{11} + a_{12}b_{21} + a_{13}b_{31}]$$

Example 7
$$[2 \quad -1 \quad 3] \begin{bmatrix} 4 \\ 7 \\ 5 \end{bmatrix} = [8 - 7 + 15] = 16$$

Before giving the general definition of the product of two matrices, we illustrate it with the following example.

Let

$$A = \begin{bmatrix} a_{11} & a_{12} \\ a_{21} & a_{22} \end{bmatrix}, \qquad B = \begin{bmatrix} b_{11} & b_{12} & b_{13} \\ b_{21} & b_{22} & b_{23} \end{bmatrix}$$

By definition, the product AB, in that order, is a matrix C such that

$$C = AB = \begin{bmatrix} a_{11}b_{11} + a_{12}b_{21} & a_{11}b_{12} + a_{12}b_{22} & a_{11}b_{13} + a_{12}b_{23} \\ a_{21}b_{11} + a_{22}b_{21} & a_{21}b_{12} + a_{22}b_{22} & a_{21}b_{13} + a_{22}b_{23} \end{bmatrix}$$

Suppose we write the matrix C in the form

$$\begin{bmatrix} c_{11} & c_{12} & c_{13} \\ c_{21} & c_{22} & c_{23} \end{bmatrix}$$

We observe that $c_{11} = a_{11}b_{11} + a_{12}b_{21}$ is simply the element in the product of the first row of A with the first column of B, *where we think of the first row of A as a row matrix and the first column of B as a column matrix*. Similarly, $c_{12} = a_{11}b_{12} + a_{12}b_{22}$ is the element in the product of the first row of A with the second column of B and $c_{13} = a_{11}b_{13} + a_{12}b_{23}$ is the element in the product of the first row of A with the third column of B. These three facts can be summarized by saying that the jth element of the first row of the product

AB is obtained by "pouring the first row of A down the jth column of B" for $j = 1, 2, 3$. Likewise, $c_{21} = a_{21}b_{11} + a_{22}b_{21}$ is obtained by "pouring the second row of A down the first column of B," $c_{22} = a_{21}b_{12} + a_{22}b_{22}$ is obtained by "pouring the second row of A down the second column of B," and $c_{23} = a_{21}b_{13} + a_{22}b_{23}$ is obtained by "pouring the second row of A down the third column of B."

In summary, we give the following more precise and more general definition.

DEFINITION 8.10. Let A be an $m \times q$ matrix and B a $q \times n$ matrix. Then the product matrix $C = AB$ is an $m \times n$ matrix whose element, c_{ij}, in the ith row and jth column is obtained by multiplying the elements of the ith row of A by the corresponding elements of the jth column of B and adding the results. Symbolically,

$$c_{ij} = a_{i1}b_{1j} + a_{i2}b_{2j} + \cdots + a_{iq}b_{qj}$$

Below are some numerical examples of matrix multiplication.

$$\begin{bmatrix} -1 & 3 \\ 2 & 1 \end{bmatrix} \begin{bmatrix} 6 & -2 \\ 1 & 5 \end{bmatrix} = \begin{bmatrix} -3 & 17 \\ 13 & 1 \end{bmatrix}$$

$$\begin{bmatrix} 1 & 0 & -1 \\ 2 & 1 & 0 \\ -1 & 3 & 1 \end{bmatrix} \begin{bmatrix} 3 & -1 & 0 \\ 0 & 2 & 1 \\ 1 & 0 & -1 \end{bmatrix} = \begin{bmatrix} 2 & -1 & 1 \\ 6 & 0 & 1 \\ -2 & 7 & 2 \end{bmatrix}$$

$$\begin{bmatrix} 2 & 1 & 3 \\ 1 & -1 & 0 \end{bmatrix} \begin{bmatrix} 3 & 0 & -1 & 0 \\ 1 & 1 & 0 & 2 \\ 0 & -1 & 3 & 1 \end{bmatrix} = \begin{bmatrix} 7 & -2 & 7 & 5 \\ 2 & -1 & -1 & -2 \end{bmatrix}$$

An obvious question at this point is that of multiplying more than two matrices together. Let A be an $m \times h$ matrix, let B be an $h \times k$ matrix, and let C be a $k \times n$ matrix. The product AB is an $m \times k$ matrix, which can be multiplied by C, giving, as a product, an $m \times n$ matrix. Thus the product $(AB)C$ exists. Similarly, the product $A(BC)$ exists; it can be shown that these two products are equal. This fact may be described by saying that matrix multiplication is associative. We define the product ABC as their common product, that is,

$$ABC = (AB)C = A(BC)$$

Given two matrices A and B, the product AB is defined if and only if the number of columns of A equals the number of rows of B. For these same two matrices, the product BA is defined if and only if the number of columns of B equals the number of rows of A. Hence for both products, AB and BA, to exist, it is necessary that both A and B be square matrices and of the same size. Thus

SEC. 8.4 Matrix Algebra

square matrices of the same size are the only matrices that can be multiplied in either order. If A and B are square matrices of the same size, it is not necessarily true that $AB = BA$. The truth of this statement is verified by the following simple example.

Let
$$A = \begin{bmatrix} 1 & 3 \\ 2 & 4 \end{bmatrix}, \quad B = \begin{bmatrix} 2 & 4 \\ 1 & 3 \end{bmatrix}$$

Then
$$AB = \begin{bmatrix} 1 & 3 \\ 2 & 4 \end{bmatrix} \begin{bmatrix} 2 & 4 \\ 1 & 3 \end{bmatrix} = \begin{bmatrix} 5 & 13 \\ 8 & 20 \end{bmatrix}$$

$$BA = \begin{bmatrix} 2 & 4 \\ 1 & 3 \end{bmatrix} \begin{bmatrix} 1 & 3 \\ 2 & 4 \end{bmatrix} = \begin{bmatrix} 10 & 22 \\ 7 & 15 \end{bmatrix}$$

and obviously $AB \neq BA$. The above discussion may be summarized in the following sentence. Matrix multiplication, where possible, is associative but not necessarily commutative.

In our discussion of addition of matrices, we found that for every matrix A there exists a matrix Z with the same dimensions as A and such that $A + Z = Z + A = A$. This additive identity matrix Z acts for matrices in the same way that the number 0 acts for numbers. Is there a multiplicative matrix that acts for matrices in the same way that the number 1 acts for numbers? The answer is yes, but such a matrix is not quite as simple as the identity matrix for addition. In the first place, the identity matrix of multiplication must be square in order to leave the dimensions of a matrix A unchanged. Thus, if A has dimensions $m \times n$, the identity matrix must have dimensions $m \times m$ if used on the left but $n \times n$ if used on the right. Thus there are two identity matrices of multiplication for each given matrix. These are known as the left identity and the right identity for multiplication. For example, consider the 2×3 matrix

$$A = \begin{bmatrix} a_{11} & a_{12} & a_{13} \\ a_{21} & a_{22} & a_{23} \end{bmatrix}$$

and let

$$I_2 = \begin{bmatrix} 1 & 0 \\ 0 & 1 \end{bmatrix}, \quad I_3 = \begin{bmatrix} 1 & 0 & 0 \\ 0 & 1 & 0 \\ 0 & 0 & 1 \end{bmatrix}$$

where the subscripts on I denote the dimension of the matrix. It is easily verified that

$$I_2 A = \begin{bmatrix} 1 & 0 \\ 0 & 1 \end{bmatrix} \begin{bmatrix} a_{11} & a_{12} & a_{13} \\ a_{21} & a_{22} & a_{23} \end{bmatrix} = \begin{bmatrix} a_{11} & a_{12} & a_{13} \\ a_{21} & a_{22} & a_{23} \end{bmatrix} = A$$

and

$$AI_3 = \begin{bmatrix} a_{11} & a_{12} & a_{13} \\ a_{21} & a_{22} & a_{23} \end{bmatrix} \begin{bmatrix} 1 & 0 & 0 \\ 0 & 1 & 0 \\ 0 & 0 & 1 \end{bmatrix} = \begin{bmatrix} a_{11} & a_{12} & a_{13} \\ a_{21} & a_{22} & a_{23} \end{bmatrix} = A$$

Hence I_2 is the left identity matrix and I_3 the right identity matrix for every 2×3 matrix. In general, let I_m represent an $m \times m$ matrix each of whose elements along the main diagonal is 1 and each of whose elements not on the main diagonal is 0, and let I_n be an $n \times n$ matrix defined in the same manner as I_m. Then I_m is the left identity matrix for multiplication and I_n the right identity matrix for every $m \times n$ matrix. If A is a square matrix, then $m = n$ and $I_m = I_n$; that is, the left identity matrix and the right identity matrix are the same.

Exercises 8.4

Perform the indicated operations in Exercises 1 through 12.

1. $\begin{bmatrix} 1 & -2 \\ 3 & -1 \end{bmatrix} + \begin{bmatrix} 2 & 3 \\ -1 & 4 \end{bmatrix}$

2. $\begin{bmatrix} 5 & 3 \\ -4 & 2 \end{bmatrix} + \begin{bmatrix} 3 & 4 \\ -1 & 6 \end{bmatrix}$

3. $\begin{bmatrix} 2 & 0 & -3 \\ 4 & 7 & 5 \end{bmatrix} + \begin{bmatrix} 8 & -2 & 6 \\ 5 & -1 & 3 \end{bmatrix}$

4. $3 \begin{bmatrix} 1 & -1 & 2 \\ -2 & 0 & 1 \end{bmatrix}$

5. $\begin{bmatrix} 3 & 2 & -1 \\ -1 & 5 & 4 \\ 7 & -2 & 3 \end{bmatrix} + \begin{bmatrix} 5 & 0 & 2 \\ -3 & 1 & 0 \\ -2 & 4 & -1 \end{bmatrix}$

6. $\begin{bmatrix} 4 & 2 \\ 3 & 1 \end{bmatrix} - \begin{bmatrix} 2 & -1 \\ 5 & 3 \end{bmatrix}$

7. $2 \begin{bmatrix} 7 & 5 \\ 4 & 8 \end{bmatrix} - 3 \begin{bmatrix} 3 & -1 \\ 2 & 6 \end{bmatrix}$

8. $\begin{bmatrix} 5 & 0 & 2 \\ 1 & 3 & -1 \\ -2 & 8 & 7 \end{bmatrix} - \begin{bmatrix} 4 & 2 & 6 \\ -3 & 7 & 0 \\ 1 & -5 & 8 \end{bmatrix}$

9. $\begin{bmatrix} 3 & -1 & 0 \\ 4 & 2 & -3 \end{bmatrix} + \begin{bmatrix} 1 & -3 & 7 \\ 5 & 2 & -1 \end{bmatrix} + \begin{bmatrix} 2 & 0 & 3 \\ -1 & 5 & 6 \end{bmatrix}$

10. $\begin{bmatrix} \sqrt{2} & 5 \\ \frac{3}{5} & -7 \end{bmatrix} + \begin{bmatrix} \sqrt{3} & \frac{5}{2} \\ -\frac{7}{11} & \sqrt{7} \end{bmatrix}$

SEC. 8.4   Matrix Algebra

11. $\begin{bmatrix} 11.215 & -7.34 \\ 2.76 & 27.951 \end{bmatrix} - \begin{bmatrix} -1.759 & -21.24 \\ 15.9 & 263 \end{bmatrix}$

12. $\begin{bmatrix} 3.719 & -22.91 \\ -7.42 & 6.5 \\ -18.25 & 12 \end{bmatrix} + \begin{bmatrix} 7.512 & \sqrt{11} & -1{,}256 \\ -91.6 & (265)^2 & 10^{10} \end{bmatrix}$

13. Consider the matrix $A = \begin{bmatrix} 5 & -1 & 4 \\ 2 & 3 & -2 \end{bmatrix}$

   a. Find the additive identity for A.   b. Find the additive inverse of A.

Find each indicated product in Exercises 14 through 25.

14. $\begin{bmatrix} 2 & 1 \\ -1 & 3 \end{bmatrix} \begin{bmatrix} 5 & -1 \\ 4 & 2 \end{bmatrix}$

15. $\begin{bmatrix} 5 & 7 \\ -3 & 2 \end{bmatrix} \begin{bmatrix} 4 & -1 \\ 2 & 3 \end{bmatrix}$

16. $\begin{bmatrix} 3 & 6 \\ 2 & -1 \end{bmatrix} \begin{bmatrix} 1 & -2 \\ 3 & 1 \end{bmatrix}$

17. $\begin{bmatrix} 7 & -4 \\ -1 & 2 \end{bmatrix} \begin{bmatrix} 1 & 5 \\ -2 & 3 \end{bmatrix}$

18. $\begin{bmatrix} 1 & 2 & 3 \\ -1 & 1 & -2 \\ 2 & -1 & 1 \end{bmatrix} \begin{bmatrix} 3 & 1 & -1 \\ 2 & 3 & 1 \\ -1 & 2 & 4 \end{bmatrix}$

19. $\begin{bmatrix} 2 & 5 & 3 \\ 1 & -2 & 1 \\ 4 & 1 & -2 \end{bmatrix} \begin{bmatrix} 1 & 3 & -2 \\ 2 & -1 & 4 \\ 1 & -2 & 3 \end{bmatrix}$

20. $\begin{bmatrix} 5 & 3 \\ 2 & 1 \end{bmatrix} \begin{bmatrix} 4 \\ 2 \end{bmatrix}$

21. $\begin{bmatrix} 3 & -2 \\ 1 & 4 \end{bmatrix} \begin{bmatrix} 7 \\ -2 \end{bmatrix}$

22. $\begin{bmatrix} 2 & -1 & 3 \\ 1 & 5 & 4 \end{bmatrix} \begin{bmatrix} 3 & -1 \\ 1 & 2 \\ -2 & 4 \end{bmatrix}$

23. $\begin{bmatrix} 1 & 2 & 5 \\ 3 & -1 & 4 \end{bmatrix} \begin{bmatrix} 1 & 2 \\ -2 & -1 \\ 3 & -4 \end{bmatrix}$

24. $\begin{bmatrix} 1 & -2 \\ -3 & 1 \\ 2 & 5 \\ 4 & 3 \end{bmatrix} \begin{bmatrix} 3 & -1 & 0 & 4 \\ 1 & 2 & 3 & -1 \end{bmatrix}$

25. $\begin{bmatrix} 5 & -1 \\ 1 & 4 \\ -2 & 0 \\ 3 & 1 \end{bmatrix} \begin{bmatrix} 2 & -1 & 0 & 3 \\ 4 & 2 & 1 & 5 \end{bmatrix}$

In Exercises 26 through 28, use a hand calculator to approximate the indicated product (round elements to five decimal places).

26. $[3.976 \quad -564.71] \begin{bmatrix} -11.297 \\ 2.56849 \end{bmatrix}$

27. $\begin{bmatrix} 3.271 & -1.35 & 2.476 \\ -1.987 & -16.21 & 48.75 \end{bmatrix} \begin{bmatrix} -9.61 & -300 \\ 5.49 & 25.71 \\ 32.67 & -36.98 \end{bmatrix}$

28. $\begin{bmatrix} -9.61 & -300 \\ 5.49 & 25.71 \\ 32.67 & -36.98 \end{bmatrix} \begin{bmatrix} 3.271 & -1.35 & 2.476 \\ -1.987 & -16.21 & 48.75 \end{bmatrix}$

8.5 THE INVERSE OF A SQUARE MATRIX

In the system of real numbers, every real number a has an additive identity and an additive inverse. Also, every real number a has a multiplicative identity, the number 1, and every number $a \neq 0$ has a multiplicative inverse. In Section 8.4 we found that every matrix A has an additive identity Z and an additive inverse $(-A)$. Similarly, we found that every $m \times n$ matrix has a left multiplicative identity matrix I_m and a right multiplicative identity matrix I_n which are equal if and only if $m = n$. Thus for each square matrix A there exists a multiplicative identity matrix I, with the same dimensions as A, each of whose elements on the main diagonal is 1, and each of whose elements not on the main diagonal is 0. Can we carry the analogy with real numbers further and define a multiplicative inverse for a square matrix? The following definition is a step in that direction.

DEFINITION 8.11. If A is a square matrix and B is another square matrix with the same dimensions and has the property that $BA = I$, where I is the identity matrix, then B is said to be the *inverse* of A. If it exists, the inverse of A will be denoted by A^{-1}.

For example, if

$$A = \begin{bmatrix} 2 & 5 \\ 1 & 3 \end{bmatrix}, \quad \text{then } A^{-1} = \begin{bmatrix} 3 & -5 \\ -1 & 2 \end{bmatrix}$$

because

$$A^{-1}A = \begin{bmatrix} 3 & -5 \\ -1 & 2 \end{bmatrix} \begin{bmatrix} 2 & 5 \\ 1 & 3 \end{bmatrix} = \begin{bmatrix} 1 & 0 \\ 0 & 1 \end{bmatrix}$$

Note that if we reverse the order of the multiplication we also get the identity matrix

$$AA^{-1} = \begin{bmatrix} 2 & 5 \\ 1 & 3 \end{bmatrix} \begin{bmatrix} 3 & -5 \\ -1 & 2 \end{bmatrix} = \begin{bmatrix} 1 & 0 \\ 0 & 1 \end{bmatrix}$$

SEC. 8.5 The Inverse of a Square Matrix

It can be shown that, in general, if the square matrix A has an inverse A^{-1}, then

$$A^{-1}A = AA^{-1} = I$$

is always true.

Although finding the inverse of a square matrix is somewhat analogous to finding the reciprocal of a real number, the analogy is not complete. It is well known that every real number except zero has a reciprocal. In the case of matrices, many nonzero square matrices do not have inverses. To see why this is true, let us develop a method for finding the inverse of the general 2×2 matrix, if it exists.

Let the general 2×2 matrix be denoted by

$$A = \begin{bmatrix} a_{11} & a_{12} \\ a_{21} & a_{22} \end{bmatrix}$$

and let the unknown inverse be denoted by

$$A^{-1} = \begin{bmatrix} b_{11} & b_{12} \\ b_{21} & b_{22} \end{bmatrix}$$

Our problem is to find four numbers b_{11}, b_{12}, b_{21}, and b_{22} such that

(21) $$A^{-1}A = \begin{bmatrix} b_{11} & b_{12} \\ b_{21} & b_{22} \end{bmatrix} \begin{bmatrix} a_{11} & a_{12} \\ a_{21} & a_{22} \end{bmatrix} = \begin{bmatrix} 1 & 0 \\ 0 & 1 \end{bmatrix}$$

By performing the indicated multiplication, Equation (21) can be written in the equivalent form

(22) $$\begin{bmatrix} a_{11}b_{11} + a_{21}b_{12} & a_{12}b_{11} + a_{22}b_{12} \\ a_{11}b_{21} + a_{21}b_{22} & a_{12}b_{21} + a_{22}b_{22} \end{bmatrix} = \begin{bmatrix} 1 & 0 \\ 0 & 1 \end{bmatrix}$$

By equating elements in corresponding positions in the two matrices in Equation (22), we obtain the following system of four linear equations in the unknowns b_{11}, b_{12}, b_{21}, and b_{22}, which is equivalent to Equation (22).

(23) $$\begin{cases} a_{11}b_{11} + a_{21}b_{12} = 1 \\ a_{12}b_{11} + a_{22}b_{12} = 0 \\ a_{11}b_{21} + a_{21}b_{22} = 0 \\ a_{12}b_{21} + a_{22}b_{22} = 1 \end{cases}$$

Our problem is now reduced to that of solving system (23) for the unknowns b_{11}, b_{12}, b_{21}, and b_{22}. Fortunately, the first two equations of this system involve only b_{11} and b_{12} and the second two equations involve only b_{21} and b_{22}, which further reduces our problem to that of solving the following two systems each consisting of two linear equations in two unknowns.

(24) $\begin{cases} a_{11}b_{11} + a_{21}b_{12} = 1 \\ a_{12}b_{11} + a_{22}b_{12} = 0 \end{cases}$

(25) $\begin{cases} a_{11}b_{21} + a_{21}b_{22} = 0 \\ a_{12}b_{21} + a_{22}b_{22} = 1 \end{cases}$

Using the method of elimination to solve systems (24) and (25), we find that

$$b_{11} = \frac{a_{22}}{a_{11}a_{22} - a_{12}a_{21}}, \quad b_{12} = \frac{-a_{12}}{a_{11}a_{22} - a_{12}a_{21}}$$

$$b_{21} = \frac{-a_{21}}{a_{11}a_{22} - a_{12}a_{21}}, \quad b_{22} = \frac{a_{11}}{a_{11}a_{22} - a_{12}a_{21}}$$

Note that the denominator in each of the four fractions just found is the same number, $a_{11}a_{22} - a_{12}a_{21}$. This number is very important, for it is the key as to whether or not the above solutions exist. It can be shown that the above solutions will exist if and only if $a_{11}a_{22} - a_{12}a_{21} \neq 0$. This number can be expressed as a second-order determinant (see Section 8.2).

$$a_{11}a_{22} - a_{12}a_{21} = \begin{vmatrix} a_{11} & a_{12} \\ a_{21} & a_{22} \end{vmatrix}$$

which is called the determinant of the matrix $A = \begin{vmatrix} a_{11} & a_{12} \\ a_{21} & a_{22} \end{vmatrix}$ and is denoted by $|A|$. It follows from the above discussion that a 2×2 matrix A has an inverse if and only if its determinant $|A| \neq 0$. If $|A| \neq 0$, the inverse A^{-1} of A is given by the formula

(26) $$A^{-1} = \begin{bmatrix} \dfrac{a_{22}}{|A|} & \dfrac{-a_{12}}{|A|} \\ \dfrac{-a_{21}}{|A|} & \dfrac{a_{11}}{|A|} \end{bmatrix}$$

In the above discussion we were concerned only with finding the inverses of 2×2 matrices. The procedure developed for such matrices can be extended to square matrices of higher order. Every $n \times n$ square matrix A has associated with it an nth-order determinant $|A|$ whose elements are the same as the elements of A. An $n \times n$ matrix is said to be of the nth *order*. We state without proof the following theorem.

THEOREM 8.4. *A square matrix A has an inverse if and only if its determinant $|A| \neq 0$.*

Finding the inverse even of a 3×3 matrix is much more difficult than finding the inverse of a 2×2 matrix, and the difficulty increases as the order of

SEC. 8.5 The Inverse of a Square Matrix

the matrix increases. A meaningful discussion of such matrices is beyond the scope of the brief introduction intended in this chapter.

Example 1 | Use Formula (26) to find the inverse of the matrix $A = \begin{bmatrix} 3 & 7 \\ 2 & 5 \end{bmatrix}$.

Solution: Here $a_{11} = 3$, $a_{12} = 7$, $a_{21} = 2$, $a_{22} = 5$, and

$$|A| = \begin{vmatrix} 3 & 7 \\ 2 & 5 \end{vmatrix} = 15 - 14 = 1$$

Applying Formula (26), we obtain

$$A^{-1} = \begin{bmatrix} 5 & -7 \\ -2 & 3 \end{bmatrix}$$

Check: $A^{-1}A = \begin{bmatrix} 5 & -7 \\ -2 & 3 \end{bmatrix} \begin{bmatrix} 3 & 7 \\ 2 & 5 \end{bmatrix} = \begin{bmatrix} 1 & 0 \\ 0 & 1 \end{bmatrix}$

The matrix $\begin{bmatrix} 1 & 2 \\ 3 & 6 \end{bmatrix}$ does not have an inverse because

$$|A| = \begin{vmatrix} 1 & 2 \\ 3 & 6 \end{vmatrix} = 6 - 6 = 0$$

Caution: Students studying matrices and determinants for the first time often confuse the terms "matrix" and "determinant." It is important to remember that a *matrix* is a rectangular array of numbers. The determinant of a matrix is a real number associated with the matrix, and this real number can be calculated by the methods given in Section 8.2 for evaluating determinants. For example,

$$A = \begin{bmatrix} 2 & -3 & 4 \\ 2 & 1 & 0 \\ 1 & 1 & -1 \end{bmatrix}$$

is a 3×3 matrix. On the other hand,

$$|A| = \begin{vmatrix} 2 & -3 & 4 \\ 3 & 1 & 0 \\ 1 & 1 & -1 \end{vmatrix} = -3$$

is the determinant of A.

Exercises 8.5

Compute the inverse of each of the following matrices.

1. $\begin{bmatrix} 7 & 4 \\ 5 & 3 \end{bmatrix}$ 2. $\begin{bmatrix} 4 & 3 \\ 1 & 1 \end{bmatrix}$ 3. $\begin{bmatrix} 2 & 8 \\ 2 & 9 \end{bmatrix}$ 4. $\begin{bmatrix} 5 & -3 \\ -4 & 2 \end{bmatrix}$

Show that each of the following matrices has no inverse.

5. $\begin{bmatrix} 2 & 6 \\ 1 & 3 \end{bmatrix}$ 6. $\begin{bmatrix} 1 & 1 \\ 1 & 1 \end{bmatrix}$ 7. $\begin{bmatrix} 8 & 12 \\ 4 & 6 \end{bmatrix}$ 8. $\begin{bmatrix} 8 & -12 \\ -6 & 9 \end{bmatrix}$

8.6 SYSTEMS OF LINEAR EQUATIONS WRITTEN AS A SINGLE MATRIX EQUATION

The definitions of equality of two matrices and of matrix multiplication make it possible to write a system of linear equations in n unknowns as a single matrix equation. To see how this is done, consider the system of two equations in two unknowns given below.

(27)
$$\begin{cases} a_{11}x + a_{12}y = b_1 \\ a_{21}x + a_{22}y = b_2 \end{cases}$$

This system can be written as the single matrix equation

(28)
$$\begin{bmatrix} a_{11} & a_{12} \\ a_{21} & a_{22} \end{bmatrix} \begin{bmatrix} x \\ y \end{bmatrix} = \begin{bmatrix} b_1 \\ b_2 \end{bmatrix}$$

To verify that this matrix equation is actually equivalent to the given system, we first perform the indicated multiplication of the two matrices on the left side of the matrix equation, giving the equation

$$\begin{bmatrix} a_{11}x + a_{12}y \\ a_{21}x + a_{22}y \end{bmatrix} = \begin{bmatrix} b_1 \\ b_2 \end{bmatrix}$$

Next we equate the elements in corresponding positions of the two matrices in the last equation and obtain the system

$$\begin{cases} a_{11}x + a_{12}y = b_1 \\ a_{21}x + a_{22}y = b_2 \end{cases}$$

which is the given system. Our verification is complete.

Equation (28) can be further abbreviated by letting $A = \begin{bmatrix} a_{11} & a_{12} \\ a_{21} & a_{22} \end{bmatrix}$,

SEC. 8.6 Systems of Linear Equations Written as a Single Matrix Equation

$X = \begin{bmatrix} x \\ y \end{bmatrix}$, and $B = \begin{bmatrix} b_1 \\ b_2 \end{bmatrix}$. With this notation, Equation (28) becomes

(29) $$AX = B$$

Equation (29) suggests a method for finding the solution of a system of two linear equations in two unknowns. To pursue this suggestion, let us suppose that the matrix A has an inverse A^{-1}. We multiply both sides of Equation (29) on the left by A^{-1}; thus

(30) $$A^{-1}(AX) = A^{-1}B$$

Since matrix multiplication is associative, $A^{-1}(AX) = (A^{-1}A)X$. Hence

$$(A^{-1}A)X = A^{-1}B$$

But $A^{-1}A = I$ and $IX = X$; therefore,

(31) $$X = A^{-1}B$$

from which the solution of the system follows immediately. Let us use this method to solve the system

(32) $$\begin{cases} 2x - 3y = 12 \\ x - y = 5 \end{cases}$$

The corresponding matrix equation is

(33) $$\begin{bmatrix} 2 & -3 \\ 1 & -1 \end{bmatrix} \begin{bmatrix} x \\ y \end{bmatrix} = \begin{bmatrix} 12 \\ 5 \end{bmatrix}$$

Using the methods of Section 8.5, we find that the inverse of $\begin{bmatrix} 2 & -3 \\ 1 & -1 \end{bmatrix}$ is $\begin{bmatrix} -1 & 3 \\ -1 & 2 \end{bmatrix}$. Multiplying the matrix equation on the left and using the associative property of matrix multiplication, we have

(34) $$\left\{ \begin{bmatrix} -1 & 3 \\ -1 & 2 \end{bmatrix} \begin{bmatrix} 2 & -3 \\ 1 & -1 \end{bmatrix} \right\} \begin{bmatrix} x \\ y \end{bmatrix} = \begin{bmatrix} -1 & 3 \\ -1 & 2 \end{bmatrix} \begin{bmatrix} 12 \\ 5 \end{bmatrix}$$

Performing the indicated multiplication gives

$$\begin{bmatrix} 1 & 0 \\ 0 & 1 \end{bmatrix} \begin{bmatrix} x \\ y \end{bmatrix} = \begin{bmatrix} 3 \\ -2 \end{bmatrix}$$

which reduces to the equation

(35) $$\begin{bmatrix} x \\ y \end{bmatrix} = \begin{bmatrix} 3 \\ -2 \end{bmatrix}$$

Equating corresponding elements of the two matrices in Equation (35), we see that $x = 3$ and $y = -2$ is the solution of that equation and hence is the solution of system (32).

Instead of a system of two equations in two unknowns, consider the following general system of n linear equations in n unknowns.

(36)
$$\begin{cases} a_{11}x_1 + a_{12}x_2 + \cdots + a_{1n}x_n = b_1 \\ a_{21}x_1 + a_{22}x_2 + \cdots + a_{2n}x_n = b_2 \\ \vdots \quad\quad \vdots \quad\quad \vdots \quad\quad \vdots \quad\quad \vdots \\ a_{n1}x_1 + a_{n2}x_2 + \cdots + a_{nn}x_n = b_n \end{cases}$$

This system can be written as the single matrix equation

(37)
$$AX = B$$

where

$$A = \begin{bmatrix} a_{11} & a_{12} & \cdots & a_{1n} \\ a_{21} & a_{22} & \cdots & a_{2n} \\ \vdots & \vdots & \vdots & \vdots \\ a_{n1} & a_{n2} & \cdots & a_{nn} \end{bmatrix}, \quad X = \begin{bmatrix} x_1 \\ x_2 \\ \vdots \\ x_n \end{bmatrix}, \quad B = \begin{bmatrix} b_1 \\ b_2 \\ \vdots \\ b_n \end{bmatrix}$$

Theoretically, assuming $|A| \neq 0$, we can solve Equation (37) by the same procedure used above to solve the system of two equations in two unknowns. The steps needed to obtain such a solution are indicated below.

$$A^{-1}(AX) = A^{-1}B$$
$$(A^{-1}A)X = A^{-1}B$$
$$IX = A^{-1}B$$
$$X = A^{-1}B$$

From this last equation, equating corresponding elements of the two matrices involved provides the solution of the given system. The big problem is, of course, that of finding the inverse of the matrix A. This is a particularly difficult problem for an $n \times n$ matrix with $n \geq 3$, and we shall not attempt to go into it here. Suffice it to say that the process of computing inverses has been the subject of much study and that methods have been found for making such computations. However, the processes for finding inverses of matrices in general delve more deeply into the subject of linear algebra than we care to in this text.

Exercises 8.6

Write each of the following systems of linear equations as a single matrix equation; multiply the matrix equation by the inverse of the coefficient matrix and thus find a solution of the system.

1. $\begin{cases} 2x + 3y = 5 \\ x + 2y = 4 \end{cases}$
2. $\begin{cases} 4x - 3y = 18 \\ x - y = 5 \end{cases}$
3. $\begin{cases} 7x + 4y = 2 \\ 5x + 3y = 1 \end{cases}$

SEC. 8.7 Systems of Linear Inequalities

4. $\begin{cases} 5x + 2y = 8 \\ 3x + y = 5 \end{cases}$
5. $\begin{cases} 3x + 2y = 11 \\ 6x + 5y = 17 \end{cases}$
6. $\begin{cases} 3x + y = 1 \\ x + y = 5 \end{cases}$

7. Verify, by matrix multiplication, that if

$$A = \begin{bmatrix} 3 & 2 & 6 \\ 1 & 1 & 2 \\ 2 & 2 & 5 \end{bmatrix} \quad \text{then } A^{-1} = \begin{bmatrix} 1 & 2 & -2 \\ -1 & 3 & 0 \\ 0 & -2 & 1 \end{bmatrix}$$

8. Write the following system of linear equations as a single matrix equation and use A^{-1} from Exercise 7 to solve the system.

$$3x + 2y + 6z = 10$$
$$x + y + 2z = 3$$
$$2x + 2y + 5z = 7$$

In Exercises 9 and 10, use a hand calculator to find approximations to the inverse of each matrix (round entries to five decimal places.)

9. $\begin{bmatrix} 7.12 & 4.53 \\ 2.15 & 6.92 \end{bmatrix}$

10. $\begin{bmatrix} 21.712 & 6.438 \\ 17.594 & -5.294 \end{bmatrix}$

8.7* SYSTEMS OF LINEAR INEQUALITIES

To solve a system of two or more linear inequalities, the best method available is the graphical method. A typical system of two linear inequalities in x and y is the following:

$$\begin{cases} a_1 x + b_1 y > r_1, & a_1 \text{ and } b_1 \text{ not both zero} \\ a_2 x + b_2 y > r_2, & a_2 \text{ and } b_2 \text{ not both zero} \end{cases}$$

where the coefficients a_1, b_1, etc., are real numbers. The solution set S of this system is the set $S = S_1 \cap S_2$, where

$$S_1 = \{(x, y) | a_1 x + b_1 y > r_1\}$$
and
$$S_2 = \{(x, y) | a_2 x + b_2 y > r_2\}$$

The graph of the solution set S_1 is a half-plane, as is the graph of the solution set S_2. The solution set S is therefore the intersection of these two half-planes. The graph of the solution set of an inequality $ax + by > r$ is an open half-plane, since it does not contain the points on the line $ax + by = r$. The graph of the solution set of $ax + by \geq r$ does contain the points on the line $ax + by = r$ and hence is a closed half-plane. Thus, in our typical system mentioned above, the graphs of S_1 and S_2 are open half-planes.

Example 1   Find the solution set of the system $\begin{cases} x - y < 2 \\ 3x + y > 2 \end{cases}$

Solution: We first draw the lines $L_1: x - y = 2$ and $L_2: 3x + y = 2$; they intersect at $(1, -1)$. The solution set S_1 of $x - y < 2$ is the left half-plane determined by L_1, while the solution set S_2 of $3x + y > 2$ is the right half-plane determined by L_2. The solution set S of the given systems is $S = S_1 \cap S_2$ and is represented graphically by the shaded portion in Figure 8.4.

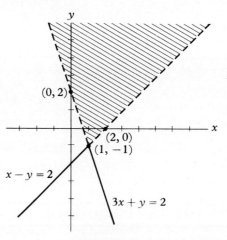

Figure 8.4

The procedure used in this example works equally well for determining the solution set of a system involving three or more linear inequalities in two variables.

Example 2   Determine the solution set of the system $\begin{cases} x - y \leq 2 \\ 3x + y \geq 2 \\ x + 3y \leq 6 \end{cases}$

Solution: Draw each of the lines $x - y = 2$, $3x + y = 2$, and $x + 3y = 6$. These lines intersect in pairs in the three points $(0, 2)$, $(1, -1)$, and $(3, 1)$, and thus determine a triangle. The graph of the desired solution set is the set of all points inside the triangle and on the sides. See Figure 8.5.

SEC. 8.7 Systems of Linear Inequalities

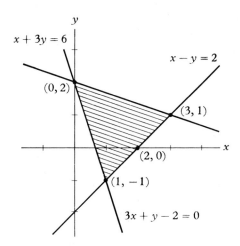

Figure 8.5

Example 3   Determine the solution set of the system $\begin{cases} 2x + y \geq 5 \\ x - y \leq 1 \\ x \geq 1 \\ x + y \leq 5 \end{cases}$.

Solution: It is easily verified that the solution set of this system has for its graph all points inside and on the sides of the quadrilateral, shaded region shown in Figure 8.6.

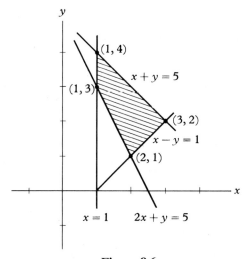

Figure 8.6

The graphs of the solution sets of the systems of linear inequalities discussed above possess a rather obvious useful common property: if any two points of the graph are joined by a line segment having the two points as endpoints, every point on the segment is also a point of the graph. A set of points having this property is called a *convex* set.

DEFINITION 8.12. A set of points is said to be *convex* if for each pair of points P_1 and P_2 which are members of the set, every point of the line segment having P_1 and P_2 as endpoints also is a member of the set.

For future reference, we state and prove the following theorem.

THEOREM 8.5. *The intersection of two convex sets is convex.*

Proof: Let S_1 and S_2 be two convex sets. If P_1 and P_2 are any two points in the set $S = S_1 \cap S_2$, then these two points are both in S_1. By definition of a convex set, all the points on the line segment having these points as endpoints are also in S_1. Similarly, since P_1 and P_2 are in S_2, all the points on this same line segment are also in S_2. Hence all the points of the segment $P_1 P_2$ belong to the set $S = S_1 \cap S_2$.

DEFINITION 8.13. The intersection of two or more closed half-planes is called a *polygonal convex set*.

The graphs of the solution sets shown in Figures 8.5 and 8.6 are typical examples of polygonal convex sets. Some applications of solution sets of systems of linear inequalities will be discussed in the next section.

Exercises 8.7

Draw the graph of the solution set of each of the following systems of linear inequalities.

1. $\begin{cases} x + y > 3 \\ y > 1 \end{cases}$
2. $\begin{cases} x \geq 2 \\ y \leq -5 \end{cases}$
3. $\begin{cases} x - 2y > 0 \\ 3x - y < 0 \end{cases}$

4. $\begin{cases} x \geq 0 \\ y \geq 0 \\ x + y \leq 5 \end{cases}$
5. $\begin{cases} x \geq 1 \\ y \leq 4 \\ x - y \leq 0 \end{cases}$
6. $\begin{cases} x - y - 3 \leq 0 \\ x + 1 \geq 0 \\ y \leq 3 \end{cases}$

7. $\begin{cases} x - 2y \geq -10 \\ 4x + 3y \geq 26 \\ 3x - 4y \leq 7 \\ x + y \leq 14 \end{cases}$
8. $\begin{cases} x - y \geq -3 \\ 2x + 3y \geq -6 \\ 2x - 5y \leq 10 \\ 3x + 5y \leq 15 \end{cases}$
9. $\begin{cases} x - 2y - 2 \geq 0 \\ 3x + y - 2 \leq 0 \\ y + 4 \geq 0 \end{cases}$

8.8* LINEAR PROGRAMMING

Suppose we are given a polygonal convex set S in the xy plane and a linear polynomial $f(x, y) = ax + by + c$, where a, b, and c are real numbers and a and b are not both zero. Suppose further that we are required to find the point or points of S at which $f(x, y)$ has its maximum (or minimum) value. Such a problem is called a *linear programming* problem in two variables. It is the purpose of this section to investigate a procedure for solving such problems.

Since a polygonal convex set is the intersection of two or more closed half-planes, such a set consists of the interior of a region together with the line segments which form its boundary. This region may, of course, be finite or infinite. If the region is finite, the boundary is a polygon, in which case we shall refer to the polygon and its interior as a *convex polygon*.

A property of polygons, well known to all who have studied high school geometry, is that a polygon with n sides has n vertices. That is, a polygon with three sides is a triangle and has three vertices, a polygon with four sides has four vertices, with five sides has five vertices, and so on. The solution set of the system of inequalities in the last example of the previous section is a convex polygon of four sides and four vertices (see Figure 8.6). The coordinates of the vertices are the ordered pairs (1, 3), (2, 1), (3, 2), and (1, 4) as shown in the figure.

We return now to the linear programming problem stated at the beginning of this section. To aid us in the solution of this problem, we need the following theorems.

THEOREM 8.6. *Let $P_1(x_1, y_1)$ and $P_2(x_2, y_2)$ be the endpoints of a line segment, and let $P(x', y')$ be any point of the segment P_1P_2. If*

$$\frac{P_1P}{P_1P_2} = r, \qquad 0 \leq r \leq 1$$

then

$$x' = x_1 + r(x_2 - x_1) \quad \text{and} \quad y' = y_1 + r(y_2 - y_1)$$

Proof: From similar triangles (Figure 8.7),

$$\frac{P_1P}{P_1P_2} = \frac{x' - x_1}{x_2 - x_1} = r \quad \text{and} \quad \frac{P_1P}{P_1P_2} = \frac{y' - y_1}{y_2 - y_1} = r$$

When simplified, these equations become

$$x' = x_1 + r(x_2 - x_1) \quad \text{and} \quad y' = y_1 + r(y_2 - y_1)$$

which proves our theorem.

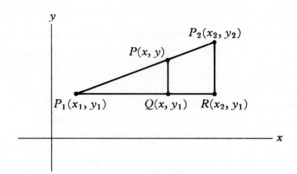

Figure 8.7

Note that if

$$r = 0, \quad P = P_1$$
$$r = \tfrac{1}{2}, \quad P \text{ is midpoint of } P_1P_2$$
$$r = 1, \quad P = P_2$$

THEOREM 8.7. *Let $P_1(x_1, y_1)$ and $P_2(x_2, y_2)$ be the endpoints, and let $P(x', y')$ be any point of a line segment P_1P_2. Denote by $f(x, y)$ the value of the linear polynomial $ax + by + c$ at (x, y), that is, $f(x, y) = ax + by + c$. If $f(x_1, y_1) = m$ and $M = f(x_2, y_2)$, where $m \leq M$, then*

$$m \leq f(x', y') \leq M$$

Proof: From Theorem 8.6 we have

$$x' = x_1 + r(x_2 - x_1) \quad \text{and} \quad y' = y_1 + r(y_2 - y_1), \quad 0 \leq r \leq 1$$

Thus

$$\begin{aligned}
f(x', y') &= ax' + by' + c \\
&= a[x_1 + r(x_2 - x_1)] + b[y_1 + r(y_2 - y_1)] + c \\
&= (1 - r)(ax_1 + by_1) + r(ax_2 + by_2) + c \\
&= (1 - r)(ax_1 + by_1 + c) + r(ax_2 + by_2 + c) \\
&= (1 - r)m + rM \\
&= m + r(M - m)
\end{aligned}$$

Since $0 \leq r \leq 1$, it is easy to verify that

$$m \leq m + r(M - m) \leq M$$

and our theorem is proved.

SEC. 8.8 Linear Programming

Example 1   Find the values of the polynomial $f(x, y) = 3x + 2y - 5$ at the vertices of the convex polygon which is the solution set of the system

$$\begin{cases} 2x + y \geq 5 \\ x - y \leq 7 \\ x \geq 1 \\ x + y \leq 5 \end{cases}$$

The solution set of this system is the convex polygon shown in Figure 8.6. The vertices are the points (1, 3), (2, 1), (3, 2), and (1, 4).

$$f(1, 3) = 3 + 6 - 5 = 4$$
$$f(2, 1) = 6 + 2 - 5 = 3$$
$$f(3, 2) = 9 + 4 - 5 = 8$$
$$f(1, 4) = 3 + 8 - 5 = 6$$

We are now ready to prove a theorem which is of vital importance in the solution of two-dimensional linear programming problems.

THEOREM 8.8. *A linear polynomial, $f(x, y) = ax + by + c$, defined on a convex polygon has its maximum (and minimum) value at a vertex of the convex polygon.*

Proof: As was stated above, a convex polygon is a convex set of finite area consisting of a polygon together with its interior. The proof is the same regardless of the number of sides the polygon may have so long as this number is finite. So, for simplicity, we shall in our proof refer to the convex polygon *ABCDE* shown in Figure 8.8. Also, we shall use the symbol $f(P)$ to denote the value of f at the point P. For example, $f(A)$ represents the value of f at the

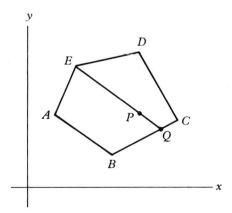

Figure 8.8

vertex A, $f(B)$ represents the value of f at vertex B, $f(C)$ represents the value of f at vertex C, and so on.

Suppose we have calculated the value of f at each of the five vertices. Let M and m denote the largest and the smallest of these five values, respectively. Next, suppose $f(E) = M$ and $f(D) = m$. Let P be any point on the convex polygon. Join vertex E, a vertex at which f has its largest value, to the point P by a line segment and continue the segment until it intersects the polygon again, say at Q. It follows from Theorem 8.7 that $f(Q)$ lies between $f(B)$ and $f(C)$. By hypothesis, $f(B)$ and $f(C)$ both lie between M and m. Hence $m \leq f(Q) \leq M$. Similarly, by Theorem 8.7, $f(Q) \leq f(P) \leq f(E)$. But $f(E) = M$ and $m \leq f(Q)$; therefore,

$$m \leq f(P) \leq M$$

Since P was any point on the convex polygon our theorem is proved.

Example 2 Find the maximum and minimum values of the polynomial $f(x, y) = 5x - 2y + 7$ over the convex polygon determined by the following system of inequalities:

$$\begin{cases} x + 3y \geq 8 \\ 2x + y \geq 6 \\ x - y \geq -3 \\ 4x - y \leq 19 \\ x + 3y \leq 21 \end{cases}$$

Solution: The solution set of the given system is the convex polygon shown in Figure 8.9. It follows from Theorem 8.8 that the maximum and minimum values of

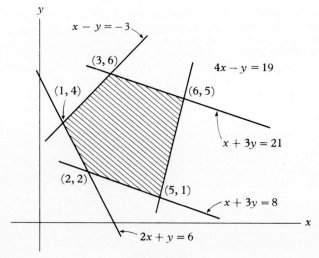

Figure 8.9

SEC. 8.8 Linear Programming

f on a convex polygon occur at vertices. Hence, to find the maximum and minimum values of $f(x, y) = 5x - 2y + 7$ over the convex polygon of Figure 8.9, we calculate the value of f at each vertex and compare as follows:

$$f(1, 4) = 5 - 8 + 7 = 4$$
$$f(2, 2) = 10 - 4 + 7 = 13$$
$$f(5, 1) = 25 - 2 + 7 = 30$$
$$f(6, 5) = 30 - 10 + 7 = 27$$
$$f(3, 6) = 15 - 12 + 7 = 10$$

which shows the maximum value of f to be 30 and the minimum value to be 4.

In the above discussions we have considered the maximum and minimum values of a linear polynomial, $f(x, y) = ax + by + c$, only over convex polygons. If the polygonal convex set S over which f is defined has an infinite area, then one and only one of the following possibilities can exist.

1. f has both a maximum value and a minimum value over S.
2. f has neither a maximum value nor a minimum value over S.
3. f has a maximum value but no minimum value over S.
4. f has a minimum value but no maximum value over S.

If f does have a maximum (or minimum) value over S, then it must occur at a point lying on the boundary of S.

Example 3 The polynomial $f(x, y) = 3x + y$ has neither a maximum value nor a minimum value over the convex set determined by the system of inequalities

$$\begin{cases} x - y \geq 0 \\ x + y \leq 0 \end{cases}$$

Solution: The solution set of the given system is the convex set S shown in Figure 8.10. To see that f has no maximum value over this convex set, note that for any

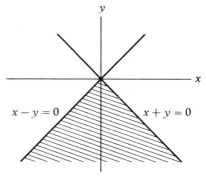

Figure 8.10

number $M > 0$, the point $(M, -M)$ belongs to S and $f(M, -M) = 3M - M = 2M > M$. Thus there is no largest value of f on S. Similarly, for any $M > 0$, the point $(-M, -M)$ belongs to S and $f(-M, -M) = -3M + M = -2M < -M$, which shows that f has no minimum value on S.

Example 4 Find the minimum value of the linear polynomial $f(x, y) = 4x + 3y + 5$ over the convex set determined by the system of inequalities

$$\begin{cases} 2x - y \geq -2 \\ 2x + 3y \geq 6 \\ x - 3y \leq 3 \end{cases}$$

Solution: We first solve the given system of inequalities and find the solution set to be the polygonal convex set shown as the shaded area of Figure 8.11. Denote this set by S. Inspection of the graph shows that S has an infinite area and two corner points, $(0, 2)$ and $(3, 0)$. We next calculate the value of f at each corner point.

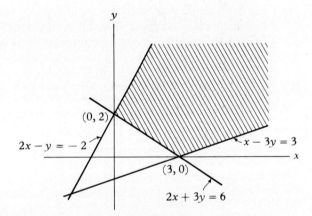

Figure 8.11

$$f(0, 2) = 0 + 6 + 5 = 11$$

$$f(3, 0) = 12 + 0 + 5 = 17$$

Thus the minimum value of f at a corner point is 11. How can we be sure that this is the minimum value of f on S, since Theorem 8.8 does not apply? Suppose (x', y') is any point of S, not a corner point. Then it follows from the second inequality of our system that $2x' + 3y' \geq 6$. Also, $x' > 0$ and $y' > 0$. Therefore,

$$f(x', y') = 4x' + 3y' + 5 > 2x' + 3y' + 5 \geq 6 + 5 = 11$$

which proves that the minimum value of f on S is 11. Obviously, f has no maximum value on S.

SEC. 8.8 Linear Programming

In the examples of linear programming problems discussed above, we have used a graphical method of solution which necessitated that we restrict ourselves to problems involving only two independent variables. Methods are available for handling problems involving more than two variables; but any worthwhile treatment of such methods would carry us too far afield, so no attempt will be made here.

In summary, we have outlined a procedure for finding the maximum (and minimum) value of a linear polynomial $f(x, y) = ax + by + c$ on a polygonal convex set determined by a system of linear inequalities in x and y. The inequalities of such a system are called *constraints*, and each point of the convex set is called a *feasible point*. A point of the set at which f has a maximum (or minimum) value is called an *optimal point*. We have found that such a maximum (or minimum) value always occurs at a boundary point of the polygonal convex set. The above procedure makes possible the solution of an important class of practical problems. We close this section with an example, a hypothetical case illustrating such a problem.

Example 5 A nutritionist in a school cafeteria wishes to ensure that the diet in the cafeteria will contain at least 12 units of Vitamin R, 5 units of Vitamin S, and 8 units of Vitamin T, for a given meal. She tests foods A and B and finds that each pound of food A contains 2 units of Vitamin R, 1 unit of Vitamin S, and 4 units of Vitamin T. On the other hand, each pound of food B contains 3 units, 1 unit, and 1 unit of Vitamins R, S, and T, respectively. Therefore, she can fulfill the minimum requirements with 6 pounds of food A or with 8 pounds of food B or with various combinations of some of each of the two foods. However, the nutritionist wishes to minimize the cost, and discovers that food A costs 60 cents per pound, while food B costs 40 cents per pound. The question then arises as to which of the possible combinations of food A and food B will fulfill the minimum requirement and yet at minimum cost.

Solution: Suppose we purchase x pounds of food A and y pounds of food B. Since we cannot purchase a negative amount of either food, $x \geq 0$ and $y \geq 0$. For a minimum of 12 units of Vitamin R, 5 units of Vitamin S, and 8 units of Vitamin T, the following inequalities must be satisfied: $2x + 3y \geq 12$, $x + y \geq 5$, and $4x + y \geq 8$. The cost C of x pounds of food A and y pounds of food B is given by the equation $C = 0.60x + 0.40y$.

Our problem reduces to that of minimizing C over the polygonal convex set determined by the following system of constraints.

$$\begin{cases} x \geq 0 \\ y \geq 0 \\ 2x + 3y \geq 12 \\ x + y \geq 5 \\ 4x + y \geq 8 \end{cases}$$

The graph of the solution set of the given systems is the polygonal set shown as the shaded area in Figure 8.12. The coordinates of the corner points are (0, 8), (1, 4), (3, 2), and (6, 0). We calculate the value of C at each corner point.

$$C(0, 8) = 0.60(0) + 0.40(8) = 3.20$$
$$C(1, 4) = 0.60(1) + 0.40(4) = 2.20$$
$$C(3, 2) = 0.60(3) + 0.40(2) = 2.60$$
$$C(6, 0) = 0.60(6) + 0.40(0) = 3.60$$

We see that $2.20 is the minimal cost. The corresponding purchases would be 1 pound of food A and 4 pounds of food B.

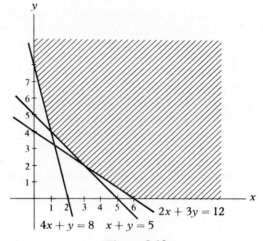

Figure 8.12

Exercises 8.8

1. Find the maximum value of the polynomial $f(x, y) = 5x - 2y + 12$ over the convex polygon determined by the following system of inequalities:

$$\begin{cases} 3x + 2y \le 12 \\ 3x - 4y \le -6 \\ 4x - y \ge -8 \end{cases}$$

2. Find the maximum value and the minimum value of $f(x, y) = 7x - 3y + 6$ over the convex closed polygon determined by the following system of inequalities:

$$\begin{cases} 2x - y \ge -2 \\ x + y \ge 5 \\ 5x + 2y \le 20 \end{cases}$$

SEC. 8.8 Linear Programming

3. Find the maximum and minimum values of $f(x, y) = -8x + 10y + 15$ on the convex polygon given by the inequalities

$$\begin{cases} y \geq -3 \\ x - y \geq -5 \\ x + y \geq -5 \\ x \leq 2 \end{cases}$$

4. Find the minimum value of $f(x, y) = 4x + 3y$ subject to the following constraints:

$$\begin{cases} x + y \geq 7 \\ x + 2y \geq 9 \\ 3x + y \geq 8 \end{cases}$$

5. A convex polygon has the points (2, 5), (3, 1), (3, 10), and (14, 8) as corner points. Find a set of inequalities which defines the convex polygon having these corner points.

6. Draw a quadrilateral and a pentagon neither of which is a convex polygon. Is every triangle a convex polygon?

7. Consider the convex polygon defined by the inequalities

$$\begin{cases} x \geq 0 \\ y \geq 0 \\ x + y \leq 7 \end{cases}$$

Find the maximum value of the polynomial $f(x, y) = 3x + 3y + 8$ over this polygon. Does f take on this maximum value at a point of the convex polygon that is not a corner point? If so, explain how this may happen.

8. Given the polynomial $f(x, y) = 20 - 2x - 3y$.
   a. Find a polygonal convex set such that over this set f has a maximum value but no minimum value.
   b. Find a polygonal convex set such that over this set f has a minimum value but no maximum value.

9. Find the maximum and minimum values, when they exist, of the polynomial $f(x, y) = 7x - 10y - 15$ over each of the convex polygons given in Exercises 7, 8, and 9 of Section 8.7.

10. A developer wants to build a ski resort on a remote mountain top. He wants to provide accommodations for at least 435 people using only two types of housing. Type I is a large A frame with a capacity of 12 persons. It requires 10 units of wood and 6 units of concrete to build, plus it requires 30 hours at a cost of $6,000 for each Type I house. Type II housing is barracks style with a capacity of 25 people. It requires 5 units of wood and 22 units of concrete to build, plus it requires 50 hours at a cost of $12,500 for each Type II house.

At least 125 units of wood and 246 units of concrete must be used. Find the number of each type of house the developer should build if he wants to (**a**) minimize construction cost and (**b**) minimize construction time.

11. There are 1,300 tons of coal to be loaded onto trucks at the processing plant. The coal must be loaded within 42 hours since the storage area will be needed again by that time. Loading the coal by conveyor belt requires 3 men who receive $5/hour. The conveyor costs $80/hour to operate and will load 60 tons/hour. A power shovel is available, but it requires 9 men who are paid $4/hour. This shovel costs $10/hour to operate and will load 10 tons/hour. The UMW contract stipulates that at least 150 hours of labor must be provided for the men during the loading of this coal. What is the minimum cost of loading the coal? Find the number of hours of operation of the conveyor and power shovel, assuming that they cannot operate simultaneously.

12. The personnel director of a large company must test the job qualifications of job applicants. At least 47 persons must be tested by one of two methods. There is a written aptitude test which can be administered or a performance test to indicate the applicant's actual abilities. It costs $11/100 to administer the aptitude test and $5/100 to administer the performance test. The time required is 10 hours/100 of the aptitude tests and 30 hours/100 of the performance. The total cost of the test administration cannot exceed $4.87 nor can the time exceed $8\frac{1}{2}$ hours.

In addition to these constraints the director must minimize the number of inaccurate appraisals. It has been discovered that 14 % of the aptitude test results and 27% of the performance test results are inaccurate. How many tests of each type should the director administer?

9

Simple Interest, Compound Interest, and Annuities

Introduction

Interest is the rental paid for the use of borrowed money. The rent depends on how much money is borrowed, the length of time the money is used, and the interest rate. The sum of money borrowed is called the *principal*. The *interest rate* is that fraction of the principal which is to be paid for its use for a specified unit period of time.

We shall consider two kinds of interest, simple interest and compound interest. If interest is computed on the original principal only, it is called *simple interest*. If the interest for a given unit period is added to the principal on which it was computed, and the interest for the next period is computed on this new principal, the sum by which the original principal has been increased is called the *compound interest* for the two unit periods.

It is common practice to express interest rates as percentages rather than as common fractions. In this connection it should be pointed out that the applications of percentage to business are countless. In fact, percentage is used in so much of the computation in the business world, particularly in commercial problems, that its importance can hardly be overemphasized. For this reason we next review briefly the notion of percentage.

9.1 PERCENTAGE

The word *percent* comes from the Latin phrase *per centum*, which means "by the hundred." Thus percentage refers to calculations in which hundredths are used as the basis of comparison. Therefore, a percent always stands for a fraction whose denominator is 100. The symbol for percent is %. Thus, 6% means $\frac{6}{100}$, 17% means $\frac{17}{100}$, 123% means $\frac{123}{100}$, etc. For example, 5% of 60 means $\frac{5}{100}$ of 60, which is 3. In making calculations involving percents, it is always necessary to express each percent as an equivalent fraction before carrying out the computation. This may be done either as a decimal fraction or a common fraction. Thus $25\% = \frac{25}{100} = 0.25 = \frac{1}{4}$. Similarly, we can express any number as a percent by multiplying it by 100 and attaching the percent sign; for example, $\frac{3}{4} = \frac{300}{4}\% = 75\%$. The following illustrative examples show how elementary algebra can often be used to solve problems involving percentage. The student is advised to study these examples very carefully before attempting to solve the problems.

Example 1

Find the number which is 7% of 280.

Solution: Let x be the desired number.

Then $\qquad x = 7\%$ of 280

That is, $\qquad x = \dfrac{7}{100}(280) = 0.07(280) = 19.6$

Example 2

What percent of 500 is 85?

Solution: Let x be the required number of percent.

Then $\qquad x\%$ of $500 = 85$

That is, $\qquad \dfrac{x}{100}(500) = 85$

or $\qquad 5x = 85$

Hence $\qquad x = 17$

Thus 85 is 17% of 500.

Example 3

If 15% of a number is 36, what is the number?

Solution: Let x be the number.

Then $\qquad 15\%$ of $x = 36$

That is, $\qquad \dfrac{15}{100}x = 36$

or $\qquad 15x = 3600$

Hence $\qquad x = 240$

Check: $\quad 15\%$ of $240 = \dfrac{15}{100}(240) = 0.15(240) = 36$

SEC. 9.1 Percentage

Example 4 In a class of 150 students, 117 received passing grades. What percent of the class failed?

Solution: Let x be the number of the percent that failed.

Then $\qquad\qquad\qquad x\%$ of $150 = 33$

That is, $\qquad\qquad\qquad \dfrac{x}{100}(150) = 33$

or $\qquad\qquad\qquad\qquad 150x = 3300$

Hence $\qquad\qquad\qquad\qquad x = 22,\qquad$ and so 22% of the class failed

Check: 22% of $150 = \dfrac{22}{100}(150) = 0.22(150) = 33$

Example 5 A man's salary for a given year was $15,680. This was 12% more than his salary for the previous year. What was his salary for the previous year?

Solution: Let x be the salary for the previous year.

Then $\qquad\qquad\qquad x + 12\%$ of $x = 15{,}680$

That is, $\qquad\qquad\qquad x + \dfrac{12}{100}x = 15{,}680$

or $\qquad\qquad\qquad\qquad 100x + 12x = 1{,}568{,}000$

which reduces to $\qquad\qquad 112x = 1{,}568{,}000$

and solving for x we get $\qquad x = 14{,}000$

As a typical example of an application of percentage to business, consider the case of a retail merchant whose business is that of buying and selling things. He may own or manage a grocery store, a department store, a hardware store, an automobile agency, or what have you, but whatever article he buys and sells he has to determine a selling price for each article. It is rather obvious that the selling price is connected in some way with the cost of the article to the merchant. It is equally obvious that, in general, if the merchant is to stay in business, the selling price must be higher than the cost. The difference between the selling price and the cost is called *markup* or *gross profit*. It is common practice to determine the markup as a certain percentage either of the cost or of the selling price. Usually the markup is based on the cost. Let S be the selling price and r the percentage markup based on the cost C. Then the markup or gross profit P is by definition

(1) $\qquad\qquad\qquad\qquad P = S - C = rC$

Therefore,

(2) $\qquad\qquad\qquad\qquad S = C + rC = C(1 + r)$

Example 6   Find the selling price of a radio that cost $30 if the percentage markup based on the cost is 40%. Also find the gross profit.

Solution: Using Equation (2) with $C = \$30$ and $r = 40\% = 0.40$, we obtain

$$S = C(1+r) = 30(1+0.40) = \$42.00$$

From Equation (1) the gross profit is

$$P = S - C = 42 - 30 = \$12.00$$

Example 7   Find the percent markup, based on the cost of an article that costs $120 and sells for $162.

Solution: From Equation (1),

$$r = \frac{S-C}{C} = \frac{162-120}{120} = 0.35$$

or

$$r = 35\%$$

As mentioned above, the markup is sometimes determined as a percent of the selling price. In this case let d represent the percentage markup based on selling price S, and let C be the cost. Then

$$S - dS = S(1-d) = C$$

or

(3) $$S = \frac{C}{1-d}$$

Example 8   Find the selling price of a piano that cost $595 if the percentage markup based on the selling price is 30%.

Solution: Using Equation (3) with $d = 30\% = 0.30$ and $C = \$595$, we get

$$S = \frac{C}{1-d} = \frac{595}{1-0.30} = \frac{595}{0.70} = \$850$$

Exercises 9.1

1. Express each of the following as a common fraction.
      a. 8%   **b.** 110%   **c.** 12.5%   **d.** 75%   **e.** $16\frac{2}{3}\%$

2. Express each of the following as a decimal fraction.
      a. 7.5%   **b.** 14%   **c.** 23%   **d.** 135%   **e.** 250%

3. Express each of the following as a percent.
      a. $\frac{1}{2}$ **b.** $\frac{3}{5}$ **c.** $\frac{2}{3}$ **d.** $\frac{3}{4}$ **e.** 1.6

SEC. 9.2   Simple Interest   441

4. Express each of the following as a percent.
   a. 0.07   b. 0.23   c. 1.24   d. 0.624   e. 1.056

5. Find a. 10% of 540, b. 6% of 385, c. 7.4% of 280, d. 18% of 875.

6. What percent of a. 60 is 5, b. 28 is 21, c. 48 is 60, d. 300 is 6?

7. a. 54 is 20% of what number?   b. 48 is 75% of what number?
   c. 6.8 is $33\frac{1}{3}$% of what number?   d. 160 is 8% of what number?

8. In a class of 40 students, 15% failed. How many students passed?

9. A piece of real estate, for taxation purposes, is evaluated at $14,500. If the tax rate is 5.4%, what is the amount of the tax?

10. If your taxable income after all deductions is $16,000 and the tax rate is 24%, what is your tax?

11. The city tax on Mr. Jones's home in a given year is $525. Find the evaluation of the home if the tax rate was 1.25% that year.

12. Find the selling price of an automobile which cost the dealer $4,750 if the percent markup based on the cost is known to be 30%.

13. A dealer sells a TV set for $891 that cost him $675. What was the percent markup based on the cost?

14. A furniture dealer wishes to sell a sofa that cost him $210 so as to make a gross profit of 35% based on the selling price. What should his selling price be?

15. A piano was sold for $1,250. If this is 15% less than the list price, what was the list price?

16. A realtor bought two lots for $2,800. He sold one lot at 9% profit, the other for 2% less. His net gain was $76. Find the cost of each lot.

17. How much water should be added to 20 gallons of alcohol, 90% pure, to reduce it to a 60% solution?

In exercises 18 through 20, round answers to five decimal places.

18. Find the number which is 19.351% of 227.972.

19. What percent of 126.283 is 251.54?

20. If 62.75% of a number is 91,381, what is the number?

9.2   SIMPLE INTEREST

By definition, *interest* is money paid for the use of borrowed money. If you borrow money from an individual, you should expect to pay this individual interest for the privilege of using his or her money. If you borrow money from a bank, savings and loan association, or other lending institution, then you will most definitely pay interest for the privilege of using the money. On the other hand, if you should lend money to an individual, then you are entitled to be paid interest for the use of your money. Finally, if you invest money in a savings account, then the institution has, in effect, borrowed money from you, and it will pay interest to you for the privilege of using your money.

Interest is always computed as a certain percent of the amount borrowed (or loaned) for one unit, or period, of time. There are three elements involved in any interest transaction: *principal, rate of interest,* and *time*. The principal is the original amount borrowed or invested. The rate of interest is the percent of the principal which is paid as interest for its use for one period of time. The time is the number of periods during which the principal is used. One year is usually taken as the unit period of time. In fact, in business circles, any quoted interest rate will be interpreted as an annual, or yearly, rate unless otherwise stated. We shall use the same interpretation in what follows. The word *amount* will be used to denote the sum of the principal and interest.

There are two kinds of interest in common use, *simple interest* and *compound interest*. If the interest is computed on the principal only, it is called *simple interest*. But if the principal is increased by adding the interest at the end of each interest period and the interest for the succeeding period is computed on the new principal each time, this type of interest is called *compound interest*. In this section we shall restrict our discussion to simple interest.

Let

P = principal in dollars
i = rate of interest in decimal form
t = time in years or fractions of years
I = total interest in dollars
S = amount in dollars of principal and interest

The following formulas involving simple interest are immediate consequences of the above definitions.

(4) $$I = Pti$$

(5) $$S = P + I = P + Pti = P(1 + ti)$$

Example 1 — Find the simple interest on $300 for 7 years at a simple interest rate of 5%. Also, find the amount.

Solution: Here $P = 300$, $t = 7$, and $i = 0.05$. Substituting in Formula (4), we obtain

$$I = Pti = 300(7)(0.05) = \$105$$

Also, $$S = P + I = 300 + 105 = \$405$$

SEC. 9.2 Simple Interest

Example 2 Find the amount of an investment of $500 for 9 months if the simple interest rate is 8%.

Solution: Here $P = 500$, $t = \dfrac{9}{12} = \dfrac{3}{4}$, and $i = 0.08$. Using Formula (5), we obtain

$$S = P(1 + ti) = 500\left[1 + \left(\dfrac{3}{4}\right)(0.08)\right] = 500(1.06) = \$530$$

Example 3 What principal will amount to $628 in 8 months at 7% simple interest?

Solution: In this problem we know that $S = 628$, $t = \dfrac{8}{12} = \dfrac{2}{3}$, and $i = 0.07$. We wish to find P. To do this, we first solve Formula (5) for P and obtain

$$P = \dfrac{S}{1 + ti}$$

Substituting the known values for S, t, and i,

$$P = \dfrac{628}{1 + \left(\dfrac{2}{3}\right)(0.07)}$$

Now, we can simply evaluate this quantity directly by using a hand calculator, obtaining

$$P = \$600$$

If a hand calculator were not available, the numerator and denominator can be multiplied by 3 to obtain

$$P = \dfrac{628}{1 + \left(\dfrac{2}{3}\right)(0.07)} = \dfrac{1{,}884}{3.14} = \$600$$

Example 4 What is the rate of simple interest if an investment of $375 amounts to $450 in $2\tfrac{1}{2}$ years?

Solution: Here $P = 375$, $t = \dfrac{5}{2}$, and $S = 450$. We wish to find i. We first find I by using Formula (5).

$$I = S - P = 450 - 375 = 75$$

Now, Formula (4) gives

$$i = \dfrac{I}{Pt} = \dfrac{75}{375\left(\dfrac{5}{2}\right)} = \dfrac{150}{1{,}875} = 0.08 = 8\%$$

Here, again, the step $\dfrac{150}{1{,}875}$ can be omitted if a hand calculator is used.

In actual practice, the majority of financial transactions where simple interest is used involve periods of time less than a year. In such transactions there have traditionally been two different methods of computing what fraction of a year is involved. The first, and oldest, is based on the assumption that one day is 1/360 of a year (that is, the assumption is that a year has 360 days). Interest computed on this basis is called *ordinary simple interest*. There are several reasons why lending institutions might have adopted the (incorrect) assumption that a year has 360 days. Perhaps the most understandable reason is that, prior to the ready availability of inexpensive calculating devices, computations were done by hand and it was simply easier to assume that each month had 30 days and that each year had 360 days. For example, the interest on a 45-day loan in this setting would involve the fraction $\frac{45}{360} = \frac{1}{8}$. Contrast this to the fraction $\frac{45}{365} = \frac{9}{73}$, which would occur in a computation based on the assumption that each year has 365 days (which is a correct assumption for nonleap years). Clearly, a computation of interest by hand would be easier using the fraction $\frac{1}{8}$ than it would using the fraction $\frac{9}{73}$. In fact, 360 has so many factors (2, 3, 4, 5, 6, 8, 9, 10, 12, 15, 18, 20, 24, 30, 36, 40, 45, 60, 72, 90, 120, 180) that many fractions with 360 as denominator reduce considerably. This list of divisors also illustrates why so many short term loans have traditionally been set up for 30, 45, 60, 90, 120, or 180 days. In fact, this tradition has persisted even among lending institutions which no longer use ordinary simple interest in their computations.

Now, with inexpensive calculating devices of all sorts available (and with expensive computers available in many cases), computations using 365 as a denominator can be just as easily performed as those using 360 as a denominator. Consequently, some banks and other lending institutions have adopted the policy of assuming that each year has 365 days. Interest computed on this basis is called *exact simple interest*.

So, what type of simple interest should you learn to apply, ordinary or exact? It turns out that you need to know both, for some lending institutions use ordinary simple interest in some cases and exact simple interest in other cases. We shall not attempt to justify this procedure, but we shall give an illustration of how it came to be in one collection of banks. At this writing, the state of Tennessee has what is called a "usury law" which sets a ceiling of 10% on the interest rate banks can charge on short-term loans of less than $25,000. Moreover, subject to interpretation in the courts, this interest ceiling is to be computed on the basis of exact simple interest. Consequently, the practice of banks in the Knoxville, Tennessee, area at this writing is to compute interest on the basis of ordinary simple interest if the quoted rate is less than 10% and on the basis of exact simple interest if the quoted rate is equal to 10%. The reason is that a 10% ordinary simple interest rate is equivalent to a 10.139% exact simple rate, and so the banks cannot legally charge 10% ordinary simple interest on such loans. Since it is important that you be able to perform computations based both on ordinary and exact simple interest, we shall be careful to say which type to use in a given example or problem. The one notable exception will be in

SEC. 9.2   Simple Interest

problems where interest is calculated for a period of $3\frac{1}{2}$ years, 8 months, $2\frac{1}{2}$ years, 6 months, and so on (see Exercises 1 through 10). In these cases, as in Examples 1 through 4, we refer to simple interest with no mention of ordinary or exact. In such problems, the difference between ordinary and exact simple interest would become evident in the dates the transactions terminate. For example, a principal of $100 would amount to $108 in one year at either ordinary or exact simple interest. The difference is that the $100 would amount to $108 in 360 days at ordinary simple interest, whereas it would take 365 days for the $100 to amount to $108 at exact simple interest.

Let I_o denote ordinary simple interest for m days at interest rate i and let I_e denote exact simple interest for m days at rate i. Then it follows from Formula (4) that

(6) $$I_o = P\left(\frac{m}{360}\right)(i) = \frac{Pmi}{360}$$

and

(7) $$I_e = P\left(\frac{m}{365}\right)(i) = \frac{Pmi}{365}$$

In problems involving simple interest, it is necessary to either calculate the exact number of days between two dates or to find the date of the day exactly m days after a given date. The accepted practices are explained in the next two examples.

Example 5 Find the ordinary and exact simple interest which would be earned on an investment of $1,800 at the rate of 6% from May 14 to July 23 of a given year.

Solution: Number of days remaining in May   17
            Number of days in June   30
            Number of days in July to be counted 23
                                                 ──
            Total number of days   70

Note that the first day, May 14, was not counted but the last day, July 23, was counted. Using Formula (6), we find that the ordinary simple interest is

$$I_o = \frac{Pmi}{360} = \frac{(1,800)(70)(0.06)}{360} = \$21.00$$

whereas Formula (7) yields the exact simple interest:

$$I_e = \frac{Pmi}{365} = \frac{(1,800)(70)(0.06)}{365} = \$20.71$$

Example 6 Suppose that you borrow $775.00 from a bank on November 18 of a given year at the rate of $9\frac{3}{4}\%$ for 127 days. On what date of the following year is the loan due, assuming that the following year is a leap year? How much must you repay the bank if the bank charges ordinary simple interest? How much must you repay if the bank charges exact simple interest?

Solution: First we find the date on which the loan becomes due.

Number of days remaining in November	12
Number of days in December	31
Number of days in January	31
Number of days in February (leap year)	29
Number of days in March to be counted	x
Total number of days	127

We see that $12 + 31 + 31 + 29 + x = 127$

$$103 + x = 127$$

$$x = 127 - 103 = 24$$

Therefore, the loan becomes due on March 24 of the following year.
Now we compute I_o and I_e as in Example 5:

$$I_o = \frac{Pmi}{360} = \frac{(775)(127)(0.0975)}{360} = \$26.66$$

and

$$I_e = \frac{Pmi}{365} = \frac{(775)(127)(0.0975)}{365} = \$26.29$$

In both cases, the computations are most easily made with a hand calculator and answers are rounded to the nearest penny. Finally, upon reading the problem again we see that it has not been solved as yet. The amount you repay if ordinary simple interest is used is

$$S = P + I_o = 775 + 26.66 = \$801.66$$

whereas the amount you repay if exact simple interest is used is

$$S = P + I_e = 775 + 26.29 = \$801.29$$

This might explain why the bank charges ordinary simple interest where possible!

As can be seen in Example 6, in many instances the amount borrowed, the length of time, and/or the interest rate will be numbers which do not lend themselves readily to long-hand computation. In such situations, some sort of calculator is an indispensable tool. Since electronic hand-held calculators are readily available on today's market at relatively low prices, we shall give a reasonable number of examples and problems for which a calculator should (if not must) be used. The student is encouraged to gain access to a calculator for the material of this chapter. For the purpose of the more complicated topics of compound interest and annuities, it is wise to obtain a calculator with the

SEC. 9.2 Simple Interest

functions y^x, $1/x$, and at least one memory. Moreover, if the student plans to study trigonometry (Chapter 11), a calculator with the trigonometric functions, *sin*, *cos*, and *tan*, would be extremely useful.

Example 7 Find the exact simple interest on $11,256.75 at 11.37% from January 11 to April 15 of a given nonleap year.

Solution: First we compute the number of days.

Number of days remaining in January	20
Number of days in February	28
Number of days in March	31
Number of days in April to be counted	15
Total number of days	94

Using Formula (7), we find

$$I_e = \frac{Pmi}{365} = \frac{(11{,}256.75)(94)(0.1137)}{365} = \$329.62$$

Such a computation would be extremely tedious and time consuming if done by hand. However, with a hand calculator it is a triviality.

Exercises 9.2

1. Find the simple interest and amount of a principal of $1,000 for $3\frac{1}{2}$ years at 8%.

2. Find the simple interest and amount of a principal of $3,600 for 8 months at $6\frac{1}{2}$%.

3. How long will it take $1,500 to yield $90 at 8% simple interest?

4. Mr. Smith repays a bank $291.20 in settlement of a loan and 6% simple interest for 8 months. How much had he borrowed 8 months earlier?

5. Find the rate of simple interest if a principal of $600 amounts to $735 in $2\frac{1}{2}$ years.

6. A certain used car can be bought for $1,000 cash or for $1,060 due in 6 months. If the second plan is accepted, what rate of simple interest does the buyer pay?

7. In how many years will $480 amount to $513 at $5\frac{1}{2}$% simple interest?

8. What is the rate of simple interest if $324 amounts to $328.05 in 3 months?

9. Mr. Smith gives a check for $576.80 in settlement of a loan of $560 with simple interest at 8%. How long did he use the money?

10. What principal will amount to $397.10 in 9 months at 6% simple interest?

11. Find the ordinary and exact simple interest on $800 at 7% from August 1 to September 20 of a given year.

12. Show that $I_e = I_o - \frac{1}{73}I_o$ and $I_o = I_e + \frac{1}{72}I_e$.

13. Find the ordinary and exact simple interest on $10,000 at 8% from May 11 to July 22 of a given year.

14. Find the ordinary and exact simple interest on $2,300 at 12% from December 20 of one year to April 11 of the following year, assuming that February has 29 days during the second year.

15. Find the exact simple interest corresponding to the ordinary simple interest $162.79.

16. Find the ordinary simple interest corresponding to the exact simple interest $96.48.

17. Find the exact simple interest on $932.58 for 175 days at 13.7%.

18. What is the exact simple interest rate if $426.75 amounts to $511.79 in 311 days?

19. Find the amount of a principal of $876.95 at 11.6% exact simple interest from March 9 to September 26 of a given year.

20. Find the exact simple interest rate if $256.17 amounts to $271.96 in 229 days.

21. What principal must be invested for 731 days at $7\frac{1}{4}$% exact simple interest in order to amount to $1,256.75?

22. How many days must $721.52 be invested at 6.259% exact simple interest to amount to (at least) $743.69? (The answer should be a positive integer m.) Exactly how much will $721.52 amount to in m days?

    Solution: Using Formula (5),

    $$I = S - P = 743.69 - 721.52$$

    and using Formula (7),

    $$I_e = \frac{Pmi}{365} = \frac{(721.52)(m)(0.06259)}{365}$$

    Setting these quantities equal and solving for m, we obtain

    $$m = \frac{(743.69 - 721.52)(365)}{(721.52)(0.06259)} \cong 179.186$$

    using a hand calculator. This means it will take more than 179 days, but less than 180 days for $721.52 to amount to $743.69. Since the answer was required to be an integer, the answer is 180 days. Finally, the exact amount $721.52 will accumulate to in 180 days at 6.259% exact simple interest is

    $$S = P + I = 721.52 + \frac{(721.52)(180)(0.06259)}{365}$$
    $$= 721.52 + 22.27$$
    $$= \$743.79$$

    rounded to the nearest cent.

23. How many days must $1,225.41 be invested at $7\frac{5}{12}$% exact simple interest to amount to (at least) $1,294.62? Exactly how much will the $1,225.41 amount to in this number of days?

SEC. 9.3   Present Value at Simple Interest   449

24. Suppose you borrow $255.00 from a friend for a period of 223 days and agree to pay interest at the rate of $8\frac{1}{2}\%$ exact simple interest. How much must you pay your friend at the end of the 223 days?

9.3 PRESENT VALUE AT SIMPLE INTEREST

Before discussing the concept of present value, we first need to acquaint ourselves with the special meanings of certain words and phrases commonly used in speaking of, or working with, financial transactions. For example, the phrase "S due in t years" or "an amount S due in t years" shall be understood to mean that a certain sum of money will have the value S t years from today. The word *debt* is a very common word in the English language, but it probably means different things to different people. In this text, the word debt shall mean a legal financial obligation, or agreement, to pay someone a certain specified sum of money on a particular date in the future. The phrase "a debt of $500 is due in 6 months" means that someone, whom we shall call the *debtor*, or *borrower*, has agreed to pay someone else, called the *lender*, an amount of $500 on a date 6 months from now. In other words, the borrower can pay the debt in full, called *canceling* the debt, by making a cash payment of $500 to the lender on the date 6 months from today. Notice that in such a phrase no mention is made of interest or an interest rate. It is common practice therefore to refer to such a debt as a *noninterest-bearing debt*. This does not mean that no interest was involved in the loan, or transaction. It means that the interest, if any, is included in the stated amount due. The date on which a debt, or loan, is due is called the *maturity date* and the value of the debt, or loan, on that date is called the *maturity value* of the debt. It should be pointed out here that the phrase "a noninterest-bearing debt of $500 is due in 6 months" means exactly the same as the phrase "a debt of $500 is due in 6 months."

The *present value*, at simple interest rate i, of an amount S due in t years is that principal P which, if invested today at the rate i, simple interest, will amount to S in t years. Since the amount S of a principal P invested at rate i, simple interest, for t years is given by the formula $S = P(1 + ti)$, it follows that the present value P at the simple interest rate i of S due in t years is

(8) $$P = \frac{S}{1 + ti} = S(1 + ti)^{-1}$$

The difference, $S - P$, between the amount S due in t years and the present value P of that amount is called the *simple discount* on S and is denoted by D_s. Thus

(9) $$D_s = S - P$$

In the business world it is assumed that money always has rental value. You and I are aware that this is certainly true to a limited extent at least. The local banks and savings and loan associations are constantly urging us through newspaper ads and TV commercials to put our money in a savings account and watch it grow. We frequently say that if money can easily be invested in a savings account, or in any manner whatsoever, at a rate i, then money is worth i, or the *going rate* for money is i. Since money always has a rental value, we shall, in the problems in this text, think of any sum mentioned as being invested at some rate of interest which may or may not be specified. For example, if through some business transaction between you and me, you agree to pay me $500 one year from today, my money is invested in that deal and is earning interest even though the interest rate is not specified. Suppose that Joe and John, twin brothers, each receive a $100 bill as a gift on their fifteenth birthday. Suppose, further, that Joe immediately invests his gift in a savings account which pays 6% interest, while John puts his bill in his father's safe. On their sixteenth birthday, Joe's investment has grown to $106, but John's bill is still worth $100. The point we are trying to make here is that for money to earn interest it must be invested. Idle money, like an idle person, earns nothing.

Example 1

Find the present value of a debt of $200 due in 1 year if the going rate for money is 7%. Also find the simple discount.

Solution: Here $S = 200$, $t = 1$, and $i = 7\% = 0.07$. Substituting these values in Formula (8), we obtain

$$P = \frac{200}{1+(0.07)} = \frac{200}{1.07} = \$186.92$$

The simple discount is $D_s = S - P = 200 - 186.92 = \$13.08$.

Example 2

A debt of $1,000 is due in 6 months. Find the value of this debt today if money is worth 8%.

Solution: Substituting $S = 1,000$, $t = \frac{1}{2}$, $i = 0.08$ in Formula (8), we get

$$P = \frac{1,000}{1 + 1/2(0.08)} = \frac{1,000}{1.04} = \$961.54$$

The debts in Examples 1 and 2 above are classified as noninterest-bearing debts since the rate of interest involved is not disclosed. In contrast, there are many debts in which the principal of the loan and the interest being charged are both specified. Such a debt, or loan, is called an *interest-bearing* debt. For any financial transaction to be legally binding in the eyes of the courts, such as borrowing or lending money, the terms of the transaction, or agreement, must be in writing. Typical forms for such agreements are illustrated in Figures 9.1 and 9.2.

SEC. 9.3 Present Value at Simple Interest

New York, N.Y., April 1, 1978

One hundred and eighty days after date I promise to pay $2,150.00 to the order of E. D. Eaves.

(signed) J. H. Carruth

Figure 9.1

$1,000   Knoxville, Tenn. May 1, 1978

Six months   after date   I   promise to pay to the order of   Ezekial Winterbottom

One thousand no/100 -------------------------dollars,

payable at   Knoxville, Tennessee .

Value received with interest at 6 percent.

John Doe

Figure 9.2

Each of these documents is called a *promissory note*. The first is noninterest bearing, whereas the second is interest bearing. In Figure 9.2, the amount borrowed is called the *face value* of the debt, or note, and is shown written in numerals in the upper left-hand corner of the note, and in words in the fourth line from the top. The date the loan was made appears in the upper right-hand corner of the note. The maturity date of the note is measured from the date shown on the note. The time connected with the debt is shown on the second line from the top and the interest rate, if any, is shown on the last line, the line above the signature. The face value of the note shown in Figure 9.2 is $1,000. The maturity value of this note is $1,030, that is, face value plus interest. In the case of a noninterest-bearing note (Figure 9.1), the face value and the maturity value are assumed to be the same and no interest rate is mentioned.

Our discussion to this point suggests that in speaking of a sum of money, to be specific, three things about the sum must be known, the *value* of the sum on a particular date, the *interest rate*, if any is involved, and the *time connected with* that sum. In solving problems involving a sum of money invested at the rate of i simple interest, Formula (5), $S = P(1 + ti)$, and Formula (8), $P = \dfrac{S}{1 + ti}$, are very important and should be memorized. For, if we know the value on the maturity date of a sum of money invested at simple interest and we know the interest rate, we can find the value of that investment on any other date by using one or the other. For example, suppose that a certain sum of money invested at the rate of i simple interest has the value $\$X$ on the maturity date. Using Formula (5), with $P = X$, we find that the sum t years after that date is $S = X(1 + ti)$. By using Formula (8), with $S = X$, we find the value of the sum, on a date t years before the date on which its value is X, to be $P = X(1 + ti)^{-1}$. Drawing a time-line diagram of the type shown in Figure 9.3(a) or 9.3(b) is often helpful in analyzing and solving problems of this kind.

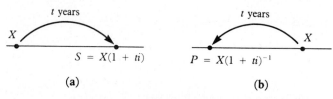

Figure 9.3

Example 3 A note bearing 5% exact simple interest, dated January 1 of a given (non-leap) year and having a face value of $800.00 is due one year later. How much should a person pay for that note on April 1 of the year in question in order to earn 7% exact simple interest on the investment?

Solution: We must first find the maturity value, $\$S$, of this note. Using Formula (5) with $P = 800$, $t = 1$, and $i = 0.05$, we find (see Figure 9.4)

$$S = P(1 + ti) = 800(1 + (1)(0.05)) = \$840.00$$

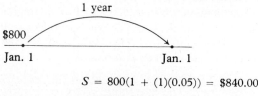

$$S = 800(1 + (1)(0.05)) = \$840.00$$

Figure 9.4

The proposed purchase date is 275 days before the maturity date (see Section 9.2). Hence, if we denote the purchase price by P and use Formula (8) with $S = 840$, $t = \dfrac{275}{365}$, and $i = 0.07$, we obtain (see Figure 9.5)

SEC. 9.3   Present Value at Simple Interest

$$P = S(1 + ti)^{-1} = 840\left[1 + \left(\frac{275}{365}\right)(0.07)\right]^{-1} = \$797.92$$

Thus, a person would pay $797.92 for the note on April 1 if the investment is to earn 7% exact simple interest.

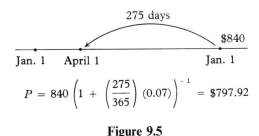

$$P = 840\left(1 + \left(\frac{275}{365}\right)(0.07)\right)^{-1} = \$797.92$$

Figure 9.5

Notice that, when the time arrow points from left to right, we use the formula $S = P(1 + ti)$, which has a $+1$ exponent implied, whereas when the time arrow points from right to left, we use the formula $P = S(1 + ti)^{-1}$, which has a -1 exponent. This is much like the situation with real numbers or coordinate axes. Moving from left to right is going in the *positive direction* (positive exponent), whereas moving from right to left is going in the *negative direction* (negative exponent). These phenomena will occur again both in dealing with compound interest and in dealing with annuities.

Example 4   A noninterest-bearing note for $1,000 is due in 8 months. Assuming that money is worth 8%, find the value of this note **a.** 4 months hence, **b.** 6 months hence, **c.** 8 months hence, and **d.** 1 year hence. (Use ordinary simple interest in your computations.)

Solution: Let P_1 and P_2 denote the value of the note 4 months hence and 6 months hence, respectively, and let S_1 and S_2 denote the value 8 months hence and 1 year hence, respectively. The time-line diagrams in Figure 9.6 illustrate parts **a**, **b**, **c**, and **d** of the given problem. Here we know the value of the note 8 months hence, that is, $S = \$1,000$.

(a) Referring to Figure 9.6(a), we see that P_1 is the value of the note 4 months before the due date. To find P_1, we use Formula (8) with $S = 1,000$, $t = \frac{1}{3}$, and $i = 0.08$. Hence

$$P_1 = 1,000\left[1 + \left(\frac{1}{3}\right)(0.08)\right]^{-1} = \frac{1,000}{1 + \frac{1}{3}(0.08)} = \frac{3,000}{3.08} = \$974.03$$

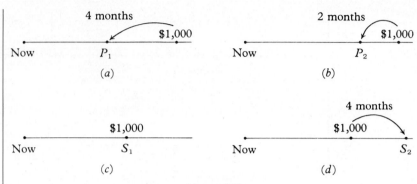

Figure 9.6

(b) To find P_2, we repeat the process used in part **a**, except in this case $t = \frac{2}{12} = \frac{1}{6}$. Thus

$$P_2 = 1{,}000\left[1 + \left(\frac{1}{6}\right)(0.08)\right]^{-1} = \frac{1{,}000}{1 + \frac{1}{6}(0.08)} = \frac{6{,}000}{6.08} = \$986.84$$

(c) Since the due date is 8 months hence, then S_1 is simply the maturity value of the note, that is, $S_1 = S = \$1{,}000$.

(d) Here S_2 is the value of the note 4 months after the due date. We use Formula (5) with $P = 1{,}000$, $t = \frac{1}{3}$, and $i = 0.08$. We obtain

$$S_2 = 1{,}000\left[1 + \left(\frac{1}{3}\right)(0.08)\right] = \$1{,}026.67$$

In most of the examples so far in this section, the amounts involved here have been whole numbers (200, 800, 1,000), the interest rates have been straightforward (5%, 7%, 8%), and the time periods have been convenient fractions of a year $(1, \frac{3}{4}, \frac{1}{3}, \frac{1}{6})$. These facts have made the computations involved rather easy. If we have access to a hand calculator, then there is no need to avoid seemingly more complicated amounts, interest rates, and periods of time. The next example illustrates this fact.

Example 5 A noninterest-bearing note of $2,456.78 is due in 48 days. Find the present value of this note if money is worth 11.53% (ordinary simple interest).

Solution: Referring to Figure 9.7 we use Formula (8) and obtain

$$P = 2{,}456.78\left[1 + (0.1153)\left(\frac{48}{360}\right)\right]^{-1}$$

SEC. 9.3   Present Value at Simple Interest

and then use a hand calculator to find

$$P = \$2419.58 \quad \text{(rounded to the nearest cent)}$$

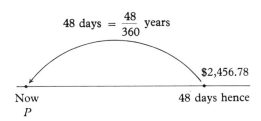

Figure 9.7

The standard practice in the banking industry until recently was to *discount* short-term loans. This practice seems to have been almost totally discontinued in this geographical area at this writing. Therefore, a discussion of this practice will not be given in this text. No matter what method a lending institution uses to compute interest charges, the Truth in Lending Act (passed by the United States Congress in 1968) requires that the equivalent simple interest rate (either ordinary or exact at the discretion of the Board of Governors of the Federal Reserve System) be quoted in writing on the note or loan disclosure statement. *The important fact is that this section and the previous one give the correct procedures for all computations involving ordinary and exact simple interest.* Any other "special types" of interest such as discount and add-on must be treated separately. In presence of the Truth in Lending Act, use of these "special types" of interest (which are almost always equivalent to much higher simple interest rates) seems to be on the decline.

The following table is placed here as a handy reference when dealing with simple interest. The formulas should be readily memorized if enough problems are solved by their use.

Simple Interest Formulas

Simple interest with t measured in years.

$$I = Pti$$

Ordinary simple interest (1 year = 360 days) for m days.

$$I_o = \frac{Pmi}{360}$$

Exact simple interest (1 year = 365 days) for m days.

$$I_e = \frac{Pmi}{365}$$

Amount at simple interest.

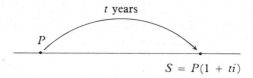

$$S = P(1 + ti)$$

Present value at simple interest.

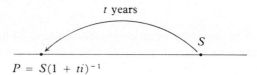

$$P = S(1 + ti)^{-1}$$

Simple discount.

$$D_s = S - P$$

Exercises 9.3

1. A non-interest bearing note for $500 is due in 90 days. Find the present value of the note based on an 8% ordinary simple interest.

2. What is the present value, based on 9% simple interest, of a $1,000 non-interest bearing debt due in 8 months? What is the simple discount?

3. A non-interest bearing note for $1,000 is due 9 months from today. Assuming that money is worth 8% simple interest, what is the value of this note **a.** 3 months from today, **b.** 6 months from today, and **c.** 12 months from today?

4. Mr. A borrows $800 from Mr. B on January 1 and gives him a note for $800 bearing 9% simple interest due in 6 months. Two months later, Mr. B sells the note to Mr. C. If B and C agree that money is worth 10% simple interest, how much does C pay B for the note? How much does the holder of the note get on the day the note is due?

5. A noninterest-bearing note for $1,200 is dated today and due 60 days later. Find the present value based on the exact simple interest rate of 8%.

6. Mr. White holds a note for $1,500, dated January 1, bearing exact simple interest at $8\frac{1}{2}\%$ and due October 1 the same year. On July 1 he sells the note to a bank whose exact simple interest rate is 10%. How much does Mr. White receive from the bank? (Assume it is not a leap year.)

SEC. 9.4 Compound Interest 457

7. A note for $600 bearing $7\frac{1}{2}\%$ simple interest is due in 8 months. What is the present value of this note at 9% simple interest?

8. To buy a piece of property, John may pay either $4,000 down and $6,000 in 6 months, or else pay $6,000 down and $4,000 in a year. If money is worth 8% simple interest, which is the better deal for John? (*Hint:* Find the present value at 8% of each alternative.)

9. Hugh Stone is to receive a trust fund of $10,000 on his 21st birthday. If money is worth $8\frac{1}{2}\%$ simple interest, find the value of this fund on his 20th birthday.

10. A note for $800 bearing exact simple interest at $7\frac{1}{2}\%$ and dated July 1 is due in 90 days. What should one pay for this note on August 30 to realize 10% ordinary simple interest on his or her investment?

11. A noninterest-bearing note of $2,456.78 is due in 48 days. Find the present value of this note if money is worth 11.5% exact simple interest.

12. If you accumulate $1,527.59 in 1,256 days at 6.37% exact simple interest, what is the total amount of interest you have earned?

9.4 COMPOUND INTEREST

It should be recalled that, in dealing with sums of money involving simple interest, the principal remains unchanged. That is, if a given principal is invested at simple interest, the interest on the investment is exactly the same each year for as long as this principal remains invested at the same rate of interest. In fact, the total interest for *n* years is *n* times the interest for 1 year. For example, the simple interest at 8% on $100 for 1 year is $8, for 7 years it is $56, and for 10 years it is $80.

Anyone that has ever had a savings account in a bank or a savings and loan association is aware that such financial institutions do not use simple interest in computing the value of a savings account. They use what is called *compound interest*. In financial transactions in which compound interest is used, the principal does not remain the same. In fact, it is increased at the end of each interest period, the increase being the interest earned during that period. When interest is added to the principal in this manner to obtain a new principal, the interest is said to be *converted*, that is, changed into principal. Another expression widely used to mean the same thing is that "the interest is compounded." In this text we shall use the terms *converted* and *compounded* to mean the same thing. If the interest is added to the principal at the end of each year, it is said to be *converted*, or *compounded*, *annually*, and the *conversion period*, or *compounding period*, is one year. If the interest is added to the principal at the end of each 6 months, the interest is said to be converted, or compounded, semiannually and the conversion period is 6 months. In fact, interest can be converted annually, semiannually, quarterly, monthly, daily, or any convenient period of time could be used; in actual practice the conversion period is usually one year or less. The total amount due, or accumulated, at the end of any conversion period is called the *compound amount* or simply the *amount* of the investment at that time. By reading newspaper ads and watching TV commercials sponsored by banks and savings and loan associations, one gets the impression that a very popular period of conversion with these institutions is quarterly, that is, every three months. However, some banks advertise that they convert interest into principal at the end of each day. The difference between the compound amount at the end of a conversion period and the original principal is called the *compound interest* for the total time interval involved in that investment.

This discussion may be clarified by the following example.

Example 1 Find the compound amount at the end of 3 years of $100 invested at 6% interest compounded semiannually.

Solution:

Original principal	$100.00
Interest for the first 6 months	3.00
Principal at the end of the first 6 months	103.00
Interest at the end of the second 6 months	3.09
Principal at the end of the second 6 months	106.09
Interest for the third 6 months	3.18
Principal at the end of the third 6 months	109.27
Interest for the fourth 6 months	3.28
Principal at the end of the fourth 6 months	112.55
Interest for the fifth 6 months	3.38
Principal at the end of the fifth 6 months	115.93
Interest for the sixth 6 months	3.48
Principal at the end of the sixth 6 months	$119.41

The interest accrued during each 6-month period is rounded to the nearest penny. Hence the compound amount at the end of 3 years, of $100 invested at 6% compounded semiannually, is $119.41. Note also that the compound interest in this case is $119.41 − $100 = $19.41.

It is interesting to compare this compound interest of $19.41 with the simple interest which would have been earned on the same investment in Example 1. The simple interest is

$$I = Pti = (100)(3)(0.06) = \$18.00$$

SEC. 9.4   Compound Interest

Hence, by investing at semiannual compound interest, an extra $1.41 interest was earned. If interest is converted monthly, then it turns out that the compound interest accrued is $19.67, which is 26¢ more than the interest earned using six months as the conversion period. If we keep the length of time at three years, the interest rate at 6%, and the principal at $100.00, decreasing the length of the conversion period will increase the interest earned. However, we shall see in Chapter 10 that no matter how short the conversion period (even if it is a microsecond) the interest earned will be less than $19.73. What really has a dramatic effect on the compound interest earned is lengthening the amount of time the principal is invested. For example, if the $100 of Example 1 were invested for 100 years, then the simple interest (at 6%) is $600.00, whereas it turns out that the compound interest (at 6% compounded semiannually) is $36,935.58! Later in this section we shall see how the latter, rather astounding, figure was obtained.

Some Terminology

Hereafter in any discussion or problem the word *interest* shall be understood to mean compound interest unless otherwise specified. Furthermore, if no conversion period is specifically stated, it shall be understood that interest is converted annually. For example, the statement "money is worth 6%" will mean 6% converted annually.

The time between two successive conversions is called the *conversion period* or *interest period*. For example, if interest is compounded quarterly, then the conversion period is three months. The number of conversion periods per year is called the *frequency of conversion*. Thus, if interest is compounded quarterly, there are 4 conversion periods per year and hence the frequency of conversion is 4.

It is common practice to express interest on an annual basis, regardless of how many times per year the interest is converted. If the conversion period is not a year, the stated annual rate is called the *nominal annual rate*. The interest rate per conversion period is found by dividing the nominal annual rate by the frequency of conversion. Thus, if interest is 8% converted quarterly, the interest rate per conversion period is 8% divided by 4, or 2% per quarter. In general, if i is the nominal annual rate (expressed as a decimal), k the frequency of conversion, and r the interest rate per conversion period (expressed as a decimal), then $r = \dfrac{i}{k}$.

A Formula for Compound Amount

In Example 1, an arithmetic step-by-step method was used for finding the compound amount. It seems clear that this is a tedious method, to say the least, and could be quite long. It is our purpose in this section to derive a general formula for finding compound amounts.

The example just mentioned suggests that the compound amount of a given principal depends on two things: the total number of conversion periods and the interest rate per conversion period. If the original principal is P, and r is the interest rate per conversion period (expressed as a decimal), the interest for the first conversion period is Pr. The principal at the end of the first interest period is $P + Pr = P(1 + r)$. For convenience, let us denote the compound amount at the end of the first conversion period by P_1, at the end of the second conversion period by P_2, at the end of the third period by P_3, and so on. In general, we denote the compound amount at the end of the nth conversion period by P_n. Thus $P_1 = P(1 + r)$. The interest for the second period is $P_1 r = P(1 + r)r$. Hence

$$P_2 = P_1 + P_1 r = P_1(1 + r) = P(1 + r)(1 + r) = P(1 + r)^2$$

Similarly,

$$P_3 = P_2 + P_2 r = P_2(1 + r) = P(1 + r)^2(1 + r) = P(1 + r)^3$$

It can be shown that the compound amount P_n at the end of n conversion periods is given by the formula $P_n = P(1 + r)^n$. It is conventional to use S to denote this compound amount, so we shall do the same and write

(10) $$S = P(1 + r)^n$$

Formula (10) gives the compound amount S of a principal of $\$P$ at the end of n conversion periods at an interest rate of r per conversion period. Figure 9.8 is a time-line diagram representation of Formula (10).

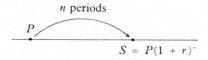

Figure 9.8

In using this formula, it must be kept in mind that n is the total number of conversion, or interest, periods and would equal the number of years the investment ran only if the interest was compounded annually, that is, if the conversion period was 1 year. Similarly, the interest rate r in the formula is not, in general, the quoted rate, which is an annual rate, but the interest rate per conversion period.

We often refer to the compound amount of a given principal as the

SEC. 9.4 Compound Interest

accumulation of that principal. Thus, to accumulate a given principal for a specified time is to find its compound amount at the end of that time. The quantity $(1+r)^n$ is called an *accumulation factor*, because multiplying any principal $P by this factor gives the compound amount of $P at the end of n conversion periods at the rate r per period. In fact, the quantity $(1+r)^n$ is itself the compound amount of $1.00 compounded for n periods at the rate r per period.

Anyone seeing the formula $S = P(1+r)^n$ for the first time might question whether its use would be preferable to the step-by-step method used in Example 1. The factor $(1+r)^n$ may look rather formidable to one inexperienced in evaluating such quantities. If we had to evaluate such a factor by a long-hand method, such as repeated multiplication or by using the binomial theorem, the calculation could be a long and tedious process, especially in a case where n is large, and we attempt to obtain the exact value of the factor expressed in decimal form. To get some idea of what is involved in such a procedure, consider the following facts. If r when expressed as a decimal contains 2 decimal places, then the exact value of $(1+r)^n$ expressed as a decimal contains $2n$ decimal places. For example, the exact value of $(1 + 0.06)^{20}$ contains 40 decimal places and $(1.06)^{80}$ expressed as a decimal contains 160 decimal places when expressed as a decimal. Similarly, if r, expressed as a decimal, contains 3 decimal places, the exact value of $(1+r)^n$, expressed as a decimal, contains $3n$ decimal places, and if r contains 4 decimal places, then $(1+r)^n$ will contain $4n$ decimal places if we insist on complete accuracy. Fortunately, to calculate the exact value of $(1+r)^n$ is, in almost all problems, not only unnecessary, but a sheer waste of time. This is due to the fact that in computing the value of S, using the formula $S = P(1+r)^n$, we always round off the result correct to two decimal places, that is, correct to cents. Certainly in any situation involving such a calculation, a value of S calculated correct to cents would be considered complete accuracy by all parties concerned. In fact, a value of S, calculated correct to within one cent, would be most acceptable. Such accuracy can be obtained by using the following simple rule.

RULE 1. In using the formula $S = P(1+r)^n$ to find the compound amount $S, at the end of n conversion periods, of a principal $P invested at r per period, use an approximate value of $(1+r)^n$ rounded off to as many decimal places as there are digits in P expressed in dollars and cents.

Thus, for a principal of $5,000.00, use a value of $(1+r)^n$ rounded off correct to 6 decimal places. Again, for a principal of $75.00, use a value of $(1+r)^n$ rounded off correct to 4 decimal places. For most of the problems in the text, we shall seldom need the value of $(1+r)^n$ correct to more than 6 decimal places, and using a hand calculator with a y^x function on it, such an approximation is easily and quickly obtained. However, tables do exist which show the value of $(1+r)^n$ and $(1+r)^{-n}$ correct to 6 decimal places for various values of r and all integral values of n and $-n$ from $n=1$ to $n=100$.

Example 2 Find the compound amount at the end of 7 years of $500 at 8% compounded quarterly.

Solution: Here $P = 500$, $n = 28$, and $r = 2\% = 0.02$. Hence $S = 500(1.02)^{28}$. Using a hand calculator, we find $(1.02)^{28} = 1.741024$ correct to 6 decimal places. Using Rule 1 we obtain, after rounding correct to cents,

$$S = 500(1.02)^{28} = 500(1.741024) = \$870.51$$

Example 3 Suppose you deposit $1,000 with a savings and loan association that pays 6% interest compounded quarterly on savings accounts. How much would you have to your credit 10 years later?

Solution: We use the formula $S = P(1 + r)^n$ with $P = \$1,000$, $n = 40$, and $r = \frac{6}{4}\%$ $= 0.015$, and obtain

$$S = 1,000(1.015)^{40} = 1,000(1.8180184) = \$1,814.02$$

This calculation was made using a hand calculator.

It is evident from the definition of compound amount and the derivation of the formula $S = P(1 + r)^n$ that this formula can be applied only to an integral number of conversion periods. Moreover, the assumption is always made that the principal is invested on a conversion date (that is, at the very beginning of a conversion period). Other circumstances certainly exist in the financial world. For example, many banks and savings and loan associations pay interest compounded quarterly on their savings accounts. However, their conversion dates are fixed as January 1, April 1, July 1, and October 1. Certainly, many persons make deposits to accounts on nonconversion dates (May 15 for example). Similarly, many withdrawals are made on nonconversion dates. Computations of interest earned in such transactions, while interesting in themselves, are complicated and often depend on the policies of the particular lending institution. Therefore, we shall not deal with such transactions.

We will deal only with investments which are made on conversion dates and we will only calculate their values on other conversion dates. Similarly, we will deal only with notes which are signed on conversion dates and mature on conversion dates.

As was mentioned in an earlier section of this chapter, when the conversion period is other than 1 year, the stated annual rate of interest is called the *nominal annual rate*. If $100 is invested at the nominal annual rate 8% converted quarterly, the amount of this investment at the end of 1 year is $100(1.02)^4 = \$108.24$. The interest actually earned on the original principal is $8.24, which represents an annual return of 8.24%. We say that the *effective annual rate* in this case is 8.24%.

Rates producing equal amounts in the same length of time on the same

SEC. 9.4 Compound Interest

principal are called *equivalent*, or *corresponding*, rates. Thus the nominal rate 8% compounded quarterly is equivalent to the effective rate 8.24%. Let j denote the effective rate equivalent to a given nominal rate i compounded k times a year. At the rate $r = \dfrac{i}{k}$, per conversion period, P amounts in 1 year to $P(1+r)^k$. At the effective rate j, P amounts in 1 year to $P(1+j)$. Since equivalent rates yield equal amounts in the same length of time,

$$P(1+j) = P(1+r)^k$$

or

$$1+j = (1+r)^k$$

whence

(11) $$j = (1+r)^k - 1$$

By use of Formula (11) we can find the effective rate j equivalent to the nominal rate $i = rk$ compounded k times a year, that is, equivalent to the rate $r = \dfrac{i}{k}$ per conversion period.

Example 4   Find the effective rate equivalent to the nominal rate 7% compounded monthly.

Solution: Here $r = \tfrac{7}{12}\%$ and $k = 12$. By substitution in Formula (11), we have

$$j = \left(1 + \frac{0.07}{12}\right)^{12} - 1 \cong 1.07229 - 1 = 0.07229$$

Hence the effective rate in this case is approximately 7.229%.

Exercises 9.4

1. Find the compound amount of $400 at the end of 10 years if interest is at 8% compounded quarterly.

2. Find the compound amount of $475 at the end of 15 years at 7% compounded quarterly.

3. If you accumulate $600 for 30 years at **a.** 6%, **b.** 7%, **c.** 8%, and **d.** 9%, what is the compound interest in each case?

4. On a boy's first birthday his grandfather deposited for him $1,000 with a trust company, which pays 8% interest compounded semiannually. How much was to the boy's credit on his eighteenth birthday?

5. Ten years ago a trust fund of $10,000 was invested at 7% compounded quarterly. Find the value of the fund now.

6. A principal of $2,000 is invested for 27 months at 7% compounded quarterly. Find the compound interest on this investment.

7. A woman borrows $600, agreeing to repay principal and interest at 7% interest compounded quarterly at the end of 3 years. Find the amount due in 3 years.

8. If $(1+r)^{100} = 1.788967$ and $(1+r)^{50} = 1.337523$, find the value of $(1+r)^{150}$ correct to five decimal places.

9. Find the value at the end of 10 years and 3 months of $3,000 invested at 6% compounded quarterly.

10. Find the amount of $800 at the end of 8 years if interest is at 5% compounded quarterly for the first 3 years and at 6% compounded semiannually for the last 5 years.

11. Find the effective rate equivalent to a nominal annual rate of 8% compounded quarterly.

12. A boy borrowed $1,000 to attend college. He was to pay no interest for four years. After that, he was to pay 6% compounded semiannually. He made no payment for $6\frac{1}{2}$ years. How much did he owe at that time?

13. Benjamin Franklin bequeathed $5,000 to the city of Boston. Assuming that the bequest was invested for 120 years at 6%, what was the investment worth at the end of that time?

14. A certain firm increases its employees' annual salary 6% each year for the first 15 years. Find the salary, for 1983, of a man who started with the firm in 1968 at an annual salary of $12,000.

15. What is the value of a $1,255 investment at the end of $11\frac{1}{2}$ years, if it earns $7\frac{1}{4}$% compounded semiannually?

9.5 PRESENT VALUE AT COMPOUND INTEREST

The *present value* at compound interest of a sum $S due at some future date is the principal $P whose compound amount on that date is $S. If the due date is n conversion periods from now and the interest rate is r percent per conversion period, then in effect $S is the compound amount of $P at the end of n conversion periods at the rate of r percent per period. Thus $S = P(1+r)^n$, or, after solving for P, we have

(12) $$P = \frac{S}{(1+r)^n} = S(1+r)^{-n}$$

The word "now," used here in a relative sense of time, refers to n periods before S is due. The time-line diagram in Figure 9.9 gives a graphical representation of Formula (12). It should be noted that the letters P and S in Formula (12) represent the value of a given obligation *at different dates*. P is the present value of the obligation, while S is the future value of the same obligation. The sum P (now) is just as good as S (n periods hence).

SEC. 9.5 Present Value at Compound Interest

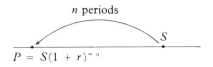

Figure 9.9

To *discount* an obligation at a compound interest rate is to find its value on some date (an integral number of periods) before the due date. The date that we choose to call the present, in dealing with problems involving money invested at compound interest, is a matter of convenience and is arbitrary. Hence, if the date on which an obligation is to be discounted is called the present, or now, then to discount the obligation is to find its present value. For this reason, the quantity $(1+r)^{-n}$ is often called the *discount factor*. Thus, to find the value of a given obligation, at a compound interest rate, n periods before the due date, we simply multiply the value on the due date by the proper discount factor. For example, suppose an obligation has the value $\$X$ on the due date and money is worth r per conversion period. It follows from Formula (12) with $S = X$ that its value on a date n periods before the due date is $P = X(1+r)^{-n}$, while its value on a date n periods after the due date (using Formula 10) with $P = X$ is $S = X(1+r)^n$. See the time-line diagrams in Figure 9.10.

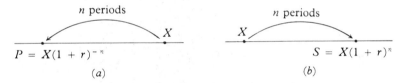

Figure 9.10

Note that the quantity $(1+r)^{-n}$ is actually the present value of $1.00 due at the end of n conversion periods with money worth r per period. The value of this factor can be easily calculated by using a hand calculator having a y^x function. To obtain, by using Formula (12), a value of P correct to within one cent, use an approximate value of $(1+r)^{-n}$ rounded off to as many decimal places as there are digits in S expressed in dollars and cents.

It must be emphasized that Formula (12) like Formula (10) is valid only for positive integral values of n.

Example 1 Find the present value of $1,000 due at the end of 3 years if money is worth 6% compounded semiannually.

Solution: First draw a time-line diagram, Figure 9.11, and then find $n = 6$ and $r = 0.03$.

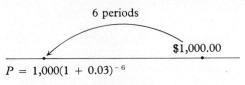

$$P = 1{,}000(1 + 0.03)^{-6}$$

Figure 9.11

Now compare Figure 9.11 with Figure 9.10(a), letting $X = \$1{,}000.00$. Since $S = X = \$1{,}000{,}000$ contains 6 dollar-and-cents digits, we read the value for $(1.03)^{-6}$ from a hand calculator, correct to 6 decimal places, and get $(1.03)^{-6} = 0.837444$. Therefore, $P = 1{,}000 \,(0.837484) = 837.48$ (correct to within 1 cent). This means that \$837.48 invested at 6% interest compounded semiannually will amount in 3 years to \$1,000, correct within 1 cent. In fact, a quick check shows that $836.48(1.03)^6 = 837.48(1.19405) = 999.993$. The difference between the future value S of an obligation and its present value P, $(S - P)$, is called the *compound discount* of S. In this problem the compound discount is

$$S - P = \$1{,}000 - \$837.48 = \$162.52$$

Example 2 A noninterest-bearing debt of \$5,000 is due in 7 years. Assuming money worth 8% compounded quarterly, find the value of this debt **a.** now, **b.** 3 years hence, **c.** 7 years hence, and **d.** 10 years hence.

Solution: In this problem we know the maturity value of the debt, $S = \$5{,}000.00$, and the maturity date is 7 years hence. Let P_1 denote the value of the debt now, let P_2 denote the value of the debt 3 years hence, and let S_1 denote the value of the debt 10 years hence. The time-line diagrams in Figure 9.12 illustrate parts **a, b, c,** and **d.** of this problem.

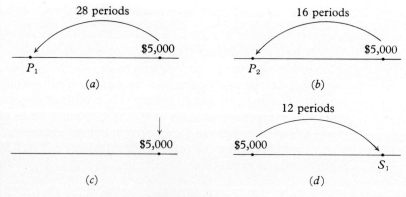

Figure 9.12

(a) In part **a** we wish to find the value of the debt now, which is 7 years before the due date. This suggests that we use Formula (12) with $S = 5{,}000$, $n = 28$, and

SEC. 9.5   Present Value at Compound Interest

$r = 0.02$. Doing this, we obtain

$$P_1 = 5{,}000(1.02)^{-28} = 5{,}000(0.574375) = \$2{,}871.88$$

A hand calculator was used to evaluate $(1.02)^{-28}$

(b) Here we wish to find the value of the debt 3 years hence, that is, 4 years before the debt is due. So again we use Formula (12) with $S = 5{,}000$, $n = 16$, and $r = 0.02$. Substituting these values in our formula, we have

$$P_2 = 5{,}000(1.02)^{-16} = 5{,}000(0.728446) = \$3{,}642.23$$

(c) Since the note is due 7 years hence, its value on that date is $5,000.

(d) In part **d** we need to find the value of the debt 10 years hence, that is, on a date 3 years after the debt is due. This means that we should use Formula (10) with $P = 5{,}000$, $n = 12$, and $r = 0.02$. Substituting these values in Formula (10) gives

$$S_1 = 5{,}000(1.02)^{12} = 5{,}000(1.268242) = \$6{,}341.21$$

Example 3   A $500 debt bearing 7% interest compounded quarterly is due 4 years from today. Find the present value of this debt if money is worth 8% compounded semiannually.

Solution: First we must find the maturity value, S, of this debt. For this we use the formula $S = P(1 + r)^n$, with $P = 500$, $n = 16$, and $r = \dfrac{0.07}{4} = 0.0175$. We get

$$S = 500(1.0175)^{16} = 500(1.31993) = \$659.96$$

Next, to find the present value, we use the formula $P = S(1 + r)^{-n}$, with $S = \$659.96$, $n = 8$, and $r = 0.04$. We obtain

$$P = 659.96(1.04)^{-8} = 659.96(0.73069) = \$482.23$$

The following table is placed here as a handy reference when dealing with compound interest. The formulas should be readily memorized if enough problems are solved by their use.

Compound Interest Formulas

Throughout, i = nominal annual interest rate, k = number of conversion periods per year, r = interest rate per conversion period, n = number of conversion periods involved, and j = effective annual interest rate corresponding to i.

Interest rate per period

$$r = \frac{i}{k}$$

Number of conversion periods

If N years are involved, then $n = Nk$.

Compound amount

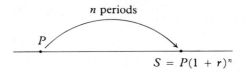

Present value at compound interest

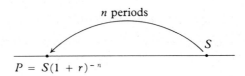

Compound discount
$$D = S - P$$

Effective annual rate
$$j = (1 + r)^k - 1$$

Exercises 9.5

1. Find the compound amount of $1,000 at the end of 20 years if it is invested at 6% compounded quarterly.

2. Find the present value of $3,200 due in 12 years if money is worth 7%.

3. Find the compound discount of an investment of $1,000 for 25 years, assuming that money is worth 6% compounded semiannually.

4. Accumulate $400 for 20 years at 7% compounded quarterly.

5. A note for $800 bearing 6% interest compounded semiannually is due in 5 years. Find the present value of the note, assuming money worth 8% is compounded quarterly.

6. Person A signs a note promising to pay B $1,000 in 5 years with interest at 6% compounded quarterly. Two years later B sells the note to C. How much does C pay B for the note if B and C agree that money is worth 8% compounded semiannually?

7. A noninterest-bearing note for $3,000 is due in 8 years. Assuming money worth 7% compounded quarterly, find the value of this note **a.** now, **b.** 4 years hence, **c.** 8 years hence, and **d.** 10 years hence.

SEC. 9.6   Equations of Value   469

8. Find the effective rate equivalent to a nominal rate of 9% compounded monthly.

9. John borrows $1,200 for his last year at college, agreeing to repay it in 5 years with interest at 7%. Three years later he wishes to settle the debt. If it is agreed that at this time money is worth 8% compounded semiannually, and he is permitted to repay the loan, how much should he pay?

The following problems are to be solved using a hand calculator having a y^x function.

10. Find the compound amount of $875 at the end of 7 years at 7% compounded semiannually.

11. Find the present value of $3,750 due in 5 years if money is worth $8\frac{1}{2}$%.

12. A certain savings and loan association advertises that it pays $7\frac{1}{2}$% interest compounded quarterly on savings accounts of $5,000 or more. What is the effective rate equivalent to this nominal rate?

13. A note for $9,575 bearing interest at $8\frac{3}{4}$% compounded quarterly is due in 8 years. Find the value of this note 5 years hence if money at that time is worth 10% compounded semiannually.

14. How much should a father place in a trust fund that pays 7% interest compounded semiannually on his son's first birthday so that the son will receive $25,000 from the fund on his 21st birthday?

9.6   EQUATIONS OF VALUE

In business and financial circles a sum of money is always considered as a growing quantity; that is, it is assumed that the sum is always invested at some rate of interest. Hence a sum of money has different values at different times, and this must always be taken into account in all transactions involving money. For example, consider a noninterest-bearing debt of $$X$ due at some specified time. Its value at the due date is, of course, $$X$. If the debt is not paid at the due date and if money is worth r per period, its value n periods *after* the due date is *more* than $$X$ and is found by accumulating $$X$ for n periods, that is, by multiplying X by $(1 + r)^n$. But its value n periods *before* the due date is less than $$X$ and is found by discounting $$X$ for n periods, that is, by multiplying X by $(1 + r)^{-n}$. Thus, if we know the value of a sum of money on any conversion date, and know the worth of money, we can find its value on any other conversion date.

A *fundamental principal of finance* is that the values of two sums of money (investments, notes, etc.) can be compared only by comparing their values on the same date. Two sums of money are said to be *equivalent on a particular date* if they have the same value on that date. For example, if money is worth 8% simple interest, then $100.00 due today and $108.00 due 1 year from today are equivalent today and they are also equivalent 1 year from today. Today both are worth $100.00, whereas 1 year from today both will be worth $108.00. Two sums of money which are equivalent at simple interest on one particular date may or may not be equivalent on another date. However, it can be shown that if two sums of money are subject to the same nominal interest rate, the same frequency of conversion, and have the same value on one particular conversion date, then they have the same value on any other conversion date. Hence, if such sums are equivalent on one conversion date, then they are equivalent on any other conversion date.

It is often desirable to exchange one set of obligations, or debts, due at various times for an equivalent set due at various other times. The most familiar situation of this sort is debt consolidation. Here a person might borrow enough money from a bank to pay off several debts on a particular day, thus replacing several debts by a single debt. Such an exchange is referred to as *commuting* the one set of obligations into an equivalent set. When we need to compare two sums of money, the important step is to focus our attention on a particular conversion date, find the value of each sum on that date, and compare these values. The date chosen is called the *focal date*.

When using compound interest, the conversion date selected as focal date in a particular problem is usually chosen in such a way that the calculations required are as easy as possible. When comparing two sets of obligations, it is necessary to first select a focal date and then find the sum of the values of each set of obligations on the focal date. An equation expressing the equivalence of two sets of obligations is called an *equation of value*. We shall always use an equation of value based on compound interest so that the equivalence of the two will be independent of the conversion date chosen as the focal date.

An equation of value is used when it is necessary, or desired, to change the dates on which the payments of a set of obligations is made. When one set of obligations is commuted into a second set, it is most important to all concerned that neither the creditor nor the debtor should be penalized by the exchange. Hence it is necessary to determine the second set of obligations so that the two sets are equivalent at some focal date. It is in this determination that an equation of value is actually used. The following examples illustrate how this is done.

Example 1

A debt of $200 due 1 year hence and another debt of $500 due 4 years hence are to be discharged, that is, paid, by a payment of $300 now and a second payment due 3 years hence. If money is worth 6% compounded semiannually, find the value of the second payment so that the two sets of payments are equivalent.

Solution: Let X denote the value of the second payment, and for ease in calculation we choose 3 years hence as the focal date. We next draw two time-line diagrams, one for each set of obligations (see Figure 9.13) to illustrate the schedule of payments for each set of obligations. We use FD in the figures to denote *focal date*.

Let S_1 denote the value of the sum of the set of old debts at the focal date, and let S_2 denote the value of the sum of the set of new debts at the focal date. Our

SEC. 9.6   Equations of Value

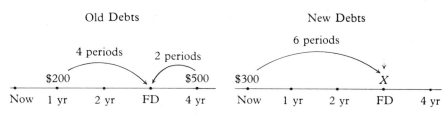

Figure 9.13

objective is to determine a value of X so that $S_1 = S_2$. Recalling that interest is at 6% compounded semiannually, it follows that

$$S_1 = 200(1.03)^4 + 500(1.03)^{-2} \quad \text{and} \quad S_2 = 300(1.03)^6 + X$$
$$= 200(1.12551) + 500(0.94260) \qquad\qquad = 300(1.19405) + X$$
$$= 225.10 + 471.30 \qquad\qquad\qquad\qquad = 358.22 + X$$
$$= \$696.40$$

The two sets of debts are equivalent if

$$X + 358.22 = 696.40$$

or if

$$X = 696.40 - 358.22 = \$338.18$$

Each of the last two equations is an equation of value.

Example 2 A debt of $1,000 due 2 years hence is to be discharged by two equal payments, one due 1 year hence and the other due 3 years hence. If money is worth 7% compounded quarterly, how much should each of these payments be?

Solution: Let X be the value of each of the two payments on the date each is due and choose 3 years hence as the focal date. The time-line diagrams in Figure 9.14 illustrate the two sets of obligations.

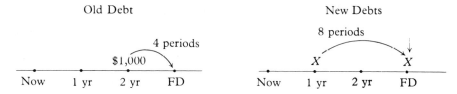

Figure 9.14

Here the interest rate per period is $r = 1\frac{3}{4}\% = 0.0175$. If S_1 denotes the value of the old debt at the focal date, and S_2 denotes the sum of the values of the new debts at the focal date, then

$$S_1 = 1{,}000(1.0175)^4 \qquad S_2 = X(1.0175)^8 + X$$
$$= 1{,}000(1.071859) \qquad\quad = X(1.148882) + X$$
$$= \$1{,}071.86 \qquad\qquad\quad = 2.148882X$$

The two sets will be equivalent if $S_1 = S_2$, that is, if

$$2.148882X = 1{,}071.86$$

or if

$$X = \frac{1{,}071.86}{2.148882} = \$498.80$$

Thus each of the two equal payments should be $498.80.

Exercises 9.6

1. A debt of $600 is due in 5 years. Given that money is worth 8% compounded quarterly, find the value of this debt **a.** now, **b.** 2 years hence, **c.** 4 years hence, **d.** 5 years hence, and **e.** 7 years hence.

2. Solve Example 1 of this section, using 1 year hence as focal date.

3. Solve Example 2 of this section, assuming the original debt is discharged by 3 equal payments made 1 year hence, 2 years hence, and 3 years hence, respectively.

4. A debt of $400 due now is to be paid in two equal installments, the first due in 1 year and the second due in 2 years. If money is worth 8%, how much should each of these payments be?

5. John Doe buys a car on the following terms: $2,500 down, $2,000 at the end of each year for 2 years. If money is worth 8% compounded semiannually, what single cash payment on the date of purchase is equivalent to these terms?

6. A debt of $200 due in 2 years and another of $400 due in 4 years are to be discharged by a payment of $100 in 1 year, $300 in 3 years, and a final payment in 5 years. Assuming money worth 8% compounded quarterly, how much should the final payment be?

7. A note for $800 with interest at 7% is due in 4 years. If money is worth 8% compounded quarterly, what single payment 2 years hence would settle this debt? (*Hint:* You must find the maturity value of the 7% note.)

8. John, aged 14, and his sister, Carolyn, aged 12, have just inherited an estate of $30,000. The will specifies that the money is to be invested at 7% compounded quarterly and that each of the heirs receive the same amount on his or her 21st birthday. Assuming that all conditions in the will are carried out, how much will each receive?

9. If money is worth 8%, what single payment 3 years hence will settle debts of $400 due in 2 years and $600 due in 5 years?

10. A debt of $197 is due 1 year from now and another of $356 is due 3 years from now. What single payment 2 years from now will discharge both debts if money is worth $9\frac{1}{2}\%$ compounded semiannually?

9.7 ANNUITIES

One of the most common experiences in our daily life is that of paying for something on the installment plan. In fact, many of the things we buy today, such as cars, TV sets, radios, furniture, houses, etc., are bought by making a down payment and agreeing to pay the balance, with interest, in equal payments at regular intervals of time. Most such payments belong to a class of payments called *annuities*. Strictly speaking, the word "annuity" implies annual payments, but we wish to use it in the succeeding sections in a more general sense. *We define an annuity as a set of equal payments made at equal intervals of time, regardless of whether the payments are made annually, quarterly, monthly, or otherwise. The important features of an annuity are that the payments are equal and that the payment intervals are of the same length. By the payment interval is meant the period between two successive payments.*

Types of Annuities

Annuities, generally, are divided into three classes, annuities certain, contingent annuities, and perpetuities.

An *annuity certain* is an annuity whose number of payments, date of first payment, and date of last payment are fixed.

A *contingent annuity* is one whose date of either the first or last payment depends on some event the date of which cannot be foretold. For example, certain life insurance policies carry the option that, when the insured reaches a stated age, he may elect to receive equal monthly payments as long as he lives; here the date of the last payment depends on when the insured dies.

A *perpetuity* is an annuity whose payments begin at a fixed date and continue forever. Dividends on preferred stock are an example of a perpetuity.

Annuities Certain

According to the definitions given above, an annuity certain is a fixed number of equal payments made at equal intervals of time, the first payment of which is made on a given fixed date. An annuity certain is called a *simple annuity* if the interest period is the same as the payment interval. Hereafter, when the word "annuity" is used in this text, it will mean a simple annuity unless otherwise stated.

It was observed earlier that, if we know the value of an obligation on a specific conversion date and also know the worth of money, we can find the value of that obligation on any other conversion date. Thus, if we think of the n payments of a given annuity as n individual payments, we can find the value of each of these payments on any conversion date. That is, we can find the value of each such payment at the beginning or end of any payment interval, assuming, of course, a mutual agreement as to the worth of money. If such a date is selected, that is, a focal date, and the value of each of the payments of an annuity is found on that date, the sum of these values is called the *value of the annuity* on that date. Stated a bit more precisely, *the value of a given annuity on any specific conversion date is that single payment which, if made on that date, is equivalent to the annuity.*

Example 1 Suppose John deposits $100 at the end of each three months in a savings account that earns 6% interest compounded quarterly. How much will he have to his credit just after he makes his sixth deposit?

Solution: These 6 equal quarterly payments form a simple annuity of 6 payments. Hence the amount John has to his credit on the date of his sixth deposit is the value of his account on that date. Let us determine this value. We observe that his first deposit draws interest for 5 quarters, the second deposit draws interest for 4 quarters, the third for 3 quarters, the fourth for 2 quarters, and the fifth draws interest for 1 quarter. The sixth deposit, being made on the date in question, draws no interest. Our calculations, keeping in mind that the interest rate is $1\frac{1}{2}\%$ per quarter, may be arranged as follows:

Value of 1st deposit on date of the 6th deposit = $100(1.015)^5 = \$107.73$
Value of 2nd deposit on date of the 6th deposit = $100(1.015)^4 = \$106.14$
Value of 3rd deposit on date of the 6th deposit = $100(1.015)^3 = \$104.57$
Value of 4th deposit on date of the 6th deposit = $100(1.015)^2 = \$103.02$
Value of 5th deposit on date of the 6th deposit = $100(1.015)^1 = \$101.50$
Value of 6th deposit on date of the 6th deposit = $100(1.015)^0 = \$100.00$

Value of the annuity on date of the 6th deposit $622.96

Thus John will have $622.96 to his credit just after he has made his sixth deposit.

The procedure used in Example 1 could become long and laborious if the number of payments were large. To shorten this task, let us develop a formula for finding the value of an annuity of n equal payments on the date the nth payment is due. In our derivation we shall use the following notation.

R = value of each equal periodic payment
r = interest rate per period or payment interval (expressed as a decimal)
n = total number of equal payments in the annuity
S_n = value of the annuity on the date the nth payment is due

Since the value of an annuity of n equal payments on the date the nth payment is due is denoted by S_n, our problem is to find a formula for S_n. To do this, we set up an equation of value, using the day the nth payment is due as focal date. The time-line diagram in Figure 9.15 may be of help in formulating our

SEC. 9.7 Annuities

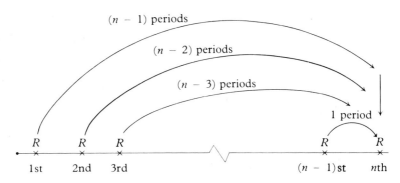

Figure 9.15

problem. Each cross mark indicates where a payment of R dollars is due. A dot indicates the end of an interest period, but where no payment is due. We see from Figure 9.15 that the first payment draws interest for $(n-1)$ periods; hence its value on the focal date is $R(1+r)^{n-1}$. The second payment draws interest for $(n-2)$ periods, and its value at the focal date is $R(1+r)^{n-2}$. Similarly, the third payment draws interest for $(n-3)$ periods, the fourth for $(n-4)$ periods, etc. The $(n-1)$st payment draws interest for only 1 period and the nth payment, being due on the focal date, draws no interest. Our equation of value is

$$S_n = R(1+r)^{n-1} + R(1+r)^{n-2} + R(1+r)^{n-3} + \cdots + R(1+r) + R$$

Factoring R out of each term in the right side, we can write

$$S_n = R[(1+r)^{n-1} + (1+r)^{n-2} + (1+r)^{n-3} + \cdots + (1+r) + 1]$$

Let us now multiply both sides of the last equation by $[(1+r)-1]$ and obtain

$$[(1+r)-1]S_n = R\{[(1+r)-1][(1+r)^{n-1} + (1+r)^{n-2} + (1+r)^{n-3} + \cdots + (1+r) + 1]\}$$

Recalling the formula (see Section 2.6)

$$x^n - y^n = (x-y)(x^{n-1} + x^{n-2}y + x^{n-3}y^2 + \cdots + xy^{n-2} + y^{n-1})$$

and letting $x = (1+r)$ and $y = 1$, we see that the product inside the braces reduces to $(1+r)^n - 1$. Hence

$$[(1+r)-1]S_n = R[(1+r)^n - 1]$$

Solving for S_n, we obtain the formula

(13) $$S_n = R\left[\frac{(1+r)^n - 1}{(1+r) - 1}\right] = R\left[\frac{(1+r)^n - 1}{r}\right]$$

The value of an annuity on the date the last payment of the annuity is due is called the *amount of the annuity*. Thus S_n is the *amount* of an annuity of n equal payments. It follows from Formula (13) that the amount of n payments of \$1 each is $\dfrac{(1+r)^n - 1}{r}$. This quantity is easily calculated for any integral value of n and any value of r if we have access to a hand calculator with a y^x function. Also, this quantity has been calculated for various values of r and n and the results have been made available in tabular form. A brief table of this type is Table VI in Appendix B. It should be pointed out that Formula (13) is only valid for integral values of n. Many texts, in discussing annuities, use the symbol $s_{\overline{n}|r}$ to denote the quantity $\dfrac{(1+r)^n - 1}{r}$. We shall not use this notation in this text.

Example 2 Find the value of an annuity of 40 semiannual payments of \$200 each on the date the 40th payment is due if money is worth 8% compounded semiannually.

Solution: Substituting into Formula (13), we get

$$S_{40} = 200\left[\frac{(1.04)^{40} - 1}{0.04}\right]$$

From Table VI we find $\dfrac{(1.04)^{40} - 1}{0.04} = 95.02552$. Hence

$$S_{40} = 200(95.02552) = \$19{,}005.10$$

Example 3 Joe and Mary Smith are depositing \$100 at the end of each month in a savings account that earns 6% compounded monthly. How much will they have in their account just after the 96th deposit?

Solution: Using Formula (13) with $R = 100$, $n = 96$, and $r = \tfrac{1}{2}\% = 0.005$, we obtain

$$S_{96} = 100\left[\frac{(1.005)^{96} - 1}{0.005}\right] = 100(122.82854) = \$12{,}282.85$$

The value of $\dfrac{(1.005)^{96} - 1}{r}$ was obtained from Table VI.

Exercises 9.7

1. What is the value of an annuity of 20 equal payments of \$50 each on the date the 20th payment is due if the payment interval is 6 months and the interest rate is 6% compounded semiannually?

SEC. 9.7 Annuities

2. Find the amount of an annuity of
   a. $600 per year for 30 years at 6%
   b. $300 each half-year for 30 years at 6% compounded semiannually
   c. $150 each 3 months for 30 years at 6% compounded quarterly

3. What is the amount of an annuity of $800 per year for 12 years if money is worth 7%?

4. Find the amount, at 7% converted monthly, of an annuity of $200 per month for
   a. 4 years, b. 8 years.

5. A woman saves $300 per quarter and invests her savings at 7% compounded quarterly. Find the amount of her savings at the end of 5 years.

6. Beginning with 1974, B deposits $500 on December 31 of each year in a savings account which earns 6%. How much will he have to his credit on January 1, 1982?

7. To provide for his daughter's college education, a man plans to deposit $500 each half-year in a fund earning 6% compounded semiannually. Assuming this plan is carried out and the first contribution is made on the daughter's 10th birthday, how much will the fund contain just after her 17th birthday?

8. Prove algebraically that

$$\frac{(1+r)^{k+m}-1}{r} = \frac{(1+r)^k-1}{r} + (1+r)^k \left[\frac{(1+r)^m-1}{r}\right]$$

9. Use the result in Exercise 8 along with Table VI to find, correct to 5 decimal places, the value of

$$\frac{(1+\tfrac{5}{12}\%)^{180}-1}{\tfrac{5}{12}\%}$$

Students with hand calculators should verify their answer.

10. A local bank makes the following statement in a TV commercial: "Deposit $50 each month for 15 years in our savings department and thereafter you can withdraw $50 per month indefinitely without decreasing the principal." By the principal is meant the amount of your annuity at the end of 15 years. Show that if you deposit $50 at the end of each month for 15 years in a savings account earning 5% interest compounded monthly, you can actually withdraw $55.68 per month thereafter indefinitely without decreasing the principal. This, of course, assumes that the interest rate remains the same.

11. If money is worth 7%, 12 annual payments of $400 each are equivalent to what single payment due at the end of 12 years?

12. A person agrees to settle a debt by making a payment of $500 at the end of each 3 months for 4 years. If money is worth 8% compounded quarterly and the first payment is due 3 months from now, what single payment now would settle the debt? (*Hint:* Find the amount of the annuity and discount this amount for 4 years at the given interest rate.)

9.8 PRESENT VALUE OF AN ANNUITY

In the previous section we derived a formula for the value, or the amount, of an annuity of n payments on the date the last, or nth, payment is due. In many problems involving the paying of an obligation by means of a series of equal payments at equal intervals of time, it is important to have a formula for finding the value of such an annuity on the date which is one period, or payment interval, before the date of the first payment. The value of an annuity of n payments, one payment period before the date the first payment is due, is called the *present value* of the annuity. We shall denote it by A_n. We found, in Section 9.7, a formula for the amount S_n of an annuity of n equal payments, of $R each, assuming money to be worth r per payment period, $S_n = R\left[\dfrac{(1+r)^n - 1}{r}\right]$. To find a formula for A_n, we only need to discount S_n for n periods at r per period, that is, $A_n = S_n(1+r)^{-n}$. Thus

$$(14) \qquad A_n = R\left[\frac{(1+r)^n - 1}{r}\right](1+r)^{-n} = R\left[\frac{1 - (1+r)^{-n}}{r}\right]$$

We observe that the quantity $\dfrac{1 - (1+r)^{-n}}{r}$ is the present value of an annuity of n payments of $1 each, assuming money to be worth r per period. The value of this quantity is easily determined for any value of r and any integral value of n by means of a hand calculator with a y^x function. However, for the benefit of those not having access to such a calculator, the values of $\dfrac{1 - (1+r)^{-n}}{r}$ have been calculated for various values of r and successive values of n and placed in convenient tables. Table VII, Appendix B, is such a table.

Formula (14) involves the four quantities A_n, R, n, and r. Theoretically, if any three of these four quantities are known, the value of the fourth can be determined. If the unknown quantity is either A_n or R, its value can be found rather quickly and easily by using Formula (14) and either a hand calculator or Table VII. It is possible to find the interest rate r of an annuity to a high degree of accuracy by methods which are beyond the scope of this text. However, interpolation in Table VI or VII gives the rate to a degree of accuracy sufficient for most practical purposes. One can also use Table VII to find n, the number of regular equal payments, R, of an annuity. A method involving the use of logarithms for finding n from Formula (14) will be discussed in the next chapter.

Example 1 A house can be bought by paying $5,000 down and $1,000 at the end of each 6 months for 20 years. What is the equivalent cash value of the house, assuming money worth 8% compounded semiannually and assuming the first $1,000 payment is due 6 months after date of purchase?

SEC. 9.8 Present Value of an Annuity

Solution: The equivalent cash value of the house is the present value of the annuity of 40 equal payments of $1,000 each plus the down payment of $5,000. Here $R = 1,000$, $r = 0.04$, and $n = 40$. Substituting these values into Formula (14), we get

$$A_{40} = 1,000 \left[\frac{1-(1.04)^{-40}}{0.04} \right]$$

From Table VII, $\dfrac{1-(1.04)^{-40}}{0.04} = 19.792774$; hence

$$A_{40} = 1,000(19.792774) = \$19,792.77$$

rounding off the answer to the nearest cent. Thus the equivalent cash value of the house is

$$\$19,792.77 + \$5,000.00 = \$24,792.77$$

Example 2 Mr. Smith buys a car, the cash price of which is $5,375. He agrees to pay $1,250 down and the balance with interest at 8% compounded quarterly in 8 equal quarterly payments, the first due three months after the date of purchase. Find the quarterly payment.

Solution: The balance after the down payment is $5,375 - \$1,250 = \$4,125$. This is the present value of the annuity of the 8 equal quarterly payments. Thus in Formula (14), $A_n = 4,125$, $n = 8$, and $r = 0.02$. We find from Table VII that $\dfrac{1-(1.02)^{-8}}{0.02} = 7.325481$. Substituting these values in Formula (14) and solving for R, we find

$$R = \frac{4,125}{7.325481} = \$563.11$$

Example 3 A loan of $2,500 with interest at 7% compounded quarterly is to be repaid by equal quarterly payments of $200 each, the first payment due in 3 months from the date of loan. Find the number of full $200 payments and the value of the final smaller payment, assuming it is made one period after the last $200 payment is made, needed in paying the loan.

Solution: Here $A_n = 2,500$, $R = 200$, and $r = 0.0175$; substituting these values into Formula (14), we have $2,500 = 200 \left[\dfrac{1-(1.0175)^{-n}}{0.0175} \right]$, from which it follows that

$$\frac{1-(1.0175)^{-n}}{0.0175} = \frac{2,500}{200} = 12.5$$

We now find n by inspecting Table VII. The values of $\dfrac{1-(1.0175)^{-n}}{0.0175}$ are found in the $1\tfrac{3}{4}\%$ column. The number 12.5 is not found, but the first entry less than 12.5 is 12.322006 and the corresponding value of n is 14. Hence 14 full $200 payments plus a final smaller payment is needed to pay the loan. To find the amount of the final smaller payment, we find the value of the original loan on the date the 14th $200 payment is due. That is, we accumulate $2,500 for 15 periods at $1\frac{3}{4}\%$ per period, getting $S = 2{,}500(1.0175)^{14} = \$3{,}187.29$. Now we find the amount of the annuity of 14 payments of $200 each, getting $S_n = 200\left[\dfrac{(1.0175)^{14}-1}{0.0175}\right]$, which reduces to $S_n = \$3{,}141.91$. The difference, $S - S_n = 3{,}187.29 - 3{,}141.91 = \$45.38$, is the amount of the original loan that is unpaid just after the 14th $200 payment is made. Hence the final payment, to be made one period after the 14th $200 payment, is $45.38 plus the interest on this amount for 1 period at $1\frac{3}{4}\%$, that is, $45.38(1.0175) = \$46.17$.

Returning again to Formula (13) for S_n, that is, $S_n = R\left[\dfrac{(1+r)^n - 1}{r}\right]$, we observe that it likewise involves four quantities, S_n, R, n, and r. When any three of these quantities are known, the fourth can be determined. But, as was mentioned in the case of Formula (14), to find an accurate value of r when the other three quantities are known is difficult. The following example illustrates how an approximate value of r may be found by using Table VI and interpolation.

Example 4 At what rate of interest, compounded quarterly, will an annuity of $200 per quarter amount to $4,682.50 in 5 years?

Solution: Since the *amount* of the annuity is given, we use the formula $S_n = R\left[\dfrac{(1+r)^n - 1}{r}\right]$. Substituting $S_n = 4{,}682.50$, $R = 200$, and $n = 20$, we have

$$4{,}682.50 = 200\left[\dfrac{(1+r)^{20} - 1}{r}\right]$$

from which it follows that

$$\dfrac{(1+r)^{20} - 1}{r} = \dfrac{4{,}682.50}{200} = 23.4125$$

Since the unknown quantity is r, it is necessary in this case to follow the line $n = 20$ in Table VI until we come to two successive values of $\dfrac{(1+r)^{20}-1}{r}$, one slightly less than 23.4125 and the other slightly greater. We record these entries and the corresponding values of r as shown in the following table. The tabulated data indicate that the desired value of r is between 0.015 and 0.0175.

Now for an increase of $23.7016 - 23.1237 = 0.5779$ in the value of $\dfrac{(1+r)^{20}-1}{r}$, there is an increase of $0.0175 - 0.015 = 0.0025$ in the value of r. Then, for an increase of $23.4125 - 23.1237 = 0.2888$ in the value of $\dfrac{(1+r)^{20}-1}{r}$, what is

SEC. 9.8 Present Value of an Annuity

$$\begin{array}{|c|c|}\hline \dfrac{(1+r)^{20}-1}{r} & r \\ \hline \end{array}$$

$$0.5779 \begin{bmatrix} 0.2888 \begin{bmatrix} 23.1237 & 0.015 \\ 23.4125 & ? \\ 23.7016 & 0.0175 \end{bmatrix} x \end{bmatrix} 0.0025$$

the corresponding increase in the value of r? Denoting this increase by x, we form the proportion

$$\frac{x}{0.0025} = \frac{0.2888}{0.5779}$$

Solving for x, we obtain

$$x = \frac{(0.2888)(0.0025)}{0.5779} = 0.00125$$

Hence the desired approximate value of r is $r = 0.015 + 0.00125 = 0.01625 = 1.625\%$. The approximate rate, compounded quarterly, is $4r = 6.5\%$. As a check, a hand calculator can be used to calculate

$$200 \left[\frac{\left(1 + \dfrac{0.065}{4}\right)^{20} - 1}{\dfrac{0.065}{4}} \right] = \$4{,}682.09$$

which indicates that our interpolated value is a very good approximation to the actual value of r.

The time-line diagrams given in the table below, together with the accompanying formulas, should prove helpful in solving the problems in the next two problem lists. Each cross mark indicates where a payment of R dollars is due. A dot indicates the end of an interest period, but where no payment is due.

Annuity Formulas

Value of an annuity on the date the last payment is due

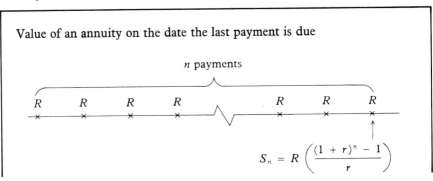

$$S_n = R \left(\frac{(1+r)^n - 1}{r} \right)$$

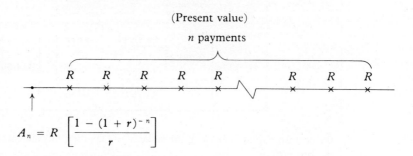

Value of an annuity one period before the first payment is due

(Present value)

n payments

$$A_n = R \left[\frac{1 - (1 + r)^{-n}}{r} \right]$$

Notice the positive exponent for S_n (left to right) and the negative exponent for A_n (right to left).

Exercises 9.8

1. Find the present value of **a.** $100 a year for 7 years at 8%, **b.** $50 each half-year for 7 years at 8% compounded semiannually, **c.** $25 each 3 months for 7 years at 8% compounded quarterly.

2. Find the present value of an annuity of $200 a month for 8 years at 7% compounded monthly.

3. Find the equivalent cash price of a house that may be purchased for $10,000 down and payments of $600 every 3 months for 10 years. Assume money to be worth 7% compounded quarterly.

4. A life insurance company settles a $1,500 claim by making 20 semiannual payments, the first due 6 months hence. What is the semiannual payment if money is worth 8% compounded semiannually?

5. A loan of $4,000, bearing interest at 8% compounded quarterly, is to be paid by payments of $400 at the end of every 3 months until the loan is paid. Find the number of full $400 payments and the amount of the concluding payment if the first $400 payment is due 3 months from the date of the loan and the concluding payment is made one period after the last $400 payment is due (see Example 3).

6. A fund is being created to pay a debt of $5,000 due in 15 years. If the fund is invested in a savings account that earns 6% interest compounded semiannually, what is the semiannual payment to be made to the fund at the end of every 6 months, if the first payment is made 6 months hence?

7. What is the cash price of a lot equivalent to $500 down and payments of $100 a month thereafter for 5 years, if money is worth 7% compounded monthly and the first $100 payment is due 1 month after the down payment?

SEC. 9.9   Amortization of a Debt   483

8. A woman borrows $10,000 with interest at 7% compounded quarterly. She agrees to pay equal installments of $500 each, beginning 3 months hence, every 3 months as long as necessary. How many full $500 payments must be paid? How much does she still owe just after the last $500 payment is made? (See illustrative Example 3.)

9. John plans to deposit $100 at the end of each month in a savings account that earns 6% compounded monthly. How long will it take him to have (at least) $4,000 to his credit? (See Exercise 22 at the end of Section 9.2.)

10. A sorority buys a chapter house for $50,000, paying $10,000 down. How long will it take to pay the balance, at 7% interest compounded monthly, in monthly installments of $500 each? What should the final payment be if it is made 1 month after the last full $500 payment?

11. At 8% interest compounded semiannually, how long will it take to pay a debt of $9,000, due now, in semiannual installments of $375 each, the first due 6 months hence? What will the final payment be if it is made 6 months after the last full $375 payment?

12. At approximately what rate converted quarterly will an annuity of $200 per quarter amount to $4,800 in 5 years? (See illustrative Example 4.) Find the value of an annuity of $200 per quarter for 5 years at the interest rate you obtained in the first part of this problem. Comparing this value with $4,800, is the approximate interest rate too high or too low?

9.9   AMORTIZATION OF A DEBT

By the amortization of an interest-bearing debt we shall mean the paying off of the debt—principal, and interest on the outstanding principal—by a series of equal payments made at equal intervals of time. These periodic payments form an annuity whose present value is the principal of the debt. This method of paying off a debt is one of the most important applications of annuities.

A common problem in connection with the amortization of a debt is to find the periodic payment when the principal of the debt, the interest rate, and the number of payments are known. A problem of this type was solved in Example 2 of the previous section, and it simply involves using the formula $A_n = R \left[\dfrac{1 - (1 + r)^{-n}}{r} \right]$ to find R when A_n, r, and n are known. However, we shall illustrate this by another example.

Example 1   A debt of $2,000, bearing 8% compounded quarterly, is to be amortized by 12 equal quarterly payments, the first due 3 months hence. Find the quarterly payment.

Solution: The 12 equal quarterly payments form an annuity whose present value is $2,000. Hence $A_n = 2,000$, $r = 0.02$, and $n = 12$. Substituting these values into

Formula (14), we get

$$2{,}000 = R\left[\frac{1-(1.02)^{-12}}{0.02}\right]$$

Solving for R and using $\dfrac{1-(1.02)^{-12}}{0.02} = 10.575341$ from Table VII, we obtain

$$R = \frac{2{,}000}{\dfrac{1-(1.02)^{-12}}{0.02}} = \frac{2{,}000}{10.57534} = \$189.12$$

At this point it is quite simple to isolate a formula for the periodic payment to amortize a debt of A_n due now at r per period in n periods if the first payment is due one period hence. It is

(15) $$R = A_n\left[\frac{r}{1-(1+r)^{-n}}\right]$$

Formula (15) can be memorized and applied directly if a hand calculator is available. It is particularly useful if you happen to be "in the market" for a home and desire to calculate the monthly payments on various homes financed for various numbers of years. It is interesting to note that, if a computation of a periodic payment yields 251.3215, then the lending institution will round this to the *next larger* penny, $251.33. The reason for this departure from the usual round-off process is that the effect obtained is a final payment which is smaller than $251.33. Had the periodic payment been rounded off to $251.32, the final payment would be larger than $251.32. Presumably for psychological reasons, the former is the accepted practice. One final word of caution: Because results of computations with money are always rounded off to cents, errors in computations naturally arise; they are called round-off errors. Discrepancies between computed values and actual values in financial computations often occur due to round-off errors, but for most computations the discrepancies are at most a few pennies.

For accounting and income tax purposes, each periodic payment is divided into two parts, a payment of interest on outstanding principal during the period and a reduction of the principal. Thus another problem of importance is that of finding the outstanding principal for any period. We illustrate a procedure for solving such a problem by the following example.

Example 2 What is the outstanding principal in Example 1 just after the eighth payment? How much of the ninth payment is applied to payment of interest and how much to reduction of principal?

Solution: The outstanding principal just after the eighth payment is the value on that date of the remaining four payments. This is simply the present value of the

SEC. 9.9 Amortization of a Debt

annuity formed by these four payments. Hence, using Formula (14),

$$A_n = 189.12 \left[\frac{1-(1.02)^{-4}}{0.02} \right] = 189.12(3.80773) = \$720.12$$

This means that immediately after the eighth payment the remainder of the debt could be paid by making an additional payment of $720.12. The interest on $720.12 for 1 period at 2% per period is $14.40. Thus $14.40 of the ninth payment is applied to the payment of interest for the ninth period. The balance, $189.12 - \$14.40 = \$174.72$, is applied to the reduction of principal.

Once again, the computation involved in Example 2 is subject to certain round-off errors. In practice, an *amortization schedule* is generated for such a loan, listing, in tabular form, the number of the payment, the portion of that payment which applies to interest, the portion of that payment which reduces the outstanding principal, and the balance due after that payment has been made. We illustrate the form of such schedules in Example 3.

Example 3   Construct an amortization schedule for repayment of the debt of Example 1, that is, repayment of $2,000 by 12 equal quarterly payments at 8% compounded quarterly.

Solution: The first step is always to compute the periodic payment. This has been done in Example 1, and the quarterly payment is $189.12. Hence the first quarterly payment is $189.12. Now, the principal on which interest is due at the end of the first quarter is $2,000.00. Hence $(2,000.00)(0.02) = \$40.00$ is the amount of the first payment which is interest. The remaining $189.12 - 40.00 = 149.12$ serves to reduce the outstanding principal. Hence the remaining balance after the first payment is $2,000 - 149.12 = \$1,850.88$. This is the principal on which interest must be paid during the second quarter. Hence the portion of the second payment which is interest is $(1,850.88)(0.02) = \$37.02$. This means $189.12 - 37.02 = \$152.10$ of the second payment will be used to reduce the outstanding balance. Hence the remaining balance after the second payment is $\$1,850.88 - 152.10 = \$1,698.78$. Continuing in this fashion, we obtain the amortization schedule exhibited in Figure 9.16. In each computation of interest we round off to the nearest penny. Ordinarily, amortization schedules are generated by computer. However, such computations can be readily performed by a person with a hand calculator if a relatively small number of payments are involved. With a programmable calculator, schedules can be generated for very large numbers of payments. For example, to amortize a home mortgage by monthly payments for 25 years involves 300 payments. It would be a rather tedious process to generate an amortization schedule for such a transaction without some sort of programmable device. Because of rounding up the computed monthly payment to the next higher cent and because of the large number of payments, the final payment is almost always smaller than the others. Finally, the total interest paid on such a transaction is rather staggering, and it must be displayed on the loan disclosure statement.

Amortization Schedule

Annual interest rate 0.08	Amount of loan $2,000.00	Number of payments 12	Periodic payment $189.12
Payment number	Interest	Principal	Remaining balance
1	40.00	149.12	1,850.88
2	37.02	152.10	1,698.78
3	33.98	155.14	1,543.64
4	30.87	158.25	1,385.39
5	27.71	161.41	1,223.98
6	24.48	164.64	1,059.34
7	21.19	167.93	891.41
8	17.83	171.29	720.12
9	14.40	174.72	545.40
10	10.91	178.21	367.19
11	7.34	181.78	185.41
12	3.71	185.41	0
Loan Totals	269.44	2,000.00	

Figure 9.16

Another type of problem that often arises is the following. A debtor can pay only a certain amount periodically. How many such payments will be required to pay off the debt? This problem is equivalent to the problem of finding n when A_n, R, and r are known. The problem in Example 3 of Section 9.8 is of this type. In general it will be found that a certain number of full payments plus a smaller payment will be needed. If such is the case, we shall assume that this final smaller payment is due one period after the last full payment.

Example 4 Suppose the debt in Example 1 above is to be amortized by quarterly payments of $100 each, the first due in three months. Find **a.** the number of full $100 payments, **b.** the outstanding principal just after the last full payment is made, and **c.** the concluding payment.

Solution: (a) $A_n = 2,000$, $R = 100$, and $r = 0.02$, and we are to find n; using the formula $A_n = R \left[\dfrac{1 - (1 + r)^{-n}}{r} \right]$. Substituting the given values, we have

$$2,000 = 100 \left[\frac{1 - (1.02)^{-n}}{0.02} \right]$$

SEC. 9.9 Amortization of a Debt 487

or
$$\frac{1-(1.02)^{-n}}{0.02} = \frac{2{,}000}{100} = 20$$

From the 2% column of Table VII, we find
$$\frac{1-(1.02)^{-25}}{0.02} = 19.523456$$

and
$$\frac{1-(1.02)^{-26}}{0.02} = 20.121036$$

It follows that 25 full payments of $100 each plus a final smaller payment will be needed. (This result can also be obtained by use of logarithms, to be discussed in Chapter 10.)

(b) The method used in Example 2 above for finding the outstanding principal does not apply here; the final, or 26th, payment is unknown as yet. However, the outstanding principal in this type of problem can be found in the following way. Find the accumulated principal, at the given interest rate, on the date the last full payment is due just as if no payments had been made. In the particular problem at hand, this is

$$S = 2{,}000(1.02)^{25} = 2{,}000(1.640606) = \$3{,}281.21$$

Next find the amount of an annuity of 25 payments of $100 each at 2% per period. We find

$$S_n = 100\left[\frac{(1.02)^{25} - 1}{0.02}\right] = 100(32.030300) = \$3{,}203.03$$

The outstanding principal just after the 25th payment is

$$S - S_n = \$3{,}281.21 - \$3{,}203.03 = \$78.18$$

Hence the concluding payment due one period later is $78.18 plus the interest on $78.18 for 1 period at 2%; that is, the concluding payment is

$$\$78.18 + \$78.18(0.02) = \$78.18 + \$1.56 = \$79.74$$

Exercises 9.9

1. A loan of $1,000, with interest at 6% payable semiannually, is to be amortized by equal payments at the end of every 6 months for 4 years. What is the semiannual payment if the first payment is due 6 months after the date of the loan?

2. In Exercise 1, what is the outstanding principal just after the fifth payment? How much of the sixth payment applies to interest and how much to reducing principal?

3. A $5,000 loan, bearing 8% interest payable quarterly, is to be amortized by quarterly payments of $200 each. Find the number of full $200 payments, the first payment being made 3 months after date of loan.

4. Suppose the debtor in Exercise 3 omits the 14th, 15th, 16th, and 17th payments. **a.** What should he pay on the day the 18th payment is due, including the 18th, to bring his payments up to schedule? **b.** What is the outstanding principal just after the 18th payment?

5. A loan of $13,000 is to be repaid, principal and interest, in equal annual installments at the end of every year for 12 years, the first due 1 year after loan, with interest at 7%. Find the amount remaining unpaid just after the eighth payment. How much of the ninth payment is applied to interest and how much to reduction of principal? Construct and fill in an amortization schedule for this transaction.

6. A certain loan is to be amortized by 20 quarterly payments of $200 each, the first due 4 years hence, and a payment of $1,000 due 9 years hence. What additional payment, along with the twelfth quarterly payment, would cancel the remainder of the debt? Interest is at 6% converted quarterly.

7. A small loan company lends $1,000 at 12% compounded monthly. It is to be repaid by 50 equal monthly payments, the first due 1 month after date of loan. **a.** Find the size of the monthly payment. **b.** What is the outstanding principal just after the 30th payment? **c.** How much of the 31st payment applies to interest and how much to reduction of principal? **d.** Construct and fill in an amortization schedule for this transaction.

8. In Exercise 7, how much of the original principal will have been repaid just after the 20th payment?

9. In Exercise 7, if the debtor omits the last 10 payments, how much would he owe on the day the 50th payment was originally due?

10. A debt of $2,000, bearing 8% interest compounded semiannually, is to be amortized by payments of $200 at the end of every six months, the first due now. Find the number of $200 payments and the concluding payment.

11. A $2,000 loan, bearing interest at 7% compounded quarterly, is to be amortized by equal quarterly payments for 5 years, the first due three months after date of loan. Find the quarterly payment. Construct and fill in an amortization schedule for this transaction.

12. To repay a $3,000 loan and 10% interest compounded semiannually, Jones will pay $200 every six months for 3 years and $400 every six months after that. Find the number of $400 payments and the concluding payment if the first $200 payment is made six months after the date of loan.

13.* A $20,000 debt is to be amortized by 20 equal annual payments, the first due now. If interest is at 7% compounded annually, what is the size of the annual payment? How much of the debt has been paid just after the tenth payment? Construct and fill in an amortization schedule for this transaction.

14.* Construct and fill in an amortization schedule for amortizing a home mortgage of $52,000, at $9\frac{1}{4}$% compounded monthly, by monthly payments for 20 years. (You will need a calculator for this problem and lots of time. A total of 240 payments are involved. Remember to round up to the next larger penny in computing the monthly payment.)

9.10 FINDING THE VALUE OF AN ANNUITY ON ANY CONVERSION DATE

We have already derived formulas for finding the value of an annuity on the date the last payment is due and on the date one period before the first payment is due. We wish now to show how we may use the two formulas to obtain the value of an annuity on any conversion date. Actually, either one of the two formulas would be sufficient for our purposes, but in certain cases one of the formulas may seem a bit easier to apply than the other. So, having both formulas at our disposal, we can use whichever we prefer.

Let us represent by a time-line diagram an annuity of n payments and indicate on it the two dates on which our known formulas are valid. In Figure 9.17 a crossmark indicates a payment date where a payment of R dollars is due, and a dot indicates the end of an interest period but where no payment is due.

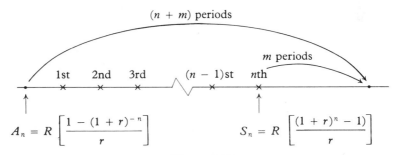

Figure 9.17

Suppose, for example, we wish to find the value V of an annuity of n payments on a date that is m periods after the date the nth payment is due. We have two choices as to how to proceed. First choice: Find the amount of the annuity, that is, the value of the annuity on the date the nth payment is due, using the formula $S_n = R\left[\dfrac{(1+r)^n - 1}{r}\right]$, and then accumulate S_n for m periods by multiplying it by $(1+r)^m$. Thus $V = S_n(1+r)^m$. Second choice: Find the value of A_n, using the formula $A_n = R\left[\dfrac{1-(1+r)^{-n}}{r}\right]$, and then accumulate A_n for $(n+m)$ periods by multiplying it by $(1+r)^{n+m}$. In this case we get $V = A_n(1+r)^{n+m}$. Due to round-off errors involved in the calculations, one is apt to find a discrepancy, usually 5 cents or less, in the two values of V obtained by the two choices.

Example 1 A contract calls for 10 annual payments of $200 each, the first due 1 year from now. If money is worth 7% compounded annually, what single payment 15 years hence could replace this contract?

First Solution: We first find the value of the contract on the date the 10th payment is due. Here $R = 200$, $n = 10$, and $r = 7\% = 0.07$. Hence, using Table VI,
$$S_n = 200\left[\frac{(1.07)^{10} - 1}{0.07}\right] = 200(13.81649) = \$2{,}763.29.$$
Next, using the formula $V = S_n(1 + r)^m$, with $S_n = 2{,}763.29$, $m = 5$, $r = 0.07$, and Table II, we obtain
$$V = 2{,}763.29(1.07)^5 = 2{,}763.29(1.402552) = \$3{,}875.66$$

Second Solution: We first find the present value of the contract using $A_n = R\left[\dfrac{1-(1+r)^{-n}}{r}\right]$, with $R = 200$, $n = 10$, and $r = 0.07$. We get
$$A_n = 200\left[\frac{1-(1.07)^{-10}}{0.07}\right] = 200(7.02358) = \$1{,}404.72$$

Next we use $V = A_n(1 + r)^{n+m}$, with $A_n = 1{,}404.72$, $n + m = 15$, and $r = 0.07$. This gives, using Table II,
$$V = 1{,}404.72(1.07)^{15} = 1{,}404.72(2.759032) = \$3{,}875.67$$

The discrepancy between the values of V found by the two different methods is only 1 cent. Either value is just as reliable as the other.

Suppose, next, that we wish to find the value V of an annuity of n payments on a date that is $(m + 1)$ periods before the date the first payment is due. This problem is very similar to the one just discussed above. Again we have two choices. We may first find the present value of the annuity A_n and discount it for m periods by multiplying A_n by $(1 + r)^{-m}$, that is, $V = A_n(1 + r)^{-m}$. Or we may first find S_n, the amount of the annuity, and discount it for $(m + n)$ periods by multiplying S_n by $(1 + r)^{-(m+n)}$, giving $V = S_n(1 + r)^{-(m+n)}$. See Figure 9.18.

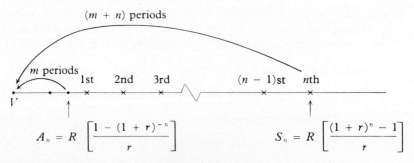

Figure 9.18

Example 2   A father has established a trust fund that will provide his daughter with 40 quarterly payments of \$1,000 each; the first payment is due on the daughter's 21st birthday. Assuming money to be worth 8% compounded quarterly, find the value of this fund on the daughter's 18th birthday.

SEC. 9.10 Finding the Value of an Annuity on Any Conversion Date

Solution: These 40 equal quarterly payments form an annuity. Let us first find the value of this annuity on the date period before the first payment is due; that is, let us find A_n. Here $R = 1{,}000$, $n = 40$, and $r = 2\% = 0.02$. Hence, using Table VII,

$$A_n = 1{,}000 \left[\frac{1 - (1.02)^{-40}}{0.02} \right] = 1{,}000(27.355479) = 27{,}355.48$$

To find the value V of this annuity on the daughter's 18th birthday, we use the formula $V = A_n(1 + r)^{-m}$, with $m = 11$ and $r = 0.02$. We obtain, using Table III,

$$V = 27{,}355.48(1.02)^{-11}$$
$$= 27{,}355.48(0.804263) = \$22{,}001.00$$

A problem, slightly different from those just discussed, that could arise in working with annuities is the following. Suppose we wish to find the value of an annuity of n equal payments on the date the kth payment is due, where k is less than n. Again we have two procedures, either of which is equally acceptable. Assume money is worth r.

1. Find $A_n = R \left[\dfrac{1 - (1 + r)^{-n}}{r} \right]$ and accumulate A_n for k periods by multiplying it by $(1 + r)^k$. If we let V denote the value of the annuity on the date the kth payment is due, then $V = A_n(1 + r)^k$.

2. Find $S_n = R \left[\dfrac{(1 + r)^n - 1}{r} \right]$ and discount S_n for $(n - k)$ periods by multiplying it by $(1 + r)^{-(n-k)}$, that is, $V = S_n(1 + r)^{-(n-k)}$.

Example 3 A contract calls for 20 semiannual payments of \$500 each, the first due 6 months hence. If money is worth 10% compounded semiannually, find the value V of this contract on the date the 12th payment is due, assuming no payments have been made.

First Solution: We find A_n with $R = 500$, $n = 20$, and $r = 5\% = 0.05$. We get, using Table VII,

$$A_n = 500 \left[\frac{1 - (1.05)^{-20}}{0.05} \right] = 500(12.46221) = \$6{,}231.11$$

Then, with $A_n = 6{,}231.11$, $m = 12$, and $r = 0.05$, we have, using Table II,

$$V = 6{,}231.11(1.05)^{12} = 6{,}231.11(1.795856) = \$11{,}190.18$$

Second Solution: $S_n = 500 \left[\dfrac{(1.05)^{20} - 1}{0.05} \right] = 500(33.06595) = \$16{,}532.98$

Then

$$V = S_n(1.05)^{-8} = 16{,}532.98(0.676839) = \$11{,}190.17$$

Again we have a 1 cent discrepancy in the value of V in the two methods. Hence, take your choice as to which method to use in a given problem.

Exercises 9.10

1. A contract calls for 15 annual payments at $100 each, the first due 4 years hence. If money is worth 7%, this annuity may be equitably replaced by what single payment due **a.** now, **b.** 3 years hence, **c.** 4 years hence, **d.** 10 years hence, **e.** 18 years hence, and **f.** 19 years hence?

2. Since 1980, John has been depositing $200 in a savings account on January 1 of each year. If he continues this practice and his savings earn 6% compounded annually, how much will he have to his credit December 31, 1990?

3. Mr. Smith owes $2,000 which he wishes to pay in 20 equal quarterly installments, the first due at once. If he is permitted to do this and it is agreed that money is worth 8% converted quarterly, what is the size of the quarterly payments?

4. A man wishes to provide for his son's college education by investing some money now in a trust fund that earns 6% converted semiannually. If the son is 14 years old now, how much should his father invest to provide him with 8 semiannual payments of $1,000 each, the first due when the son reaches 18 years of age, 4 years hence?

5. A house is leased for 5 years at $800 payable annually in advance. Find the cash equivalent of this lease if money is worth 6%.

6. A television set is bought at an installment price of $40 a month for 18 months, the first payment due at the time of purchase. If interest at 12% converted monthly is charged, what should be the cash price of this set?

7. A sum of $10,000 is deposited with a trust company that pays 7% interest converted quarterly. This is to provide a quarterly income of $500, the first due in 5 years. How long will the payments continue?

8. Instead of a cash settlement of $5,000, the beneficiary of a life insurance policy elects to receive 40 equal quarterly payments, the first due three months hence. If the company allows 4% interest, converted quarterly, what will each payment be?

9. A young couple wishes to accumulate $8,000 to use as a down payment, 5 years hence, on a house. How much should they deposit at the beginning of every 6 months for 5 years? Interest on the savings is 6% converted semiannually.

10. A contract calls for 20 quarterly payments of $200 each, the first due three months hence. If, instead of following this schedule, a single payment of $1,000 is made 1 year hence and another single payment of $2,000 is made 3 years hence, what final payment 5 years hence will close the transaction? Assume money worth 8% converted quarterly.

11. What equal quarterly payments, made each quarter for 10 years (40 payments), will pay for a $40,000 house if the first payment is made on the day of purchase and interest is at 8% converted quarterly?

12. An annuity of $200 every two months, the first payment due January 1, 1979, and the last payment due January 1, 1989, is equivalent at 6% converted bimonthly to what single payment due **a.** July 1, 1978, **b.** November 1, 1978, **c.** January 1, 1979, **d.** January 1, 1981, and **e.** January 1, 1989?

SEC. 9.10   Finding the Value of an Annuity on Any Conversion Date   493

13. A county owes $100,000 which it wishes to pay in five equal annual payments, the first due 1 year hence. Find the annual payment if the debt bears 6% interest converted annually.

14. Mr. Jones owes a debt due in 5 years, the maturity value of which is $2,000. He wishes to pay this debt in 10 equal semiannual installments, the first due in 6 months. If money is worth 8% converted semiannually, find the size of these semiannual payments.

15. Find the value of an annuity of 20 monthly payments of $250 at 18% compounded monthly:
    a. One month before the 1st payment is due
    b. Six months before the 1st payment is due
    c. Twelve months before the 1st payment is due
    d. On the date the last payment is due
    e. One year after the last payment is due
    f. Eighteen months after the last payment is due
    g. On the date the 7th payment is due
    h. On the date the 11th payment is due
    i. On the date the 17th payment is due

16. Answer each of the parts of Exercise 1 for an annuity of 360 monthly payments of $456.31 at $8\frac{3}{4}$% compounded monthly. (You will need a calculator.)

17. Find the value of an annuity of 37 quarterly payments of $321.56 at 12% compounded quarterly:
    a. Six months before the 1st payment is due
    b. One year before the 1st payment is due
    c. Nine months after the last payment is due
    d. Three years after the last payment is due
    e. On the date the 36th payment is due
    f. Twenty-seven months after the 1st payment is due

18. Answer each of the parts of Exercise 2 for an annuity of 44 quarterly payments at $1,398.27 at $7\frac{5}{8}$% compounded quarterly. (You will need a calculator.)

10

Exponential and Logarithmic Functions

Introduction

In Chapter 9 we learned that money invested at compound interest is a growing quantity. Although the percentage rate remains constant, the total value of the investment grows faster and faster. In fact, the total rate of growth, measured in dollars per year, during any interest period, is a fixed percent of the amount accumulated at the beginning of that period. Thus, for the same nominal rate of interest, the greater the frequency of conversion, the greater is the total rate of growth of the investment.

    For example, if money is invested at 6% converted annually, the total rate of growth of the investment (number of dollars per year) at any instant during the year is 6% of the value of the investment at the beginning of the year. If the interest is compounded semiannually, the total rate of growth of the investment at any instant will be 6% of the accumulation at the beginning of the half-year containing the instant in question. Similarly, if the conversion is monthly or daily or hourly or, in fact, of any frequency, the total rate of growth of the investment during any conversion period is 6% of the compound amount at the beginning of that conversion period. The total rate of growth in each of these cases may be considered the average total rate of growth over an interval, the interval being the conversion period in question. Hence, to obtain an instantaneous total rate of growth, we need to use a limiting process; that is, we need to find the limit approached by the average total rate of growth over a conversion period as the length of the conversion period is made to approach zero or, what is the same thing, as the frequency of conversion is increased indefinitely. The result of increasing the frequency indefinitely is called *compounding continuously*. We next consider the effect of compounding continuously.

10.1 COMPOUNDING CONTINUOUSLY AND EXPONENTIAL FUNCTIONS

In Section 9.5, we found that the compound amount of original principal of P at the end of n conversion periods at the interest rate of r per period is given by the formula $S = P(1 + r)^n$. If, instead of being the interest rate per period, r is the nominal rate and if k is the frequency of conversion, that is, the interest is compounded k times per year, then the interest rate per period is $\frac{r}{k}$. The formula for compound amount at the end of n periods then becomes $S = P\left(1 + \frac{r}{k}\right)^n$. Hence, at the end of 1 year, $n = k$ and the compound amount is

$$S = P\left(1 + \frac{r}{k}\right)^k$$

Let us first consider the special case in which $P = 1$ and $r = 100\% = 1$. Our previous equation then becomes

$$S = \left(1 + \frac{1}{k}\right)^k$$

We are now concerned with two questions regarding this formula. First, does

$$\lim_{k \to \infty} \left(1 + \frac{1}{k}\right)^k$$

exist? That is, does $\left(1 + \frac{1}{k}\right)^k$ approach some real number as k becomes large? Second, if so, what is this limit? We wish now to give some evidence in support of the contention that the limit does exist and also to obtain some notion of its value. Let us expand $\left(1 + \frac{1}{k}\right)^k$, using the method for expanding a binomial given in Chapter 2. We get

$$\left(1 + \frac{1}{k}\right)^k = 1^k + k(1^{k-1})\left(\frac{1}{k}\right) + \frac{k(k-1)}{2}(1^{k-2})\left(\frac{1}{k}\right)^2$$
$$+ \frac{k(k-1)(k-2)}{2 \cdot 3}(1^{k-3})\left(\frac{1}{k}\right)^3 + \cdots$$
$$+ \frac{k(k-1)(k-2)\cdots[k-(k-1)]}{2 \cdot 3 \cdot 4 \cdots (k-2)(k-1)k}\left(\frac{1}{k}\right)^k$$

whose right side we may simplify so that we have

$$\left(1+\frac{1}{k}\right)^k = 2 + \frac{1-\frac{1}{k}}{2} + \frac{\left(1-\frac{1}{k}\right)\left(1-\frac{2}{k}\right)}{2\cdot 3} + \cdots$$

$$+ \frac{\left(1-\frac{1}{k}\right)\left(1-\frac{2}{k}\right)\cdots\left[1-\frac{(k-1)}{k}\right]}{2\cdot 3\cdot 4 \cdots (k-2)(k-1)k}$$

We observe from this equation that each term of the expansion is positive and increases with k; hence $\left(1+\frac{1}{k}\right)^k$ increases as k increases. Furthermore, $\left(1+\frac{1}{k}\right)^k = 2$ when $k = 1$, and each of the factors $\left(1-\frac{1}{k}\right)$, $\left(1-\frac{2}{k}\right)$, ..., $\left[1-\frac{(k-1)}{k}\right]$ is positive and less than 1. Therefore, if we replace each factor by 1, we can write

$$\left(1+\frac{1}{k}\right)^k < 2 + \frac{1}{2} + \left(\frac{1}{2}\right)\left(\frac{1}{3}\right) + \left(\frac{1}{2}\right)\left(\frac{1}{3}\right)\left(\frac{1}{4}\right)$$

$$+ \cdots + \left[\left(\frac{1}{2}\right)\left(\frac{1}{3}\right)\left(\frac{1}{4}\right)\cdots\left(\frac{1}{k}\right)\right]$$

whose right side is, in turn, less than the following sum:

$$2 + \frac{1}{2} + \left(\frac{1}{2}\right)\left(\frac{1}{2}\right) + \left(\frac{1}{2}\right)\left(\frac{1}{2}\right)\left(\frac{1}{2}\right) + \cdots + \left[\left(\frac{1}{2}\right)\left(\frac{1}{2}\right)\left(\frac{1}{2}\right)\cdots\left(\frac{1}{2}\right)\right]$$

obtained by replacing each fractional factor of the inequality by $\frac{1}{2}$. This sum can be written

$$2 + \frac{1}{2} + \frac{1}{4} + \frac{1}{8} + \cdots + \frac{1}{2^{k-1}}$$

which can be shown to be less than 3, no matter how large k may become. Hence, for k any positive integer,

$$\left(1+\frac{1}{k}\right)^k < 3$$

We have now observed the following three facts:

$\left(1+\frac{1}{k}\right)^k = 2$ when $k = 1$

$\left(1+\frac{1}{k}\right)^k$ increases as k increases

SEC. 10.1 Compounding Continuously and Exponential Functions

$$\left(1 + \frac{1}{k}\right)^k < 3 \text{ for } k \text{ any positive integer}$$

Therefore, it seems reasonable to conclude that $\lim_{k \to \infty} \left(1 + \frac{1}{k}\right)^k$ does exist and that the value of the limit is a number greater than 2 and less than or equal to 3. It can be proved that this is indeed true. In fact, it can be shown that

(1) $$\lim_{k \to \infty} \left(1 + \frac{1}{k}\right)^k = e$$

where e is an irrational number, whose value correct to five decimal places is

$$e = 2.71828$$

and which, like the number π, plays an important role in mathematics. In our discussion of this limit, k was assumed to be a positive integer. This restriction is, however, not necessary. The same limit e exists if k is allowed to increase numerically through either positive or negative real values; in fact,

$$\lim_{|x| \to \infty} \left(1 + \frac{1}{x}\right)^x = e$$

Returning to Equation (1) above, this means that \$1 invested at 100% interest compounded continuously will amount, at the end of 1 year, to $e = \$2.72$, rounded off to the nearest cent.

We wish now to find the amount at the end of t years of any original principal P invested at the nominal interest rate r compounded continuously. It is easily shown that the compound amount in t years of P dollars invested at the nominal rate r compounded k times per year is

(2) $$S = P\left(1 + \frac{r}{k}\right)^{kt}$$

and to find the amount when the interest is compounded continuously, we need to find the limit

$$\lim_{k \to \infty} P\left(1 + \frac{r}{k}\right)^{kt}$$

To do this, let $k = rx$; then $\frac{r}{k} = \frac{1}{x}$ and $kt = rtx$. Therefore,

$$P\left(1 + \frac{r}{k}\right)^{kt} = P\left(1 + \frac{1}{x}\right)^{rtx} = P\left[\left(1 + \frac{1}{x}\right)^x\right]^{rt}$$

Since r is to remain fixed, x must become infinite as k becomes infinite, and thus we have

$$\lim_{k\to\infty} P\left(1+\frac{r}{k}\right)^{kt} = \lim_{x\to\infty} P\left[\left(1+\frac{1}{x}\right)^{x}\right]^{rt}$$
$$= P\left[\lim_{x\to\infty}\left(1+\frac{1}{x}\right)^{x}\right]^{rt} = P(e)^{rt}$$

That is,

(3) $$S = Pe^{rt}$$

It is important to note that Equation (2) is strictly correct only at the end of any conversion period, that is, only for rational values of t such that the product kt is a positive integer. On the other hand, Equation (3) is strictly correct for all positive real values of t because in this case the interest is being compounded continuously, that is, at each instant, which means that the compound amount of the investment is also growing continuously and that its value S, at any instant t, is given by the formula

$$S = Pe^{rt}$$

Let us now take a closer look at Equation (2), that is, at the equation

$$S = P\left(1+\frac{r}{k}\right)^{kt}$$

keeping in mind the meaning of the letters involved. For each permissible value of k, S is a function of t. Consider a fixed value of t and let ΔS be the increase in S over the interval from t to $t+\Delta t$, where $\Delta t = \frac{1}{k}$. Then

$$\Delta S = P\left(1+\frac{r}{k}\right)^{k(t+1/k)} - P\left(1+\frac{r}{k}\right)^{kt}$$
$$= P\left(1+\frac{r}{k}\right)^{kt}\left(1+\frac{r}{k}-1\right) = \frac{rP}{k}\left(1+\frac{r}{k}\right)^{kt}$$

Since $\Delta t = \frac{1}{k}$, we can divide the left side of the last equation by Δt and the right side by $\frac{1}{k}$, and obtain

(4) $$\frac{\Delta S}{\Delta t} = \frac{\frac{rP}{k}\left(1+\frac{r}{k}\right)^{kt}}{\frac{1}{k}} = rP\left(1+\frac{r}{k}\right)^{kt}$$

Equation (4) states that the *average rate of growth over any conversion period*, of an investment accumulating at a nominal interest rate r compounded k times

SEC. 10.1   Compounding Continuously and Exponential Functions   499

per year, is equal to r times the compound amount of that investment at the beginning of the interest period involved. Note also that this equation is valid for all permissible values of k which include values as large as we please. As k increases, Δt decreases; that is, the length of the interest period approaches zero, and the average rate, expressed by Equation (4), approaches a limiting value. In fact, it follows from Equation (4) and the discussion leading to Equation (3) that

$$(5) \qquad \lim_{\Delta t \to 0} \frac{\Delta S}{\Delta t} = \lim_{k \to \infty} \left[rP\left(1 + \frac{r}{k}\right)^{kt} \right] = rPe^{rt}$$

which, in the language of Chapters 5 and 6, is an instantaneous rate. Furthermore, as k increases the compound amount $P\left(1 + \frac{r}{k}\right)^{kt}$ at the beginning of an interest period of an investment whose interest is compounded k times per year approaches the compound amount Pe^{rt} at that instant of that same investment but whose interest is compounded continuously. That is,

$$(6) \qquad \lim_{k \to \infty} \left[P\left(1 + \frac{r}{k}\right)^{kt} \right] = Pe^{rt}$$

After comparing Equations (4), (5), and (6), it seems reasonable to assume that the instantaneous rate of change of the compound amount Pe^{rt} of an investment with interest compounded continuously should be the limit of the average rate of that investment, with interest compounded k times per year as the interval over which the average rate is calculated is made to approach zero, that is, as $k \to \infty$. In symbols, this assumption is that if

$$S = Pe^{rt}$$

then

$$D_t S = D_t(Pe^{rt}) = \lim_{k \to \infty} \left[rP\left(1 + \frac{r}{k}\right)^{kt} \right] = rPe^{rt}$$

or

$$(7) \qquad D_t S = rPe^{rt} = rS$$

This assumption can be shown to be correct.

Examples. If $S = 100e^{0.04t}$, then

$$D_t S = 0.04 S = 4e^{0.04t}$$

If $y = 5e^{3x}$, then

$$D_x y = 3y = 15e^{3x}$$

It follows from Equation (7) and the definition of an antiderivative that if

$$D_t S = rS$$

then

$$S = ce^{rt}$$

where c is the constant of integration. Note that when $t = 0$, then $S = ce^0 = c$; therefore, c is the value of S at $t = 0$. Thus, if any quantity Q increases with time in such a way that its instantaneous rate of change per year (or per hour or per any convenient unit of time) is a constant percentage of the value of Q at that instant, that is, if

(8) $$D_t Q = rQ, \quad r > 0$$

then Q obeys a law of the form

(9) $$Q = Q_0 e^{rt}$$

where Q_0 is the value of Q when $t = 0$, r is the constant percentage rate, and t is the time measured in some suitable unit. If, in Equation (8), r is negative, then Q decreases at a constant percentage rate.

One would seldom expect a bank or a building and loan association to convert one's interest continuously, but many quantities in nature do grow in just such a manner. Some examples are the number of bacteria in a culture, the number of cells in the human body, the weight of a growing organism, or even the total human population, which actually increases almost continuously at a fairly steady percentage rate. In fact, the quantities in nature that grow in such a way are so numerous that Equation (9) is often referred to as the *law of growth*. It is also sometimes called the *compound interest law* or, abbreviated, the CIL.

Example 1   The number of bacteria N in a culture increased at an instantaneous rate per hour, constantly equal to 30% of N. If at time $t = 0$, $N = 500$, find a formula which expresses N as a function of t and use this formula to find N when $t = 5$.

Solution: Here we use the law of growth with $r = 30\% = 0.30$ and obtain

$$D_t N = 0.30 N$$

Integrating, $$N = ce^{0.3t}$$

We are given that $N = 500$ when $t = 0$; hence

$$500 = ce^0 = c$$

and therefore the desired formula is

$$N = 500 e^{0.3t}$$

when $t = 5$, $$N = 500 e^{1.5}$$

Now, by either using a hand calculator with the function e^x (or "inverse ln") or by using Table IV, we obtain

$$N = 500(4.4817)$$
$$= 2,241 \text{ (approx.)}$$

SEC. 10.1 Compounding Continuously and Exponential Functions

All radioactive substances disintegrate at a rate that is constantly equal to some percentage of the amount of the substance remaining; thus all such substances follow the law of growth, but with r negative.

Example 2 Radium decomposes at a time rate per 100 years which at any instant is about 4% of the amount remaining. Write a formula for the amount of the original 1,000 milligrams remaining after t centuries. Also, find the number of milligrams remaining after 1,000 years.

Solution: Here we again use the law of growth with $r = -4\% = -0.04$; hence

$$D_t Q = -0.04 Q$$

Integrating,

$$Q = ce^{-0.04t}$$

We are given that when $t = 0$, $Q = 1{,}000$ milligrams; hence

$$1{,}000 = ce^0 = c$$

The required formula is

$$Q = 1{,}000 e^{-0.04t}$$

After 1,000 years, $t = 10$, and when $t = 10$,

$$Q = 1{,}000 e^{-0.4} = 1{,}000(0.67032)$$
$$= 670 \text{ (approx.)}$$

Exponential Functions

In previous chapters we have discussed functions that could be expressed as a sum of terms of the type kx^n. Any term of this type is called a *power function*, which is characterized by the property that a variable base x is raised to a constant power n. The law of growth introduces a new kind of function, of the form e^x, called an *exponential function*. Here the base is a constant and the exponent is a variable. In general, an exponential function is a constant base, not necessarily e, raised to a variable power. Functions like e^x, 2^x, 3^{x^2}, a^t, 10^{2t}, etc., are exponential functions. For the present we shall restrict our discussion of exponential functions to the case in which the constant base is e. To solve problems involving exponential functions, we need to be able to evaluate these functions. Such evaluations are readily performed by using a hand calculator which has the function key y^x. In case the base is e, many calculators have the special key e^x (or "inverse ln" of x), and in case the base is 10, many calculators have the special key 10^x (or "inverse log" of x). Exponential functions play such an important role in mathematics that the values of e^x and e^{-x} have been computed for many values of x and made available in the form of a table, as was done for accumulation factors, discount factors, amount of annuity of $1, and present value of $1. Table IV, Appendix B, gives the approximate values of e^x

and e^{-x} for values of x from $x = 0$ to $x = 10$ for each successive hundredth. By interpolation this can be extended to each successive thousandth.

Example 3 Use Table IV to evaluate $e^{1.74}$ and $e^{-1.74}$.

Solution: Here $x = 1.74$. We find 1.74 in the x column and read the value of $e^{1.74}$ in the e^x column, and we find $e^{1.74} = 5.6973$. Similarly, we find from the e^{-x} column that $e^{-1.74} = 0.17552$.

Of course, we could use a hand calculator to evaluate these quantities directly, without the use of tables.

Table IV suggests the fact that there exists a value of e^x for each real value of x. In order to approximate the value of e^x for values of x which are not in Table IV (without the use of a hand calculator), we must use either straight-line interpolation alone or a combination of interpolation and the laws of exponents listed below. As always, we shall illustrate how this is done by examples.

The Laws of Exponents

LAW 1. $e^{x_1} e^{x_2} = e^{x_1 + x_2}$

LAW 2. $\dfrac{e^{x_1}}{e^{x_2}} = e^{x_1 - x_2}$

LAW 3. $(e^x)^n = e^{nx}$

LAW 4. $e^{-x} = \dfrac{1}{e^x}$

LAW 5. $e^0 = 1$

Since e itself is a positive number greater than 1, the number e^x is greater than zero for all real values of x, whether positive, negative, or zero.

Example 4 Use Table IV to approximate $e^{1.159}$.

Solution: Using interpolation, we arrange our work as follows:

$$0.010 \begin{cases} 0.009 \begin{cases} x & e^x \\ 1.15 & 3.1582 \\ 1.159 & ? \end{cases} d \\ 1.16 & 3.1899 \end{cases} 0.0317$$

SEC. 10.1 Compounding Continuously and Exponential Functions

$$\frac{d}{0.0317} = \frac{0.009}{0.010} = 0.9$$

$$d = (0.9)(0.0317) = 0.02856$$

Hence $e^{1.159}$ is approximately equal to

$$3.1582 + 0.02856 = 3.18676$$

Example 5   Use Table IV to approximate $e^{11.25}$.

Solution: Since 11.25 is not in Table IV, we use Law 1 to write

$$\begin{aligned}
e^{11.25} &= e^{10+1.25} \\
&= (e^{10})(e^{1.25}) \\
&= (22{,}026)(3.4903) \\
&= 76{,}877.35 \quad \text{(rounded to two decimal places)}
\end{aligned}$$

Example 6   Use Table IV to approximate e^{-12}.

Solution: Using Law 1, we write

$$\begin{aligned}
e^{-12} &= e^{-10-2} \\
&= (e^{-10})(e^{-2}) \\
&= (0.0000454)(0.13534) \\
&= 0.000006144 \quad \text{(rounded to four significant figures)}
\end{aligned}$$

Exercises 10.1

1. From Table IV, find the approximate value of each of the following, using straight-line interpolation, where necessary. Check your answers by using a hand calculator to approximate the value of the number in each case.
   a. $e^{0.07}$ b. $e^{-0.05}$ c. $e^{7.82}$ d. $e^{9.95}$ e. $e^{-9.85}$
   f. $e^{2.345}$ g. $e^{-1.634}$

2. Assuming that the number of bacteria in a certain culture grows at a rate per hour that is constantly equal to 20% of the number then present, find a formula for the number present after t hours if $N = 1{,}000$ when $t = 0$. Also find the number present when $t = 10$ and when $t = 20$.

3. a. Find a formula for the amount after t years of $2,000 with interest at 6% converted continuously. How much is this after 10 years?

   b. What would the amount be after 10 years if the interest is at 6% converted semiannually?

4. Radium decomposes at a rate per century that is constantly equal to 3.8% of the quantity remaining at that time. How much will be left after 6,000 years from an original 400 milligrams?

5. A certain house cost $20,000 to build. If it depreciates at an instantaneous rate per year that is constantly equal to 5% of the value V of the house at that instant, what is the value of the house at the end of 10 years?

6. The population of a certain city in 1930 was 100,000. Assuming that its total rate of growth per year is at each instant 5% of the population at that time, find the expected population in 1960; in 1970; in 1980; in 1990; in 2000.

7. Given $D_t S = 0.03S$ and $S = 100$ when $t = 0$, write a formula for S in terms of t. What is S when $t = 75$?

8. An electric current dies out according to the law $D_t i = -100i$. Find i in terms of t if $i = 20$ when $t = 0$.

9. Given $D_x y = 2y$ and $y = 40$ when $x = 0$, express y as a function of x. Find the value of y when $x = 3$; when $x = -\frac{1}{2}$.

10.2 NATURAL LOGARITHMS DEFINED

We now wish to define the inverse function of e^x, which we shall call the natural logarithm of x and denote by $\ln x$.

The *natural logarithm* of x for all positive real values of x is defined as that number y such that $x = e^y$; that is,

$$\ln x = y \quad \text{if and only if} \quad x = e^y$$

These last two equations shall be taken as equivalent statements. According to this definition, the natural logarithm of x is an exponent, the power to which the number e must be raised to produce x. For this reason e is said to be the base of natural logarithms. Later we shall define logarithms to other bases.

It follows from the definition of $\ln x$ that

(1) $\quad\quad\quad\quad\quad\quad\quad \ln 1 = 0 \quad\quad \text{because } e^0 = 1$

(2) $\quad\quad\quad\quad\quad\quad\quad \ln e = 1 \quad\quad \text{because } e^1 = e$

or in general for all real numbers a,

(3) $\quad\quad\quad\quad\quad\quad\quad \ln e^a = a \quad\quad \text{because } e^a = e^a$

A hand calculator with the function key ln can be used to obtain excellent approximations to $\ln x$ for all positive real numbers x. Notice that, since e^y is always positive, $\ln x$ is only defined for positive real numbers x. Moreover, in case one does not have access to an appropriate hand calculator, and since the function $\ln x$ is of such importance, its values have been tabulated. Table V, Appendix B, gives the natural logarithms of real numbers between 1 and 9.99 in steps of 0.01, correct to four decimal places. With the use of Table V and interpolation, one can approximate $\ln N$ if N is known and use four decimal places in the approximation. For this reason the table is called a four-place table.

SEC. 10.2 Natural Logarithms Defined

The Use of a Natural-Logarithm Table

If a number N is one of the numbers from 1.00 to 9.99, its natural logarithm can be read directly from Table V. Of course, a hand calculator with the function key $\ln x$ can be used to obtain the desired value also. In either case, the values obtained are approximations to, and not the exact values of, the number involved.

Example 1 Find an approximation to $\ln 3.76$.

Solution: In Table V look for 3.7 in the column headed N; then move across the table horizontally to the column headed 0.06. The number there is 1.3244. Thus $\ln 3.76 \cong 1.3244$.

As usual, we could use a hand calculator to find this approximation without the use of tables. In fact, a higher degree of accuracy can be obtained, for instance, $\ln 3.76 \cong 1.32441896$.

Example 2 Find an approximation to $\ln 6.438$.

Solution: Here, if we desire to use Table V, we must interpolate. We may arrange our work as follows.

$$0.010 \left\{ 0.008 \left\{ \begin{array}{cc} N & \ln N \\ 6.430 & 1.8610 \\ 6.438 & ? \\ 6.440 & 1.8625 \end{array} \right\} d \right\} 0.0015$$

$$\frac{d}{0.0015} = \frac{0.008}{0.010} = 0.8$$

$$d = 0.8(0.0015) = 0.00120$$

Hence $\ln 6.438 \cong 1.8610 + 0.0012 = 1.8622$.

Clearly, using a hand calculator with the function key $\ln x$ makes this problem much easier. We simply key in the number 6.438 and then hit the ln key. With the use of hand calculators, in dealing with exponents, logarithms, and trigonometric functions (Chapter 11), interpolation will likely become a thing of the past. There remain many functions, however, for which interpolation is important, and so we continue to emphasize the ability to interpolate between values given in a table.

It is also important to be able to find N if $\ln N$ is known. To do this, we can either reverse the procedure used above by finding (an approximation to) $\ln N$ in Table V and reading off N, or we can use the fact that if $\ln N = y$, then $N = e^y$. Knowing this fact, we simply find N by evaluating $e^{\ln N}$ (using a hand calculator or Table IV), where $\ln N$ is known. We shall illustrate with examples.

Example 3 Given $\ln N = 1.4493$, find N.

Solution: The simplest way to find (that is, approximate) N is to evaluate $e^{\ln N} = e^{1.4493}$ by using a hand calculator. We find $N = e^{\ln N} = e^{1.4493} \cong 4.26013$. If we desire to use Table V, we look for 1.4493 in the body of the table. We now move across horizontally and to the left to the entry in the column headed N. The value found there is 4.26; that is, $N = 4.26$. The number 4.26 is called the *antilogarithm* of 1.4493. In other words, 4.26 is (an approximation to) the number whose natural logarithm is 1.4493. So, to find N when its logarithm is given is to find the antilogarithm of the given logarithm.

Example 4 Given $\ln N = 1.6872$, find N.

Solution: Once again, with a hand calculator we simply find $N = e^{\ln N} = e^{1.6872} \cong 5.40433$ (rounded to five decimal places). If we desire to use Table V, since 1.6872 does not appear in the table, interpolation is required. We look for the two numbers in the table nearest to 1.6872 and between which our number lies. Again we arrange our work in tabular form:

$$
\begin{array}{cc}
N & \ln N \\
\end{array}
$$

$$
0.010 \left\{ d \left\{ \begin{array}{l} 5.400 \\ ? \\ 5.410 \end{array} \right. \quad \left. \begin{array}{l} 1.6864 \\ 1.6872 \\ 1.6883 \end{array} \right\} 0.0008 \right\} 0.0019
$$

$$\frac{d}{0.010} = \frac{0.0008}{0.0019} = \frac{8}{19}$$

$$d = \frac{8}{19}(0.010) = 0.004 \quad \text{(to three decimal places)}$$

Therefore, our approximation by using Table V and interpolation is $N = 5.400 + 0.004 = 5.404$ (rounded to three decimal places).

Example 5 Use natural logarithms to solve the equation $200e^{2x} = 988$ for x.

Solution: We first solve the equation for the exponential term and obtain

$$e^{2x} = \frac{988}{200} = 4.94$$

If two numbers are equal, their natural logarithms are equal; hence

$$\ln e^{2x} = \ln 4.94$$

It follows from Equation (3) that $\ln e^{2x} = 2x$. Therefore,

$$2x = \ln 4.94$$

Using either a hand calculator or Table V, we get

$$2x \cong 1.5974$$

$$x \cong 0.7987$$

SEC. 10.3 The Laws of Natural Logarithms

Exercises 10.2

1. Approximate the natural logarithms of the following:
   a. 3.06   **b.** 5.72   **c.** 7.88   **d.** 6.734   **e.** 4.006   **f.** 8.296

2. Approximate the numbers whose natural logarithms are:
   a. 1.1725   **b.** 1.5872   **c.** 2.1883   **d.** 1.5357   **e.** 2.1662   **f.** 1.8554

3. By using natural logarithms, approximate the value of x for each of the following:
   a. $e^x = 4.78$ **b.** $e^{6x} = 7.84$ **c.** $e^{0.05x} = 5.32$
   d. $e^{-x} = 3.75$ (*Hint:* $-x = \ln 3.75$)   **e.** $e^{-0.06x} = 2.47$

4. Approximate x in each of the following, after first solving for the exponential function:
   a. $536 = 100e^{10x}$ **b.** $476 = 200e^{0.6x}$ **c.** $0.564 = 0.08e^{0.04x}$
   d. $748 = 125e^{-2x}$

10.3 THE LAWS OF NATURAL LOGARITHMS

In working with natural logarithms it is convenient to have available the following laws, which are a direct consequence of the definition of a natural logarithm.

LAW 1. The natural logarithm of the product of two or more factors is equal to the sum of the natural logarithms of the factors; that is,

$$\ln(MN \cdots Q) = \ln M + \ln N + \cdots + \ln Q$$

Proof: Let $m = \ln M$, $n = \ln N$, ..., and $q = \ln Q$; in exponential form, these are $M = e^m$, $N = e^n$, ..., and $Q = e^q$. Multiplication gives $MN \cdots Q = e^m \cdot e^n \cdots e^q = e^{m+n+\cdots+q}$. By the definition of a natural logarithm,

$$\ln(MN \cdots Q) = m + n + \cdots + q = \ln M + \ln N + \cdots + \ln Q$$

LAW 2. The natural logarithm of a fraction is equal to the natural logarithm of the numerator minus the natural logarithm of the denominator; that is,

$$\ln \frac{M}{N} = \ln M - \ln N$$

Proof: Let $x = \ln M$ and $y = \ln N$; in exponential form this is written $M = e^x$ and $N = e^y$. Division gives $\dfrac{M}{N} = \dfrac{e^x}{e^y} = e^{x-y}$. By definition,

$$\ln \frac{M}{N} = x - y = \ln M - \ln N$$

LAW 3. The natural logarithm of the nth power of a number is equal to n times the natural logarithm of the number; that is,

$$\ln M^n = n \ln M$$

Proof: Let $x = \ln M$; then $M = e^x$. Raise both sides to the nth power and get $M^n = (e^x)^n = e^{nx}$. By definition,

$$\ln M^n = nx = n \ln M$$

By using Laws 1 and 3, we are able to find the natural logarithm of any four-digit number, regardless of whether it falls within our table or not. If a hand calculator is not available, we can use Laws 1 and 3 to approximate the natural logarithm of any positive real number, regardless of whether it falls within our table or not. To do this, let us first note that any number can be expressed as the product of a number, lying on the interval 1 to 10, and the number 10 raised to some integral power. This integral power of 10 will be positive, zero, or negative, depending on whether the given number is greater than 10, lies on the interval from 1 to 10, or is less than 1. For example,

$$475 = (4.75)(10^2) \qquad 0.8974 = (8.974)(10^{-1})$$

$$0.0065 = (6.5)(10^{-3}) \qquad 350{,}000 = (3.5)(10^5)$$

Expressing a number in this form is called expressing the number in *scientific notation*.

Example 1 Use Table V to approximate $\ln 284$.

Solution: In scientific notation, $284 = (2.84)(10^2)$,

Hence $\qquad\qquad\qquad \ln 284 = \ln (2.84)(10^2)$
By Law 1, $\qquad\qquad\qquad\quad = \ln 2.84 + \ln 10^2$
By Law 3, $\qquad\qquad\qquad\quad = \ln 2.84 + 2 \ln 10$
and, from Table V, $\qquad\quad \cong 1.0438 + 4.6052$
and, adding, $\qquad\qquad\qquad = 5.6490$

Example 2 Use Table V to approximate $\ln 0.763$.

Solution: Proceeding as in the previous example, we may write

$$\ln 0.763 = \ln (7.63)(10^{-1})$$
$$= \ln 7.63 + \ln (10^{-1})$$
$$= \ln 7.63 - \ln 10$$
$$\cong 2.0321 - 2.3026$$
$$= -0.2705$$

The next two examples illustrate how one might approximate N if $\ln N$ is known, is not in Table V, and we have access only to Table V (no hand

SEC. 10.3 The Laws of Natural Logarithms

calculator and no Table of Exponents). If we did have access to a table of exponents, we would simply evaluate $e^{\ln N}$ to find N. The primary purpose of these two examples is to illustrate how one can use properties of logarithms (or exponents) to solve certain problems in different ways. Similar examples could be given where we are given e^x and are asked to solve for x.

Example 3 Given $\ln N = 9.0780$, approximate N.

Solution: The number 9.0780 is outside the range of Table V, since the largest logarithm occurring there is 2.3016. In such a case we subtract from $\ln N$ the smallest multiple of $\ln 10$ that will give a number in the table. In this case, $3 \ln 10 = 6.9078$ subtracted from 9.0780 does give a number in our table, while $2 \ln 10 = 4.6052$ is too small. We proceed as follows:

$$\ln N - 3 \ln 10 = 9.0780 - 6.9078 = 2.1702$$

$$\ln N - \ln 1{,}000 = 2.1702$$

$$\ln \frac{N}{1{,}000} = 2.1702$$

From Table V we find the antilogarithm of 2.1702 to be 8.76, and therefore

$$\frac{N}{1{,}000} \cong 8.76 \quad \text{or} \quad N \cong 8{,}760$$

Example 4 Given $\ln N = -2.3183$, approximate N.

Solution: The number -2.3183, being negative, is not in our table of logarithms, so to obtain a number that falls within the table we add to $\ln N$ the smallest multiple of $\ln 10$ necessary to produce a positive number as sum. In the case at hand, $2 \ln 10 = \ln 100 = 4.6052$ will be sufficient. We proceed as follows:

$$\ln N + \ln 100 = -2.3183 + 4.6052$$

$$\ln 100N = 2.2869$$

To find the antilogarithm of 2.2869, we use Table V and interpolation. For convenience, let $x = 100N$:

$$0.01 \left\{ d \left\{ \begin{array}{l} 9.84 \\ ? \\ 9.85 \end{array} \right. \begin{array}{l} \ln x \\ 2.2865 \\ 2.2869 \\ 2.2875 \end{array} \left. \begin{array}{l} 0.0004 \end{array} \right\} 0.0010 \right.$$

$$\frac{d}{0.01} = \frac{0.0004}{0.0010} = \frac{4}{10}$$

$$d = 0.004$$

$$x \cong 9.84 + 0.004 = 9.844$$

Therefore, $100N \cong 9.844$ and $N \cong 0.09844$.

Example 5 Use natural logarithms to approximate $(3.57)^{4.36}$.

Solution: If we have access to a hand calculator with the function key y^x, we simply compute an approximation to the desired number directly:

$$(3.57)^{4.36} = 256.8227 \quad (rounded\ to\ four\ decimal\ places)$$

If we desire to make the approximation by using tables, for convenience let $N = (3.57)^{4.36}$. If two numbers are equal, their natural logarithms are equal; hence

$$\ln N = \ln(3.57)^{4.36} = 4.36 \ln 3.57$$

In Table V we find $\ln 3.57 = 1.2726$; therefore,

$$\ln N = 4.36(1.2726) = 5.5485$$

Since 5.5485 is not in Table V, we subtract $\ln 100$ from each side of the last equation; thus

$$\ln N - \ln 100 = 5.5485 - \ln 100$$

Simplifying, we get

$$\ln \frac{N}{100} = 5.5485 - 4.6052 = 0.9433$$

Using Table V and interpolation, we obtain

$$\frac{N}{100} = 2.568$$

Hence $\qquad N = 256.8 \quad (approx.)$

The use of logarithms is particularly helpful in solving problems in one unknown when that unknown occurs as an exponent. We illustrate this in the following examples.

Example 6 Approximately how long would it take an investment of $800 to amount to $2,000 if the investment earns 8% interest compounded semiannually?

Solution: Here we use the formula for compound amount derived in Chapter 9: $S = P(1 + r)^n$, with $S = 2,000$, $P = 800$, and $r = 0.04$. Substituting these values into our formula, we obtain the equation

$$2{,}000 = 800(1.04)^n$$

Solving for the exponential term, we get

$$(1.04)^n = \frac{2{,}000}{800} = 2.5$$

Equating the natural logarithms of the two sides gives

$$n \ln 1.04 = \ln 2.5$$

SEC. 10.3   The Laws of Natural Logarithms

Hence
$$n = \frac{\ln 2.5}{\ln 1.04} = \frac{0.9163}{0.0392} = 23.375$$

The logarithms of 2.5 and 1.04 were read from Table V. (As always, we could use a hand calculator.) Thus it would take slightly more than 23 six-months periods, or $11\frac{1}{2}$ years, for the original investment of $800 to amount to $2,000 at the given rate of interest. We must keep in mind that the formula for compound amount is only valid for integral values of n. We therefore interpret our result, $n = 23.375$, as indicating that the compound amount at the end of 23 periods would be less than $2,000 and at the end of 24 periods would be more than $2,000. This is easily verified as follows

$$800(1.04)^{23} = 800(2.46472) = \$1{,}971.78$$

$$800(1.04)^{24} = 800(2.56330) = \$2{,}050.64$$

Example 7   A debt of $3,720, bearing 7% interest compounded semiannually, is to be amortized by payments of $300 at the end of each 6 months, the first payment due 6 months hence. Find the number of full $300 payments.

Solution: In Chapter 9 we derived a formula for the present value of an annuity of n equal payments, $A_n = R\left[\dfrac{1 - (1+r)^{-n}}{r}\right]$. In our problem, $3,720 is the present value of an annuity of n payments of $300 each with interest at $r = 3\frac{1}{2}\% = 0.035$ per period. Substituting these values into the present value formula, we get

$$3{,}720 = 300\left[\frac{1 - (1.035)^{-n}}{0.035}\right]$$

which when simplified can be written

$$\frac{3{,}720(0.035)}{300} = 1 - (1.035)^{-n}$$

or
$$0.434 = 1 - (1.035)^{-n}$$

Solving for the exponential term, we obtain

$$(1.035)^{-n} = 1 - 0.434 = 0.566$$

Equating the natural logarithms of the two sides gives

$$-n \ln 1.035 = \ln 0.566$$

or
$$-n = \frac{\ln 0.566}{\ln 1.035} = \frac{-0.5692}{0.0344}$$

Finally,
$$n = \frac{0.5692}{0.0344} \cong 16.5465$$

Hence there should be 16 full $300 payments and a single smaller payment.

For convenience, both the laws of exponents and the laws of natural logarithms are repeated here in tabular form.

Laws of Exponents and Natural Logarithms

Exponents

LAW 1. $e^x \cdot e^y = e^{x+y}$ LAW 2. $\dfrac{e^x}{e^y} = e^{x-y}$

LAW 3. $(e^x)^y = e^{xy}$ LAW 4. $e^{-x} = \dfrac{1}{e^x}$

LAW 5. $e^0 = 1$

Natural logarithms

LAW 1. $\ln xy = \ln x + \ln y$ LAW 2. $\ln \dfrac{x}{y} = \ln x - \ln y$

LAW 3. $\ln x^n = n \cdot \ln x$

Exercises 10.3

1. Approximate the natural logarithm of each of the following numbers: **a.** 532, **b.** 6,740, **c.** 0.485, **d.** 0.0879, **e.** 47,500, **f.** 0.003864, **g.** 962.7.

2. Approximate the numbers whose natural logarithms are the following: **a.** 5.9296, **b.** 9.7757, **c.** 6.8748, **d.** -0.2446, **e.** -2.3820, **f.** -1, **g.** 5.

3. Approximate the value of x given $x = 476e^{1.3246}$. (*Hint:* $\ln x = \ln 476 + 1.3246$.)

4. Approximate the value of x given $x = 850e^{-1.4652}$.

5. Approximate x, given $682 = 394e^{10x}$. (*Hint:* Show that $10x = \ln 6.82 - \ln 3.94$.)

6. The number of bacteria in a certain culture increases at an instantaneous rate per hour that is equal at any instant to 15% of the number in the culture at that instant. Starting with an original number N_0, in how many hours will the number be doubled?

7. Use natural logarithms to approximate each of the following.

   a. $\dfrac{(47.3)(86.5)}{237}$ **b.** $(2.78)^{4.2}$ **c.** $\sqrt[5]{534}$

8. Use natural logarithms to approximate x, given $(2.5)^x = 48$.

9. At 8% interest compounded semiannually, approximately how long will it take to pay a debt of $9,000, due now, in semiannual installments of $375, the first due six months hence? Use natural logarithms.

10. How many quarterly payments of $250 each are necessary to accumulate (at least) $10,000, if interest is at 8% compounded quarterly?

SEC. 10.4   The Derivatives of $e^{f(x)}$ and $\ln f(x)$

11. If $\ln 2 \simeq 0.6932$, $\ln 3 \simeq 1.0986$, and $\ln 10 \simeq 2.3026$, approximate the value of each of the following by using the laws of natural logarithms.

   a. $\ln 16$ b. $\ln 12$ c. $\ln \sqrt{8}$ d. $\ln 20$

   e. $\ln 200$ f. $\ln 5$ g. $\ln \frac{4}{9}$ h. $\ln \sqrt[3]{9}$

   i. $\ln 13.5$ j. $\ln 0.125$ k. $\ln 72$ l. $\ln 1{,}000$
   m. $\ln 0.018$ n. $\ln \sqrt[3]{24}$ o. $\ln 0.25$ p. $\ln \sqrt{54}$

10.4 THE DERIVATIVES OF $e^{f(x)}$ AND $\ln f(x)$

In Section 10.1 we observed that if $S = Pe^{rt}$, then $D_t S = rPe^{rt}$. For the special case $P = 1$ and $r = 1$, we have, if $S = e^t$, then $D_t S = e^t$; that is, $D_t[e^t] = e^t$. We observe the interesting result that the function e^t is a function of t whose derivative with respect to t is the function itself. Since the letter used to represent a variable is a matter of personal choice, it follows that $D_x[e^x] = e^x$, $D_y[e^y] = e^y$, $D_u[e^u] = e^u$, $D_v[e^v] = e^v$, and so on, regardless of the letter used to represent the independent variable. Hence, if

$$y = e^u \quad \text{and} \quad u = f(x)$$

where f is a differentiable function of x, then

$$D_u y = e^u$$

and applying the Chain Rule (see Section 6.7), $D_x y = D_u y \cdot D_x u$, we obtain the formula

(1) $\qquad D_x y = D_u[e^u] D_x u = e^u D_x[f(x)]$

Thus, if $y = e^{f(x)}$, where f is a differentiable function of x, then

(2) $\qquad D_x y = D_x[e^{f(x)}] = e^{f(x)} \cdot D_x[f(x)]$

In words, Formula (2) states that the derivative, with respect to an independent variable of e raised to a power which is a differentiable function of that variable is equal to e raised to that same power multiplied by the derivative of that power.

Example 1   If $y = e^{3x^2}$, find $D_x y$.

Solution: Here we use Formula (2) with $f(x) = 3x^2$ and $D_x[f(x)] = 6x$; then

$$D_x y = e^{3x^2} \cdot 6x = 6xe^{3x^2}$$

Example 2 If $y = \dfrac{4}{e^{7x}}$, find $D_x y$.

Solution: We can write $y = \dfrac{4}{e^{7x}} = 4e^{-7x}$; hence

$$D_x y = D_x[4e^{-7x}] = 4D_x[e^{-7x}]$$

Again we use Formula (2), this time with $f(x) = (-7x)$ and $D_x[f(x)] = -7$ and obtain $D_x y = 4e^{-7x}(-7) = -28e^{-7x} = -\dfrac{28}{e^{7x}}$.

Example 3 If $y = 10e^{\sqrt{x}}$, find $D_x y$.

Solution: Here $f(x) = \sqrt{x} = x^{1/2}$ and $D_x[f(x)] = \dfrac{1}{2}x^{-1/2} = \dfrac{1}{2\sqrt{x}}$. Now applying Formula (2), we have

$$D_x y = D_x[10e^{\sqrt{x}}] = 10D_x[e^{\sqrt{x}}] = 10D_x[e^{x^{1/2}}] = 10e^{x^{1/2}} \cdot \dfrac{1}{2\sqrt{x}} = \dfrac{5e^{\sqrt{x}}}{\sqrt{x}}$$

Example 4 If $y = x^2 e^{5x}$, find $D_x y$.

Solution: We need to recall the formula for the derivative of a product of two functions, $D_x[f(x)g(x)] = f(x)D_x[g(x)] + g(x)D_x[f(x)]$, which was derived in Section 5.8. Using this formula and Formula (2) of this section, we find

$$D_x y = D_x[x^2 e^{5x}] = x^2 D_x[e^{5x}] + e^{5x} \cdot D_x[x^2] = x^2(5e^{5x}) + e^{5x}(2x)$$
$$= xe^{5x}(5x + 2)$$

To find a formula for the derivative of $\ln f(x)$, we proceed as follows. Let $y = \ln f(x)$; recalling the definition in $\ln f(x)$, we may write the equation $y = \ln f(x)$ in the equivalent form

$$f(x) = e^y$$

Since $e^y > 0$ for all real values of y, $\ln f(x)$ is defined only for values of x such that $f(x) > 0$. Differentiating with respect to x and using the Chain Rule, we get

$$D_x[f(x)] = e^y D_x y$$

Solving for $D_x y$, we obtain the desired formula:

(3) $$D_x y = \dfrac{D_x[f(x)]}{e^y} = \dfrac{D_x[f(x)]}{f(x)} \quad \text{or} \quad D_x[\ln f(x)] = \dfrac{D_x[f(x)]}{f(x)}$$

In words, Formula (3) states that the derivative of the natural logarithm of a function of a single independent variable with respect to that variable is equal to the derivative of that function divided by the function itself.

SEC. 10.4 The Derivatives of $e^{f(x)}$ and ln $f(x)$

Example 5 If $y = \ln x$, find $D_x y$.

Solution: Using Formula (3) with $f(x) = x$ and $D_x[f(x)] = 1$, we have

$$D_x y = \frac{1}{x}$$

Example 6 If $y = \ln(3x^2 - 7x + 5)$, find $D_x y$.

Solution: Here $f(x) = 3x^2 - 7x + 5$, and $D_x[f(x)] = 6x - 7$; hence

$$D_x y = \frac{6x - 7}{3x^2 - 7x + 5}$$

Often, before attempting to differentiate a logarithmic function such as ln $f(x)$, it is possible, by using one or more of the laws of natural logarithms discussed in Section 10.3, to simplify the logarithmic function and thereby make the process of differentiating the function much simpler and easier. This type of procedure is illustrated in the following examples.

Example 7 If $y = \ln \dfrac{x^2 - 1}{x^2 + 1}$, find $D_x y$.

Solution: Using Law 2 of natural logarithms, we can write

$$y = \ln(x^2 - 1) - \ln(x^2 + 1)$$

Differentiating term by term gives

$$D_x y = \frac{2x}{x^2 - 1} - \frac{2x}{x^2 + 1} = \frac{2(x)(x^2 + 1) - 2x(x^2 - 1)}{(x^2 - 1)(x^2 + 1)} = \frac{4x}{x^4 - 1}$$

Example 8 If $y = \ln \sqrt{\dfrac{2x + 5}{2x - 1}}$, find $D_x y$.

Solution: First we write $y = \ln \left(\dfrac{2x + 5}{2x - 1} \right)^{1/2}$ and apply Law 3, getting $y = \dfrac{1}{2} \ln \dfrac{2x + 5}{2x - 1}$. Now apply Law 2, and we have

$$y = \frac{1}{2}[\ln(2x + 5) - \ln(2x - 1)]$$

Hence

$$D_x y = \frac{1}{2}\left[\frac{2}{2x + 5} - \frac{2}{2x - 1}\right] = \frac{1}{2}\left[\frac{2(2x - 1) - 2(2x + 5)}{(2x + 5)(2x - 1)}\right]$$

$$= -\frac{6}{(2x + 5)(2x - 1)}$$

Example 9 | If $y = 6 \ln 7\sqrt{x^3}$, find $D_x y$.

Solution: Simplifying, we can write
$$y = 6 \ln 7x^{3/2} = 6[\ln 7 + \ln x^{3/2}]$$
$$= 6\left[\ln 7 + \frac{3}{2}\ln x\right]$$

Therefore, $\quad D_x y = 6\left[0 + \frac{3}{2} \cdot \frac{1}{x}\right] = \frac{9}{x}$

Exercises 10.4

Find $D_x y$ in each of the following problems.

1. $y = 5e^{2x}$
2. $y = 10e^{-4x}$
3. $y = \dfrac{7}{e^{3x}}$

4. $y = 10e^{x^3}$
5. $y = xe^x$
6. $y = x^2 e^{-x}$

7. $y = \ln(3x)$
8. $y = \ln x^7$
9. $y = \ln 6x^5$

10. $y = 4 \ln x^{3/2}$
11. $y = 6 \ln(8\sqrt[3]{x^2})$
12. $y = \ln \dfrac{3x+5}{2x-3}$

13. $y = \ln \sqrt{x^2 - 4x + 7}$
14. $y = \ln(x\sqrt{x^2 - 1})$
15. $y = (\ln x)^4$

16. $y = e^x \ln x$
17. $y = \ln(4x - 3)^{3/2}$
18. $y = \ln \dfrac{x}{3x+2}$

19. $y = \ln x^4$
20. $y = \ln \sqrt{\dfrac{x^2}{x^2 + 1}}$
21. $y = \dfrac{1}{2}(e^x + e^{-x})$

22. $y = x \ln x$
23. $y = \dfrac{\ln x}{x}$
24. $y = 2x - \dfrac{1}{2}e^{2x}$

25. $y = e^{(2x - 1/2e^{2x})}$
26. $y = x \ln x - x$
27. $y = \dfrac{1}{2} \ln \dfrac{1+x}{1-x}$

28. $y = \dfrac{e^x - e^{-x}}{e^x + e^{-x}}$
29. $y = \dfrac{\ln x}{e^x}$
30. $y = \ln(xe^x)$

10.5 THE INTEGRALS $\int e^{f(x)} f'(x)\, dx$ AND $\int \dfrac{f'(x)\, dx}{f(x)}$

First, let us recall the definition of an integral. In Chapter 7 we defined the integral, $\int f(x)\, dx$, by the equation

$$\int f(x)\, dx = F(x) + C$$

where F is a function such that $F'(x) = f(x)$ and C is a constant, called a constant of integration. In other words, an integral is an antiderivative.

In the previous section we showed that $D_x e^{f(x)} = e^{f(x)} \cdot f'(x)$; it follows from the above definition of an integral that

(1) $$\int e^{f(x)} f'(x)\, dx = e^{f(x)} + C$$

Example 1 Evaluate the integral $\int e^{x^2} 2x\, dx$.

Solution: Comparing with Formula (1), $f(x) = x^2$ and $f'(x) = 2x$; hence

$$\int e^{x^2} 2x\, dx = e^{x^2} + C$$

Example 2 Evaluate the integral $\int e^{5x}\, dx$.

Solution: Here $f(x) = 5x$ and $f'(x) = 5$. Our integral is not in the form $\int e^{f(x)} \cdot f'(x)\, dx$ but by making use of the theorem that, if k is any constant, except zero, then $\int kf(x)\, dx = k\int f(x)\, dx$, we can write

$$\int e^{5x}\, dx = \frac{1}{5}\int e^{5x} 5\, dx = \frac{1}{5} e^{5x} + C$$

In the last section we showed that $D_x[\ln f(x)] = \dfrac{f'(x)}{f(x)}$ and pointed out $\ln f(x)$ is defined only for values of x such that $f(x) > 0$. Now applying our definition of an integral, we see that

(2) $$\int \frac{f'(x)}{f(x)}\, dx = \ln f(x) + C, \qquad \text{for } f(x) > 0$$

If, however, $f(x) < 0$, then $-f(x) > 0$, and we can write

(3) $$\int \frac{f'(x)}{f(x)}\, dx = \int \frac{-f'(x)}{-f(x)}\, dx = \ln(-f(x)) + C, \qquad \text{for } f(x) < 0$$

In Chapter 2 we defined the absolute value $|u|$ of a number u as follows: $|u| = u$ if u is positive or zero and $|u| = -u$ if u is negative. With this absolute value notation, we can now combine Equations (2) and (3) into the general formula

(4)
$$\int \frac{f'(x)}{f(x)} dx = \ln |f(x)| + C$$

Example 3

Evaluate the integral $\int \frac{2x}{x^2 + 1} dx$.

Solution: Here $f(x) = x^2 + 1$ and $f'(x) = 2x$. Hence it follows from Formula (4) that

$$\int \frac{2x}{x^2 + 1} dx = \ln (x^2 + 1) + C$$

Note that the absolute value sign is not necessary here because $x^2 + 1$ is never negative.

Example 4

Evaluate the integral $\int \frac{dx}{3x - 5}$.

Solution: In this case $f(x) = 3x - 5$ and $f'(x) = 3$. To put our integral in the form needed to use Formula (4), we write

$$\int \frac{dx}{3x - 5} = \frac{1}{3} \int \frac{3 dx}{3x - 5} = \frac{1}{3} \ln |3x - 5| + C$$

Example 5

Evaluate the integral $\int \frac{x \, dx}{4 - x^2}$.

Solution: Here $f(x) = 4 - x^2$ and $f'(x) = -2x$. So we write

$$\int \frac{x \, dx}{4 - x^2} = -\frac{1}{2} \int \frac{-2x \, dx}{4 - x^2} = -\frac{1}{2} \ln |4 - x^2| + C$$

In Chapter 7, it was shown that if f is continuous on a closed interval $[a, b]$ and if $F'(x) = f(x)$ for all x on $[a, b]$, then $\int_a^b f(x) \, dx = F(x)|_a^b = F(b) - F(a)$.

Example 6

Evaluate the definite integral $\int_2^{10} \frac{40}{x} dx$.

Solution:
$$\int_2^{10} \frac{40}{x} dx = 40 \int_2^{10} \frac{1}{x} dx = 40 \ln |x| \Big]_2^{10} = 40[\ln 10 - \ln 2]$$
$$= 40 \ln \frac{10}{2} = 40 \ln 5 = 64.38 \quad \text{(rounded to 2 decimal places)}$$

SEC. 10.5 The Integrals $\int e^{f(x)}f'(x)\,dx$ and $\int [f'(x)\,dx/f(x)]$

Example 7   Evaluate the definite integral $\int_{-8}^{-4} \frac{1}{x}\,dx$.

Solution: $\int_{-8}^{-4} \frac{1}{x}\,dx = \ln|x|\Big]_{-8}^{-4} = \ln|-4| - \ln|-8| = \ln 4 - \ln 8$
$$= -\ln 8 + \ln 4 = -[\ln 8 - \ln 4] = -\ln 2 = -0.6932$$

Example 8   Evaluate the definite integral $\int_{1}^{2} e^{x^2} 2x\,dx$.

Solution: $\int_{1}^{2} e^{x^2} 2x\,dx = e^{x^2}\Big]_{1}^{2} = e^4 - e^1 = 51.8799$ (rounded to 4 decimal places)

For convenience, the formulas for differentiation and integration of exponential and logarithmic functions are repeated here in tabular form.

$$D_x[e^{f(x)}] = e^{f(x)} D_x[f(x)]$$

$$D_x[\ln |f(x)|] = \frac{1}{f(x)} D_x[f(x)]$$

$$\int e^{f(x)} f'(x)\,dx = e^{f(x)} + C$$

$$\int \frac{f'(x)}{f(x)}\,dx = \ln |f(x)| + C$$

Exercises 10.5

Evaluate each of the following integrals:

1. $\int e^{3x}\,dx$

2. $\int e^{x^2} 2x\,dx$

3. $\int \frac{2\,dx}{e^{5x}}$

4. $\int xe^{2-x^2}\,dx$

5. $\int \frac{2x^2 + 3x + 5}{x}\,dx$

6. $\int \frac{dx}{x+1}$

7. $\int \frac{3\,dx}{2x-3}$

8. $\int \frac{e^x\,dx}{4+e^x}$

9. $\int \frac{x\,dx}{x^2+1}$

10. $\int \frac{dx}{5-3x}$

11. $\int \frac{x\,dx}{4x^2+1}$

12. $\int \frac{2x-5}{x}\,dx$

13. $\int \frac{x^2}{5-x^3}\,dx$

14. $\int (\ln x)^4 \frac{1}{x}\,dx$

15. $\int \frac{dx}{(2x+1)^2}$

Evaluate each of the following definite integrals.

16. $\int_0^3 e^x \, dx$

17. $\int_{-1}^2 e^{2x} \, dx$

18. $\int_1^4 \frac{dx}{x}$

19. $\int_0^3 \frac{x \, dx}{x^2 + 1}$

20. $\int_0^{\ln 2} e^{3x} \, dx$

21. $\int_{-7}^{-1} \frac{dx}{x - 1}$

22. $\int_0^1 \frac{x^2 \, dx}{2 - x^3}$

10.6* LOGARITHMS TO OTHER BASES

In Section 10.2 we defined the natural logarithm of any positive real number x to be the power to which e must be raised to give x, that is, $y = \ln x$, if and only if $x = e^y$. The number e is called the base of natural logarithms. We shall use this same idea to define logarithms to bases other than e.

Let a be a positive real number, $a \neq 1$, and let N be any positive real number. We define the *logarithm* of N to the base a as the power to which a must be raised to give N. In symbols

$$\log_a N = x \quad \text{if and only if} \quad N = a^x$$

These two equations may be taken as equivalent statements. The meaning of a^x was given in Section 2.3. The exponential a^x possesses the same general properties as the exponential e^x, that is:

LAW 1. $a^{x_1} \cdot a^{x_2} = a^{x_1 + x_2}$

LAW 2. $\dfrac{a^{x_1}}{a^{x_2}} = a^{x_1 - x_2}$

LAW 3. $(a^x)^n = a^{nx}$

LAW 4. $a^{-x} = \dfrac{1}{a^x}$

LAW 5. $a^0 = 1$

By using these properties of a^x, it can be shown that $\log_a N$ obeys analogous laws to those obeyed by $\ln N$; these are the following.

LAW 1. $\log_a MN = \log_a M + \log_a N$

LAW 2. $\log_a \dfrac{M}{N} = \log_a M - \log_a N$

SEC. 10.6 Logarithms to Other Bases

LAW 3. $\log_a M^N = N \log_a M$

The precise meaning of the definition of $\log_a N$ may be clarified by the following examples.

Example 1 $\log_5 25 = 2$ since $25 = 5^2$

Example 2 $\log_{10} 1{,}000 = 3$ since $1{,}000 = 10^3$

Example 3 $\log_9 3 = \frac{1}{2}$ since $3 = 9^{1/2}$

Example 4 $\log_{10} 0.01 = -2$ since $0.01 = 10^{-2}$

Example 5 $\log_7 1 = 0$ since $1 = 7^0$

It should be observed that since $a^0 = 1$, $\log_a 1 = 0$, and since $a^1 = a$, $\log_a a = 1$.

The bases of logarithms of most practical importance are the bases $a = e$ and $a = 10$. Logarithms to the base 10 are called *common logarithms*. Up until fairly recently, common logarithms were very useful for the purpose of making numerical computations because our number system was the base 10. However, since the invention of electronic computers and hand calculators, they are not used as often now. One advantage common logarithms have over natural logarithms is that they give a person more of a feel for the magnitude of numbers being dealt with. For example, if $\log_{10} N = 5.21$, then, since $N = 10^{5.21} = (10^{0.21})(10^5)$, we see that N is something between 10^5 and 10^6, that is, between 10,000 and 100,000. Similarly, if $10^x = 456$, then, since

$$x = \log_{10} 456 = \log_{10} (10^2)(4.56)$$
$$= \log_{10} 10^2 + \log_{10} 4.56$$
$$= 2 + \log_{10} 4.56$$

we see that x lies somewhere between 2 and 3. In fact, the number to the left of the decimal point (called the characteristic) of a logarithm to the base 10 yields immediately the largest power of 10 which is less than or equal to the number N whose logarithm we have taken. Some texts spend considerable time on common logarithms. Our attitude here is that natural logarithms are more prevalent in scientific fields, and persons desiring to use common logarithms will likely have access to a hand calculator. Therefore, we shall not go into a detailed study of common logarithms.

A rather simple relation exists between $\log_a N$ and $\ln N$. We determine this relation as follows. By definition,

$$\log_a N = x \quad \text{if and only if} \quad N = a^x$$

Hence $$\ln N = x \ln a \quad \text{or} \quad x = \frac{\ln N}{\ln a}$$

That is,

(1) $$\log_a N = \frac{\ln N}{\ln a}$$

We shall end this section, for the sake of completeness, by developing formulas for $D_x[a^{f(x)}]$, $D_x[\log_a f(x)]$, and $\int a^{f(x)} f'(x)\, dx$, where f is assumed to be a differentiable function of x.

Let $y = a^{f(x)}$, then

$$\ln y = f(x) \ln a$$

Differentiating both sides of this last equation with respect to x, we obtain

$$\frac{1}{y} D_x y = D_x[f(x)] \ln a \quad \text{since } a \text{ is a constant}$$

Hence $$D_x y = y D_x[f(x)] \ln a = a^{f(x)} D_x[f(x)] \ln a$$

or

(2) $$D_x[a^{f(x)}] = a^{f(x)} D_x[f(x)] \ln a$$

Note that if in Formula (2) we let $a = e$, it reduces to the formula for $D_x[e^{f(x)}]$ derived in Section 10.4.

Next, it follows from Formula (1) that

$$\log_a f(x) = \frac{\ln f(x)}{\ln a}$$

Therefore,

$$D_x[\log_a f(x)] = D_x\left[\frac{\ln f(x)}{\ln a}\right] = \frac{1}{\ln a} D_x[\ln f(x)]$$
$$= \frac{1}{\ln a} \frac{D_x[f(x)]}{f(x)}$$

or

(3) $$D_x[\log_a f(x)] = \frac{D_x[f(x)]}{f(x) \ln a}$$

Using Formula (2) and the definition of an integral, we observe that

$$\int a^{f(x)} f'(x)\, dx = \frac{1}{\ln a} \int a^{f(x)} f'(x) \ln a\, dx = \frac{1}{\ln a} a^{f(x)} + C$$

SEC. 10.6 Logarithms to Other Bases

or simply

(4) $$\int a^{f(x)} f'(x)\, dx = \frac{a^{f(x)}}{\ln a} + C$$

Example 6 If $y = 7^{x^3}$, find $D_x y$.

Solution: Here we use Formula (2) with $a = 7$, $f(x) = x^3$, and $D_x[f(x)] = 3x^2$. Hence

$$D_x y = 7^{x^3}(3x^2) \ln 7 = 3x^2 7^{x^3} \ln 7$$

Example 7 Approximate $\log_4 26.7$.

Solution: It follows from Formula (1) that

$$\log_4 26.7 = \frac{\ln 26.7}{\ln 4} \cong \frac{3.28466}{1.38629} \cong 2.36939$$

Example 8 If $y = \log_{10}(x^2 + 3x)$, find $D_x y$.

Solution: Using Formula (3) with $a = 10$, $f(x) = x^2 + 3x$, and $D_x[f(x)] = 2x + 3$, we obtain

$$D_x y \cong \frac{2x + 3}{(x^2 + 3x)(2.3026)} \cong \frac{0.43429(2x + 2)}{x^2 + 3x}$$

Example 9 Evaluate the integral $\int 3^{5x}\, dx$.

Solution: $$\int 3^{5x}\, dx = \frac{1}{5 \ln 3} \int 3^{5x} \cdot 5 \ln 3\, dx = \frac{3^{5x}}{5 \ln 3} + C$$

For convenience, we repeat the formulas for differentiating and integrating exponential and logarithmic functions to other bases in tabular form.

$$D_x[a^{f(x)}] = a^{f(x)} D_x[f(x)] \ln a$$

$$D_x[\ln_a |f(x)|] = \frac{D_x[f(x)]}{f(x) \ln a}$$

$$\int a^{f(x)} f'(x)\, dx = \frac{1}{\ln a} \cdot a^{f(x)} + C$$

Exercises 10.6

Use the definition of $\log_a N$ to find the value of each of the following logarithms.

1. $\log_8 64$
2. $\log_{25} 5$
3. $\log_{10} 0.001$
4. $\log_2 16$
5. $\log_3 \dfrac{1}{27}$
6. $\log_5 1$

Find the value of x for which the following statements are true.

7. $\log_3 81 = x$
8. $\log_4 x = 3$
9. $\log_x 32 = 5$
10. $\log_x 16 = 4$
11. $\log_5 x = -2$
12. $\log_x 1{,}000 = \dfrac{3}{4}$
13. $\log_x 0.001 = -3$
14. $\log_8 x = 0$
15. $\log_8 x = -\dfrac{4}{3}$

Approximate, by using Formula (1), the value of each of the following logarithms.

16. $\log_7 4$
17. $\log_3 6$
18. $\log_5 17.5$
19. $\log_8 42$
20. $\log_3 5.4$
21. $\log_{12} 347$

Find $D_x y$ in each of the following problems.

22. $y = 5^{x^2}$
23. $y = 7^{3x}$
24. $y = \log_{10} \dfrac{3x+5}{2x-7}$
25. $y = 10^{(3x+2)}$
26. $y = \dfrac{4}{3^{2x}}$
27. $y = \log_{10} e^{3x}$
28. $y = 8^{\ln x}$
29. $y = \log_4 (x^2 + 1)$
30. $y = (\log_5 x)^4$

Evaluate each of the following integrals.

31. $\displaystyle\int 10^{2x}\, dx$
32. $\displaystyle\int x^{2-x^2}\, dx$
33. $\displaystyle\int_0^1 7^{2t}\, dt$

Write each of the following expressions as a single logarithm, assuming all logarithms are to the base a.

34. $\log P + \log Q + \log R$
35. $\log x + 2 \log y$
36. $2 \log x - \log y$
37. $2 \log x + 3 \log y - 5 \log z$
38. $\log x + \dfrac{1}{2} \log (x^2 + 1) - 2 \log (2x + 3)$
39. $3 \log x + 2 \log y - \log z - \dfrac{1}{2} \log w$
40. $\dfrac{1}{3}(\log x + \log y - 2 \log z)$

11

Trigonometric Functions

11.1 ANGLES AND THEIR MEASURES

Consider a line L in the plane and let O and A be two distinct points on L (see Figure 11.1). The set of points consisting of O, A, and all the points on L between O and A is called a *line segment*. Note that a line segment has two endpoints. The set of points consisting of all the points on the segment OA and all points P on L such that A is between O and P is called a *ray*. The point O is called the *endpoint*, or *initial point*, of the ray. All other points of the ray are called *interior points*. The ray with endpoint O and interior point A is denoted by the symbol **OA**. The first letter mentioned in the symbol for a ray is always the endpoint. A ray has one and only one endpoint. The set of all points on a ray, excluding the endpoint, is called a *half-line*. The angle AOB is defined as the union of two rays **OA** and **OB** with a common initial point O called the vertex of the angle. How are such angles formed or generated? For our purposes in the study of trigonometric functions, we shall say that angle AOB has been generated when a ray, having its endpoint at O, is rotated about O from an initial position **OA** to a terminal position **OB**. The amount of rotation is called the *magnitude of the angle*. Obviously, such a rotation could take place in either of two directions, clockwise or counterclockwise. It is conventional to consider an angle as positive when the rotation is counterclockwise and as negative when the

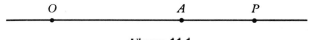

Figure 11.1

rotation is clockwise. In drawing an angle, the direction of rotation should be indicated with a curved arrow to prevent any confusion. See Figure 11.2.

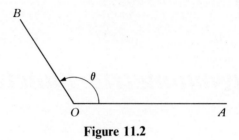

Figure 11.2

Since the magnitude or measure of an angle is the amount of rotation necessary to move the initial side into the position of the terminal side, a rather natural unit of measure for angles is a complete rotation, or *one revolution*. See Figure 11.3. A revolution is too large a unit for most practical purposes. It is common practice, therefore, to use $\frac{1}{360}$ of a revolution as a unit of measure for angles. This unit is called a *degree*. To obtain still smaller units, we define *one minute* as $\frac{1}{60}$ of a degree and *one second* as $\frac{1}{60}$ of a minute. In symbols, these become

$$1° = \frac{1}{360} \text{ rev} \quad \text{or} \quad 360° = 1 \text{ rev}, \qquad 60' = 1° \quad \text{and} \quad 60'' = 1'$$

where the symbols °, ′, and ″ signify degrees, minutes, and seconds, respectively.

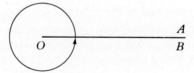

Figure 11.3

If an angle has a magnitude of 90° (that is, $\frac{1}{4}$ rev), it is called a *right angle*, and its sides are said to be *perpendicular* to each other. Angles less than 90° are called *acute angles*. Angles greater than 90° and less than 180° are called *obtuse angles*. If the magnitude of an angle is 180°, the angle is called a *straight angle*. See Figure 11.4.

For many types of problems in geometry and physics the most convenient units of measure for angles are degrees, minutes, and seconds. However, for some purposes another unit of measure called a *radian* is more convenient to

SEC. 11.1   Angles and Their Measures   527

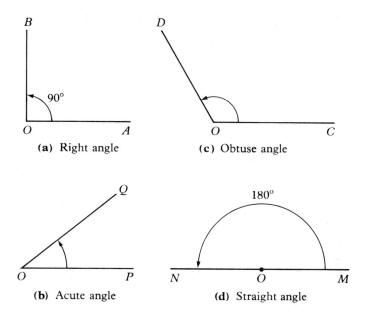

Figure 11.4

use. A radian is $\dfrac{1}{2\pi}$ of a revolution, that is,

$$2\pi(r) = 1 \text{ rev}, \qquad 1(r) = \frac{1}{2\pi} \text{ rev}$$

Recalling that the formula for the length, C, of the circumference of a circle of radius r is $C = 2\pi r$, or $C/2\pi = r$, we see that an equivalent definition of a radian is the following. A *radian* is the measure of the central angle in a circle which subtends an arc equal in length to the radius of the circle. Since 1 rev = 360°, it follows that

$$2\pi(r) = 360°, \qquad \pi(r) = 180°, \qquad 1(r) = \left(\frac{180}{\pi}\right)° \cong 57° \; 17' \; 44.8''$$

$$1° = \frac{\pi(r)}{180} \cong 0.01745(r)$$

In the case of angles measured in the units of degrees, minutes, and seconds, we have the standard symbols °, ′, and ″ to represent these units, but there is no standard symbol in common use to represent a radian. It is conventional therefore to use no symbol at all on the measure of angles if the unit is a radian. Thus, when no symbol is indicated on the measure of an angle, we shall understand the unit of measure to be a radian.

One simple example of a case where radian measure is more convenient to use than degree measure is the formula for the length of the arc of a circle subtending a given central angle of the circle. From geometry we know that in any circle the length of an arc subtending a central angle is proportional to the central angle (see Figure 11.5). If the central angle is doubled, the arc is doubled, etc. In fact, if the central angle is multiplied by any constant, the length of the arc subtending the angle is multiplied by that same constant. It follows from the definition of a degree that the length of the arc subtending a central angle of $1°$ in a circle of radius r is $\frac{1}{360}$ of the circumference of the circle, that is, $\frac{1}{360}(2\pi r) = \pi r/180$. Hence the length S of the arc subtending a central angle of $\theta°$ is $S = \pi r \theta/180$. By definition of a radian, the length of the arc subtending a central angle of 1 radian in a circle of radius r is r. Hence the length, S, of the arc subtending a central angle of θ radians is simply $S = r\theta$. In each of these two formulas for arc length, S is measured in the same units as the radius r.

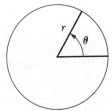

Figure 11.5

It turns out, as we shall see, that the formulas for the derivatives and integrals of functions involving angles are simplest if the angles are measured in radians. This assumes of course that the variables involved are angles, which in turn may be functions of other variables. For example, suppose we mark a point P on the circumference of a wheel, of radius r feet, at the end of a particular spoke in the wheel, and then start the wheel turning on its axle. As the wheel turns, the spoke in question generates an angle θ measured from some given fixed position. It seems clear that the angle so generated is a function of the time t elapsed since the wheel started turning. The rate at which the measure of θ changes with t is given by $D_t\theta$ and is called the *angular velocity* of the marked spoke. As θ increases, so does the distance, s feet, traveled by the point P on the circumference of the wheel. The distance s is given by the formula $s = r\theta$ if θ is measured in radians. Since r, the radius of the wheel, is constant, the linear velocity, $D_t s$, of the point P is related to the angular velocity of the spoke by the equation

(1) $$D_t s = r D_t \theta$$

if θ is measured in radians and is a differentiable function of t. This suggests that

SEC. 11.2   Triangles

the linear velocity of a point P moving on the circumference of a circle with center at 0 and radius r is equal to the radius of the circle times the angular velocity of the line OP.

We can differentiate Equation (1) and obtain

(2) $$D_t^2 s = r D_t^2 \theta$$

which states that the linear acceleration of a point P on the circumference of a circle is equal to the radius of the circle multiplied by the angular acceleration of the line joining the center of the circle to the point P, if θ is measured in radians. The interpretation of the angular acceleration as the instantaneous rate of change of angular velocity should be apparent.

Exercises 11.1

1. Convert to radians expressed in terms of π each of the following angles.
   a. 180°   b. 90°   c. 45°   d. 150°   e. 210°   f. 240°
   g. 135°   h. 270°   i. 60°   j. 315   k. 225°   l. −30°

2. Use a hand calculator to convert the following angles to radians.
   a. 36°   b. 68.2°   c. 110°   d. 140.6°   e. 48.8°
   f. 230°   g. 142°18′32″

3. Convert to degrees each of the following angles
   a. $\dfrac{\pi}{4}$ b. $\dfrac{\pi}{3}$ c. $\dfrac{5\pi}{6}$ d. $\dfrac{7\pi}{4}$ e. $\dfrac{3\pi}{2}$ f. $-\dfrac{5\pi}{4}$
   g. $\dfrac{\pi}{12}$ h. $\pi/9°$

4. Convert to degrees, minutes and seconds each of the following angles.
   a. 1.4   b. 0.6   c. 2.34   d. 0.42   e. 3.2   f. 1.67
   g. 1.5708   h. 4.76

11.2   TRIANGLES

If three points, not in a straight line, are joined in succession by line segments, the figure formed is called a *triangle*. The three points are called the *vertices* of the triangle, each point is a vertex, and the three line segments joining the vertices are called the *sides* of the triangle. Each vertex and the two sides meeting at that vertex form an angle of the triangle. Every triangle has three sides and three angles which are often referred to as the six parts of the triangle.

A very important theorem from plane geometry states that the sum of the measures of the three angles of any triangle is 180° (or π radians). We shall assume this theorem to be true and in so doing we are using Euclidean geometry.

A triangle one of whose angles is a right angle (90°) is called a *right triangle*, a triangle having two equal sides is called an *isosceles triangle*, and a triangle having three equal sides is called an *equilateral triangle*. It can be shown that an equilateral triangle is also *equiangular*, and conversely an equiangular triangle is equilateral. A triangle that does not contain a right angle is called an *oblique triangle*.

Suppose we set up a rectangular coordinate system in the plane, using the same scale on both the x axis and the y axis. If $P(x, y)$ is any point in the plane, where x and y are the coordinates of P, referred to our coordinate system, and r is the distance from P to the origin, it follows from the distance formula, $d = \sqrt{(x_2 - x_1)^2 + (y_2 - y_1)^2}$, that $r = \sqrt{x^2 + y^2}$. Hence r is always positive or zero and is zero only when P is at the origin.

Example 1 Find the distance of the point $(-4, -3)$ from the origin.

Solution: Here $x = -4$ and $y = -3$; therefore,

$$r = \sqrt{(-4)^2 + (-3)^2} = \sqrt{16 + 9} = \sqrt{25} = 5$$

Example 2 Find the distance of the point $(-3, 2)$ from the origin.

Solution: $r = \sqrt{(-3)^2 + (2)^2} = \sqrt{9 + 4} = \sqrt{13}$

Note here that we cannot express the exact value of r in decimal form.

We have now associated with each point in the plane three numbers, its abscissa x, its ordinate y, and its distance from the origin r. If for a particular point we know any two of these three numbers, and we also know the quadrant in which the point lies when it is not on one of the axes, we can use the relation $r^2 = x^2 + y^2$, $r > 0$, to find the third of these three numbers. If the point is on the x axis, then $|x| = r$ and $y = 0$. If it is on the y axis, $|y| = r$ and $x = 0$. If $r = 0$, then $x = y = 0$.

Example 3 If P is a point in the fourth quadrant, $x = 4$ and $r = 5$, find y.

Solution: Since P is in the fourth quadrant, we know that y must be negative. Using the relation $x^2 + y^2 = r^2$, with $x = 4$ and $r = 5$, we have

$$4^2 + y^2 = 5^2 \quad \text{or} \quad y^2 = 25 - 16 = 9$$

Therefore, $y = -\sqrt{9} = -3$.

SEC. 11.2 Triangles

Exercises 11.2

1. Use the relation $r = \sqrt{x^2 + y^2}$ to find the distance r of each of the following points from the origin.
   a. $P(3, 4)$ b. $P(-4, 3)$ c. $P(12, 5)$ d. $P(24, 7)$
   e. $P(-5, 12)$ f. $P(2, 3)$ g. $P(6, 2)$ h. $P(-7, -24)$

2. Use the relation $x^2 + y^2 = r^2$ to find the missing coordinate in each of the following, where QI means quadrant I, QII means quadrant II, QIII means quadrant III, and QIV means quadrant IV.
   a. $x = -4, r = 13$, point in QIII
   b. $y = 15, r = 17$, point in QI
   c. $x = -12, r = 13$, point in QIII
   d. $y = -7, v = 25$, point in QIV
   e. $x = -4, r = 4$, point on a coordinate axis
   f. $x = 2, r = \sqrt{13}$, point in QI

3. Find the third angle in a triangle if two of its angles are those given.
   a. 46° and 87°   b. 37° 14′ 20″ and 68° 26′ 44″   c. $\dfrac{\pi}{8}$ and $\dfrac{\pi}{3}$

4. Set up a rectangular coordinate system in the xy plane and draw the following angles with each having its vertex at the origin of your coordinate system and its initial side along the positive x axis.
   a. 135°   b. 60°   c. 90°   d. $\dfrac{\pi}{6}$ e. $-120°$ f. $-\dfrac{\pi}{4}$
   g. $\dfrac{7\pi}{6}$ h. 450°

5. A point P is moving along the circumference of a circle with center at 0, as shown in the figure, and a radius of 10 inches in such a way that the angle θ generated by the line OP, t seconds after starting, is given by the formula $\theta = 30t^2 - 2t^3$. Find at the instant $t = 4$ the angular velocity and the angular acceleration of the line OP and the speed of the point P.

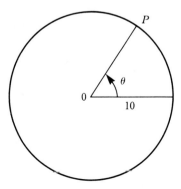

6. A wheel 28 inches in diameter revolves at a rate of 60 revolutions per minute. How fast is a point on the circumference of the wheel moving? $\left(\text{Use the approximation } \pi \cong \frac{22}{7}.\right)$

7. In a circle of diameter 20 inches, find the lengths of the arcs subtended by a central angle of
   a. 2 radians **b.** 1.84 radians **c.** 60° **d.** 135°

8. How large is the central angle in a circle of radius 100 inches that subtends an arc of 30 inches?

11.3 THE TRIGONOMETRIC FUNCTIONS

An angle will be said to be in standard position if its vertex is at the origin of a rectangular coordinate system and its initial side coincides with the positive x axis. It is said to be in a certain *quadrant* if it is in standard position and its terminal side lies in that quadrant. For example, if drawn in standard position, 110° is in the second quadrant, 315° is in the fourth quadrant, and $-100°$ is in the third quadrant.

An angle that is in standard position and whose terminal side coincides with one of the coordinate axes is called a *quadrantal angle*. Examples of quadrantal angles are 90°, 180°, 270°, $-90°$, $-450°$, etc.

Two or more angles are said to be *coterminal* if they are in standard position and have the same terminal side. For example, 0° and 360° in standard position are coterminal angles; so are 120°, 480°; and also $\frac{7\pi}{6}$, $\frac{-5\pi}{6}$, and $\frac{19\pi}{6}$. See Figure 11.6.

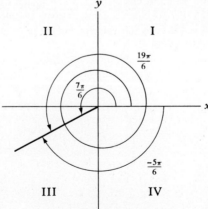

Figure 11.6

SEC. 11.3   The Trigonometric Functions

Consider an angle θ in standard position relative to a given coordinate system, Figure 11.7. Now select an arbitrary but fixed point P, not the origin, on the terminal side of θ, and let the coordinates of P be represented by (x, y). Let the distance of the point P from the origin be represented by the letter r. We shall assume that r is always positive. It follows from the right triangle relation

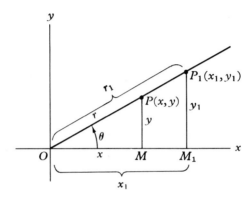

Figure 11.7

that $x^2 + y^2 = r^2$. We have thus associated with the angle θ in standard position three numbers, x, y, and r. These three numbers determine six possible ratios, $\dfrac{y}{r}, \dfrac{x}{r}, \dfrac{y}{x}, \dfrac{r}{x}, \dfrac{r}{y}$, and $\dfrac{x}{y}$. It is easy to show that the values of these six ratios are independent of the choice of the fixed point P. To show this, select a second point, say $P_1(x_1, y_1)$, in Figure 11.7 on the terminal side of θ and distinct from P. Denote by r_1 the distance of P_1 from the origin. The right triangles OMP and OM_1P_1 have an acute angle in common and hence are similar. It follows that the ratios of their corresponding sides are equal. That is,

$$\frac{y}{r} = \frac{y_1}{r_1}, \quad \frac{x}{r} = \frac{x_1}{r_1}, \quad \frac{y}{x} = \frac{y_1}{x_1},$$

$$\frac{x}{y} = \frac{x_1}{y_1}, \quad \frac{r}{x} = \frac{r_1}{x_1}, \quad \frac{r}{y} = \frac{r_1}{y_1}.$$

Thus, when θ is given, each of these ratios that exists can be determined. Furthermore, if θ is changed, the value of each of these ratios will in general be changed. That is, each of the six ratios determined by the coordinates x, y, and r of a point P on the terminal side of θ is a function of θ. To emphasize this fact and also as a matter of convenience in computation, each of these ratios is given a special name which associates it with the specific angle involved. The names and definitions of the six functions of θ are

$$\sin \theta = \frac{y}{r}, \qquad \csc \theta = \frac{r}{y}$$

$$\cos \theta = \frac{x}{r}, \qquad \sec \theta = \frac{r}{x}$$

$$\tan \theta = \frac{y}{x}, \qquad \cot \theta = \frac{x}{y}$$

where sin, cos, tan, cot, sec, and csc are abbreviations for sine, cosine, tangent, cotangent, secant, and cosecant, respectively. These six functions are called the *six trigonometric functions of angle θ*.

In the above definitions of the six trigonometric functions of an angle θ, it is understood that θ represents an arbitrary angle in the set of all possible angles, with the added condition that it be an angle in standard position relative to some rectangular coordinate system. The definitions of the trigonometric functions consist of the six possible ratios determined by the three numbers x, y, and r, where x and y are the coordinates (x, y) of an arbitrary point P, not the origin, on the terminal side of angle θ. The number r is the distance of the point P from the origin. Since P cannot be the origin, r is always positive. In fact x, y, and r also satisfy the equation $r = \sqrt{x^2 + y^2}$. Since r cannot be zero, $\sin \theta$ and $\cos \theta$ are defined for all possible angles. However, if θ is a quadrantal angle, then either x or y must be zero, for in this case the terminal side of θ must coincide with one of the coordinate axes. Since two of the six ratios have x as a denominator, the two functions defined by these ratios, $\tan \theta$ and $\sec \theta$, will fail to exist if the terminal side of θ coincides with the y axis, that is, if $\theta = 90° + n(180°)$ for some integer n. Stated briefly, $\tan \theta$ and $\sec \theta$ are not defined if $\theta = \frac{\pi}{2} + n\pi$ (n is an integer). Similarly, two of the ratios have y as a denominator, and hence the two functions, $\cot \theta$ and $\csc \theta$, defined by these ratios do not exist if $\theta = n\pi$ (n an integer).

Example 1 | Find the trigonometric functions, that exist, of $\frac{\pi}{2} = 90°$.

Solution: First we place our angle in standard position relative to a rectangular coordinate system. See Figure 11.8. Next we select a point on the terminal side of $90°$; for convenience we choose $P(0, 1)$. With this choice of P we have $x = 0$, $y = 1$, and $r = 1$. Applying the above definitions, we obtain

$$\sin 90° = \sin \frac{\pi}{2} = \frac{y}{r} = \frac{1}{1} = 1$$

$$\cos 90° = \cos \frac{\pi}{2} = \frac{x}{r} = \frac{0}{1} = 0$$

$$\tan 90° = \tan \frac{\pi}{2} = \frac{y}{x} = \frac{1}{0} \quad \text{(not defined)}$$

$$\csc 90° = \csc \frac{\pi}{2} = \frac{r}{y} = \frac{1}{1} = 1$$

$$\sec 90° = \sec \frac{\pi}{2} = \frac{r}{x} = \frac{1}{0} \quad \text{(not defined)}$$

$$\cot 90° = \cot \frac{\pi}{2} = \frac{x}{y} = \frac{0}{1} = 0$$

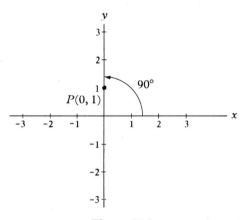

Figure 11.8

Trigonometry is a detailed study of the various relations that exist among the six trigonometric functions and the applications of these functions to problems in geometry and other fields. For example, one very useful set of relations that exists among the trigonometric functions is the set of reciprocal relations. The six ratios used above consist of three pairs of reciprocals, $\frac{y}{r}$ and $\frac{r}{y}$, $\frac{x}{r}$ and $\frac{r}{x}$, $\frac{y}{x}$ and $\frac{x}{y}$; that is, $\csc \theta = \frac{1}{\sin \theta}$, $\sec \theta = \frac{1}{\cos \theta}$, and $\cot \theta = \frac{1}{\tan \theta}$. These will be discussed in more detail in Section 11.11.

Example 2   Given $P(-12, 5)$, a point on the terminal side of the angle θ (see Figure 11.9 for one possibility for θ), find the six trigonometric functions of θ.

Solution: $r^2 = (-12)^2 + (5)^2 = 144 + 25 = 169; r = 13$

$$\sin \theta = \frac{5}{13}, \quad \csc \theta = \frac{13}{5}$$

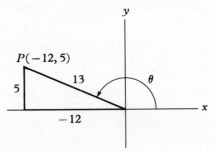

$$\cos\theta = -\frac{12}{13}, \qquad \sec\theta = -\frac{13}{12}$$

$$\tan\theta = -\frac{5}{12}, \qquad \cot\theta = -\frac{12}{5}$$

Figure 11.9

For convenience, the definitions of the trigonometric functions of an acute angle θ are repeated here in tabular form.

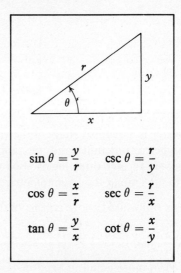

$$\sin\theta = \frac{y}{r} \qquad \csc\theta = \frac{r}{y}$$

$$\cos\theta = \frac{x}{r} \qquad \sec\theta = \frac{r}{x}$$

$$\tan\theta = \frac{y}{x} \qquad \cot\theta = \frac{x}{y}$$

Exercises 11.3

1. Place each of the following angles in standard position relative to a rectangular coordinate system and draw a curved arrow to indicate the direction of rotation. For each of these angles, find two other angles, one positive and one negative, which are coterminal with the given angle.
   a. 60° **b.** 135° **c.** 240° **d.** 540° **e.** 180°

SEC. 11.4   Related Angles

2. Find the values of the trigonometric functions, that exist, of each of the following quadrantal angles.
   a. $0°$ b. π radians   c. $270°$ d. $\dfrac{\pi}{2}$ radians

3. Each of the following given points is on the terminal side of the positive angle θ in standard position. Find the six trigonometric functions of the angle θ associated with each of these points.
   a. $P(1, 1)$ b. $P(4, 3)$ c. $P(-12, 5)$ d. $P(-1, 2)$
   e. $P(-8, -15)$ f. $P(5, -12)$ g. $P(15, -8)$ h. $P(3, -2)$
   i. $P(-\sqrt{5}, 2)$

4. An angle θ is in standard position and the point $P(1, 1)$ is on its terminal side.
   a. Find the smallest positive angle θ satisfying these conditions.
   b. Find the six trigonometric functions of the angle found in part a.
   c. Find a negative angle θ satisfying the given conditions.
   d. What are the six trigonometric functions of the angle found in part c?

5. In which quadrant is θ if:
   a. $\sin\theta$ is positive and $\cos\theta$ is negative?
   b. $\sin\theta$ is negative and $\tan\theta$ is positive?
   c. $\sin\theta$ and $\cos\theta$ are both negative?
   d. $\cos\theta$ and $\tan\theta$ are both negative?

6. An angle θ is in standard position and a point P on its terminal side is such that
   a. $x = -8$ and $r = 17$ b. $y = 24$ and $r = 25$
   Find the six trigonometric functions of θ in case a and case b.

7. Find the values of the other five trigonometric functions of an angle θ in standard position if:
   a. $\tan\theta = -3/4$ and θ is in QIV   b. $\sin\theta = 2/3$ and θ is in QI
   c. $\cos\theta = -12/13$ and θ is in QII   d. $\sin\theta = -15/17$ and θ is in QIII
   e. $\sec\theta = -5$ and θ lies in QII
   (*Hint:* Find the coordinates of a point P on the terminal side of θ in each case.)

8. Which of the following are impossible and why?
   a. $\cos\theta = 2$ b. $\csc\theta = \tfrac{1}{2}$ c. $\tan\theta = 1{,}000{,}000$
   d. $\cot\theta = 0.00001$ e. $\sin\theta = -5$ f. $\sec\theta = 0.999$

11.4 RELATED ANGLES

The *related angle* of a given angle θ is the positive acute angle determined by the terminal side of θ and the x axis when θ is in standard position; for example, $30°$ is the related angle of $150°$, $\dfrac{\pi}{3}$ is the related angle of $\dfrac{4\pi}{3}$, and $\dfrac{\pi}{5}$ is the related angle of $\frac{-4\pi}{5}$. We can now state an important theorem which we shall refer to as *the related-angle theorem.*

THEOREM 11.1 (RELATED-ANGLE THEOREM). *The absolute value of any trigonometric function of an angle θ is equal to the same function of its related angle.*

Proof: Let θ be any angle and let θ_1 be its related angle. Draw θ and θ_1 in standard position (Figure 11.10). Select on the terminal side of θ a point P with coordinates (x, y) and a corresponding value of r. Now locate the point

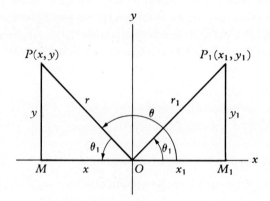

Figure 11.10

$P_1(x_1, y_1)$ on the terminal side of θ_1 such that $r_1 = OP_1 = OP = r$; the two right triangles OMP and OM_1P_1 are congruent. Therefore, since x_1 and y_1 are nonnegative,

$$|x| = x_1, \qquad |y| = y_1, \qquad |r| = r_1$$

From the definitions in Section 11.3 and the relations given above, we have

$$|\sin \theta| = \frac{|y|}{|r|} = \frac{y_1}{r} = \sin \theta_1$$

$$|\cos \theta| = \frac{|x|}{|r|} = \frac{x_1}{r} = \cos \theta_1$$

$$|\tan \theta| = \frac{|y|}{|x|} = \frac{y_1}{x_1} = \tan \theta_1$$

the other three functions being represented in a similar manner. This completes our proof.

SEC. 11.4 Related Angles

The theorem permits the following symbolic statement:

any function of $\theta = \pm$ the same function of θ_1

where θ is any angle and θ_1 is the related angle of θ. The proper sign on the right side here is determined by the quadrant in which the given angle θ lies and by the function involved. To determine the proper sign, one needs to recall the definition of the function in question and the algebraic signs that x and y have in each quadrant and to remember that r is always positive.

For example, x and y are both positive for every point in quadrant I, x is negative and y is positive in quadrant II, x and y are both negative in quadrant III, and x is positive and y negative in quadrant IV. Suppose we wish to know the algebraic sign for each of the functions sin 140°, cos 140°, and tan 140°. In standard position, 140° is an angle in quadrant II and, for any point (x, y) in quadrant II, x is negative, y is positive, and r is positive. So we have

$$\sin 140° = \frac{y}{r} = \frac{+}{+} = +, \qquad \cos 140° = \frac{-}{+} = -$$

$$\tan 140° = \frac{y}{x} = \frac{+}{-} = -$$

The importance of our theorem lies in the fact that it enables us to express any function of an angle that is not an acute angle in terms of the same function of an acute angle. In particular, it enables us to determine the following relations between the functions of an angle θ and its negative:

$$\sin(-\theta) = -\sin\theta, \qquad \csc(-\theta) = -\csc\theta$$
$$\cos(-\theta) = +\cos\theta, \qquad \sec(-\theta) = +\sec\theta$$
$$\tan(-\theta) = -\tan\theta, \qquad \cot(-\theta) = -\cot\theta$$

Example 1   Express the sine, cosine, and tangent of 200° in terms of the related angle.

Solution: 200°, in standard position, is an angle in quadrant III with a related angle of 20°. Consequently, sin 200° = −sin 20°, cos 200° = −cos 20°, tan 200° = tan 20°.

Example 2   Express the sine, cosine, and tangent of $\dfrac{-2\pi}{3}$ in terms of $\dfrac{2\pi}{3}$.

Solution: $\sin\dfrac{-2\pi}{3} = -\sin\dfrac{2\pi}{3}$, $\cos\dfrac{-2\pi}{3} = \cos\dfrac{2\pi}{3}$, $\tan\dfrac{-2\pi}{3} = -\tan\dfrac{2\pi}{3}$

Example 3   Given θ an angle in the third quadrant such that sec $\theta = -2$, find the other functions of θ.

Solution: Since $\sec\theta = -2 = \dfrac{2}{-1}$, we may let $r = 2$ and $x = -1$ (note that r is always positive, and therefore the negative sign must be associated with x). By the relation $x^2 + y^2 = r^2$, we have $y^2 = 3$. In the third quadrant, y is negative and hence $y = -\sqrt{3}$ (see Figure 11.11). The other functions can now be written at once:

$$\sin\theta = -\frac{\sqrt{3}}{2}, \qquad \cos\theta = -\frac{1}{2}, \qquad \tan\theta = +\sqrt{3}$$

$$\cot\theta = +\left(\frac{1}{\sqrt{3}}\right), \qquad \csc\theta = -\frac{2}{\sqrt{3}}$$

It should be pointed out that any choice of r and x such that $\dfrac{r}{x} = -2$ would serve equally well and would yield the same results.

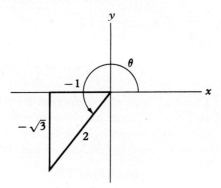

Figure 11.11

Exercises 11.4

1. Find the related angle of each of the following angles.

   a. $120°$ **b.** $310°$ **c.** $225°$ **d.** $140°$ **e.** $-220°$ **f.** $440°$ **g.** $\dfrac{8\pi}{15}$

2. List each of the functions that are negative in the second quadrant, in the third quadrant, and in the fourth quadrant. In how many lists does each function occur?

3. Find the other five functions in each of the following cases.

   a. $\sin\theta = \dfrac{2}{3}$, θ in QI   **b.** $\cos\theta = -\dfrac{4}{5}$, θ in QII   **c.** $\tan\theta = \dfrac{3}{4}$, θ in QIII

   d. $\sin\theta = -\dfrac{3}{5}$, θ in QIV   **e.** $\sec\theta = -5$, θ in QII

SEC. 11.5   The Trigonometric Functions of Acute Angles

4. Find two different angles, one positive and one negative, each of which has the given angle as its related angle.
   a. 40°   b. 54°   c. 72° 40′   d. 85°

5. Express the sine, cosine, and tangent of each of the following angles in terms of its related angle.
   a. 140°   b. 250°   c. 315°   d. 400°   e. −200°
   f. 780°   g. 1450°   h. −100°   i. 1375°   j. $\dfrac{11\pi}{6}$

6. By making use of the related angle theorem, find the values of the sine, cosine, and tangent of each of the following angles.
   a. 120°   b. 210°   c. 135°   d. 240°   e. 405°

11.5   THE TRIGONOMETRIC FUNCTIONS OF ACUTE ANGLES

Theorem 11.1 enables us to find the six trigonometric functions of any angle θ provided we know the corresponding six trigonometric functions of its related angle. According to the definition, a related angle to a given angle is always a positive acute angle. Thus, if we know the values of the trigonometric functions of all acute angles, then we can obtain the values of the trigonometric functions of all angles. For this reason we shall turn our attention to the problem of determining the trigonometric functions of positive acute angles.

If θ is a positive acute angle, it is always possible to construct a right triangle having θ as one of its angles. Let θ be a given positive acute angle. We now construct a right triangle having θ as one of its angles. See Figure 11.12. For convenience we shall adopt the following notation. We shall denote the vertices of our triangle by the capital letters A, B, and C, the right angle being at vertex C. It is common practice to denote the angle at a vertex of such a triangle by the letter at that vertex; that is, the angle at vertex A is called angle A, the angle at

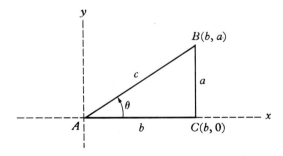

Figure 11.12

vertex B is called angle B, and the angle at vertex C is called angle C. If we follow this practice here, then angle θ and angle A are the same angle. The lowercase letters a, b, and c will be used to denote the length of the sides that are opposite the vertices A, B, and C, respectively. Thus the side opposite vertex A has length a, the side opposite vertex B has length b, and the side opposite vertex C has length c. Hence c is the length of the hypotenuse, a is the length of the side opposite angle A, and b is the length of the side adjacent to angle A. In the case of a right triangle, the hypotenuse is never referred to as a side but always as the hypotenuse of the triangle.

Let us now construct a rectangular coordinate system in the plane of right triangle ABC such that the origin is at A and C is a point on the positive x axis with coordinates $(b, 0)$. Vertex B will be a point in the first quadrant, with coordinates (b, a). See Figure 11.12. Since angle A and angle θ are the same angle, it follows from the definitions of the trigonometric functions given in Section 11.3 that

$$\sin A = \sin \theta = \frac{a}{c}, \qquad \csc A = \csc \theta = \frac{c}{a}$$

(3) $$\cos A = \cos \theta = \frac{b}{c}, \qquad \sec A = \sec \theta = \frac{c}{b}$$

$$\tan A = \tan \theta = \frac{a}{b}, \qquad \cot A = \cot \theta = \frac{b}{a}$$

Since angle θ can represent an acute angle in any right triangle, the above results show that for any acute angle θ lying in a right triangle,

$$\sin \theta = \frac{\text{length of side opposite } \theta}{\text{length of the hypotenuse}}, \qquad \csc \theta = \frac{\text{hyp.}}{\text{opp. side}}$$

(4) $$\cos \theta = \frac{\text{length of side adjacent to } \theta}{\text{length of the hypotenuse}}, \qquad \sec \theta = \frac{\text{hyp.}}{\text{adj. side}}$$

$$\tan \theta = \frac{\text{length of side opposite } \theta}{\text{length of adjacent side}}, \qquad \cot \theta = \frac{\text{adj. side}}{\text{opp. side}}$$

These special definitions of the trigonometric functions of acute angles lying in right triangles are so important that they should be carefully memorized.

Two positive acute angles are called *complementary* if the sum of their measure is 90°. Some very useful relations exist between functions of complementary angles. To see that this is true, we observe that any two such angles can be constructed as the two acute angles A and B of a right triangle ABC as shown in Figure 11.13.

SEC. 11.5 The Trigonometric Functions of Acute Angles

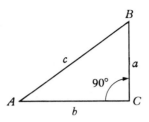

Figure 11.13

Applying the above special definition to angle A and angle B, we obtain

(5)
$$\sin A = \frac{a}{c} = \cos B, \qquad \csc A = \frac{c}{a} = \sec B$$
$$\cos A = \frac{b}{c} = \sin B, \qquad \sec A = \frac{c}{b} = \csc B$$
$$\tan A = \frac{a}{b} = \cot B, \qquad \cot A = \frac{b}{a} = \tan B$$

As you have probably already observed, the names of the six trigonometric functions come in pairs, sine and cosine, tangent and cotangent, and secant and cosecant. The two functions comprising any one of these pairs are called cofunctions. Equation (5) states the following important general principle regarding cofunctions: *Any trigonometric function of a positive acute angle is equal to the cofunction of its complementary angle.*

This principle plays an important role in the construction of tables of values of trigonometric functions, as we shall see later. Some examples of this principle are

$$\sin 40° = \cos 50°$$
$$\cot 10° = \tan 80°$$
$$\sec 27° = \csc 63°$$

That these names come in pairs is no accident. It was planned that way. For example, "cosine" is short for "complement's sine," "cotangent" is an abbreviation for "complement's tangent," etc.

The following more general theorem can be shown to be true.

THEOREM 11.2. *If A and B are any angles whose algebraic sum is $90°$, then*

$$\sin A = \cos B, \qquad \csc A = \sec B$$
$$\cos A = \sin B, \qquad \sec A = \csc B$$
$$\tan A = \cot B, \qquad \cot A = \tan B$$

The Trigonometric Functions of 30°, 45°, and 60°

We recall that an equilateral triangle is one whose sides are equal and whose angles are also equal. Since the sum of the measures of the angles in any triangle is 180°, it follows that each angle of an equilateral triangle has measure 60°. Consider an equilateral triangle each of whose sides has length $2a$ units (see Figure 11.14.) From a theorem in geometry we know that a perpendicular line dropped from a vertex of an equilateral triangle to the opposite side bisects the angle at that vertex and it also bisects the opposite side, thereby dividing the given triangle into two equal right triangles, as shown in Figure 11.14. This

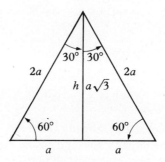

Figure 11.14

shows that in any right triangle whose acute angles are 30° and 60°, respectively, the length of the side opposite the 30° angle is always equal in length to one-half the length of the hypotenuse. In Figure 11.14, let h represent the length of the perpendicular shown. Now applying the Pythagorean Theorem to either of the right triangles involved, we obtain

$$h^2 + a^2 = (2a)^2 = 4a^2$$

or
$$h^2 = 3a^2$$

and hence
$$h = a\sqrt{3}$$

This shows that in any right triangle whose acute angles are 30° and 60°, the length of the side opposite the 60° angle is equal to $\sqrt{3}$ times the length of the side opposite the 30° angle. To calculate the trigonometric functions of 30° and 60°, we prefer to use the simplest right triangle possible. To obtain such a triangle, we let $a = 1$ in the triangle of Figure 11.14 and obtain the right triangle shown in Figure 11.15. Now applying the special definitions (4) to the acute

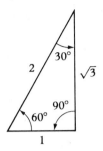

Figure 11.15

angles in this right triangle, we have

$$\sin 30° = \cos 60° = \frac{1}{2}$$

$$\cos 30° = \sin 60° = \frac{\sqrt{3}}{2}$$

$$\tan 30° = \cot 60° = \frac{1}{\sqrt{3}} = \frac{\sqrt{3}}{3}$$

$$\cot 30° = \tan 60° = \sqrt{3}$$

$$\sec 30° = \csc 60° = \frac{2}{\sqrt{3}} = \frac{2\sqrt{3}}{3}$$

$$\csc 30° = \sec 60° = 2$$

To calculate the trigonometric functions of 45°, any isosceles right **triangle would serve the purpose, but the most convenient one to use is one whose equal sides are each 1 unit in length. See Figure 11.16. It follows from definitions (4) that**

$$\sin 45° = \cos 45° = \frac{1}{\sqrt{2}} = \frac{\sqrt{2}}{2}$$

$$\tan 45° = \cot 45° = 1$$

$$\sec 45° = \csc 45° = \sqrt{2}$$

Having calculated the trigonometric functions of each of the special **angles** 30°, 45°, and 60°, we can now, by using the Related-Angle Theorem 11.1, find **the six** trigonometric functions of any angle whose related angle is one of these **special** angles.

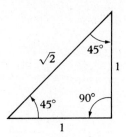

Figure 11.16

Example 1   Find the sine, cosine, and tangent of 120°.

Solution: The related angle of 120° is 60°. Since 120° is in QII, its cosine and tangent are negative. Hence

$$\sin 120° = \sin 60° = \frac{\sqrt{3}}{2}$$

$$\cos 120° = -\cos 60° = -\frac{1}{2}$$

$$\tan 120° = -\tan 60° = -\sqrt{3}$$

Example 2   Find the sine, cosine, and tangent of 225°.

Solution: Since 225° = 180° + 45°, the related angle is 45°. Also, 225° is in QIII, and hence the sine and cosine of 225° are negative, but the tangent of this angle is positive. Therefore,

$$\sin 225° = -\sin 45° = -\frac{\sqrt{2}}{2}$$

$$\cos 225° = -\cos 45° = -\frac{\sqrt{2}}{2}$$

$$\tan 225° = \tan 45° = 1$$

Exercises 11.5

1. A right triangle whose acute angles are 30° and 60° has a hypotenuse whose length is 10 units.

   a. Find the lengths of the other two sides.

   b. Use this triangle to find the six trigonometric functions of 30° and of 60°.

2. With your book closed, draw a suitable isosceles right triangle and use it to find the six trigonometric functions of 45°.

SEC. 11.6   Tables of Trigonometric Functions

3. Use definitions (4) and the triangle shown below to find the six trigonometric functions of angle B in that triangle.

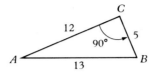

4. Read from the right triangle shown below the six trigonometric functions of angle θ in that triangle.

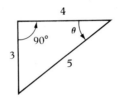

By making use of the values of the trigonometric functions of 30°, 45°, and 60°, calculated in this section, and the Related-Angle Theorem, find the sine, cosine, tangent, and cotangent of the angles given in the following problems.

5. 210°   **6.** 135°   **7.** 240°   **8.** 150°   **9.** $-120°$

10. 300°   **11.** 315°   **12.** $\dfrac{-5\pi}{6}$ **13.** $\dfrac{2\pi}{3}$ **14.** 390°

15. $-45°$ **16.** 780°   **17.** 495°   **18.** $-315°$

19. Using the values of the trigonometric functions of 30°, 45°, and 60° and the Related-Angle Theorem, evaluate the following:
  a. $\sin 90° - \sec 135°$
  b. $\csc 315° + \tan 210°$
  c. $\sin 120° + \cos 210° + \tan 300°$
  d. $(\sin 330°)(\cos 60°)(\tan 135°)$

20. Using the values of the trigonometric functions of 30°, 45°, and 60° and the Related-Angle Theorem, verify that the following statements are true.
  a. $\sin^2 30° = 1 - \cos^2 30°$ **b.** $\sin 240° = 2(\sin 120°)(\cos 120°)$

11.6 TABLES OF TRIGONOMETRIC FUNCTIONS

Trigonometric functions are used in so many different types of problems that tables of their values have been compiled and are easily available to anyone needing them. Moreover, many hand calculators have the function keys *sin, cos,* and *tan.* Using such a calculator alleviates the need for tables of trigonometric functions, as well as the tedious interpolation process which inevitably accompanies their use. Many calculators with trigonometric functions on them can do calculations in either degrees or radians. Some only handle degrees and, presumably, some only handle radians. The instructions for a particular calculator must be consulted to determine the capabilities and the recommended computational procedures. The most common type of trigonometric table gives, correct to five significant digits, the values of the sine, cosine, tangent, and cotangent of each angle from $0°$ to $90°$ at intervals of $1'$. Table I in Appendix B is such a table, and a small portion of it is reproduced here.

Notice how the table utilizes the fundamental principle regarding cofunctions of complementary angles stated in Section 11.5. The degrees, from $0°$ to $44°$, are at the tops of the pages and the minutes, from $0'$ to $60'$ (reading down), are listed in the first column on each page; the degrees, from $45°$ to $89°$, are at the bottom of the pages and the minutes, from $0'$ to $60'$ (reading up), are listed in the last column of each page. Thus the degrees at the top of a page and any minutes in the first column are the complement of the degrees at the bottom and the minutes immediately across in the last column. For example, $35°17'$ is the complement of $54°43'$. Note how the remaining four columns are labeled and that the label at the top of a column is the cofunction of the label at the bottom.

To see how the table is used, let us find $\sin 35°18'$. Since $35°$ occurs at the top of the table, we move down the first column to $18'$ and across to the entry in the column labeled "sin" at the top. Thus $\sin 35°18' = 0.57786$. To find $\tan 54°41'$, we move up the last column to $41'$ and read the entry in the column labeled "tan" at the bottom. Thus $\tan 54°41' = 1.4115$. We can also use such a table to find an acute angle if any one of the four functions, sine, cosine, tangent, and cotangent, of the given angle is known. As always, all values obtained are only approximations.

$35°$

′	sin	tan	cot	cos	′
0	0.57358	0.70021	1.4281	0.81915	60
1	0.57381	0.70064	1.4273	0.81899	59
2	0.57405	0.70107	1.4264	0.81882	58
3	0.57429	0.70151	1.4255	0.81865	57
—					—
—					—
17	0.57762	0.70760	1.4132	0.81631	43
18	0.57786	0.70804	1.4124	0.81614	42
19	0.57810	0.70848	1.4115	0.81597	41
—					—
59	0.58755	0.72610	1.3772	0.80919	1
60	0.58779	0.72654	1.3764	0.80902	0
	cos	cot	tan	sin	′

$54°$

SEC. 11.6   Tables of Trigonometric Functions

Example 1   If A is an acute angle and $\sin A = 0.57405$, find A.

Solution: We search for this number in either of the sine columns of our table in which it might occur. If it occurs in the column labeled "sin" at the top, we read the angle in degrees from the top of the table and the correct minutes from the minutes column on the left. If it occurs in the column labeled "sin" at the bottom, we read the degrees at the bottom and the minutes from the minutes column on the right. Since 0.57405 is found in the sine column at the top, we read $A = 35°2'$.

By using straight-line interpolation, it is possible to find from a table of trigonometric functions the functions of angles that lie between consecutive entries in the table. Again, these are approximations only.

Example 2   Find $\tan 35°18.4'$.

Solution: We arrange our work in tabular form thus:

$$
1'\left\{0.4'\left\{\begin{array}{ll} \text{angle} & \text{tan} \\ 35°18' & 0.70804 \\ 35°18.4' & ? \\ 35°19' & 0.70848 \end{array}\right\}d\right\}0.00044
$$

$$\frac{d}{0.00044} = \frac{0.4}{1}$$

$$d = 0.000176 \cong 0.00018 \quad \text{(rounded off to five places)}$$

$$\tan 35°18.4' \cong 0.70804 + 0.00018 = 0.70822$$

Example 3   Given that A is an acute angle and $\sin A = 0.81621$, use interpolation to approximate A. Check the answer with a hand calculator.

Solution: Arranging our work in tabular form, we have

$$
1'\left\{d\left\{\begin{array}{ll} \text{angle} & \text{sin} \\ 54°42' & 0.81614 \\ ? & 0.81621 \\ 54°43' & 0.81631 \end{array}\right.\right.\}0.00007\}0.00017
$$

$$\frac{d}{1} = \frac{0.00007}{0.00017} = \frac{7}{17} = 0.41$$

$$d \cong 0.4 \quad \text{(rounding off to tenths)}$$

$$A \cong 54°42' + 0.4' = 54°42.4'$$

We want to use a hand calculator now to check the result. This will be done by taking $\sin 54°42.4'$ to see if we obtain 0.81621. First we must convert 42.4' to degrees. This is done by the process of *cancellation of units* as follows:

$$(42.4 \text{ minutes}) \cdot \left(\frac{1}{60} \text{ degrees/minute}\right) \cong 0.7067 \text{ degrees}$$

or

$$\frac{42.4 \text{ minutes}}{60 \text{ minutes/degree}} \cong 0.7067 \text{ degrees}$$

rounding to four decimal places in each case. Hence $54°42.4' \cong 54.7067°$. Using a hand calculator, we obtain

$$\sin 54.7067° \cong 0.81621 \quad \textit{(rounded to five decimal places)}$$

thus checking our answer.

Exercises 11.6

Use a hand calculator to approximate the value of each of the following functions.

1. $\sin 26°$
2. $\tan 47°$
3. $\cos 40.7°$
4. $\tan 34.31°$
5. $\cot 52°12'$
6. $\sin 65°30'$
7. $\cos 25°54'$
8. $\tan 78°$
9. $\cos 4°3'$
10. $\cot 36°48'$
11. $\sin 32°24'$
12. $\tan 28°6'$
13. $\sin 5,236$
14. $\cos 3.14159$
15. $\tan 2.56$
16. $\cot 2.84$

Assuming A is an acute angle, use a hand calculator to find an approximate value of A, correct to the nearest minute, in each of the following problems.

17. $\sin A = 0.53411$
18. $\tan A = 0.32042$
19. $\cos A = 0.90948$
20. $\cot A = 1.7045$
21. $\sin A = 0.85264$
22. $\tan A = 1.5070$
23. $\cos A = 0.52175$
24. $\cot A = 0.77848$
25. $\sin A = 0.97176$
26. $\tan A = 4.4676$
27. $\cos A = 0.21104$
28. $\sin A = 0.70711$
29. $\sin A = 0.64536$
30. $\tan A = 0.73160$
31. $\cos A = 0.83651$

11.7 SOLUTION OF RIGHT TRIANGLES

Every triangle is said to have six parts—three sides and three angles. If one of the angles is a right angle, 90°, it is called a right triangle. It can be shown that a right triangle is completely determined if in addition to the right angle the magnitudes of two other parts of the triangle are given, provided that at least one of these known magnitudes is the length of a side or the length of the hypotenuse. This means that there exists one and only one right triangle with two of its parts having the given magnitudes. For example, there exists one and only one right triangle having a hypotenuse 10 units long and a side 6 units long,

SEC. 11.7 Solution of Right Triangles

or having a side 8 units long and an acute angle with measure 37°. In fact, if we know that a given triangle is a right triangle and in addition we are given the magnitudes of two other of its parts at least one of which is not an angle, we can by the use of trigonometry calculate the magnitudes of each of the three parts not given. All that is needed to make these calculations are the special definitions (5) and a table of values of the trigonometric functions or an appropriate hand calculator. Making such calculations is called *solving the triangle*. We shall use the notation for labeling the parts of our right triangle as was used in Section 11.5. See Figure 11.12. The conventions that are important to note are that C is always the right angle and c is always the hypotenuse. The actual procedures used in solving right triangles are demonstrated in the following examples.

Example 1 Solve the right triangle ABC in which $c = 100$ and $A = 35°18'$.

Solution: For the given right triangle, we know $C = 90°$, $A = 35°18'$, and the hypotenuse $c = 100$. The unknown parts are angle B, side a, and side b. Since angles A and B are complementary,

$$B = 90° - A = 90° - 35°18' = 54°42'$$

To find side a, we form the ratio of a and c, since $c = 100$ is given, and obtain $\dfrac{a}{c}$. It follows from definitions (5) that

$$\frac{a}{c} = \sin A \quad \text{or} \quad \frac{a}{c} = \cos B$$

Since both A and B are known, we could use either of these two equations to find a. We choose the first and write

$$\frac{a}{c} = \sin A \quad \text{or} \quad a = c \sin A$$

Hence
$$a = 100 \sin 35°18'$$

From Table I or using a hand calculator, $\sin 35°18' \cong 0.57786$. Therefore,

$$a \cong 100(0.57786) = 57.786$$

Similarly,
$$\frac{b}{c} = \cos A \quad \text{or} \quad \frac{b}{c} = \sin B$$

Again using the first of these equations, we rewrite it in the form $b = c \cos A$. Substituting $c = 100$ and $A = 35°18'$, we get

$$b = 100 \cos 35°18' \cong 100(0.81614) = 81.614$$

Note that once we have found a we could have used the Pythagorean Theorem to find b, a computation which is easily performed on a hand calculator having the function key $\sqrt{x}$.

Example 2 Solve the right triangle ABC in which $c = 98.46$ and $a = 50.52$.

Solution: Here the unknown parts are the two acute angles A and B and side b. We start by finding angle A (we could just as easily find angle B). To do this, we write

$$\sin A = \frac{a}{c} = \frac{50.52}{98.46} \cong 0.51312 \quad \text{(to five decimal places)}$$

Using Table I and interpolating, or using a hand calculator, we find

$$A \cong 30°52.3'$$

Hence, $\quad B = 90° - A \cong 90° - 30°52.3' = 59°7.7'$

To obtain side b, we write

$$\frac{b}{c} = \sin B \quad \text{or} \quad b = c \sin B$$

Substituting the known values of c and B, we get

$$b \cong 98.46 \sin 57°7.7' \cong 98.46(0.83989) \cong 82.696$$

Exercises 11.7

Solve each of the following right triangles. Hand calculators may be used if available.

1. $A = 26°10'$, $a = 120$
2. $B = 54°18'$, $a = 75$
3. $c = 40$, $A = 30°$
4. $c = 200$, $a = 148.6$
5. $a = 96.44$, $b = 64.32$
6. $A = 41°20'$, $c = 140$
7. $b = 10$, $c = 14.14$
8. $a = 0.0648$, $c = 0.9724$
9. $B = 57°32'$, $c = 0.482$
10. $a = 0.6478$, $c = 2$
11. $A = 30°$, $c = 196.8$
12. $A = 54°12'$, $b = 40$
13. $B = 44°35'$, $c = 1{,}000$
14. $B = 45°$, $a = 28.74$
15. $A = 72°30'$, $b = 150$
16. $a = 486$, $b = 600$
17. $a = 32$, $c = 64$
18. $a = 6$, $b = 8$
19. $B = 58°14'$, $a = 250$
20. $B = 60°$, $b = 112$

21. If we denote the area of the right triangle ABC by K, then $K = \frac{1}{2}ab$. Use this formula to find the area of each of the triangles in the odd-numbered problems above.

11.8 ANGLES OF ELEVATION AND ANGLES OF DEPRESSION

An *angle of elevation* is the angle, measured from the horizontal, through which an observer would have to elevate his normal line of sight, EH, in order to see an object O situated above EH. In Figure 11.17, angle E is the angle of elevation.

An *angle of depression* is the angle, measured from the horizontal, through

SEC. 11.8 Angles of Elevation and Angles of Depression

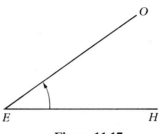

Figure 11.17

which an observer would have to depress his normal line of sight, *DH*, in order to see an object *O* situated below *DH*. In Figure 11.18, angle *D* is the angle of depression.

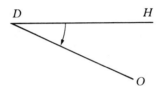

Figure 11.18

Example 1 Find the length of the shadow cast by a tree 50 feet tall at the instant that the angle of elevation of the sun is 45°. See Figure 11.19.

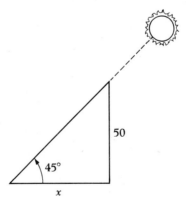

Figure 11.19

Solution: Let *x* be the length of the shadow; from Figure 11.19 we can write

$$\frac{x}{50} = \cot 45°$$

Hence

$$x = 50 \cot 45° = 50(1) = 50 \text{ ft}$$

Example 2 As seen from the top of a tower, the angle of depression of an object on the level ground 100 feet from the base of the tower is 28°46′. Find the height of the tower.

Solution: Let h be the height of the tower; from Figure 11.20 we have

$$\frac{h}{100} = \tan 28°46'$$

From Table I, or a hand calculator, $\tan 28°46' \cong 0.54900$; hence $h = 54.9$ feet.

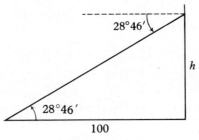

Figure 11.20

Exercises 11.8

1. A ladder 30 feet long has one end on the ground and the other end against a wall. If the ladder makes an angle of 68° with the level ground, how far up the wall does it reach?

2. A tree 48 feet tall is casting a shadow 32 feet long. What is the angle of elevation of the sun at that moment?

3. From the top of a vertical cliff 80 feet high on one bank of a river, the angle of depression of a point on the opposite bank directly across the river is 20°. How wide is the river at this point?

4. At the instant that the angle of elevation of the sun is 48°, a boy standing erect casts a shadow 4.9 feet in length. How tall is the boy?

5. A boy flying a kite has let out 220 feet of cord. The kite is directly above a spot 140 feet distant from the boy. Assuming that the cord is taut (and neglecting the height of the boy), find the height of the kite and its angle of elevation.

6. The angle of depression of a boat, as seen from a vertical cliff 225 feet high, is 30°. Find the distance of the boat from the foot of the cliff.

7. A flagpole on the school grounds stands 78 feet high. If a student whose eye level is $5\frac{1}{2}$ feet above the ground is standing in a spot where his angle of elevation to the top of the pole is 45°, how far is he standing from the base of the pole?

8. A surveyor marks off a right-angle corner of a rectangular house foundation. In sighting on the diagonally opposite corner of the foundation, he finds that his line of sight has moved through an angle of 30°. Determine the length of the short side of the foundation, if the length of the long side is 53.2 feet.

SEC. 11.9 Projections, Vectors, and Components of Vectors

9. A balloon is 430 feet above one end of a bridge. The angle of depression of the other end of the bridge from the balloon is 60°. How long is the bridge?

10. A lighthouse stands 120 feet above sea level at high tide. From the top of the lighthouse, the angle of depression of a buoy is 40°30′ at high tide and 43°10′ at low tide. Determine the height of the tide.

11. From a point on a bridge 75 feet above the water, the angles of depression of two boats, on the river and in the same straight line with a point on the water directly below the observer, are 16°20′ and 24°40′. Find the distance between the boats.

12. From a window in a building across a level street from a skyscraper, the angle of elevation of the top of the skyscraper is 56°40′ and the angle of depression of the base is 37°26′. The window is 75 feet above the street level. Find the height of the skyscraper and the width of the street.

13. It takes the earth 6 hours to rotate through 90°. Assuming this rotation uniform, each degree of elevation of the sum will correspond to $\frac{6}{90}$ of an hour, or 4 minutes. A vertical stick 16.4 inches high casts a horizontal shadow 8 inches long. What time is it if the sun rose at 6:00 A.M. and will be directly overhead at noon?

14. From a point on the ground and 100 feet from the base of a building, the angles of elevation of the bottom and top of a flagpole standing on top of the building are 24°10′ and 33°32′, respectively. Find the length of the flagpole.

15. Two straight highways, A and B, intersect at a point O, and the smallest angle between them is 58°. A service station is located on highway A, 500 yards from the intersection. Where is the point on highway B that is nearest to the service station, and how near is it?

16. The sides of an isosceles triangle are 20, 20, and 16. Find the magnitudes of the angles to the nearest minute. (*Hint:* Use a hand calculator.)

11.9 PROJECTIONS, VECTORS, AND COMPONENTS OF VECTORS

Before discussing vectors, we need to introduce the notion of projections. This we shall do next.

The *projection of a point P* on a line L is the point Q of the intersection of L and the line drawn through P perpendicular to L; see Figure 11.21. If P is on L, P and Q will coincide and P projects into itself.

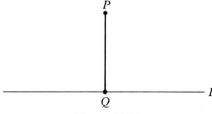

Figure 11.21

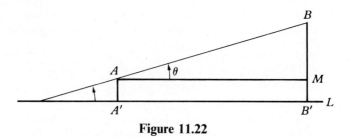

Figure 11.22

The *projection of a line segment AB* on a line L is the segment $A'B'$, where A' is the projection of A on L and B' is the projection of B on L; see Figure 11.22. If the segment AB lies on a line perpendicular to L, A' and B' will coincide and the projection of AB on L is a single point. Let $|AB|$ denote the length of the line segment AB. If the line AB is parallel to L, then $|A'B'| = |AB|$. If AB is neither perpendicular nor parallel to L, it will intersect L at some acute angle θ. Through A, draw a line parallel to L and let it intersect BB' in a point M. Now

$$|A'B'| = |AM| = |AB| \cos \theta$$

which proves the following theorem.

THEOREM 11.3. *The length of the projection of a line segment AB on a line L is equal to the length of AB times the cosine of the angle between AB and L.*

Vectors

In the study of physics and mechanics, two common types of quantities are encountered. The first is a quantity that has magnitude but no direction; positive and negative numbers, speed, temperature, volume, mass, etc., are of this kind. Such a quantity is called a *scalar*. The second is a quantity that has both magnitude and direction; velocities, forces, and displacements are of this type. Such a quantity is called a *vector quantity*.

It is convenient to represent a vector quantity by a straight arrow whose length represents the magnitude of the quantity and whose direction indicates the direction of the quantity. Such a directed line segment is called a *vector*. We shall use lowercase bold letters to designate vectors.

We shall denote the length of a vector a by the notation $|\mathbf{a}|$.

Two vectors are said to be equal if they are of the same length, are parallel, and point in the same direction. In Figure 11.23, $\mathbf{a} = \mathbf{b}$.

If two vectors have the same length, are parallel, but point in opposite directions, they are said to be the negatives of each other. In Figure 11.24, $\mathbf{c} = -\mathbf{d}$ and $\mathbf{d} = -\mathbf{c}$.

Hence a vector can be moved from one position to another without being changed if its direction and magnitude are kept unchanged.

SEC. 11.9 Projections, Vectors, and Components of Vectors

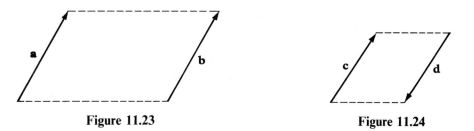

Figure 11.23 Figure 11.24

Consider two vectors **a** and **b** and a point P. Place the initial end of vector **a** at the point P. Move vector **b** until its initial end coincides with the terminal end of vector **a**. Draw the vector **c** with initial end at P and terminal end at the terminal end of **b**, Figure 11.25 (left). Vector **c** is called the *resultant*, or *sum*, of the vectors **a** and **b** and is written $\mathbf{c} = \mathbf{a} + \mathbf{b}$.

If we reverse the process by starting with vector **b**, placing its initial end at P, and moving vector **a** until its initial end coincides with the terminal end of **b**, we find the same vector **c** as a resultant, Figure 11.25 (right). Thus $\mathbf{c} = \mathbf{a} + \mathbf{b} = \mathbf{b} + \mathbf{a}$.

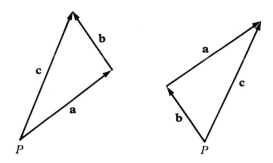

Figure 11.25

If **a** and **b** are parallel vectors, the resultant **c** is parallel to **a** and to **b** (Figure 11.26).

Now consider the case in which **a** and **b** are not parallel, and place the initial end of **a** at P and the initial end of **b** at P. Complete the parallelogram having **a** and **b** as adjacent sides. The diagonal from P to the opposite vertex determines a vector **c**, and again $\mathbf{c} = \mathbf{a} + \mathbf{b}$, Figure 11.27. Thus the sum of two vectors may be found graphically by drawing a triangle, as in Figure 11.25, or by drawing a parallelogram, as in Figure 11.27.

In a similar manner, three or more vectors may be added graphically by finding the resultant of two of the vectors, adding a third vector to this resultant, then adding a fourth vector to the resultant of the first three, and so on. The order in which vectors are added does not change the sum.

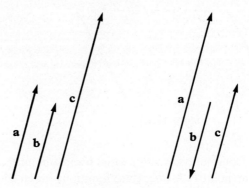

Figure 11.26

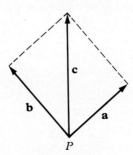

Figure 11.27

Components of Vectors

The concept of the *component* of a vector **a** along a line L is illustrated in Figure 11.28 (where **b** represents the component of **a** along L). Notice that the magnitude of **b** is the length of the projection of the line segment determined by **a** onto L. If θ is the angle between L and the direction of **a** (that is, any line determined by **a**; see Figure 11.28), then $|\mathbf{b}| = |\mathbf{a}| \cos \theta$.

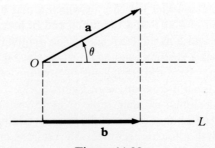

Figure 11.28

SEC. 11.9 Projections, Vectors, and Components of Vectors

Two components of a vector that are often useful are the horizontal component and the vertical component. If **a** is a vector whose direction makes an angle θ with the horizontal, and $\mathbf{a}_h$ and $\mathbf{a}_v$ are its horizontal and vertical components, then by definition $|\mathbf{a}_h| = |\mathbf{a}| \cos \theta$ and $|\mathbf{a}_v| = |\mathbf{a}| \sin \theta$. Moreover, $\mathbf{a} = \mathbf{a}_h + \mathbf{a}_v$; see Figure 11.29.

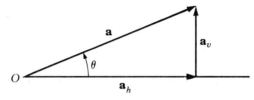

Figure 11.29

Example 1 A force of 50 pounds acts in a direction making an angle of 30° with the horizontal. Find the magnitude of the horizontal and vertical components of the force.

Solution: From Figure 11.29, we have

$$|\mathbf{a}_h| = 50 \cos 30° = 25\sqrt{3} \text{ lb}$$

$$|\mathbf{a}_v| = 50 \sin 30° = 25 \text{ lb}$$

Example 2 A force of 40 pounds and a force of 60 pounds act on the same point O in directions that are at right angles with each other. Find the resultant of these two forces and the angle it makes with the direction of the 60-pound force.

Solution: Complete the rectangle determined by the vectors representing the given forces; see Figure 11.30. The diagonal of this rectangle, with O as initial point,

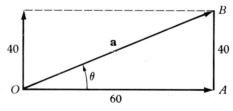

Figure 11.30

is a vector **a** which represents the resultant of the given forces. Let θ represent the angle the resultant makes with the 60-pound force. Then from the figure we have

$$\tan \theta = \frac{40}{60} = \frac{2}{3} \cong 0.66667$$

which gives $\theta \cong 33°41'$ correct to the nearest minute.

Next let $R = |\mathbf{a}|$ represent the magnitude of the resultant. Referring again to the figure, we find

$$\frac{40}{R} = \sin \theta$$

$$40 = R \sin \theta$$

$$R = \frac{40}{\sin \theta} \cong \frac{40}{\sin 33°41'} \cong \frac{40}{0.5546} \cong 72.11$$

Exercises 11.9

1. Find the horizontal and vertical projections of a line segment 40 inches long and inclined at an angle of 32° with the horizontal.

2. A line segment inclined 60° with the horizontal has a horizontal projection of 20 feet. Find the length of the segment.

3. A line segment 50 feet long has a vertical projection 30 feet long. Find the angle the line makes with the horizontal.

4. If a ship is sailing 30° north of east at the rate of 20 miles per hour, what are the components of its velocity eastward and northward?

5. A force of 100 pounds and a force of 50 pounds act at a common point O and at right angles to each other. Find the resultant of these two forces and the angle this resultant makes with the 100-pound force.

6. A 4,000-pound car is parked on a hill inclined at an angle of 10° with the horizontal. What force must the brakes overcome to keep the car from rolling down the hill? With what force is the car pressing against the hill?

7. A cake of ice weighing 100 pounds is held on a smooth plane, inclined 20° with the horizontal, by a single force acting along the plane. Find the magnitude of the force.

8. A lawnmower weighing 30 pounds is pushed on level ground with a force of 40 pounds directed along the handle, which is inclined 32° with the ground. How large is the force that produces the forward motion? With what total force do the wheels press against the ground?

11.10 OBLIQUE TRIANGLES

An oblique triangle was defined earlier in this chapter as one not containing a right angle. Thus the ratio of two sides of an oblique triangle does not represent a function of one of the angles of the triangle. Hence the method used in solving right triangles does not apply here. In fact, to solve oblique triangles we shall need two additional formulas, the law of sines and the law of cosines.

SEC. 11.10 Oblique Triangles

LAW OF SINES. In any triangle the sides are proportional to the sines of the opposite angles or, in symbols,

$$\frac{a}{\sin A} = \frac{b}{\sin B} = \frac{c}{\sin C}$$

Case 1: All angles are acute. Let h be the length of the perpendicular CD from the vertex C to the side AB; see Figure 11.31. From the right triangles ACD and BCD, we find

$$\sin A = \frac{h}{b} \quad \text{and} \quad \sin B = \frac{h}{a}$$

$$h = b \sin A \quad \text{and} \quad h = a \sin B$$

Equating these two values for h, we have

$$a \sin B = b \sin A$$

After dividing by $\sin A \sin B$, we get

$$\frac{a}{\sin A} = \frac{b}{\sin B}$$

In a similar manner we can show that

$$\frac{a}{\sin A} = \frac{c}{\sin C}$$

Consequently, our theorem is proved.

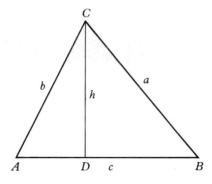

Figure 11.31

Case 2: One angle is obtuse. Suppose that angle B is obtuse. Let h be the length of the perpendicular CD drawn from the vertex C to the side AB extended; see Figure 11.32. From the right triangles ADC and BDC we have the

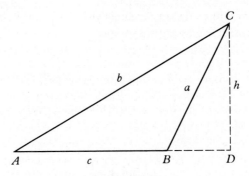

Figure 11.32

two following relations:

$$\sin A = \frac{h}{b} \quad \text{or} \quad h = b \sin A$$

and

$$\sin (CBD) = \frac{h}{a} \quad \text{or} \quad h = a \sin (CBD)$$

But angle CBD is the angle related to the obtuse angle B; thus

$$\sin (CBD) = \sin B$$

Hence
$$h = a \sin B$$

Again, if we equate the two values found for h, $a \sin B = b \sin A$, or

$$\frac{a}{\sin A} = \frac{b}{\sin B}$$

In a similar manner

$$\frac{c}{\sin C} = \frac{b}{\sin B}$$

Consequently, our theorem is proved.

It is shown in plane geometry that if one side and any other two parts of a triangle are given, there is in general one and only one triangle so determined. (There is one exception to this general law which will be explained in a moment.) There are four ways of combining one side of a triangle with two other parts:

Case 1. Given one side and two angles.
Case 2. Given two sides and the angle opposite one of them.
Case 3. Given two sides and the included angle.
Case 4. Given three sides.

SEC. 11.10 Oblique Triangles

The law of sines is equivalent to the three formulas

$$\frac{a}{\sin A} = \frac{b}{\sin B}, \quad \frac{a}{\sin A} = \frac{c}{\sin C}, \quad \frac{b}{\sin B} = \frac{c}{\sin C}$$

It is obvious from an inspection of these three formulas that the law of sines is adequate for solving oblique triangles when the given information is either that in Case 1 or Case 2. The exceptional case mentioned earlier comes under Case 2. When the given angle is acute and the given side opposite this angle is shorter than the other given side, there may be two, one, or no solutions. If the given angle is obtuse and the given side opposite this angle is shorter than the other given side, then there is no solution. These last two statements are easily verified by drawing a figure for each. We shall not consider problems of the ambiguous type here.

Example 1   Given $A = 70°$, $B = 30°$, and $b = 40$, find angle C, side a, and side c.

Solution: $A + B + C = 180°$

$$C = 180° - (A + B) = 180° - 100° = 80°$$

From the law of sines we have

$$\frac{a}{\sin A} = \frac{b}{\sin B} \quad \text{and} \quad \frac{c}{\sin C} = \frac{b}{\sin B}$$

or

$$a = \frac{b \sin A}{\sin B} \quad \text{and} \quad c = \frac{b \sin C}{\sin B}$$

Substituting the values given for A, B, and b and the value found for C, we obtain the following:

$$a = \frac{40 \sin 70°}{\sin 30°} \cong \frac{40(0.93969)}{0.5} \quad \text{and} \quad c = \frac{40 \sin 80°}{\sin 30°} \cong \frac{40(0.98481)}{0.5}$$

$$= 80(0.93969) \qquad\qquad\qquad\qquad = 80(0.98481)$$

$$\cong 75.18 \qquad\qquad\qquad\qquad\qquad \cong 78.78$$

We shall assume that all data given in triangle problems are exact.

LAW OF COSINES.   The square of any side of a triangle is equal to the sum of the squares of the other two sides minus twice the product of those sides and the cosine of the angle between them or, in symbols,

$$a^2 = b^2 + c^2 - 2bc \cos A$$
$$b^2 = a^2 + c^2 - 2ac \cos B$$
$$c^2 = a^2 + b^2 - 2ab \cos C$$

Proof: Consider the triangle ABC of Figure 11.33. We shall follow the same convention regarding notation as was used in the discussion of right

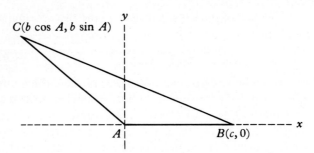

Figure 11.33

triangles, with the exception that no one of the angles is necessarily a right angle. In the plane of triangle ABC, construct a rectangular coordinate system with the origin at vertex A and the positive x axis passing through B (Figure 11.33). Relative to this coordinate system, the coordinates of A, B, and C, respectively, are $(0, 0)$, $(c, 0)$, and $(b \cos A, b \sin A)$. Using the distance formula to calculate $d(B, C)$, noting that according to our notation $d(B, C) = a$, and using the fact that $\cos^2 A + \sin^2 A = 1$ (see Section 11.11), we obtain

$$a^2 = (b \cos A - c)^2 + (b \sin A - 0)^2$$
$$= b^2(\cos^2 A + \sin^2 A) - 2bc \cos A + c^2$$
$$= b^2 + c^2 - 2bc \cos A$$

By rearranging the letters A, B, C, a, b, and c on our triangle, we find that the equations for b^2 and c^2 follow.

Example 2 Given $a = 14$, $b = 8$, $C = 40°$, find side c.

Solution: To find side c, we use

$$c^2 = a^2 + b^2 - 2ab \cos C$$
$$= 196 + 64 - 2(14)(8) \cos 40°$$
$$= 260 - 224 \cos 40°$$
$$\cong 260 - 224(0.7660) \cong 88.42$$
$$c \cong 9.40$$

Example 3 Given $a = 20$, $b = 14$, and $c = 10$, find angle A.

Solution: To find angle A, we use $a^2 = b^2 + c^2 - 2bc \cos A$ and solve for $\cos A$. The result is

$$\cos A = \frac{b^2 + c^2 - a^2}{2bc} = \frac{196 + 100 - 400}{280} \cong -0.37143$$

If A_1 is the related angle, then $\cos A_1 = 0.37143$ and $A_1 \cong 68°12'$; therefore, $A \cong 180° - 68°12' = 111°48'$.

SEC. 11.11 Basic Trigonometric Identities

For convenience, the law of sines and the law of cosines are repeated here in tabular form.

$$\text{Law of Sines:} \quad \frac{a}{\sin A} = \frac{b}{\sin B} = \frac{c}{\sin C}$$

$$\text{Law of Cosines:} \begin{cases} a^2 = b^2 + c^2 - 2bc \cos A \\ b^2 = a^2 + c^2 - 2ac \cos B \\ c^2 = a^2 + b^2 - 2ab \cos C \end{cases}$$

Exercises 11.10

Find the unknown parts of the given triangles in Exercises 1 through 10.

1. $A = 72°, C = 64°10', c = 20$
2. $A = 100°, B = 34°45', b = 50$
3. $B = 80°, C = 70°, a = 100$
4. $A = 60°, a = 10, b = 8$
5. $A = 110°, C = 40°, b = 20$
6. $A = 120°, a = 30, b = 20$
7. $b = 30, c = 40, A = 60°$
8. $a = 6, b = 7, c = 8$
9. $a = 20, b = 24, C = 40°$
10. $a = 60, c = 50, B = 110°$

11. A force of 60 pounds and a force of 100 pounds act at the same point on an object. If the angle between the two forces is 60°, what are the direction and magnitude of the resultant?

12. Two points A and B are separated by a pond. If the distance from A to a third point C is 316 yards, the distance from C to B is 346 yards, and the angle ACB is 41°15′, how far is it from A to B?

13. How long is the chord subtending an angle of 40° at the center of a circle of radius 15 feet?

14. Prove that the area of any triangle is equal to one-half the product of any two sides and the sine of the angle between those sides.

15. Two adjacent sides of a parallelogram are 16 feet and 25 feet, and the angle between them is 50°. Find the length of each diagonal.

11.11 BASIC TRIGONOMETRIC IDENTITIES

In Chapter 3 we defined algebraic identities and discussed their importance in simplifying algebraic expressions, reducing fractions to simplest form, solving equations, etc. In this section we shall define a trigonometric identity and list eight such identities, which are called the *basic identities of trigonometry*.

Let θ represent an arbitrary angle from the set of all possible angles. An equation involving two or more trigonometric functions of θ is a trigonometric identity if the equation holds, that is, is true for all values of θ for which each function in the equation is defined.

The eight basic trigonometric identities are

$$[\text{I}] \ \csc \theta = \frac{1}{\sin \theta} \qquad\qquad [\text{V}] \ 1 + \tan^2 \theta = \sec^2 \theta$$

$$[\text{II}] \ \sec \theta = \frac{1}{\cos \theta} \qquad\qquad [\text{VI}] \ 1 + \cot^2 \theta = \csc^2 \theta$$

$$[\text{III}] \ \cot \theta = \frac{1}{\tan \theta} \qquad\qquad [\text{VII}] \ \tan \theta = \frac{\sin \theta}{\cos \theta}$$

$$[\text{IV}] \ \sin^2 \theta + \cos^2 \theta = 1 \qquad [\text{VIII}] \ \cot \theta = \frac{\cos \theta}{\sin \theta}$$

These eight identities are easily proved by using the definitions of the six trigonometric functions and the fact that the numbers x, y, and r appearing in the definitions always satisfy the relation $x^2 + y^2 = r^2$. For example, to prove Identity [I] we use the definitions of $\csc \theta$ and $\sin \theta$ as follows.

$$\csc \theta = \frac{r}{y} = \frac{1}{y/r} = \frac{1}{\sin \theta} \qquad \text{provided } y \neq 0$$

which shows that Identity [I] is true for all values of θ except those for which $\sin \theta = 0$. Identities [II] and [III] can be proved in a similar manner. Identities [I], [II], and [III] are often called the *reciprocal relations*. To prove Identity [IV], we start with the relation

$$x^2 + y^2 = r^2$$

Since r is never zero, we can divide by r^2 and obtain

$$\frac{x^2}{r^2} + \frac{y^2}{r^2} = \frac{r^2}{r^2} = 1$$

or

$$\left(\frac{x}{r}\right)^2 + \left(\frac{y}{r}\right)^2 = 1$$

By definition $\frac{x}{r} = \cos \theta$ and $\frac{y}{r} = \sin \theta$; hence

$$(\cos \theta)^2 + (\sin \theta)^2 = 1 \qquad (\textit{for all values of } \theta)$$

or

$$\sin^2 \theta + \cos^2 \theta = 1$$

where $\sin^2 \theta$ and $\cos^2 \theta$ are accepted shorter ways of writing $(\sin \theta)^2$ and $(\cos \theta)^2$. This short form will be used hereafter in writing powers of trigonometric

SEC. 11.11   Basic Trigonometric Identities

functions. Similarly, if $x \neq 0$, we can divide $x^2 + y^2 = r^2$ by x^2 and get

$$\frac{x^2}{x^2} + \frac{y^2}{x^2} = \frac{r^2}{x^2}$$

or

$$1 + \left(\frac{y}{x}\right)^2 = \left(\frac{r}{x}\right)^2$$

By definition, $\frac{y}{x} = \tan \theta$ and $\frac{r}{x} = \sec \theta$; therefore,

$$1 + \tan^2 \theta = \sec^2 \theta$$

for all values of θ for which $\tan \theta$ and $\sec \theta$ are defined. This proves Identity [V]. To prove Identity [VI], divide $x^2 + y^2 = r^2$ by y^2, $y \neq 0$. Identities [IV], [V], and [VI] are often called the *squared relations*. Identity [VII] is proved as follows.

$$\tan \theta = \frac{y}{x} = \frac{y/r}{x/r} = \frac{\sin \theta}{\cos \theta}$$

which holds for all values of θ for which $\tan \theta$ is defined. Identity [VIII] can be proved in a similar manner. Identities [VII] and [VIII] are called the *quotient relations*.

These basic identities are invaluable in many problems involving trigonometric functions. The following examples illustrate some important applications.

Example 1   Simplify $\dfrac{\sin^2 \theta}{1 + \cos \theta}$.

Solution: It follows from Identity [IV] that $\sin^2 \theta = 1 - \cos^2 \theta$. Hence

$$\frac{\sin^2 \theta}{1 + \cos \theta} = \frac{1 - \cos^2 \theta}{1 + \cos \theta}$$

But $1 - \cos^2 \theta = (1 - \cos \theta)(1 + \cos \theta)$; therefore,

$$\frac{\sin^2 \theta}{1 + \cos \theta} = \frac{(1 - \cos \theta)(1 + \cos \theta)}{1 + \cos \theta}$$

$$= 1 - \cos \theta$$

Example 2   If $\sin \theta = \dfrac{4}{5}$ and θ is in QII, use the basic identities to find the values of the other five functions.

Solution: From Identity [IV], $\cos^2 \theta + \sin^2 \theta = 1$ or $\cos^2 \theta = 1 - \sin^2 \theta$. Extracting square roots, we get

$$\cos \theta = \pm \sqrt{1 - \sin^2 \theta}$$

Since the cosine of an angle in the second quadrant is negative, we choose the negative sign and, substituting $\frac{4}{5}$ for $\sin\theta$, we obtain

$$\cos\theta = -\sqrt{1-\left(\frac{4}{5}\right)^2} = -\frac{3}{5}$$

From Identity [VII], $\quad \tan\theta = \dfrac{\sin\theta}{\cos\theta} = \dfrac{\frac{4}{5}}{-\frac{3}{5}} = -\dfrac{4}{3}$

From Identity [III], $\quad \cot\theta = \dfrac{1}{\tan\theta} = \dfrac{1}{-\frac{4}{3}} = -\dfrac{3}{4}$

From Identity [II], $\quad \sec\theta = \dfrac{1}{\cos\theta} = \dfrac{1}{-\frac{3}{5}} = -\dfrac{5}{3}$

From Identity [I], $\quad \csc\theta = \dfrac{1}{\sin\theta} = \dfrac{1}{\frac{4}{5}} = \dfrac{5}{4}$

An inspection of the basic identities reveals that the functions occurring most often are $\sin\theta$ and $\cos\theta$. For example, Identities [I], [II], [VII], and [VIII] express each of the other functions in terms of $\sin\theta$ and $\cos\theta$. This suggests the following "golden rule" for simplifying complicated trigonometric expressions: When in doubt how to proceed, write the expression in terms of $\sin\theta$ and $\cos\theta$ and simplify.

Example 3 Simplify the expression $\dfrac{\cos\theta - \sin\theta}{\cot\theta - 1}$.

Solution: Our solution will consist of three steps: (a) writing the given expression in terms of $\sin\theta$ and $\cos\theta$, (b) multiplying numerator and denominator of the derived expression by $\sin\theta$, and (c) dividing numerator and denominator of the fraction resulting from step (b) by $\cos\theta - \sin\theta$. These steps are shown below.

(a) $\quad \dfrac{\cos\theta - \sin\theta}{\cot\theta - 1} = \dfrac{\cos\theta - \sin\theta}{\dfrac{\cos\theta}{\sin\theta} - 1}$

(b) $\quad = \dfrac{\sin\theta(\cos\theta - \sin\theta)}{\sin\theta\left(\dfrac{\cos\theta}{\sin\theta} - 1\right)}$

SEC. 11.11 Basic Trigonometric Identities

$$\text{or} \qquad = \frac{\sin\theta(\cos\theta - \sin\theta)}{\cos\theta - \sin\theta}$$

(c) $\qquad = \sin\theta$

For convenience, the eight basic trigonometric identities are repeated here in tabular form.

Eight Basic Trigonometric Identities

$$[\text{I}] \quad \csc\theta = \frac{1}{\sin\theta}$$

$$[\text{II}] \quad \sec\theta = \frac{1}{\cos\theta}$$

$$[\text{III}] \quad \cot\theta = \frac{1}{\tan\theta}$$

$$[\text{IV}] \quad \sin^2\theta + \cos^2\theta = 1$$
$$[\text{V}] \quad 1 + \tan^2\theta = \sec^2\theta$$
$$[\text{VI}] \quad 1 + \cot^2\theta = \csc^2\theta$$

$$[\text{VII}] \quad \tan\theta = \frac{\sin\theta}{\cos\theta}$$

$$[\text{VIII}] \quad \cot\theta = \frac{\cos\theta}{\sin\theta}$$

Exercises 11.11

1. Prove Identity [II].
2. Prove Identity [III].
3. Prove Identity [VI].
4. Prove Identity [VIII].
5. Use the basic identities to find the values of the other five functions if $\cos\theta = -\frac{12}{13}$ and θ is an angle in the third quadrant.
6. Express each of the other five functions in terms of $\sin\theta$.

Use the basic identities to simplify each trigonometric expression in Exercises 7 through 26.

7. $\sin\theta \csc\theta$
8. $\cos\theta \tan\theta$

9. $\dfrac{\cot\theta}{\csc\theta}$ 10. $\dfrac{\cos^2\theta}{1-\sin\theta}$

11. $\csc\theta - \cot\theta\cos\theta$ 12. $\sin\theta(\cot\theta + \csc\theta)$

13. $\dfrac{\sin\theta\sec\theta}{\tan\theta + \cot\theta}$ 14. $\dfrac{1+\tan\theta}{1+\cot\theta}$

15. $\dfrac{\cos\theta - \sin\theta}{\tan\theta - 1}$ 16. $\dfrac{\sec\theta - \cos\theta}{\tan\theta}$

17. $\dfrac{\sin\theta - \cos\theta}{\tan\theta - 1}$ 18. $\dfrac{\sec^2\theta - \tan^2\theta}{\csc\theta}$

19. $\cos^4\theta + 2\sin^2\theta\cos^2\theta + \sin^4\theta$ 20. $\sec^2\theta(1-\sin^2\theta)$

21. $\sin\theta\cos^2\theta(\tan\theta + \cot\theta)$ 22. $(1-\sin^2\theta)(1+\tan^2\theta)$

23. $\dfrac{\cos\theta}{1+\sin\theta} + \dfrac{1+\sin\theta}{\cos\theta}$ 24. $\dfrac{\tan\theta + \sin\theta}{\cot\theta + \csc\theta}$

25. $\dfrac{\tan\theta + \cot\theta}{\sec\theta}$ 26. $\dfrac{\sin^3\theta - \cos^3\theta}{\sin\theta - \cos\theta}$

27. Prove the following identity by transforming the left side into the right side.

$$\dfrac{\cos\theta + \cot\theta}{1+\sin\theta} = \cot\theta$$

28. Use a hand calculator to find the value of the following expression after first simplifying it as much as possible.

$$\dfrac{\cot 37°(\sec^2 37° - 1)}{\cos 37° + \tan 37° \sin 37°}$$

11.12* FUNCTIONS OF $(A+B)$, $(A-B)$, AND $2A$

It is a simple matter to show that in general

$$\sin(A+B) \neq \sin A + \sin B$$

or that

$$\sin 2A \neq 2\sin A$$

For example, $\sin 30° = \dfrac{1}{2}$, $\sin 60° = \dfrac{\sqrt{3}}{2}$, and $\sin 90° = 1$. Since $30° + 60° = 90°$

SEC. 11.12   Functions of $(A+B)$, $(A-B)$, and $2A$

and $\dfrac{1}{2} + \dfrac{\sqrt{3}}{2} \neq 1$, then

$$\sin(30° + 60°) = \sin 90° \neq \sin 30° + \sin 60°$$

Similarly, since $60° = 2 \cdot 30°$, and $2 \cdot \dfrac{1}{2} = 1 \neq \dfrac{\sqrt{3}}{2}$, then

$$\sin 60° = \sin 2 \cdot 30° \neq 2 \sin 30°$$

We may therefore ask if it is possible to express $\sin(A+B)$, or in fact any function of $(A+B)$ or $(A-B)$, in terms of functions of A and functions of B? To answer such questions, we prove the following identities. We start by proving that if A and B are any two angles then

(6) $\qquad\qquad \cos(A-B) = \cos A \cos B + \sin A \sin B$

Since $\cos(-\theta) = \cos\theta$ and $\sin(-\theta) = -\sin\theta$ for all angles θ, if A and B are interchanged in Equation (6), both sides of the equation remain unchanged. We may therefore assume $A > B$. Also, since $\sin(\theta \pm 2\pi) = \sin\theta$ and $\cos(\theta \pm 2\pi) = \cos\theta$ are identities, we may assume without loss of generality that $0 \leq A \leq 2\pi$ and $0 \leq B \leq 2\pi$.

Draw a unit circle on a rectangular coordinate system and place angles A and B in standard position on this same system (see Figure 11.34). Let $P_1(x_1, y_1)$ be the point of intersection of the terminal side of B and the unit circle, and let $P_2(x_2, y_2)$ be the point of intersection of the terminal side of A and the circle. By definition,

$$\cos B = \frac{x_1}{1} = x_1, \qquad \cos A = \frac{x_2}{1} = x_2$$

$$\sin B = \frac{y_1}{1} = y_1, \qquad \sin A = \frac{y_2}{1} = y_2$$

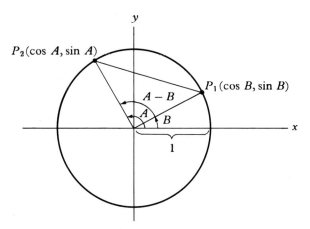

Figure 11.34

Hence $(x_1, y_1) = (\cos B, \sin B)$ and $(x_2, y_2) = (\cos A, \sin A)$, as shown in the figure. Applying the Law of Cosines,

$$(P_1 P_2)^2 = 1^2 + 1^2 - 2 \cdot 1 \cdot 1 \cos(A - B)$$
$$= 2 - 2 \cos(A - B)$$

Using the distance formula to find $(P_1 P_2)^2$, we get

$$(P_1 P_2)^2 = (\cos A - \cos B)^2 + (\sin A - \sin B)^2$$
$$= \cos^2 A - 2 \cos A \cos B + \cos^2 B + \sin^2 A$$
$$\quad - 2 \sin A \sin B + \sin^2 B$$
$$= (\cos^2 A + \sin^2 A) + (\cos^2 B + \sin^2 B)$$
$$\quad - 2 \cos A \cos B - 2 \sin A \sin B$$
$$= 2 - 2 \cos A \cos B - 2 \sin A \sin B$$

Equating the two expressions for $(P_1 P_2)^2$ and simplifying, we obtain

$$\cos(A - B) = \cos A \cos B + \sin A \sin B$$

and our identity is proved.

Since Equation (6) holds for all angles A and B, we can replace B by $(-B)$, which gives

$$\cos[A - (-B)] = \cos A \cos(-B) + \sin A \sin(-B)$$

But $\cos(-B) = \cos B$ and $\sin(-B) = -\sin B$; hence

(7) $$\cos(A + B) = \cos A \cos B - \sin A \sin B$$

Example 1

If $\sin A = \dfrac{12}{13}$, A in quadrant II

and $\cos B = \dfrac{4}{5}$, B in quadrant IV

find $\cos(A - B)$ and $\cos(A + B)$.

Solution: Before we can calculate $\cos(A - B)$ and $\cos(A + B)$, we need to calculate $\cos A$ and $\sin B$. To do this, place A and B in standard position on a coordinate system as shown in Figure 11.35. From the figure we see that $\cos A = -\dfrac{5}{13}$ and $\sin B = -\dfrac{3}{5}$.

We now substitute the known values for $\cos A$, $\sin A$, $\cos B$, and $\sin B$ into Equation (6), giving

$$\cos(A - B) = \left(-\frac{5}{13}\right)\left(\frac{4}{5}\right) + \left(\frac{12}{13}\right)\left(-\frac{3}{5}\right)$$
$$= -\frac{20}{65} - \frac{36}{65} = -\frac{56}{65}$$

SEC. 11.12   Functions of $(A+B)$, $(A-B)$, and $2A$

To find $\cos(A+B)$, we use Equation (7) and find that

$$\cos(A+B) = -\frac{20}{65} + \frac{36}{65} = \frac{16}{65}$$

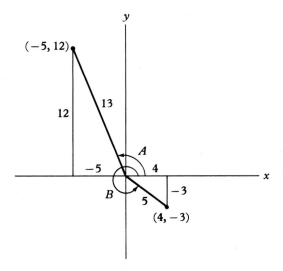

Figure 11.35

The angles $\frac{\pi}{2} - \theta$ and θ are complementary angles, and hence for any angle θ (Theorem 11.2),

$$\cos\left(\frac{\pi}{2} - \theta\right) = \sin\theta$$

$$\sin\left(\frac{\pi}{2} - \theta\right) = \cos\theta$$

To obtain a formula for $\sin(A+B)$, we proceed as follows:

$$\sin(A+B) = \cos\left[\frac{\pi}{2} - (A+B)\right]$$
$$= \cos\left[\left(\frac{\pi}{2} - A\right) - B\right]$$
$$= \cos\left(\frac{\pi}{2} - A\right)\cos B + \sin\left(\frac{\pi}{2} - A\right)\sin B$$

and finally

(8) $\qquad\qquad \sin(A+B) = \sin A \cos B + \cos A \sin B$

since $\cos\left(\dfrac{\pi}{2} - A\right) = \sin A$ and $\sin\left(\dfrac{\pi}{2} - A\right) = \cos A$. If in Equation (8) we replace B by $-B$, we obtain

(9) $$\sin(A - B) = \sin A \cos B - \cos A \sin B$$

The details are left to the student.

We can now obtain a formula for $\tan(A - B)$ in the following way.

$$\tan(A - B) = \dfrac{\sin(A - B)}{\cos(A - B)} \quad (Identity \text{ [IV]}, Section\ 11.11)$$

$$= \dfrac{\sin A \cos B - \cos A \sin B}{\cos A \cos B + \sin A \sin B} \quad (expanding)$$

$$\tan(A - B) = \dfrac{\dfrac{\sin A \cos B}{\cos A \cos B} - \dfrac{\cos A \sin B}{\cos A \cos B}}{\dfrac{\cos A \cos B}{\cos A \cos B} + \dfrac{\sin A \sin B}{\cos A \cos B}} \quad (divide\ by\ \cos A \cos B)$$

Then

(10) $$\tan(A - B) = \dfrac{\tan A - \tan B}{1 + \tan A \tan B}$$

Replacing B by $-B$ and recalling that $\tan(-B) = -\tan B$, Equation (10) becomes

(11) $$\tan(A + B) = \dfrac{\tan A + \tan B}{1 - \tan A \tan B}$$

It should be emphasized that Identities (10) and (11) are valid only for those angles for which the functions involved are defined and for which the denominators are not zero.

DEFINITION 11.1. The inclination of a line L is the angle ϕ measured counterclockwise from the x axis to the line. The angle ϕ satisfies the inequality

$$0° \leq \phi < 180°$$

and $\phi = 0$ if and only if L is parallel to the x axis.

It is obvious from Figure 11.36 that the slope m of line L is equal to the tangent of the angle of inclination ϕ of L. That is,

$$m = \dfrac{y_2 - y_1}{x_2 - x_1} = \tan \phi$$

SEC. 11.12 Functions of $(A+B)$, $(A-B)$, and $2A$

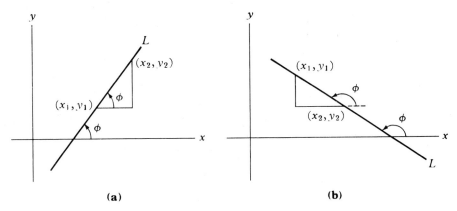

Figure 11.36

Let L_1 and L_2 be two lines, with inclination angles ϕ_1 and ϕ_2, respectively, which intersect at P, and let θ be the angle of intersection of L_1 and L_2 measured from L_1 counterclockwise around to L_2 (see Figure 11.37). Then $\theta = \phi_2 - \phi_1$ and

$$\tan \theta = \tan(\phi_2 - \phi_1) = \frac{\tan \phi_2 - \tan \phi_1}{1 + \tan \phi_2 \tan \phi_1}$$

If we denote the slopes of L_1 and L_2 by m_1 and m_2, the last equation can be written in the form

$$\tan \theta = \frac{m_2 - m_1}{1 + m_1 m_2}$$

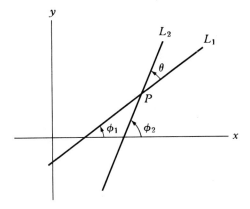

Figure 11.37

Example 2 Find the tangent of the acute angle θ between L_1 and L_2 if L_1: $2x - y = 3$ and L_2: $3x + y = 7$.

Solution: Here $m_1 = 2$ and $m_2 = -3$; hence

$$\tan \theta = \frac{m_2 - m_1}{1 + m_1 m_2} = \frac{-3 - 2}{1 - 6} = 1$$

so $\tan \theta = 1$ and $\theta = 45°$.

Double-Angle Formulas

If in Equation (7) we let $B = A$, it becomes

(12) $$\cos 2A = \cos^2 A - \sin^2 A$$

Similarly, by setting $B = A$ in Equation (8), we obtain

(13) $$\sin 2A = 2 \sin A \cos A$$

Formulas (12) and (13) are called double-angle formulas because the angle exhibited on the left is double the angle exhibited on the right.

For convenience, we gather together the more commonly used identities which have been developed in this section in tabular form.

Some Commonly Used Identities

$$\sin (A \pm B) = \sin A \cos B \pm \cos A \sin B$$

$$\cos (A \pm B) = \cos A \cos B \mp \sin A \sin B$$

$$\tan (A \pm B) = \frac{\tan A \pm \tan B}{1 \mp \tan A \tan B}$$

$$\sin 2A = 2 \sin A \cos A$$

$$\cos 2A = \cos^2 A - \sin^2 A$$

$$\tan 2A = \frac{2 \tan A}{1 - \tan^2 A}$$

Here, all "top signs" in a given formula yield an identity, and similarly, all "bottom signs" in a given formula yield an identity.

SEC. 11.12   Functions of $(A+B)$, $(A-B)$, and $2A$

Exercises 11.12

1. If $\cos A = -\frac{3}{5}$, A in quadrant II, and $\cos B = \frac{15}{17}$, B in quadrant I, find:
   a. $\sin(A+B)$ b. $\sin(A-B)$ c. $\cos(A+B)$ d. $\cos(A-B)$
   e. $\tan(A+B)$ f. $\tan(A-B)$ g. $\cos 2A$ h. $\sin 2B$
   i. $\cos 2B$

2. If $\tan A = \frac{1}{2}$, $\tan B = \frac{1}{3}$, and A and B are both acute angles, find $\tan(A+B)$ and thus show that $A + B = 45°$.

3. Simplify each of the following expressions and then find its value.
   a. $\cos 40° \cos 20° - \sin 40° \sin 20°$ b. $\sin 70° \cos 40° - \cos 70° \sin 40°$
   c. $\cos^2 15° - \sin^2 15°$ d. $2 \sin 15° \cos 15°$
   e. $\cos 75° \cos 15° + \sin 75° \sin 15°$ f. $\sin 40° \cos 50° + \cos 40° \sin 50°$
   g. $\dfrac{\tan 25° + \tan 20°}{1 - \tan 25° \tan 20°}$ h. $\dfrac{\tan 200° - \tan 65°}{1 + \tan 200° \tan 65°}$

4. By regarding 3θ as $2\theta + \theta$ and using Formulas (12) and (13), show that
$$\sin 3\theta = 3 \sin \theta - 4 \sin^3 \theta$$

5. Find the tangent of the acute angle of intersection of each given pair of lines.
   a. $3x + y = 7$, $x - y = 3$ b. $2x + 3y = 8$, $2x + y = 4$

6. Express each of the following as a function of θ only.
   a. $\cos(90° + \theta)$ b. $\sin(180° - \theta)$ c. $\sin(270° - \theta)$

7. If $\sin A = -\frac{4}{5}$, $\cos B = -\frac{5}{13}$, and A and B are in the same quadrant, find $\sin(A+B)$ and $\cos(A-B)$.

8. The angle of intersection of two intersecting curves is defined as the angle of intersection of the tangents to the two curves drawn at the point of intersection. Using this definition, find the tangent of the angle of intersection of the parabolas $y = x^2$ and $y^2 = x$ at the point $(1, 1)$.

9. Using the identities $\cos^2 \theta + \sin^2 \theta = 1$ and $\cos^2 \theta - \sin^2 \theta = \cos 2\theta$, show that
$$\sin^2 \theta = \frac{1 - \cos 2\theta}{2}, \quad \cos^2 \theta = \frac{1 + \cos 2\theta}{2}$$

10. Identify each given statement as true or false and give a reason. Do not use tables.
    a. $\cos^2 18° = \sin^2 18° + \cos 36°$ b. $\cos^2\left(\dfrac{\theta}{2}\right) + \sin^2\left(\dfrac{\theta}{2}\right) = \dfrac{1}{2}$
    c. $\cos 80° = 1 - 2 \sin^2 40°$ d. $\sin^2 70° + \sin^2 20° = 1$

11.13* GRAPHS OF THE TRIGONOMETRIC FUNCTIONS

A better understanding of the nature and the properties of the trigonometric functions may be obtained by the study of their graphs. Before discussing these graphs, we need to call attention to a very important property possessed by each of the trigonometric functions. These functions belong to a very special class of functions called *periodic functions*. We now define a periodic function.

DEFINITION 11.2. A function f is *periodic* if and only if $f(x+p)=f(x)$ for every x in the domain of f. The smallest $p > 0$ for which $f(x+p)=f(x)$ is called the *fundamental period* of f. If F is any trigonometric function and x is any angle, for which F is defined, it follows from the definition of F that $F(x+2n\pi) = F(x)$, where x is measured in radians, or $F(x° + n360°) = F(x°)$, where x is measured in degrees. The fundamental period of sin x, cos x, csc x, and sec x is 2π radians (or 360°). The fundamental period for the tangent and cotangent is π radians (or 180°). It is their periodicity that make the trigonometric functions so extremely useful, especially in electrical engineering, physics, and applied mathematics.

We now construct the graph of the function defined by the equation $y = \sin x$. Since $\sin(x + 2n\pi) = \sin x$ $(n = 0, \pm 1, \pm 2, ...)$, we shall draw the graph for $0 \le x \le 2\pi$ only. Since $y = f(x) = \sin x$ is a periodic function of fundamental period 2π, its graph outside the interval $[0, 2\pi]$ is easily obtained. We first construct the table shown in Figure 11.38 by calculating values of sin x at intervals of $\pi/6$ units, starting with $x = 0$. Next we plot these points on a rectangular coordinate system and draw a smooth curve through them as shown in Figure 11.39. The degree measure of the angle x is shown in the graph along with the value in radians. In the remainder of this chapter it will be assumed that all angles are measured in radians unless otherwise specified.

The graph of $y = A \sin x$ is just the same as the graph of $y = \sin x$ except each value of y is multiplied by the same constant, A. The range of values for y is from $-A$ to $+A$; the maximum height of the values is $+A$, and the minimum height is $-A$. This constant A is called the *amplitude*. For example, the graph of $y = 5 \sin x$ has an amplitude of 5.

Similarly, the graph of $y = \sin Bx$ resembles the graph of $y = \sin x$ except it will repeat itself B times as often; that is, the period of Bx is $2\pi/B$. For example, the graph of $y = 3 \sin 2x$ has an amplitude of 3 and a period of π.

Since $\cos x = \sin\left(\dfrac{\pi}{2} - x\right)$ for all values of x, we see that the values of cos x are the same as the values of sin x but for different values of x. In fact, the graph of $y = \cos x$ is the same as the graph of $y = \sin x$ but shifted $\dfrac{\pi}{2}$ units to the left. See Figure 11.40.

x	$y = \sin x$
0	0.00
$\dfrac{\pi}{6}$	0.50
$\dfrac{\pi}{3}$	0.87
$\dfrac{\pi}{2}$	1.00
$\dfrac{2\pi}{3}$	0.87
$\dfrac{5\pi}{6}$	0.50
π	0.00
$\dfrac{7\pi}{6}$	-0.50
$\dfrac{4\pi}{3}$	-0.87
$\dfrac{3\pi}{2}$	-1.00
$\dfrac{5\pi}{3}$	-0.87
$\dfrac{11\pi}{3}$	-0.50
2π	-0.00

Figure 11.38

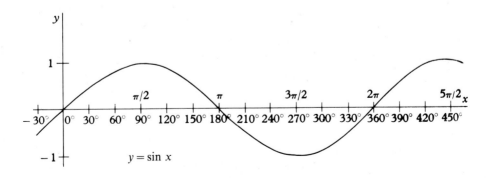

Figure 11.39

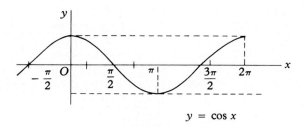

$y = \cos x$

Figure 11.40

The graphs of sin x and cos x show that each of these functions is continuous for all values of x. This means that $\lim_{x \to x_0} \sin x = \sin x_0$ and $\lim_{x \to x_0} \cos x = \cos x_0$. For example, $\lim_{x \to 0} \sin x = \sin 0 = 0$, and $\lim_{x \to 0} \cos x = \cos 0 = 1$.

Before drawing the graph of $y = \tan x$, we make the following observations. First we note that $\tan x$ is not defined at $x = \pm \dfrac{\pi}{2}, \pm \dfrac{3\pi}{2}, \pm \dfrac{5\pi}{2}$, and so on, for all odd multiples of $\dfrac{\pi}{2}$. Also, $\tan x = 0$ at $x = n\pi$ ($n = 0, \pm 1, \pm 2, \pm 3, \ldots$).

As was mentioned earlier, tan x has a fundamental period of π. This can be shown to be true as follows:

$$\tan(x + \pi) = \frac{\tan x + \tan \pi}{1 - \tan x \tan \pi} = \tan x \quad \text{since } \tan \pi = 0$$

It follows from the definition of tan x and the fact that $\tan(-x) = -\tan x$ that, as x increases from $-\dfrac{\pi}{2}$ to $x = \dfrac{\pi}{2}$, tan x increases from $-\infty$ to $+\infty$. These values are repeated in each successive interval of length π. We shall draw the graph in the interval at $\left(-\dfrac{\pi}{2}, \dfrac{\pi}{2}\right)$. The graph outside this interval is easily obtained. By using Table I or a hand calculator, we construct the table shown in Figure 11.41 and then plot the points from this table. The graph of $y = \tan x$ is shown in Figure 11.42. Actually, neither $\tan -\dfrac{\pi}{2}$ nor $\tan \dfrac{\pi}{2}$ is defined, but thinking of their values as $-\infty$ and $+\infty$, respectively, is helpful on occasion.

x	$\left(-\dfrac{\pi}{2}\right)$	$-\dfrac{\pi}{3}$	$-\dfrac{\pi}{4}$	$-\dfrac{\pi}{6}$	0	$\dfrac{\pi}{6}$	$\dfrac{\pi}{4}$	$\dfrac{\pi}{3}$	$\left(\dfrac{\pi}{2}\right)$
y	$(-\infty)$	-1.7	-1	-0.58	0	0.58	1	1.7	(∞)

Figure 11.41

SEC. 11.13 Graphs of the Trigonometric Functions

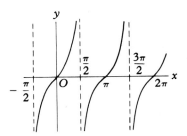

Figure 11.42

Therefore, we include these values in the table, placing them in parentheses. The portion of the complete graph of $y = \tan x$ shown in Figure 11.42 indicates that the function $\tan x$ is continuous in any interval in which it is defined. The graphs of the other three trigonometric functions are shown in Figure 11.43.

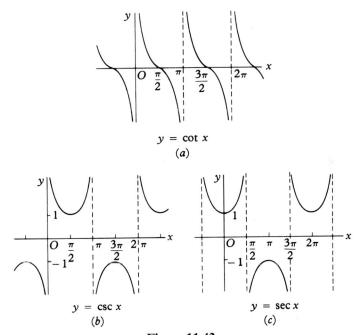

Figure 11.43

Exercises 11.13

Sketch the graph of each of the following equations.

1. $y = 2 \cos \frac{1}{2} x$ **2.** $y = 3 \sin 2x$ **3.** $y = 2 \sec \frac{1}{2} x$

4. $y = 3 \csc 2x$ 5. $y = \tan \frac{1}{2} x$ 6. $y = \cot 2x$

7. $y = \cos\left(x + \frac{\pi}{6}\right)$ 8. $y = \sin\left(x - \frac{\pi}{4}\right)$

11.14 DIFFERENTIATION AND INTEGRATION OF TRIGONOMETRIC FUNCTIONS

Before we attempt to develop formulas for differentiating the trigonometric functions, we shall first establish two important limit theorems.

THEOREM 11.4. *Suppose that $f(x) < h(x) < g(x)$ for all values of x in some interval containing c. Suppose further*

$$\lim_{x \to c} f(x) = \lim_{x \to c} g(x) = L$$

Then
$$\lim_{x \to c} h(x) = L$$

Proof: It follows from the definition of a limit that for each positive number d there corresponds a positive number k such that

$$L - d < f(x) < L + d \quad \text{and} \quad L - d < g(x) < L + d$$

when $|x - c| < k$. This implies that

$$L - d < h(x) < L + d \quad \text{when } |x - c| < k$$

This last statement is equivalent to the statement

$$|h(x) - L| < d \quad \text{when } |x - c| < k$$

Hence, by Definition 5.3,
$$\lim_{x \to c} h(x) = L$$

It is rather easy to see why this theorem is sometimes referred to as the squeeze theorem.

THEOREM 11.5. *If θ is an angle measured in radians, then*

(14)
$$\lim_{\theta \to 0} \frac{\sin \theta}{\theta} = 1$$

Proof: Since we are interested in $\lim \frac{\sin \theta}{\theta}$ as θ approaches zero, we may assume that $|\theta| < \frac{\pi}{2}$. First let us suppose that $\theta > 0$ and that it is the central

SEC. 11.14 Differentiation and Integration of Trigonometric Functions

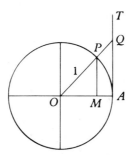

Figure 11.44

angle in a circle of radius $r = 1$ (Figure 11.44). In the figure, OA and OP are the sides of angle θ; AT is tangent to the circle at A and intersects the line OP at Q. Referring to Figure 11.44, we observe that

(15) $\qquad$ Area $\triangle OMP <$ area sector $OAP <$ area $\triangle OAQ$

These areas may be expressed in terms of θ as follows,

$$\text{Area } \triangle OMP = \frac{1}{2}|OM|\cdot|MP| = \frac{1}{2}\cos\theta\sin\theta$$

$$\text{Area sector } OAP = \frac{1}{2}r^2\theta = \frac{1}{2}\theta$$

$$\text{Area } \triangle OAQ = \frac{1}{2}|OA|\cdot|AQ| = \frac{1}{2}\tan\theta = \frac{1}{2}\frac{\sin\theta}{\cos\theta}$$

Substituting these results into Formula (15), we get

$$\frac{1}{2}\cos\theta\sin\theta < \frac{1}{2}\theta < \frac{1}{2}\frac{\sin\theta}{\cos\theta}, \qquad 0 < \theta < \frac{\pi}{2}$$

Since $\sin\theta$ is positive, we may divide all three terms by $\frac{1}{2}\sin\theta$ and obtain

(16) $\qquad \cos\theta < \dfrac{\theta}{\sin\theta} < \dfrac{1}{\cos\theta}$

Next we take the reciprocal of each term in Formula (16), which requires that the inequality symbols be reversed, obtaining

(17) $\qquad \dfrac{1}{\cos\theta} > \dfrac{\sin\theta}{\theta} > \cos\theta, \qquad 0 < \theta < \dfrac{\pi}{2}$

It follows that

(18) $\qquad \lim\limits_{\theta \to 0}\cos\theta \leq \lim\limits_{\theta \to 0}\dfrac{\sin\theta}{\theta} \leq \lim\limits_{\theta \to 0}\dfrac{1}{\cos\theta}$

It was pointed out in Section 11.13 that $\cos \theta$ is a continuous function and hence $\lim_{\theta \to 0} \cos \theta = 1$. Since $\lim_{\theta \to 0} \dfrac{1}{\cos \theta} = 1$, we may apply Theorem 11.4 to Formula (18) and conclude that

$$(19) \qquad \lim_{\theta \to 0} \frac{\sin \theta}{\theta} = 1$$

if θ approaches zero through positive values.

Although the argument leading to the result (19) is based on the assumption that θ is positive, the same limit is obtained if θ approaches zero through negative values. For, suppose θ is negative and $-\theta = \beta$, where β is positive. Then

$$\frac{\sin \theta}{\theta} = \frac{\sin(-\beta)}{-\beta} = \frac{-\sin \beta}{-\beta} = \frac{\sin \beta}{\beta}$$

Hence
$$\lim_{\theta \to 0} \frac{\sin \theta}{\theta} = \lim_{\beta \to 0} \frac{\sin \beta}{\beta} = 1$$

and our theorem is proved.

It is easily shown that if θ is measured in degrees instead of radians, then

$$(20) \qquad \lim_{\theta \to 0} \frac{\sin \theta^\circ}{\theta} = \frac{\pi}{180}$$

By contrasting the values of the limits in Formulas (14) and (20), we shall discover below why it is desirable to use radian measure in all calculus operations involving the trigonometric functions.

Example 1 | Use Theorem 11.5 to evaluate $\lim_{\theta \to 0} \dfrac{\sin 3\theta}{\theta}$.

Solution: Let $3\theta = x$; then $\theta = \dfrac{x}{3}$ and as $\theta \to 0$, $x \to 0$. Hence

$$\lim_{\theta \to 0} \frac{\sin 3\theta}{\theta} = \lim_{x \to 0} \frac{\sin x}{\frac{1}{3}x} = 3 \lim_{x \to 0} \frac{\sin x}{x} = 3 \cdot 1 = 3$$

Example 2 | Evaluate $\lim_{\theta \to 0} \dfrac{1 - \cos \theta}{\theta^2}$.

Solution: If $|\theta| < \pi$,

$$1 - \cos \theta = \frac{(1 - \cos \theta)(1 + \cos \theta)}{1 + \cos \theta} = \frac{\sin^2 \theta}{1 + \cos \theta}$$

SEC. 11.14 Differentiation and Integration of Trigonometric Functions

Hence

$$\lim_{\theta \to 0} \frac{1-\cos\theta}{\theta^2} = \lim_{\theta \to 0}\left(\frac{\sin^2\theta}{\theta^2} \cdot \frac{1}{1+\cos\theta}\right)$$

$$= \left(\lim_{\theta \to 0}\frac{\sin\theta}{\theta}\right)^2 \left(\lim_{\theta \to 0}\frac{1}{1+\cos\theta}\right) = 1^2 \cdot \frac{1}{1+1} = \frac{1}{2}$$

We now use the definition of a derivative to find $D_x y$ if $y = \sin x$, with x measured in radians.

$$D_x y = \lim_{h \to 0} \frac{\sin(x+h) - \sin x}{h}$$

Expanding $\sin(x+h)$, we can write

$$D_x y = \lim_{h \to 0} \frac{\sin x \cos h + \cos x \sin h - \sin x}{h}$$

$$= \lim_{h \to 0} \cos x \frac{\sin h}{h} - \lim_{h \to 0} \sin x \frac{1-\cos h}{h}$$

Using the identity $\cos 2A = \cos^2 A - \sin^2 A = 1 - 2\sin^2 A$, and letting $A = \frac{h}{2}$, we obtain the identity $1 - \cos h = 2\sin^2 \frac{h}{2}$. Making this substitution in the last equation, we have

$$D_x y = \lim_{h \to 0} \cos x \frac{\sin h}{h} - \lim_{h \to 0}\left[(\sin x)\left(\frac{2\sin^2 \frac{h}{2}}{h}\right)\right]$$

$$= \cos x \lim_{h \to 0} \frac{\sin h}{h} - \sin x \lim_{h \to 0} \sin \frac{h}{2} \lim_{h \to 0} \frac{\sin \frac{h}{2}}{\frac{h}{2}}$$

$$= (\cos x)(1) - (\sin x)(0)(1) = \cos x$$

Thus we have established the formula

(21) $\qquad D_x[\sin x] = \cos x \qquad$ (if x is measured in radians)

To generalize this result, suppose $y = \sin f(x)$, where f is a differentiable function of x. Using the chain rule, we obtain

(22) $\quad D_x y = \cos[f(x)] \cdot D_x f(x) \quad$ or $\quad D_x[\sin f(x)] = [\cos f(x)] \cdot D_x[f(x)]$

This last formula may be stated in words thus: the derivative of the sine of an angle, which is a function of a single independent variable, with respect to that variable is equal to the cosine of that same angle multiplied by the derivative of the angle.

Example 3 Find $D_x y$ if $y = \sin 5x$.

Solution: Here $f(x) = 5x$ and $D_x[f(x)] = 5$; hence
$$D_x y = (\cos 5x)(5) = 5 \cos 5x$$

Example 4 Find $D_x y$ if $y = 4 \sin(3x^2 - 7)$.

Solution: $D_x y = 4 \cos(3x^2 - 7) D_x[3x^2 - 7] = 4[\cos(3x^2 - 7)](6x)$
$$= 24x \cos(3x^2 - 7)$$

Example 5 Find $D_x y$ if $y = \sin^4(2x) = (\sin 2x)^4$.

Solution: We need to recall the formula $D_x[F(x)^n] = n[F(x)]^{n-1} \cdot D_x[F(x)]$.
$$D_x y = 4 \sin^3(2x) D_x[\sin 2x]$$
$$= 4 \sin^3(2x)(\cos 2x)(D_x[2x])$$
$$= 4(\sin^3 2x)(\cos 2x)(2)$$
$$= 8 \sin^3 2x \cos 2x$$

We could use the definition of a derivative to find $D_x(\cos x)$, but it is much easier to use Formula (21). First we write the identity

$$\cos x = \sin\left(\frac{\pi}{2} - x\right)$$

Then

$$D_x[\cos x] = D_x\left[\sin\left(\frac{\pi}{2} - x\right)\right]$$
$$= \cos\left(\frac{\pi}{2} - x\right) D_x\left[\frac{\pi}{2} - x\right] = -\cos\left(\frac{\pi}{2} - x\right)$$
$$= -\sin x$$

or

(23) $$D_x[\cos x] = -\sin x$$

which generalizes to the formula

(24) $$D_x[\cos f(x)] = [-\sin f(x)] \cdot D_x[f(x)]$$

Example 6 Find $D_x y$ if $y = \cos 7x$.

Solution: Using Formula (24), we obtain
$$D_x y = D_x[\cos 7x] = (-\sin 7x) D_x[7x]$$
$$= -7 \sin 7x$$

SEC. 11.14   Differentiation and Integration of Trigonometric Functions

Example 7 Find $D_x y$ if $y = 5 \cos^3 7x = 5[\cos 7x]^3$.

Solution: $D_x y = 3 \cdot 5 [\cos 7x]^2 \cdot D_x [\cos 7x]$
$= 15 [\cos 7x]^2 [-(\sin 7x)(7)]$
$= -105 \cos^2 7x \sin 7x$

Example 8 Find $D_x y$ if $y = \cos(e^x)$.

Solution: $D_x y = -\sin(e^x) \cdot D_x [e^x] = -\sin(e^x) \cdot e^x$
$= -e^x \sin e^x$

To find $D_x[\tan x]$, we take advantage of the fact that $\tan x = \dfrac{\sin x}{\cos x}$ and use the formula for the derivative of a fraction:

$$D_x \left[\frac{f(x)}{g(x)} \right] = \frac{g(x) \cdot D_x [f(x)] - f(x) \cdot D_x [g(x)]}{(g(x))^2}$$

Hence

$$D_x[\tan x] = D_x \left[\frac{\sin x}{\cos x} \right] = \frac{\cos x D_x[\sin x] - \sin x D_x[\cos x]}{\cos^2 x}$$
$$= \frac{\cos x \cdot \cos x - \sin x \cdot (-\sin x)}{\cos^2 x}$$
$$= \frac{\cos^2 x + \sin^2 x}{\cos^2 x} = \frac{1}{\cos^2 x} = \sec^2 x$$

(25) $$D_x[\tan x] = \sec^2 x$$

Generalizing, using the chain rule, we have

(26) $$D_x[\tan f(x)] = \sec^2 f(x) \cdot D_x[f(x)]$$

Example 9 Find $D_x y$ if $y = \tan 3x$.

Solution: Using Formula (26), we get
$D_x y = D_x[\tan 3x] = \sec^2 3x \cdot D_x[3x]$
$= 3 \sec^2 3x$

Example 10 Find $D_x y$ if $y = \tan \sqrt{x}$.

Solution: $D_x y = D_x[\tan \sqrt{x}] = \sec^2 \sqrt{x} \cdot D_x[\sqrt{x}] = \sec^2 \sqrt{x} D_x[x^{1/2}]$
$= \sec^2 \sqrt{x} \cdot \left(\dfrac{1}{2} x^{-1/2} \right) = \dfrac{\sec^2 \sqrt{x}}{2\sqrt{x}}$

Example 11 Find $D_x y$ if $y = \tan^5(x^2) = [\tan(x^2)]^5$.

Solution: $D_x y = 5[\tan(x^2)]^4 D_x[\tan(x^2)]$
$= 5[\tan(x^2)]^4 [\sec^2(x^2)] D_x[x^2]$
$= 5 \tan^4 x^2 \sec^2 x^2 \cdot (2x)$
$= 10x \tan^4 x^2 \sec^2 x^2$

By applying the formulas for the derivative of sin x, cos x, and the derivative of a fraction, as was done in finding the derivative of tan x, to the identities

$$\cot x = \frac{\cos x}{\sin x}, \qquad \sec x = \frac{1}{\cos x}, \qquad \csc x = \frac{1}{\sin x}$$

we obtain the following formulas:

(27) $\qquad D_x[\cot x] = -\csc^2 x$

(28) $\qquad D_x[\sec x] = \sec x \tan x$

(29) $\qquad D_x[\csc x] = -\csc x \cot x$

These are readily generalized to the formulas

(30) $\qquad D_x[\cot f(x)] = -[\csc^2 f(x)] D_x[f(x)]$

(31) $\qquad D_x[\sec f(x)] = [\sec f(x) \tan f(x)] D_x[f(x)]$

(32) $\qquad D_x[\csc f(x)] = -[\csc f(x) \cot f(x)] D_x[f(x)]$

This completes our list of derivative formulas for the derivatives of the six trigonometric functions. The proofs of Formulas (27), (28), and (29) and their generalizations are left to the student in the exercises.

Integrals of the Trigonometric Functions

Since $D_x[\sin x] = \cos x$ and $D_x[\cos x] = -\sin x$, it follows from the definition of an integral that

(33) $\qquad \int \cos x\, dx = \sin x + C \quad \text{and} \quad \int \sin x\, dx = -\cos x + C$

or in general

(34) $\qquad \int \cos f(x) \cdot f'(x)\, dx = \sin f(x) + C$

(35) $\qquad \int \sin f(x) f'(x)\, dx = -\cos f(x) + C$

Similarly, since $D_x[\tan f(x)] = \sec^2 f(x) \cdot D_x[f(x)]$ and $D_x[\cot f(x)] = -\csc^2 f(x) \cdot D_x[f(x)]$, we obtain the following integrals:

(36) $\qquad \int \sec^2 f(x) \cdot f'(x)\, dx = \tan f(x) + C$

and

(37) $\qquad \int \csc^2 f(x) \cdot f'(x)\, dx = -\cot f(x) + C$

SEC. 11.14 Differentiation and Integration of Trigonometric Functions

Example 12 Evaluate the integral $\int \sin 5x \, dx$.

Solution: $\int \sin 5x \, dx = \dfrac{1}{5} \int \sin 5x \, 5dx = -\dfrac{1}{5} \cos 5x + C$

Example 13 $\int \cos(3x+2) \, dx = \dfrac{1}{3} \int \cos(3x+2) \cdot 3 \, dx = \dfrac{1}{3} \sin(3x+2) + C$

Example 14 $\int \sec^2 2x \, dx = \dfrac{1}{2} \int \sec^2 2x \cdot 2 \, dx = \dfrac{1}{2} \tan 2x + C$

Example 15 $\int \dfrac{\csc^2 \sqrt{x}}{\sqrt{x}} \, dx = 2 \int \csc^2 \sqrt{x} \cdot \dfrac{1}{2\sqrt{x}} \, dx = -2 \cot \sqrt{x} + C$

To evaluate the integral $\int \tan x \, dx$, we need to recall the integral formula

$$\int \dfrac{f'(x) \, dx}{f(x)} = \ln |f(x)| + C$$

To use this formula, we write

(38) $\quad \int \tan x \, dx = \int \dfrac{\sin x}{\cos x} \, dx = -\int \dfrac{-\sin x}{\cos x} \, dx = -\ln |\cos x| + C$

Similarly,

(39) $\quad \int \cot x \, dx = \int \dfrac{\cos x \, dx}{\sin x} = \ln |\sin x| + C$

In more general form, the last two integrals become

(40) $\quad \int \tan f(x) \cdot f'(x) \, dx = \int \dfrac{\sin f(x) \cdot f'(x) \, dx}{\cos f(x)} = -\ln |\cos f(x)| + C$

(41) $\quad \int \cot f(x) \cdot f'(x) \, dx = \int \dfrac{\cos f(x) \cdot f'(x) \, dx}{\sin f(x)} = \ln |\sin f(x)| + C$

We observe that

$$D_x[\sec f(x) + \tan f(x)] = [\sec f(x) \tan f(x) + \sec^2 f(x)] f'(x)$$

and that

$$\sec f(x) = \dfrac{\sec f(x)[\sec f(x) + \tan f(x)]}{\sec f(x) + \tan f(x)}$$
$$= \dfrac{\sec f(x) \tan f(x) + \sec^2 f(x)}{\sec f(x) + \tan f(x)}$$

Hence

(42) $\displaystyle\int \sec f(x) \cdot f'(x)\, dx = \int \frac{[\sec f(x) \tan f(x) + \sec^2 f(x)] \cdot f'(x)\, dx}{\sec f(x) + \tan f(x)}$
$= \ln |\sec f(x) + \tan f(x)| + C$

Using a similar argument, it can be shown that

(43) $\displaystyle\int \csc f(x) \cdot f'(x)\, dx = \ln |\csc f(x) - \cot f(x)| + C$

Example 16 $\displaystyle\int \tan 2x\, dx = \frac{1}{2}\int \tan 2x \cdot 2\, dx = -\frac{1}{2} \ln |\cos 2x| + C$

Example 17 $\displaystyle\int \sec 3x\, dx = \frac{1}{3}\int \sec 3x \cdot 3\, dx = \frac{1}{3} \ln |\sec 3x + \tan 3x| + C$

The integral formula

$$\int [f(x)^n] f'(x) = \frac{[f(x)]^{n+1}}{n+1} + C, \qquad n \neq -1$$

may be used to evaluate certain integrals involving powers of the trigonometric functions, as illustrated by the following examples.

Example 18 Evaluate the integral $\int \sin^4 3x \cos 3x\, dx$.

Solution: Let $f(x) = \sin 3x$; then $f'(x) = 3 \cos 3x$. Hence

$\displaystyle\int (\sin 3x)^4 (\cos 3x)\, dx = \frac{1}{3}\int (\sin 3x)^4 (3 \cos 3x\, dx) = \frac{1}{3} \frac{(\sin 3x)^5}{5} + C$
$= \frac{1}{15} \sin^5 3x + C$

Example 19 Evaluate the integral $\int \sin^3 2x\, dx$.

Solution: $\sin^3 2x = (\sin^2 2x)(\sin 2x) = (1 - \cos^2 2x) \sin 2x$. Hence

$\displaystyle\int \sin^3 2x\, dx = \int (1 - \cos^2 2x) \sin 2x\, dx$
$= \int \sin 2x\, dx - \int \cos^2 2x \sin 2x\, dx$
$= \frac{1}{2}\int \sin 2x\, 2dx + \frac{1}{2}\int (\cos 2x)^2 (-2 \sin 2x)\, dx$
$= -\frac{1}{2} \cos 2x + \frac{1}{6}(\cos 2x)^3 + C$
$= -\frac{1}{2} \cos 2x + \frac{1}{6} \cos^3 2x + C$

SEC. 11.14 Differentiation and Integration of Trigonometric Functions

For convenience, we gather the formulas for derivatives and integrals of trigonometric functions together in tabular form.

Derivatives and Integrals of Trigonometric Functions

$$D_x[\sin f(x)] = \cos f(x) \cdot D_x[f(x)]$$
$$D_x[\cos f(x)] = -\sin f(x) \cdot D_x[f(x)]$$
$$D_x[\tan f(x)] = \sec^2 f(x) \cdot D_x[f(x)]$$
$$D_x[\cot f(x)] = -\csc^2 f(x) \cdot D_x[f(x)]$$
$$D_x[\sec f(x)] = \sec f(x) \cdot \tan f(x) \cdot D_x[f(x)]$$
$$D_x[\csc f(x)] = -\csc f(x) \cdot \cot f(x) \cdot D_x[f(x)]$$
$$\int \sin f(x) f'(x)\, dx = -\cos f(x) + C$$
$$\int \cos f(x) f'(x)\, dx = \sin f(x) + C$$
$$\int \tan f(x) f'(x)\, dx = -\ln |\cos f(x)| + C$$
$$\int \cot f(x) f'(x)\, dx = \ln |\sin f(x)| + C$$
$$\int \sec f(x) f'(x)\, dx = \ln |\sec f(x) + \tan f(x)| + C$$
$$\int \csc f(x) f'(x)\, dx = \ln |\csc f(x) - \cot f(x)| + C$$

Exercises 11.14

1. Evaluate each of the following limits by making use of Theorem 11.5 with appropriate trigonometric identities and theorems on limits.

   a. $\lim\limits_{x \to 0} \dfrac{\sin 3x}{x}$ b. $\lim\limits_{x \to 0} \dfrac{\tan x}{x}$ c. $\lim\limits_{x \to 0} \dfrac{\sin x}{5x}$ d. $\lim\limits_{x \to 0} \dfrac{\sin x}{2x^2 + 3x}$

In each of Exercises 2 through 15, find $D_x y$.

2. $y = 2 \sin 7x$ 3. $y = \cos(3x + 7)$ 4. $y = \sin(x^3 + 5)$

5. $y = \dfrac{1 - \cos^2 x}{\sin x}$ 6. $y = \dfrac{\cos^2 x}{1 - \sin x}$ 7. $y = \sin^4 3x$

8. $y = \cos^5 2x$
9. $y = e^x \sin x$
10. $y = \dfrac{1}{\cos 2x}$

11. $y = \tan(2x+1)$
12. $y = \cos^2 3x + \sin^2 3x$
13. $y = \ln(\sec x)$
14. $y = \csc 3x$
15. $y = e^{2x} \cos 3x$

Evaluate the following integrals.

16. $\displaystyle\int \sin 5x \, dx$
17. $\displaystyle\int \cos(2x+3) \, dx$
18. $\displaystyle\int 7 \cos 4x \, dx$

19. $\displaystyle\int \sec^2 3x \, dx$
20. $\displaystyle\int \tan^2 x \, dx$
21. $\displaystyle\int \cot 7x \, dx$

22. $\displaystyle\int \dfrac{3 \, dx}{\sin^2 x}$
23. $\displaystyle\int \dfrac{\cos x \, dx}{\sqrt{3 + \sin x}}$
24. $\displaystyle\int \sin x \cos^3 x \, dx$

25. $\displaystyle\int \tan^3 x \sec^2 x \, dx$

12

Probability and Statistics

Introduction

The word *probability* is used loosely in our everyday conversation, and we know only vaguely what it means. We talk of the chance that Tennessee will win a football game or that it will rain tomorrow. We are interested in a future event, of which the outcome is uncertain and about which we want to make a kind of prediction. In a mathematical discussion of probability we are more precise in that we try to present conditions under which we can make sensible numerical statements about uncertainties and to present methods of calculating numerical values of probabilities.

12.1 PROBABILITY

All mathematical probabilities are measured on a scale that extends from 0 to 1. The value 0 is assigned to any event that certainly will not occur; that is, the probability is 0 that the Earth and Mars will collide tomorrow. The value 1 is assigned to events that are certain to occur; that is, the probability is 1 that the sun will rise tomorrow. Events that may occur or may not occur are assigned fractional probabilities between 0 and 1.

Let us consider one of the simplest phenomena of chance, the outcome of the toss of a coin. Here there are two possible outcomes, a head or a tail. It is certain that a head or a tail will occur on any one toss of the coin. If it be agreed that these two outcomes are equally likely, the total probability of 1 must be equally divided between them. Thus we assign $\frac{1}{2}$ to each as its probability.

Another simple example is the roll of a die. An ideal die is a perfect cube with six faces having one to six dots. If we roll it once, the set of possible outcomes is $S = \{1, 2, 3, 4, 5, 6\}$. It is certain that one of these will occur, and if we agree to assign equal probabilities to the six numbers, then each gets, as its equal share, $\frac{1}{6}$ of the total probability.

Many times we want the probability of an event E which consists of several elementary outcomes. Thus the event "a die shows an even number of dots" corresponds to the subset $\{2, 4, 6\}$ of the original set S, and we naturally add the probabilities assigned to the elementary outcomes in the subset to obtain $\frac{3}{6}$ as the probability that the die will turn up even.

It should be noted that the set $S = \{1, 2, 3, 4, 5, 6\}$ of possible outcomes of the roll of a true die is not unique. We might, for instance, use the sets $S_1 = \{$odd number, even number$\}$ or $S_2 = \{$number is 2 or less, number is 3 or more$\}$. Each of these sets provides an *exhaustive* set of *mutually exclusive* outcomes for the experiment. The two outcomes in S_1 would be considered equally likely, and we would assign probability $\frac{1}{2}$ to each. For S_2, however, the outcomes are not equally likely, and it seems natural to assign the following probabilities:

$$\text{Pr (number is 2 or less)} = \tfrac{1}{3}$$

$$\text{Pr (number is 3 or more)} = \tfrac{2}{3}$$

DEFINITION 12.1. If an event E can happen in m cases out of a total of n possible cases, which are all considered to be equally likely, then the *probability* of the event E is defined as m/n. The notation for this is $\Pr(E) = m/n$.

This definition is perfectly clear-cut and agrees with our intuitive notion of chance in the simplest situations. However, it is difficult or impossible to apply in more complicated situations, when we cannot make a simple enumeration of cases that are equally likely. In such situations we appeal to observational evidence for probabilities. Suppose, for example, that a coin is tossed a large number of times. The probability prediction is that heads will turn up one half of the time. That is, we can think of probability as predicting the relative frequency of occurrence of some event. It does not predict the outcome of any one trial or the order of the outcomes of several trials, but only what will happen in the long run or on the average, in a large number of trials. This leads us to assume that the relative frequency of occurrence of an event in an actual long-run series of trials will approximate the probability and that, the larger the number of trials, the better the approximation generally. For example, from mortality tables compiled in past years, we find that, of 100,000 new-born white American males, about 92,300 are alive at age 20. This allows us to estimate the probability that a new-born white American male will live to be 20 years old as 0.923. Insurance companies use such empirically determined relative frequencies in calculating life insurance premiums.

SEC. 12.1 Probability

Exercises 12.1

1. What is the probability that a letter of the alphabet picked at random (by pure chance) will be a vowel?

2. If a card is drawn at random (by pure chance) from a pack of playing cards, what is the probability that it will be a heart? That it will be an honor card? That it will not be a diamond?

3. If two coins are tossed, what is the probability that both will show heads? That one will show heads and one tails?

4. Construct a table showing all the possible results when three coins are tossed. What is the probability of at least two heads? Of at most two heads?

5. In a roll of one die, what is the probability that the number of dots obtained will not exceed four? Will be at least four?

6. List the set of 36 equally likely, mutually exclusive, outcomes when two dice, one red and one green, are rolled.
   a. What is the probability of obtaining exactly one 6? Of not obtaining a 6? Of obtaining at least one 6?
   b. What is the probability that both dice show the same face?

7. Suppose you are only interested in the total number of dots that show when two dice are rolled. Use the material of Exercise 6 to compile a set of outcomes and probabilities pertinent to this problem.
   a. What is the probability of obtaining a total of 6 dots? A total of at most 6 dots?
   b. What total number of dots has the greatest probability?

8. The following mortality table shows the number of survivors, to various ages, among 100,000 new-born white American males.

Age x	Survivors to age x
0	100,000
10	93,601
20	92,293
30	90,092
40	86,880
50	80,521
60	67,787

   a. Estimate the probability that a new-born infant in this class will live to be 50 years old.
   b. Estimate the probability that a 20-year-old in this class will live to be 50 years old.

9. The probability that an individual will die from a particular operation is 0.01. A doctor has performed this operation 99 times in succession with no resulting deaths. Does this mean that the next such operation he performs will be fatal to the patient?

12.2 PERMUTATIONS AND COMBINATIONS

Many probability calculations involve questions similar to the following. If a coin is tossed n times (or if n separate coins are tossed), in how many distinguishable ways can we get exactly m heads and $n - m$ tails? If n is small, the question may be answered easily by listing all possible arrangements of n heads and tails, as in Exercises 3 and 4 of Section 12.1, and then simply counting those arrangements that contain m heads and $n - m$ tails. When n is large, a complete listing of the possible arrangements becomes onerous to say the least. The answer to the question is obtainable without a listing, by means of a general formula for the number of arrangements of n objects that are not all distinct. Before we derive the formula, however, we shall consider some simpler problems in the enumeration of arrangements.

Permutations

We shall consider first the number of ways in which we can select and arrange two objects if we have five distinct objects from which to choose: If the objects are designated by the letters A, B, C, D, and E, we have the following possibilities:

AB	BA	CA	DA	EA
AC	BC	CB	DB	EB
AD	BD	CD	DC	EC
AE	BE	CE	DE	ED

Here we consider AB and BA to be separate arrangements because the objects appear in different orders. Arrangements such as these, in which the order of the objects and the objects themselves are both distinguishing elements, are called *permutations*. Specifically, we have enumerated all possible permutations of the five objects A, B, C, D, and E, using exactly two at one time. The number of such permutations will be denoted by $P(5, 2)$, and here we see that $P(5, 2) = 20$. In general, $P(n, r)$ denotes the number of permutations which may be formed by using precisely r objects if n objects are available.

Observe in the example above that we have five choices for the first object, and for each of these five choices there are four choices for the second object. Thus in this example we have $P(5, 2) = 5 \cdot 4 = 20$.

In general, in the computation of $P(n, r)$ we have n choices for the first object. After the first object has been chosen, there are $n - 1$ choices available for the second object, and each of these $n - 1$ choices may be associated with each of the n choices of the first object. Thus there are $n(n - 1)$ choices of the first two objects. There are $n - 2$ choices available for the third object, and each of these choices may be associated with each of the choices for the first two objects. Thus there are $n(n - 1)(n - 2)$ choices for the first three objects. This line of reasoning

SEC. 12.2 Permutations and Combinations

may be continued down to the $n - r + 1$ available choices for the rth object and to the final formula for the permutations of r objects when n are available:

$$P(n, r) = n(n - 1)(n - 2) \cdots (n - r + 1)$$

Expressions similar to this will occur frequently in the remaining discussion; consequently, it will be convenient to have a more concise notation. We define n *factorial*, written $n!$, where n is a positive integer, to be the product of the first n positive integers:

$$1! = 1$$
$$2! = 1 \cdot 2$$
$$3! = 1 \cdot 2 \cdot 3$$
$$\vdots$$
$$n! = 1 \cdot 2 \cdot 3 \cdots (n - 1)n$$

Note: This list is completed with the special definition $0! = 1$.

Now we may rewrite our formula by multiplying it by the fraction $(n - r)!/(n - r)!$, which puts it in a much more convenient form to remember:

$$P(n, r) = \frac{n(n - 1) \cdots (n - r + 1)(n - r) \cdots 2 \cdot 1}{(n - r)(n - r - 1) \cdots 2 \cdot 1} = \frac{n!}{(n - r)!}$$

Example 1 How many different itineraries can be made out if each is to include 3 points of interest from a selected list of 9 points of interest? (Two itineraries that include the same points of interest will be considered different if they include these 3 points in a different order.)

Solution: Here $n = 9$ and $r = 3$. The order of arrangement is important, so the number of itineraries is the number of permutations:

$$P(9, 3) = \frac{9!}{(9 - 3)!} = \frac{9!}{6!} = \frac{9 \cdot 8 \cdot 7 \cdot \cancel{6} \cdot \cancel{5} \cdot \cancel{4} \cdot \cancel{3} \cdot \cancel{2} \cdot \cancel{1}}{\cancel{6} \cdot \cancel{5} \cdot \cancel{4} \cdot \cancel{3} \cdot \cancel{2} \cdot \cancel{1}} = 504$$

In some instances one or more of the objects being arranged may be indistinguishable, and the interchanges of these objects may be discounted in the enumeration of the number of permutations.

Example 2 How many distinct 9-letter arrangements may be made from the letters of the word *Tennessee*?

Solution: There are 9 letters and all 9 are to be used, but it is not true that $9!$ distinguishable arrangements are possible. The 4 letters e may be interchanged without affecting the apparent arrangement. Since, for each apparent arrangement, there are $P(4, 4) = 24$ ways of interchanging the letters e without affecting anything else, the number $9!$ counts each arrangement of them 24 times. By the same argument, each of the pairs of letters n and s gives twice as many apparent arrangements as are exactly distinguishable. Thus the number of distinct arrangements is

$$\frac{P(9,9)}{P(4,4)P(2,2)P(2,2)} = \frac{9!}{4!2!2!} = 3{,}780$$

In general, the formula for computing the number of distinguishable permutations of n objects of which a certain group, i in number, are alike and of another group, j in number, are alike, and so on, is

$$\frac{n!}{i!j!\cdots}$$

Combinations

Consider again our first example above. We might have requested the number of itineraries possible in which the only requirement was that the three points of interest were to be distinct. In this case the order in which the three points were visited would be of no importance; the distinguishing feature here is the specific choice of three points. We wish the number of combinations of nine points, three of which are to be chosen each time. We shall denote this by $C(9, 3)$. The distinction between permutations and combinations is a matter of order and of whether it makes any difference.

In counting permutations, two different arrangements of the same set of objects are considered to be different, as are two sets of objects some of which are different or all of which are different. In counting combinations, we distinguish only sets whose objects are some or all different. A rearrangement of the same set of objects does not produce a different combination. We determined that $P(9, 3) = 504$, but to determine $C(9, 3)$ we observe that for each combination of three things, all of which are to be used, there are $P(3, 3) = 6$ ways of permuting them. Hence we have

$$P(3, 3)C(9, 3) = P(9, 3)$$

or

$$C(9, 3) = \frac{P(9, 3)}{P(3, 3)}$$

thus

$$C(9, 3) = \frac{9!}{(9-3)!} \div \frac{3!}{(3-3)!}$$

or

$$C(9, 3) = \frac{9!}{6!3!} = 84$$

In general, we observe in similar fashion that if $C(n, r)$ is the number of

SEC. 12.2 Permutations and Combinations

combinations of n things, r of which are to be used each time, then

$$P(r, r)C(n, r) = P(n, r)$$

or

$$C(n, r) = \frac{P(n, r)}{P(r, r)}$$

and this reduces to

$$C(n, r) = \frac{n!}{(n-r)!r!}$$

It should be noted that the formulas for $C(n, r)$ and for the number of permutations of n objects, of which r are indistinguishable from each other and the remaining $n - r$ are also indistinguishable from each other, are identical. The identity of the formulas does not identify the two concepts.

Example 3 Find the number of ways in which a committee of 4 can be chosen from a class of 11. (Here the order of choice makes no difference.)

Solution: $C(11, 4) = \dfrac{11!}{7!4!} = \dfrac{11 \cdot 10 \cdot 9 \cdot 8 \cdot 7!}{7! \cdot 4 \cdot 3 \cdot 2 \cdot 1} = 330$

Example 4 Suppose that the class in the example above is composed of 7 men and 4 women and that the committee is to be chosen by lot. What is the probability that the committee will be composed of 2 men and 2 women?

Solution: The total number n of committees possible is again 330. For the committee to be composed of 2 men and 2 women, we note that there are $C(7, 2) = 21$ ways of selecting the men and $C(4, 2) = 6$ ways of selecting the women. The favorable committee may then be chosen in $m = 21 \cdot 6 = 126$ ways. By the definition of the probability of an event, the probability required here is $126/330 = 21/55$.

Example 5 Five cards are to be dealt from the standard deck of 52 cards. What is the probability that all 5 are spades?

Solution: Since there are 13 spades, the number m of possible 5-card hands, all of which are spades, is $C(13, 5)$. Similarly, the number n of 5-card hands which can be dealt from the entire deck is $C(52, 5)$. Hence

$$\Pr(E) = \frac{C(13, 5)}{C(52, 5)} = \frac{33}{66{,}640}$$

Exercises 12.2

Exercises 1 through 10 are problems in permutations.

1. Evaluate $P(5, 5)$ and $P(5, 3)$.

2. How many 7-letter arrangements may be formed from the letters in *Alabama*?

3. How many 4-letter arrangements may be formed from the letters in *New York*?

4. How many 11-letter arrangements may be formed from the letters in *Mississippi*?

5. How many 4-digit numbers with distinct digits can be written by use of the digits 1, 2, 3, 4, and 5 if the number is to be an even number?

6. How many 3-letter "words" are possible with our usual alphabet if the only requirement for a "word" is that it have at least 1 of the 5 vowels? How many such "words" are possible if none is to contain any single letter more than once?

7. A certain airline has 10 terminals. If a ticket shows only the terminals of departure and destination, how many different tickets may be written for this airline?

8. If one each of the tickets of different types for the airline in Exercise 7 is placed in a hat and then one is drawn from the hat at random, what is the probability that a specific one of the 10 terminals will be named on that ticket?

9. In how many ways can 10 coins be arranged in a row so that 4 heads and 6 tails show up?

10. If 10 equally balanced coins are tossed, what is the probability of obtaining 4 heads and 6 tails?

11. Evaluate $C(9, 3)$ and $C(9, 6)$. How do these values compare? Why?

12. Prove that $C(n, r) = C(n, n - r)$ if n is a positive integer and r is a nonnegative integer $(n > r)$.

13. A committee of 3 men and 2 women is to be appointed from a group of 12 men and 10 women. How many distinct possibilities are there?

14. A certain military unit has 48 privates on the roster. How many different lists of 5 privates each can be made up for KP duty?

15. A baseball team has 18 men on the roster as follows: 2 catchers, 7 pitchers, 5 infielders other than pitcher or catcher, and 4 outfielders. How many different lineups, without regard to batting order, can be fielded? How many different batting orders are possible?

16. What is the probability that 2 cards drawn at random from a deck of 52 cards will both be black?

17. What is the probability that 4 cards drawn at random from a standard deck will be the same suit?

18. A box contains 15 small plastic balls all alike except for color; 9 are white and the rest are red. If 3 balls are drawn simultaneously and at random, what is the probability that they will all be red? That there will be 1 red and 2 whites?

19. A purchaser agrees to buy a lot of 100 items if a careful examination of 10 of the items selected at random reveals no defects. If, unknown to the purchaser, the lot of 100 contains 2 defective items, what is the probability that he will buy the lot?

20. A box contains 50 good and 5 defective screws. If 10 screws are used, what is the probability that none is defective?

21. Suppose 3 bad light bulbs get mixed up with 12 good ones, and that you start testing the bulbs one by one. What is the probability that you will find 2 defective among the first 6 tested?

22. There are 3 roads between towns A and B, and a salesman wants to go from A to B and return. How many different routes are possible? Explain the conditions under which each of the following answers would be correct: 9, 6, 3.

12.3 THE BINOMIAL EXPANSION; PROBABILITY RULES

The use of permutations enables us to establish the binomial expansion of $(a+b)^n$, where n is a positive integer; this expansion was introduced in Section 2.8. We need notice only that in forming the product of n factors, each of which is $a+b$, we shall choose either an a or a b from each factor, but not both. Thus the terms must be of the type

$$C \cdot a^r \cdot b^{n-r}$$

where r is an integer. The coefficient C is the number of distinguishable orders of arrangement of r factors a and $n-r$ factors b. The identity of the formula for this number of permutations and the formula for $C(n, r)$ has already been noted. It is customary to take advantage of that identity for purposes of notation. Thus the term that involves $a^r b^{n-r}$ will be written as

(1) $$C(n, r) \cdot a^r \cdot b^{n-r}$$

The entire expansion is obtained by summing all such terms, r taking on the integral values from 0 to n inclusive. Thus

(2) $$\begin{aligned}(a+b)^n = C(n, n) \cdot a^n &+ C(n, n-1) \cdot a^{n-1} b \\ &+ C(n, n-2) \cdot a^{n-2} b^2 + \cdots \\ &+ C(n, 1) \cdot a b^{n-1} + C(n, 0) \cdot b^n\end{aligned}$$

Note that, since $C(n, r) = C(n, n-r)$, the coefficients in Equation (2) appear symmetrically from the ends of the expansion.

The student should review the examples and exercises on the binomial expansion given in Section 2.8.

The binomial expansion finds application in probability theory in the calculation of the probabilities of repeated trials, such as in tossing a coin many times or tossing many equally balanced coins.

Example 1 | What is the probability that a random throw of n coins will show x heads and $n-x$ tails for any one of the possible values $x = 0, 1, 2, \ldots, n$?

Solution: There are 2 ways of throwing each coin and so 2^n ways of throwing the n coins. Out of these 2^n ways, the number of ways yielding x heads and $n-x$ tails is the number of permutations of x heads and $n-x$ tails. This number is equal in value to $C(n, x)$. Thus the required probability is

$$(3) \qquad \frac{C(n, x)}{2^n} = C(n, x)\left(\frac{1}{2}\right)^x \left(\frac{1}{2}\right)^{n-x}$$

Notice that the right side of Equation (3) is the special case of Equation (1) obtained by setting $a = b = \frac{1}{2}$ and $r = x$. The example displays a general principle in the calculation of probabilities for repeated trials.

Principle: If an experiment can lead to only two possible results, the success of an event E or the failure of that event, and if the probabilities that the event succeed or fail in a single trial of the experiment are p and $q = 1 - p$, respectively, then the probability that the event E will succeed in x and fail in $n - x$ trials out of n independent trials of the experiment is

$$(4) \qquad C(n, x) \cdot p^x \cdot q^{n-x}$$

We have used implicitly three basic rules from probability theory in the discussion thus far. These will now be stated without proof.

RULE 1. If E and F denote events that are mutually exclusive (that is, that cannot occur simultaneously), then the probability that one of E or F will occur is the sum of the separate probabilities of E and F. In symbols, $\Pr(E \text{ or } F) = \Pr(E) + \Pr(F)$.

RULE 2. If E is any event and if "not E" denotes the failure of E to occur, then $\Pr(E) = 1 - \Pr(\text{not } E)$.

This rule is a corollary of the first one, since the events E and "not E" are mutually exclusive and one or the other is certain to occur.

RULE 3. If E and F denote events that are independent of each other (that is, the occurrence of either one does not affect the probability of occurrence of the other), then the probability that E and F will both occur is the product of the probabilities $\Pr(E)$ and $\Pr(F)$ of the single events E and F, each calculated without reference to whether or not the other event has or has not occurred. In symbols, $\Pr(E \text{ and } F) = \Pr(E) \cdot \Pr(F)$.

Example 2 | The game of chuck-a-luck is often played at small carnivals. A player pays a nickel to play. Three dice are rolled. If any "sixes" appear up, the player gets his nickel back and as bonus he gets an extra nickel for each "six" that appears. What is the probability that a player will win one or more nickels?

SEC. 12.3 The Binomial Expansion; Probability Rules

Solution: There are 3 mutually exclusive outcomes of the role of the dice that will win for the player: 3 "sixes," 2 "sixes" and anything else on the third die, and 1 "six" and anything else on the other 2 dice. To calculate the probabilities of these three cases, think of the dice as numbered: first die, second die, and third die. Let E, F, and G, respectively, denote the possible events described above.

For E to occur, all 3 dice must show "sixes." The probability of a six on any one die is $\frac{1}{6}$ and the dice are independent so, by Rule 3, $\Pr(E) = (\frac{1}{6})^3 = \frac{1}{216}$.

The event F can occur in three mutually exclusive ways, to be denoted by F_1, F_2, and F_3, where the subscript denotes in each case the number of the die that does not show a "six." Again the probability of any one "six" is $\frac{1}{6}$, while the probability of not a "six" is $\frac{5}{6}$. By Rule 3,

$$\Pr(F_1) = \tfrac{5}{6} \cdot \tfrac{1}{6} \cdot \tfrac{1}{6} = \tfrac{5}{216}$$

$$\Pr(F_2) = \tfrac{1}{6} \cdot \tfrac{5}{6} \cdot \tfrac{1}{6} = \tfrac{5}{216}$$

$$\Pr(F_3) = \tfrac{1}{6} \cdot \tfrac{1}{6} \cdot \tfrac{5}{6} = \tfrac{5}{216}$$

By Rule 1, then, $\Pr(F) = \Pr(F_1) + \Pr(F_2) + \Pr(F_3) = \frac{15}{216}$.

In a similar way, the event G may be broken into three mutually exclusive cases, G_1, G_2, and G_3, where the subscript denotes in each case the number of the die that shows a "six." As above,

$$\Pr(G_1) = \tfrac{1}{6} \cdot \tfrac{5}{6} \cdot \tfrac{5}{6} = \tfrac{25}{216}$$

$$\Pr(G_2) = \tfrac{5}{6} \cdot \tfrac{1}{6} \cdot \tfrac{5}{6} = \tfrac{25}{216}$$

$$\Pr(G_3) = \tfrac{5}{6} \cdot \tfrac{5}{6} \cdot \tfrac{1}{6} = \tfrac{25}{216}$$

and $\Pr(G) = \Pr(G_1) + \Pr(G_2) + \Pr(G_3) = \frac{75}{216}$

Finally, by Rule 1, $\Pr(\text{player wins}) = \Pr(E) + \Pr(F) + \Pr(G) = \frac{91}{216}$.

The meaning of independence in Rule 3 is further illustrated by the following example, in which the trials are not independent.

Example 3 A bag contains 15 marbles of the same size. Of these, 10 are white and 5 are red. Marbles are to be drawn at random from the bag one at a time until 3 have been withdrawn. What is the probability that the 3 marbles drawn will be red?

Solution: For the first marble drawn there are $n_1 = 15$ from which to draw and of these $m_1 = 5$ are red. Thus the probability that the first marble drawn will be red is $\frac{5}{15}$. After the first marble has been drawn, there remain $n_2 = 14$ marbles, of which $m_2 = 4$ are red for the second draw. Thus the probability that the second marble drawn will be red is $\frac{4}{14}$. Similarly, after the first 2 red marbles have been withdrawn, the probability that the third marble drawn will be red is $\frac{3}{13}$.

To obtain the probability that all three marbles will be red, we still multiply together the three single trial probabilities, just as in Rule 3, to get the result $\frac{1}{3} \cdot \frac{2}{7} \cdot \frac{3}{13} = \frac{2}{91}$. The departure from Rule 3 occurred in that the single trial probabilities were not calculated independently of each other.

Note that the same final result in this example could have been obtained by the use of a ratio of combinations, $C(5, 3)/C(15, 3) = \frac{2}{91}$. Compare the last two examples of Section 12.2. Thus this last method takes into account automatically the lack of independence of successive trials. It should not be used when trials are independent.

Exercises 12.3

1. What is the probability that a 5-card hand containing 3 aces and 2 kings will be dealt from a standard 52-card deck?

2. What is the probability that a penny will land with heads showing exactly 5 times if it is tossed 8 times?

3. What is the probability that if a pair of dice is thrown the sum of the dots will be at least 7? Exactly 7?

4. John has 3 white shirts and 4 colored shirts, while Joe has 2 white shirts and 6 colored shirts. What is the probability that both John and Joe wear white shirts the same day, assuming choices are random? What is the probability, assuming they each wear a clean shirt each day and have no 24-hour laundry service, that they both wear white shirts 2 days in succession?

5. What is the probability that a "six" will appear exactly 3 times if a die is thrown 6 times?

6. If it rains, on the average, 20% of the days in July, what is the probability that 3 randomly specified days in July will be dry?

7. Eighty percent of a certain tribe are right-handed and 60% have brown eyes. What is the probability that a randomly selected member of the tribe will be left-handed and have brown eyes?

8. If 90% of the items manufactured by a certain company pass inspection, what is the probability that 8 chosen at random out of a shipment of 10 will pass inspection?

9. A case of 100 toy autos contains 10 defective toys. If 5 are selected at random and shipped to a store, what is the probability that the store will receive at least 1 defective toy?

10. Assume that on the average 1 telephone number out of 5 called between 4 and 5 P.M. on weekdays in a certain city is busy. What is the probability that if 10 randomly selected telephone numbers are called during this hour, not more than 2 of them will be busy?

11. If 12 coins are tossed, what is the probability that at least 2 heads show?

12. Each of the 50 states has two senators. What is the probability that Tennessee will be represented on a committee of 50 chosen at random from the 100 senators?

13. Suppose there are 2 nickels and 2 dimes in each of 2 boxes. Find the probability of drawing 2 dimes when 1 coin is drawn at random from each box, and when all the coins are put in 1 box and then 2 coins are drawn at random.

SEC. 12.4   Mathematical Expectation

14. An urn contains 4 white, 5 red, and 6 black balls. Another urn contains 5 white, 6 red, and 7 black balls. One ball is selected from each urn. What is the probability that they will be of the same color?

12.4 MATHEMATICAL EXPECTATION

In any experiment or game in which numerical values are attached to each of the mutually exclusive events in an exhaustive set, the sum, over that set of events, of the products of the values of the events and the probabilities of the events is called the *mathematical expectation* of the values (briefly, the "expectation" or the "expected value") of the experiment or game.

Example 1   In a toss of 3 coins, let the value of each possible outcome be the number x of heads that show on the coins. Find the expectation of a toss.

Solution: The possible values of x are 0, 1, 2, and 3, and these values may be used to denote also the mutually exclusive set of events in this case. Thus

$$\text{Expectation} = 0 \cdot \Pr(x=0) + 1 \cdot \Pr(x=1) + 2 \cdot \Pr(x=2) + 3 \cdot \Pr(x=3)$$
$$= 0 \cdot (\tfrac{1}{2})^3 + 1 \cdot 3 \cdot (\tfrac{1}{2})^3 + 2 \cdot 3 \cdot (\tfrac{1}{2})^3 + 3 \cdot 1 \cdot (\tfrac{1}{2})^3$$
$$= \tfrac{12}{8} = 1.5$$

In a gambling game, the attached values for which the expectation is desired are usually the amounts of money to be gained or lost (taken to be positive if there is a gain, negative if there is a loss, zero if there is neither gain nor loss) from the occurrence of the various outcomes of the game. A game is called a *fair game* if the player's expectation from the game is zero. By way of illustration, let us return to Example 2 of Section 12.3 in which the game chuck-a-luck was described.

Example 2   Is chuck-a-luck a fair game?

Solution: We found that the total probability that a player will win is $\tfrac{91}{216}$. Thus the probability of not winning is $\tfrac{125}{216}$, and the value attached to this event is -5 since the player loses the nickel that he paid. The event E has the value 15, F has the value 10, and G has the value 5, since the player has gains of those amounts associated with those events. The player's expectation from the game is

$$15 \cdot \Pr(E) + 10 \cdot \Pr(F) + 5 \cdot \Pr(G) - 5 \cdot \Pr(\text{not winning})$$
$$= \tfrac{15}{216} + \tfrac{150}{216} + \tfrac{375}{216} - \tfrac{625}{216} = -\tfrac{85}{216}$$

This is approximately -0.4 cent; the game is not a fair game.

Example 3   John and Joe frequently have their coffee breaks together. It is their habit to match pennies to see which one will pay for the coffee each time. Coffee costs them 10 cents per cup. Is this a fair game?

Solution: On those occasions that John matches Joe, the game and its results may be tabulated as follows:

John's coin	Joe's coin	Who pays
H	H	Joe
T	T	Joe
H	T	John
T	H	John

The probability that John will pay is $\frac{1}{2}$, and in this case he pays 20 cents and receives a 10-cent cup of coffee; the net value to him is -10. The probability that John will not pay is also $\frac{1}{2}$, and in this case he receives a free cup of coffee; the net value to him is 10. Thus his expectation from the game is zero. Clearly, the situation is the same if Joe matches John. This is a fair game for John. It is also a fair game for Joe.

The expectation of value of an experiment or game is interpretable physically as the average of the values obtained from the experiment or game from a long series of repeated runs of the experiment or plays of the game.

Exercises 12.4

1. Find the expected value of the sum of the number of dots that show up on the roll of a pair of dice.

2. A game is played on the basis of a roll of 1 die. The player receives 2 pennies for each dot that shows up on the die. What is a fair price to play this game?

3. A certain lottery has 100,000 tickets numbered serially. A ticket costs its buyer $2.00. The tickets numbered 500, 8,325, and 92,350 pay off $10,000 each to their purchasers. What is a purchaser's expectation of value from a single ticket in this lottery?

4. If the probability of death from an airplane accident of any individual passenger on a one-way air trip is 10^{-5}, what is a fair price for an individual one-way-air-trip accident-insurance policy for $25,000? Ignore the costs and profits of running the insurance company.

5. If 5,000 lottery tickets are sold at $1 each on a $2,000 car, what is the expected gain of a person who buys 1 ticket?

6. A bag contains 7 white and 5 red balls. Three are withdrawn successively without replacement. What is the expected number of red balls drawn?

7. One purse contains 10 pennies and 4 dimes. Another purse contains 5 pennies and 6 dimes. One coin is drawn at random from each purse. What is the expected total draw?

SEC. 12.5 Probability Distributions

8. A game is played in which 2 coins are tossed. If both coins turn up heads, a player collects $1. If both coins turn up tails, the player collects $2. If the coins do not land the same way, the player pays $0.50. What is the expectation for the player of this game?

9. A card is drawn at random from a deck. If an ace is drawn, the drawer is credited with 15 points; if a king or queen, 12 points; and if a jack, 8 points. If a number card is drawn, the drawer loses 5 points. Find the expected number of points to be scored if this is done 20 times, with the card drawn being returned to the deck each time.

12.5 PROBABILITY DISTRIBUTIONS

It has been remarked earlier in the text that one of the roles of mathematics in applications is to provide models for observed relationships. The theory of probability provides models for the prediction of the relative frequencies of events arising from random experiments. In this section we shall give examples of several widely used models.

Example 1 This is an example from industrial quality control. Suppose that, under the well-controlled conditions of a modern assembly line, 10% of all finished items are defective; that is, the probability of a defective item is 0.1. What is the probability of 0, 1, 2, 3 defectives in a sample of 3 items drawn at random for testing?

Solution: The experiment in this case consists of drawing one item at random from the production line and noting whether it is good (G) or defective (D). The experiment is performed 3 times. If we assume that the 3 trials are independent of each other and that $\Pr(G) = 0.9$ and $\Pr(D) = 0.1$ for each trial, then Equation (4) of Section 12.3 can be applied in finding the required probabilities. Thus, if x is the number of defectives in the sample of 3, then

$$\Pr(x = x_i) = C(3, x_i)(0.1)^{x_i}(0.9)^{3-x_i} \quad \{x_i\} = \{0, 1, 2, 3\}$$

It is convenient to present the results in tabular form:

No. of defectives, x_i	0	1	2	3
$\Pr(x = x_i)$	0.729	0.243	0.027	0.001

We have listed a mutually exclusive and exhaustive set of outcomes with associated probabilities. The ordered pairs $[x_i, \Pr(x_i)]$ define a function called the probability function for this experiment; see Figure 12.1.

Since the formula for the probability function is one term of a binomial expansion, we may say that we have an example of the *binomial probability distribution*. It serves as a model for many practical situations.

The physical interpretation in this example is that, if we were to draw a great many 3-item samples for testing, about 73% of the samples would contain no defective, 24% would contain 1 defective, and only 3% would contain as

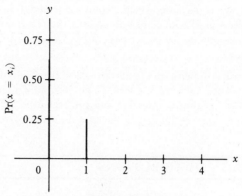

Figure 12.1

many as 2 defectives. If we do obtain as improbable a result as a sample with 2 or 3 defectives, we reason that there may be some trouble in the control of the production process that has caused $\Pr(D)$ to be larger than 0.1.

Thus far our discussion has been limited to cases in which the number of possible outcomes is finite. In many chance experiments the outcome is a measurement such as length, weight, or time, and the set of all possible outcomes is associated with a continuous variable. Our previous definition of probability then will not apply. We shall give an intuitive geometric approach to this type of problem (a rigorous treatment would require mathematical techniques that are beyond the scope of this text).

Example 2

Suppose a traffic signal has a 90-second cycle showing green for 50, amber for 10, and red for 30 seconds. What is the probability that a motorist will "hit" the green signal, assuming that he approaches the signal under chance conditions?

Solution: The time at which the motorist reaches the signal, measured from the start of the green signal, may be represented by the line interval of 0 to 90. The probability that he reaches the signal at some time in the cycle is 1 and will be represented by a rectangle of height $\frac{1}{90}$ and area 1, as shown in Figure 12.2. The

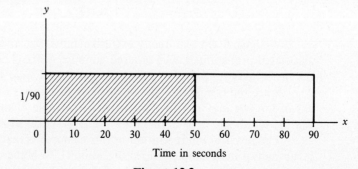

Figure 12.2

SEC. 12.5 Probability Distributions

probability that he "hits" the green light is, then,

$$\frac{\text{area of shaded rectangle}}{\text{area of large rectangle}} = \frac{\frac{50}{90}}{\frac{90}{90}} = \frac{5}{9}$$

Example 3 Experience in a certain factory has shown that an exponential function is a satisfactory probability model for the times required to serve mechanics at a tool crib. To be specific, suppose that areas under the curve $y = e^{-x}$, where x is time in minutes, can be used to represent the probability that the service time is of a given duration. What is the probability that the service time will be 1 minute or less? That it will be 3 minutes or less?

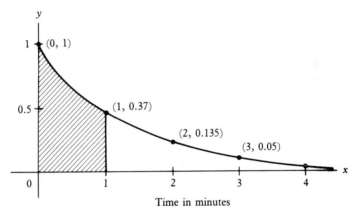

Figure 12.3

Solution: The graph of $y = e^{-x}$ is given in Figure 12.3, in which are used several values from Table IV of Appendix B. The probability that the service time will be 1 minute or less will be the ratio of the shaded area to the total area, or

$$\frac{\int_0^1 e^{-x}\, dx}{\int_0^\infty e^{-x}\, dx}$$

We state without proof that the denominator is 1, so the required probability (see Section 10.8) is

$$\int_0^1 e^{-x}\, dx = -e^{-x}\Big]_0^1 = 1 - e^{-1} = 1 - 0.36788 = 0.63212$$

Approximately 63% of the time calls at the tool crib will require 1 minute or less, provided the probability model adequately describes the real situation.

We leave it to the reader to show that, in the example just given, about 95% of all calls will require 3 minutes or less. Note that probability is represented by the *area* under a curve and not by the height of y in each of the last two examples. In general, probabilities associated with a continuous variable x are defined in such a way that the probability that x takes on a value between a and b is the area under a curve over the interval a to b. Since the area over a single point is 0, the probability that a continuous chance variable x will take on a specific value is defined as 0.

The most important probability distribution is the *standard normal distribution* specified by the curve whose equation is

(1) $$y = \frac{1}{\sqrt{2\pi}} e^{-x^2/2}$$

The table below gives the values of y for a few values of x. Further values could be computed by using the table of exponentials.

x	0.0	±0.5	±1.0	±1.5	±2.0	±2.5	±3.0	±3.5	±4.0
y	0.3989	0.3521	0.2420	0.1295	0.0540	0.0175	0.0044	0.0009	0.0001

If we plot these points and join them with a smooth curve, we obtain the well-known bell-shaped curve of Figure 12.4. This curve has the following properties,

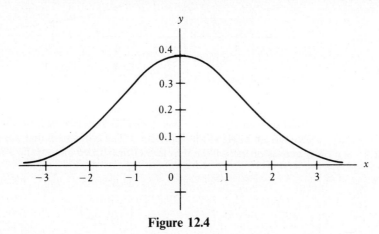

Figure 12.4

some of which are obvious and some of which require advanced calculus for proof:

1. It is symmetric about the y axis.
2. It is always above the x axis, extends indefinitely to the left and to the right, and approaches the x axis as its distance from the y axis increases.
3. The area bounded by the curve and the x axis is 1.
4. The area under the curve between $x = a$ and $x = b$ can be interpreted as the probability that a value of x selected at random from a set of numbers which has this distribution will have a value between a and b.

SEC. 12.5 Probability Distributions

The area under the standard normal curve over the interval $x = a$ to $x = b$ can be obtained by evaluating the integral

$$\text{(2)} \qquad \frac{1}{\sqrt{2\pi}} \int_a^b e^{-x^2/2} \, dx$$

although, unfortunately, this integral is one that cannot be evaluated by ordinary means. We shall, however, give a table of values for the integral $I(a)$, the area from $x = 0$ to $x = a$,

$$\text{(3)} \qquad I(a) = \frac{1}{\sqrt{2\pi}} \int_0^a e^{-x^2/2} \, dx$$

and from the table we shall be able to evaluate Equation (2) for any values of a and b. To do this, we note that

$$\text{(4)} \qquad \frac{1}{\sqrt{2\pi}} \int_a^b e^{-x^2/2} \, dx = \frac{1}{\sqrt{2\pi}} \int_0^b e^{-x^2/2} \, dx - \frac{1}{\sqrt{2\pi}} \int_0^a e^{-x^2/2} \, dx$$

and

$$\text{(5)} \qquad \frac{1}{\sqrt{2\pi}} \int_{-a}^0 e^{-x^2/2} \, dx = \frac{1}{\sqrt{2\pi}} \int_0^a e^{-x^2/2} \, dx$$

The accompanying table, gives the values of $I(a)$ for values of a from 0 to 4. Since the curve is so very close to the x axis at $x = 4$ and continues to approach the x axis, the area to the right of $x = 4$ is negligible.

Areas under the Standard Normal Curve

a	$I(a)$	a	$I(a)$	a	$I(a)$
0.0	0.0000	1.4	0.4192	2.8	0.4974
0.1	0.0398	1.5	0.4332	2.9	0.4981
0.2	0.0793	1.6	0.4452	3.0	0.4987
0.3	0.1179	1.7	0.4554	3.1	0.4990
0.4	0.1554	1.8	0.4641	3.2	0.4993
0.5	0.1915	1.9	0.4713	3.3	0.4995
0.6	0.2257	2.0	0.4772	3.4	0.4997
0.7	0.2580	2.1	0.4821	3.5	0.4998
0.8	0.2881	2.2	0.4861	3.6	0.4998
0.9	0.3159	2.3	0.4893	3.7	0.4999
1.0	0.3413	2.4	0.4918	3.8	0.4999
1.1	0.3643	2.5	0.4938	3.9	0.5000
1.2	0.3849	2.6	0.4953	4.0	0.5000
1.3	0.4032	2.7	0.4965		

Example 4   Given a set of numbers that has the standard normal distribution, find the probability that an element selected at random has a value between -0.4 and $+1.2$.

Solution: This probability is the area under the curve from $x = -0.4$ to $x = 1.2$, or the area from $x = -0.4$ to $x = 0$ plus the area from $x = 0$ to $x = 1.2$. By symmetry, the area from $x = -0.4$ to $x = 0$ is equal to the area from $x = 0$ to $x = 0.4$. Therefore, the area is

$$I(0.4) + I(1.2) = 0.1554 + 0.3849 = 0.5403$$

and so the required probability is 0.5403. Alternatively, using Equations (4) and (5), we obtain

$$\frac{1}{\sqrt{2\pi}} \int_{-0.4}^{1.2} e^{-x^2/2}\, dx = \frac{1}{\sqrt{2\pi}} \int_{0}^{1.2} e^{-x^2/2}\, dx - \frac{1}{\sqrt{2\pi}} \int_{0}^{-0.4} e^{-x^2/2}\, dx$$

$$= \frac{1}{\sqrt{2\pi}} \int_{0}^{1.2} e^{-x^2/2}\, dx + \frac{1}{\sqrt{2\pi}} \int_{-0.4}^{0} e^{-x^2/2}\, dx$$

$$= \frac{1}{\sqrt{2\pi}} \int_{0}^{1.2} e^{-x^2/2}\, dx + \frac{1}{\sqrt{2\pi}} \int_{0}^{0.4} e^{-x^2/2}\, dx$$

$$= 0.3849 + 0.1554 = 0.5403$$

Example 5   Given a set of numbers that has the standard normal distribution, find the probability that an element selected at random has a value between $x = 0.8$ and $x = 2.2$.

Solution: This probability is the area under the curve from $x = 0.8$ to $x = 2.2$, or the area under the curve from $x = 0$ to $x = 2.2$ minus the area under the curve from $x = 0$ to $x = 0.8$. Hence the required probability is

$$I(2.2) - I(0.8) = 0.4861 - 0.2881 = 0.1980$$

This distribution is a satisfactory model for a wide variety of practical situations. For example, heights of men, intelligence test scores, and diameters of bolts often have approximately normal distributions. We shall discuss some of these applications in Section 12.8, but first it will be necessary to develop some additional theory.

Exercises 12.5

1. A certain machine produces 5% defective items in the long run. What is the probability distribution of x, the number of defectives in random samples of 2 items each? Exhibit your answer as a table and as a formula.

2. Evaluate $\dfrac{1}{\sqrt{2\pi}} \displaystyle\int_{-2.3}^{1.8} e^{-x^2/2}\, dx$ 3. Evaluate $\dfrac{1}{\sqrt{2\pi}} \displaystyle\int_{2.2}^{2.1} e^{-x^2/2}\, dx$

4. Evaluate $\dfrac{1}{\sqrt{2\pi}} \displaystyle\int_{3.2}^{4.0} e^{-x^2/2}\, dx$ **5.** Evaluate $\dfrac{1}{\sqrt{2\pi}} \displaystyle\int_{3.5}^{1,000,000} e^{-x^2/2}\, dx$

12.6 MEAN VALUE, VARIANCE, AND STANDARD DEVIATION

It is often useful to have measures that summarize the information in a probability distribution. The three most useful measures are the mean value (which we have studied in Section 12.4 as mathematical expectation), the variance, and the standard deviation. We shall give definitions for the case of distributions in which the variable takes only a finite number of values; generalizations require material from the integral calculus, beyond the scope of this text.

DEFINITION 12.2. If an experiment can result in the set of numbers $\{x_i\} = \{x_1, x_2, \ldots, x_n\}$ with probabilities $\Pr(x_i)$, respectively, then the *expected value* of x is

(1) $$E(x) = x_1 \Pr(x_1) + x_2 \Pr(x_2) + \cdots + x_n \Pr(x_n)$$

In the special case in which each of the x_i has probability $1/n$, we obtain the usual formula for the arithmetic mean:

(2) $$E(x) = \frac{x_1 + x_2 + \cdots + x_n}{n}$$

$E(x)$ is often denoted by the symbol μ (the Greek letter for m, the first letter of the word "mean"). We now state some theorems about expected values. The theorems hold in general, but we shall limit the discussion in this section to the simple case in which $\Pr(x_i) = 1/n$ for all x_i.

THEOREM 12.1. *If the set $\{x_i - a\}$ is formed from the set $\{x_i\}$ by the subtraction of a from each element, then $E(x - a) = E(x) - a$.*

Proof: $\quad E(x - a) = \dfrac{(x_1 - a) + (x_2 - a) + \cdots + (x_n - a)}{n}$

$\qquad\qquad\quad = \dfrac{x_1 + x_2 + \cdots + x_n - na}{n}$

$\qquad\qquad\quad = \dfrac{x_1 + x_2 + \cdots + x_n}{n} - \dfrac{na}{a}$

$\qquad\qquad\quad = E(x) - a$

THEOREM 12.2. *If the set $\{cx_i\}$ is formed from the set $\{x_i\}$ by the multiplication of each element by c, then $E(cx) = cE(x)$.*

Proof:
$$E(cx) = \frac{cx_1 + cx_2 + \cdots + cx_n}{n}$$
$$= c\frac{x_1 + x_2 + \cdots + x_n}{n} = cE(x)$$

These two theorems are sometimes useful in reducing the amount of arithmetic required for calculating mean values.

Example 1 Find the arithmetic mean of the set $\{x_i\} = \{245, 251, 248, 253, 255\}$.

Solution 1: By definition,
$$E(x) = \frac{245 + 251 + 248 + 253 + 255}{5}$$
$$= \tfrac{1252}{5} = 250.4$$

Solution 2: Subtract 250 from each element to obtain the set $\{-5, 1, -2, 3, 5\}$, whose mean is $\tfrac{2}{5} = 0.4$. By Theorem 12.1,
$$E(x - 250) = E(x) - 250$$
$$E(x) = E(x - 250) + 250 = 250.4$$

Example 2 Find the arithmetic mean of the set $\{x_i\} = \{0.04, 0.07, 0.08, 0.05\}$.

Solution: Multiply each element of the set by 100 to obtain the set $\{4, 7, 8, 5\}$, whose mean is $\tfrac{24}{4} = 6$. By Theorem 12.2,
$$E(100x) = 100E(x)$$
$$E(x) = \frac{E(100x)}{100} = \frac{6}{100} = 0.06$$

Note that the mean value of a set is not necessarily an element of the set. The "usefulness" of the mean value of a set of numbers is generally greater if the numbers of the set do not deviate a great deal from the mean. The variance can be used as a measure of this deviation.

DEFINITION 12.3. The *variance* $V(x)$ of the set $\{x_i\}$ with mean value $E(x) = \mu$ is the *expected value* of the set $\{(x_i - \mu)^2\}$ whose elements are the squares of the differences between the corresponding elements of the original set and the mean of that set. Thus

(3) $\quad V(x) = (x_1 - \mu)^2 \Pr(x_1) + (x_2 - \mu)^2 \Pr(x_2) + \cdots + (x_n - \mu)^2 \Pr(x_n)$

If $\Pr(x_i) = 1/n$ for each x_i, this reduces to

(4) $$V(x) = \frac{(x_1 - \mu)^2 + (x_2 - \mu)^2 + \cdots + (x_n - \mu)^2}{n}$$

SEC. 12.6 Mean Value, Variance and Standard Deviation

Note that the variance is nonnegative. If all the x_i are equal and hence equal to μ, the variance is zero. If the x_i do not differ much from μ, the variance will be small, but if the deviations $x_i - \mu$ are large, then the variance will be large.

Example 3

Compute the variance of the set $\{x_i\} = \{2, 4, 0, 3, -1\}$.

Solution: $\mu = \frac{8}{5} = 1.6$. We have $\{x_i - \mu\} = \{0.4, 2.4, -1.6, 1.4, -2.6\}$. Hence the variance is the arithmetic mean of the set

$$\{(x_i - \mu)^2\} = \{0.16, 5.76, 2.56, 1.96, 6.76\}$$

$$V(x) = \frac{17.20}{5} = 3.44$$

The variance has several interesting properties, which are stated in the following theorems.

THEOREM 12.3. *If μ and $V(x)$ are the mean and variance, respectively, of the set $\{x_i\}$ and if $E(x^2)$ is the mean of the set $\{x_i^2\}$, then $V(x) = E(x^2) - \mu^2$.*

Proof:
$$V(x) = \frac{(x_1 - \mu)^2 + (x_2 - \mu)^2 + \cdots + (x_n - \mu)^2}{n}$$

$$= \frac{x_1^2 + x_2^2 + \cdots + x_n^2 - 2\mu(x_1 + x_2 + \cdots + x_n) + n\mu^2}{n}$$

But
$$\frac{x_1^2 + x_2^2 + \cdots + x_n^2}{n} = E(x^2)$$

and
$$\frac{x_1 + x_2 + \cdots + x_n}{n} = \mu, \quad \text{by definition}$$

Therefore, $V(x) = E(x^2) - 2\mu \cdot \mu + \mu^2 = E(x^2) - \mu^2$.

Thus the variance of a set of numbers may be calculated by taking the mean of the squares minus the square of the mean of the numbers. This method is usually simpler.

Example 4

Compute the variance of the set $\{x_i\}$ of the last example by the method given above.

Solution: Given the set $\{x_i\} = \{2, 4, 0, 3, -1\}$, the set $\{x_i^2\} = \{4, 16, 0, 9, 1\}$ and $E(x^2) = \frac{30}{5} = 6$. Moreover, $\mu = 1.6$, and so

$$V(x) = E(x^2) - \mu^2 = 6 - (1.6)^2 = 6 - 2.56 = 3.44$$

as before.

THEOREM 12.4. *If the set $\{cx_i\}$ is formed from the set $\{x_i\}$ by multiplying each element of the set by the constant c, then $V(cx) = c^2 V(x)$.*

Proof: By the preceding theorem, we know that
$$V(cx) = E(c^2 x^2) - [E(cx)]^2$$
But, by Theorem 12.2, we have
$$\begin{aligned} V(cx) &= c^2 E(x^2) - [cE(x)]^2 \\ &= c^2[E(x^2) - \mu^2] \\ &= c^2 V(x) \end{aligned}$$

The arithmetic mean of a set and deviations from the mean are measured in the same units as the elements of the set, but the variance is measured in square units. Thus, if the elements of a set are measurements in inches, the mean will be given in inches but the variance will be in inches squared. To obtain a measure of variability measured in original units, it is natural to take the square root of the variance.

DEFINITION 12.4. The *standard deviation* of a set $\{x_i\}$, denoted by $\sigma(x)$ (read "sigma of x"), is the positive square root of the variance of the set. Thus $\sigma(x) = \sqrt{V(x)}$.

The standard deviation is frequently called the *root mean square*, since it is the square root of the mean of the squares of the deviations from the mean.

We are now ready to show how to construct a standard set, a concept which will be required in the last section of the chapter when we deal with some practical applications of the normal distribution.

DEFINITION 12.5. A set will be called a *standard set* if the mean of the set is zero and the standard deviation is 1.

THEOREM 12.5. *If the set $\{x_i\}$ has mean μ and standard deviation $\sigma \neq 0$, then the set $\{(x_i - \mu)/\sigma\}$ is a standard set.*

Proof: By Theorems 12.1 and 12.2,
$$E\left(\frac{x-\mu}{\sigma}\right) = E\left(\frac{x}{\sigma} - \frac{\mu}{\sigma}\right) = E\left(\frac{x}{\sigma}\right) - \frac{\mu}{\sigma}$$
$$= \frac{1}{\sigma} E(x) - \frac{\mu}{\sigma} = \frac{\mu}{\sigma} - \frac{\mu}{\sigma} = 0$$

By Theorem 12.3,
$$V\left(\frac{x-\mu}{\sigma}\right) = E\left(\frac{x-\mu}{\sigma}\right)^2 - \left[E\left(\frac{x-\mu}{\sigma}\right)\right]^2$$

By Theorem 12.4,

$$E\left(\frac{x-\mu}{\sigma}\right)^2 = \frac{1}{\sigma^2}E(x-\mu)^2$$

$$V\left(\frac{x-\mu}{\sigma}\right) = \frac{1}{\sigma^2}E(x-\mu)^2 - 0 = \frac{1}{\sigma^2}\cdot\sigma^2 = 1$$

It follows that the standard deviation of the set $\{(x_i - \mu)/\sigma\}$ is also 1. Thus a standard set may be constructed from any set by subtracting the mean from each element and dividing the differences by the standard deviation.

Example 5 Construct a standard set from the set $\{4, 0, 3, 4, 1, 6\}$.

Solution: We first calculate $E(x) = \mu = \frac{18}{6} = 3$ and $E(x^2) = \frac{78}{6} = 13$, so that

$$V(x) = E(x^2) - \mu^2 = 13 - 9 = 4, \quad \text{by Theorem 12.3}$$

We also have $\sigma(x) = \sqrt{V(x)} = 2$. Now construct the set $\{x - \mu\} = \{1, -3, 0, 1, -2, 3\}$, whose mean is 0. Finally, construct the set

$$\left\{\frac{x-\mu}{\sigma}\right\} = \{\tfrac{1}{2}, -\tfrac{3}{2}, 0, \tfrac{1}{2}, -1, \tfrac{3}{2}\} = \{0.5, -1.5, 0, 0.5, -1, 1.5\}$$

This is the required standard set.

Check: The mean of this set is 0 and the variance is

$$\frac{0.25 + 2.25 + 0 + 0.25 + 1 + 2.25}{6} = 1$$

Therefore, the standard deviation is also 1.

Exercises 12.6

Find the mean, variance, and standard deviation of each of the following sets.

1. $\{4, 4.2, 5, 5.6, 7, 9\}$
2. $\{7, 3, 5, 4, 1, 9, 8, 7, 5, 4\}$
3. $\{7, 8, 9, 9, 10, 10, 11, 12, 14\}$
4. $\{2, 3, 6, 7, 10, 13, 17, 18, 20, 24\}$
5. Obtain a standard set corresponding to $\{4, 6, 7, 5, 2, 6, 6, 4, 4, 6\}$. Check your result.
6. Obtain a standard set corresponding to $\{5, 10, 13, 7, 17, 8\}$. Check your result.

12.7 POPULATIONS AND SAMPLES

The observations that are recorded when an experiment is performed usually will be only a subset of all the possible observations under the given conditions. We say that we have a *sample* from a given *population*. In the theory of statistics it is assumed that the population can be adequately described by some probability distribution and that the data of the sample, when properly drawn, can be used to make inferences about the characteristics of this distribution, such as its form, mean, and standard deviation.

Since the data in a sample represent only a subset of the population, it is necessary to introduce some new notation for the measures calculated from the sample data. Suppose that an experiment yields a sample of n observations with values $x_1, x_2, \ldots, x_k$ and respective frequencies $f_1, f_2, \ldots, f_k$. Of course, $f_1 + f_2 + \cdots + f_k = n$, and the relative frequency of occurrence of any x_i is f_i/n. The information in the sample may be summarized by the *sample mean* $\bar{x}$, the *sample variance* s^2, and the *sample standard deviation* s. Formulas for these three quantities are

$$(1) \qquad \bar{x} = \frac{x_1 f_1 + x_2 f_2 + \cdots + x_k f_k}{n}$$

$$(2) \qquad s^2 = \frac{(x_1 - \bar{x})^2 f_1 + (x_2 - \bar{x})^2 f_2 + \cdots + (x_k - \bar{x})^2 f_k}{n}$$

$$(3) \qquad s = \sqrt{\text{sample variance } s^2}$$

When the respective frequencies f_i are all equal to unity, Equations (1) and (2) simplify to

$$(1') \qquad \bar{x} = \frac{x_1 + x_2 + \cdots + x_n}{n}$$

$$(2') \qquad s^2 = \frac{(x_1 - \bar{x})^2 + (x_2 - \bar{x})^2 + \cdots + (x_n - \bar{x})^2}{n}$$

A convenient computing formula, algebraically equivalent to Equation (2), is

$$(2'') \qquad s^2 = \frac{x_1^2 f_1 + x_2^2 f_2 + \cdots + x_k^2 f_k}{n} - \bar{x}^2$$

Note that we are here using n to denote the number of observations in the sample, not the total number of possible outcomes. Theorems 12.1 through 12.5 may be restated in terms of $\bar{x}$, s^2, and s with only a change of notation, and are not repeated here.

We have already remarked the relationship between relative frequency and probability when n is large. It is also true that, for large samples drawn at random, $\bar{x}$, s^2, and s will usually be good approximations of $E(x) = \mu$, $\sigma^2(x)$, $\sigma(x)$, respectively.

Example 1 Three pennies were tossed 100 times, with the following results:

No. of heads, x_i	0	1	2	3
Frequency, f_i	19	31	41	9

SEC. 12.7 Populations and Samples

Calculate the relative frequencies, the mean, the variance, and the standard deviation of the sample data. Compare the results with the corresponding quantities for the theoretical frequency distribution.

Solution: The necessary calculations are presented below in tabular form in order to display the analogies between the distribution of the sample data and the probability distribution which serves as the population model.

Sample Data

x_i	f_i	f_i/n	$x_i f_i$	$x_i - \bar{x}$	$(x_i - \bar{x})^2$	$(x_i - \bar{x})^2 f_i$
0	19	0.19	0	-1.4	1.96	37.24
1	31	0.31	31	-0.4	0.16	4.96
2	41	0.41	82	0.6	0.36	14.76
3	9	0.09	27	1.6	2.56	23.04
$n = 100$		1.00	140			80.00

$\bar{x} = \frac{140}{100} = 1.4$, $\quad s^2 = \frac{80}{100} = 0.8$, $\quad s = 0.894$

Probability Distribution

x_i	$\Pr(x_i)$	$x_i \Pr(x_i)$	$x_i - \mu$	$(x_i - \mu)^2$	$(x_i - \mu)^2 \Pr(x_i)$
0	0.125	0	-1.5	2.25	0.28125
1	0.375	0.375	-0.5	0.25	0.09375
2	0.375	0.750	0.5	0.25	0.09375
3	0.125	0.375	1.5	2.25	0.28125
	1.000	$\mu = 1.500$			$\sigma^2 = 0.75000$
					$\sigma = 0.866$

Exercises 12.7

1. Compute s^2 and σ^2 in the example of this section, using the computing formulas for these quantities.

2. Prove that the following equation holds for the simple case in which $f_i = 1$ for all i. Start with Equation (2) of this section.
$$s^2 = \frac{x_1^2 + x_2^2 + \cdots + x_n^2}{n} - \bar{x}^2$$

3. **a.** Roll a die 100 times and record your results as a frequency distribution. Compute the mean, variance, and standard deviation for your data.
   b. Find the expected value, variance, and standard deviation of the probability distribution associated with the roll of a true die. Compare these results with the experimental results just obtained.

12.8 NORMAL PROBABILITY DISTRIBUTIONS

Large sets of observations arising in a wide variety of fields of study often can be described by the probability distribution drawn in Figure 12.5. This is called the normal probability distribution with mean μ and standard deviation σ. Actually, there are infinitely many normal distributions, because there are infinitely many possible values of μ and σ. The specific normal distribution that we studied in Section 12.5 is called the standard normal distribution because its mean is 0 and its standard deviation is 1. Fortunately, questions about any normal distribution can be reduced to questions about the standard normal distribution by the standardization technique described in Section 12.5. Consider the points $\mu - 2\sigma$, $\mu - \sigma$, μ, $\mu + \sigma$, and $\mu + 2\sigma$ on the x axis, which are labeled in the figure. These would become the elements $-2, -1, 0, 1, 2$ of the standard set $\{(x - \mu)/\sigma\}$ and, since $\sigma = 1$ for this set, Figure 12.5 would become Figure 12.4 (Section 12.5) by a mere relabeling of the axes. Thus the table of the same section can be used to find probabilities for any normal distribution.

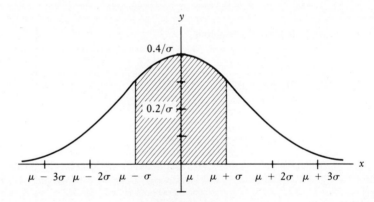

Figure 12.5

Example 1 — Given a set of numbers that is normally distributed with mean μ and standard deviation σ, what is the probability that an element of the set has a value between $\mu - \sigma$ and $\mu + \sigma$?

Solution: We have remarked that $\mu - \sigma$ and $\mu + \sigma$ correspond to the elements -1 and $+1$ of the standard set; then the required probability is the shaded area in Figure 12.5 or the corresponding area under the standard normal curve (Figure 12.4). This is

$$\frac{1}{\sqrt{2\pi}} \int_{-1}^{1} e^{-x^2/2}\, dx = 2I(1) = 2 \cdot 0.3413 = 0.6826$$

Thus, for any normal distribution, approximately 68% of the values will deviate from the mean by 1 standard deviation or less. Similarly, it can be shown

SEC. 12.8 Normal Probability Distributions

that approximately 95% of the numbers will have values within 2 standard deviations of the mean.

Example 2 Measurements of 400 metal rods selected at random from the output of a certain production process gave the following distribution for the rod diameters, measured to the nearest hundredth of an inch; present the data of this sample graphically.

Diameter	0.33	0.34	0.35	0.36	0.37	0.38	0.39	0.40	0.41
Frequency	1	0	4	54	139	146	50	4	2

Solution: Since the diameters are measured to the nearest hundredth of an inch, a reading of 0.33 may represent a true value anywhere between 0.325 and 0.335. The one observation at 0.33 is therefore pictured as a rectangle, its base running from 0.325 to 0.335 and its height being 1. The other frequencies are treated similarly. The resulting graph is called a *histogram* and is the usual method of graphing such data; see Figure 12.6.

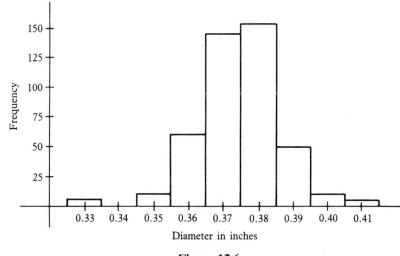

Figure 12.6

Here we have used areas of rectangles to correspond to frequencies just as we used areas under curves to represent probabilities. The shape of the histogram of the sample data helps us to select an appropriate model for the population distribution. In this case it seems reasonable to assume that the diameters of all rods manufactured by this process will be normally distributed.

Example 3 Assuming that the variation in the diameters of the rods in the example above is normally distributed and that the sample mean and standard deviation, $\bar{x}$ and s, respectively, are good approximations of μ and σ (a reasonable assumption, since the sample is large), what is the probability that a given rod has a diameter of at least 0.355 inch? How many rods should be purchased to obtain at least 1,000 rods with diameters of 0.355 inch or more?

Solution: Calculations in which the formulas of Section 12.6 are used give a mean of 0.375 inch and a standard deviation of 0.01 inch for the 400 diameters. Therefore, the deviation in which we are interested, $0.355 - 0.375 = -0.02$ inch, is equal to -2 standard deviations. The required probability is the area to the right of -2 under the standard normal curve, or $I(2) + 0.5000 = 0.4772 + 0.5000 = 0.9772$. In other words, we would expect 97.72% of the rods to have a diameter at least as large as 0.355 inch. Hence, out of N rods, we would expect $0.9772N$ rods of the desired size. Since we want at least 1,000, we must have $0.9772N \geq 1,000$. This requires that $N \geq 1023.3$, or 1,024 rods.

Example 4

A manufacturer of light bulbs would like to claim that 90% of bulbs of a certain type will have a life of at least T hours when used under standard conditions. Five hundred of these bulbs were selected at random and burned until all failed. The distribution of the times to failure, rounded to the nearest 50 hours, is given in the following table. Use these data to estimate the time T which the manufacturer can use in his claim.

Time to failure in hours	150	200	250	300	350	400	450	500	550
Number of bulbs	1	4	67	153	184	83	6	0	2

Solution: It will first be necessary to calculate the mean and variance of the data. The calculations are arranged below in the usual tabular form. In the column headed f, place the number of bulbs having the life given in the column headed x. In the third column, place $f_i x_i$, the number in the f column multiplied by the corresponding number in the x column. In the fourth column, place $f_i x_i^2$, the number in the fx column multiplied by the corresponding number in the x column. The totals of the f, fx, and fx^2 columns provide the necessary sums for computing the mean and variance by Equations (1) and (2″) of Section 12.7.

f	x	fx	fx^2
1	150	150	22,500
4	200	800	160,000
67	250	16,750	4,187,500
153	300	45,900	13,770,000
184	350	64,400	22,540,000
83	400	33,200	13,280,000
6	450	2,700	1,215,000
0	500	0	0
2	550	1,100	605,000
500		165,000	55,780,000

SEC. 12.8 Normal Probability Distributions

$$\bar{x} = \frac{165{,}000}{500} = 330 \text{ hours}$$

$$s^2 = \frac{55{,}780{,}000}{500} - (330)^2 = 111{,}560 - 108{,}900$$

$$= 2{,}660(\text{hour})^2$$

$$s = \sqrt{2660} - 51.6 \text{ hours}$$

We are now ready to solve the problem of the estimation of T. We shall assume that the distribution of the lives (x) of all bulbs of the special type is normal and that μ and σ are approximated closely by $\bar{x} = 330$ and $s = 51.6$, respectively. We require the value $x = T$ such that the probability of a value of x greater than T is 0.90. This is equivalent to asking for that value c of the standard normal distribution which has 90% of the area to the right of it. It is clear that c is negative and such that

$$\frac{1}{\sqrt{2\pi}} \int_c^\infty e^{-x^2/2}\, dx = \frac{1}{\sqrt{2\pi}} \int_c^0 e^{-x^2/2}\, dx + 0.5 = 0.90$$

$$\frac{1}{\sqrt{2\pi}} \int_c^0 e^{-x^2/2}\, dx = 0.4$$

Reference to the table in Section 12.5 shows that c is between -1.2 and -1.3, and interpolation yields the more accurate value $c = -1.28$. Thus the required value T deviates from the mean by -1.28 standard deviations, or $(T - 330)/51.6 = -1.28$ and $T = 264$ hours.

The last two examples are illustrations of statistical inference. In each we have used the incomplete information contained in a sample to make **predictions** about a larger set, the population. There are, of course, risks of **error in** such procedures, but modern statistical methods based on a firm **foundation of** probability theory enable us to make useful estimates and decisions **and to** assess the reliability of the results. The usefulness of the methods is indicated by their widespread acceptance in such fields as agriculture, medicine, psychology, the physical sciences, marketing, economics, and management science.

Exercises 12.8

1. Given a set of numbers that is normally distributed, with mean 50 and standard deviation 4, find the probability that an element selected at random from the set will have a value between 48 and 54; will have a value less than 56.

2. Scores on a certain college entrance test can be assumed to be normally distributed, with mean 200 and standard deviation 75.
   a. What percentage of the scores is between 50 and 350? What percentage is greater than 275?

b. College A will admit any applicant who makes a score of at least 125 on the test. How many will be expected to be admitted out of 600 applicants?

c. College B will admit any applicant who makes a score in the upper 40%. Find the minimum score necessary for admission to this college.

3. Consider the data on rod diameters given in Example 2.

   a. Show that $\bar{x} = 0.375$ inch and $s = 0.01$ inch.

   b. Suppose that rods with diameters between 0.355 inch and 0.395 inch can be sold at a profit of 5¢ each. Rods under 0.355 inch in diameter must be scrapped, with a resulting loss of 2¢ per rod. Rods over 0.395 inch in diameter can be reworked to specifications and then sold for a profit of 3¢ each. Estimate the expected profit per rod, based on this sample of 400 rods.

4. A group of men was measured for height; the following table records the results.

Height in inches	62	63	64	65	66	67	68	69
Number of men with that height	1	2	0	3	6	15	70	130

Height in inches	70	71	72	73	74	75	76	77
Number of men with that height	360	410	250	110	35	10	7	1

Assuming the population to be normal, what is the probability that a man picked at random will be between 68 and 70 inches tall?

5. Ten coins were tossed 1,000 times, and the following results were obtained.

Number of heads	0	1	2	3	4	5	6	7	8	9	10
Number of occurrences	2	10	44	107	209	252	211	110	47	7	1

Assuming the population distribution to be normal, what is the probability that at least 8 coins will be alike on a given throw?

6. A company manufactures steel plates. A random sample was tested for thickness with the following results.

Thickness in inches	1.97	1.98	1.99	2.00	2.01	2.02	2.03
Number of plates	2	23	212	571	253	36	3

A customer wants 2,000 plates but specifies that the thickness must be at least 1.98 inches and not more than 2.02 inches. How many plates should be shipped to satisfy the requirements?

Appendix A

Logic and Sets

Introduction

In developing any mathematical theory, which is the prime function of a textbook in mathematics, one has to decide where to start. In an elementary text such as this the usual procedure is to start by defining some words or terms that are to be used and by listing certain statements that are to be assumed true. By the word "statements," we mean declarative sentences. These statements are called axioms and may or may not seem true intuitively. From these assumed truths or axioms, we should be able to deduce many other truths by a mental process called logical reasoning. These deduced truths are usually stated as conclusions of propositions called theorems. We affirm the conclusion of a theorem by *proving* the *theorem*. How does one reason logically? How does one know when an argument is valid?

Many discussions in elementary mathematics can often be greatly simplified and at the same time be made more meaningful to the student by the use of the language and concept of sets. In fact, the concept of sets is one of the important fundamental concepts in modern mathematics and it permeates many branches of mathematics.

In an attempt to shed some light on the two questions raised in the first paragraph above, and in order to be able to use the language and concept of sets whenever needed in the text, a brief discussion of logic and sets is given in this appendix.

A.1 LOGIC, PROPOSITIONS, AND CONNECTIVES

In our routine daily lives most of us must make certain decisions based on our ability to reason. To be sure, many of these decisions are made more or less subconsciously and without any thought or realization as to whether we are or whether we are not reasoning correctly. In mathematics we must reason correctly if the conclusions based on our reasoning are to be valid.

How can one be sure that he or she is reasoning correctly? This is where formal logic comes to our aid. Logicians have very carefully worked out formal rules designed to help us draw valid conclusions from a given set of statements which are assumed to be true. Care must be taken to deal only with declarative statements which are either true or false, but not both. We shall refer to such statements as *propositions*. Thus, *a statement is a proposition if and only if one of the terms "true" or "false" can be meaningfully applied to it*. If a proposition is true, we say that it has truth value T (true), and if a proposition is false, we say that it has truth value F (false). For a statement to be a proposition, it is not necessary that we know which of the two truth values actually applies to the statement. The important thing to know is that one and only one of the truth values does apply. For example, the student is probably familiar with one of the standard processes for obtaining a decimal approximation to the square root of a number that is not a perfect square. If such a process is repeated over and over, a result may be obtained that contains as many decimal places as is desired, even a thousand or more decimal places if we were willing to waste a sufficient amount of time in such an activity. Thus the statement "the digit in the 100th decimal place in a decimal approximation to $\sqrt{2}$ that contains more than 100 decimal places is 7" is a proposition, because either the statement is true or the statement is false even though we do not know which. Other examples of propositions are the following.

(a) Dallas is the capital of Texas.
(b) 7 is a prime number.
(c) If a triangle is equilateral, it is equiangular.
(d) $8 + 5 = 12$.
(e) All rectangles are squares.

It is obvious that propositions (a), (d), and (e) are all false; hence each has truth value F. Propositions (b) and (c) are both true; hence the truth value of each is T.

The student has learned from the study of arithmetic that when two or more numbers are combined using one or more of the fundamental operations, addition, subtraction, multiplication, and division, the result is in general a number whose value is determined by the kinds of operations used and by the values of the numbers involved in the combination. Likewise, two or more propositions may be combined to form a new proposition whose truth value

A.1 Logic, Propositions, and Connectives

may be determined from the truth values of the propositions from which the new one is formed. In arithmetic, combinations may be formed by using one of the fundamental operations. In the case of propositions, combinations are formed by the use of certain words or terms called connectives. The most common connectives, the ones which we shall discuss here, are the words *and*, *or*, *if–then*, *if and only if*, and *not*. Two definitions are now in order.

DEFINITION A.1. A proposition that contains no connectives is called a *simple proposition*.

DEFINITION A.2. A proposition that contains one or more connectives is called a *compound proposition*.

As an example of how the connectives listed above may be used to form compound propositions from simple propositions, let us consider the two simple propositions (**a**) "a horse is a four-legged animal" and (**b**) "2 times 3 is 5." From (**a**) and (**b**) we can construct the new compound propositions: "a horse is a four-legged animal and 2 times 3 is 5," "a horse is a four-legged animal or 2 times 3 is 5," "if a horse is a four-legged animal, then 2 times 3 is 5," "a horse is a four-legged animal if and only if 2 times 3 is 5." Finally, from the proposition "2 times 3 is 5" we can form the new proposition "2 times 3 is not 5." It should be observed that in the last case (1) we used only one of the two given propositions in forming our new proposition and (2) the new proposition is a denial of the old. We say that the new proposition is the *negative* of the old or a *negation* of the old. In other words, the connective word *not* applied to a given proposition gives us a new proposition whose truth value is the opposite of the truth value of the given proposition. For example, the proposition "2 times 3 is 5" obviously is false, hence has truth value F, whereas the proposition "2 times 3 is not 5" has truth value T. To even include the word *not* as a connective may seem arbitrary, but suffice it to say that it is most important in mathematics to be able to form a proposition that denies or negates a given proposition. Negation will be discussed more fully in a later section.

In the succeeding sections we shall consider each of the connectives one at a time and define the truth values of compound propositions involving them. However, before we start our discussion of each of the connectives, some general remarks are in order. First, we want to emphasize the fact that in symbolic logic we are not concerned with the meaning of propositions but only with their truth values. This being the case, the truth value of a compound proposition is determined by the truth values of its components. To simplify our discussion, suppose that we have a compound proposition formed from the two unspecified propositions p and q. Obviously, proposition p is either true or false. Likewise, proposition q is either true or false. Thus the only possible combinations of truth values for p and q are the following:

(a) p true and q true,
(b) p true and q false,
(c) p false and q true,
(d) p false and q false.

In the following subsections we shall define the truth value for each of the compound propositions "p and q," "p or q," and "if p, then q" in each of the four cases (a), (b), (c), and (d) listed above. Also, we shall define the truth value for "not p" in each of the two cases

(i) p is true,
(ii) p is false.

Conjunctions

The compound proposition "p and q," denoted symbolically as $p \wedge q$, is called the *conjunction* of p and q. It should be intuitively obvious that the conjunction $p \wedge q$ is true when and only when both p and q are true. But to be more explicit, we define the truth values of the conjunction as shown in the following table, called a truth table. See Figure A.1.

	Conjunction	
p	q	$p \wedge q$
T	T	T
T	F	F
F	T	F
F	F	F

Figure A.1

Disjunctions

The compound proposition "p or q," denoted symbolically as $p \vee q$, is called the *inclusive disjunction* of p and q and is false when and only when both p and q are false. In ordinary language, the compound statement "p or q" is often used to mean p is true or q is true but not both. Such a disjunction is called an *exclusive* disjunction and denoted symbolically as $p \veebar q$. Unless otherwise stated,

A.1 Logic, Propositions, and Connectives

Inclusive Disjunction				Exclusive Disjunction		
p	q	$p \vee q$		p	q	$p \underline{\vee} q$
T	T	T		T	T	F
T	F	T		T	F	T
F	T	T		F	T	T
F	F	F		F	F	F

Figure A.2   **Figure A.3**

disjunctions used in this text should be interpreted as inclusive disjunctions. The truth tables for the two types of disjunction are shown in Figures A.2 and A.3.

A typical example of an inclusive disjunction as used in mathematics is the following.

Example 1   If p and q are the propositions

p: 2 is greater than 0

q: 2 is greater than 1

then both p and q have truth value T. Therefore, $p \vee q$ has truth value T, whereas $p \underline{\vee} q$ has truth value F.

The following is an example which illustrates the fact that, if $p \underline{\vee} q$ has truth value T, then $p \vee q$ must also have truth value T.

Example 2   Suppose two basketball teams, Team A and Team B, are scheduled to play each other in a game with the understanding that overtimes will be used if necessary to prevent the game from ending in a tie. If p and q are the propositions

p: Team A will win the game

q: Team B will win the game

then both $p \underline{\vee} q$ and $p \vee q$ have truth value T.

Negations

As was pointed out earlier, it is often desirable to form a proposition that negates, or denies, a given proposition. To be able to form such a proposition, it is necessary that we state precisely what is meant by the negation of a proposition. Suppose p is a given proposition. By the negation of p is meant the proposition, denoted symbolically as $\sim p$ and read "not p," such that

(a) if p is true, then $\sim p$ is false,
(b) if p is false, then $\sim p$ is true.

The truth table is shown in Figure A.4. Often there are several equivalent ways of forming the negation of a given proposition. For example, let p be the

Negation

p	$\sim p$
T	F
F	T

Figure A.4

proposition "the weather is hot." The negation $\sim p$ could take one of the following forms.

(a) The weather is not hot.
(b) It is false that the weather is hot.
(c) The weather is cool.

It should be noted that one needs to exercise a certain amount of care in forming the negation of a proposition. To see that this is true, let p be the proposition "all men are honest." Now consider the following propositions as possibilities for $\sim p$.

(a) All men are not honest.
(b) Not all men are honest.
(c) Some men are dishonest.
(d) All men are dishonest.
(e) It is false that all men are honest.

Which of the propositions (a), (b), (c), (d), and (e) is a negation of the given proposition p? In making our decision we must use the definition of the negation of a proposition: if p is true, then $\sim p$ is false, and if p is false, then $\sim p$ is true. Let us now apply this test to propositions (a), (b), (c), (d), and (e), keeping in mind that the statement "All men are honest" means that each and every man is honest.

Suppose p is true; then obviously (a), (b), (c), (d), and (e) are all false. Next suppose p is false. This means that at least one man is not honest. That is, p can be false without each and every man being dishonest. Thus, if p is false, then

A.1 Logic, Propositions, and Connectives

propositions **(b)**, **(c)**, and **(e)** are definitely true. But propositions **(a)** and **(d)** are not necessarily true. We therefore conclude that each of the propositions **(b)**, **(c)**, and **(e)** is an acceptable negation of p, whereas neither of the propositions **(a)** nor **(d)** is an acceptable negation of p.

In a later subsection the symbol used above to denote negation, $\sim$, will be used to denote that two sets are equipollent, but this double usage will not cause any confusion. Also, negation is sometimes denoted by the symbol $/$, called a *slash*. For example, to negate the statement "$3 = 5$," we would write "$3 \neq 5$."

Implications

It often happens that instead of making an outright assertion we feel that it is necessary or at least desirable to qualify a statement by inserting a condition. Most everyone has made, or heard, or read such statements as "If I go to Nashville tomorrow, I will go by plane;" "If $x = y$, then $x^2 = y^2$;" "If the diagonals of a parallelogram are equal, then it is a rectangle." Each of these statements is a compound proposition formed from two simple propositions by use of the connective "if–then," called the *conditional connective*.

Suppose p and q are propositions. The compound proposition "if p, then q," denoted symbolically as $p \to q$, is called an *implication*. The truth values of the implication are defined in the truth table of Figure A.5.

The alert student, upon careful examination of Figure A.5, may question the last two entries in the third column. The first two entries in the third column seem quite natural and obvious. That is, if p and q are both true, then $p \to q$ is obviously true. Likewise, if p is true and q is false, it is equally obvious that $p \to q$ is false. But what about the cases where p is false? On first thought, we might be inclined to leave these cases undefined. However, to do so would violate our basic principle that a proposition is true or it is false. Thus, to carry out this principle, logicians had to make a decision; they decided to define the truth value of $p \to q$ as true when p is false regardless of the truth value of q. Although such a

Implication

p	q	$p \to q$
T	T	T
T	F	F
F	T	T
F	F	T

Figure A.5

definition may seem quite arbitrary, it may be justified, at least in part, in the following manner. Suppose a husband says to his wife, "If I get paid today, I will take you to a movie this evening." If he gets paid and takes his wife to a movie, then certainly he told the truth. If he gets paid and does not take his wife to a movie, then his statement to her was false. The conclusion reached in each of the last two sentences is in agreement with the first two entries in the third column of Figure A.5. Next, suppose the husband does not get paid that day. Notice that in this case his statement to his wife does not commit him to any type of action. Thus, whether he takes his wife to a movie or not, we could hardly say his statement was false; if we are impelled to classify it as true or false, the better choice seems to be to define the statement as true, as was done in Figure A.5.

It should be observed that the order in which p and q occur in the implication is important; that is, the implication $p \to q$ does not have the same meaning as the implication $q \to p$. A more precise way of saying this is made possible with the following definition.

DEFINITION A.3. Two compound propositions are said to be *logically equivalent* if they have the same truth values in every possible case.

A comparison of the truth tables given in Figures A.5 and A.6 shows that the third columns are not identical. Hence, according to Definition A.3, the implications $p \to q$ and $q \to p$ are not logically equivalent.

Implication

q	p	$q \to p$
T	T	T
F	T	T
T	F	F
F	F	T

Figure A.6

Associated with each implication $p \to q$ are three other implications, defined as follows.

DEFINITION A.4. The implication $q \to p$ is called the *converse* of the implication $p \to q$.

DEFINITION A.5. The implication $\sim q \to \sim p$ is called the *contrapositive* of the implication $p \to q$.

A.1 Logic, Propositions, and Connectives

DEFINITION A.6. The implication $\sim p \to \sim q$ is called the *inverse* of the implication $p \to q$.

Example 3

Implication: If a parallelogram is a square, then it is a rectangle.
Converse: If a parallelogram is a rectangle, then it is a square.
Contrapositive: If a parallelogram is not a rectangle, then it is not a square.
Inverse: If a parallelogram is not a square, then it is not a rectangle.

It can be shown by comparing truth tables that any implication and its contrapositive are logically equivalent.

In Figures A.1 through A.5 are shown the truth tables for the basic connectives when they are used singly to form compound statements. It is also possible to use two or more of these basic connectives to form a more complicated compound proposition. For example, $\sim p \wedge q$, $\sim(p \wedge q)$, $(p \wedge q) \to p$, and $[p \wedge (p \to q)] \to q$ are all compound propositions. They are to be read "from the inside out"; that is, the quantities inside the innermost parentheses are first grouped together, then these parentheses are grouped together, and so on. Each compound proposition has a truth table which can be constructed in a step-by-step process. The following examples show this step-by-step method for constructing truth tables.

Example 4

Consider the compound proposition $\sim p \wedge q$. We start by placing in the first two columns the four possible pairs of truth values for the simple propositions p and q, as shown in Figure A.7. In Step 2 we write in the proposition in question, leaving enough space between symbols so that we can fill in the columns below. Then we fill in the truth values of $\sim p$ and q in the columns below these symbols. See Figure A.8.

p	q
T	T
T	F
F	T
F	F

Step No. 1

Figure A.7

p	q	$\sim p$	$\wedge$	q
T	T	F		T
T	F	F		F
F	T	T		T
F	F	T		F

Step No. 2

Figure A.8

Finally, we fill in the column under the conjunction symbol, which gives us the truth value of the given compound proposition in each pair of possible cases. This final truth table is shown in Figure A.9. Column 4 in Figure A.9 gives the truth value of the proposition $\sim p \wedge q$ for each of the four possible pairs of truth values for p and q.

p	q	$\sim p \land q$
T	T	F F T
T	F	F F F
F	T	T T T
F	F	T F F

Step No. 3

Figure A.9

Example 5   For our second example we construct the truth table for the compound proposition $[p \land (p \to q)] \to q$. The truth table, together with the numbers indicating the order in which the columns are filled, is shown in Figure A.10.

p	q	$[p \land (p \to q)] \to q$
T	T	T T T T T T T
T	F	T F T F F T F
F	T	F F F T T T T
F	F	F F F T F T F

Step No.   1 3   1 2 1   4 1

Figure A.10

The truth values of the compound proposition in question are always found in the last column filled in the construction of the truth table. For example, in the truth table shown in Figure A.10 the column labeled No. 4 gives the truth values of the compound proposition $[p \land (p \to q)] \to q$. Note that this particular proposition has the truth value T for each possible pair of truth values of the simple propositions p and q; that is, the compound proposition is true regardless of the truth values of p and q. Such a compound proposition is called a *tautology*.

Exercises A.1

1. Assume the following statements are compound propositions. Give their simple components. Denoting the first component as p and the second as q, express each compound proposition in symbolic form.

A.1 Logic, Propositions, and Connectives

    a. Mathematics and physics are difficult subjects.
    b. The goods will be shipped by train or by truck.
    c. If $2 + 3 = 5$, then $2 + 3 + 4 = 9$.
    d. If $x^2 = 4$, then $x = 2$.
    e. English is easy but history is boring. (*Note:* We shall assume the connective *but* to mean the same as the connective *and*.)
    f. If I do not study I will fail my math course.

2. Write the following propositions in symbolic form, letting p be "wages are high" and q be "food prices are high."
    a. Wages are low and food prices are high.
    b. Wages are low and food prices are low.
    c. Wages are high and food prices are low.
    d. Wages are low or food prices are high.
    e. Neither wages nor food prices are high.
    f. It is not true that wages and food prices are both low.

3. Assuming that wages and food prices are both high, which of the compound statements in Exercise 2 are true?

4. Write the following propositions in symbolic form, letting p be "the government is expanding" and q be "taxes are increasing."
    a. The government is shrinking.
    b. The government is expanding and taxes are increasing.
    c. Taxes are increasing or the government is expanding.
    d. If the government is expanding, then taxes are increasing.

5. Form the conjunction, disjunction, and two implications for each given pair of propositions.
    a. John is smart. Tom is rich.
    b. Five is a counting number. $3 + 2 = 5$.
    c. Five is a counting number. $3 + 2 = 6$.
    d. Bill does not do his homework. Bill calls Mary.
    e. Six is not a prime number. Three times five does not equal fifteen.

6. Discuss the truth values of the implications called for in Exercise 5.

7. Form the negation of the following propositions.
    a. The square of any number is greater than the number itself.
    b. $5 + 3 = 7$.
    c. It is false that Bob and Joe are cousins.
    d. Some men are wealthy.
    e. All students who graduate from college get good jobs.
    f. All boys are not good in mathematics.
    g. Not all boys are good in mathematics.

8. Write the converse, the contrapositive, and the inverse of each of the following implications.
   a. If he is my father, then I am not his brother.
   b. If 2 is greater than 5, then 2 is greater than 4.
   c. If Tom is rich, then Joan will marry him.
   d. If $a = b$, then $a^2 = b^2$.
   e. If $a^2 + b^2$ is positive, then $a + b$ is positive.
   f. If Bill and Joe are not brothers, then their wives are not sisters.

9. By comparing truth tables, show that any implication and its contrapositive are logically equivalent.

10. By using truth tables, show that the converse and the inverse of any implication are logically equivalent.

11. Let p be "the triangle ABC is isosceles" and let q be "the angle C is a right angle." Write in words a translation for each of the following.
   a. $p \wedge q$ b. $p \wedge \sim q$ c. $\sim p \wedge \sim q$ d. $\sim p \vee q$
   e. $p \rightarrow q$ f. $q \rightarrow p$ g. $\sim(p \wedge q)$ h. $\sim(p \vee q)$
   i. $\sim p \rightarrow q$ j. $\sim(\sim p \vee \sim q)$

12. By using truth tables, verify that the following table is correct.

	Given Proposition	Negation
Conjunction	$p \wedge q$	$(\sim p) \vee (\sim q)$
Disjunction	$p \vee q$	$(\sim p) \wedge (\sim q)$
Implication	$p \rightarrow q$	$p \wedge (\sim q)$

13. By referring to the table in Exercise 12, form the negation of the following compound propositions.
   a. John is sick and Harry is hungry.
   b. Alice or Mary will be here on time.
   c. If 7 is a prime, then $2 + 3 = 6$.
   d. Only if $x = 3$ does $x + 5 = 7$.
   e. The polygon is a square or it is not a rectangle.

14. Which connective (conjunction, disjunction, negation, or implication) is contained in each of the following compound propositions?
   a. San Francisco is in California or Boston is in Massachusetts.
   b. I will go fishing if it doesn't rain.
   c. The stock market is not on an upswing.
   d. Jury duty is interesting as well as time consuming.
   e. The government is expanding to meet the needs of an expanding government.

A.2 Bi-Implications and Theorems

15. Construct a truth table for each of the following compound propositions.
    a. $p \vee \sim q$ **b.** $\sim(p \wedge q)$ **c.** $(p \vee \sim q) \wedge \sim p$ **d.** $p \rightarrow (p \vee q)$
    e. $\sim(\sim p)$ **f.** $\sim(p \wedge \sim p)$ **g.** $(p \vee q) \wedge \sim p$ **h.** $(p \rightarrow q) \wedge \sim q$

16. Show that each of the following compound propositions is a *tautology*.
    a. $[(p \rightarrow q) \wedge (q \rightarrow r)] \rightarrow (p \rightarrow r)$
    b. $\sim(\sim p) \rightarrow p$
    c. $[(p \wedge q) \rightarrow r] \rightarrow [p \rightarrow (q \rightarrow r)]$
    d. $(p \wedge q) \rightarrow p$
    e. $[p \wedge (q \vee r)] \rightarrow [(p \wedge q) \vee (p \wedge r)]$
    f. $[p \vee (q \wedge r)] \rightarrow [(p \vee q) \wedge (p \vee r)]$

17. Suppose p, q, and r are three simple propositions. Complete the truth table shown below.

p	q	r	$p \wedge r$	$q \wedge r$	$p \vee r$	$q \vee r$	$p \wedge (q \wedge r)$	$p \wedge (q \vee r)$	$p \vee (q \wedge r)$
T	T	T							
T	T	F							
T	F	T							
T	F	F							
F	T	T							
F	T	F							
F	F	T							
F	F	F							

18. Construct truth tables for the following compound propositions.
    a. $p \rightarrow (q \vee r)$ **b.** $(p \vee \sim q) \wedge r$

19. The truth table for a proposition formed from two simple propositions has four rows, and the truth table for a proposition formed from three simple propositions has eight rows. How many rows would the truth table for a proposition formed from four simple propositions have? For n?

A.2 BI-IMPLICATIONS AND THEOREMS

The fifth and last connective which we shall discuss here is the one consisting of the words "if and only if." Given any two propositions p and q, the compound proposition "p if and only if q" (denoted symbolically as $p \leftrightarrow q$) is called the *bi-implication* of p and q. This proposition is true when and only when p and q are both true or both false. It follows that the bi-implication $p \leftrightarrow q$ is true when, and only when, p and q are logically equivalent. Some texts call the bi-implication

$p \leftrightarrow q$ an "equivalence." This convention is not adopted here in order to avoid the use of the same word to describe several different phenomena. The word "equivalence" is reserved for the notions of logical equivalence of propositions (Definition A.3), equivalence of open statements (Definition A.26), and equivalence of equations and inequalities (Chapter 3). The proposition "$p \leftrightarrow q$" is the conjunction of the two propositions "$p \rightarrow q$" and "$q \rightarrow p$," that is, the conjunction of an implication and its converse. The truth values of the bi-implication $p \leftrightarrow q$ are given in the truth table shown in Figure A.11.

Bi-implication

p	q	$p \leftrightarrow q$
T	T	T
T	F	F
F	T	F
F	F	T

Figure A.11

The importance of the role of implications and bi-implications in mathematics is based on the following facts. The foundation of mathematics consists of a set of undefined terms, a set of definitions, and a set of propositions, called *axioms*, which are assumed to be true. The superstructure, or framework, consists of a set of implications and bi-implications, called *theorems*, which are true either as a direct or an indirect consequence of the axioms previously assumed to be true.

If the implication $p \rightarrow q$ is a theorem, then we say any of the following:

1. If p is true, then q is true.
2. p is true only if q is true.
3. p is a *sufficient condition* for q.
4. q is a *necessary condition* for p.
5. p is the *hypothesis* of the theorem.
6. q is the *conclusion* of the theorem.

Similarly, if the bi-implication $p \leftrightarrow q$ is a theorem, then we say any of the following:

1. p is true if and only if q is true.
2. q is true if and only if p is true.
3. p is a *necessary and sufficient condition* for q.
4. q is a *necessary and sufficient condition* for p.

A.2 Bi-Implications and Theorems

In mathematics we are primarily interested in two things: (1) establishing that certain propositions are true, and (2) using the fact that certain propositions are true to enable us to solve certain problems.

As noted earlier, we ordinarily refer to a true proposition as a *theorem*. Hence, when we write

THEOREM. *If $a = 1$, then $a^2 = 1$.*

we mean that the proposition "If $a = 1$, then $a^2 = 1$" is true and is capable of being proved. A careful and orderly presentation of logical facts which lead to the conclusion that, in fact, the proposition in question is true is called a *proof* of the theorem. A number of theorems and their proofs are presented in the text. As a final word about terminology, we write "Prove $p \to q$" or "Show $p \to q$" as a shorthand for "Prove $p \to q$ is true" or "Show $p \to q$ is true."

It should be observed here that a theorem of the form "$p \leftrightarrow q$" is indeed the combination of two theorems into one, and to prove such a theorem, we must prove each of the two theorems from which it was formed. For example, suppose a and b are numbers, and we consider the theorem "$a^2 = b^2$ if and only if $a = b$ or $a = -b$." This theorem may be reformulated as the two theorems:

1. If $a^2 = b^2$, then $a = b$ or $a = -b$.
2. If $a = b$ or $a = -b$, then $a^2 = b^2$.

To prove the original theorem, we must prove both Theorem 1 and Theorem 2. It should be noted that Theorem 2 is the converse of Theorem 1, and vice versa.

In the text we encounter many theorems of the form "$p \leftrightarrow q$." It is therefore most important that you recognize the form and also the general procedure used in the proof of such a theorem. This general procedure consists of two steps and can be summarized briefly as follows.

To prove: *p if and only if q*.

Proof: Step 1. Assume p is true. Prove q is true.
Step 2. Assume q is true. Prove p is true.

Exercises A.2

1. Use truth tables to show that the propositions "$p \leftrightarrow q$" and "$(p \to q) \land (q \to p)$" are logically equivalent.
2. Show that $p \leftrightarrow q$ and $q \leftrightarrow p$ are logically equivalent.
3. Use truth tables to show that the negation of $p \leftrightarrow q$ is $p \leftrightarrow (\sim q)$ or $(\sim p) \leftrightarrow q$.
4. Form the negation of the following propositions.
   a. A parallelogram is a rectangle if and only if its diagonals are equal.

b. A necessary and sufficient condition that a parallelogram be a rectangle is that a vertex angle be a right angle.
c. A necessary and sufficient condition that $x^2 - 3x = 0$ is that $x = 0$ or $x = 3$.
d. A necessary condition that x^2 be greater than 4 is that x be greater than 2.
e. A sufficient condition that an integer be an even integer is that it be a multiple of 4.

5. A husband said to his wife, "I will buy you a new car only if I get a raise." He got the raise but did not buy his wife the car. Was she logically justified in thinking her husband had lied to her?

6. A notice sent to all members of a sandlot softball team had printed on it, "The softball game on Friday will not be played only if the weather is not fair."
a. State in other words precisely what this statement says and only what it says.
b. It was fair Friday. Should there have been a game?
c. There was a game. What was the weather like?

7. By appropriately identifying p and q with the simple propositions from which the compound proposition is formed, write each of the following compound propositions in the form $p \rightarrow q$ or $p \leftrightarrow q$, whichever applies.
a. $7x + 3 = 10$ only if $x = 1$.
b. A sufficient condition that $x^2 = 9$ is that $x = 3$.
c. A necessary condition that a triangle be a right triangle is that no angle be greater than $90°$.
d. A polygon is a parallelogram if and only if a pair of opposite sides are both parallel and equal.
e. A necessary and sufficient condition that a triangle be equiangular is that it be equilateral.

8. Smith and Brown are running for Mayor. Let p be "Smith is smart," let q be "Brown is stupid," and let r be "Smith will win the election." For each of the following propositions find a symbolic form, and then construct a truth table.
a. If Smith is smart and Brown is stupid, Smith will win the election.
b. Smith will win the election if and only if either he is smart or Brown is stupid.

A.3 SETS, ELEMENTS, AND MEMBERSHIP

Few words in the English language are more common than the word *set* as it applies to a collection of *things*. In an attempt to be as general and as inclusive as possible, we shall use the word *element* to mean the most general notion associated with such words as *things, entities, objects,* and the like. That is, an element may be any definite distinct object of our perception or of our thought. For example, an element could be a book, a house, a number, a theorem in geometry, a mathematical symbol, a word, a paragraph, a thought, a sound, a color, or a day of the week.

In general, we shall think of a *set* as any collection of elements, and we shall always think of elements as members of sets. The elements in the collection which form a particular set are said to *belong* to that set or to be *members* of that set. An element which belongs to a set is said to be *contained* in the set. For example, consider the set consisting of the vowels in the alphabet. The elements are *a, e, i, o*, and *u*. The letter *a* is a member of this set, or it belongs to this set. The set called the set of vowels contains the letters *a, e, i, o, u*.

As has been suggested above, the notion of a set, or of belonging to a set, is common knowledge gained from experience. It is not at all uncommon in our daily conversation to use or hear such phrases as *a set of books, a set of dishes, a set of tools, a set of boys,* and so forth. In fact, almost everyone is interested in some kind of set. The two-year-old child is interested in his set of toys. The history student may be interested in a particular set of events or in a set of historical dates. The teacher of elementary arithmetic is concerned with the set of counting numbers. The student of mathematics is interested in many different sets—the set of real numbers, a set of points, a set of axioms, a set of formulas, a set of operations, to name just a few. However, to be useful in mathematics, it is necessary that a set be well defined. That is, we must be able to say of a given set and a given element that the element does belong to the set or that it does not belong to the set. This implies that a set is determined by some rule or property which completely characterizes the set. For example, the set consisting of the capitals of the 50 states in the United States is a well-defined set because it is possible to say of any element that it does or does not belong to the set. Nashville is a member of the set; Knoxville is not a member of the set. On the other hand, the set consisting of the five happiest persons in Tennessee is not well defined, because it is impossible to say exactly who belongs to this set. Thus, in constructing a well-defined set, it is necessary to describe its elements so precisely that there is no question about what elements the set contains. Methods of doing this will be discussed in the following subsection.

Methods of Constructing Sets

In our daily life we find it convenient, if not absolutely necessary, to give names to everything around us—people, food, our possessions, our emotions, and things in general. This obviously makes it much easier for us to identify, describe, or single out any object or concept under consideration in a conversation, discussion, or discourse. Hence it seems natural, when discussing sets, to give particular sets names and then to refer to them by name. It is more or less conventional to use capital letters as names of sets and small letters as names of elements. This convention will be followed here except in special cases, where other names are self-explanatory and more appropriate. We shall also use braces, { }, to denote sets, and the symbols within the braces either will be a list of the names of the elements of the set or will describe the properties of the elements of the set. The symbol $\in$ will be used to indicate that an element belongs to a particular set. For example, if S is a set and a is an element, we write $a \in S$, which is read "a belongs to S" or "a is a member of S." If we wish to indicate that an element d does not belong to a set S, we write $d \notin S$, which is read "d does not belong to S" or "d is not a member of S." Note that the slash used in the last symbol is in keeping with the familiar use of the slash to denote unequal quantities, as when we write $a \neq b$ (read "a is not equal to b"). In general, when it is used "over" other symbols, we shall interpret the slash as a denial. Other uses of the slash will be indicated later.

As mentioned above, for a set to be useful in mathematics it must be described so clearly that everyone interested in the set will know exactly what elements it contains. What method or methods may one use to describe a set precisely? An obvious way is, where possible, to list the names of all the elements in the set. For example, let A be the set consisting of the names of three persons: George Washington, Abraham Lincoln, and Woodrow Wilson. We write this in abbreviated form:

$$A = \{\text{George Washington, Abraham Lincoln, Woodrow Wilson}\}$$

Some other examples are:

$$B = \{\text{Knoxville, Nashville, Chattanooga, Memphis}\}$$
$$C = \{\text{Tom, Dick, Harry}\}$$
$$H = \{a, b, c, d, e\}$$
$$J = \{\text{boy, dog, gun}\}$$
$$K = \{1, 2, 3, 4, \ldots, 99\}$$

This method is in general not practical if the number of elements in the set to be defined is large unless the first few elements in the set, together with the last element if the set has a fixed number of elements, are such that the composition of the set is completely determined. Set K above is an example of such a set. Notice that in defining set K we used a shortcut, an ellipsis or three dots. These three dots indicate that we are to include in set K all the counting numbers from 1 to 99 inclusive. Another method of defining the set K is given by the following notation:

$$K = \{x | x \text{ is a counting number less than } 100\}$$

which is read "K is the set consisting of all the elements x such that x is a counting number less than 100." Note the interpretation of the vertical line, which is read "such that." Another very important point to be emphasized regarding the latter method of defining the set K is the meaning of the letter x. Here the letter x is used not as the name of a particular element of the set but to represent an arbitrary element of the set; it is called a variable.

A.3 Sets, Elements, and Membership

DEFINITION A.7. A letter used to represent an arbitrary element of a set (containing more than one element) is called a *variable*. If a set contains only a single element, the letter used to represent this element is called a *constant*.

Consider each of the following sets.

1. $R = \{c|c$ is an even counting number less than $4\}$ is read "R is the set consisting of all the elements c such that c is an even counting number less than 4." Since there is only one even counting number less than 4, that is, 2, set R contains only one element. It follows from Definition A.7 that c is a constant. It would be correct to write $c = 2$.
2. $T = \{x|x$ is the name of a state in the United States$\}$ is read "T is the set consisting of all the elements x such that x is the name of a state in the United States." In this case the set T contains 50 elements; that is, the variable x has 50 different possible values. One such value is $x =$ Tennessee.
3. $N = \{n|n$ is a counting number$\}$ is read "N is the set consisting of all the elements n such that n is a counting number." Another method that is often used for describing this set is $N = \{1, 2, 3, 4, 5, \ldots\}$. The three dots in this case are placed last in the braces and may be read "and so on without end." This is interpreted to mean that if the counting numbers are arranged in natural order, there is no last, or largest, such number. We say that the set N contains infinitely many elements, or, more briefly, N is an infinite set.
4. $S = \{p|p$ is a real number which is both greater than 0 and less than 1$\}$ is read "S is the set consisting of all the elements p such that p is a real number which is both greater than 0 and less than 1." For example, $\frac{1}{2}$ is a member of S. This is another example of an infinite set.

DEFINITION A.8. A set of elements is said to be *finite* if there exists a fixed counting number, denoted by M, such that the number of elements in the set is M. A set that is not finite is said to be *infinite*.

Sets A, B, C, H, J, and K defined above are finite sets. Set A contains 3 elements; hence in this case $M = 3$. For set B, $M = 4$; for set C, $M = 3$; for set H, $M = 5$; for set J, $M = 3$; for set K, $M = 99$. Likewise, sets R and T are finite, while N and S are infinite.

The Universal Set and the Empty Set

Consider the following sets: $A = \{x|x$ is a student in the University of Tennessee$\}$, $B = \{x|x$ is a freshman in the University of Tennessee$\}$, $C = \{x|x$ is a student enrolled in mathematics at the University of Tennessee$\}$, and $D = \{x|x$ is an engineering student in the University of Tennessee$\}$. Here set A is a type of overall set which contains all the elements with which one might be concerned in any discussion involving one or more of the sets B, C, and D. Such a fixed overall set as A, when used in a particular discussion, is called a universal set for that discussion.

DEFINITION A.9. In discussing particular sets, we may assume or have specified some fixed overall set of elements to which all elements under consideration must belong. Such a fixed set is called the *universal set* for that discussion. A universal set is always denoted by the capital letter U.

It should be emphasized here that the universal set for a particular discussion is usually determined by the nature of the discussion and the elements involved. In a particular case, we may choose the universal set to suit our purpose or convenience. The basic use of the universal set is one of restriction; that is, it is designed to keep a discussion within the desired bounds. This is especially true in the case of mathematical discussion. For example, if $B = \{x|x$ is a counting number less than $20\}$ is being considered by a class in fourth-grade arithmetic, the universal set would most probably be the set of all counting numbers. However, for a class of college freshmen, the universal set might be the set of all real numbers.

We have already had an example of a set that contains only one element. Is it possible to have a set that has no elements? Consider the set consisting of all the counting numbers less than 5, each of which is exactly divisible by 7. Since the counting numbers less than 5 are the numbers 1, 2, 3, 4 and none of these is exactly divisible by 7, our set contains no elements. It is an empty set. Using formal logic, it can be shown that there is only one empty set and hence we can refer to it as *the* empty set. If such a set at first seems a bit artificial, suffice it to say that for purposes of complete generality it is convenient and desirable to define the empty set and to include it in our general discussion of sets. The empty set is also called the *null* set. In a certain sense the empty set plays a role somewhat analogous to that played by the number zero in the real number system.

DEFINITION A.10. By the *empty* or *null* set, denoted by the symbol $\emptyset$, is meant the set that contains no elements. The null set is sometimes denoted by empty braces thus: $\{\ \}$. To indicate that a set A is not the empty set, we would write $A \neq \emptyset$.

Example 1 | The set of all the students in your class who are over 10 feet tall is the empty set.

Exercises A.3

1. Label each of the following sets as being finite, infinite, or the empty set. For the sets in **a**, **b**, and **c**, find the number of elements M in each.
    a. The set consisting of all the letters in the English alphabet.
    b. The set consisting of all the senators in the United States Congress.

A.3 Sets, Elements, and Membership

    c. The set consisting of all the even counting numbers less than 20.
    d. The set of men now living who fought in the War of 1812.
    e. The set consisting of all the people now living in the world. Is this set well defined?
    f. The set of stars in the Milky Way.
    g. The set of all the even counting numbers less than 20 each of which is exactly divisible by 7.
    h. The set of all the counting numbers each of which is exactly divisible by 1,379.

2. Give an example, not given in the text, of each of the following kinds of sets.
    a. A finite set   b. An infinite set
    c. An empty set   d. A set that is not well defined

3. For each of the following sets, write the set in set notation using braces and within the braces list the elements of the set.
    a. The set consisting of the names of all the states in the United States whose names begin with M.
    b. The set of all counting numbers which lie between 2 and 11.
    c. The set of all counting numbers less than 50, each of which is the square of a counting number.
    d. The set of all counting numbers less than 10, each of which is the square root of a counting number.
    e. The set consisting of the first name of each member of your immediate family.
    f. The set consisting of the names of the states bordering Tennessee.

4. Describe in words each of the following sets.
    a. $A = \{\text{Alabama, Alaska, Arizona, Arkansas}\}$
    b. $B = \{1, 2, 3, \ldots, 30\}$
    c. $C = \{x \mid x + 3 = 5\}$
    d. $D = \{2, 4, 6, 8, 10, \ldots\}$
    e. $E = \{y \mid y \text{ is an even counting number less than } 20\}$
    f. $F = \{y \in E \mid y \text{ is exactly divisible by } 3\}$
    g. $G = \{x \mid x \text{ is the name of a baseball team which belongs to the American League}\}$

5. The sets mentioned below are defined in Exercise 4 above. Label each of the following statements as T (true) or F (false).
    a. $17 \in B$ b. $1478 \in D$ c. $20 \in E$ d. $\text{Argentina} \notin A$
    e. $2 \in C$ f. $\text{Atlanta Braves} \in G$ g. $15 \in F$ h. $\text{New York Yankees} \notin G$

6. For each of the following sets, give a possible universal set that could be useful in a discussion involving the elements of the set.
    a. $L = \{x \mid x \text{ is a city in Texas with population over } 100{,}000\}$
    b. $M = \{x \mid x \text{ is a vowel in the English alphabet}\}$
    c. $P = \{y \mid y \text{ is a woman over 40 years of age}\}$
    d. $Q = \{x \mid x \text{ is a teacher of mathematics}\}$
    e. $R = \{x \mid x \text{ is a college freshman}\}$

A.4 SUBSETS

Consider the two sets $A = \{1, 2, 3\}$ and $B = \{1, 2, 3, 4, 5\}$. Every element of A is also an element of B, but there are elements of B that do not belong to A. We say that A is a *proper subset* of B.

DEFINITION A.11. A set A is said to be a *proper subset* of a set B when every element of A is also an element of B and there is at least one element of B that does not belong to A. This relation is denoted by the special symbol $A \subset B$, which is read "A is a proper subset of B."

It may happen that set A and set B are identical; that is, every element of A belongs to B and every element of B belongs to A; in this case we say that set A is *equal* to set B.

DEFINITION A.12. Two sets, A and B, are said to be *equal* if, and only if, they have exactly the same members. We indicate this relation by writing $A = B$.

The set A is said to be a *subset* of B if either $A \subset B$ or $A = B$. We usually combine these two symbols into the single symbol $A \subseteq B$, and read "A is a subset of B." This relation is convenient when we do not have enough information to be more specific. For example, suppose that A is the set consisting of all the boys in your math class and B is the set consisting of all the students in your math class; without further information, one is certainly correct in writing $A \subseteq B$. Let us now write this definition in symbolic notation.

DEFINITION A.13. A is a *subset* of B if, and only if, for every element a such that $a \in A$ it is also true that $a \in B$. We indicate this relation by writing $A \subseteq B$. Note that Definition A.12 can now be restated in the simple form, $A = B$ if and only if $A \subseteq B$ and $B \subseteq A$.

The empty set, or null set, is considered to be a subset of every set. Similarly, every set is a subset of itself; that is, $A \subseteq A$.

From the set $A = \{1, 2, 3\}$, let us form all the possible subsets. First, there is the empty set, $\{\ \}$. Next, the subsets with only one element are $\{1\}, \{2\}, \{3\}$. Next, the subsets with two elements are $\{1, 2\}, \{1, 3\}, \{2, 3\}$. Then, last, we have the set itself, $\{1, 2, 3\}$. A quick count shows that $A = \{1, 2, 3\}$ has exactly 2^3, or 8, subsets. It can be shown that for every set of n elements there are exactly 2^n subsets.

Next consider the two sets $A = \{a, c, e, g\}$ and $B = \{a, b, e, f, h\}$. Some elements of A also belong to B, but there are elements of A not contained in B. There are also elements in B which do not belong to A. We say that the sets A and B are *overlapping*.

A.4 Subsets

DEFINITION A.14. Two sets are said to be *overlapping* if, and only if, they have at least one element in common but neither is a subset of the other.

Another possible relation between two sets is that they have no elements in common: $A = \{$Chicago, Cleveland, Dallas$\}$ and $B = \{$Alabama, Georgia, Tennessee$\}$ are two such sets, and they are said to be *disjoint*.

DEFINITION A.15. Two sets are said to be *disjoint* if, and only if, (a) both are nonempty and (b) they have no elements in common.

In the above remarks we have noted that two sets may be related in any one of four principal ways. In fact, any two sets must be related in one and only one of the four ways: (a) the two sets are equal, (b) one set is a proper subset of the other, (c) the two sets overlap, (d) the two sets are disjoint.

Venn Diagrams

In studying sets and subsets it is often helpful to have a pictorial representation of them. A very convenient diagram is the Venn diagram, named after the English logician John Venn (1834–1883). A Venn diagram uses regions bounded by suitable closed curves, usually circles, to graphically represent sets. The usual procedure is to draw a rectangle and to let the region inside the rectangle represent the universal set U and then to draw circular regions inside the rectangle to represent each subset of U under consideration. For example, suppose Professor Smith has one chemistry class of about 100 students; some have blue eyes, some have brown eyes, and some have eyes that are neither blue nor brown. Suppose further we are discussing the students of this class who have blue eyes and those who have brown eyes. For purposes of our discussion, we could define the following three sets.

$U = \{x|x$ is a student in Professor Smith's chemistry class$\}$

$A = \{x \in U|x$ has blue eyes$\}$

$B = \{y \in U|y$ has brown eyes$\}$

It is obvious that $A \subset U$, $B \subset U$, and sets A and B are disjoint. The relationships among the sets U, A, and B can be made readily apparent and emphasized by means of the Venn diagram shown in Figure A.12. Observe that the name of each subset is placed inside the region representing that subset. Also, since A and B are disjoint, the regions representing them must not overlap. We shall agree that, if a region appears in a Venn diagram with a letter inside of it, then the set represented by that letter is not the empty set. An important feature of any Venn diagram representing a universal set and its subsets is that the diagram shows at a glance which subsets are disjoint, which are overlapping,

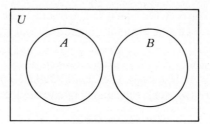

Figure A.12

and which are proper subsets of others. As a further illustration, consider Figure A.13, which shows the Venn diagram of the following universal set and subsets.

$U = \{x | x \text{ is a student in the University of Tennessee}\}$

$A = \{x \in U | x \text{ is a freshman}\}$

$B = \{x \in U | x \text{ is enrolled in mathematics}\}$

$C = \{x \in U | x \text{ is studying engineering}\}$

$D = \{x \in U | x \text{ is a senior law student}\}$

$E = \{x \in D | x \text{ is female}\}$

The Venn diagram in Figure A.13 emphasizes at a glance the following facts regarding students at the University of Tennessee: (a) some freshmen, but not all, are enrolled in mathematics; (b) some freshmen, but not all, are studying engineering; (c) some engineering students who are not freshmen are enrolled in mathematics; (d) some students are enrolled in mathematics who are neither freshmen nor studying engineering; (e) some freshmen are enrolled in mathematics and are studying engineering (see shaded region); (f) no senior law student is a freshman, is enrolled in mathematics, or is studying engineering; (g) there is at least one female senior law student and not all senior law students are female.

Venn diagrams have their greatest value as an aid in the discussion of sets, like those represented in Figure A.13, where the number of elements involved is

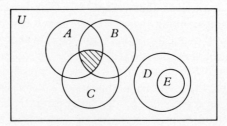

Figure A.13

A.4 Subsets

too large for a listing to be practical or sets where such a listing would be impossible. In such a case it is convenient, but not necessary, to think of the "points" which make up the region representing a set as the elements of the set. It should be emphasized, however, that no significance is to be attached to the relative sizes of the regions used to represent subsets of a given universal set. In fact, it must be clearly understood that Venn diagrams are strictly schematic representations and nothing more.

The following example illustrates another use of Venn diagrams. Suppose that at the beginning of the fall semester in a certain high school the freshman class consists of 100 students. Suppose further that each beginning freshman is required to take at least one of three subjects: mathematics, general science, and Latin. After the fall registration was completed, the class rolls revealed the following statistics regarding the freshman class:

60 were taking mathematics
44 were taking general science
30 were taking Latin
15 were taking both general science and Latin
 6 were taking mathematics and general science but not Latin
10 were taking mathematics, general science, and Latin

Several questions concerning this freshman class might be asked. For example, how many of the class enrolled in only one of the three subjects mentioned above? Or how many enrolled in at least two of these subjects? To answer these and other such questions, we shall use Venn diagrams and the information given above. We start by defining four specific sets.

$$U = \{\text{all students in the class}\}$$

$$L = \{\text{all students in the class taking Latin}\}$$

$$M = \{\text{all students in the class taking mathematics}\}$$

$$S = \{\text{all students in the class taking general science}\}$$

Since the sets L, M, and S are overlapping sets, the circular regions used to represent them must likewise overlap. Hence our Venn diagram consists of a rectangle enclosing three intersecting circles as shown in Figure A.14. We label the three circular regions L, M, and S to correspond respectively to the three sets L (Latin), M (mathematics), and S (general science), as shown. We observe that these three intersecting circles determine seven distinct regions each of which represents a fixed number of students corresponding to that region. We shall now determine the number of students represented by each of the seven regions; as such a number is determined, it will be written in the appropriate region of our Venn diagram. Although one diagram would suffice, we shall do this in slow motion, using a series of six diagrams to indicate our line of reasoning at each

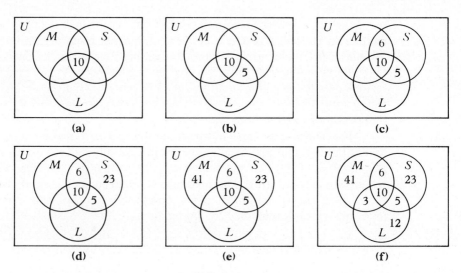

Figure A.14

step. We will work "from the inside out." Since 10 members of the class enrolled in all three of the given subjects, we place the number 10 in the region common to all three circles; see Figure A.14(**a**). According to the given information, 15 were taking both Latin and general science, but 10 of these have already been placed. This leaves 5 for the region inside regions L and S but outside region M; see Figure A.14(**b**). Also, 6 students were taking mathematics and general science but not Latin, so we place a 6 in the region common only to regions M and S; see Figure A.14(**c**). A total of 44 of the class were taking general science. An inspection of Figure A.14(**c**) shows that 21 of these have already been placed, leaving 23 who were taking general science but not mathematics or Latin; see Figure A.14(**d**). We next observe that 30 members of the class were taking Latin and 44 general science. However, 15 of the 74 were taking both; hence 59 of the 100 members of the class were taking Latin or general science or both. Each freshman was required to take at least one of the three subjects. Thus 41 were taking mathematics and not Latin or general science; see Figure A.14(**e**). We have now determined the number of students represented by each of five of the seven regions under consideration. The last two are easily found. For of the 60 students taking mathematics, 57 have already been placed. This leaves 3 for the region common only to circles L and M. Finally, of the 30 taking Latin, 18 are now placed, leaving 12 for the region inside circle L but outside both circles M and S; see Figure A.14(**f**).

The answers to the two questions asked in the early part of our example are readily obtained from the Venn diagram in Figure A.14(**f**). We see that 76 students were taking only one of the required subjects; hence 24 were taking at least two.

Exercises A.4

1. Let the universal set U consist of all the people now living in the world. Next, let A be the set of all people living in Russia, B the set of all people living in the United States, and C the set of all people living in Tennessee. Draw a Venn diagram representing the sets U, A, B, and C.

2. Let $U = \{$all quadrilaterals$\}$, $R = \{$all rectangles$\}$, $S = \{$all squares$\}$, and $P = \{$all parallelograms$\}$. Draw a Venn diagram representing U, R, S, and P.

3. For the sets in Exercise 2, label each of these statements T (true) or F (false).
   a. $S \subset U$ b. $R \subset P$ c. $S \subseteq R$ d. $P \subset U$ e. $S \subset P$ f. $P \subset R$
   g. $R \subseteq U$ h. $R = P$ i. $R = S$

4. Write all the subsets of the set $A = \{1, 2, 3, 4\}$. How many subsets of A are there?

5. Construct a Venn diagram to represent the following subsets of the set of all triangles: all right triangles, all isosceles triangles, and all equilateral triangles.

6. Suppose that X, Y, and Z are sets such that $X \subset Y$ and $Y \subset Z$. Prove that $X \subset Z$.

7. A total of 50 boys enrolled in a summer camp. Each boy was required to participate in at least one of three sports: swimming, tennis, and softball. After registration, it was found that

   22 had chosen swimming
   20 had chosen tennis
   25 had chosen softball
    5 had chosen both swimming and tennis
    4 had chosen swimming and softball but not tennis
    3 had chosen all three sports

   Let $U = \{$all boys enrolled in the camp$\}$, $B = \{$all boys who chose softball$\}$, $S = \{$all boys who chose swimming$\}$, and $T = \{$all boys who chose tennis$\}$. Use this information to place numbers in the appropriate places in a Venn diagram similar to Figure A.14.

8. Referring to Exercise 7 and using the information obtainable from your Venn diagram, answer the following questions.
   a. How many boys chose swimming and softball but not tennis?
   b. How many chose swimming and softball?
   c. How many chose at least two of the three sports?
   d. How many chose tennis only?
   e. How many chose softball only?

9. If in Exercise 7 we change from 4 to 6 the number who chose swimming and softball but not tennis, what change will result in the numbers on your Venn diagram?

10. In the illustrative example at the end of this section, change the number enrolled in Latin and general science to 14, the number enrolled in mathematics and general

science but not Latin from 6 to 10, and change the number enrolled in all three from 10 to 6. Now use this information to place numbers in the appropriate regions of a Venn diagram.

11. A, B, and C are three sets such that $a \in A$, $b \in B$, $A \subset C$, and $B \subset C$.
   a. By drawing Venn diagrams, exhibit as many different possible relations as you can which could exist among the sets A, B, and C, yet still preserve the given relations.
   b. Does $a \in C$?
   c. Does $b \in C$?
   d. Could $a \in B$?
   e. Could there exist an element $c \in C$ such that $c \notin A$ and $c \notin B$?

A.5 SET OPERATIONS

In the previous sections we have seen how two sets may be related. In this section we wish to introduce three operations that may be performed on sets. These operations may be regarded as set-forming operations, for the result of performing each of these operations is to form a set.

The first of these operations is that of combining into one set all of the elements of two sets. The operation is called the *union* of the two sets.

DEFINITION A.16. The *union* of two sets A and B is the set of all elements belonging to A or B or to both A and B. The operation of union is represented by the symbol $A \cup B$ and is read "A union B" or "A cup B" or "the union of A and B."

Although union was defined as an operation on two sets, the notion can be extended to any number of sets simply by combining them two at a time. That is, we define

$$A \cup B \cup C = (A \cup B) \cup C$$

and then define

$$A \cup B \cup C \cup D = (A \cup B \cup C) \cup D, \text{ etc.}$$

The final result in each case is one set.

It follows from the definition of the union of two sets that the order in which the sets are considered does not matter. That is,

1. $A \cup B = B \cup A$
2. $(A \cup B) \cup C = A \cup (B \cup C)$

Property (1) is called *the commutative law for the union of two sets* and Property (2) is called *the associative law for the union of three sets*.

A.5 Set Operations

Example 1 Let $G = \{1, 2, 3\}$, $H = \{3, 4, 5\}$, and $K = \{7, 8\}$. Find (a) $G \cup H$, (b) $H \cup K$, and (c) $G \cup H \cup K$.

$$G \cup H = \{1, 2, 3, 4, 5\}$$
$$H \cup K = \{3, 4, 5, 7, 8\}$$
$$G \cup H \cup K = \{1, 2, 3, 4, 5, 7, 8\}$$

The shaded area in each of the Venn diagrams in Figure A.15 represents the set $A \cup B$ in the three important cases, that A and B are overlapping sets, that $B \subset A$, and that A and B are disjoint.

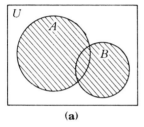
(a)

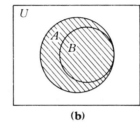
(b)

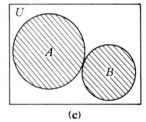
(c)

Figure A.15

The second set-forming operation we wish to consider is that of the *intersection* of two sets.

DEFINITION A.17. The *intersection* of two sets A and B is the set of all the elements which are in both A and B. We denote the operation of intersection by the symbol $\cap$ and write $A \cap B$, which is read "the intersection of A and B" or "A cap B."

The commutative and associative laws are valid also for this operation. That is,

$$A \cap B = B \cap A$$
$$(A \cap B) \cap C = A \cap (B \cap C)$$

The set $A \cap B$ is represented by the shaded area in the Venn diagrams in Figure A.16.

It often happens in mathematical discussions that it is desirable to talk about the set which is called the *complement* of a set S relative to a universal set U. Such a set is defined as follows.

DEFINITION A.18. By the *complement* of a set S relative to a universal set U is meant the set $\bar{S}$ containing all the elements in U not contained in S.

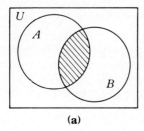 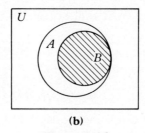 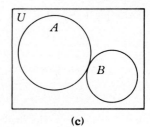

(a)   (b)   (c)

Figure A.16

Example 2

Suppose that S is the set of all the students in your school that are enrolled in at least one mathematics course. Suppose further that the universal set U is the set of all students enrolled in your school. Then $\bar{S}$ is the set consisting of all the students enrolled in your school who are not taking mathematics.

The Venn diagram in Figure A.17 shows the relation between any set S and its complement $\bar{S}$ relative to a universal set U. The shaded area represents $\bar{S}$ and the rectangle represents U.

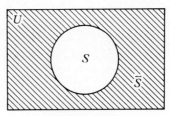

Figure A.17

Exercises A.5

1. Show that
   a. If $A = B$, then $A \cup B = A \cap B = A$.
   b. If $A \subset B$, then $A \cup B = B$ and $A \cap B = A$.
   c. If A is any set and $\emptyset$ is the empty set, then $A \cup \emptyset = A$ and $A \cap \emptyset = \emptyset$.

2. Two subsets A and B of a universal set U are related as shown in the Venn diagram of Figure A.18. Use a separate copy of this Venn diagram and shading to represent each of the following sets.
   a. $A \cap B$ b. $\bar{A}$ c. $\bar{B}$ d. $\overline{A \cup B}$ e. $\bar{A} \cup \bar{B}$

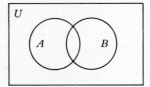

Figure A.18

A.5   Set Operations

3. If the universal set $U = \{a, b, c, d, e, f, g, h\}$ and $A = \{a, c, g, h\}$ and $B = \{c, d, e, g\}$, find each of the following sets.
   a. $A \cap B$ b. $A \cup B$ c. $\bar{A}$ d. $\bar{B}$ e. $\bar{A} \cup B$ f. $A \cup \bar{B}$
   g. $\bar{A} \cap B$ h. $A \cap \bar{B}$ i. $\bar{A} \cup \bar{B}$ j. $\overline{A \cup B}$ k. $\bar{A} \cap \bar{B}$ l. $\overline{A \cap B}$

4. If the universal set $U = \{x \mid x \text{ is a counting number less than } 10\}$ and $A = \{1, 2, 3, 4, 5\}$, $B = \{2, 5, 7, 9\}$, and $C = \{3, 5, 6, 7, 8\}$, find each of the following sets.
   a. $A \cup B$ b. $A \cup C$ c. $B \cup C$
   d. $A \cap C$ e. $B \cap C$ f. $(A \cap B) \cup C$
   g. $A \cup (B \cap C)$ h. $(A \cup B) \cap (A \cup C)$ i. $A \cap (B \cup C)$
   j. $(A \cap B) \cup (A \cap C)$ k. $\bar{A} \cap B$ l. $\bar{A} \cup B$
   m. $A \cup \bar{C}$ n. $\bar{B} \cup C$ o. $\overline{B \cap C}$
   p. $\bar{A} \cap (B \cup C)$ q. $(A \cap B) \cap \bar{C}$ r. $(B \cap C) \cup \overline{(A \cap C)}$

5. Three subsets A, B, and C of a universal U are related as shown in Figure A.19. Use a separate copy of this Venn diagram and shading to represent each of the following sets.
   a. $A \cup C$ b. $B \cap C$ c. $(A \cap B) \cap C$ d. $A \cup (B \cap C)$
   e. $(A \cup B) \cap (A \cup C)$ f. $\bar{A} \cap (B \cup C)$ g. $\bar{B} \cup (A \cap C)$ h. $A \cup \overline{(B \cup C)}$

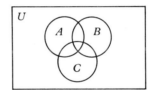

Figure A.19

Compare the Venn diagrams of the sets in parts **d** and **e**. What conclusion do you make?

6. Repeat parts **d** and **e** of Exercise 5 for the following two cases.
   Case 1. A, B, and C are related as shown in Figure A.20.

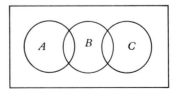

Figure A.20

Case 2. A, B, and C are related as shown in Figure A.21.
What relation between $A \cup (B \cap C)$ and $(A \cup B) \cap (A \cup C)$ is suggested by the Venn diagrams of the two sets in Case 1 and Case 2?

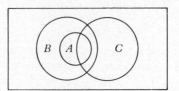

Figure A.21

7. There are many possible ways in which three sets A, B, and C might be related. Three such possibilities are exhibited in Figures A.19, A.20, and A.21. It was shown in Exercises 5 and 6, by use of Venn diagrams, that for the three cases considered

$$A \cup (B \cap C) = (A \cup B) \cap (A \cup C)$$

This equation states a law called the *distributive law of union over intersection*, which can easily be proved without the use of Venn diagrams to be true in general. For convenience in making such a proof, let $S = A \cup (B \cap C)$ and $T = (A \cup B) \cap (A \cup C)$. First prove that, if $x \in S$, then $x \in T$, and hence $S \subseteq T$. Do this by considering two cases:

Case 1. $x \in S$ and $x \in A$; hence $x \in A \cup B$ and $x \in A \cup C$. That is, $x \in T$. Why?

Case 2. $x \in S$ and $x \notin A$; therefore, $x \in B \cap C$. That is, $x \in B$ and $x \in C$; hence $x \in A \cup B$ and $x \in A \cup C$. Thus $x \in T$. Why?

Conclusion: $S \subseteq T$.

Next, prove that if $y \in T$, then $y \in S$ and hence $T \subseteq S$. Again consider two cases.

Case 1. $y \in T$ and $y \in A$.

Case 2. $y \in T$ and $y \notin A$.

Complete the argument as above.

8. Using the method of Exercise 7, prove the *distributive law of intersection over union*.

$$A \cap (B \cup C) = (A \cap B) \cup (A \cap C)$$

9. DeMorgan's laws:

    a. $\overline{(A \cup B)} = \bar{A} \cap \bar{B}$ **b.** $\overline{(A \cap B)} = \bar{A} \cup \bar{B}$

Use the method of Exercise 7 to prove DeMorgan's laws.

10. The difference, $A - B$, between two sets A and B is defined as the set consisting of all the elements of A that do not belong to B. In symbols, $A - B = \{x | x \in A, x \notin B\}$. Represent the set $A - B$ by means of a Venn diagram and shading for each of the following cases.

    a. A and B are overlapping sets   **b.** A and B are disjoint   **c.** $B \subset A$

    d. $A \subset B$ **e.** $A = B$

11. If $A = \{a, b, c, d, e\}$, $B = \{b, d, f, h, j\}$, and $C = \{a, c, f, g, h\}$, find

    a. $A - B$ **b.** $A - C$ **c.** $B - C$ **d.** $B - A$

    e. $C - B$ **f.** $(A \cup C) - B$

A.5 Set Operations

12. Let $N(A)$ denote the number of elements in set A. For example, if $A = \{1, 3, 4, 7, 9, 11\}$, $N(A) = 6$, and if $B = \{x|x \text{ is a U.S. senator}\}$, $N(B) = 100$.
    a. Give an example of two sets A and B such that $N(A \cup B) = N(A) + N(B)$.
    b. Give an example of two sets A and B such that $N(A \cup B) \neq N(A) + N(B)$.
    c. Give an example of two sets A and B such that $N(A - B) = N(A) - N(B)$.

13. If A and B are any two overlapping sets, use a Venn diagram to show that $N(A \cup B) = N(A) + N(B) - N(A \cap B)$. Would this be true for any two sets?

14. A survey of the homes in a new subdivision showed that every family subscribed to at least one of the two local daily papers. In fact, 25 families subscribed to the morning paper, 30 families subscribed to the evening paper, and 10 families subscribed to both. How many families live in the subdivision?

15. If A and B are any two overlapping subsets of a universal set U, find
    a. $N(A \cup B)$ if $N(A \cap B) = 6$, $N(A) = 13$, and $N(B) = 16$
    b. $N(A \cap B)$ if $N(B - A) = 25$ and $N(B) = 37$
    c. $N(B - A)$ if $N(A \cup B) = 23$ and $N(A) = 18$
    d. $N(B)$ if $N(A \cap B) = 3$, $N(A \cup B) = 19$, and $N(A) = 8$
    e. $N(U)$ if $N(A - B) = 7$, $N(B) = 15$, and $N(\bar{A} \cap \bar{B}) = 4$

16. In an adult education class of 25 students, one stormy evening it was observed that 14 wore raincoats, 9 carried umbrellas, and 3 persons had both. How many persons in this class had neither a raincoat nor an umbrella that evening?

17. Let A, B, and C be any three overlapping subsets of a universal set U. Use a Venn diagram to show that

$$N(A \cup B \cup C) = N(A) + N(B) + N(C) - N(A \cap B) \\ - N(A \cap C) - N(B \cap C) + N(A \cap B \cap C)$$

Is this result valid for any three sets in general?

18. Suppose that in a certain school every member of the freshman class had to take at least one of three subjects: English, mathematics, and a foreign language. Suppose further that the school records revealed that

    41 were taking English
    36 were taking mathematics
    18 were taking a foreign language
    13 were taking English and mathematics
    9 were taking English and a foreign language
    8 were taking mathematics and a foreign language
    5 were taking all three subjects

    How many members were in the freshman class?

19. A supermarket employee working on a consumer-purchasing study submitted the following report on the purchases of 40 shoppers. Thirty-two purchased groceries, 23 purchased meats, 17 purchased housewares, 19 purchased groceries and meats, 7

purchased meats and housewares, 14 purchased groceries and housewares, and 6 purchased items in all three categories. How many of the shoppers in the survey purchased from none of these three categories?

20. In a survey of 100 male seniors at a college regarding participation in intramural sports, it was found that

25 had participated in baseball
29 had participated in football
7 had participated in baseball and football
9 had participated in football and basketball
12 had participated in basketball and baseball
5 had participated in baseball, football, and basketball
38 had not participated in any of these three sports

a. How many participated in basketball?
b. How many participated in exactly two of these sports?
c. How many participated in baseball only?
d. How many participated in football but not in baseball?

A.6 THE CARTESIAN PRODUCT OF TWO SETS

In the previous section we discussed some operations which we called set-forming operations. The result of each operation was a set; that is, $A \cup B$ is a set, $A \cap B$ is a set, and $\bar{A}$ is a set, assuming, of course, a universal set U. If the universal set is determined, then the operation of union, intersection, or complement performed on subsets of U always gives rise to a set which is a subset of U. For example, if $C = A \cup B$, then the elements of C are also elements of U. Furthermore, these operations were applied to the sets rather than to the elements in the sets.

In this section we shall show how two sets A and B may give rise to a third set each of whose elements is a pair, (a, b), of elements, such that $a \in A$ and $b \in B$. We shall refer to the pair (a, b) as an *ordered pair* because, in general, the pair (a, b) will not be the same as the pair (b, a). In other words, the order in which the pair is written may and generally does affect whatever the pair represents. We are familiar with this fact from elementary arithmetic; the number resulting from combining a pair of integers may depend on the order in which you combine the integers. For example, an ordinary fraction may be thought of as a number pair, say $\frac{2}{3}$. We could even use the above notation by agreeing that the two symbols $(2, 3)$ and $\frac{2}{3}$ mean the same thing, that is $(2, 3) = \frac{2}{3}$. Now certainly, if we change the order in the number pair, we get a different fraction for $(3, 2) = \frac{3}{2} \neq \frac{2}{3}$. So you see, if we define fractions as number pairs, we have to define them as ordered number pairs, lest we get into difficulty.

In referring to an ordered pair, say (x, y), we call x the *first element* of the

A.6 The Cartesian Product of Two Sets

pair and y the *second element* of the pair. This is easy to remember, for the element written first is the first element, assuming of course that in writing we use the natural order in which we write words, from left to right.

Now suppose we have the two sets $A = \{1, 2, 3\}$ and $B = \{2, 4\}$. How many simple fractions can be formed from these two sets if the numerator must be an element of A and the denominator must be an element of B? Obviously, there are as many different fractions as there are possible number pairs, if the first number in each pair has to be from A and the second number from B. Writing down all the possible pairs as described, we get $(1, 2), (1, 4), (2, 2), (2, 4), (3, 2)$, and $(3, 4)$. Hence there are exactly six different such fractions possible. We now consider these ordered pairs as being the elements of a set, say C. We may write

$$C = \{(1, 2), (1, 4), (2, 2), (2, 4), (3, 2), (3, 4)\}$$

This new set C, which we have formed from A and B, is called the *Cartesian product of A and B*. It is customary, instead of giving this new set a name, to denote it by $A \times B$, read "A cross B." This leads us to the following definition.

DEFINITION A.19. By the *Cartesian product* of two nonempty sets A and B is meant the set of all ordered pairs (a, b) such that $a \in A$ and $b \in B$. In symbols, this becomes $A \times B = \{(a, b) | a \in A \text{ and } b \in B\}$.

It should be noted that the definition just given for the Cartesian product of two sets does not specify any relation between the two sets forming the product. The two sets could be the same or they could be different. Also, the elements of the two sets need have no relation. However, the real usefulness in mathematics of such a product will be found in cases in which the elements of both sets are related. For example, we shall be interested in the case in which the elements of both sets are real numbers.

Example 1 Let $A = \{1, 2\}$; then $A \times A = \{(1, 1), (1, 2), (2, 1), (2, 2)\}$.

Example 2 Let $A = \{a, b, c\}$ and $B = \{d, e\}$; then

$$A \times B = \{(a, d), (a, e), (b, d), (b, e), (c, d), (c, e)\}$$
$$B \times A = \{(d, a), (d, b), (d, c), (e, a), (e, b), (e, c)\}$$

A very convenient way of representing graphically a Cartesian product is illustrated in Figure A.22. The Cartesian product is $S \times T$, where $S = \{a, b, c\}$ and $T = \{d, e\}$. By definition,

$$S \times T = \{(x, y) | x \in S \text{ and } y \in T\}$$

To obtain the graph shown in Figure A.22, two mutually perpendicular lines are drawn. Just as a matter of convenience, one is drawn horizontally and the other vertically. The horizontal line is labeled at the far right with an x and is referred

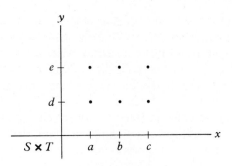

Figure A.22

to as the x axis. The vertical line is labeled with a y and is referred to as the y axis. Three arbitrary but distinct points are chosen on the x axis to represent the elements of the set S; one represents the element a, one the element b, and one the element c. Which point is associated with what element of S is of no particular significance, except that one point of the x axis is associated with only one element of S. In a similar manner, two distinct points are chosen on the y axis to represent the two elements d and e of the set T. Each of the three points on the x axis is labeled with the name of the element it represents, and so is each of the two points on the y axis. Next, to represent the element (a, d) of the Cartesian product $S \times T$, a point is placed vertically above a and on the same level as d. Similarly, to represent the element (a, e), a point is placed vertically above a and on the same level with e. This scheme is repeated until six points, representing the six ordered pair elements of $S \times T$, are located. Points arranged in this manner are called *lattice points*.

To read the graph, which is to find the ordered pairs associated with a particular point, one reverses the procedure just described. The first element of the ordered pair represented by a particular lattice point is the element on the x axis vertically below the given point. The second element of the ordered pair is the element on the y axis and on the same level as the given point.

A graph of the type just described can, of course, be constructed for any Cartesian product, regardless of the kind of elements involved. However, the lattice-type graphs that are of most interest to mathematicians are those in which the elements x and y of the given sets are real numbers and, hence, the elements of the Cartesian product are ordered pairs, (x, y), of real numbers. Then it is customary to select an arbitrary unit of measure for measuring distances and to associate with each number $x \in S$ a point on the x axis that is x units to the right of the y axis, assuming x a positive number, and to associate with each number $y \in T$ a point on the y axis that is y units above the x axis if y is a positive number. The number elements x and y of the ordered pair (x, y) are then called the coordinates of the lattice point with which the number pair is associated. The lattice point associated with the number pair $(3, 2)$ is located three units to the

A.6 The Cartesian Product of Two Sets

right of the y axis and two units above the x axis. One should notice that such a scheme would associate with (2, 3) a point different from the one associated with (3, 2). Hence, in considering a pair of elements (x, y), the order is as important as the elements themselves. Graphs of Cartesian products whose elements are real numbers were discussed in Chapter 1.

Example 3 Construct the graph of the Cartesian product $A \times B = \{(x, y) | x \in A \text{ and } y \in B\}$, where $A = \{1, 2, 3, 4, 5\}$ and $B = \{1, 2, 3, 4\}$. See Figure A.23.

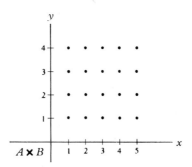

Figure A.23

Exercises A.6

1. Let $A = \{3, 4, 5\}$ and $B = \{2, 7\}$.
   a. Find $A \times B$, $A \times A$, $B \times B$, and $B \times A$.
   b. How many elements are in each set of part **a**?

2. Let A and B be finite sets and let $N(A)$ and $N(B)$ denote the number of elements in A and B, respectively. If $N(A) = p$ and $N(B) = q$, what is $N(A \times B)$ in terms of p and q?

3. If one coin is tossed, it will fall heads (H) or tails (T). If two coins are tossed, what is the set of possible outcomes, expressed as a Cartesian product.

4. An ideal die is a perfect cube with six faces. One face has one dot, a second face has two dots, a third face three dots, and so on, the sixth face having six dots. Associate with each face a number corresponding to the number of dots on the face and let S be a set consisting of the numbers $S = \{1, 2, 3, 4, 5, 6\}$. If two dice are rolled, $S \times S$ is the set of possible outcomes. Draw a lattice graph representing $S \times S$.

5. Let A and B be two sets. Discuss the possibility that $A \times B = B \times A$.

6. The school cafeteria serves three kinds of salads: tossed, head lettuce, and tomato. There is also a choice of three kinds of dressing: Thousand Island, French, and mayonnaise. Treating the salads as the elements of a set A and the three dressings as elements of a second set B, draw a lattice graph representing $A \times B$. Label one or two of the lattice points with the kind of salad they represent.

7. Figure A.23 represents $A \times B$, where $A = \{1, 2, 3, 4, 5\}$ and $B = \{1, 2, 3, 4\}$. List the elements of each of the following subsets of $A \times B$.
   a. $C = \{(x, y) | (x, y) \in A \times B \text{ and } x \text{ is less than } 4\}$
   b. $D = \{(x, y) | (x, y) \in A \times B \text{ and } x \text{ is 1 and } y \text{ is even}\}$
   c. $E = \{(x, y) | (x, y) \in A \times B \text{ and } x = y\}$
   d. $F = \{(x, y) | (x, y) \in A \times B \text{ and } x + y = 4\}$
   e. $G = \{(x, y) | (x, y) \in A \times B \text{ and } xy = 30\}$
   f. $H = \{(x, y) | (x, y) \in A \times B \text{ and } x \text{ is greater than } 3, y \text{ is less than } 3\}$
8. Referring to Exercise 4, list the elements in each of the following subsets of $S \times S$.
   a. $M = \{(a, b) | (a, b) \in S \times S \text{ and } a + b = 7\}$
   b. $N = \{(a, b) | (a, b) \in S \times S \text{ and } a + b \text{ is at least } 7\}$
   c. $Q = \{(a, b) | (a, b) \in S \times S \text{ and } a + b \neq 8\}$
   d. $K = \{(a, b) | (a, b) \in S \times S \text{ and } a \neq 6, b \neq 6\}$
   e. $L = \{(a, b) | (a, b) \in S \times S \text{ and } a = 6, b \neq 6\}$
   f. $J = \{(a, b) | (a, b) \in S \times S \text{ and } a = 6 \text{ or } b = 6, \text{ but not both}\}$
9. If $A = \{\text{John, Paul}\}$ and $B = \{\text{Alice, Mary}\}$, what are a. all the elements of $A \times B$ and b. all the subsets of $A \times B$?

A.7 EQUIPOLLENT SETS

Consider the sets $A = \{1, 2, 3\}$ and $B = \{a, b, c\}$. Obviously, the only property that these sets have in common is that they have the same number of elements, 3. Since A and B are disjoint, they are not equal in the sense of our definition of the equality of two sets; that is, they are not identical. However, the sets A and B are essentially indistinguishable if our only interest is the number of elements in each set. One word which describes this phenomenon is *equipollence*. Hence we say that the sets A and B are *equipollent*. The words equivalence and equivalent are not used in this situation in order to avoid the use of the same word to describe several different phenomena. The words equivalence and equivalent will be reserved for the notions of logical equivalence of propositions (Definition A.3), equivalence of open statements (Definition A.6), and equivalence of equations and inequalities (Chapter 3).

Before we make a precise definition of equipollent sets, let us consider another example. Let S be the set of students in Professor Smith's chemistry class and let T be the set of chairs in his classroom. At the first meeting of the class, Professor Smith does not need to count either students or chairs to decide whether or not he has the same number of students as chairs. He simply asks the students to be seated. If every chair is occupied and no student is left standing, his conclusion is that the number of students is the same as the number of chairs. The mental process used in arriving at this conclusion was one of pairing students and chairs. In mathematical language we say that a correspondence has

A.7 Equipollent Sets

been set up between students and chairs. To each student there corresponds one and only one chair, the chair occupied by the student. Conversely, to each chair there corresponds one and only one student, the student occupying the chair. This type of pairing is called *one-to-one correspondence*.

DEFINITION A.20. Two sets A and B are said to be in *one-to-one correspondence* when it is possible to find a pairing of the elements of A and B such that each element of A corresponds to one and only one element of B, and each element of B corresponds to one and only one element of A.

We are now ready to define *equipollent* sets.

DEFINITION A.21. Two sets A and B are said to be *equipollent* when the elements of A and the elements of B can be placed in one-to-one correspondence. We denote this relation by the symbol $A \sim B$, read "A is equipollent to B."

In the example given above, if Professor Smith finds that he has exactly the same number of students as chairs, then $S \sim T$.

It should be pointed out that the common property of two or more equipollent sets is that they each contain the same number of elements. That is, two finite sets A and B are equipollent if the number of elements in A is the same as the number of elements in B. Hence, if $A = B$, then $A \sim B$. However, the converse is not necessarily true. A may be equipollent to B when A is not equal to B. For example, the sets $A = \{1, 2, 3\}$ and $B = \{a, b, c\}$ are equipollent but not equal.

A very important aspect of our definition of equipollent sets is that it gives us a means of comparing the number of elements in each of two sets when each set contains infinitely many elements. As a matter of fact, when we say that one infinite set has the same number of elements as a second infinite set, we simply mean that the two sets are equipollent.

Example 1

Let $N = \{1, 2, 3, \ldots\}$ and $E = \{2, 4, 6, \ldots\}$. It is easy to show that $N \sim E$. We need to show that the elements of N and E can be placed in one-to-one correspondence. To do this, let x represent an arbitrary element of N. Then certainly there is an element $y \in E$ such that $y = 2x$. Now, if we pair with each $x \in N$ the number $2x \in E$, then to each element of N there corresponds one and only one element of E, and to each element of E there corresponds one and only one element of N. Thus, by definition, the elements of N and E have been placed in one-to-one correspondence. Hence $N \sim E$. We often indicate this particular equipollence by saying that there are as many even counting numbers as there are counting numbers. The pairing of the element of N with the elements of E described above can be exhibited as follows:

$$\begin{array}{ccccccc} 1 & 2 & 3 & 4 & \ldots & x & \ldots \\ \updownarrow & \updownarrow & \updownarrow & \updownarrow & & \updownarrow & \\ 2 & 4 & 6 & 8 & \ldots & 2x & \ldots \end{array}$$

Example 2 Let us indicate a proof that a line segment $\frac{1}{2}$ inch long has the same number of points as a line segment 1 inch long. First we construct the two segments, one above the other, as shown in Figure A.24. Denote the endpoints of the 1-inch segment by A and B and the endpoints of the $\frac{1}{2}$-inch segment by A' and B'. Now let S be the set consisting of points A and B and all points between them on segment AB, and let T be the set consisting of points A' and B' and all points between them on segment $A'B'$. The two segments have the same number of points if $S \sim T$. To show this is true, draw lines AA' and BB', and let O be their point of intersection. The elements of S and T can be put into a one-to-one correspondence in the following manner: Let C be an arbitrary point of set S. Pair with each such point C of set S the point C' of set T which is obtained as the point of intersection of lines OC and $A'B'$. Similarly, pair with each point D' of set T the point of intersection of lines OD' and AB, which we denote by D. Thus to each point C of set S there corresponds, by this procedure, one and only one point of set T, that is, C', and conversely to each point D' of T there corresponds one and only one point of set S, that is, D. Thus a one-to-one correspondence has been established between the elements of the sets S and T, and hence $S \sim T$. That is, there are as many points in T as in S.

If A, B and C are sets, it follows from the definition of equipollent sets that each of the following statements is true.

1. $A \sim A$; each set is equipollent to itself.
2. If $A \sim B$, then $B \sim A$.
3. If $A \sim B$ and $B \sim C$, then $A \sim C$.

These three relations may be stated in words: the property of equipollence of sets is reflexive, symmetric, and transitive.

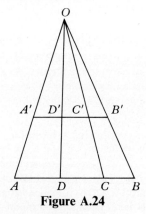

Figure A.24

Exercises A.7

1. Determine which of the following pairs of sets are equal and which are equipollent but not equal.

    a. $A = \{x | x \text{ is a letter in the word "Tennessee"}\}$ and $B = \{y | y \text{ is a letter in the word "nest"}\}$

b. $C = \{\text{Washington, Jefferson, Lincoln}\}$ and $D = \{w, j, l\}$
   c. $E = \{1, 2, 3, \ldots, 30\}$ and $F = \{3, 6, 9, \ldots, 90\}$
   d. $G = \{x | x + 1 = 4\}$ and $H = \{x | x - 2 = 1\}$
   e. $I = \{\text{the vowels in the English alphabet}\}$ and $J = \{\text{the letters in the word "balloon"}\}$

2. If $S = \{1, 2, 3, 4, 5\}$, write four subsets of S that are equipollent but not equal.

3. If $N = \{1, 2, 3, 4, 5, \ldots\}$ and $T = \{5, 10, 15, 20, 25, \ldots\}$, show that $N \sim T$. Is the statement $T \subset N$ true or false? One characteristic property of an infinite set A is that it has a proper subset B which is equipollent to A.

4. If $E = \{2, 4, 6, 8, \ldots\}$, find a proper subset of E that is equipollent to E.

5. List all the possible ways in which the elements of the two sets $\{1, 2, 3\}$ and $\{5, 6, 7\}$ can be paired in a one-to-one correspondence.

6. Suggest a method of showing that the semicircular arc ACB and the diameter AB in Figure A.25 have the same number of points.

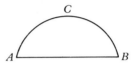

Figure A.25

7. Give examples of the following pairs of sets:
   a. Two sets that are overlapping
   b. Two sets, one of which is a proper subset of the other
   c. Two sets that are disjoint
   d. Two sets that are equal
   e. Two sets that are equipollent but not equal

A.8* COUNTABLE SETS

In one of the examples of Section A.7 it was shown that the set $N = \{1, 2, 3, 4, \ldots\}$ and the set $E = \{2, 4, 6, 8, \ldots\}$ are equipollent. That is, it was shown that the elements of set N can be put into a one-to-one correspondence with the elements of set E. The set E is obviously a proper subset of N. Here we have the surprising result of the elements of a set being put into a one-to-one correspondence with the elements of a proper subset of itself. This is not a rare case. In fact, every infinite set can be put into one-to-one correspondence with *some* proper subset of itself. This is one of the characteristic properties of an infinite set and is often used in defining such a set.

It is easy to see how such sets as $K = \{51, 52, 53, 54, \ldots\}$, $L = \{1,000, 2,000, 3,000, 4,000, \ldots\}$, and $M = \{7, 14, 21, 28, \ldots\}$ can be put into a one-to-one correspondence with the set N of counting numbers. For example, the correspondence between N and K could be shown as follows:

$$1 \leftrightarrow 51,\ 2 \leftrightarrow 52,\ 3 \leftrightarrow 53,\ 4 \leftrightarrow 54,\ \ldots,\ n \leftrightarrow n + 50,\ \ldots$$

For the correspondence between N and M, we could use $1 \leftrightarrow 7,\ 2 \leftrightarrow 14,\ 3 \leftrightarrow 21,\ 4 \leftrightarrow 28,\ \ldots,\ n \leftrightarrow 7n,\ \ldots$. Each of the sets K, L, and M is a proper subset of N. Let us now consider the set

$$P = \{-5,\ -4,\ -3,\ -2,\ -1,\ 0,\ 1,\ 2,\ 3,\ 4,\ \ldots\}$$

In this case $N \subset P$ and $N \sim P$. To show this equipollence between N and P, we could pair the elements thus: $1 \leftrightarrow -5,\ 2 \leftrightarrow -4,\ 3 \leftrightarrow -3,\ 4 \leftrightarrow -2,\ \ldots,\ n \leftrightarrow n - 6,\ \ldots$. Note here that the relation $n \leftrightarrow n - 6$ serves as a formula for the correspondence which may be used to determine quickly the number in P that is paired with a specific number in N, or vice versa. For example, since $735 \in N$, the number in P that is paired with 735 is $735 - 6$, or 729.

As a slightly more complicated case, let us show that the set

$$I = \{\ldots,\ -4,\ -3,\ -2,\ -1,\ 0,\ 1,\ 2,\ 3,\ 4,\ \ldots\}$$

can be put into a one-to-one correspondence with set N. One method is the following.

1	2	3	4	5	6	7	8	9	10	11	$\cdots$	$2k$	$2k+1$	$\cdots$
$\updownarrow$	$\updownarrow$	$\updownarrow$	$\updownarrow$	$\updownarrow$	$\updownarrow$	$\updownarrow$	$\updownarrow$	$\updownarrow$	$\updownarrow$	$\updownarrow$		$\updownarrow$	$\updownarrow$	
0	-1	1	-2	2	-3	3	-4	4	-5	5	$\cdots$	$-k$	k	$\cdots$

In this case we really need two formulas to express the correspondence between the elements of N and I. If $n \in N$ and $n = 2k$, where k is a counting number, then $-k \in I$ and $2k \leftrightarrow -k$; but if $n \in N$ and $n = 2k + 1$, where k is either zero or a counting number, then $k \in I$ and $2k + 1 \leftrightarrow k$. It should be noted that since n is a counting number, to say that $n = 2k$ simply means that n is an even counting number. To say that $n = 2k + 1$ means that n is an odd counting number and that $n = 1$ if and only if $k = 0$.

The above discussion suggests that many infinite sets possess this common property: each can be put into a one-to-one correspondence with the set N of counting numbers. Any infinite set having this property is said to be a *countably infinite set*.

DEFINITION A.22. A set whose elements can be put into a one-to-one correspondence with the elements of the set of counting numbers is called a *countably infinite set*.

DEFINITION A.23. A set is said to be countable if (a) it is the empty set, or (b) it is a finite set, or (c) it is a countably infinite set.

We may ask, are all infinite sets countable? That is, are all infinite sets countably infinite? The answer is no. To justify this answer we shall show that

A.8 Countable Sets

the set S consisting of all the real numbers lying between 0 and 1 is not a countably infinite set. To make our proof, we shall need to make use of the fact that every real number between 0 and 1 can be written uniquely as a nonending decimal if we agree to write fractions like 0.5, 0.37, 0.854, etc., as 0.4999 ..., 0.369999 ..., 0.8539999 ..., etc. We shall use an indirect method of proof. That is, we shall assume that set S is a countably infinite set and show that this assumption leads us to a contradiction. If S is a countably infinite set, then its elements can be put into a one-to-one correspondence with the elements of N; and let us suppose that such a correspondence has been set up. Furthermore, let us denote by r_1 the number in S that corresponds to 1, denote by r_2 the number in S that corresponds to 2, denote by r_3 the number in S that corresponds to 3, and so on. Thus $\{r_1, r_2, r_3, \ldots, r_n, \ldots\} \subseteq S$. Next, we express r_1, r_2, r_3, etc., as nonending decimal fractions and denote them as follows.

$$1 \leftrightarrow r_1 = 0.a_{11}a_{12}a_{13}a_{14} \cdots a_{1n} \cdots$$
$$2 \leftrightarrow r_2 = 0.a_{21}a_{22}a_{23}a_{24} \cdots a_{2n} \cdots$$
$$3 \leftrightarrow r_3 = 0.a_{31}a_{32}a_{33}a_{34} \cdots a_{3n} \cdots$$
$$\vdots$$
$$n \leftrightarrow r_n = 0.a_{n1}a_{n2}a_{n3}a_{n4} \cdots a_{nn} \cdots$$
$$\vdots$$

Here each symbol of the form a_{ij} represents some one of the digits 0, 1, 2, 3, 4, 5, 6, 7, 8, 9. If our assumption that set S is countable is true, then each number between 0 and 1 must be one of the numbers $r_1, r_2, r_3, \ldots$ listed above. We shall now show that there is a number which lies between 0 and 1 and which does not occur in the above list and hence does not correspond to one of the counting numbers. For example, consider the number $s = 0.b_1 b_2 b_3 \ldots b_n \ldots$, where each of the symbols b_1, b_2, b_3, and so on, represents some one of the digits 0, 1, 2, 3, 4, 5, 6, 7, 8, 9 and is determined as follows. If $a_{11} \neq 1$, then $b_1 = 1$; if $a_{11} = 1$, then $b_1 = 2$; if $a_{22} \neq 1$, then $b_2 = 1$; if $a_{22} = 1$, then $b_2 = 2$; if $a_{33} \neq 1$, then $b_3 = 1$; if $a_{33} = 1$, then $b_3 = 2$; or in general $b_k = 1$ if $a_{kk} \neq 1$, but $b_k = 2$ if $a_{kk} = 1$. Obviously, $s \in S$. Also, $s \neq r_1$ since $b_1 \neq a_{11}$, $s \neq r_2$ since $b_2 \neq a_{22}$, $s \neq r_3$ since $b_3 \neq a_{33}$, and so on. Therefore, s does not correspond to one of the counting numbers and hence the correspondence is not one-to-one, a contradiction of our assumption that set S is countable. We must conclude therefore that not all infinite sets are countable.

A.9* OPEN STATEMENTS AND QUANTIFIERS

In Section A.1 we discussed a special kind of statement, which we called a proposition. It will be recalled that a proposition was defined as a statement for which it could be said that the statement is true or the statement is false, but not both. Then in Section A.3 we defined a variable as a letter used to represent an arbitrary element of some given or implied set. In this section we want to define a second special kind of statement called an *open* statement. As we shall see, open statements are statements involving one or more variables, and it is this fact which makes them important in mathematics. For example, in mathematics we often need to write and use such statements as:

1. $2x + 4 = 10$
2. $x^2 > 4$ (Read: "x^2 is greater than 4.")
3. $(x - 2)(x - 5) = 0$
4. $2x < 6$ (Read: "$2x$ is less than 6.")
5. $x + y = 5$

As it stands, none of the five statements listed above is a proposition, because it is not possible to say of any statement that it is true or that it is false. However, in each case if a specific number is substituted for x, or in statement 5 if a number is substituted for x and a number for y, then the statement becomes a proposition.

DEFINITION A.24. An *open statement* is a statement which contains one or more variables and which becomes a proposition when each variable is replaced by one of its possible values.

Example 1 Let $N = \{1, 2, 3, 4, \ldots\}$ and suppose $x \in N$. Then the statement

$$2x + 4 = 10$$

is an open statement. For, first, the statement contains the variable x and, second, if x is replaced by any specific element of N, the statement becomes a proposition. In fact, $2x + 4 = 10$ is a true statement when x is replaced by 3, and the statement is false when x is replaced by any other specific number from the set N.

In our discussion of propositions we used the letters $p, q, r, \ldots$ to designate unspecified propositions. It seems quite natural, therefore, to designate unspecified open statements involving the variable x by p_x, q_x, r_x, etc. Also, open statements involving two variables x and y will be denoted $p_{x,y}, q_{x,y}, r_{x,y}$, etc., and so on for open statements containing three or more variables. We now come to the important notion of the *truth set* of an open statement.

A.9 Open Statements and Quantifiers

DEFINITION A.25. If p_x is an open statement which contains the variable x, then the *truth set* of p is the set of elements a for which p_a is a true proposition. Similarly, the *truth set* of an open statement $p_{x,y}$, which contains the variables x and y, is the set of ordered pairs (a, b) for which $p_{a,b}$ is a true proposition and so on for open statements containing three or more variables.

In keeping with the notation developed above, we shall use corresponding capital letters to designate truth sets of open statements. That is, we shall designate by the capital letter P the set of elements a for which p_a is a true proposition, by the capital letter Q the set of all elements a for which q_a is a true proposition, and so on. A similar convention is adopted for open statements containing two or more variables.

Example 2 Let p_x be the statement "$(x-1)(x-3) = 0$." Then
$$P = \{x | (x-1)(x-3) = 0\} = \{1, 3\}$$

Example 3 Let p_x be the statement "$x + 5 = 7$" and let q_x be the statement "$5x = 10$." Then
$$P = \{x | x + 5 = 7\} = \{2\}$$
and
$$Q = \{x | 5x = 10\} = \{2\}$$

Observe that in Example 3 the truth sets of p_x and q_x are equal. Whenever this is the case, it must be true that if a is a possible value of x, then p_a is true when and only when q_a is true. That is, for each possible value a of the variable x, the bi-implication $p_a \leftrightarrow q_a$ is true (see Section A.2). In such a case we say that p_x and q_x are *equivalent open statements*. Similar remarks apply to open statements containing two or more variables.

DEFINITION A.26. Two open statements are said to be *equivalent open statements* if their truth sets are equal.

Remark: All the statements given above as examples of open statements are either statements commonly known as "equations" or are statements known as "inequalities." In fact, our chief concern with open statements in the text is with those which are either equations or inequalities. In Example 3 we saw that the open statement "$x + 5 = 7$" was equivalent to the open statement "$5x = 10$." Hence we say that the *equation* $x + 5 = 7$ is equivalent to the *equation* $5x = 10$. Of course, this simply means that the truth sets of the equations are equal.
In the text, when dealing with open statements which are equations or inequalities, we refer to their truth sets as *solution sets*, since a number a is in the truth set when and only when it is a *solution* to the equation or inequality. A considerable emphasis in the text will be on "solving" equations and inequalities for which the role played by equivalent open statements (called equivalent equations or equivalent inequalities, as the case may be) is a very important one.

Quantifiers

A statement that involves a variable is not necessarily an open statement. It may be a proposition. For example, in a course in high school algebra where it is generally understood that, unless otherwise specified, the universal set is the set consisting of all the real numbers, statements similar to the following are quite commonplace.

For all x, $3x + 4x = 7x$.
For every x, $x^2 + 1 > 0$.
For each integer x, $2x$ is an even integer.

Obviously, each of the above statements is true and hence each is a proposition. In mathematics it often happens that we need to express the fact that a certain property is true for each and every element a in some suitably restricted set. This is usually done by use of one of the phrases "for all x," "for every x," or "for each x." We shall regard these three phrases as having the same meaning. Each is called a *universal quantifier*, and each is symbolized by the common symbol "$\forall_x$." By using this quantifier symbol, the examples listed above can be written

$$\forall_x[3x + 4x = 7x]$$

$$\forall_x[x^2 + 1 > 0]$$

$$\forall_x[x \text{ an integer implies } 2x \text{ is an even integer}]$$

The symbol $\forall_x$ is most often read "for all x," but either of the phrases "for every x" or "for each x" would be equally appropriate.

Another type of quantifier that is most useful in mathematics is one illustrated by such statements as:

There exists an x such that $x + 2 = 5$.
For some x, $x^2 - 5x + 6 = 0$.
There is at least one integer x such that $x/x \neq 1$.

Again, each of the above statements is true (taking $x = 0$ for the third one). Here we have used the three phrases "there exists an x," "for some x," and "there is at least one x" to express one central fact, that there exists at least one element x which belongs to an implied universal set and for which our statement is true. That is, these three phrases are regarded as having the same meaning. They are called *existential quantifiers*, and each is symbolized by the common symbol $\exists_x$. By using this symbol the above three statements can be written

$$\exists_x[x + 2 = 5]$$

$$\exists_x[x^2 - 5x + 6 = 0]$$

$$\exists_x[x \text{ an integer}, x/x \neq 1]$$

A.9 Open Statements and Quantifiers

The symbol $\exists_x$ is most commonly read "there exists an x," but it could just as well be read "for some x" or "there is at least one x."

An important problem in connection with propositions involving quantifiers is that of writing the negation of such a statement. This problem is easily solved provided we keep in mind the definition of the negation of a proposition: the negation of a proposition p is a proposition $\sim p$ such that

(a) If p is true, then $\sim p$ is false.
(b) If p is false, then $\sim p$ is true.

Let x be an element of a universal set U and let p_x be an open statement whose truth set is P; then $P \subseteq U$. Furthermore, the truth set of $\sim p_x$ is $\bar{P}$, the complement of P. Now consider the proposition $\forall_x(p_x)$. If $\forall_x(p_x)$ is true, then $P = U$, $\bar{P} = \emptyset$, and hence $\exists_x(\sim p_x)$ is false. On the other hand, if $\forall_x(p_x)$ is false, then $P \neq U$, $\bar{P} \neq \emptyset$, and therefore $\exists_x(\sim p_x)$ is true. In other words, a correct negation of $\forall_x(p_x)$ is $\exists_x(\sim p_x)$. Similarly, suppose the proposition $\exists_x(p_x)$ is true; then $\bar{P} \neq U$, $P \neq \emptyset$, and thus $\forall_x(\sim p_x)$ is false. Next, suppose $\exists_x(p_x)$ is false; then $P = \emptyset$, $\bar{P} = U$, and accordingly $\forall_x(\sim p_x)$ is true. It follows then from the definition of negation that the negation of $\exists_x(p_x)$ is $\forall_x(\sim p_x)$. In summary, we have shown that the negations of propositions involving quantifiers are given by the following:

$\sim \forall_x(p_x)$ is logically equivalent to (and may be replaced by) $\exists_x(\sim p_x)$

$\sim \exists_x(p_x)$ is logically equivalent to (and may be replaced by) $\forall_x(\sim p_x)$

Example 4 Write the negation of $\forall_x[x^2 + 1 > 0]$.

Solution: The negation is $\exists_x[x^2 + 1 \leq 0]$.

Example 5 Write the negation of $\exists_x[x + 2 = 5]$.

Solution: The negation is $\forall_x[x + 2 \neq 5]$.

The reader should observe that the statements being negated in Examples 4 and 5 are true, and so the negation of each statement is false.

Exercises A.9

1. Find the truth set for each given open statement.
   a. x is a prime less than 10
   b. $2x + 3 = x + 5$
   c. $x + 4 = 0$
   d. x is a U.S. senator
   e. x is a baseball team in the National League

2. Write four open statements, which are not equations, involving a variable x and in each case state a universal set to which x belongs.

3. Replace each given open statement with an equivalent one containing the phrase "for all x." (The given statements are not necessarily true.)
   a. For each prime number x greater than 2, $x + 1$ is an even integer.
   b. For every real number x, $x^2 - 4x + 7$ is positive.
   c. If $x > 2$, then $x^2 > 4$.
   d. No man is 8 feet tall.
   e. The square of every odd integer is an odd integer.

4. By using quantifiers, express the negation of each statement in Exercise 3.

5. For each of the following open statements, write an open statement equivalent to it. (See definition of equivalent open statements, Definition A.26.)
   a. $2x + 3x = 10$ b. $x^2 - 3x + 2 = 0$ c. $2x + 5 = 2x + 7$
   d. $3x < 6$

6. Express each of the following open statements using the quantifier "there exists an x." (The given statements are not necessarily true.)
   a. For some prime number x, $x + 1$ is a prime number.
   b. Some men are rich.
   c. There is at least one number x such that $x^2 = 1$.
   d. In the set of all triangles in the plane, some are isosceles.
   e. For some number x, $1 - 17x - 5x^2$ is positive.

7. By using quantifiers, write the negation of each statement in Exercise 6.

8. Show that each of the following sets is countable.
   a. $A = \{7, 14, 21, 28, \ldots\}$
   b. $B = \{\frac{1}{1}, \frac{1}{2}, \frac{1}{3}, \frac{1}{4}, \ldots\}$
   c. $C = \{\frac{1}{2}, \frac{2}{3}, \frac{3}{4}, \frac{4}{5}, \ldots\}$
   d. $D = \{\ldots, -4, -3, -2, -1, 0, 1, 2\}$
   e. $E = \{4, 9, 14, 19, 24, 29, \ldots\}$
   f. $F = \{1, \frac{1}{2}, \frac{1}{4}, \frac{1}{8}, \frac{1}{16}, \ldots\}$

Appendix B

Tables

Table I, Natural Trigonometric Functions, and Table IV, Exponential Functions, are reproduced from *Rinehart Mathematical Tables, Formulas, and Curves*, enlarged edition, compiled by Harold D. Larsen, copyright 1948, 1953 by Harold D. Larsen. Reprinted by permission of Holt, Rinehart & Winston, Inc.

Table I Natural Trigonometric Functions

0°

′	Sin	Tan	Ctn	Cos	′
0	.00000	.00000	—	1.0000	60
1	.00029	.00029	3437.7	1.0000	59
2	.00058	.00058	1718.9	1.0000	58
3	.00087	.00087	1145.9	1.0000	57
4	.00116	.00116	859.44	1.0000	56
5	.00145	.00145	687.55	1.0000	55
6	.00175	.00175	572.96	1.0000	54
7	.00204	.00204	491.11	1.0000	53
8	.00233	.00233	429.72	1.0000	52
9	.00262	.00262	381.97	1.0000	51
10	.00291	.00291	343.77	1.0000	50
11	.00320	.00320	312.52	.99999	49
12	.00349	.00349	286.48	.99999	48
13	.00378	.00378	264.44	.99999	47
14	.00407	.00407	245.55	.99999	46
15	.00436	.00436	229.18	.99999	45
16	.00465	.00465	214.86	.99999	44
17	.00495	.00495	202.22	.99999	43
18	.00524	.00524	190.98	.99999	42
19	.00553	.00553	180.93	.99998	41
20	.00582	.00582	171.89	.99998	40
21	.00611	.00611	163.70	.99998	39
22	.00640	.00640	156.26	.99998	38
23	.00669	.00669	149.47	.99998	37
24	.00698	.00698	143.24	.99998	36
25	.00727	.00727	137.51	.99997	35
26	.00756	.00756	132.22	.99997	34
27	.00785	.00785	127.32	.99997	33
28	.00814	.00815	122.77	.99997	32
29	.00844	.00844	118.54	.99996	31
30	.00873	.00873	114.59	.99996	30
31	.00902	.00902	110.89	.99996	29
32	.00931	.00931	107.43	.99996	28
33	.00960	.00960	104.17	.99995	27
34	.00989	.00989	101.11	.99995	26
35	.01018	.01018	98.218	.99995	25
36	.01047	.01047	95.489	.99995	24
37	.01076	.01076	92.908	.99994	23
38	.01105	.01105	90.463	.99994	22
39	.01134	.01135	88.144	.99994	21
40	.01164	.01164	85.940	.99993	20
41	.01193	.01193	83.844	.99993	19
42	.01222	.01222	81.847	.99993	18
43	.01251	.01251	79.943	.99992	17
44	.01280	.01280	78.126	.99992	16
45	.01309	.01309	76.390	.99991	15
46	.01338	.01338	74.729	.99991	14
47	.01367	.01367	73.139	.99991	13
48	.01396	.01396	71.615	.99990	12
49	.01425	.01425	70.153	.99990	11
50	.01454	.01455	68.750	.99989	10
51	.01483	.01484	67.402	.99989	9
52	.01513	.01513	66.105	.99989	8
53	.01542	.01542	64.858	.99988	7
54	.01571	.01571	63.657	.99988	6
55	.01600	.01600	62.499	.99987	5
56	.01629	.01629	61.383	.99987	4
57	.01658	.01658	60.306	.99986	3
58	.01687	.01687	59.266	.99986	2
59	.01716	.01716	58.261	.99985	1
60	.01745	.01746	57.290	.99985	0
′	Cos	Ctn	Tan	Sin	′

89°

1°

′	Sin	Tan	Ctn	Cos	′
0	.01745	.01746	57.290	.99985	60
1	.01774	.01775	56.351	.99984	59
2	.01803	.01804	55.442	.99984	58
3	.01832	.01833	54.561	.99983	57
4	.01862	.01862	53.709	.99983	56
5	.01891	.01891	52.882	.99982	55
6	.01920	.01920	52.081	.99982	54
7	.01949	.01949	51.303	.99981	53
8	.01978	.01978	50.549	.99980	52
9	.02007	.02007	49.816	.99980	51
10	.02036	.02036	49.104	.99979	50
11	.02065	.02066	48.412	.99979	49
12	.02094	.02095	47.740	.99978	48
13	.02123	.02124	47.085	.99977	47
14	.02152	.02153	46.449	.99977	46
15	.02181	.02182	45.829	.99976	45
16	.02211	.02211	45.226	.99976	44
17	.02240	.02240	44.639	.99975	43
18	.02269	.02269	44.066	.99974	42
19	.02298	.02298	43.508	.99974	41
20	.02327	.02328	42.964	.99973	40
21	.02356	.02357	42.433	.99972	39
22	.02385	.02386	41.916	.99972	38
23	.02414	.02415	41.411	.99971	37
24	.02443	.02444	40.917	.99970	36
25	.02472	.02473	40.436	.99969	35
26	.02501	.02502	39.965	.99969	34
27	.02530	.02531	39.506	.99968	33
28	.02560	.02560	39.057	.99967	32
29	.02589	.02589	38.618	.99966	31
30	.02618	.02619	38.188	.99966	30
31	.02647	.02648	37.769	.99965	29
32	.02676	.02677	37.358	.99964	28
33	.02705	.02706	36.956	.99963	27
34	.02734	.02735	36.563	.99963	26
35	.02763	.02764	36.178	.99962	25
36	.02792	.02793	35.801	.99961	24
37	.02821	.02822	35.431	.99960	23
38	.02850	.02851	35.070	.99959	22
39	.02879	.02881	34.715	.99959	21
40	.02908	.02910	34.368	.99958	20
41	.02938	.02939	34.027	.99957	19
42	.02967	.02968	33.694	.99956	18
43	.02996	.02997	33.366	.99955	17
44	.03025	.03026	33.045	.99954	16
45	.03054	.03055	32.730	.99953	15
46	.03083	.03084	32.421	.99952	14
47	.03112	.03114	32.118	.99952	13
48	.03141	.03143	31.821	.99951	12
49	.03170	.03172	31.528	.99950	11
50	.03199	.03201	31.242	.99949	10
51	.03228	.03230	30.960	.99948	9
52	.03257	.03259	30.683	.99947	8
53	.03286	.03288	30.412	.99946	7
54	.03316	.03317	30.145	.99945	6
55	.03345	.03346	29.882	.99944	5
56	.03374	.03376	29.624	.99943	4
57	.03403	.03405	29.371	.99942	3
58	.03432	.03434	29.122	.99941	2
59	.03461	.03463	28.877	.99940	1
60	.03490	.03492	28.636	.99939	0
′	Cos	Ctn	Tan	Sin	′

88°

Table I Natural Trigonometric Functions

2°

′	Sin	Tan	Ctn	Cos	′
0	.03490	.03492	28.636	.99939	60
1	.03519	.03521	28.399	.99938	59
2	.03548	.03550	28.166	.99937	58
3	.03577	.03579	27.937	.99936	57
4	.03606	.03609	27.712	.99935	56
5	.03635	.03638	27.490	.99934	55
6	.03664	.03667	27.271	.99933	54
7	.03693	.03696	27.057	.99932	53
8	.03723	.03725	26.845	.99931	52
9	.03752	.03754	26.637	.99930	51
10	.03781	.03783	26.432	.99929	50
11	.03810	.03812	26.230	.99927	49
12	.03839	.03842	26.031	.99926	48
13	.03868	.03871	25.835	.99925	47
14	.03897	.03900	25.642	.99924	46
15	.03926	.03929	25.452	.99923	45
16	.03955	.03958	25.264	.99922	44
17	.03984	.03987	25.080	.99921	43
18	.04013	.04016	24.898	.99919	42
19	.04042	.04046	24.719	.99918	41
20	.04071	.04075	24.542	.99917	40
21	.04100	.04104	24.368	.99916	39
22	.04129	.04133	24.196	.99915	38
23	.04159	.04162	24.026	.99913	37
24	.04188	.04191	23.859	.99912	36
25	.04217	.04220	23.695	.99911	35
26	.04246	.04250	23.532	.99910	34
27	.04275	.04279	23.372	.99909	33
28	.04304	.04308	23.214	.99907	32
29	.04333	.04337	23.058	.99906	31
30	.04362	.04366	22.904	.99905	30
31	.04391	.04395	22.752	.99904	29
32	.04420	.04424	22.602	.99902	28
33	.04449	.04454	22.454	.99901	27
34	.04478	.04483	22.308	.99900	26
35	.04507	.04512	22.164	.99898	25
36	.04536	.04541	22.022	.99897	24
37	.04565	.04570	21.881	.99896	23
38	.04594	.04599	21.743	.99894	22
39	.04623	.04628	21.606	.99893	21
40	.04653	.04658	21.470	.99892	20
41	.04682	.04687	21.337	.99890	19
42	.04711	.04716	21.205	.99889	18
43	.04740	.04745	21.075	.99888	17
44	.04769	.04774	20.946	.99886	16
45	.04798	.04803	20.819	.99885	15
46	.04827	.04833	20.693	.99883	14
47	.04856	.04862	20.569	.99882	13
48	.04885	.04891	20.446	.99881	12
49	.04914	.04920	20.325	.99879	11
50	.04943	.04949	20.206	.99878	10
51	.04972	.04978	20.087	.99876	9
52	.05001	.05007	19.970	.99875	8
53	.05030	.05037	19.855	.99873	7
54	.05059	.05066	19.740	.99872	6
55	.05088	.05095	19.627	.99870	5
56	.05117	.05124	19.516	.99869	4
57	.05146	.05153	19.405	.99867	3
58	.05175	.05182	19.296	.99866	2
59	.05205	.05212	19.188	.99864	1
60	.05234	.05241	19.081	.99863	0
′	Cos	Ctn	Tan	Sin	′

87°

3°

′	Sin	Tan	Ctn	Cos	′
0	.05234	.05241	19.081	.99863	60
1	.05263	.05270	18.976	.99861	59
2	.05292	.05299	18.871	.99860	58
3	.05321	.05328	18.768	.99858	57
4	.05350	.05357	18.666	.99857	56
5	.05379	.05387	18.564	.99855	55
6	.05408	.05416	18.464	.99854	54
7	.05437	.05445	18.366	.99852	53
8	.05466	.05474	18.268	.99851	52
9	.05495	.05503	18.171	.99849	51
10	.05524	.05533	18.075	.99847	50
11	.05553	.05562	17.980	.99846	49
12	.05582	.05591	17.886	.99844	48
13	.05611	.05620	17.793	.99842	47
14	.05640	.05649	17.702	.99841	46
15	.05669	.05678	17.611	.99839	45
16	.05698	.05708	17.521	.99838	44
17	.05727	.05737	17.431	.99836	43
18	.05756	.05766	17.343	.99834	42
19	.05785	.05795	17.256	.99833	41
20	.05814	.05824	17.169	.99831	40
21	.05844	.05854	17.084	.99829	39
22	.05873	.05883	16.999	.99827	38
23	.05902	.05912	16.915	.99826	37
24	.05931	.05941	16.832	.99824	36
25	.05960	.05970	16.750	.99822	35
26	.05989	.05999	16.668	.99821	34
27	.06018	.06029	16.587	.99819	33
28	.06047	.06058	16.507	.99817	32
29	.06076	.06087	16.428	.99815	31
30	.06105	.06116	16.350	.99813	30
31	.06134	.06145	16.272	.99812	29
32	.06163	.06175	16.195	.99810	28
33	.06192	.06204	16.119	.99808	27
34	.06221	.06233	16.043	.99806	26
35	.06250	.06262	15.969	.99804	25
36	.06279	.06291	15.895	.99803	24
37	.06308	.06321	15.821	.99801	23
38	.06337	.06350	15.748	.99799	22
39	.06366	.06379	15.676	.99797	21
40	.06395	.06408	15.605	.99795	20
41	.06424	.06438	15.534	.99793	19
42	.06453	.06467	15.464	.99792	18
43	.06482	.06496	15.394	.99790	17
44	.06511	.06525	15.325	.99788	16
45	.06540	.06554	15.257	.99786	15
46	.06569	.06584	15.189	.99784	14
47	.06598	.06613	15.122	.99782	13
48	.06627	.06642	15.056	.99780	12
49	.06656	.06671	14.990	.99778	11
50	.06685	.06700	14.924	.99776	10
51	.06714	.06730	14.860	.99774	9
52	.06743	.06759	14.795	.99772	8
53	.06773	.06788	14.732	.99770	7
54	.06802	.06817	14.669	.99768	6
55	.06831	.06847	14.606	.99766	5
56	.06860	.06876	14.544	.99764	4
57	.06889	.06905	14.482	.99762	3
58	.06918	.06934	14.421	.99760	2
59	.06947	.06963	14.361	.99758	1
60	.06976	.06993	14.301	.99756	0
′	Cos	Ctn	Tan	Sin	′

86°

Table I Natural Trigonometric Functions

4°

′	Sin	Tan	Ctn	Cos	′
0	.06976	.06993	14.301	.99756	60
1	.07005	.07022	14.241	.99754	59
2	.07034	.07051	14.182	.99752	58
3	.07063	.07080	14.124	.99750	57
4	.07092	.07110	14.065	.99748	56
5	.07121	.07139	14.008	.99746	55
6	.07150	.07168	13.951	.99744	54
7	.07179	.07197	13.894	.99742	53
8	.07208	.07227	13.838	.99740	52
9	.07237	.07256	13.782	.99738	51
10	.07266	.07285	13.727	.99736	50
11	.07295	.07314	13.672	.99734	49
12	.07324	.07344	13.617	.99731	48
13	.07353	.07373	13.563	.99729	47
14	.07382	.07402	13.510	.99727	46
15	.07411	.07431	13.457	.99725	45
16	.07440	.07461	13.404	.99723	44
17	.07469	.07490	13.352	.99721	43
18	.07498	.07519	13.300	.99719	42
19	.07527	.07548	13.248	.99716	41
20	.07556	.07578	13.197	.99714	40
21	.07585	.07607	13.146	.99712	39
22	.07614	.07636	13.096	.99710	38
23	.07643	.07665	13.046	.99708	37
24	.07672	.07695	12.996	.99705	36
25	.07701	.07724	12.947	.99703	35
26	.07730	.07753	12.898	.99701	34
27	.07759	.07782	12.850	.99699	33
28	.07788	.07812	12.801	.99696	32
29	.07817	.07841	12.754	.99694	31
30	.07846	.07870	12.706	.99692	30
31	.07875	.07899	12.659	.99689	29
32	.07904	.07929	12.612	.99687	28
33	.07933	.07958	12.566	.99685	27
34	.07962	.07987	12.520	.99683	26
35	.07991	.08017	12.474	.99680	25
36	.08020	.08046	12.429	.99678	24
37	.08049	.08075	12.384	.99676	23
38	.08078	.08104	12.339	.99673	22
39	.08107	.08134	12.295	.99671	21
40	.08136	.08163	12.251	.99668	20
41	.08165	.08192	12.207	.99666	19
42	.08194	.08221	12.163	.99664	18
43	.08223	.08251	12.120	.99661	17
44	.08252	.08280	12.077	.99659	16
45	.08281	.08309	12.035	.99657	15
46	.08310	.08339	11.992	.99654	14
47	.08339	.08368	11.950	.99652	13
48	.08368	.08397	11.909	.99649	12
49	.08397	.08427	11.867	.99647	11
50	.08426	.08456	11.826	.99644	10
51	.08455	.08485	11.785	.99642	9
52	.08484	.08514	11.745	.99639	8
53	.08513	.08544	11.705	.99637	7
54	.08542	.08573	11.664	.99635	6
55	.08571	.08602	11.625	.99632	5
56	.08600	.08632	11.585	.99630	4
57	.08629	.08661	11.546	.99627	3
58	.08658	.08690	11.507	.99625	2
59	.08687	.08720	11.468	.99622	1
60	.08716	.08749	11.430	.99619	0
′	Cos	Ctn	Tan	Sin	′

85°

5°

′	Sin	Tan	Ctn	Cos	′
0	.08716	.08749	11.430	.99619	60
1	.08745	.08778	11.392	.99617	59
2	.08774	.08807	11.354	.99614	58
3	.08803	.08837	11.316	.99612	57
4	.08831	.08866	11.279	.99609	56
5	.08860	.08895	11.242	.99607	55
6	.08889	.08925	11.205	.99604	54
7	.08918	.08954	11.168	.99602	53
8	.08947	.08983	11.132	.99599	52
9	.08976	.09013	11.095	.99596	51
10	.09005	.09042	11.059	.99594	50
11	.09034	.09071	11.024	.99591	49
12	.09063	.09101	10.988	.99588	48
13	.09092	.09130	10.953	.99586	47
14	.09121	.09159	10.918	.99583	46
15	.09150	.09189	10.883	.99580	45
16	.09179	.09218	10.848	.99578	44
17	.09208	.09247	10.814	.99575	43
18	.09237	.09277	10.780	.99572	42
19	.09266	.09306	10.746	.99570	41
20	.09295	.09335	10.712	.99567	40
21	.09324	.09365	10.678	.99564	39
22	.09353	.09394	10.645	.99562	38
23	.09382	.09423	10.612	.99559	37
24	.09411	.09453	10.579	.99556	36
25	.09440	.09482	10.546	.99553	35
26	.09469	.09511	10.514	.99551	34
27	.09498	.09541	10.481	.99548	33
28	.09527	.09570	10.449	.99545	32
29	.09556	.09600	10.417	.99542	31
30	.09585	.09629	10.385	.99540	30
31	.09614	.09658	10.354	.99537	29
32	.09642	.09688	10.322	.99534	28
33	.09671	.09717	10.291	.99531	27
34	.09700	.09746	10.260	.99528	26
35	.09729	.09776	10.229	.99526	25
36	.09758	.09805	10.199	.99523	24
37	.09787	.09834	10.168	.99520	23
38	.09816	.09864	10.138	.99517	22
39	.09845	.09893	10.108	.99514	21
40	.09874	.09923	10.078	.99511	20
41	.09903	.09952	10.048	.99508	19
42	.09932	.09981	10.019	.99506	18
43	.09961	.10011	9.9893	.99503	17
44	.09990	.10040	9.9601	.99500	16
45	.10019	.10069	9.9310	.99497	15
46	.10048	.10099	9.9021	.99494	14
47	.10077	.10128	9.8734	.99491	13
48	.10106	.10158	9.8448	.99488	12
49	.10135	.10187	9.8164	.99485	11
50	.10164	.10216	9.7882	.99482	10
51	.10192	.10246	9.7601	.99479	9
52	.10221	.10275	9.7322	.99476	8
53	.10250	.10305	9.7044	.99473	7
54	.10279	.10334	9.6768	.99470	6
55	.10308	.10363	9.6493	.99467	5
56	.10337	.10393	9.6220	.99464	4
57	.10366	.10422	9.5949	.99461	3
58	.10395	.10452	9.5679	.99458	2
59	.10424	.10481	9.5411	.99455	1
60	.10453	.10510	9.5144	.99452	0
′	Cos	Ctn	Tan	Sin	′

84°

Table I Natural Trigonometric Functions

6°

′	Sin	Tan	Ctn	Cos	′
0	.10453	.10510	9.5144	.99452	60
1	.10482	.10540	9.4878	.99449	59
2	.10511	.10569	9.4614	.99446	58
3	.10540	.10599	9.4352	.99443	57
4	.10569	.10628	9.4090	.99440	56
5	.10597	.10657	9.3831	.99437	55
6	.10626	.10687	9.3572	.99434	54
7	.10655	.10716	9.3315	.99431	53
8	.10684	.10746	9.3060	.99428	52
9	.10713	.10775	9.2806	.99424	51
10	.10742	.10805	9.2553	.99421	50
11	.10771	.10834	9.2302	.99418	49
12	.10800	.10863	9.2052	.99415	48
13	.10829	.10893	9.1803	.99412	47
14	.10858	.10922	9.1555	.99409	46
15	.10887	.10952	9.1309	.99406	45
16	.10916	.10981	9.1065	.99402	44
17	.10945	.11011	9.0821	.99399	43
18	.10973	.11040	9.0579	.99396	42
19	.11002	.11070	9.0338	.99393	41
20	.11031	.11099	9.0098	.99390	40
21	.11060	.11128	8.9860	.99386	39
22	.11089	.11158	8.9623	.99383	38
23	.11118	.11187	8.9387	.99380	37
24	.11147	.11217	8.9152	.99377	36
25	.11176	.11246	8.8919	.99374	35
26	.11205	.11276	8.8686	.99370	34
27	.11234	.11305	8.8455	.99367	33
28	.11263	.11335	8.8225	.99364	32
29	.11291	.11364	8.7996	.99360	31
30	.11320	.11394	8.7769	.99357	30
31	.11349	.11423	8.7542	.99354	29
32	.11378	.11452	8.7317	.99351	28
33	.11407	.11482	8.7093	.99347	27
34	.11436	.11511	8.6870	.99344	26
35	.11465	.11541	8.6648	.99341	25
36	.11494	.11570	8.6427	.99337	24
37	.11523	.11600	8.6208	.99334	23
38	.11552	.11629	8.5989	.99331	22
39	.11580	.11659	8.5772	.99327	21
40	.11609	.11688	8.5555	.99324	20
41	.11638	.11718	8.5340	.99320	19
42	.11667	.11747	8.5126	.99317	18
43	.11696	.11777	8.4913	.99314	17
44	.11725	.11806	8.4701	.99310	16
45	.11754	.11836	8.4490	.99307	15
46	.11783	.11865	8.4280	.99303	14
47	.11812	.11895	8.4071	.99300	13
48	.11840	.11924	8.3863	.99297	12
49	.11869	.11954	8.3656	.99293	11
50	.11898	.11983	8.3450	.99290	10
51	.11927	.12013	8.3245	.99286	9
52	.11956	.12042	8.3041	.99283	8
53	.11985	.12072	8.2838	.99279	7
54	.12014	.12101	8.2636	.99276	6
55	.12043	.12131	8.2434	.99272	5
56	.12071	.12160	8.2234	.99269	4
57	.12100	.12190	8.2035	.99265	3
58	.12129	.12219	8.1837	.99262	2
59	.12158	.12249	8.1640	.99258	1
60	.12187	.12278	8.1443	.99255	0
′	Cos	Ctn	Tan	Sin	′

83°

7°

′	Sin	Tan	Ctn	Cos	′
0	.12187	.12278	8.1443	.99255	60
1	.12216	.12308	8.1248	.99251	59
2	.12245	.12338	8.1054	.99248	58
3	.12274	.12367	8.0860	.99244	57
4	.12302	.12397	8.0667	.99240	56
5	.12331	.12426	8.0476	.99237	55
6	.12360	.12456	8.0285	.99233	54
7	.12389	.12485	8.0095	.99230	53
8	.12418	.12515	7.9906	.99226	52
9	.12447	.12544	7.9718	.99222	51
10	.12476	.12574	7.9530	.99219	50
11	.12504	.12603	7.9344	.99215	49
12	.12533	.12633	7.9158	.99211	48
13	.12562	.12662	7.8973	.99208	47
14	.12591	.12692	7.8789	.99204	46
15	.12620	.12722	7.8606	.99200	45
16	.12649	.12751	7.8424	.99197	44
17	.12678	.12781	7.8243	.99193	43
18	.12706	.12810	7.8062	.99189	42
19	.12735	.12840	7.7882	.99186	41
20	.12764	.12869	7.7704	.99182	40
21	.12793	.12899	7.7525	.99178	39
22	.12822	.12929	7.7348	.99175	38
23	.12851	.12958	7.7171	.99171	37
24	.12880	.12988	7.6996	.99167	36
25	.12908	.13017	7.6821	.99163	35
26	.12937	.13047	7.6647	.99160	34
27	.12966	.13076	7.6473	.99156	33
28	.12995	.13106	7.6301	.99152	32
29	.13024	.13136	7.6129	.99148	31
30	.13053	.13165	7.5958	.99144	30
31	.13081	.13195	7.5787	.99141	29
32	.13110	.13224	7.5618	.99137	28
33	.13139	.13254	7.5449	.99133	27
34	.13168	.13284	7.5281	.99129	26
35	.13197	.13313	7.5113	.99125	25
36	.13226	.13343	7.4947	.99122	24
37	.13254	.13372	7.4781	.99118	23
38	.13283	.13402	7.4615	.99114	22
39	.13312	.13432	7.4451	.99110	21
40	.13341	.13461	7.4287	.99106	20
41	.13370	.13491	7.4124	.99102	19
42	.13399	.13521	7.3962	.99098	18
43	.13427	.13550	7.3800	.99094	17
44	.13456	.13580	7.3639	.99091	16
45	.13485	.13609	7.3479	.99087	15
46	.13514	.13639	7.3319	.99083	14
47	.13543	.13669	7.3160	.99079	13
48	.13572	.13698	7.3002	.99075	12
49	.13600	.13728	7.2844	.99071	11
50	.13629	.13758	7.2687	.99067	10
51	.13658	.13787	7.2531	.99063	9
52	.13687	.13817	7.2375	.99059	8
53	.13716	.13846	7.2220	.99055	7
54	.13744	.13876	7.2066	.99051	6
55	.13773	.13906	7.1912	.99047	5
56	.13802	.13935	7.1759	.99043	4
57	.13831	.13965	7.1607	.99039	3
58	.13860	.13995	7.1455	.99035	2
59	.13889	.14024	7.1304	.99031	1
60	.13917	.14054	7.1154	.99027	0
′	Cos	Ctn	Tan	Sin	′

82°

8°

′	Sin	Tan	Ctn	Cos	′
0	.13917	.14054	7.1154	.99027	60
1	.13946	.14084	7.1004	.99023	59
2	.13975	.14113	7.0855	.99019	58
3	.14004	.14143	7.0706	.99015	57
4	.14033	.14173	7.0558	.99011	56
5	.14061	.14202	7.0410	.99006	55
6	.14090	.14232	7.0264	.99002	54
7	.14119	.14262	7.0117	.98998	53
8	.14148	.14291	6.9972	.98994	52
9	.14177	.14321	6.9827	.98990	51
10	.14205	.14351	6.9682	.98986	50
11	.14234	.14381	6.9538	.98982	49
12	.14263	.14410	6.9395	.98978	48
13	.14292	.14440	6.9252	.98973	47
14	.14320	.14470	6.9110	.98969	46
15	.14349	.14499	6.8969	.98965	45
16	.14378	.14529	6.8828	.98961	44
17	.14407	.14559	6.8687	.98957	43
18	.14436	.14588	6.8548	.98953	42
19	.14464	.14618	6.8408	.98948	41
20	.14493	.14648	6.8269	.98944	40
21	.14522	.14678	6.8131	.98940	39
22	.14551	.14707	6.7994	.98936	38
23	.14580	.14737	6.7856	.98931	37
24	.14608	.14767	6.7720	.98927	36
25	.14637	.14796	6.7584	.98923	35
26	.14666	.14826	6.7448	.98919	34
27	.14695	.14856	6.7313	.98914	33
28	.14723	.14886	6.7179	.98910	32
29	.14752	.14915	6.7045	.98906	31
30	.14781	.14945	6.6912	.98902	30
31	.14810	.14975	6.6779	.98897	29
32	.14838	.15005	6.6646	.98893	28
33	.14867	.15034	6.6514	.98889	27
34	.14896	.15064	6.6383	.98884	26
35	.14925	.15094	6.6252	.98880	25
36	.14954	.15124	6.6122	.98876	24
37	.14982	.15153	6.5992	.98871	23
38	.15011	.15183	6.5863	.98867	22
39	.15040	.15213	6.5734	.98863	21
40	.15069	.15243	6.5606	.98858	20
41	.15097	.15272	6.5478	.98854	19
42	.15126	.15302	6.5350	.98849	18
43	.15155	.15332	6.5223	.98845	17
44	.15184	.15362	6.5097	.98841	16
45	.15212	.15391	6.4971	.98836	15
46	.15241	.15421	6.4846	.98832	14
47	.15270	.15451	6.4721	.98827	13
48	.15299	.15481	6.4596	.98823	12
49	.15327	.15511	6.4472	.98818	11
50	.15356	.15540	6.4348	.98814	10
51	.15385	.15570	6.4225	.98809	9
52	.15414	.15600	6.4103	.98805	8
53	.15442	.15630	6.3980	.98800	7
54	.15471	.15660	6.3859	.98796	6
55	.15500	.15689	6.3737	.98791	5
56	.15529	.15719	6.3617	.98787	4
57	.15557	.15749	6.3496	.98782	3
58	.15586	.15779	6.3376	.98778	2
59	.15615	.15809	6.3257	.98773	1
60	.15643	.15838	6.3138	.98769	0
′	Cos	Ctn	Tan	Sin	′

81°

9°

′	Sin	Tan	Ctn	Cos	′
0	.15643	.15838	6.3138	.98769	60
1	.15672	.15868	6.3019	.98764	59
2	.15701	.15898	6.2901	.98760	58
3	.15730	.15928	6.2783	.98755	57
4	.15758	.15958	6.2666	.98751	56
5	.15787	.15988	6.2549	.98746	55
6	.15816	.16017	6.2432	.98741	54
7	.15845	.16047	6.2316	.98737	53
8	.15873	.16077	6.2200	.98732	52
9	.15902	.16107	6.2085	.98728	51
10	.15931	.16137	6.1970	.98723	50
11	.15959	.16167	6.1856	.98718	49
12	.15988	.16196	6.1742	.98714	48
13	.16017	.16226	6.1628	.98709	47
14	.16046	.16256	6.1515	.98704	46
15	.16074	.16286	6.1402	.98700	45
16	.16103	.16316	6.1290	.98695	44
17	.16132	.16346	6.1178	.98690	43
18	.16160	.16376	6.1066	.98686	42
19	.16189	.16405	6.0955	.98681	41
20	.16218	.16435	6.0844	.98676	40
21	.16246	.16465	6.0734	.98671	39
22	.16275	.16495	6.0624	.98667	38
23	.16304	.16525	6.0514	.98662	37
24	.16333	.16555	6.0405	.98657	36
25	.16361	.16585	6.0296	.98652	35
26	.16390	.16615	6.0188	.98648	34
27	.16419	.16645	6.0080	.98643	33
28	.16447	.16674	5.9972	.98638	32
29	.16476	.16704	5.9865	.98633	31
30	.16505	.16734	5.9758	.98629	30
31	.16533	.16764	5.9651	.98624	29
32	.16562	.16794	5.9545	.98619	28
33	.16591	.16824	5.9439	.98614	27
34	.16620	.16854	5.9333	.98609	26
35	.16648	.16884	5.9228	.98604	25
36	.16677	.16914	5.9124	.98600	24
37	.16706	.16944	5.9019	.98595	23
38	.16734	.16974	5.8915	.98590	22
39	.16763	.17004	5.8811	.98585	21
40	.16792	.17033	5.8708	.98580	20
41	.16820	.17063	5.8605	.98575	19
42	.16849	.17093	5.8502	.98570	18
43	.16878	.17123	5.8400	.98565	17
44	.16906	.17153	5.8298	.98561	16
45	.16935	.17183	5.8197	.98556	15
46	.16964	.17213	5.8095	.98551	14
47	.16992	.17243	5.7994	.98546	13
48	.17021	.17273	5.7894	.98541	12
49	.17050	.17303	5.7794	.98536	11
50	.17078	.17333	5.7694	.98531	10
51	.17107	.17363	5.7594	.98526	9
52	.17136	.17393	5.7495	.98521	8
53	.17164	.17423	5.7396	.98516	7
54	.17193	.17453	5.7297	.98511	6
55	.17222	.17483	5.7199	.98506	5
56	.17250	.17513	5.7101	.98501	4
57	.17279	.17543	5.7004	.98496	3
58	.17308	.17573	5.6906	.98491	2
59	.17336	.17603	5.6809	.98486	1
60	.17365	.17633	5.6713	.98481	0
′	Cos	Ctn	Tan	Sin	′

80°

Table I   Natural Trigonometric Functions

10°

′	Sin	Tan	Ctn	Cos	′
0	.17365	.17633	5.6713	.98481	60
1	.17393	.17663	5.6617	.98476	59
2	.17422	.17693	5.6521	.98471	58
3	.17451	.17723	5.6425	.98466	57
4	.17479	.17753	5.6329	.98461	56
5	.17508	.17783	5.6234	.98455	55
6	.17537	.17813	5.6140	.98450	54
7	.17565	.17843	5.6045	.98445	53
8	.17594	.17873	5.5951	.98440	52
9	.17623	.17903	5.5857	.98435	51
10	.17651	17933	5.5764	.98430	50
11	.17680	.17963	5.5671	.98425	49
12	.17708	.17993	5.5578	.98420	48
13	.17737	.18023	5.5485	.98414	47
14	.17766	.18053	5.5393	.98409	46
15	.17794	.18083	5.5301	.98404	45
16	.17823	.18113	5.5209	.98399	44
17	.17852	.18143	5.5118	.98394	43
18	.17880	.18173	5.5026	.98389	42
19	.17909	.18203	5.4936	.98383	41
20	.17937	.18233	5.4845	.98378	40
21	.17966	.18263	5.4755	.98373	39
22	.17995	.18293	5.4665	.98368	38
23	.18023	.18323	5.4575	.98362	37
24	.18052	.18353	5.4486	.98357	36
25	.18081	.18384	5.4397	.98352	35
26	.18109	.18414	5.4308	.98347	34
27	.18138	.18444	5.4219	.98341	33
28	.18166	.18474	5.4131	.98336	32
29	.18195	.18504	5.4043	.98331	31
30	.18224	.18534	5.3955	.98325	30
31	.18252	.18564	5.3868	.98320	29
32	.18281	.18594	5.3781	.98315	28
33	.18309	.18624	5.3694	.98310	27
34	.18338	.18654	5.3607	.98304	26
35	.18367	.18684	5.3521	.98299	25
36	.18395	.18714	5.3435	.98294	24
37	.18424	.18745	5.3349	.98288	23
38	.18452	.18775	5.3263	.98283	22
39	.18481	.18805	5.3178	.98277	21
40	.18509	.18835	5.3093	.98272	20
41	.18538	.18865	5.3008	.98267	19
42	.18567	.18895	5.2924	.98261	18
43	.18595	.18925	5.2839	.98256	17
44	.18624	.18955	5.2755	.98250	16
45	.18652	.18986	5.2672	.98245	15
46	.18681	.19016	5.2588	.98240	14
47	.18710	.19046	5.2505	.98234	13
48	.18738	.19076	5.2422	.98229	12
49	.18767	.19106	5.2339	.98223	11
50	.18795	.19136	5.2257	.93218	10
51	.18824	.19166	5.2174	.98212	9
52	.18852	.19197	5.2092	.98207	8
53	.18881	.19227	5.2011	.98201	7
54	.18910	.19257	5.1929	.98196	6
55	.18938	.19287	5.1848	.98190	5
56	.18967	.19317	5.1767	.98185	4
57	.18995	.19347	5.1686	.98179	3
58	.19024	.19378	5.1606	.98174	2
59	.19052	.19408	5.1526	.98168	1
60	.19081	.19438	5.1446	.98163	0
′	Cos	Ctn	Tan	Sin	′

79°

11°

′	Sin	Tan	Ctn	Cos	′
0	.19081	.19438	5.1446	.98163	60
1	.19109	.19468	5.1366	.98157	59
2	.19138	.19498	5.1286	.98152	58
3	.19167	.19529	5.1207	.98146	57
4	.19195	.19559	5.1128	.98140	56
5	.19224	.19589	5.1049	.98135	55
6	.19252	.19619	5.0970	.98129	54
7	.19281	.19649	5.0892	.98124	53
8	.19309	.19680	5.0814	.98118	52
9	.19338	.19710	5.0736	.98112	51
10	.19366	.19740	5.0658	.98107	50
11	.19395	.19770	5.0581	.98101	49
12	.19423	.19801	5.0504	.98096	48
13	.19452	.19831	5.0427	.98090	47
14	.19481	.19861	5.0350	.98084	46
15	.19509	.19891	5.0273	.98079	45
16	.19538	.19921	5.0197	.98073	44
17	.19566	.19952	5.0121	.98067	43
18	.19595	.19982	5.0045	.98061	42
19	.19623	.20012	4.9969	.98056	41
20	.19652	.20042	4.9894	.98050	40
21	.19680	.20073	4.9819	.98044	39
22	.19709	.20103	4.9744	.98039	38
23	.19737	.20133	4.9669	.98033	37
24	.19766	.20164	4.9594	.98027	36
25	.19794	.20194	4.9520	.98021	35
26	.19823	.20224	4.9446	.98016	34
27	.19851	.20254	4.9372	.98010	33
28	.19880	.20285	4.9298	.98004	32
29	.19908	.20315	4.9225	.97998	31
30	.19937	.20345	4.9152	.97992	30
31	.19965	.20376	4.9078	.97987	29
32	.19994	.20406	4.9006	.97981	28
33	.20022	.20436	4.8933	.97975	27
34	.20051	.20466	4.8860	.97969	26
35	.20079	.20497	4.8788	.97963	25
36	.20108	.20527	4.8716	.97958	24
37	.20136	.20557	4.8644	.97952	23
38	.20165	.20588	4.8573	.97946	22
39	.20193	.20618	4.8501	.97940	21
40	.20222	.20648	4.8430	.97934	20
41	.20250	.20679	4.8359	.97928	19
42	.20279	.20709	4.8288	.97922	18
43	.20307	.20739	4.8218	.97916	17
44	.20336	.20770	4.8147	.97910	16
45	20364	.20800	4.8077	.97905	15
46	.20393	.20830	4.8007	.97899	14
47	.20421	.20861	4.7937	.97893	13
48	.20450	.20891	4.7867	.97887	12
49	.20478	.20921	4.7798	.97881	11
50	.20507	.20952	4.7729	.97875	10
51	.20535	.20982	4.7659	.97869	9
52	.20563	.21013	4.7591	.97863	8
53	.20592	.21043	4.7522	.97857	7
54	.20620	.21073	4.7453	.97851	6
55	.20649	.21104	4.7385	.97845	5
56	.20677	.21134	4.7317	.97839	4
57	.20706	.21164	4.7249	.97833	3
58	.20734	.21195	4.7181	.97827	2
59	.20763	.21225	4.7114	.97821	1
60	.20791	.21256	4.7046	.97815	0
′	Cos	Ctn	Tan	Sin	′

78°

Table I Natural Trigonometric Functions

12°

′	Sin	Tan	Ctn	Cos	′
0	.20791	.21256	4.7046	.97815	60
1	.20820	.21286	4.6979	.97809	59
2	.20848	.21316	4.6912	.97803	58
3	.20877	.21347	4.6845	.97797	57
4	.20905	.21377	4.6779	.97791	56
5	.20933	.21408	4.6712	.97784	55
6	.20962	.21438	4.6646	.97778	54
7	.20990	.21469	4.6580	.97772	53
8	.21019	.21499	4.6514	.97766	52
9	.21047	.21529	4.6448	.97760	51
10	.21076	.21560	4.6382	.97754	50
11	.21104	.21590	4.6317	.97748	49
12	.21132	.21621	4.6252	.97742	48
13	.21161	.21651	4.6187	.97735	47
14	.21189	.21682	4.6122	.97729	46
15	.21218	.21712	4.6057	.97723	45
16	.21246	.21743	4.5993	.97717	44
17	.21275	.21773	4.5928	.97711	43
18	.21303	.21804	4.5864	.97705	42
19	.21331	.21834	4.5800	.97698	41
20	.21360	.21864	4.5736	.97692	40
21	.21388	.21895	4.5673	.97686	39
22	.21417	.21925	4.5609	.97680	38
23	.21445	.21956	4.5546	.97673	37
24	.21474	.21986	4.5483	.97667	36
25	.21502	.22017	4.5420	.97661	35
26	.21530	.22047	4.5357	.97655	34
27	.21559	.22078	4.5294	.97648	33
28	.21587	.22108	4.5232	.97642	32
29	.21616	.22139	4.5169	.97636	31
30	.21644	.22169	4.5107	.97630	30
31	.21672	.22200	4.5045	.97623	29
32	.21701	.22231	4.4983	.97617	28
33	.21729	.22261	4.4922	.97611	27
34	.21758	.22292	4.4860	.97604	26
35	.21786	.22322	4.4799	.97598	25
36	.21814	.22353	4.4737	.97592	24
37	.21843	.22383	4.4676	.97585	23
38	.21871	.22414	4.4615	.97579	22
39	.21899	.22444	4.4555	.97573	21
40	.21928	.22475	4.4494	.97566	20
41	.21956	.22505	4.4434	.97560	19
42	.21985	.22536	4.4373	.97553	18
43	.22013	.22567	4.4313	.97547	17
44	.22041	.22597	4.4253	.97541	16
45	.22070	.22628	4.4194	.97534	15
46	.22098	.22658	4.4134	.97528	14
47	.22126	.22689	4.4075	.97521	13
48	.22155	.22719	4.4015	.97515	12
49	.22183	.22750	4.3956	.97508	11
50	.22212	.22781	4.3897	.97502	10
51	.22240	.22811	4.3838	.97496	9
52	.22268	.22842	4.3779	.97489	8
53	.22297	.22872	4.3721	.97483	7
54	.22325	.22903	4.3662	.97476	6
55	.22353	.22934	4.3604	.97470	5
56	.22382	.22964	4.3546	.97463	4
57	.22410	.22995	4.3488	.97457	3
58	.22438	.23026	4.3430	.97450	2
59	.22467	.23056	4.3372	.97444	1
60	.22495	.23087	4.3315	.97437	0
′	Cos	Ctn	Tan	Sin	′

77°

13°

′	Sin	Tan	Ctn	Cos	′
0	.22495	.23087	4.3315	.97437	60
1	.22523	.23117	4.3257	.97430	59
2	.22552	.23148	4.3200	.97424	58
3	.22580	.23179	4.3143	.97417	57
4	.22608	.23209	4.3086	.97411	56
5	.22637	.23240	4.3029	.97404	55
6	.22665	.23271	4.2972	.97398	54
7	.22693	.23301	4.2916	.97391	53
8	.22722	.23332	4.2859	.97384	52
9	.22750	.23363	4.2803	.97378	51
10	.22778	.23393	4.2747	.97371	50
11	.22807	.23424	4.2691	.97365	49
12	.22835	.23455	4.2635	.97358	48
13	.22863	.23485	4.2580	.97351	47
14	.22892	.23516	4.2524	.97345	46
15	.22920	.23547	4.2468	.97338	45
16	.22948	.23578	4.2413	.97331	44
17	.22977	.23608	4.2358	.97325	43
18	.23005	.23639	4.2303	.97318	42
19	.23033	.23670	4.2248	.97311	41
20	.23062	.23700	4.2193	.97304	40
21	.23090	.23731	4.2139	.97298	39
22	.23118	.23762	4.2084	.97291	38
23	.23146	.23793	4.2030	.97284	37
24	.23175	.23823	4.1976	.97278	36
25	.23203	.23854	4.1922	.97271	35
26	.23231	.23885	4.1868	.97264	34
27	.23260	.23916	4.1814	.97257	33
28	.23288	.23946	4.1760	.97251	32
29	.23316	.23977	4.1706	.97244	31
30	.23345	.24008	4.1653	.97237	30
31	.23373	.24039	4.1600	.97230	29
32	.23401	.24069	4.1547	.97223	28
33	.23429	.24100	4.1493	.97217	27
34	.23458	.24131	4.1441	.97210	26
35	.25486	.24162	4.1388	.97203	25
36	.23514	.24193	4.1335	.97196	24
37	.23542	.24223	4.1282	.97189	23
38	.23571	.24254	4.1230	.97182	22
39	.23599	.24285	4.1178	.97176	21
40	.23627	.24316	4.1126	.97169	20
41	.23656	.24347	4.1074	.97162	19
42	.23684	.24377	4.1022	.97155	18
43	.23712	.24408	4.0970	.97148	17
44	.23740	.24439	4.0918	.97141	16
45	.23769	.24470	4.0867	.97134	15
46	.23797	.24501	4.0815	.97127	14
47	.23825	.24532	4.0764	.97120	13
48	.23853	.24562	4.0713	.97113	12
49	.23882	.24593	4.0662	.97106	11
50	.23910	.24624	4.0611	.97100	10
51	.23938	.24655	4.0560	.97093	9
52	.23966	.24686	4.0509	.97086	8
53	.23995	.24717	4.0459	.97079	7
54	.24023	.24747	4.0408	.97072	6
55	.24051	.24778	4.0358	.97065	5
56	.24079	.24809	4.0308	.97058	4
57	.24108	.24840	4.0257	.97051	3
58	.24136	.24871	4.0207	.97044	2
59	.24164	.24902	4.0158	.97037	1
60	.24192	.24933	4.0108	.97030	0
′	Cos	Ctn	Tan	Sin	′

76°

Table I   Natural Trigonometric Functions   681

14°

′	Sin	Tan	Ctn	Cos	′
0	.24192	.24933	4.0108	.97030	60
1	.24220	.24964	4.0058	.97023	59
2	.24249	.24995	4.0009	.97015	58
3	.24277	.25026	3.9959	.97008	57
4	.24305	.25056	3.9910	.97001	56
5	.24333	.25087	3.9861	.96994	55
6	.24362	.25118	3.9812	.96987	54
7	.24390	.25149	3.9763	.96980	53
8	.24418	.25180	3.9714	.96973	52
9	.24446	.25211	3.9665	.96966	51
10	.24474	.25242	3.9617	.96959	50
11	.24503	.25273	3.9568	.96952	49
12	.24531	.25304	3.9520	.96945	48
13	.24559	.25335	3.9471	.96937	47
14	.24587	.25366	3.9423	.96930	46
15	.24615	.25397	3.9375	.96923	45
16	.24644	.25428	3.9327	.96916	44
17	.24672	.25459	3.9279	.96909	43
18	.24700	.25490	3.9232	.96902	42
19	.24728	.25521	3.9184	.96894	41
20	.24756	.25552	3.9136	.96887	40
21	.24784	.25583	3.9089	.96880	39
22	.24813	.25614	3.9042	.96873	38
23	.24841	.25645	3.8995	.96866	37
24	.24869	.25676	3.8947	.96858	36
25	.24897	.25707	3.8900	.96851	35
26	.24925	.25738	3.8854	.96844	34
27	.24954	.25769	3.8807	.96837	33
28	.24982	.25800	3.8760	.96829	32
29	.25010	.25831	3.8714	.96822	31
30	.25038	.25862	3.8667	.96815	30
31	.25066	.25893	3.8621	.96807	29
32	.25094	.25924	3.8575	.96800	28
33	.25122	.25955	3.8528	.96793	27
34	.25151	.25986	3.8482	.96786	26
35	.25179	.26017	3.8436	.96778	25
36	.25207	.26048	3.8391	.96771	24
37	.25235	.26079	3.8345	.96764	23
38	.25263	.26110	3.8299	.96756	22
39	.25291	.26141	3.8254	.96749	21
40	.25320	.26172	3.8208	.96742	20
41	.25348	.26203	3.8163	.96734	19
42	.25376	.26235	3.8118	.96727	18
43	.25404	.26266	3.8073	.96719	17
44	.25432	.26297	3.8028	.96712	16
45	.25460	.26328	3.7983	.96705	15
46	.25488	.26359	3.7938	.96697	14
47	.25516	.26390	3.7893	.96690	13
48	.25545	.26421	3.7848	.96682	12
49	.25573	.26452	3.7804	.96675	11
50	.25601	.26483	3.7760	.96667	10
51	.25629	.26515	3.7715	.96660	9
52	.25657	.26546	3.7671	.96653	8
53	.25685	.26577	3.7627	.96645	7
54	.25713	.26608	3.7583	.96638	6
55	.25741	.26639	3.7539	.96630	5
56	.25769	.26670	3.7495	.96623	4
57	.25798	.26701	3.7451	.96615	3
58	.25826	.26733	3.7408	.96608	2
59	.25854	.26764	3.7364	.96600	1
60	.25882	.26795	3.7321	.96593	0
′	Cos	Ctn	Tan	Sin	′

75°

15°

′	Sin	Tan	Ctn	Cos	′
0	.25882	.26795	3.7321	.96593	60
1	.25910	.26826	3.7277	.96585	59
2	.25938	.26857	3.7234	.96578	58
3	.25966	.26888	3.7191	.96570	57
4	.25994	.26920	3.7148	.96562	56
5	.26022	.26951	3.7105	.96555	55
6	.26050	.26982	3.7062	.96547	54
7	.26079	.27013	3.7019	.96540	53
8	.26107	.27044	3.6976	.96532	52′
9	.26135	.27076	3.6933	.96524	51
10	.26163	.27107	3.6891	.96517	50
11	.26191	.27138	3.6848	.96509	49
12	.26219	.27169	3.6806	.96502	48
13	.26247	.27201	3.6764	.96494	47
14	.26275	.27232	3.6722	.96486	46
15	.26303	.27263	3.6680	.96479	45
16	.26331	.27294	3.6638	.96471	44
17	.26359	.27326	3.6596	.96463	43
18	.26387	.27357	3.6554	.96456	42
19	.26415	.27388	3.6512	.96448	41
20	.26443	.27419	3.6470	.96440	40
21	.26471	.27451	3.6429	.96433	39
22	.26500	.27482	3.6387	.96425	38
23	.26528	.27513	3.6346	.96417	37
24	.26556	.27545	3.6305	.96410	36
25	.26584	.27576	3.6264	.96402	35
26	.26612	.27607	3.6222	.96394	34
27	.26640	.27638	3.6181	.96386	33
28	.26668	.27670	3.6140	.96379	32
29	.26696	.27701	3.6100	.96371	31
30	.26724	.27732	3.6059	.96363	30
31	.26752	.27764	3.6018	.96355	29
32	.26780	.27795	3.5978	.96347	28
33	.26808	.27826	3.5937	.96340	27
34	.26836	.27858	3.5897	.96332	26
35	.26864	.27889	3.5856	.96324	25
36	.26892	.27921	3.5816	.96316	24
37	.26920	.27952	3.5776	.96308	23
38	.26948	.27983	3.5736	.96301	22
39	.26976	.28015	3.5696	.96293	21
40	.27004	.28046	3.5656	.96285	20
41	.27032	.28077	3.5616	.96277	19
42	.27060	.28109	3.5576	.96269	18
43	.27088	.28140	3.5536	.96261	17
44	.27116	.28172	3.5497	.96253	16
45	.27144	.28203	3.5457	.96246	15
46	.27172	.28234	3.5418	.96238	14
47	.27200	.28266	3.5379	.96230	13
48	.27228	.28297	3.5339	.96222	12
49	.27256	.28329	3.5300	.96214	11
50	.27284	.28360	3.5261	.96206	10
51	.27312	.28391	3.5222	.96198	9
52	.27340	.28423	3.5183	.96190	8
53	.27368	.28454	3.5144	.96182	7
54	.27396	.28486	3.5105	.96174	6
55	.27424	.28517	3.5067	.96166	5
56	.27452	.28549	3.5028	.96158	4
57	.27480	.28580	3.4989	.96150	3
58	.27508	.28612	3.4951	.96142	2
59	.27536	.28643	3.4912	.96134	1
60	.27564	.28675	3.4874	.96126	0
′	Cos	Ctn	Tan	Sin	′

74°

16°

′	Sin	Tan	Ctn	Cos	′
0	.27564	.28675	3.4874	.96126	60
1	.27592	.28706	3.4836	.96118	59
2	.27620	.28738	3.4798	.96110	58
3	.27648	.28769	3.4760	.96102	57
4	.27676	.28801	3.4722	.96094	56
5	.27704	.28832	3.4684	.96086	55
6	.27731	.28864	3.4646	.96078	54
7	.27759	.28895	3.4608	.96070	53
8	.27787	.28927	3.4570	.96062	52
9	.27815	.28958	3.4533	.96054	51
10	.27843	.28990	3.4495	.96046	50
11	.27871	.29021	3.4458	.96037	49
12	.27899	.29053	3.4420	.96029	48
13	.27927	.29084	3.4383	.96021	47
14	.27955	.29116	3.4346	.96013	46
15	.27983	.29147	3.4308	.96005	45
16	.28011	.29179	3.4271	.95997	44
17	.28039	.29210	3.4234	.95989	43
18	.28067	.29242	3.4197	.95981	42
19	.28095	.29274	3.4160	.95972	41
20	.28123	.29305	3.4124	.95964	40
21	.28150	.29337	3.4087	.95956	39
22	.28178	.29368	3.4050	.95948	38
23	.28206	.29400	3.4014	.95940	37
24	.28234	.29432	3.3977	.95931	36
25	.28262	.29463	3.3941	.95923	35
26	.28290	.29495	3.3904	.95915	34
27	.28318	.29526	3.3868	.95907	33
28	.28346	.29558	3.3832	.95898	32
29	.28374	.29590	3.3796	.95890	31
30	.28402	.29621	3.3759	.95882	30
31	.28429	.29653	3.3723	.95874	29
32	.28457	.29685	3.3687	.95865	28
33	.28485	.29716	3.3652	.95857	27
34	.28513	.29748	3.3616	.95849	26
35	.28541	.29780	3.3580	.95841	25
36	.28569	.29811	3.3544	.95832	24
37	.28597	.29843	3.3509	.95824	23
38	.28625	.29875	3.3473	.95816	22
39	.28652	.29906	3.3438	.95807	21
40	.28680	.29938	3.3402	.95799	20
41	.28708	.29970	3.3367	.95791	19
42	.28736	.30001	3.3332	.95782	18
43	.28764	.30033	3.3297	.95774	17
44	.28792	.30065	3.3261	.95766	16
45	.28820	.30097	3.3226	.95757	15
46	.28847	.30128	3.3191	.95749	14
47	.28875	.30160	3.3156	.95740	13
48	.28903	.30192	3.3122	.95732	12
49	.28931	.30224	3.3087	.95724	11
50	.28959	.30255	3.3052	.95715	10
51	.28987	.30287	3.3017	.95707	9
52	.29015	.30319	3.2983	.95698	8
53	.29042	.30351	3.2948	.95690	7
54	.29070	.30382	3.2914	.95681	6
55	.29098	.30414	3.2879	.95673	5
56	.29126	.30446	3.2845	.95664	4
57	.29154	.30478	3.2811	.95656	3
58	.29182	.30509	3.2777	.95647	2
59	.29209	.30541	3.2743	.95639	1
60	.29237	.30573	3.2709	.95630	0
′	Cos	Ctn	Tan	Sin	′

73°

17°

′	Sin	Tan	Ctn	Cos	′
0	.29237	.30573	3.2709	.95630	60
1	.29265	.30605	3.2675	.95622	59
2	.29293	.30637	3.2641	.95613	58
3	.29321	.30669	3.2607	.95605	57
4	.29348	.30700	3.2573	.95596	56
5	.29376	.30732	3.2539	.95588	55
6	.29404	.30764	3.2506	.95579	54
7	.29432	.30796	3.2472	.95571	53
8	.29460	.30828	3.2438	.95562	52
9	.29487	.30860	3.2405	.95554	51
10	.29515	.30891	3.2371	.95545	50
11	.29543	.30923	3.2338	.95536	49
12	.29571	.30955	3.2305	.95528	48
13	.29599	.30987	3.2272	.95519	47
14	.29626	.31019	3.2238	.95511	46
15	.29654	.31051	3.2205	.95502	45
16	.29682	.31083	3.2172	.95493	44
17	.29710	.31115	3.2139	.95485	43
18	.29737	.31147	3.2106	.95476	42
19	.29765	.31178	3.2073	.95467	41
20	.29793	.31210	3.2041	.95459	40
21	.29821	.31242	3.2008	.95450	39
22	.29849	.31274	3.1975	.95441	38
23	.29876	.31306	3.1943	.95433	37
24	.29904	.31338	3.1910	.95424	36
25	.29932	.31370	3.1878	.95415	35
26	.29960	.31402	3.1845	.95407	34
27	.29987	.31434	3.1813	.95398	33
28	.30015	.31466	3.1780	.95389	32
29	.30043	.31498	3.1748	.95380	31
30	.30071	.31530	3.1716	.95372	30
31	.30098	.31562	3.1684	.95363	29
32	.30126	.31594	3.1652	.95354	28
33	.30154	.31626	3.1620	.95345	27
34	.30182	.31658	3.1588	.95337	26
35	.30209	.31690	3.1556	.95328	25
36	.30237	.31722	3.1524	.95319	24
37	.30265	.31754	3.1492	.95310	23
38	.30292	.31786	3.1460	.95301	22
39	.30320	.31818	3.1429	.95293	21
40	.30348	.31850	3.1397	.95284	20
41	.30376	.31882	3.1366	.95275	19
42	.30403	.31914	3.1334	.95266	18
43	.30431	.31946	3.1303	.95257	17
44	.30459	.31978	3.1271	.95248	16
45	.30486	.32010	3.1240	.95240	15
46	.30514	.32042	3.1209	.95231	14
47	.30542	.32074	3.1178	.95222	13
48	.30570	.32106	3.1146	.95213	12
49	.30597	.32139	3.1115	.95204	11
50	.30625	.32171	3.1084	.95195	10
51	.30653	.32203	3.1053	.95186	9
52	.30680	.32235	3.1022	.95177	8
53	.30708	.32267	3.0991	.95168	7
54	.30736	.32299	3.0961	.95159	6
55	.30763	.32331	3.0930	.95150	5
56	.30791	.32363	3.0899	.95142	4
57	.30819	.32396	3.0868	.95133	3
58	.30846	.32428	3.0838	.95124	2
59	.30874	.32460	3.0807	.95115	1
60	.30902	.32492	3.0777	.95106	0
′	Cos	Ctn	Tan	Sin	′

72°

Table I   Natural Trigonometric Functions

18°

′	Sin	Tan	Ctn	Cos	′
0	.30902	.32492	3.0777	.95106	60
1	.30929	.32524	3.0746	.95097	59
2	.30957	.32556	3.0716	.95088	58
3	.30985	.32588	3.0686	.95079	57
4	.31012	.32621	3.0655	.95070	56
5	.31040	.32653	3.0625	.95061	55
6	.31068	.32685	3.0595	.95052	54
7	.31095	.32717	3.0565	.95043	53
8	.31123	.32749	3.0535	.95033	52
9	.31151	.32782	3.0505	.95024	51
10	.31178	.32814	3.0475	.95015	50
11	.31206	.32846	3.0445	.95006	49
12	.31233	.32878	3.0415	.94997	48
13	.31261	.32911	3.0385	.94988	47
14	.31289	.32943	3.0356	.94979	46
15	.31316	.32975	3.0326	.94970	45
16	.31344	.33007	3.0296	.94961	44
17	.31372	.33040	3.0267	.94952	43
18	.31399	.33072	3.0237	.94943	42
19	.31427	.33104	3.0208	.94933	41
20	.31454	.33136	3.0178	.94924	40
21	.31482	.33169	3.0149	.94915	39
22	.31510	.33201	3.0120	.94906	38
23	.31537	.33233	3.0090	.94897	37
24	.31565	.33266	3.0061	.94888	36
25	.31593	.33298	3.0032	.94878	35
26	.31620	.33330	3.0003	.94869	34
27	.31648	.33363	2.9974	.94860	33
28	.31675	.33395	2.9945	.94851	32
29	.31703	.33427	2.9916	.94842	31
30	.31730	.33460	2.9887	.94832	30
31	.31758	.33492	2.9858	.94823	29
32	.31786	.33524	2.9829	.94814	28
33	.31813	.33557	2.9800	.94805	27
34	.31841	.33589	2.9772	.94795	26
35	.31868	.33621	2.9743	.94786	25
36	.31896	.33654	2.9714	.94777	24
37	.31923	.33686	2.9686	.94768	23
38	.31951	.33718	2.9657	.94758	22
39	.31979	.33751	2.9629	.94749	21
40	.32006	.33783	2.9600	.94740	20
41	.32034	.33816	2.9572	.94730	19
42	.32061	.33848	2.9544	.94721	18
43	.32089	.33881	2.9515	.94712	17
44	.32116	.33913	2.9487	.94702	16
45	.32144	.33945	2.9459	.94693	15
46	.32171	.33978	2.9431	.94684	14
47	.32199	.34010	2.9403	.94674	13
48	.32227	.34043	2.9375	.94665	12
49	.32254	.34075	2.9347	.94656	11
50	.32282	.34108	2.9319	.94646	10
51	.32309	.34140	2.9291	.94637	9
52	.32337	.34173	2.9263	.94627	8
53	.32364	.34205	2.9235	.94618	7
54	.32392	.34238	2.9208	.94609	6
55	.32419	.34270	2.9180	.94599	5
56	.32447	.34303	2.9152	.94590	4
57	.32474	.34335	2.9125	.94580	3
58	.32502	.34368	2.9097	.94571	2
59	.32529	.34400	2.9070	.94561	1
60	.32557	.34433	2.9042	.94552	0
′	Cos	Ctn	Tan	Sin	′

71°

19°

′	Sin	Tan	Ctn	Cos	′
0	.32557	.34433	2.9042	.94552	60
1	.32584	.34465	2.9015	.94542	59
2	.32612	.34498	2.8987	.94533	58
3	.32639	.34530	2.8960	.94523	57
4	.32667	.34563	2.8933	.94514	56
5	.32694	.34596	2.8905	.94504	55
6	.32722	.34628	2.8878	.94495	54
7	.32749	.34661	2.8851	.94485	53
8	.32777	.34693	2.8824	.94476	52
9	.32804	.34726	2.8797	.94466	51
10	.32832	.34758	2.8770	.94457	50
11	.32859	.34791	2.8743	.94447	49
12	.32887	.34824	2.8716	.94438	48
13	.32914	.34856	2.8689	.94428	47
14	.32942	.34889	2.8662	.94418	46
15	.32969	.34922	2.8636	.94409	45
16	.32997	.34954	2.8609	.94399	44
17	.33024	.34987	2.8582	.94390	43
18	.33051	.35020	2.8556	.94380	42
19	.33079	.35052	2.8529	.94370	41
20	.33106	.35085	2.8502	.94361	40
21	.33134	.35118	2.8476	.94351	39
22	.33161	.35150	2.8449	.94342	38
23	.33189	.35183	2.8423	.94332	37
24	.33216	.35216	2.8397	.94322	36
25	.33244	.35248	2.8370	.94313	35
26	.33271	.35281	2.8344	.94303	34
27	.33298	.35314	2.8318	.94293	33
28	.33326	.35346	2.8291	.94284	32
29	.33353	.35379	2.8265	.94274	31
30	.33381	.35412	2.8239	.94264	30
31	.33408	.35445	2.8213	.94254	29
32	.33436	.35477	2.8187	.94245	28
33	.33463	.35510	2.8161	.94235	27
34	.33490	.35543	2.8135	.94225	26
35	.33518	.35576	2.8109	.94215	25
36	.33545	.35608	2.8083	.94206	24
37	.33573	.35641	2.8057	.94196	23
38	.33600	.35674	2.8032	.94186	22
39	.33627	.35707	2.8006	.94176	21
40	.33655	.35740	2.7980	.94167	20
41	.33682	.35772	2.7955	.94157	19
42	.33710	.35805	2.7929	.94147	18
43	.33737	.35838	2.7903	.94137	17
44	.33764	.35871	2.7878	.94127	16
45	.33792	.35904	2.7852	.94118	15
46	.33819	.35937	2.7827	.94108	14
47	.33846	.35969	2.7801	.94098	13
48	.33874	.36002	2.7776	.94088	12
49	.33901	.36035	2.7751	.94078	11
50	.33929	.36068	2.7725	.94068	10
51	.33956	.36101	2.7700	.94058	9
52	.33983	.36134	2.7675	.94049	8
53	.34011	.36167	2.7650	.94039	7
54	.34038	.36199	2.7625	.94029	6
55	.34065	.36232	2.7600	.94019	5
56	.34093	.36265	2.7575	.94009	4
57	.34120	.36298	2.7550	.93999	3
58	.34147	.36331	2.7525	.93989	2
59	.34175	.36364	2.7500	.93979	1
60	.34202	.36397	2.7475	.93969	0
′	Cos	Ctn	Tan	Sin	′

70°

Table I Natural Trigonometric Functions

20°

′	Sin	Tan	Ctn	Cos	′
0	.34202	.36397	2.7475	.93969	60
1	.34229	.36430	2.7450	.93959	59
2	.34257	.36463	2.7425	.93949	58
3	.34284	.36496	2.7400	.93939	57
4	.34311	.36529	2.7376	.93929	56
5	.34339	.36562	2.7351	.93919	55
6	.34366	.36595	2.7326	.93909	54
7	.34393	.36628	2.7302	.93899	53
8	.34421	.36661	2.7277	.93889	52
9	.34448	.36694	2.7253	.93879	51
10	.34475	.36727	2.7228	.93869	50
11	.34503	.36760	2.7204	.93859	49
12	.34530	.36793	2.7179	.93849	48
13	.34557	.36826	2.7155	.93839	47
14	.34584	.36859	2.7130	.93829	46
15	.34612	.36892	2.7106	.93819	45
16	.34639	.36925	2.7082	.93809	44
17	.34666	.36958	2.7058	.93799	43
18	.34694	.36991	2.7034	.93789	42
19	.34721	.37024	2.7009	.93779	41
20	.34748	.37057	2.6985	.93769	40
21	.34775	.37090	2.6961	.93759	39
22	.34803	.37123	2.6937	.93748	38
23	.34830	.37157	2.6913	.93738	37
24	.34857	.37190	2.6889	.93728	36
25	.34884	.37223	2.6865	.93718	35
26	.34912	.37256	2.6841	.93708	34
27	.34939	.37289	2.6818	.93698	33
28	.34966	.37322	2.6794	.93688	32
29	.34993	.37355	2.6770	.93677	31
30	.35021	.37388	2.6746	.93667	30
31	.35048	.37422	2.6723	.93657	29
32	.35075	.37455	2.6699	.93647	28
33	.35102	.37488	2.6675	.93637	27
34	.35130	.37521	2.6652	.93626	26
35	.35157	.37554	2.6628	.93616	25
36	.35184	.37588	2.6605	.93606	24
37	.35211	.37621	2.6581	.93596	23
38	.35239	.37654	2.6558	.93585	22
39	.35266	.37687	2.6534	.93575	21
40	.35293	.37720	2.6511	.93565	20
41	.35320	.37754	2.6488	.93555	19
42	.35347	.37787	2.6464	.93544	18
43	.35375	.37820	2.6441	.93534	17
44	.35402	.37853	2.6418	.93524	16
45	.35429	.37887	2.6395	.93514	15
46	.35456	.37920	2.6371	.93503	14
47	.35484	.37953	2.6348	.93493	13
48	.35511	.37986	2.6325	.93483	12
49	.35538	.38020	2.6302	.93472	11
50	.35565	.38053	2.6279	.93462	10
51	.35592	.38086	2.6256	.93452	9
52	.35619	.38120	2.6233	.93441	8
53	.35647	.38153	2.6210	.93431	7
54	.35674	.38186	2.6187	.93420	6
55	.35701	.38220	2.6165	.93410	5
56	.35728	.38253	2.6142	.93400	4
57	.35755	.38286	2.6119	.93389	3
58	.35782	.38320	2.6096	.93379	2
59	.35810	.38353	2.6074	.93368	1
60	.35837	.38386	2.6051	.93358	0
′	Cos	Ctn	Tan	Sin	′

69°

21°

′	Sin	Tan	Ctn	Cos	′
0	.35837	.38386	2.6051	.93358	60
1	.35864	.38420	2.6028	.93348	59
2	.35891	.38453	2.6006	.93337	58
3	.35918	.38487	2.5983	.93327	57
4	.35945	.38520	2.5961	.93316	56
5	.35973	.38553	2.5938	.93306	55
6	.36000	.38587	2.5916	.93295	54
7	.36027	.38620	2.5893	.93285	53
8	.36054	.38654	2.5871	.93274	52
9	.36081	.38687	2.5848	.93264	51
10	.36108	.38721	2.5826	.93253	50
11	.36135	.38754	2.5804	.93243	49
12	.36162	.38787	2.5782	.93232	48
13	.36190	.38821	2.5759	.93222	47
14	.36217	.38854	2.5737	.93211	46
15	.36244	.38888	2.5715	.93201	45
16	.36271	.38921	2.5693	.93190	44
17	.36298	.38955	2.5671	.93180	43
18	.36325	.38988	2.5649	.93169	42
19	.36352	.39022	2.5627	.93159	41
20	.36379	.39055	2.5605	.93148	40
21	.36406	.39089	2.5583	.93137	39
22	.36434	.39122	2.5561	.93127	38
23	.36461	.39156	2.5539	.93116	37
24	.36488	.39190	2.5517	.93106	36
25	.36515	.39223	2.5495	.93095	35
26	.36542	.39257	2.5473	.93084	34
27	.36569	.39290	2.5452	.93074	33
28	.36596	.39324	2.5430	.93063	32
29	.36623	.39357	2.5408	.93052	31
30	.36650	.39391	2.5386	.93042	30
31	.36677	.39425	2.5365	.93031	29
32	.36704	.39458	2.5343	.93020	28
33	.36731	.39492	2.5322	.93010	27
34	.36758	.39526	2.5300	.92999	26
35	.36785	.39559	2.5279	.92988	25
36	.36812	.39593	2.5257	.92978	24
37	.36839	.39626	2.5236	.92967	23
38	.36867	.39660	2.5214	.92956	22
39	.36894	.39694	2.5193	.92945	21
40	.36921	.39727	2.5172	.92935	20
41	.36948	.39761	2.5150	.92924	19
42	.36975	.39795	2.5129	.92913	18
43	.37002	.39829	2.5108	.92902	17
44	.37029	.39862	2.5086	.92892	16
45	.37056	.39896	2.5065	.92881	15
46	.37083	.39930	2.5044	.92870	14
47	.37110	.39963	2.5023	.92859	13
48	.37137	.39997	2.5002	.92849	12
49	.37164	.40031	2.4981	.92838	11
50	.37191	.40065	2.4960	.92827	10
51	.37218	.40098	2.4939	.92816	9
52	.37245	.40132	2.4918	.92805	8
53	.37272	.40166	2.4897	.92794	7
54	.37299	.40200	2.4876	.92784	6
55	.37326	.40234	2.4855	.92773	5
56	.37353	.40267	2.4834	.92762	4
57	.37380	.40301	2.4813	.92751	3
58	.37407	.40335	2.4792	.92740	2
59	.37434	.40369	2.4772	.92729	1
60	.37461	.40403	2.4751	.92718	0
′	Cos	Ctn	Tan	Sin	′

68°

Table I Natural Trigonometric Functions

22°

′	Sin	Tan	Ctn	Cos	′
0	.37461	.40403	2.4751	.92718	60
1	.37488	.40436	2.4730	.92707	59
2	.37515	.40470	2.4709	.92697	58
3	.37542	.40504	2.4689	.92686	57
4	.37569	.40538	2.4668	.92675	56
5	.37595	.40572	2.4648	.92664	55
6	.37622	.40606	2.4627	.92653	54
7	.37649	.40640	2.4606	.92642	53
8	.37676	.40674	2.4586	.92631	52
9	.37703	.40707	2.4566	.92620	51
10	.37730	.40741	2.4545	.92609	50
11	.37757	.40775	2.4525	.92598	49
12	.37784	.40809	2.4504	.92587	48
13	.37811	.40843	2.4484	.92576	47
14	.37838	.40877	2.4464	.92565	46
15	.37865	.40911	2.4443	.92554	45
16	.37892	.40945	2.4423	.92543	44
17	.37919	.40979	2.4403	.92532	43
18	.37946	.41013	2.4383	.92521	42
19	.37973	.41047	2.4362	.92510	41
20	.37999	.41081	2.4342	.92499	40
21	.38026	.41115	2.4322	.92488	39
22	.38053	.41149	2.4302	.92477	38
23	.38080	.41183	2.4282	.92466	37
24	.38107	.41217	2.4262	.92455	36
25	.38134	.41251	2.4242	.92444	35
26	.38161	.41285	2.4222	.92432	34
27	.38188	.41319	2.4202	.92421	33
28	.38215	.41353	2.4182	.92410	32
29	.38241	.41387	2.4162	.92399	31
30	.38268	.41421	2.4142	.92388	30
31	.38295	.41455	2.4122	.92377	29
32	.38322	.41490	2.4102	.92366	28
33	.38349	.41524	2.4083	.92355	27
34	.38376	.41558	2.4063	.92343	26
35	.38403	.41592	2.4043	.92332	25
36	.38430	.41626	2.4023	.92321	24
37	.38456	.41660	2.4004	.92310	23
38	.38483	.41694	2.3984	.92299	22
39	.38510	.41728	2.3964	.92287	21
40	.38537	.41763	2.3945	.92276	20
41	.38564	.41797	2.3925	.92265	19
42	.38591	.41831	2.3906	.92254	18
43	.38617	.41865	2.3886	.92243	17
44	.38644	.41899	2.3867	.92231	16
45	.38671	.41933	2.3847	.92220	15
46	.38698	.41968	2.3828	.92209	14
47	.38725	.42002	2.3808	.92198	13
48	.38752	.42036	2.3789	.92186	12
49	.38778	.42070	2.3770	.92175	11
50	.38805	.42105	2.3750	.92164	10
51	.38832	.42139	2.3731	.92152	9
52	.38859	.42173	2.3712	.92141	8
53	.38886	.42207	2.3693	.92130	7
54	.38912	.42242	2.3673	.92119	6
55	.38939	.42276	2.3654	.92107	5
56	.38966	.42310	2.3635	.92096	4
57	.38993	.42345	2.3616	.92085	3
58	.39020	.42379	2.3597	.92073	2
59	.39046	.42413	2.3578	.92062	1
60	.39073	.42447	2.3559	.92050	0
′	Cos	Ctn	Tan	Sin	′

67°

23°

′	Sin	Tan	Ctn	Cos	′
0	.39073	.42447	2.3559	.92050	60
1	.39100	.42482	2.3539	.92039	59
2	.39127	.42516	2.3520	.92028	58
3	.39153	.42551	2.3501	.92016	57
4	.39180	.42585	2.3483	.92005	56
5	.39207	.42619	2.3464	.91994	55
6	.39234	.42654	2.3445	.91982	54
7	.39260	.42688	2.3426	.91971	53
8	.39287	.42722	2.3407	.91959	52
9	.39314	.42757	2.3388	.91948	51
10	.39341	.42791	2.3369	.91936	50
11	.39367	.42826	2.3351	.91925	49
12	.39394	.42860	2.3332	.91914	48
13	.39421	.42894	2.3313	.91902	47
14	.39448	.42929	2.3294	.91891	46
15	.39474	.42963	2.3276	.91879	45
16	.39501	.42998	2.3257	.91868	44
17	.39528	.43032	2.3238	.91856	43
18	.39555	.43067	2.3220	.91845	42
19	.39581	.43101	2.3201	.91833	41
20	.39608	.43136	2.3183	.91822	40
21	.39635	.43170	2.3164	.91810	39
22	.39661	.43205	2.3146	.91799	38
23	.39688	.43239	2.3127	.91787	37
24	.39715	.43274	2.3109	.91775	36
25	.39741	.43308	2.3090	.91764	35
26	.39768	.43343	2.3072	.91752	34
27	.39795	.43378	2.3053	.91741	33
28	.39822	.43412	2.3035	.91729	32
29	.39848	.43447	2.3017	.91718	31
30	.39875	.43481	2.2998	.91706	30
31	.39902	.43516	2.2980	.91694	29
32	.39928	.43550	2.2962	.91683	28
33	.39955	.43585	2.2944	.91671	27
34	.39982	.43620	2.2925	.91660	26
35	.40008	.43654	2.2907	.91648	25
36	.40035	.43689	2.2889	.91636	24
37	.40062	.43724	2.2871	.91625	23
38	.40088	.43758	2.2853	.91613	22
39	.40115	.43793	2.2835	.91601	21
40	.40141	.43828	2.2817	.91590	20
41	.40168	.43862	2.2799	.91578	19
42	.40195	.43897	2.2781	.91566	18
43	.40221	.43932	2.2763	.91555	17
44	.40248	.43966	2.2745	.91543	16
45	.40275	.44001	2.2727	.91531	15
46	.40301	.44036	2.2709	.91519	14
47	.40328	.44071	2.2691	.91508	13
48	.40355	.44105	2.2673	.91496	12
49	.40381	.44140	2.2655	.91484	11
50	.40408	.44175	2.2637	.91472	10
51	.40434	.44210	2.2620	.91461	9
52	.40461	.44244	2.2602	.91449	8
53	.40488	.44279	2.2584	.91437	7
54	.40514	.44314	2.2566	.91425	6
55	.40541	.44349	2.2549	.91414	5
56	.40567	.44384	2.2531	.91402	4
57	.40594	.44418	2.2513	.91390	3
58	.40621	.44453	2.2496	.91378	2
59	.40647	.44488	2.2478	.91366	1
60	.40674	.44523	2.2460	.91355	0
′	Cos	Ctn	Tan	Sin	′

66°

Table I Natural Trigonometric Functions

24°

′	Sin	Tan	Ctn	Cos	′
0	.40674	.44523	2.2460	.91355	60
1	.40700	.44558	2.2443	.91343	59
2	.40727	.44593	2.2425	.91331	58
3	.40753	.44627	2.2408	.91319	57
4	.40780	.44662	2.2390	.91307	56
5	.40806	.44697	2.2373	.91295	55
6	.40833	.44732	2.2355	.91283	54
7	.40860	.44767	2.2338	.91272	53
8	.40886	.44802	2.2320	.91260	52
9	.40913	.44837	2.2303	.91248	51
10	.40939	.44872	2.2286	.91236	50
11	.40966	.44907	2.2268	.91224	49
12	.40992	.44942	2.2251	.91212	48
13	.41019	.44977	2.2234	.91200	47
14	.41045	.45012	2.2216	.91188	46
15	.41072	.45047	2.2199	.91176	45
16	.41098	.45082	2.2182	.91164	44
17	.41125	.45117	2.2165	.91152	43
18	.41151	.45152	2.2148	.91140	42
19	.41178	.45187	2.2130	.91128	41
20	.41204	.45222	2.2113	.91116	40
21	.41231	.45257	2.2096	.91104	39
22	.41257	.45292	2.2079	.91092	38
23	.41284	.45327	2.2062	.91080	37
24	.41310	.45362	2.2045	.91068	36
25	.41337	.45397	2.2028	.91056	35
26	.41363	.45432	2.2011	.91044	34
27	.41390	.45467	2.1994	.91032	33
28	.41416	.45502	2.1977	.91020	32
29	.41443	.45538	2.1960	.91008	31
30	.41469	.45573	2.1943	.90996	30
31	.41496	.45608	2.1926	.90984	29
32	.41522	.45643	2.1909	.90972	28
33	.41549	.45678	2.1892	.90960	27
34	.41575	.45713	2.1876	.90948	26
35	.41602	.45748	2.1859	.90936	25
36	.41628	.45784	2.1842	.90924	24
37	.41655	.45819	2.1825	.90911	23
38	.41681	.45854	2.1808	.90899	22
39	.41707	.45889	2.1792	.90887	21
40	.41734	.45924	2.1775	.90875	20
41	.41760	.45960	2.1758	.90863	19
42	.41787	.45995	2.1742	.90851	18
43	.41813	.46030	2.1725	.90839	17
44	.41840	.46065	2.1708	.90826	16
45	.41866	.46101	2.1692	.90814	15
46	.41892	.46136	2.1675	.90802	14
47	.41919	.46171	2.1659	.90790	13
48	.41945	.46206	2.1642	.90778	12
49	.41972	.46242	2.1625	.90766	11
50	.41998	.46277	2.1609	.90753	10
51	.42024	.46312	2.1592	.90741	9
52	.42051	.46348	2.1576	.90729	8
53	.42077	.46383	2.1560	.90717	7
54	.42104	.46418	2.1543	.90704	6
55	.42130	.46454	2.1527	.90692	5
56	.42156	.46489	2.1510	.90680	4
57	.42183	.46525	2.1494	.90668	3
58	.42209	.46560	2.1478	.90655	2
59	.42235	.46595	2.1461	.90643	1
60	.42262	.46631	2.1445	.90631	0
′	Cos	Ctn	Tan	Sin	′

65°

25°

′	Sin	Tan	Ctn	Cos	′
0	.42262	.46631	2.1445	.90631	60
1	.42288	.46666	2.1429	.90618	59
2	.42315	.46702	2.1413	.90606	58
3	.42341	.46737	2.1396	.90594	57
4	.42367	.46772	2.1380	.90582	56
5	.42394	.46808	2.1364	.90569	55
6	.42420	.46843	2.1348	.90557	54
7	.42446	.46879	2.1332	.90545	53
8	.42473	.46914	2.1315	.90532	52
9	.42499	.46950	2.1299	.90520	51
10	.42525	.46985	2.1283	.90507	50
11	.42552	.47021	2.1267	.90495	49
12	.42578	.47056	2.1251	.90483	48
13	.42604	.47092	2.1235	.90470	47
14	.42631	.47128	2.1219	.90458	46
15	.42657	.47163	2.1203	.90446	45
16	.42683	.47199	2.1187	.90433	44
17	.42709	.47234	2.1171	.90421	43
18	.42736	.47270	2.1155	.90408	42
19	.42762	.47305	2.1139	.90396	41
20	.42788	.47341	2.1123	.90383	40
21	.42815	.47377	2.1107	.90371	39
22	.42841	.47412	2.1092	.90358	38
23	.42867	.47448	2.1076	.90346	37
24	.42894	.47483	2.1060	.90334	36
25	.42920	.47519	2.1044	.90321	35
26	.42946	.47555	2.1028	.90309	34
27	.42972	.47590	2.1013	.90296	33
28	.42999	.47626	2.0997	.90284	32
29	.43025	.47662	2.0981	.90271	31
30	.43051	.47698	2.0965	.90259	30
31	.43077	.47733	2.0950	.90246	29
32	.43104	.47769	2.0934	.90233	28
33	.43130	.47805	2.0918	.90221	27
34	.43156	.47840	2.0903	.90208	26
35	.43182	.47876	2.0887	.90196	25
36	.43209	.47912	2.0872	.90183	24
37	.43235	.47948	2.0856	.90171	23
38	.43261	.47984	2.0840	.90158	22
39	.43287	.48019	2.0825	.90146	21
40	.43313	.48055	2.0809	.90133	20
41	.43340	.48091	2.0794	.90120	19
42	.43366	.48127	2.0778	.90108	18
43	.43392	.48163	2.0763	.90095	17
44	.43418	.48198	2.0748	.90082	16
45	.43445	.48234	2.0732	.90070	15
46	.43471	.48270	2.0717	.90057	14
47	.43497	.48306	2.0701	.90045	13
48	.43523	.48342	2.0686	.90032	12
49	.43549	.48378	2.0671	.90019	11
50	.43575	.48414	2.0655	.90007	10
51	.43602	.48450	2.0640	.89994	9
52	.43628	.48486	2.0625	.89981	8
53	.43654	.48521	2.0609	.89968	7
54	.43680	.48557	2.0594	.89956	6
55	.43706	.48593	2.0579	.89943	5
56	.43733	.48629	2.0564	.89930	4
57	.43759	.48665	2.0549	.89918	3
58	.43785	.48701	2.0533	.89905	2
59	.43811	.48737	2.0518	.89892	1
60	.43837	.48773	2.0503	.89879	0
′	Cos	Ctn	Tan	Sin	′

64°

Table I Natural Trigonometric Functions

26°

′	Sin	Tan	Ctn	Cos	′
0	.43837	.48773	2.0503	.89879	60
1	.43863	.48809	2.0488	.89867	59
2	.43889	.48845	2.0473	.89854	58
3	.43916	.48881	2.0458	.89841	57
4	.43942	.48917	2.0443	.89828	56
5	.43968	.48953	2.0428	.89816	55
6	.43994	.48989	2.0413	.89803	54
7	.44020	.49026	2.0398	.89790	53
8	.44046	.49062	2.0383	.89777	52
9	.44072	.49098	2.0368	.89764	51
10	.44098	.49134	2.0353	.89752	50
11	.44124	.49170	2.0338	.89739	49
12	.44151	.49206	2.0323	.89726	48
13	.44177	.49242	2.0308	.89713	47
14	.44203	.49278	2.0293	.89700	46
15	.44229	.49315	2.0278	.89687	45
16	.44255	.49351	2.0263	.89674	44
17	.44281	.49387	2.0248	.89662	43
18	.44307	.49423	2.0233	.89649	42
19	.44333	.49459	2.0219	.89636	41
20	.44359	.49495	2.0204	.89623	40
21	.44385	.49532	2.0189	.89610	39
22	.44411	.49568	2.0174	.89597	38
23	.44437	.49604	2.0160	.89584	37
24	.44464	.49640	2.0145	.89571	36
25	.44490	.49677	2.0130	.89558	35
26	.44516	.49713	2.0115	.89545	34
27	.44542	.49749	2.0101	.89532	33
28	.44568	.49786	2.0086	.89519	32
29	.44594	.49822	2.0072	.89506	31
30	.44620	.49858	2.0057	.89493	30
31	.44646	.49894	2.0042	.89480	29
32	.44672	.49931	2.0028	.89467	28
33	.44698	.49967	2.0013	.89454	27
34	.44724	.50004	1.9999	.89441	26
35	.44750	.50040	1.9984	.89428	25
36	.44776	.50076	1.9970	.89415	24
37	.44802	.50113	1.9955	.89402	23
38	.44828	.50149	1.9941	.89389	22
39	.44854	.50185	1.9926	.89376	21
40	.44880	.50222	1.9912	.89363	20
41	.44906	.50258	1.9897	.89350	19
42	.44932	.50295	1.9883	.89337	18
43	.44958	.50331	1.9868	.89324	17
44	.44984	.50368	1.9854	.89311	16
45	.45010	.50404	1.9840	.89298	15
46	.45036	.50441	1.9825	.89285	14
47	.45062	.50477	1.9811	.89272	13
48	.45088	.50514	1.9797	.89259	12
49	.45114	.50550	1.9782	.89245	11
50	.45140	.50587	1.9768	.89232	10
51	.45166	.50623	1.9754	.89219	9
52	.45192	.50660	1.9740	.89206	8
53	.45218	.50696	1.9725	.89193	7
54	.45243	.50733	1.9711	.89180	6
55	.45269	.50769	1.9697	.89167	5
56	.45295	.50806	1.9683	.89153	4
57	.45321	.50843	1.9669	.89140	3
58	.45347	.50879	1.9654	.89127	2
59	.45373	.50916	1.9640	.89114	1
60	.45399	.50953	1.9626	.89101	0
′	Cos	Ctn	Tan	Sin	′

63°

27°

′	Sin	Tan	Ctn	Cos	′
0	.45399	.50953	1.9626	.89101	60
1	.45425	.50989	1.9612	.89087	59
2	.45451	.51026	1.9598	.89074	58
3	.45477	.51063	1.9584	.89061	57
4	.45503	.51099	1.9570	.89048	56
5	.45529	.51136	1.9556	.89035	55
6	.45554	.51173	1.9542	.89021	54
7	.45580	.51209	1.9528	.89008	53
8	.45606	.51246	1.9514	.88995	52
9	.45632	.51283	1.9500	.88981	51
10	.45658	.51319	1.9486	.88968	50
11	.45684	.51356	1.9472	.88955	49
12	.45710	.51393	1.9458	.88942	48
13	.45736	.51430	1.9444	.88928	47
14	.45762	.51467	1.9430	.88915	46
15	.45787	.51503	1.9416	.88902	45
16	.45813	.51540	1.9402	.88888	44
17	.45839	.51577	1.9388	.88875	43
18	.45865	.51614	1.9375	.88862	42
19	.45891	.51651	1.9361	.88848	41
20	.45917	.51688	1.9347	.88835	40
21	.45942	.51724	1.9333	.88822	39
22	.45968	.51761	1.9319	.88808	38
23	.45994	.51798	1.9306	.88795	37
24	.46020	.51835	1.9292	.88782	36
25	.46046	.51872	1.9278	.88768	35
26	.46072	.51909	1.9265	.88755	34
27	.46097	.51946	1.9251	.88741	33
28	.46123	.51983	1.9237	.88728	32
29	.46149	.52020	1.9223	.88715	31
30	.46175	.52057	1.9210	.88701	30
31	.46201	.52094	1.9196	.88688	29
32	.46226	.52131	1.9183	.88674	28
33	.46252	.52168	1.9169	.88661	27
34	.46278	.52205	1.9155	.88647	26
35	.46304	.52242	1.9142	.88634	25
36	.46330	.52279	1.9128	.88620	24
37	.46355	.52316	1.9115	.88607	23
38	.46381	.52353	1.9101	.88593	22
39	.46407	.52390	1.9088	.88580	21
40	.46433	.52427	1.9074	.88566	20
41	.46458	.52464	1.9061	.88553	19
42	.46484	.52501	1.9047	.88539	18
43	.46510	.52538	1.9034	.88526	17
44	.46536	.52575	1.9020	.88512	16
45	.46561	.52613	1.9007	.88499	15
46	.46587	.52650	1.8993	.88485	14
47	.46613	.52687	1.8980	.88472	13
48	.46639	.52724	1.8967	.88458	12
49	.46664	.52761	1.8953	.88445	11
50	.46690	.52798	1.8940	.88431	10
51	.46716	.52836	1.8927	.88417	9
52	.46742	.52873	1.8913	.88404	8
53	.46767	.52910	1.8900	.88390	7
54	.46793	.52947	1.8887	.88377	6
55	.46819	.52985	1.8873	.88363	5
56	.46844	.53022	1.8860	.88349	4
57	.46870	.53059	1.8847	.88336	3
58	.46896	.53096	1.8834	.88322	2
59	.46921	.53134	1.8820	.88308	1
60	.46947	.53171	1.8807	.88295	0
′	Cos	Ctn	Tan	Sin	′

62°

Table I Natural Trigonometric Functions

28°

′	Sin	Tan	Ctn	Cos	′
0	.46947	.53171	1.8807	.88295	60
1	.46973	.53208	1.8794	.88281	59
2	.46999	.53246	1.8781	.88267	58
3	.47024	.53283	1.8768	.88254	57
4	.47050	.53320	1.8755	.88240	56
5	.47076	.53358	1.8741	.88226	55
6	.47101	.53395	1.8728	.88213	54
7	.47127	.53432	1.8715	.88199	53
8	.47153	.53470	1.8702	.88185	52
9	.47178	.53507	1.8689	.88172	51
10	.47204	.53545	1.8676	.88158	50
11	.47229	.53582	1.8663	.88144	49
12	.47255	.53620	1.8650	.88130	48
13	.47281	.53657	1.8637	.88117	47
14	.47306	.53694	1.8624	.88103	46
15	.47332	.53732	1.8611	.88089	45
16	.47358	.53769	1.8598	.88075	44
17	.47383	.53807	1.8585	.88062	43
18	.47409	.53844	1.8572	.88048	42
19	.47434	.53882	1.8559	.88034	41
20	.47460	.53920	1.8546	.88020	40
21	.47486	.53957	1.8533	.88006	39
22	.47511	.53995	1.8520	.87993	38
23	.47537	.54032	1.8507	.87979	37
24	.47562	.54070	1.8495	.87965	36
25	.47588	.54107	1.8482	.87951	35
26	.47614	.54145	1.8469	.87937	34
27	.47639	.54183	1.8456	.87923	33
28	.47665	.54220	1.8443	.87909	32
29	.47690	.54258	1.8430	.87896	31
30	.47716	.54296	1.8418	.87882	30
31	.47741	.54333	1.8405	.87868	29
32	.47767	.54371	1.8392	.87854	28
33	.47793	.54409	1.8379	.87840	27
34	.47818	.54446	1.8367	.87826	26
35	.47844	.54484	1.8354	.87812	25
36	.47869	.54522	1.8341	.87798	24
37	.47895	.54560	1.8329	.87784	23
38	.47920	.54597	1.8316	.87770	22
39	.47946	.54635	1.8303	.87756	21
40	.47971	.54673	1.8291	.87743	20
41	.47997	.54711	1.8278	.87729	19
42	.48022	.54748	1.8265	.87715	18
43	.48048	.54786	1.8253	.87701	17
44	.48073	.54824	1.8240	.87687	16
45	.48099	.54862	1.8228	.87673	15
46	.48124	.54900	1.8215	.87659	14
47	.48150	.54938	1.8202	.87645	13
48	.48175	.54975	1.8190	.87631	12
49	.48201	.55013	1.8177	.87617	11
50	.48226	.55051	1.8165	.87603	10
51	.48252	.55089	1.8152	.87589	9
52	.48277	.55127	1.8140	.87575	8
53	.48303	.55165	1.8127	.87561	7
54	.48328	.55203	1.8115	.87546	6
55	.48354	.55241	1.8103	.87532	5
56	.48379	.55279	1.8090	.87518	4
57	.48405	.55317	1.8078	.87504	3
58	.48430	.55355	1.8065	.87490	2
59	.48456	.55393	1.8053	.87476	1
60	.48481	.55431	1.8040	.87462	0
′	Cos	Ctn	Tan	Sin	′

61°

29°

′	Sin	Tan	Ctn	Cos	′
0	.48481	.55431	1.8040	.87462	60
1	.48506	.55469	1.8028	.87448	59
2	.48532	.55507	1.8016	.87434	58
3	.48557	.55545	1.8003	.87420	57
4	.48583	.55583	1.7991	.87406	56
5	.48608	.55621	1.7979	.87391	55
6	.48634	.55659	1.7966	.87377	54
7	.48659	.55697	1.7954	.87363	53
8	.48684	.55736	1.7942	.87349	52
9	.48710	.55774	1.7930	.87335	51
10	.48735	.55812	1.7917	.87321	50
11	.48761	.55850	1.7905	.87306	49
12	.48786	.55888	1.7893	.87292	48
13	.48811	.55926	1.7881	.87278	47
14	.48837	.55964	1.7868	.87264	46
15	.48862	.56003	1.7856	.87250	45
16	.48888	.56041	1.7844	.87235	44
17	.48913	.56079	1.7832	.87221	43
18	.48938	.56117	1.7820	.87207	42
19	.48964	.56156	1.7808	.87193	41
20	.48989	.56194	1.7796	.87178	40
21	.49014	.56232	1.7783	.87164	39
22	.49040	.56270	1.7771	.87150	38
23	.49065	.56309	1.7759	.87136	37
24	.49090	.56347	1.7747	.87121	36
25	.49116	.56385	1.7735	.87107	35
26	.49141	.56424	1.7723	.87093	34
27	.49166	.56462	1.7711	.87079	33
28	.49192	.56501	1.7699	.87064	32
29	.49217	.56539	1.7687	.87050	31
30	.49242	.56577	1.7675	.87036	30
31	.49268	.56616	1.7663	.87021	29
32	.49293	.56654	1.7651	.87007	28
33	.49318	.56693	1.7639	.86993	27
34	.49344	.56731	1.7627	.86978	26
35	.49369	.56769	1.7615	.86964	25
36	.49394	.56808	1.7603	.86949	24
37	.49419	.56846	1.7591	.86935	23
38	.49445	.56885	1.7579	.86921	22
39	.49470	.56923	1.7567	.86906	21
40	.49495	.56962	1.7556	.86892	20
41	.49521	.57000	1.7544	.86878	19
42	.49546	.57039	1.7532	.86863	18
43	.49571	.57078	1.7520	.86849	17
44	.49596	.57116	1.7508	.86834	16
45	.49622	.57155	1.7496	.86820	15
46	.49647	.57193	1.7485	.86805	14
47	.49672	.57232	1.7473	.86791	13
48	.49697	.57271	1.7461	.86777	12
49	.49723	.57309	1.7449	.86762	11
50	.49748	.57348	1.7437	.86748	10
51	.49773	.57386	1.7426	.86733	9
52	.49798	.57425	1.7414	.86719	8
53	.49824	.57464	1.7402	.86704	7
54	.49849	.57503	1.7391	.86690	6
55	.49874	.57541	1.7379	.86675	5
56	.49899	.57580	1.7367	.86661	4
57	.49924	.57619	1.7355	.86646	3
58	.49950	.57657	1.7344	.86632	2
59	.49975	.57696	1.7332	.86617	1
60	.50000	.57735	1.7321	.86603	0
′	Cos	Ctn	Tan	Sin	′

60°

Table I Natural Trigonometric Functions

30°

′	Sin	Tan	Ctn	Cos	′
0	.50000	.57735	1.7321	.86603	60
1	.50025	.57774	1.7309	.86588	59
2	.50050	.57813	1.7297	.86573	58
3	.50076	.57851	1.7286	.86559	57
4	.50101	.57890	1.7274	.86544	56
5	.50126	.57929	1.7262	.86530	55
6	.50151	.57968	1.7251	.86515	54
7	.50176	.58007	1.7239	.86501	53
8	.50201	.58046	1.7228	.86486	52
9	.50227	.58085	1.7216	.86471	51
10	.50252	.58124	1.7205	.86457	50
11	.50277	.58162	1.7193	.86442	49
12	.50302	.58201	1.7182	.86427	48
13	.50327	.58240	1.7170	.86413	47
14	.50352	.58279	1.7159	.86398	46
15	.50377	.58318	1.7147	.86384	45
16	.50403	.58357	1.7136	.86369	44
17	.50428	.58396	1.7124	.86354	43
18	.50453	.58435	1.7113	.86340	42
19	.50477	.58474	1.7102	.86325	41
20	.50503	.58513	1.7090	.86310	40
21	.50528	.58552	1.7079	.86295	39
22	.50553	.58591	1.7067	.86281	38
23	.50578	.58631	1.7056	.86266	37
24	.50603	.58670	1.7045	.86251	36
25	.50628	.58709	1.7033	.86237	35
26	.50654	.58748	1.7022	.86222	34
27	.50679	.58787	1.7011	.86207	33
28	.50704	.58826	1.6999	.86192	32
29	.50729	.58865	1.6988	.86178	31
30	.50754	.58905	1.6977	.86163	30
31	.50779	.58944	1.6965	.86148	29
32	.50804	.58983	1.6954	.86133	28
33	.50829	.59022	1.6943	.86119	27
34	.50854	.59061	1.6932	.86104	26
35	.50879	.59101	1.6920	.86089	25
36	.50904	.59140	1.6909	.86074	24
37	.50929	.59179	1.6898	.86059	23
38	.50954	.59218	1.6887	.86045	22
39	.50979	.59258	1.6875	.86030	21
40	.51004	.59297	1.6864	.86015	20
41	.51029	.59336	1.6853	.86000	19
42	.51054	.59376	1.6842	.85985	18
43	.51079	.59415	1.6831	.85970	17
44	.51104	.59454	1.6820	.85956	16
45	.51129	.59494	1.6808	.85941	15
46	.51154	.59533	1.6797	.85926	14
47	.51179	.59573	1.6786	.85911	13
48	.51204	.59612	1.6775	.85896	12
49	.51229	.59651	1.6764	.85881	11
50	.51254	.59691	1.6753	.85866	10
51	.51279	.59730	1.6742	.85851	9
52	.51304	.59770	1.6731	.85836	8
53	.51329	.59809	1.6720	.85821	7
54	.51354	.59849	1.6709	.85806	6
55	.51379	.59888	1.6698	.85792	5
56	.51404	.59928	1.6687	.85777	4
57	.51429	.59967	1.6676	.85762	3
58	.51454	.60007	1.6665	.85747	2
59	.51479	.60046	1.6654	.85732	1
60	.51504	.60086	1.6643	.85717	0
′	Cos	Ctn	Tan	Sin	′

59°

31°

′	Sin	Tan	Ctn	Cos	′
0	.51504	.60086	1.6643	.85717	60
1	.51529	.60126	1.6632	.85702	59
2	.51554	.60165	1.6621	.85687	58
3	.51579	.60205	1.6610	.85672	57
4	.51604	.60245	1.6599	.85657	56
5	.51628	.60284	1.6588	.85642	55
6	.51653	.60324	1.6577	.85627	54
7	.51678	.60364	1.6566	.85612	53
8	.51703	.60403	1.6555	.85597	52
9	.51728	.60443	1.6545	.85582	51
10	.51753	.60483	1.6534	.85567	50
11	.51778	.60522	1.6523	.85551	49
12	.51803	.60562	1.6512	.85536	48
13	.51828	.60602	1.6501	.85521	47
14	.51852	.60642	1.6490	.85506	46
15	.51877	.60681	1.6479	.85491	45
16	.51902	.60721	1.6469	.85476	44
17	.51927	.60761	1.6458	.85461	43
18	.51952	.60801	1.6447	.85446	42
19	.51977	.60841	1.6436	.85431	41
20	.52002	.60881	1.6426	.85416	40
21	.52026	.60921	1.6415	.85401	39
22	.52051	.60960	1.6404	.85385	38
23	.52076	.61000	1.6393	.85370	37
24	.52101	.61040	1.6383	.85355	36
25	.52126	.61080	1.6372	.85340	35
26	.52151	.61120	1.6361	.85325	34
27	.52175	.61160	1.6351	.85310	33
28	.52200	.61200	1.6340	.85294	32
29	.52225	.61240	1.6329	.85279	31
30	.52250	.61280	1.6319	.85264	30
31	.52275	.61320	1.6308	.85249	29
32	.52299	.61360	1.6297	.85234	28
33	.52324	.61400	1.6287	.85218	27
34	.52349	.61440	1.6276	.85203	26
35	.52374	.61480	1.6265	.85188	25
36	.52399	.61520	1.6255	.85173	24
37	.52423	.61561	1.6244	.85157	23
38	.52448	.61601	1.6234	.85142	22
39	.52473	.61641	1.6223	.85127	21
40	.52498	.61681	1.6212	.85112	20
41	.52522	.61721	1.6202	.85096	19
42	.52547	.61761	1.6191	.85081	18
43	.52572	.61801	1.6181	.85066	17
44	.52597	.61842	1.6170	.85051	16
45	.52621	.61882	1.6160	.85035	15
46	.52646	.61922	1.6149	.85020	14
47	.52671	.61962	1.6139	.85005	13
48	.52696	.62003	1.6128	.84989	12
49	.52720	.62043	1.6118	.84974	11
50	.52745	.62083	1.6107	.84959	10
51	.52770	.62124	1.6097	.84943	9
52	.52794	.62164	1.6087	.84928	8
53	.52819	.62204	1.6076	.84913	7
54	.52844	.62245	1.6066	.84897	6
55	.52869	.62285	1.6055	.84882	5
56	.52893	.62325	1.6045	.84866	4
57	.52918	.62366	1.6034	.84851	3
58	.52943	.62406	1.6024	.84836	2
59	.52967	.62446	1.6014	.84820	1
60	.52992	.62487	1.6003	.84805	0
′	Cos	Ctn	Tan	Sin	′

58°

Table I Natural Trigonometric Functions

32°

'	Sin	Tan	Ctn	Cos	'
0	.52992	.62487	1.6003	.84805	60
1	.53017	.62527	1.5993	.84789	59
2	.53041	.62568	1.5983	.84774	58
3	.53066	.62608	1.5972	.84759	57
4	.53091	.62649	1.5962	.84743	56
5	.53115	.62689	1.5952	.84728	55
6	.53140	.62730	1.5941	.84712	54
7	.53164	.62770	1.5931	.84697	53
8	.53189	.62811	1.5921	.84681	52
9	.53214	.62852	1.5911	.84666	51
10	.53238	.62892	1.5900	.84650	50
11	.53263	.62933	1.5890	.84635	49
12	.53288	.62973	1.5880	.84619	48
13	.53312	.63014	1.5869	.84604	47
14	.53337	.63055	1.5859	.84588	46
15	.53361	.63095	1.5849	.84573	45
16	.53386	.63136	1.5839	.84557	44
17	.53411	.63177	1.5829	.84542	43
18	.53435	.63217	1.5818	.84526	42
19	.53460	.63258	1.5808	.84511	41
20	.53484	.63299	1.5798	.84495	40
21	.53509	.63340	1.5788	.84480	39
22	.53534	.63380	1.5778	.84464	38
23	.53558	.63421	1.5768	.84448	37
24	.53583	.63462	1.5757	.84433	36
25	.53607	.63503	1.5747	.84417	35
26	.53632	.63544	1.5737	.84402	34
27	.53656	.63584	1.5727	.84386	33
28	.53681	.63625	1.5717	.84370	32
29	.53705	.63666	1.5707	.84355	31
30	.53730	.63707	1.5697	.84339	30
31	.53754	.63748	1.5687	.84324	29
32	.53779	.63789	1.5677	.84308	28
33	.53804	.63830	1.5667	.84292	27
34	.53828	.63871	1.5657	.84277	26
35	.53853	.63912	1.5647	.84261	25
36	.53877	.63953	1.5637	.84245	24
37	.53902	.63994	1.5627	.84230	23
38	.53926	.64035	1.5617	.84214	22
39	.53951	.64076	1.5607	.84198	21
40	.53975	.64117	1.5597	.84182	20
41	.54000	.64158	1.5587	.84167	19
42	.54024	.64199	1.5577	.84151	18
43	.54049	.64240	1.5567	.84135	17
44	.54073	.64281	1.5557	.84120	16
45	.54097	.64322	1.5547	.84104	15
46	.54122	.64363	1.5537	.84088	14
47	.54146	.64404	1.5527	.84072	13
48	.54171	.64446	1.5517	.84057	12
49	.54195	.64487	1.5507	.84041	11
50	.54220	.64528	1.5497	.84025	10
51	.54244	.64569	1.5487	.84009	9
52	.54269	.64610	1.5477	.83994	8
53	.54293	.64652	1.5468	.83978	7
54	.54317	.64693	1.5458	.83962	6
55	.54342	.64734	1.5448	.83946	5
56	.54366	.64775	1.5438	.83930	4
57	.54391	.64817	1.5428	.83915	3
58	.54415	.64858	1.5418	.83899	2
59	.54440	.64899	1.5408	.83883	1
60	.54464	.64941	1.5399	.83867	0
'	Cos	Ctn	Tan	Sin	'

57°

33°

'	Sin	Tan	Ctn	Cos	'
0	.54464	.64941	1.5399	.83867	60
1	.54488	.64982	1.5389	.83851	59
2	.54513	.65024	1.5379	.83835	58
3	.54537	.65065	1.5369	.83819	57
4	.54561	.65106	1.5359	.83804	56
5	.54586	.65148	1.5350	.83788	55
6	.54610	.65189	1.5340	.83772	54
7	.54635	.65231	1.5330	.83756	53
8	.54659	.65272	1.5320	.83740	52
9	.54683	.65314	1.5311	.83724	51
10	.54708	.65355	1.5301	.83708	50
11	.54732	.65397	1.5291	.83692	49
12	.54756	.65438	1.5282	.83676	48
13	.54781	.65480	1.5272	.83660	47
14	.54805	.65521	1.5262	.83645	46
15	.54829	.65563	1.5253	.83629	45
16	.54854	.65604	1.5243	.83613	44
17	.54878	.65646	1.5233	.83597	43
18	.54902	.65688	1.5224	.83581	42
19	.54927	.65729	1.5214	.83565	41
20	.54951	.65771	1.5204	.83549	40
21	.54975	.65813	1.5195	.83533	39
22	.54999	.65854	1.5185	.83517	38
23	.55024	.65896	1.5175	.83501	37
24	.55048	.65938	1.5166	.83485	36
25	.55072	.65980	1.5156	.83469	35
26	.55097	.66021	1.5147	.83453	34
27	.55121	.66063	1.5137	.83437	33
28	.55145	.66105	1.5127	.83421	32
29	.55169	.66147	1.5118	.83405	31
30	.55194	.66189	1.5108	.83389	30
31	.55218	.66230	1.5099	.83373	29
32	.55242	.66272	1.5089	.83356	28
33	.55266	.66314	1.5080	.83340	27
34	.55291	.66356	1.5070	.83324	26
35	.55315	.66398	1.5061	.83308	25
36	.55339	.66440	1.5051	.83292	24
37	.55363	.66482	1.5042	.83276	23
38	.55388	.66524	1.5032	.83260	22
39	.55412	.66566	1.5023	.83244	21
40	.55436	.66608	1.5013	.83228	20
41	.55460	.66650	1.5004	.83212	19
42	.55484	.66692	1.4994	.83195	18
43	.55509	.66734	1.4985	.83179	17
44	.55533	.66776	1.4975	.83163	16
45	.55557	.66818	1.4966	.83147	15
46	.55581	.66860	1.4957	.83131	14
47	.55605	.66902	1.4947	.83115	13
48	.55630	.66944	1.4938	.83098	12
49	.55654	.66986	1.4928	.83082	11
50	.55678	.67028	1.4919	.83066	10
51	.55702	.67071	1.4910	.83050	9
52	.55726	.67113	1.4900	.83034	8
53	.55750	.67155	1.4891	.83017	7
54	.55775	.67197	1.4882	.83001	6
55	.55799	.67239	1.4872	.82985	5
56	.55823	.67282	1.4863	.82969	4
57	.55847	.67324	1.4854	.82953	3
58	.55871	.67366	1.4844	.82936	2
59	.55895	.67409	1.4835	.82920	1
60	.55919	.67451	1.4826	.82904	0
'	Cos	Ctn	Tan	Sin	'

56°

Table I Natural Trigonometric Functions

34°

′	Sin	Tan	Ctn	Cos	′
0	.55919	.67451	1.4826	.82904	**60**
1	.55943	.67493	1.4816	.82887	59
2	.55968	.67536	1.4807	.82871	58
3	.55992	.67578	1.4798	.82855	57
4	.56016	.67620	1.4788	.82839	56
5	.56040	.67663	1.4779	.82822	**55**
6	.56064	.67705	1.4770	.82806	54
7	.56088	.67748	1.4761	.82790	53
8	.56112	.67790	1.4751	.82773	52
9	.56136	.67832	1.4742	.82757	51
10	.56160	.67875	1.4733	.82741	**50**
11	.56184	.67917	1.4724	.82724	49
12	.56208	.67960	1.4715	.82708	48
13	.56232	.68002	1.4705	.82692	47
14	.56256	.68045	1.4696	.82675	46
15	.56280	.68088	1.4687	.82659	**45**
16	.56305	.68130	1.4678	.82643	44
17	.56329	.68173	1.4669	.82626	43
18	.56353	.68215	1.4659	.82610	42
19	.56377	.68258	1.4650	.82593	41
20	.56401	.68301	1.4641	.82577	**40**
21	.56425	.68343	1.4632	.82561	39
22	.56449	.68386	1.4623	.82544	38
23	.56473	.68429	1.4614	.82528	37
24	.56497	.68471	1.4605	.82511	36
25	.56521	.68514	1.4596	.82495	**35**
26	.56545	.68557	1.4586	.82478	34
27	.56569	.68600	1.4577	.82462	33
28	.56593	.68642	1.4568	.82446	32
29	.56617	.68685	1.4559	.82429	31
30	.56641	.68728	1.4550	.82413	**30**
31	.56665	.68771	1.4541	.82396	29
32	.56689	.68814	1.4532	.82380	28
33	.56713	.68857	1.4523	.82363	27
34	.56736	.68900	1.4514	.82347	26
35	.56760	.68942	1.4505	.82330	**25**
36	.56784	.68985	1.4496	.82314	24
37	.56808	.69028	1.4487	.82297	23
38	.56832	.69071	1.4478	.82281	22
39	.56856	.69114	1.4469	.82264	21
40	.56880	.69157	1.4460	.82248	**20**
41	.56904	.69200	1.4451	.82231	19
42	.56928	.69243	1.4442	.82214	18
43	.56952	.69286	1.4433	.82198	17
44	.56976	.69329	1.4424	.82181	16
45	.57000	.69372	1.4415	.82165	**15**
46	.57024	.69416	1.4406	.82148	14
47	.57047	.69459	1.4397	.82132	13
48	.57071	.69502	1.4388	.82115	12
49	.57095	.69545	1.4379	.82098	11
50	.57119	.69588	1.4370	.82082	**10**
51	.57143	.69631	1.4361	.82065	9
52	.57167	.69675	1.4352	.82048	8
53	.57191	.69718	1.4344	.82032	7
54	.57215	.69761	1.4335	.82015	6
55	.57238	.69804	1.4326	.81999	**5**
56	.57262	.69847	1.4317	.81982	4
57	.57286	.69891	1.4308	.81965	3
58	.57310	.69934	1.4299	.81949	2
59	.57334	.69977	1.4290	.81932	1
60	.57358	.70021	1.4281	.81915	**0**
′	Cos	Ctn	Tan	Sin	′

55°

35°

′	Sin	Tan	Ctn	Cos	′
0	.57358	.70021	1.4281	.81915	**60**
1	.57381	.70064	1.4273	.81899	59
2	.57405	.70107	1.4264	.81882	58
3	.57429	.70151	1.4255	.81865	57
4	.57453	.70194	1.4246	.81848	56
5	.57477	.70238	1.4237	.81832	**55**
6	.57501	.70281	1.4229	.81815	54
7	.57524	.70325	1.4220	.81798	53
8	.57548	.70368	1.4211	.81782	52
9	.57572	.70412	1.4202	.81765	51
10	.57596	.70455	1.4193	.81748	**50**
11	.57619	.70499	1.4185	.81731	49
12	.57643	.70542	1.4176	.81714	48
13	.57667	.70586	1.4167	.81698	47
14	.57691	.70629	1.4158	.81681	46
15	.57715	.70673	1.4150	.81664	**45**
16	.57738	.70717	1.4141	.81647	44
17	.57762	.70760	1.4132	.81631	43
18	.57786	.70804	1.4124	.81614	42
19	.57810	.70848	1.4115	.81597	41
20	.57833	.70891	1.4106	.81580	**40**
21	.57857	.70935	1.4097	.81563	39
22	.57881	.70979	1.4089	.81546	38
23	.57904	.71023	1.4080	.81530	37
24	.57928	.71066	1.4071	.81513	36
25	.57952	.71110	1.4063	.81496	**35**
26	.57976	.71154	1.4054	.81479	34
27	.57999	.71198	1.4045	.81462	33
28	.58023	.71242	1.4037	.81445	32
29	.58047	.71285	1.4028	.81428	31
30	.58070	.71329	1.4019	.81412	**30**
31	.58094	.71373	1.4011	.81395	29
32	.58118	.71417	1.4002	.81378	28
33	.58141	.71461	1.3994	.81361	27
34	.58165	.71505	1.3985	.81344	26
35	.58189	.71549	1.3976	.81327	**25**
36	.58212	.71593	1.3968	.81310	24
37	.58236	.71637	1.3959	.81293	23
38	.58260	.71681	1.3951	.81276	22
39	.58283	.71725	1.3942	.81259	21
40	.58307	.71769	1.3934	.81242	**20**
41	.58330	.71813	1.3925	.81225	19
42	.58354	.71857	1.3916	.81208	18
43	.58378	.71901	1.3908	.81191	17
44	.58401	.71946	1.3899	.81174	16
45	.58425	.71990	1.3891	.81157	**15**
46	.58449	.72034	1.3882	.81140	14
47	.58472	.72078	1.3874	.81123	13
48	.58496	.72122	1.3865	.81106	12
49	.58519	.72167	1.3857	.81089	11
50	.58543	.72211	1.3848	.81072	**10**
51	.58567	.72255	1.3840	.81055	9
52	.58590	.72299	1.3831	.81038	8
53	.58614	.72344	1.3823	.81021	7
54	.58637	.72388	1.3814	.81004	6
55	.58661	.72432	1.3806	.80987	**5**
56	.58684	.72477	1.3798	.80970	4
57	.58708	.72521	1.3789	.80953	3
58	.58731	.72565	1.3781	.80936	2
59	.58755	.72610	1.3772	.80919	1
60	.58779	.72654	1.3764	.80902	**0**
′	Cos	Ctn	Tan	Sin	′

54°

Table I Natural Trigonometric Functions

36°

′	Sin	Tan	Ctn	Cos	′
0	.58779	.72654	1.3764	.80902	60
1	.58802	.72699	1.3755	.80885	59
2	.58826	.72743	1.3747	.80867	58
3	.58849	.72788	1.3739	.80850	57
4	.58873	.72832	1.3730	.80833	56
5	.58896	.72877	1.3722	.80816	55
6	.58920	.72921	1.3713	.80799	54
7	.58943	.72966	1.3705	.80782	53
8	.58967	.73010	1.3697	.80765	52
9	.58990	.73055	1.3688	.80748	51
10	.59014	.73100	1.3680	.80730	50
11	.59037	.73144	1.3672	.80713	49
12	.59061	.73189	1.3663	.80696	48
13	.59084	.73234	1.3655	.80679	47
14	.59108	.73278	1.3647	.80662	46
15	.59131	.73323	1.3638	.80644	45
16	.59154	.73368	1.3630	.80627	44
17	.59178	.73413	1.3622	.80610	43
18	.59201	.73457	1.3613	.80593	42
19	.59225	.73502	1.3605	.80576	41
20	.59248	.73547	1.3597	.80558	40
21	.59272	.73592	1.3588	.80541	39
22	.59295	.73637	1.3580	.80524	38
23	.59318	.73681	1.3572	.80507	37
24	.59342	.73726	1.3564	.80489	36
25	.59365	.73771	1.3555	.80472	35
26	.59389	.73816	1.3547	.80455	34
27	.59412	.73861	1.3539	.80438	33
28	.59436	.73906	1.3531	.80420	32
29	.59459	.73951	1.3522	.80403	31
30	.59482	.73996	1.3514	.80386	30
31	.59506	.74041	1.3506	.80368	29
32	.59529	.74086	1.3498	.80351	28
33	.59552	.74131	1.3490	.80334	27
34	.59576	.74176	1.3481	.80316	26
35	.59599	.74221	1.3473	.80299	25
36	.59622	.74267	1.3465	.80282	24
37	.59646	.74312	1.3457	.80264	23
38	.59669	.74357	1.3449	.80247	22
39	.59693	.74402	1.3440	.80230	21
40	.59716	.74447	1.3432	.80212	20
41	.59739	.74492	1.3424	.80195	19
42	.59763	.74538	1.3416	.80178	18
43	.59786	.74583	1.3408	.80160	17
44	.59809	.74628	1.3400	.80143	16
45	.59832	.74674	1.3392	.80125	15
46	.59856	.74719	1.3384	.80108	14
47	.59879	.74764	1.3375	.80091	13
48	.59902	.74810	1.3367	.80073	12
49	.59926	.74855	1.3359	.80056	11
50	.59949	.74900	1.3351	.80038	10
51	.59972	.74946	1.3343	.80021	9
52	.59995	.74991	1.3335	.80003	8
53	.60019	.75037	1.3327	.79986	7
54	.60042	.75082	1.3319	.79968	6
55	.60065	.75128	1.3311	.79951	5
56	.60089	.75173	1.3303	.79934	4
57	.60112	.75219	1.3295	.79916	3
58	.60135	.75264	1.3287	.79899	2
59	.60158	.75310	1.3278	.79881	1
60	.60182	.75355	1.3270	.79864	0
′	Cos	Ctn	Tan	Sin	′

53°

37°

′	Sin	Tan	Ctn	Cos	′
0	.60182	.75355	1.3270	.79864	60
1	.60205	.75401	1.3262	.79846	59
2	.60228	.75447	1.3254	.79829	58
3	.60251	.75492	1.3246	.79811	57
4	.60274	.75538	1.3238	.79793	56
5	.60298	.75584	1.3230	.79776	55
6	.60321	.75629	1.3222	.79758	54
7	.60344	.75675	1.3214	.79741	53
8	.60367	.75721	1.3206	.79723	52
9	.60390	.75767	1.3198	.79706	51
10	.60414	.75812	1.3190	.79688	50
11	.60437	.75858	1.3182	.79671	49
12	.60460	.75904	1.3175	.79653	48
13	.60483	.75950	1.3167	.79635	47
14	.60506	.75996	1.3159	.79618	46
15	.60529	.76042	1.3151	.79600	45
16	.60553	.76088	1.3143	.79583	44
17	.60576	.76134	1.3135	.79565	43
18	.60599	.76180	1.3127	.79547	42
19	.60622	.76226	1.3119	.79530	41
20	.60645	.76272	1.3111	.79512	40
21	.60668	.76318	1.3103	.79494	39
22	.60691	.76364	1.3095	.79477	38
23	.60714	.76410	1.3087	.79459	37
24	.60738	.76456	1.3079	.79441	36
25	.60761	.76502	1.3072	.79424	35
26	.60784	.76548	1.3064	.79406	34
27	.60807	.76594	1.3056	.79388	33
28	.60830	.76640	1.3048	.79371	32
29	.60853	.76686	1.3040	.79353	31
30	.60876	.76733	1.3032	.79335	30
31	.60899	.76779	1.3024	.79318	29
32	.60922	.76825	1.3017	.79300	28
33	.60945	.76871	1.3009	.79282	27
34	.60968	.76918	1.3001	.79264	26
35	.60991	.76964	1.2993	.79247	25
36	.61015	.77010	1.2985	.79229	24
37	.61038	.77057	1.2977	.79211	23
38	.61061	.77103	1.2970	.79193	22
39	.61084	.77149	1.2962	.79176	21
40	.61107	.77196	1.2954	.79158	20
41	.61130	.77242	1.2946	.79140	19
42	.61153	.77289	1.2938	.79122	18
43	.61176	.77335	1.2931	.79105	17
44	.61199	.77382	1.2923	.79087	16
45	.61222	.77428	1.2915	.79069	15
46	.61245	.77475	1.2907	.79051	14
47	.61268	.77521	1.2900	.79033	13
48	.61291	.77568	1.2892	.79016	12
49	.61314	.77615	1.2884	.78998	11
50	.61337	.77661	1.2876	.78980	10
51	.61360	.77708	1.2869	.78962	9
52	.61383	.77754	1.2861	.78944	8
53	.61406	.77801	1.2853	.78926	7
54	.61429	.77848	1.2846	.78908	6
55	.61451	.77895	1.2838	.78891	5
56	.61474	.77941	1.2830	.78873	4
57	.61497	.77988	1.2822	.78855	3
58	.61520	.78035	1.2815	.78837	2
59	.61543	.78082	1.2807	.78819	1
60	.61566	.78129	1.2799	.78801	0
′	Cos	Ctn	Tan	Sin	′

52°

Table I Natural Trigonometric Functions

38°

′	Sin	Tan	Ctn	Cos	′
0	.61566	.78129	1.2799	.78801	60
1	.61589	.78175	1.2792	.78783	59
2	.61612	.78222	1.2784	.78765	58
3	.61635	.78269	1.2776	.78747	57
4	.61658	.78316	1.2769	.78729	56
5	.61681	.78363	1.2761	.78711	55
6	.61704	.78410	1.2753	.78694	54
7	.61726	.78457	1.2746	.78676	53
8	.61749	.78504	1.2738	.78658	52
9	.61772	.78551	1.2731	.78640	51
10	.61795	.78598	1.2723	.78622	50
11	.61818	.78645	1.2715	.78604	49
12	.61841	.78692	1.2708	.78586	48
13	.61864	.78739	1.2700	.78568	47
14	.61887	.78786	1.2693	.78550	46
15	.61909	.78834	1.2685	.78532	45
16	.61932	.78881	1.2677	.78514	44
17	.61955	.78928	1.2670	.78496	43
18	.61978	.78975	1.2662	.78478	42
19	.62001	.79022	1.2655	.78460	41
20	.62024	.79070	1.2647	.78442	40
21	.62046	.79117	1.2640	.78424	39
22	.62069	.79164	1.2632	.78405	38
23	.62092	.79212	1.2624	.78387	37
24	.62115	.79259	1.2617	.78369	36
25	.62138	.79306	1.2609	.78351	35
26	.62160	.79354	1.2602	.78333	34
27	.62183	.79401	1.2594	.78315	33
28	.62206	.79449	1.2587	.78297	32
29	.62229	.79496	1.2579	.78279	31
30	.62251	.79544	1.2572	.78261	30
31	.62274	.79591	1.2564	.78243	29
32	.62297	.79639	1.2557	.78225	28
33	.62320	.79686	1.2549	.78206	27
34	.62342	.79734	1.2542	.78188	26
35	.62365	.79781	1.2534	.78170	25
36	.62388	.79829	1.2527	.78152	24
37	.62411	.79877	1.2519	.78134	23
38	.62433	.79924	1.2512	.78116	22
39	.62456	.79972	1.2504	.78098	21
40	.62479	.80020	1.2497	.78079	20
41	.62502	.80067	1.2489	.78061	19
42	.62524	.80115	1.2482	.78043	18
43	.62547	.80163	1.2475	.78025	17
44	.62570	.80211	1.2467	.78007	16
45	.62592	.80258	1.2460	.77988	15
46	.62615	.80306	1.2452	.77970	14
47	.62638	.80354	1.2445	.77952	13
48	.62660	.80402	1.2437	.77934	12
49	.62683	.80450	1.2430	.77916	11
50	.62706	.80498	1.2423	.77897	10
51	.62728	.80546	1.2415	.77879	9
52	.62751	.80594	1.2408	.77861	8
53	.62774	.80642	1.2401	.77843	7
54	.62796	.80690	1.2393	.77824	6
55	.62819	.80738	1.2386	.77806	5
56	.62842	.80786	1.2378	.77788	4
57	.62864	.80834	1.2371	.77769	3
58	.62887	.80882	1.2364	.77751	2
59	.62909	.80930	1.2356	.77733	1
60	.62932	.80978	1.2349	.77715	0
′	Cos	Ctn	Tan	Sin	′

51°

39°

′	Sin	Tan	Ctn	Cos	′
0	.62932	.80978	1.2349	.77715	60
1	.62955	.81027	1.2342	.77696	59
2	.62977	.81075	1.2334	.77678	58
3	.63000	.81123	1.2327	.77660	57
4	.63022	.81171	1.2320	.77641	56
5	.63045	.81220	1.2312	.77623	55
6	.63068	.81268	1.2305	.77605	54
7	.63090	.81316	1.2298	.77586	53
8	.63113	.81364	1.2290	.77568	52
9	.63135	.81413	1.2283	.77550	51
10	.63158	.81461	1.2276	.77531	50
11	.63180	.81510	1.2268	.77513	49
12	.63203	.81558	1.2261	.77494	48
13	.63225	.81606	1.2254	.77476	47
14	.63248	.81655	1.2247	.77458	46
15	.63271	.81703	1.2239	.77439	45
16	.63293	.81752	1.2232	.77421	44
17	.63316	.81800	1.2225	.77402	43
18	.63338	.81849	1.2218	.77384	42
19	.63361	.81898	1.2210	.77366	41
20	.63383	.81946	1.2203	.77347	40
21	.63406	.81995	1.2196	.77329	39
22	.63428	.82044	1.2189	.77310	38
23	.63451	.82092	1.2181	.77292	37
24	.63473	.82141	1.2174	.77273	36
25	.63496	.82190	1.2167	.77255	35
26	.63518	.82238	1.2160	.77236	34
27	.63540	.82287	1.2153	.77218	33
28	.63563	.82336	1.2145	.77199	32
29	.63585	.82385	1.2138	.77181	31
30	.63608	.82434	1.2131	.77162	30
31	.63630	.82483	1.2124	.77144	29
32	.63653	.82531	1.2117	.77125	28
33	.63675	.82580	1.2109	.77107	27
34	.63698	.82629	1.2102	.77088	26
35	.63720	.82678	1.2095	.77070	25
36	.63742	.82727	1.2088	.77051	24
37	.63765	.82776	1.2081	.77033	23
38	.63787	.82825	1.2074	.77014	22
39	.63810	.82874	1.2066	.76996	21
40	.63832	.82923	1.2059	.76977	20
41	.63854	.82972	1.2052	.76959	19
42	.63877	.83022	1.2045	.76940	18
43	.63899	.83071	1.2038	.76921	17
44	.63922	.83120	1.2031	.76903	16
45	.63944	.83169	1.2024	.76884	15
46	.63966	.83218	1.2017	.76866	14
47	.63989	.83268	1.2009	.76847	13
48	.64011	.83317	1.2002	.76828	12
49	.64033	.83366	1.1995	.76810	11
50	.64056	.83415	1.1988	.76791	10
51	.64078	.83465	1.1981	.76772	9
52	.64100	.83514	1.1974	.76754	8
53	.64123	.83564	1.1967	.76735	7
54	.64145	.83613	1.1960	.76717	6
55	.64167	.83662	1.1953	.76698	5
56	.64190	.83712	1.1946	.76679	4
57	.64212	.83761	1.1939	.76661	3
58	.64234	.83811	1.1932	.76642	2
59	.64256	.83860	1.1925	.76623	1
60	.64279	.83910	1.1918	.76604	0
′	Cos	Ctn	Tan	Sin	′

50°

Table I   Natural Trigonometric Functions

40°

′	Sin	Tan	Ctn	Cos	′
0	.64279	.83910	1.1918	.76604	60
1	.64301	.83960	1.1910	.76586	59
2	.64323	.84009	1.1903	.76567	58
3	.64346	.84059	1.1896	.76548	57
4	.64368	.84108	1.1889	.76530	56
5	.64390	.84158	1.1882	.76511	55
6	.64412	.84208	1.1875	.76492	54
7	.64435	.84258	1.1868	.76473	53
8	.64457	.84307	1.1861	.76455	52
9	.64479	.84357	1.1854	.76436	51
10	.64501	.84407	1.1847	.76417	50
11	.64524	.84457	1.1840	.76398	49
12	.64546	.84507	1.1833	.76380	48
13	.64568	.84556	1.1826	.76361	47
14	.64590	.84606	1.1819	.76342	46
15	.64612	.84656	1.1812	.76323	45
16	.64635	.84706	1.1806	.76304	44
17	.64657	.84756	1.1799	.76286	43
18	.64679	.84806	1.1792	.76267	42
19	.64701	.84856	1.1785	.76248	41
20	.64723	.84906	1.1778	.76229	40
21	.64746	.84956	1.1771	.76210	39
22	.64768	.85006	1.1764	.76192	38
23	.64790	.85057	1.1757	.76173	37
24	.64812	.85107	1.1750	.76154	36
25	.64834	.85157	1.1743	.76135	35
26	.64856	.85207	1.1736	.76116	34
27	.64878	.85257	1.1729	.76097	33
28	.64901	.85308	1.1722	.76078	32
29	.64923	.85358	1.1715	.76059	31
30	.64945	.85408	1.1708	.76041	30
31	.64967	.85458	1.1702	.76022	29
32	.64989	.85509	1.1695	.76003	28
33	.65011	.85559	1.1688	.75984	27
34	.65033	.85609	1.1681	.75965	26
35	.65055	.85660	1.1674	.75946	25
36	.65077	.85710	1.1667	.75927	24
37	.65100	.85761	1.1660	.75908	23
38	.65122	.85811	1.1653	.75889	22
39	.65144	.85862	1.1647	.75870	21
40	.65166	.85912	1.1640	.75851	20
41	.65188	.85963	1.1633	.75832	19
42	.65210	.86014	1.1626	.75813	18
43	.65232	.86064	1.1619	.75794	17
44	.65254	.86115	1.1612	.75775	16
45	.65276	.86166	1.1606	.75756	15
46	.65298	.86216	1.1599	.75738	14
47	.65320	.86267	1.1592	.75719	13
48	.65342	.86318	1.1585	.75700	12
49	.65364	.86368	1.1578	.75680	11
50	.65386	.86419	1.1571	.75661	10
51	.65408	.86470	1.1565	.75642	9
52	.65430	.86521	1.1558	.75623	8
53	.65452	.86572	1.1551	.75604	7
54	.65474	.86623	1.1544	.75585	6
55	.65496	.86674	1.1538	.75566	5
56	.65518	.86725	1.1531	.75547	4
57	.65540	.86776	1.1524	.75528	3
58	.65562	.86827	1.1517	.75509	2
59	.65584	.86878	1.1510	.75490	1
60	.65606	.86929	1.1504	.75471	0
′	Cos	Ctn	Tan	Sin	′

49°

41°

′	Sin	Tan	Ctn	Cos	′
0	.65606	.86929	1.1504	.75471	60
1	.65628	.86980	1.1497	.75452	59
2	.65650	.87031	1.1490	.75433	58
3	.65672	.87082	1.1483	.75414	57
4	.65694	.87133	1.1477	.75395	56
5	.65716	.87184	1.1470	.75375	55
6	.65738	.87236	1.1463	.75356	54
7	.65759	.87287	1.1456	.75337	53
8	.65781	.87338	1.1450	.75318	52
9	.65803	.87389	1.1443	.75299	51
10	.65825	.87441	1.1436	.75280	50
11	.65847	.87492	1.1430	.75261	49
12	.65869	.87543	1.1423	.75241	48
13	.65891	.87595	1.1416	.75222	47
14	.65913	.87646	1.1410	.75203	46
15	.65935	.87698	1.1403	.75184	45
16	.65956	.87749	1.1396	.75165	44
17	.65978	.87801	1.1389	.75146	43
18	.66000	.87852	1.1383	.75126	42
19	.66022	.87904	1.1376	.75107	41
20	.66044	.87955	1.1369	.75088	40
21	.66066	.88007	1.1363	.75069	39
22	.66088	.88059	1.1356	.75050	38
23	.66109	.88110	1.1349	.75030	37
24	.66131	.88162	1.1343	.75011	36
25	.66153	.88214	1.1336	.74992	35
26	.66175	.88265	1.1329	.74973	34
27	.66197	.88317	1.1323	.74953	33
28	.66218	.88369	1.1316	.74934	32
29	.66240	.88421	1.1310	.74915	31
30	.66262	.88473	1.1303	.74896	30
31	.66284	.88524	1.1296	.74876	29
32	.66306	.88576	1.1290	.74857	28
33	.66327	.88628	1.1283	.74838	27
34	.66349	.88680	1.1276	.74818	26
35	.66371	.88732	1.1270	.74799	25
36	.66393	.88784	1.1263	.74780	24
37	.66414	.88836	1.1257	.74760	23
38	.66436	.88888	1.1250	.74741	22
39	.66458	.88940	1.1243	.74722	21
40	.66480	.88992	1.1237	.74703	20
41	.66501	.89045	1.1230	.74683	19
42	.66523	.89097	1.1224	.74664	18
43	.66545	.89149	1.1217	.74644	17
44	.66566	.89201	1.1211	.74625	16
45	.66588	.89253	1.1204	.74606	15
46	.66610	.89306	1.1197	.74586	14
47	.66632	.89358	1.1191	.74567	13
48	.66653	.89410	1.1184	.74548	12
49	.66675	.89463	1.1178	.74528	11
50	.66697	.89515	1.1171	.74509	10
51	.66718	.89567	1.1165	.74489	9
52	.66740	.89620	1.1158	.74470	8
53	.66762	.89672	1.1152	.74451	7
54	.66783	.89725	1.1145	.74431	6
55	.66805	.89777	1.1139	.74412	5
56	.66827	.89830	1.1132	.74392	4
57	.66848	.89883	1.1126	.74373	3
58	.66870	.89935	1.1119	.74353	2
59	.66891	.89988	1.1113	.74334	1
60	.66913	.90040	1.1106	.74314	0
′	Cos	Ctn	Tan	Sin	′

48°

Table I Natural Trigonometric Functions

42°

′	Sin	Tan	Ctn	Cos	′
0	.66913	.90040	1.1106	.74314	60
1	.66935	.90093	1.1100	.74295	59
2	.66956	.90146	1.1093	.74276	58
3	.66978	.90199	1.1087	.74256	57
4	.66999	.90251	1.1080	.74237	56
5	.67021	.90304	1.1074	.74217	55
6	.67043	.90357	1.1067	.74198	54
7	.67064	.90410	1.1061	.74178	53
8	.67086	.90463	1.1054	.74159	52
9	.67107	.90516	1.1048	.74139	51
10	.67129	.90569	1.1041	.74120	50
11	.67151	.90621	1.1035	.74100	49
12	.67172	.90674	1.1028	.74080	48
13	.67194	.90727	1.1022	.74061	47
14	.67215	.90781	1.1016	.74041	46
15	.67237	.90834	1.1009	.74022	45
16	.67258	.90887	1.1003	.74002	44
17	.67280	.90940	1.0996	.73983	43
18	.67301	.90993	1.0990	.73963	42
19	.67323	.91046	1.0983	.73944	41
20	.67344	.91099	1.0977	.73924	40
21	.67366	.91153	1.0971	.73904	39
22	.67387	.91206	1.0964	.73885	38
23	.67409	.91259	1.0958	.73865	37
24	.67430	.91313	1.0951	.73846	36
25	.67452	.91366	1.0945	.73826	35
26	.67473	.91419	1.0939	.73806	34
27	.67495	.91473	1.0932	.73787	33
28	.67516	.91526	1.0926	.73767	32
29	.67538	.91580	1.0919	.73747	31
30	.67559	.91633	1.0913	.73728	30
31	.67580	.91687	1.0907	.73708	29
32	.67602	.91740	1.0900	.73688	28
33	.67623	.91794	1.0894	.73669	27
34	.67645	.91847	1.0888	.73649	26
35	.67666	.91901	1.0881	.73629	25
36	.67688	.91955	1.0875	.73610	24
37	.67709	.92008	1.0869	.73590	23
38	.67730	.92062	1.0862	.73570	22
39	.67752	.92116	1.0856	.73551	21
40	.67773	.92170	1.0850	.73531	20
41	.67795	.92224	1.0843	.73511	19
42	.67816	.92277	1.0837	.73491	18
43	.67837	.92331	1.0831	.73472	17
44	.67859	.92385	1.0824	.73452	16
45	.67880	.92439	1.0818	.73432	15
46	.67901	.92493	1.0812	.73413	14
47	.67923	.92547	1.0805	.73393	13
48	.67944	.92601	1.0799	.73373	12
49	.67965	.92655	1.0793	.73353	11
50	.67987	.92709	1.0786	.73333	10
51	.68008	.92763	1.0780	.73314	9
52	.68029	.92817	1.0774	.73294	8
53	.68051	.92872	1.0768	.73274	7
54	.68072	.92926	1.0761	.73254	6
55	.68093	.92980	1.0755	.73234	5
56	.68115	.93034	1.0749	.73215	4
57	.68136	.93088	1.0742	.73195	3
58	.68157	.93143	1.0736	.73175	2
59	.68179	.93197	1.0730	.73155	1
60	.68200	.93252	1.0724	.73135	0
′	Cos	Ctn	Tan	Sin	′

47°

43°

′	Sin	Tan	Ctn	Cos	′
0	.68200	.93252	1.0724	.73135	60
1	.68221	.93306	1.0717	.73116	59
2	.68242	.93360	1.0711	.73096	58
3	.68264	.93415	1.0705	.73076	57
4	.68285	.93469	1.0699	.73056	56
5	.68306	.93524	1.0692	.73036	55
6	.68327	.93578	1.0686	.73016	54
7	.68349	.93633	1.0680	.72996	53
8	.68370	.93688	1.0674	.72976	52
9	.68391	.93742	1.0668	.72957	51
10	.68412	.93797	1.0661	.72937	50
11	.68434	.93852	1.0655	.72917	49
12	.68455	.93906	1.0649	.72897	48
13	.68476	.93961	1.0643	.72877	47
14	.68497	.94016	1.0637	.72857	46
15	.68518	.94071	1.0630	.72837	45
16	.68539	.94125	1.0624	.72817	44
17	.68561	.94180	1.0618	.72797	43
18	.68582	.94235	1.0612	.72777	42
19	.68603	.94290	1.0606	.72757	41
20	.68624	.94345	1.0599	.72737	40
21	.68645	.94400	1.0593	.72717	39
22	.68666	.94455	1.0587	.72697	38
23	.68688	.94510	1.0581	.72677	37
24	.68709	.94565	1.0575	.72657	36
25	.68730	.94620	1.0569	.72637	35
26	.68751	.94676	1.0562	.72617	34
27	.68772	.94731	1.0556	.72597	33
28	.68793	.94786	1.0550	.72577	32
29	.68814	.94841	1.0544	.72557	31
30	.68835	.94896	1.0538	.72537	30
31	.68857	.94952	1.0532	.72517	29
32	.68878	.95007	1.0526	.72497	28
33	.68899	.95062	1.0519	.72477	27
34	.68920	.95118	1.0513	.72457	26
35	.68941	.95173	1.0507	.72437	25
36	.68962	.95229	1.0501	.72417	24
37	.68983	.95284	1.0495	.72397	23
38	.69004	.95340	1.0489	.72377	22
39	.69025	.95395	1.0483	.72357	21
40	.69046	.95451	1.0477	.72337	20
41	.69067	.95506	1.0470	.72317	19
42	.69088	.95562	1.0464	.72297	18
43	.69109	.95618	1.0458	.72277	17
44	.69130	.95673	1.0452	.72257	16
45	.69151	.95729	1.0446	.72236	15
46	.69172	.95785	1.0440	.72216	14
47	.69193	.95841	1.0434	.72196	13
48	.69214	.95897	1.0428	.72176	12
49	.69235	.95952	1.0422	.72156	11
50	.69256	.96008	1.0416	.72136	10
51	.69277	.96064	1.0410	.72116	9
52	.69298	.96120	1.0404	.72095	8
53	.69319	.96176	1.0398	.72075	7
54	.69340	.96232	1.0392	.72055	6
55	.69361	.96288	1.0385	.72035	5
56	.69382	.96344	1.0379	.72015	4
57	.69403	.96400	1.0373	.71995	3
58	.69424	.96457	1.0367	.71974	2
59	.69445	.96513	1.0361	.71954	1
60	.69466	.96569	1.0355	.71934	0
′	Cos	Ctn	Tan	Sin	′

46°

44°

′	Sin	Tan	Ctn	Cos	′
0	.69466	.96569	1.0355	.71934	60
1	.69487	.96625	1.0349	.71914	59
2	.69508	.96681	1.0343	.71894	58
3	.69529	.96738	1.0337	.71873	57
4	.69549	.96794	1.0331	.71853	56
5	.69570	.96850	1.0325	.71833	55
6	.69591	.96907	1.0319	.71813	54
7	.69612	.96963	1.0313	.71792	53
8	.69633	.97020	1.0307	.71772	52
9	.69654	.97076	1.0301	.71752	51
10	.69675	.97133	1.0295	.71732	50
11	.69696	.97189	1.0289	.71711	49
12	.69717	.97246	1.0283	.71691	48
13	.69737	.97302	1.0277	.71671	47
14	.69758	.97359	1.0271	.71650	46
15	.69779	.97416	1.0265	.71630	45
16	.69800	.97472	1.0259	.71610	44
17	.69821	.97529	1.0253	.71590	43
18	.69842	.97586	1.0247	.71569	42
19	.69862	.97643	1.0241	.71549	41
20	.69883	.97700	1.0235	.71529	40
21	.69904	.97756	1.0230	.71508	39
22	.69925	.97813	1.0224	.71488	38
23	.69946	.97870	1.0218	.71468	37
24	.69966	.97927	1.0212	.71447	36
25	.69987	.97984	1.0206	.71427	35
26	.70008	.98041	1.0200	.71407	34
27	.70029	.98098	1.0194	.71386	33
28	.70049	.98155	1.0188	.71366	32
29	.70070	.98213	1.0182	.71345	31
30	.70091	.98270	1.0176	.71325	30
31	.70112	.98327	1.0170	.71305	29
32	.70132	.98384	1.0164	.71284	28
33	.70153	.98441	1.0158	.71264	27
34	.70174	.98499	1.0152	.71243	26
35	.70195	.98556	1.0147	.71223	25
36	.70215	.98613	1.0141	.71203	24
37	.70236	.98671	1.0135	.71182	23
38	.70257	.98728	1.0129	.71162	22
39	.70277	.98786	1.0123	.71141	21
40	.70298	.98843	1.0117	.71121	20
41	.70319	.98901	1.0111	.71100	19
42	.70339	.98958	1.0105	.71080	18
43	.70360	.99016	1.0099	.71059	17
44	.70381	.99073	1.0094	.71039	16
45	.70401	.99131	1.0088	.71019	15
46	.70422	.99189	1.0082	.70998	14
47	.70443	.99247	1.0076	.70978	13
48	.70463	.99304	1.0070	.70957	12
49	.70484	.99362	1.0064	.70937	11
50	.70505	.99420	1.0058	.70916	10
51	.70525	.99478	1.0052	.70896	9
52	.70546	.99536	1.0047	.70875	8
53	.70567	.99594	1.0041	.70855	7
54	.70587	.99652	1.0035	.70834	6
55	.70608	.99710	1.0029	.70813	5
56	.70628	.99768	1.0023	.70793	4
57	.70649	.99826	1.0017	.70772	3
58	.70670	.99884	1.0012	.70752	2
59	.70690	.99942	1.0006	.70731	1
60	.70711	1.0000	1.0000	.70711	0
′	Cos	Ctn	Tan	Sin	′

45°

Table II   Compound Amount

$$(1+r)^n$$

n	$\frac{5}{12}\%$	$\frac{1}{2}\%$	$\frac{7}{12}\%$	$\frac{3}{4}\%$	1%
1	1.004 167	1.005 000	1.005 833	1.007 500	1.010 000
2	1.008 351	1.010 025	1.011 701	1.015 056	1.020 100
3	1.012 552	1.015 075	1.017 602	1.022 669	1.030 301
4	1.016 771	1.020 151	1.023 538	1.030 339	1.040 604
5	1.021 008	1.025 251	1.029 509	1.038 067	1.051 010
6	1.025 262	1.030 378	1.035 514	1.045 852	1.061 520
7	1.029 534	1.035 529	1.041 555	1.053 696	1.072 135
8	1.033 824	1.040 707	1.047 631	1.061 599	1.082 857
9	1.038 131	1.045 911	1.053 742	1.069 561	1.093 685
10	1.042 457	1.051 140	1.059 889	1.077 583	1.104 622
11	1.046 800	1.056 396	1.066 071	1.085 664	1.115 668
12	1.051 162	1.061 678	1.072 290	1.093 807	1.126 825
13	1.055 542	1.066 986	1.078 545	1.102 010	1.138 093
14	1.059 940	1.072 321	1.084 837	1.110 276	1.149 474
15	1.064 356	1.077 683	1.091 165	1.118 603	1.160 969
16	1.068 791	1.083 071	1.097 530	1.126 992	1.172 579
17	1.073 244	1.088 487	1.103 932	1.135 445	1.184 304
18	1.077 716	1.093 929	1.110 372	1.143 960	1.196 147
19	1.082 207	1.099 399	1.116 849	1.152 540	1.208 109
20	1.086 716	1.104 896	1.123 364	1.161 184	1.220 190
21	1 091 244	1.110 420	1 129 917	1.169 893	1.232 392
22	1.095 791	1.115 972	1.136 508	1.178 667	1.244 716
23	1.100 357	1.121 552	1.143 138	1.187 507	1.257 163
24	1.104 941	1.127 160	1.149 806	1.196 414	1.269 735
25	1 109 545	1.132 796	1.156 513	1.205 387	1.282 432
26	1.114 168	1.138 460	1.163 260	1.214 427	1.295 256
27	1.118 811	1.144 152	1.170 045	1.223 535	1.308 209
28	1.123 472	1.149 873	1.176 870	1.232 712	1.321 291
29	1.128 154	1.155 622	1.183 736	1.241 957	1.334 504
30	1.132 854	1.161 400	1.190 641	1.251 272	1.347 849
31	1 137 574	1.167 207	1.197 586	1.260 656	1.361 327
32	1.142 314	1.173 043	1 204 572	1.270 111	1.374 941
33	1.147 074	1.178 908	1.211 599	1.279 637	1.388 690
34	1 151 853	1.184 803	1.218 666	1.289 234	1.402 577
35	1.156 653	1.190 727	1.225 775	1.298 904	1.416 603
36	1.161 472	1.196 681	1.232 926	1.308 645	1.430 769
37	1.166 312	1.202 664	1.240 118	1.318 460	1.445 076
38	1.171 171	1.208 677	1.247 352	1.328 349	1.459 527
39	1.176 051	1.214 721	1.254 628	1.338 311	1.474 123
40	1.180 951	1.220 794	1.261 947	1.348 349	1.488 864
41	1 185 872	1.226 898	1.269 308	1.358 461	1.503 752
42	1.190 813	1.233 033	1.276 712	1.368 650	1.518 790
43	1 195 775	1.239 198	1.284 160	1.378 915	1.533 978
44	1 200 757	1.245 394	1.291 651	1.389 256	1.549 318
45	1.205 760	1.251 621	1.299 185	1.399 676	1.564 811
46	1.210 784	1.257 879	1.306 764	1.410 173	1.580 459
47	1 215 829	1.264 168	1.314 387	1.420 750	1.596 263
48	1 220 895	1.270 489	1.322 054	1.431 405	1.612 226
49	1 225 982	1.276 842	1.329 766	1.442 141	1.628 348
50	1.231 091	1.283 226	1.337 523	1.452 957	1.644 632

Table II   Compound Amount

$$(1+r)^n$$

n	$\frac{5}{12}$%	$\frac{1}{2}$%	$\frac{7}{12}$%	$\frac{3}{4}$%	1%
51	1.236 220	1.289 642	1.345 325	1.463 854	1.661 078
52	1.241 371	1.296 090	1.353 173	1.474 833	1.677 689
53	1.246 544	1.302 571	1.361 066	1.485 894	1.694 466
54	1.251 737	1.309 083	1.369 006	1.497 038	1.711 410
55	1.256 953	1.315 629	1.376 992	1.508 266	1.728 525
56	1.262 190	1.322 207	1.385 024	1.519 578	1.745 810
57	1.267 449	1.328 818	1.393 103	1.530 975	1.763 268
58	1.272 730	1.335 462	1.401 230	1.542 457	1.780 901
59	1.278 034	1.342 139	1.409 404	1.554 026	1.798 710
60	1.283 359	1.348 850	1.417 625	1.565 681	1.816 697
61	1.288 706	1.355 594	1.425 895	1.577 424	1.834 864
62	1.294 076	1.362 372	1.434 212	1.589 254	1.853 212
63	1.299 468	1.369 184	1.442 579	1.601 174	1.871 744
64	1.304 882	1.376 030	1.450 994	1.613 183	1.890 462
65	1.310 319	1.382 910	1.459 458	1.625 281	1.909 366
66	1.315 779	1.389 825	1.467 971	1.637 471	1.928 460
67	1.321 261	1.396 774	1.476 535	1.649 752	1.947 745
68	1.326 766	1.403 758	1.485 148	1.662 125	1.967 222
69	1.332 295	1.410 777	1.493 811	1.674 591	1.986 894
70	1.337 846	1.417 831	1.502 525	1.687 151	2.006 763
71	1.343 420	1.424 920	1.511 290	1.699 804	2.026 831
72	1.349 018	1.432 044	1.520 106	1.712 553	2.047 099
73	1.354 639	1.439 204	1.528 973	1.725 397	2.067 570
74	1.360 283	1.446 401	1.537 892	1.738 337	2.088 246
75	1.365 951	1.453 633	1.546 863	1.751 375	2.109 128
76	1.371 642	1.460 901	1.555 886	1.764 510	2.130 220
77	1.377 357	1.468 205	1.564 962	1.777 744	2.151 522
78	1.383 096	1.475 546	1.574 091	1.791 077	2.173 037
79	1.388 859	1.482 924	1.583 273	1.804 510	2.194 768
80	1.394 646	1.490 339	1.592 509	1.818 044	2.216 715
81	1.400 457	1.497 790	1.601 799	1.831 679	2.238 882
82	1.406 293	1.505 279	1.611 143	1.845 417	2.261 271
83	1.412 152	1.512 806	1.620 541	1.859 258	2.283 884
84	1.418 036	1.520 370	1.629 994	1.873 202	2.306 723
85	1.423 945	1.527 971	1.639 502	1.887 251	2.329 790
86	1.429 878	1.535 611	1.649 066	1.901 405	2.353 088
87	1.435 835	1.543 289	1.658 686	1.915 666	2.376 619
88	1.441 818	1.551 006	1.668 361	1.930 033	2.400 385
89	1.447 826	1.558 761	1.678 093	1.944 509	2.424 389
90	1.453 858	1.566 555	1.687 882	1.959 092	2.448 633
91	1.459 916	1.574 387	1.697 728	1.973 786	2.473 119
92	1.465 999	1.582 259	1.707 632	1.988 589	2.497 850
93	1.472 107	1.590 171	1.717 593	2.003 503	2.522 829
94	1.478 241	1.598 122	1.727 612	2.018 530	2.548 057
95	1.484 400	1.606 112	1.737 690	2.033 669	2.573 538
96	1.490 585	1.614 143	1.747 826	2.048 921	2.599 273
97	1.496 796	1.622 213	1.758 022	2.064 288	2.625 266
98	1.503 033	1.630 324	1.768 277	2.079 770	2.651 518
99	1.509 296	1.638 776	1.778 592	2.095 369	2.678 033
100	1.515 584	1.646 668	1.788 967	2.111 084	2.704 814

Table II   **Compound Amount**

$$(1 + r)^n$$

n	$1\tfrac{1}{8}\%$	$1\tfrac{1}{4}\%$	$1\tfrac{1}{2}\%$	$1\tfrac{3}{4}\%$	2%
1	1.011 250	1.012 500	1.015 000	1.017 500	1.020 000
2	1.022 627	1.025 156	1.030 225	1.035 306	1.040 400
3	1.034 131	1.037 971	1.045 678	1.053 424	1.061 208
4	1.045 765	1.050 945	1.061 364	1.071 859	1.082 432
5	1.057 530	1.064 082	1.077 284	1.090 617	1.104 081
6	1.069 427	1.077 383	1.093 443	1.109 702	1.126 162
7	1.081 458	1.090 850	1.109 845	1.129 122	1.148 686
8	1.093 625	1.104 486	1.126 493	1.148 882	1.171 659
9	1.105 928	1.118 292	1.143 390	1.168 987	1.195 093
10	1.118 370	1.132 271	1.160 541	1.189 444	1.218 994
11	1.130 951	1.146 424	1.177 949	1.210 260	1.243 374
12	1.143 674	1.160 755	1.195 618	1.231 439	1.268 242
13	1.156 541	1.175 264	1.213 552	1.252 990	1.293 607
14	1.169 552	1.189 955	1.231 756	1.274 917	1.319 479
15	1.182 709	1.204 829	1.250 232	1.297 228	1.345 868
16	1.196 015	1.219 890	1.268 986	1.319 929	1.372 786
17	1.209 470	1.235 138	1.288 020	1.343 028	1.400 241
18	1.223 077	1.250 577	1.307 341	1.366 531	1.428 246
19	1.236 836	1.266 210	1.326 951	1.390 445	1.456 811
20	1.250 751	1.282 037	1.346 855	1.414 778	1.485 947
21	1.264 821	1.298 063	1.367 058	1.439 537	1.515 666
22	1.279 051	1.314 288	1.387 564	1.464 729	1.545 980
23	1.293 440	1.330 717	1.408 377	1.490 361	1.576 899
24	1.307 991	1.347 351	1.429 503	1.516 443	1.608 437
25	1.322 706	1.364 193	1.450 945	1.542 981	1.640 606
26	1.337 587	1.381 245	1.472 710	1.569 983	1.673 418
27	1.352 634	1.398 511	1.494 800	1.597 457	1.706 886
28	1.367 852	1.415 992	1.517 222	1.625 413	1.741 024
29	1.383 240	1.433 692	1.539 981	1.653 858	1.775 845
30	1.398 801	1.451 613	1.563 080	1.682 800	1.811 362
31	1.414 538	1.469 759	1.586 526	1.712 249	1.847 589
32	1.430 451	1.488 131	1.610 324	1.742 213	1.884 541
33	1.446 544	1.506 732	1.634 479	1.772 702	1.922 231
34	1.462 818	1.525 566	1.658 996	1.803 725	1.960 676
35	1.479 274	1.544 636	1.683 881	1.835 290	1.999 890
36	1.495 916	1.563 944	1.709 140	1.867 407	2.039 887
37	1.512 745	1.583 493	1.734 777	1.900 087	2.080 685
38	1.529 764	1.603 287	1.760 798	1.933 338	2.122 299
39	1.546 973	1.623 328	1.787 210	1.967 172	2.164 745
40	1.564 377	1.643 619	1.814 018	2.001 597	2.208 040
41	1.581 976	1.664 165	1.841 229	2.036 625	2.252 200
42	1.599 773	1.684 967	1.868 847	2.072 266	2.297 244
43	1.617 771	1.706 029	1.896 880	2.108 531	2.343 189
44	1.635 971	1.727 354	1.925 333	2.145 430	2.390 053
45	1.654 375	1.748 946	1.954 213	2.182 975	2.437 854
46	1.672 987	1.770 808	1.983 526	2.221 177	2.486 611
47	1.691 808	1.792 943	2.013 279	2.260 048	2.536 344
48	1.710 841	1.815 355	2.043 478	2.299 599	2.587 070
49	1.730 088	1.838 047	2.074 130	2.339 842	2.638 812
50	1.749 552	1.861 022	2.105 242	2.380 789	2.691 588

Table II Compound Amount

$$(1+r)^n$$

n	$1\frac{1}{8}\%$	$1\frac{1}{4}\%$	$1\frac{1}{2}\%$	$1\frac{3}{4}\%$	2%
51	1.769 234	1.884 285	2.136 821	2.422 453	2.745 420
52	1.789 138	1.907 839	2.168 873	2.464 846	2.800 328
53	1.809 266	1.931 687	2.201 406	2.507 980	2.856 335
54	1.829 620	1.955 833	2.234 428	2.551 870	2.913 461
55	1.850 203	1.980 281	2.267 944	2.596 528	2.971 731
56	1.871 018	2.005 034	2.301 963	2.641 967	3.031 165
57	1.892 067	2.030 097	2.336 493	2.688 202	3.091 789
58	1.913 353	2.055 473	2.371 540	2.735 245	3.153 624
59	1.934 878	2.081 167	2.407 113	2.783 112	3.216 697
60	1.956 645	2.107 181	2.443 220	2.831 816	3.281 031
61	1.978 657	2.133 521	2.479 868	2.881 373	3.346 651
62	2.000 917	2.160 190	2.517 066	2.931 797	3.413 584
63	2.023 428	2.187 193	2.554 822	2.983 104	3.481 856
64	2.046 191	2.214 532	2.593 144	3.035 308	3.551 493
65	2.069 211	2.242 214	2.632 042	3.088 426	3.622 523
66	2.092 489	2.270 242	2.671 522	3.142 473	3.694 974
67	2.116 030	2.298 620	2.711 595	3.197 466	3.768 873
68	2.139 835	2.327 353	2.752 269	3.253 422	3.844 251
69	2.163 908	2.356 444	2.793 553	3.310 357	3.921 136
70	2.188 252	2.385 900	2.835 456	3.368 288	3.999 558
71	2.212 870	2.415 724	2.877 988	3.427 233	4.079 549
72	2.237 765	2.445 920	2.921 158	3.487 210	4.161 140
73	2.262 940	2.476 494	2.964 975	3.548 236	4.244 363
74	2.288 398	2.507 450	3.009 450	3.610 330	4.329 250
75	2.314 142	2.538 794	3.054 592	3.673 511	4.415 835
76	2.340 177	2.570 529	3.100 411	3.737 797	4.504 152
77	2.366 504	2.602 660	3.146 917	3.803 209	4.594 235
78	2.393 127	2.635 193	3.194 120	3.869 765	4.686 120
79	2.420 049	2.668 133	3.242 032	3.937 486	4.779 842
80	2.447 275	2.701 485	3.290 663	4.006 392	4.875 439
81	2.474 807	2.735 254	3.340 023	4.076 504	4.972 948
82	2.502 648	2.769 444	3.390 123	4.147 843	5.072 407
83	2.530 803	2.804 062	3.440 975	4.220 430	5.173 855
84	2.559 275	2.839 113	3.492 590	4.294 287	5.277 332
85	2.588 067	2.874 602	3.544 978	4.369 437	5.382 879
86	2.617 182	2.910 534	3.598 153	4.445 903	5.490 536
87	2.646 626	2.946 916	3.652 125	4.523 706	5.600 347
88	2.676 400	2.983 753	3.706 907	4.602 871	5.712 354
89	2.706 510	3.021 049	3.762 511	4.683 421	5.826 601
90	2.736 958	3.058 813	3.818 949	4.765 381	5.943 133
91	2.767 749	3.097 048	3.876 233	4.848 775	6.061 996
92	2.798 886	3.135 761	3.934 376	4.933 629	6.183 236
93	2.830 373	3.174 958	3.993 392	5.019 967	6.306 900
94	2.862 215	3.214 645	4.053 293	5.107 816	6.433 038
95	2.894 415	3.254 828	4.114 092	5.197 203	6.561 699
96	2.926 977	3.295 513	4.175 804	5.288 154	6.692 933
97	2.959 906	3.336 707	4.238 441	5.380 697	6.826 792
98	2.993 205	3.378 416	4.302 017	5.474 859	6.963 328
99	3.026 878	3.420 646	4.366 547	5.570 669	7.102 594
100	3.060 930	3.463 404	4.432 046	5.668 156	7.244 646

Table II Compound Amount

$$(1 + r)^n$$

n	3%	4%	5%	6%	7%
1	1.030 000	1.040 000	1.050 000	1.060 000	1.070 000
2	1.060 900	1.081 600	1.102 500	1.123 600	1.144 900
3	1.092 727	1.124 864	1.157 625	1.191 016	1.225 043
4	1.125 509	1.169 859	1.215 506	1.262 477	1.310 796
5	1.159 274	1.216 653	1.276 282	1.338 226	1.402 552
6	1.194 052	1.265 319	1.340 096	1.418 519	1.500 730
7	1.229 874	1.315 932	1.407 100	1.503 630	1.605 781
8	1.266 770	1.368 569	1.477 455	1.593 848	1.718 186
9	1.304 773	1.423 312	1.551 328	1.689 479	1.838 459
10	1.343 916	1.480 244	1.628 895	1.790 848	1.967 151
11	1.384 234	1.539 454	1.710 339	1.898 299	2.104 852
12	1.425 761	1.601 032	1.795 856	2.012 196	2.252 192
13	1.468 534	1.665 074	1.885 649	2.132 928	2.409 845
14	1.512 590	1.731 676	1.979 932	2.260 904	2.578 534
15	1.557 967	1.800 944	2.078 928	2.396 558	2.759 032
16	1.604 706	1.872 981	2.182 875	2.540 352	2.952 164
17	1.652 848	1.947 900	2.292 018	2.692 773	3.158 815
18	1.702 433	2.025 817	2.406 619	2.854 339	3.379 932
19	1.753 506	2.106 849	2.526 950	3.025 600	3.616 528
20	1.806 111	2.191 123	2.653 298	3.207 135	3.869 684
21	1.860 295	2.278 768	2.785 963	3.399 564	4.140 562
22	1.916 103	2.369 919	2.925 261	3.603 537	4.430 402
23	1.973 587	2.464 716	3.071 524	3.819 750	4.740 530
24	2.032 794	2.563 304	3.225 100	4.048 935	5.072 367
25	2.093 778	2.665 836	3.386 355	4.291 871	5.427 433
26	2.156 591	2.772 470	3.555 673	4.549 383	5 807 353
27	2.221 289	2.883 369	3.733 456	4.822 346	6.213 868
28	2.287 928	2.998 703	3.920 129	5.111 687	6.648 838
29	2.356 566	3.118 651	4.116 136	5.418 388	7.114 257
30	2.427 262	3.243 398	4.321 942	5.743 491	7.612 255
31	2.500 080	3.373 133	4.538 039	6.088 101	8.145 113
32	2.575 083	3.508 059	4.764 941	6.453 387	8 715 271
33	2.652 335	3.648 381	5 003 189	6.840 590	9 325 340
34	2.731 905	3.794 316	5.253 348	7.251 025	9.978 114
35	2.813 862	3.946 089	5 516 015	7.686 087	10 676 581
36	2.898 278	4.103 933	5.791 816	8.147 252	11.423 942
37	2.985 227	4.268 090	6.081 407	8.636 087	12 223 618
38	3.074 783	4.438 813	6.385 477	9.154 252	13.079 271
39	3.167 027	4.616 366	6.704 751	9.703 507	13.994 820
40	3.262 038	4.801 021	7.039 989	10.285 718	14.974 458
41	3.359 899	4.993 061	7.391 988	10.902 861	16.022 670
42	3.460 696	5.192 784	7.761 588	11.557 033	17.144 257
43	3.564 517	5.400 495	8.149 667	12.250 455	18.344 355
44	3.671 452	5.616 515	8.557 150	12.985 482	19.628 460
45	3.781 596	5.841 176	8.985 008	13.764 611	21 002 452
46	3.895 044	6.074 823	9.434 258	14 590 487	22.472 623
47	4.011 895	6.317 816	9.905 971	15.465 917	24.045 707
48	4.132 252	6.570 528	10.401 270	16.393 872	25.728 907
49	4.256 219	6.833 349	10.921 333	17.377 504	27.529 930
50	4.383 906	7.106 683	11.467 400	18.420 154	29.457 025

Table II   Compound Amount

$$(1+r)^n$$

n	3%	4%	5%	6%	7%
51	4.515 423	7.390 951	12.040 770	19.525 364	31.519 017
52	4.650 886	7.686 589	12.642 808	20.696 885	33.725 348
53	4.790 412	7.994 052	13.274 949	21.938 698	36.086 122
54	4.934 125	8.313 814	13.938 696	23.255 020	38.612 151
55	5.082 149	8.646 367	14.635 631	24.650 322	41.315 001
56	5.234 613	8.992 222	15.367 412	26.129 341	44.207 052
57	5.391 651	9.351 910	16.135 783	27.697 101	47.301 545
58	5.553 401	9.725 987	16.942 572	29.358 927	50.612 653
59	5.720 003	10.115 026	17.789 701	31.120 463	54.155 539
60	5.891 603	10.519 627	18.679 186	32.987 691	57.946 427
61	6.068 351	10.940 413	19.613 145	34.966 952	62.002 677
62	6.250 402	11.378 029	20.593 802	37.064 969	66.342 864
63	6.437 914	11.833 150	21.623 493	39.288 868	70.986 865
64	6.631 051	12.306 476	22.704 667	41.646 200	75.955 945
65	6.829 983	12.798 735	23.839 901	44.144 972	81.272 861
66	7.034 882	13.310 685	25.031 896	46.793 670	86.961 962
67	7.245 929	13.843 112	26.283 490	49.601 290	93.049 299
68	7.463 307	14.396 836	27.597 665	52.577 368	99.562 750
69	7.687 206	14.972 710	28.977 548	55.732 010	106.532 142
70	7.917 822	15.571 618	30.426 426	59.075 930	113.989 392
71	8.155 357	16.194 483	31.947 747	62.620 486	121.968 650
72	8.400 017	16.842 262	33.545 134	66.377 715	130.506 455
73	8.652 018	17.515 953	35.222 391	70.360 378	139.641 907
74	8.911 578	18.216 591	36.983 510	74.582 001	149.416 840
75	9.178 926	18.945 255	38.832 686	79.056 921	159.876 019
76	9.454 293	19.703 065	40.774 320	83.800 336	171.067 341
77	9.737 922	20.491 187	42.813 036	88.828 356	183.042 055
78	10.030 060	21.310 835	44.953 688	94.158 058	195.854 998
79	10.330 962	22.163 268	47.201 372	99.807 541	209.564 848
80	10.640 891	23.049 799	49.561 441	105.795 993	224.234 388
81	10.960 117	23.971 791	52.039 531	112.143 753	239.930 795
82	11.288 921	24.930 663	54.641 489	118.872 378	256.725 950
83	11.627 588	25.927 889	57.373 563	126.004 721	274.696 767
84	11.976 416	26.965 005	60.242 241	133.565 004	293.925 541
85	12.335 709	28.043 605	63.254 353	141.578 904	314.500 328
86	12.705 780	29.165 349	66.417 071	150.073 639	336.515 351
87	13.086 953	30.331 963	69.737 925	159.078 057	360.071 426
88	13.479 562	31.545 242	73.224 821	168.622 741	385.276 426
89	13.883 949	32.807 051	76.886 062	178.740 105	412.245 776
90	14.300 467	34.119 333	80.730 365	189.464 511	441.102 980
91	14.729 481	35.484 107	84.766 883	200.832 382	471.980 188
92	15.171 366	36.903 471	89.005 227	212.882 325	505.018 802
93	15.626 507	38.379 610	93.455 489	225.655 264	540.370 118
94	16.095 302	39.914 794	98.128 263	239.194 580	578.196 026
95	16.578 161	41.511 386	103.034 676	253.546 255	618.669 748
96	17.075 506	43.171 841	108.186 410	268.759 030	661.976 630
97	17.587 771	44.898 715	113.595 731	284.884 572	708.314 994
98	18.115 404	46.694 664	119.275 517	301.977 646	757.897 044
99	18.658 866	48.562 450	125.239 293	320.096 305	810.949 837
100	19.218 632	50.504 948	131.501 258	339.302 084	867.716 326

Table II   Compound Amount

$$(1+r)^{-n}$$

n	$\frac{5}{12}\%$	$\frac{1}{2}\%$	$\frac{7}{12}\%$	$\frac{3}{4}\%$	1%
1	0.995 851	0.995 025	0.994 200	0.992 556	0.990 099
2	0.991 718	0.990 075	0.984 435	0.985 167	0.980 296
3	0 987 603	0.985 149	0.982 702	0 977 833	0.970 590
4	0.983 506	0.980 248	0.977 003	0.970 554	0.960 980
5	0.979 425	0.975 371	0.971 337	0.963 329	0.951 466
6	0.975 361	0.970 518	0.965 704	0 956 158	0.942 045
7	0.971 313	0.965 690	0.960 103	0.949 040	0.932 718
8	0.967 283	0.960 885	0.954 535	0.941 975	0.923 483
9	0.963 269	0.956 105	0.948 999	0.934 963	0.914 340
10	0.959 272	0.951 348	0.943 495	0.928 003	0.905 287
11	0.955 292	0.946 615	0.938 024	0 921 095	0 896 324
12	0.951 328	0.941 905	0.932 583	0 914 238	0.887 449
13	0.947 381	0.937 219	0.927 175	0.907 432	0 878 663
14	0.943 450	0.932 556	0.921 798	0.900 677	0.869 963
15	0.939 535	0.927 917	0.916 452	0.893 973	0.861 349
16	0.935 637	0.923 300	0.911 137	0.887 318	0.852 821
17	0.931 754	0.918 707	0.905 853	0 880 712	0.844 377
18	0.927 888	0.914 136	0.900 599	0.874 156	0 836 017
19	0.924 038	0.909 588	0.895 376	0.867 649	0.827 740
20	0.920 204	0.905 063	0.890 183	0.861 190	0.819 544
21	0.916 385	0.900 560	0.885 021	0.854 779	0 811 430
22	0.912 583	0.896 080	0.879 888	0.848 416	0.803 396
23	0 908 796	0.891 622	0.874 785	0.842 100	0 795 442
24	0.905 025	0 887 186	0.869 712	0.835 831	0 787 566
25	0 901 270	0.882 772	0.864 668	0.829 609	0.779 768
26	0.897 530	0.878 380	0.859 653	0.823 434	0 772 048
27	0.893 806	0.874 010	0.854 668	0.817 304	0.764 404
28	0.890 097	0.869 662	0.849 711	0.811 220	0.756 836
29	0.886 404	0.865 335	0.844 783	0.805 181	0.749 342
30	0.882 726	0.861 030	0.839 884	0.799 187	0.741 923
31	0.879 063	0.856 746	0.835 013	0.793 238	0 734 577
32	0.875 416	0.852 484	0.830 170	0.787 333	0.727 304
33	0.871 783	0.848 242	0.825 356	0.781 472	0.720 103
34	0.868 166	0.844 022	0.820 569	0.775 654	0.712 973
35	0.864 564	0.839 823	0.815 810	0.769 880	0 705 914
36	0.860 976	0.835 645	0.811 079	0.764 149	0.698 925
37	0.857 404	0.831 487	0.806 375	0.758 461	0.692 005
38	0.853 846	0.827 351	0.801 699	0.752 814	0 685 153
39	0.850 303	0.823 235	0.797 049	0.747 210	0.678 370
40	0.846 775	0.819 139	0.792 427	0.741 648	0.671 653
41	0.843 261	0.815 064	0.787 831	0.736 127	0.665 003
42	0.839 762	0.811 009	0.783 262	0.730 647	0.658 419
43	0.836 278	0.806 974	0.778 719	0.725 208	0.651 900
44	0.832 808	0.802 959	0.774 203	0.719 810	0.645 445
45	0.829 352	0.798 964	0.769 713	0.714 451	0.639 055
46	0.825 911	0.794 989	0.765 249	0.709 133	0.632 728
47	0.822 484	0.791 034	0.760 811	0.703 854	0.626 463
48	0.819 071	0.787 098	0.756 399	0.698 614	0.620 260
49	0.815 672	0.783 182	0.752 012	0.693 414	0.614 119
50	0.812 288	0.779 286	0.747 651	0 688 252	0.608 039

Table II Compound Amount

$$(1+r)^{-n}$$

n	$\frac{5}{12}\%$	$\frac{1}{2}\%$	$\frac{7}{12}\%$	$\frac{3}{4}\%$	1%
51	0.808 917	0.775 409	0.743 315	0.683 128	0.602 019
52	0.805 561	0.771 551	0.739 004	0.678 043	0.596 058
53	0 802 218	0.767 713	0.734 718	0.672 995	0.590 156
54	0.798 890	0.763 893	0.730 457	0.667 986	0.584 313
55	0.795 575	0.760 093	0.726 221	0.663 013	0.578 528
56	0.792 274	0.756 311	0.722 009	0.658 077	0.572 800
57	0.788 986	0.752 548	0.717 822	0.653 178	0.567 129
58	0.785 712	0.748 804	0.713 659	0.648 316	0.561 514
59	0.782 452	0.745 079	0.709 520	0.643 490	0.555 954
60	0.779 205	0.741 372	0.705 405	0.638 700	0.550 450
61	0.775 972	0.737 684	0.701 314	0.633 945	0.545 000
62	0.772 752	0.734 014	0.697 247	0.629 226	0.539 604
63	0 769 546	0.730 362	0.693 203	0.624 542	0.534 261
64	0.766 353	0.726 728	0.689 183	0.619 893	0.528 971
65	0.763 173	0.723 113	0.685 186	0.615 278	0.523 734
66	0 760 006	0.719 515	0.681 212	0.610 698	0.518 548
67	0.756 853	0.715 935	0.677 262	0.606 152	0.513 414
68	0 753 712	0.712 374	0.673 334	0.601 639	0.508 331
69	0.750 585	0.708 829	0.669 429	0.597 161	0.503 298
70	0.747 470	0.705 303	0.665 546	0.592 715	0.498 315
71	0.744 369	0.701 794	0.661 687	0.588 303	0.493 381
72	0.741 280	0.698 302	0.657 849	0.583 924	0.488 496
73	0 738 204	0.694 828	0.654 034	0.579 577	0.483 659
74	0 735 141	0.691 371	0.650 241	0.575 262	0.478 871
75	0.732 091	0.687 932	0.646 470	0.570 980	0.474 129
76	0 729 053	0.684 509	0.642 721	0.566 730	0.469 435
77	0 726 028	0.681 104	0.638 993	0.562 511	0.464 787
78	0 723 015	0.677 715	0.635 287	0.558 323	0.460 185
79	0 720 015	0.674 343	0.631 603	0.554 167	0.455 629
80	0.717 028	0.670 988	0.627 940	0.550 042	0.451 118
81	0 714 052	0.667 650	0.624 298	0.545 947	0.446 651
82	0 711 090	0.664 329	0.620 678	0.541 883	0.442 229
83	0 708 139	0.661 023	0 617 078	0.537 849	0.437 851
84	0 705 201	0.657 735	0.613 499	0.533 845	0.433 515
85	0 702 275	0.654 462	0.609 941	0.529 871	0.429 223
86	0.699 361	0 651 206	0.606 404	0.525 927	0.424 974
87	0.696 459	0.647 967	0.602 887	0.522 012	0.420 766
88	0.693 569	0.644 743	0.599 391	0.518 126	0.416 600
89	0.690 691	0.641 535	0.595 914	0.514 269	0.412 475
90	0.687 825	0.638 344	0.592 458	0.510 440	0.408 391
91	0.684 971	0.635 168	0.589 022	0.506 641	0.404 348
92	0.682 129	0.632 008	0.585 606	0.502 869	0.400 344
93	0.679 298	0.628 863	0.582 210	0.499 126	0.396 380
94	0.676 480	0.625 735	0.578 834	0.495 410	0.392 456
95	0 673 673	0.622 622	0.575 477	0.491 722	0.388 570
96	0 670 877	0.619 524	0.572 139	0.488 062	0.384 723
97	0 668 094	0.616 442	0.568 821	0.484 428	0.380 914
98	0 665 321	0.613 375	0.565 522	0.480 822	0.377 142
99	0 662 561	0.610 323	0.562 242	0.477 243	0.373 408
100	0.659 812	0.607 287	0.558 982	0.473 690	0.369 711

Table III Present Value

$$(1 + r)^{-n}$$

n	$1\frac{1}{8}\%$	$1\frac{1}{4}\%$	$1\frac{1}{2}\%$	$1\frac{3}{4}\%$	2%
1	0.988 875	0.987 654	0.985 222	0.982 801	0.980 392
2	0.977 874	0.975 461	0.970 662	0.965 898	0.961 169
3	0.966 995	0.963 418	0.956 317	0.949 285	0.942 322
4	0.956 238	0.951 524	0.942 184	0.932 959	0.923 845
5	0.945 600	0.939 777	0.928 260	0.916 913	0.905 731
6	0.935 080	0.928 175	0.914 542	0.901 143	0.887 971
7	0.924 677	0.916 716	0.901 027	0.885 644	0.870 560
8	0.914 391	0.905 398	0.887 711	0.870 412	0.853 490
9	0.904 218	0.894 221	0.874 592	0.855 441	0.836 755
10	0.894 159	0.883 181	0.861 667	0.840 729	0.820 348
11	0.884 211	0.872 277	0.848 933	0.826 269	0.804 263
12	0.874 375	0.861 509	0.836 387	0.812 058	0.788 493
13	0.864 647	0.850 873	0.824 027	0.798 091	0.773 033
14	0.855 028	0.840 368	0.811 849	0.784 365	0.757 875
15	0.845 516	0.829 993	0.799 852	0.770 875	0.743 015
16	0.836 110	0.819 746	0.788 031	0.757 616	0.728 446
17	0.826 808	0.809 626	0.776 385	0.744 586	0.714 163
18	0.817 610	0.799 631	0.764 912	0.731 780	0.700 159
19	0.808 515	0.789 759	0.753 607	0.719 194	0.686 431
20	0.799 520	0.780 009	0.742 470	0.706 825	0.672 971
21	0.790 625	0.770 379	0.731 498	0.694 668	0.659 776
22	0.781 830	0.760 868	0.720 688	0.682 720	0.646 839
23	0.773 132	0.751 475	0.710 037	0.670 978	0.634 156
24	0.764 531	0.742 197	0.699 544	0.659 438	0.621 721
25	0.756 026	0.733 034	0.689 206	0.648 096	0.609 531
26	0.747 615	0.723 984	0.679 021	0.636 950	0.597 579
27	0.739 298	0.715 046	0.668 986	0.625 995	0.585 862
28	0.731 073	0.706 219	0.659 099	0.615 228	0.574 375
29	0.722 940	0.697 500	0.649 359	0.604 647	0.563 112
30	0.714 898	0.688 889	0.639 762	0.594 248	0.552 071
31	0.706 945	0.680 384	0.630 308	0.584 027	0.541 246
32	0.699 080	0.671 984	0.620 993	0.573 982	0.530 633
33	0.691 303	0.663 688	0.611 816	0.564 111	0.520 229
34	0.683 612	0.655 494	0.602 774	0.554 408	0.510 028
35	0.676 007	0.647 402	0.593 866	0.544 873	0.500 028
36	0.668 487	0.639 409	0.585 090	0.535 502	0.490 223
37	0.661 050	0.631 515	0.576 443	0.526 292	0.480 611
38	0.653 696	0.623 719	0.567 924	0.517 240	0.471 187
39	0.646 424	0.616 019	0.559 531	0.508 344	0.461 948
40	0.639 232	0.608 413	0.551 262	0.499 601	0.452 890
41	0.632 121	0.600 902	0.543 116	0.491 008	0.444 010
42	0.625 089	0.593 484	0.535 089	0.482 563	0.435 304
43	0.618 135	0.586 157	0.527 182	0.474 264	0.426 769
44	0.611 258	0.578 920	0.519 391	0.466 107	0.418 401
45	0.604 458	0.571 773	0.511 715	0.458 090	0.410 197
46	0.597 733	0.564 714	0.504 153	0.450 212	0.402 154
47	0.591 084	0.557 742	0.496 702	0.442 469	0.394 268
48	0.584 508	0.550 856	0.489 362	0.434 858	0.386 538
49	0.578 005	0.544 056	0.482 130	0.427 379	0.378 958
50	0.571 575	0.537 339	0.475 005	0.420 029	0.371 528

Table III Present Value

$$(1+r)^{-n}$$

n	$1\frac{1}{8}\%$	$1\frac{1}{4}\%$	$1\frac{1}{2}\%$	$1\frac{3}{4}\%$	2%
51	0.565 216	0.530 705	0.467 985	0.412 805	0.364 243
52	0.558 928	0.524 153	0.461 069	0.405 705	0.357 101
53	0.552 710	0.517 682	0.454 255	0.398 727	0.350 099
54	0.546 562	0.511 291	0.447 542	0.391 869	0.343 234
55	0.540 481	0.504 979	0.440 928	0.385 130	0.336 504
56	0.534 468	0.498 745	0.434 412	0.378 506	0.329 906
57	0.528 523	0.492 587	0.427 992	0.371 996	0.323 437
58	0.522 643	0.486 506	0.421 667	0.365 598	0.317 095
59	0.516 829	0.480 500	0.415 435	0.359 310	0.310 878
60	0.511 079	0.474 568	0.409 296	0.353 130	0.304 782
61	0.505 393	0.468 709	0.403 247	0.347 057	0.298 806
62	0.499 771	0.462 922	0.397 288	0.341 088	0.292 947
63	0.494 211	0.457 207	0.391 417	0.335 221	0.287 203
64	0.488 713	0.451 563	0.385 632	0.329 456	0.281 572
65	0.483 276	0.445 988	0.379 933	0.323 790	0.276 051
66	0.477 900	0.440 482	0.374 318	0.318 221	0.270 638
67	0.472 583	0.435 044	0.368 787	0.312 748	0.265 331
68	0.467 326	0.429 673	0.363 337	0.307 369	0.260 129
69	0.462 127	0.424 368	0.357 967	0.302 082	0.255 028
70	0.456 986	0.419 129	0.352 677	0.296 887	0.250 028
71	0.451 902	0.413 955	0.347 465	0.291 781	0.245 125
72	0.446 874	0.408 844	0.342 330	0.286 762	0.240 319
73	0.441 903	0.403 797	0.337 271	0.281 830	0.235 607
74	0.436 987	0.398 811	0.332 287	0.276 983	0.230 987
75	0.432 126	0.393 888	0.327 376	0.272 219	0.226 458
76	0.427 318	0.389 025	0.322 538	0.267 537	0.222 017
77	0.422 564	0.384 222	0.317 771	0.262 936	0.217 664
78	0.417 863	0.379 479	0.313 075	0.258 414	0.213 396
79	0.413 215	0.374 794	0.308 448	0.253 969	0.209 212
80	0.408 618	0.370 167	0.303 890	0.249 601	0.205 110
81	0.404 072	0.365 597	0.299 399	0.245 308	0.201 088
82	0.399 577	0.361 083	0.294 975	0.241 089	0.197 145
83	0.395 131	0.356 625	0.290 615	0.236 943	0.193 279
84	0.390 736	0.352 223	0.286 321	0.232 868	0.189 490
85	0.386 389	0.347 874	0.282 089	0.228 862	0.185 774
86	0.382 090	0.343 580	0.277 920	0.224 926	0.182 132
87	0.377 840	0.339 338	0.273 813	0.221 058	0.178 560
88	0.373 636	0.335 148	0.269 767	0.217 256	0.175 059
89	0.369 480	0.331 011	0.265 780	0.213 519	0.171 627
90	0.365 369	0.326 924	0.261 852	0.209 847	0.168 261
91	0.361 304	0.322 888	0.257 982	0.206 238	0.164 962
92	0.357 285	0.318 902	0.254 170	0.202 691	0.161 728
93	0.353 310	0.314 965	0.250 414	0.199 204	0.158 556
94	0.349 380	0.311 076	0.246 713	0.195 778	0.155 448
95	0.345 493	0.307 236	0.243 067	0.192 411	0.152 400
96	0.341 649	0.303 443	0.239 475	0.189 102	0.149 411
97	0.337 849	0.299 697	0.235 936	0.185 850	0.146 482
98	0.334 090	0.295 997	0.232 449	0.182 653	0.143 609
99	0.330 373	0.292 342	0.229 014	0.179 512	0.140 794
100	0.326 698	0.288 733	0.225 629	0.176 424	0.138 033

Table III   **Present Value**

$$(1+r)^{-n}$$

n	3%	4%	5%	6%	7%
1	0.970 874	0.961 538	0.952 381	0.943 396	0.934 579
2	0.942 596	0.924 556	0.907 029	0.889 996	0.873 439
3	0.915 142	0.888 996	0.863 838	0.839 619	0.816 298
4	0.888 487	0.854 804	0.822 702	0.792 094	0.762 895
5	0.862 609	0.821 927	0.783 526	0.747 258	0.712 986
6	0.837 484	0.790 315	0.746 215	0.704 961	0.666 342
7	0.813 092	0.759 918	0.710 681	0.665 057	0.622 750
8	0 789 409	0.730 690	0.676 839	0.627 412	0.582 009
9	0.766 417	0.702 587	0.644 609	0.591 898	0.543 934
10	0.744 094	0.675 564	0.613 913	0.558 395	0.508 349
11	0.722 421	0.649 581	0.584 679	0.526 788	0.475 093
12	0.701 380	0.624 597	0.556 837	0.496 969	0.444 012
13	0.680 951	0.600 574	0.530 321	0.468 839	0.414 964
14	0.661 118	0.577 475	0.505 068	0.442 301	0.387 817
15	0.641 862	0.555 265	0.481 017	0.417 265	0.362 446
16	0.623 167	0.533 908	0.458 112	0.393 646	0.338 735
17	0.605 016	0.513 373	0.436 297	0.371 364	0.316 574
18	0.587 395	0.493 628	0.415 521	0.350 344	0.295 864
19	0.570 286	0.474 642	0.395 734	0.330 513	0.276 508
20	0.553 676	0.456 387	0.376 889	0.311 805	0.258 419
21	0.537 549	0.438 834	0.358 942	0.294 155	0.241 513
22	0.521 893	0.421 955	0.341 850	0.277 505	0.225 713
23	0.506 692	0.405 726	0.325 571	0.261 797	0.210 947
24	0.491 934	0.390 121	0.310 068	0.246 979	0.197 147
25	0.477 606	0.375 117	0.295 303	0.232 999	0.184 249
26	0.463 695	0.360 689	0.281 241	0.219 810	0.172 195
27	0.450 189	0.346 817	0.267 848	0.207 368	0.160 930
28	0.437 077	0.333 477	0.255 094	0.195 630	0.150 402
29	0.424 346	0.320 651	0.242 946	0.184 557	0.140 563
30	0.411 987	0.308 319	0.231 377	0.174 110	0.131 367
31	0.399 987	0.296 460	0.220 359	0.164 255	0.122 773
32	0.388 337	0.285 058	0.209 866	0.154 957	0.114 741
33	0.377 026	0.274 094	0.199 873	0.146 186	0.107 235
34	0.366 045	0.263 552	0.190 355	0.137 912	0.100 219
35	0.355 383	0.253 415	0.181 291	0.130 105	0.093 663
36	0.345 032	0.243 669	0.172 657	0.122 741	0.087 535
37	0.334 983	0.234 297	0.164 436	0.115 793	0.081 809
38	0.325 226	0.225 285	0.156 605	0.109 239	0.076 457
39	0.315 754	0.216 621	0.149 148	0.103 056	0.071 455
40	0.306 557	0.208 289	0.142 046	0.097 222	0.066 780
41	0.297 628	0.200 278	0.135 282	0.091 719	0.062 412
42	0.288 959	0.192 575	0.128 840	0.086 527	0.058 329
43	0.280 543	0.185 168	0.122 704	0.081 630	0.054 513
44	0.272 372	0.178 046	0.116 861	0.077 009	0.050 946
45	0.264 439	0.171 198	0.111 297	0.072 650	0.047 613
46	0.256 737	0.164 614	0.105 997	0.068 538	0.044 499
47	0.249 259	0.158 283	0.100 949	0.064 658	0.041 587
48	0.241 999	0.152 195	0.096 142	0.060 998	0.038 867
49	0.234 950	0.146 341	0.091 564	0.057 546	0.036 324
50	0.228 107	0.140 713	0.087 204	0.054 288	0.033 945

Table III Present Value

$$(1+r)^{-n}$$

n	3%	4%	5%	6%	7%
51	0.221 463	0.135 301	0.083 051	0.051 215	0.031 727
52	0.215 013	0.130 097	0.079 096	0.048 316	0.029 651
53	0.208 750	0.125 093	0.075 330	0.045 582	0.027 711
54	0.202 670	0.120 282	0.071 743	0.043 001	0.025 899
55	0.196 767	0.115 656	0.068 326	0.040 567	0.024 204
56	0.191 036	0.111 207	0.065 073	0.038 271	0.022 621
57	0.185 472	0.106 930	0.061 974	0.036 105	0.021 141
58	0.180 070	0.102 817	0.059 023	0.034 061	0.019 758
59	0.174 825	0.098 863	0.056 212	0.032 133	0.018 465
60	0.169 733	0.095 060	0.053 536	0.030 314	0.017 257
61	0.164 789	0.091 404	0.050 986	0.028 598	0.016 128
62	0.159 990	0.087 889	0.048 558	0.026 980	0.015 073
63	0.155 330	0.084 508	0.046 246	0.025 453	0.014 087
64	0.150 806	0.081 258	0.044 044	0.024 012	0.013 166
65	0.146 413	0.078 133	0.041 946	0.022 653	0.012 304
66	0.142 149	0.075 128	0.039 949	0.021 370	0.011 499
67	0.138 009	0.072 238	0.038 047	0.020 161	0.010 747
68	0.133 989	0.069 460	0.036 235	0.019 020	0.010 044
69	0.130 086	0.066 788	0.034 509	0.017 943	0.009 387
70	0.126 297	0.064 219	0.032 866	0.016 927	0.008 773
71	0.122 619	0.061 749	0.031 301	0.015 969	0.008 199
72	0.119 047	0.059 374	0.029 811	0.015 065	0.007 662
73	0.115 580	0.057 091	0.028 391	0.014 213	0.007 161
74	0.112 214	0.054 895	0.027 039	0.013 408	0.006 693
75	0.108 945	0.052 784	0.025 752	0.012 649	0.006 255
76	0.105 772	0.050 754	0.024 525	0.011 933	0.005 846
77	0.102 691	0.048 801	0.023 357	0.011 258	0.005 463
78	0.099 700	0.046 924	0.022 245	0.010 620	0.005 106
79	0.096 796	0.045 120	0.021 186	0.010 019	0.004 772
80	0.093 977	0.043 384	0.020 177	0.009 452	0.004 460
81	0.091 240	0.041 716	0.019 216	0.008 917	0.004 168
82	0.088 582	0.040 111	0.018 301	0.008 412	0.003 895
83	0.086 002	0.038 569	0.017 430	0.007 936	0.003 640
84	0.083 497	0.037 085	0.016 600	0.007 487	0.003 402
85	0.081 065	0.035 659	0.015 809	0.007 063	0.003 180
86	0.078 704	0.034 287	0.015 056	0.006 663	0.002 972
87	0 076 412	0.032 969	0.014 339	0.006 286	0 002 777
88	0.074 186	0.031 701	0.013 657	0.005 930	0.002 596
89	0.072 026	0.030 481	0 013 006	0.005 595	0.002 426
90	0.069 928	0.029 309	0.012 387	0.005 278	0.002 267
91	0.067 891	0.028 182	0.011 797	0.004 979	0.002 119
92	0.065 914	0.027 098	0.011 235	0.004 697	0.001 980
93	0.063 994	0.026 056	0.010 700	0.004 432	0.001 851
94	0.062 130	0.025 053	0.010 191	0.004 181	0.001 730
95	0.060 320	0.024 090	0.009 705	0.003 944	0.001 616
96	0.058 563	0.023 163	0.009 243	0.003 721	0.001 511
97	0.056 858	0.022 272	0.008 803	0.003 510	0.001 412
98	0.055 202	0.021 416	0.008 384	0 003 312	0.001 319
99	0.053 594	0.020 592	0.007 985	0.003 124	0.001 233
100	0.052 033	0.019 800	0.007 604	0.002 947	0.001 152

Table IV Exponential Functions

x	e^x	$\text{Log}_{10} e^x$	e^{-x}	x	e^x	$\text{Log}_{10} e^x$	e^{-x}
0.00	1.0000	.00 000	1.00 000	**0.50**	1.6487	.21 715	.60 653
0.01	1.0101	.00 434	0.99 005	0.51	1.6653	.22 149	.60 050
0.02	1.0202	.00 869	.98 020	0.52	1.6820	.22 583	.59 452
0.03	1.0305	.01 303	.97 045	0.53	1.6989	.23 018	.58 860
0.04	1.0408	.01 737	.96 079	0.54	1.7160	.23 452	.58 275
0.05	1.0513	.02 171	.95 123	**0.55**	1.7333	.23 886	.57 695
0.06	1.0618	.02 606	.94 176	0.56	1.7507	.24 320	.57 121
0.07	1.0725	.03 040	.93 239	0.57	1.7683	.24 755	.56 553
0.08	1.0833	.03 474	.92 312	0.58	1.7860	.25 189	.55 990
0.09	1.0942	.03 909	.91 393	0.59	1.8040	.25 623	.55 433
0.10	1.1052	.04 343	.90 484	**0.60**	1.8221	.26 058	.54 881
0.11	1.1163	.04 777	.89 583	0.61	1.8404	.26 492	.54 335
0.12	1.1275	.05 212	.88 692	0.62	1.8589	.26 926	.53 794
0.13	1.1388	.05 646	.87 810	0.63	1.8776	.27 361	.53 259
0.14	1.1503	.06 080	.86 936	0.64	1.8965	.27 795	.52 729
0.15	1.1618	.06 514	.86 071	**0.65**	1.9155	.28 229	.52 205
0.16	1.1735	.06 949	.85 214	0.66	1.9348	.28 663	.51 685
0.17	1.1853	.07 383	.84 366	0.67	1.9542	.29 098	.51 171
0.18	1.1972	.07 817	.83 527	0.68	1.9739	.29 532	.50 662
0.19	1.2092	.08 252	.82 696	0.69	1.9937	.29 966	.50 158
0.20	1.2214	.08 686	.81 873	**0.70**	2.0138	.30 401	.49 659
0.21	1.2337	.09 120	.81 058	0.71	2.0340	.30 835	.49 164
0.22	1.2461	.09 554	.80 252	0.72	2.0544	.31 269	.48 675
0.23	1.2586	.09 989	.79 453	0.73	2.0751	.31 703	.48 191
0.24	1.2712	.10 423	.78 663	0.74	2.0959	.32 138	.47 711
0.25	1.2840	.10 857	.77 880	**0.75**	2.1170	.32 572	.47 237
0.26	1.2969	.11 292	.77 105	0.76	2.1383	.33 006	.46 767
0.27	1.3100	.11 726	.76 338	0.77	2.1598	.33 441	.46 301
0.28	1.3231	.12 160	.75 578	0.78	2.1815	.33 875	.45 841
0.29	1.3364	.12 595	.74 826	0.79	2.2034	.34 309	.45 384
0.30	1.3499	.13 029	.74 082	**0.80**	2.2255	.34 744	.44 933
0.31	1.3634	.13 463	.73 345	0.81	2.2479	.35 178	.44 486
0.32	1.3771	.13 897	.72 615	0.82	2.2705	.35 612	.44 043
0.33	1.3910	.14 332	.71 892	0.83	2.2933	.36 046	.43 605
0.34	1.4049	.14 766	.71 177	0.84	2.3164	.36 481	.43 171
0.35	1.4191	.15 200	.70 469	**0.85**	2.3396	.36 915	.42 741
0.36	1.4333	.15 635	.69 768	0.86	2.3632	.37 349	.42 316
0.37	1.4477	.16 069	.69 073	0.87	2.3869	.37 784	.41 895
0.38	1.4623	.16 503	.68 386	0.88	2.4109	.38 218	.41 478
0.39	1.4770	.16 937	.67 706	0.89	2.4351	.38 652	.41 066
0.40	1.4918	.17 372	.67 032	**0.90**	2.4596	.39 087	.40 657
0.41	1.5068	.17 806	.66 365	0.91	2.4843	.39 521	.40 252
0.42	1.5220	.18 240	.65 705	0.92	2.5093	.39 955	.39 852
0.43	1.5373	.18 675	.65 051	0.93	2.5345	.40 389	.39 455
0.44	1.5527	.19 109	.64 404	0.94	2.5600	.40 824	.39 063
0.45	1.5683	.19 543	.63 763	**0.95**	2.5857	.41 258	.38 674
0.46	1.5841	.19 978	.63 128	0.96	2.6117	.41 692	.38 289
0.47	1.6000	.20 412	.62 500	0.97	2.6379	.42 127	.37 908
0.48	1.6161	.20 846	.61 878	0.98	2.6645	.42 561	.37 531
0.49	1.6323	.21 280	.61 263	0.99	2.6912	.42 995	.37 158
0.50	1.6487	.21 715	.60 653	**1.00**	2.7183	.43 429	.36 788
x	e^x	$\text{Log}_{10} e^x$	e^{-x}	x	e^x	$\text{Log}_{10} e^x$	e^{-x}

Table IV Exponential Functions

x	e^x	$\text{Log}_{10}\,e^x$	e^{-x}	x	e^x	$\text{Log}_{10}\,e^x$	e^{-x}
1.00	2.7183	.43 429	.36 788	**1.50**	4.4817	.65 144	.22 313
1.01	2.7456	.43 864	.36 422	1.51	4.5267	.65 578	.22 091
1.02	2.7732	.44 298	.36 059	1.52	4.5722	.66 013	.21 871
1.03	2.8011	.44 732	.35 701	1.53	4.6182	.66 447	.21 654
1.04	2.8292	.45 167	.35 345	1.54	4.6646	.66 881	.21 438
1.05	2.8577	.45 601	.34 994	**1.55**	4.7115	.67 316	.21 225
1.06	2.8864	.46 035	.34 646	1.56	4.7588	.67 750	.21 014
1.07	2.9154	.46 470	.34 301	1.57	4.8066	.68 184	.20 805
1.08	2.9447	.46 904	.33 960	1.58	4.8550	.68 619	.20 598
1.09	2.9743	.47 338	.33 622	1.59	4.9037	.69 053	.20 393
1.10	3.0042	.47 772	.33 287	**1.60**	4.9530	.69 487	.20 190
1.11	3.0344	.48 207	.32 956	1.61	5.0028	.69 921	.19 989
1.12	3.0649	.48 641	.32 628	1.62	5.0531	.70 356	.19 790
1.13	3.0957	.49 075	.32 303	1.63	5.1039	.70 790	.19 593
1.14	3.1268	.49 510	.31 982	1.64	5.1552	.71 224	.19 398
1.15	3.1582	.49 944	.31 664	**1.65**	5.2070	.71 659	.19 205
1.16	3.1899	.50 378	.31 349	1.66	5.2593	.72 093	.19 014
1.17	3.2220	.50 812	.31 037	1.67	5.3122	.72 527	.18 825
1.18	3.2544	.51 247	.30 728	1.68	5.3656	.72 961	.18 637
1.19	3.2871	.51 681	.30 422	1.69	5.4195	.73 396	.18 452
1.20	3.3201	.52 115	.30 119	**1.70**	5.4739	.73 830	.18 268
1.21	3.3535	.52 550	.29 820	1.71	5.5290	.74 264	.18 087
1.22	3.3872	.52 984	.29 523	1.72	5.5845	.74 699	.17 907
1.23	3.4212	.53 418	.29 229	1.73	5.6407	.75 133	.17 728
1.24	3.4556	.53 853	.28 938	1.74	5.6973	.75 567	.17 552
1.25	3.4903	.54 287	.28 650	**1.75**	5.7546	.76 002	.17 377
1.26	3.5254	.54 721	.28 365	1.76	5.8124	.76 436	.17 204
1.27	3.5609	.55 155	.28 083	1.77	5.8709	.76 870	.17 033
1.28	3.5966	.55 590	.27 804	1.78	5.9299	.77 304	.16 864
1.29	3.6328	.56 024	.27 527	1.79	5.9895	.77 739	.16 696
1.30	3.6693	.56 458	.27 253	**1.80**	6.0496	.78 173	.16 530
1.31	3.7062	.56 893	.26 982	1.81	6.1104	.78 607	.16 365
1.32	3.7434	.57 327	.26 714	1.82	6.1719	.79 042	.16 203
1.33	3.7810	.57 761	.26 448	1.83	6.2339	.79 476	.16 041
1.34	3.8190	.58 195	.26 185	1.84	6.2965	.79 910	.15 882
1.35	3.8574	.58 630	.25 924	**1.85**	6.3598	.80 344	.15 724
1.36	3.8962	.59 064	.25 666	1.86	6.4237	.80 779	.15 567
1.37	3.9354	.59 498	.25 411	1.87	6.4883	.81 213	.15 412
1.38	3.9749	.59 933	.25 158	1.88	6.5535	.81 647	.15 259
1.39	4.0149	.60 367	.24 908	1.89	6.6194	.82 082	.15 107
1.40	4.0552	.60 801	.24 660	**1.90**	6.6859	.82 516	.14 957
1.41	4.0960	.61 236	.24 414	1.91	6.7531	.82 950	.14 808
1.42	4.1371	.61 670	.24 171	1.92	6.8210	.83 385	.14 661
1.43	4.1787	.62 104	.23 931	1.93	6.8895	.83 819	.14 515
1.44	4.2207	.62 538	.23 693	1.94	6.9588	.84 253	.14 370
1.45	4.2631	.62 973	.23 457	**1.95**	7.0287	.84 687	.14 227
1.46	4.3060	.63 407	.23 224	1.96	7.0993	.85 122	.14 086
1.47	4.3492	.63 841	.22 993	1.97	7.1707	.85 556	.13 946
1.48	4.3929	.64 276	.22 764	1.98	7.2427	.85 990	.13 807
1.49	4.4371	.64 710	.22 537	1.99	7.3155	.86 425	.13 670
1.50	4.4817	.65 144	.22 313	**2.00**	7.3891	.86 859	.13 534
x	e^x	$\text{Log}_{10}\,e^x$	e^{-x}	x	e^x	$\text{Log}_{10}\,e^x$	e^{-x}

Table IV **Exponential Functions**

x	e^x	$\text{Log}_{10} e^x$	e^{-x}	x	e^x	$\text{Log}_{10} e^x$	e^{-x}
2.00	7.3891	.86 859	.13 534	**2.50**	12.182	1.08 574	.082 085
2.01	7.4633	.87 293	.13 399	2.51	12.305	1.09 008	.081 268
2.02	7.5383	.87 727	.13 266	2.52	12.429	1.09 442	.080 460
2.03	7.6141	.88 162	.13 134	2.53	12.554	1.09 877	.079 659
2.04	7.6906	.88 596	.13 003	2.54	12.680	1.10 311	.078 866
2.05	7.7679	.89 030	.12 873	**2.55**	12.807	1.10 745	.078 082
2.06	7.8460	.89 465	.12 745	2.56	12.936	1.11 179	.077 305
2.07	7.9248	.89 899	.12 619	2.57	13.066	1.11 614	.076 536
2.08	8.0045	.90 333	.12 493	2.58	13.197	1.12 048	.075 774
2.09	8.0849	.90 768	.12 369	2.59	13.330	1.12 482	.075 020
2.10	8.1662	.91 202	.12 246	**2.60**	13.464	1.12 917	.074 274
2.11	8.2482	.91 636	.12 124	2.61	13.599	1.13 351	.073 535
2.12	8.3311	.92 070	.12 003	2.62	13.736	1.13 785	.072 803
2.13	8.4149	.92 505	.11 884	2.63	13.874	1.14 219	.072 078
2.14	8.4994	.92 939	.11 765	2.64	14.013	1.14 654	.071 361
2.15	8.5849	.93 373	.11 648	**2.65**	14.154	1.15 088	.070 651
2.16	8.6711	.93 808	.11 533	2.66	14.296	1.15 522	.069 948
2.17	8.7583	.94 242	.11 418	2.67	14.440	1.15 957	.069 252
2.18	8.8463	.94 676	.11 304	2.68	14.585	1.16 391	.068 563
2.19	8.9352	.95 110	.11 192	2.69	14.732	1.16 825	.067 881
2.20	9.0250	.95 545	.11 080	**2.70**	14.880	1.17 260	.067 206
2.21	9.1157	.95 979	.10 970	2.71	15.029	1.17 694	.066 537
2.22	9.2073	.96 413	.10 861	2.72	15.180	1.18 128	.065 875
2.23	9.2999	.96 848	.10 753	2.73	15.333	1.18 562	.065 219
2.24	9.3933	.97 282	.10 646	2.74	15.487	1.18 997	.064 570
2.25	9.4877	.97 716	.10 540	**2.75**	15.643	1.19 431	.063 928
2.26	9.5831	.98 151	.10 435	2.76	15.800	1.19 865	.063 292
2.27	9.6794	.98 585	.10 331	2.77	15.959	1.20 300	.062 662
2.28	9.7767	.99 019	.10 228	2.78	16.119	1.20 734	.062 039
2.29	9.8749	.99 453	.10 127	2.79	16.281	1.21 168	.061 421
2.30	9.9742	.99 888	.10 026	**2.80**	16.445	1.21 602	.060 810
2.31	10.074	1.00 322	.09 9261	2.81	16.610	1.22 037	.060 205
2.32	10.176	1.00 756	.09 8274	2.82	16.777	1.22 471	.059 606
2.33	10.278	1.01 191	.09 7296	2.83	16.945	1.22 905	.059 013
2.34	10.381	1.01 625	.09 6328	2.84	17.116	1.23 340	.058 426
2.35	10.486	1.02 059	.09 5369	**2.85**	17.288	1.23 774	.057 844
2.36	10.591	1.02 493	.09 4420	2.86	17.462	1.24 208	.057 269
2.37	10.697	1.02 928	.09 3481	2.87	17.637	1.24 643	.056 699
2.38	10.805	1.03 362	.09 2551	2.88	17.814	1.25 077	.056 135
2.39	10.913	1.03 796	.09 1630	2.89	17.993	1.25 511	.055 576
2.40	11.023	1.04 231	.09 0718	**2.90**	18.174	1.25 945	.055 023
2.41	11.134	1.04 665	.08 9815	2.91	18.357	1.26 380	.054 476
2.42	11.246	1.05 099	.08 8922	2.92	18.541	1.26 814	.053 934
2.43	11.359	1.05 534	.08 8037	2.93	18.728	1.27 248	.053 397
2.44	11.473	1.05 968	.08 7161	2.94	18.916	1.27 683	.052 866
2.45	11.588	1.06 402	.08 6294	**2.95**	19.106	1.28 117	.052 340
2.46	11.705	1.06 836	.08 5435	2.96	19.298	1.28 551	.051 819
2.47	11.822	1.07 271	.08 4585	2.97	19.492	1.28 985	.051 303
2.48	11.941	1.07 705	.08 3743	2.98	19.688	1.29 420	.050 793
2.49	12.061	1.08 139	.08 2910	2.99	19.886	1.29 854	.050 287
2.50	12.182	1.08 574	.08 2085	**3.00**	20.086	1.30 288	.049 787
x	e^x	$\text{Log}_{10} e^x$	e^{-x}	x	e^x	$\text{Log}_{10} e^x$	e^{-x}

Table IV Exponential Functions

x	e^x	$\text{Log}_{10} e^x$	e^{-x}	x	e^x	$\text{Log}_{10} e^x$	e^{-x}
3.00	20.086	1.30 288	.04 9787	**3.50**	33.115	1.52 003	.030 197
3.01	20.287	1.30 723	.04 9292	3.51	33.448	1.52 437	.029 897
3.02	20.491	1.31 157	.04 8801	3.52	33.784	1.52 872	.029 599
3.03	20.697	1.31 591	.04 8316	3.53	34.124	1.53 306	.029 305
3.04	20.905	1.32 026	.04 7835	3.54	34.467	1.53 740	.029 013
3.05	21.115	1.32 460	.04 7359	**3.55**	34.813	1.54 175	.028 725
3.06	21.328	1.32 894	.04 6888	3.56	35.163	1.54 609	.028 439
3.07	21.542	1.33 328	.04 6421	3.57	35.517	1.55 043	.028 156
3.08	21.758	1.33 763	.04 5959	3.58	35.874	1.55 477	.027 876
3.09	21.977	1.34 197	.04 5502	3.59	36.234	1.55 912	.027 598
3.10	22.198	1.34 631	.04 5049	**3.60**	36.598	1.56 346	.027 324
3.11	22.421	1.35 066	.04 4601	3.61	36.966	1.56 780	.027 052
3.12	22.646	1.35 500	.04 4157	3.62	37.338	1.57 215	.026 783
3.13	22.874	1.35 934	.04 3718	3.63	37.713	1.57 649	.026 516
3.14	23.104	1.36 368	.04 3283	3.64	38.092	1.58 083	.026 252
3.15	23.336	1.36 803	.04 2852	**3.65**	38.475	1.58 517	.025 991
3.16	23.571	1.37 237	.04 2426	3.66	38.861	1.58 952	.025 733
3.17	23.807	1.37 671	.04 2004	3.67	39.252	1.59 386	.025 476
3.18	24.047	1.38 106	.04 1586	3.68	39.646	1.59 820	.025 223
3.19	24.288	1.38 540	.04 1172	3.69	40.045	1.60 255	.024 972
3.20	24.533	1.38 974	.04 0762	**3.70**	40.447	1.60 689	.024 724
3.21	24.779	1.39 409	.04 0357	3.71	40.854	1.61 123	.024 478
3.22	25.028	1.39 843	.03 9955	3.72	41.264	1.61 558	.024 234
3.23	25.280	1.40 277	.03 9557	3.73	41.679	1.61 992	.023 993
3.24	25.534	1.40 711	.03 9164	3.74	42.098	1.62 426	.023 754
3.25	25.790	1.41 146	.03 8774	**3.75**	42.521	1.62 860	.023 518
3.26	26.050	1.41 580	.03 8388	3.76	42.948	1.63 295	.023 284
3.27	26.311	1.42 014	.03 8006	3.77	43.380	1.63 729	.023 052
3.28	26.576	1.42 449	.03 7628	3.78	43.816	1.64 163	.022 823
3.29	26.843	1.42 883	.03 7254	3.79	44.256	1.64 598	.022 596
3.30	27.113	1.43 317	.03 6883	**3.80**	44.701	1.65 032	.022 371
3.31	27.385	1.43 751	.03 6516	3.81	45.150	1.65 466	.022 148
3.32	27.660	1.44 186	.03 6153	3.82	45.604	1.65 900	.021 928
3.33	27.938	1.44 620	.03 5793	3.83	46.063	1.66 335	.021 710
3.34	28.219	1.45 054	.03 5437	3.84	46.525	1.66 769	.021 494
3.35	28.503	1.45 489	.03 5084	**3.85**	46.993	1.67 203	.021 280
3.36	28.789	1.45 923	.03 4735	3.86	47.465	1.67 638	.021 068
3.37	29.079	1.46 357	.03 4390	3.87	47.942	1.68 072	.020 858
3.38	29.371	1.46 792	.03 4047	3.88	48.424	1.68 506	.020 651
3.39	29.666	1.47 226	.03 3709	3.89	48.911	1.68 941	.020 445
3.40	29.964	1.47 660	.03 3373	**3.90**	49.402	1.69 375	.020 242
3.41	30.265	1.48 094	.03 3041	3.91	49.899	1.69 809	.020 041
3.42	30.569	1.48 529	.03 2712	3.92	50.400	1.70 243	.019 841
3.43	30.877	1.48 963	.03 2387	3.93	50.907	1.70 678	.019 644
3.44	31.187	1.49 397	.03 2065	3.94	51.419	1.71 112	.019 448
3.45	31.500	1.49 832	.03 1746	**3.95**	51.935	1.71 546	.019 255
3.46	31.817	1.50 266	.03 1430	3.96	52.457	1.71 981	.019 063
3.47	32.137	1.50 700	.03 1117	3.97	52.985	1.72 415	.018 873
3.48	32.460	1.51 134	.03 0807	3.98	53.517	1.72 849	.018 686
3.49	32.786	1.51 569	.03 0501	3.99	54.055	1.73 283	.018 500
3.50	33.115	1.52 003	.03 0197	**4.00**	54.598	1.73 718	.018 316
x	e^x	$\text{Log}_{10} e^x$	e^{-x}	x	e^x	$\text{Log}_{10} e^x$	e^{-x}

Table IV Exponential Functions

x	e^x	$\text{Log}_{10}\, e^x$	e^{-x}	x	e^x	$\text{Log}_{10}\, e^x$	e^{-x}
4.00	54.598	1.73 718	.01 8316	**4.50**	90.017	1.95 433	.011 109
4.01	55.147	1.74 152	.01 8133	4.51	90.922	1.95 867	.010 998
4.02	55.701	1.74 586	.01 7953	4.52	91.836	1.96 301	.010 889
4.03	56.261	1.75 021	.01 7774	4.53	92.759	1.96 735	.010 781
4.04	56.826	1.75 455	.01 7597	4.54	93.691	1.97 170	.010 673
4.05	57.397	1.75 889	.01 7422	**4.55**	94.632	1 97 604	.010 567
4.06	57.974	1.76 324	.01 7249	4.56	95.583	1.98 038	.010 462
4.07	58.557	1.76 758	.01 7077	4.57	96.544	1.98 473	.010 358
4.08	59.145	1.77 192	.01 6907	4.58	97.514	1.98 907	.010 255
4.09	59.740	1.77 626	.01 6739	4.59	98.494	1.99 341	.010 153
4.10	60.340	1.78 061	.01 6573	**4.60**	99.484	1.99 775	.010 052
4.11	60.947	1.78 495	.01 6408	4.61	100.48	2.00 210	.009 9518
4.12	61.559	1.78 929	.01 6245	4.62	101.49	2.00 644	.009 8528
4.13	62.178	1.79 364	.01 6083	4.63	102.51	2.01 078	.009 7548
4.14	62.803	1.79 798	.01 5923	4.64	103.54	2.01 513	.009 6577
4.15	63.434	1.80 232	.01 5764	**4.65**	104.58	2.01 947	.009 5616
4.16	64.072	1.80 667	.01 5608	4.66	105.64	2.02 381	.009 4665
4.17	64.715	1.81 101	.01 5452	4.67	106.70	2.02 816	.009 3723
4.18	65.366	1.81 535	.01 5299	4.68	107.77	2.03 250	.009 2790
4.19	66.023	1.81 969	.01 5146	4.69	108.85	2.03 684	.009 1867
4.20	66.686	1.82 404	.01 4996	**4.70**	109.95	2.04 118	.009 0953
4.21	67.357	1.82 838	.01 4846	4.71	111.05	2.04 553	.009 0048
4.22	68.033	1.83 272	.01 4699	4.72	112.17	2.04 987	.008 9152
4.23	68.717	1.83 707	.01 4552	4.73	113.30	2.05 421	.008 8265
4.24	69.408	1.84 141	.01 4408	4.74	114.43	2.05 856	.008 7386
4.25	70.105	1.84 575	.01 4264	**4.75**	115.58	2.06 290	.008 6517
4.26	70.810	1.85 009	.01 4122	4.76	116.75	2.06 724	.008 5656
4.27	71.522	1.85 444	.01 3982	4.77	117.92	2.07 158	.008 4804
4.28	72.240	1.85 878	.01 3843	4.78	119.10	2.07 593	.008 3960
4.29	72.966	1.86 312	.01 3705	4.79	120.30	2.08 027	.008 3125
4.30	73.700	1.86 747	.01 3569	**4.80**	121.51	2.08 461	.008 2297
4.31	74.440	1.87 181	.01 3434	4.81	122.73	2.08 896	.008 1479
4.32	75.189	1.87 615	.01 3300	4.82	123.97	2.09 330	.008 0668
4.33	75.944	1.88 050	.01 3168	4.83	125.21	2.09 764	.007 9865
4.34	76.708	1.88 484	.01 3037	4.84	126.47	2.10 199	.007 9071
4.35	77.478	1.88 918	.01 2907	**4.85**	127.74	2.10 633	.007 8284
4.36	78.257	1.89 352	.01 2778	4.86	129.02	2.11 067	.007 7505
4.37	79.044	1.89 787	.01 2651	4.87	130.32	2.11 501	.007 6734
4.38	79.838	1.90 221	.01 2525	4.88	131.63	2.11 936	.007 5970
4.39	80.640	1.90 655	.01 2401	4.89	132.95	2.12 370	.007 5214
4.40	81.451	1.91 090	.01 2277	**4.90**	134.29	2.12 804	.007 4466
4.41	82.269	1.91 524	.01 2155	4.91	135.64	2.13 239	.007 3725
4.42	83.096	1.91 958	.01 2034	4.92	137.00	2.13 673	.007 2991
4.43	83.931	1.92 392	.01 1914	4.93	138.38	2.14 107	.007 2265
4.44	84.775	1.92 827	.01 1796	4.94	139.77	2.14 541	.007 1546
4.45	85.627	1.93 261	.01 1679	**4.95**	141.17	2.14 976	.007 0834
4.46	86.488	1.93 695	.01 1562	4.96	142.59	2.15 410	.007 0129
4.47	87.357	1.94 130	.01 1447	4.97	144.03	2.15 844	.006 9431
4.48	88.235	1.94 564	.01 1333	4.98	145.47	2.16 279	.006 8741
4.49	89.121	1.94 998	.01 1221	4.99	146.94	2.16 713	.006 8057
4.50	90.017	1.95 433	.01 1109	**5.00**	148.41	2.17 147	.006 7379
x	e^x	$\text{Log}_{10}\, e^x$	e^{-x}	x	e^x	$\text{Log}_{10}\, e^x$	e^{-x}

Table IV Exponential Functions

x	e^x	$\text{Log}_{10} e^x$	e^{-x}	x	e^x	$\text{Log}_{10} e^x$	e^{-x}
5.00	148.41	2.17 147	.00 67379	**7.50**	1 808.0	3.25 721	.000 5531
5.05	156.02	2.19 319	.00 64093	7.55	1 900.7	3.27 892	.000 5261
5.10	164.02	2.21 490	.00 60967	7.60	1 998.2	3.30 064	.000 5005
5.15	172.43	2.23 662	.00 57994	7.65	2 100.6	3.32 235	.000 4760
5.20	181.27	2.25 833	.00 55166	7.70	2 208.3	3.34 407	.000 4528
5.25	190.57	2.28 005	.00 52475	**7.75**	2 321.6	3.36 578	.000 4307
5.30	200.34	2.30 176	.00 49916	7.80	2 440.6	3.38 750	.000 4097
5.35	210.61	2.32 348	.00 47482	7.85	2 565.7	3.40 921	.000 3898
5.40	221.41	2.34 519	.00 45166	7.90	2 697.3	3.43 093	.000 3707
5.45	232.76	2.36 690	.00 42963	7.95	2 835.6	3.45 264	.000 3527
5.50	244.69	2.38 862	.00 40868	**8.00**	2 981.0	3.47 436	.000 3355
5.55	257.24	2.41 033	.00 38875	8.05	3 133.8	3.49 607	.000 3191
5.60	270.43	2.43 205	.00 36979	8.10	3 294.5	3.51 779	.000 3035
5.65	284.29	2.45 376	.00 35175	8.15	3 463.4	3.53 950	.000 2887
5.70	298.87	2.47 548	.00 33460	8.20	3 641.0	3.56 121	.000 2747
5.75	314.19	2.49 719	.00 31828	**8.25**	3 827.6	3.58 293	.000 2613
5.80	330.30	2.51 891	.00 30276	8.30	4 023.9	3.60 464	.000 2485
5.85	347.23	2.54 062	.00 28799	8.35	4 230.2	3.62 636	.000 2364
5.90	365.04	2.56 234	.00 27394	8.40	4 447.1	3.64 807	.000 2249
5.95	383.75	2.58 405	.00 26058	8.45	4 675.1	3.66 979	.000 2139
6.00	403.43	2.60 577	.00 24788	**8.50**	4 914.8	3.69 150	.000 2035
6.05	424.11	2.62 748	.00 23579	8.55	5 166.8	3.71 322	.000 1935
6.10	445.86	2.64 920	.00 22429	8.60	5 431.7	3.73 493	.000 1841
6.15	468.72	2.67 091	.00 21335	8.65	5 710.1	3.75 665	.000 1751
6.20	492.75	2.69 263	.00 20294	8.70	6 002.9	3.77 836	.000 1666
6.25	518.01	2.71 434	.00 19305	**8.75**	6 310.7	3.80 008	.000 1585
6.30	544.57	2.73 606	.00 18363	8.80	6 634.2	3.82 179	.000 1507
6.35	572.49	2.75 777	.00 17467	8.85	6 974.4	3.84 351	.000 1434
6.40	601.85	2.77 948	.00 16616	8.90	7 332.0	3.86 522	.000 1364
6.45	632.70	2.80 120	.00 15805	8.95	7 707.9	3.88 694	.000 1297
6.50	665.14	2.82 291	.00 15034	**9.00**	8 103.1	3.90 865	.000 1234
6.55	699.24	2.84 463	.00 14301	9.05	8 518.5	3.93 037	.000 1174
6.60	735.10	2.86 634	.00 13604	9.10	8 955.3	3.95 208	.000 1117
6.65	772.78	2.88 806	.00 12940	9.15	9 414.4	3.97 379	.000 1062
6.70	812.41	2.90 977	.00 12309	9.20	9 897.1	3.99 551	.000 1010
6.75	854.06	2.93 149	.00 11709	**9.25**	10 405	4.01 722	.000 0961
6.80	897.85	2.95 320	.00 11138	9.30	10 938	4.03 894	.000 0914
6.85	943.88	2.97 492	.00 10595	9.35	11 499	4.06 065	.000 0870
6.90	992.27	2.99 663	.00 10078	9.40	12 088	4.08 237	.000 0827
6.95	1 043.1	3.01 835	.00 09586	9.45	12 708	4.10 408	.000 0787
7.00	1 096.6	3.04 006	.00 09119	**9.50**	13 360	4.12 580	.000 0749
7.05	1 152.9	3.06 178	.00 08674	9.55	14 045	4.14 751	.000 0712
7.10	1 212.0	3.08 349	.00 08251	9.60	14 765	4.16 923	.000 0677
7.15	1 274.1	3.10 521	.00 07849	9.65	15 522	4.19 094	.000 0644
7.20	1 339.4	3.12 692	.00 07466	9.70	16 318	4.21 266	.000 0613
7.25	1 408.1	3.14 863	.00 07102	**9.75**	17 154	4.23 437	.000 0583
7.30	1 480.3	3.17 035	.00 06755	9.80	18 034	4.25 609	.000 0555
7.35	1 556.2	3.19 206	.00 06426	9.85	18 958	4.27 780	.000 0527
7.40	1 636.0	3.21 378	.00 06113	9.90	19 930	4.29 952	.000 0502
7.45	1 719.9	3.23 549	.00 05814	9.95	20 952	4.32 123	.000 0477
7.50	1 808.0	3.25 721	.00 05531	**10.00**	22 026	4.34 294	.000 0454
x	e^x	$\text{Log}_{10} e^x$	e^{-x}	x	e^x	$\text{Log}_{10} e^x$	e^{-x}

Table V Natural Logarithms

	.00	.01	.02	.03	.04	.05	.06	.07	.08	.09
1.0	0.0000	0.0100	0.0198	0.0296	0.0392	0.0488	0.0583	0.0677	0.0770	0.0862
1.1	.0953	.1044	.1133	.1222	.1310	.1398	.1484	.1570	.1655	.1740
1.2	.1823	.1906	.1989	.2070	.2151	.2231	.2311	.2390	.2469	.2546
1.3	.2624	.2700	.2776	.2852	.2927	.3001	.3075	.3148	.3221	.3293
1.4	.3365	.3436	.3507	.3577	.3646	.3716	.3784	.3853	.3920	.3988
1.5	.4055	.4121	.4187	.4253	.4318	.4383	.4447	.4511	.4574	.4637
1.6	.4700	.4762	.4824	.4886	.4947	.5008	.5068	.5128	.5188	.5247
1.7	.5306	.5365	.5423	.5481	.5539	.5596	.5653	.5710	.5766	.5822
1.8	.5878	.5933	.5988	.6043	.6098	.6152	.6206	.6259	.6313	.6366
1.9	.6419	.6471	.6523	.6575	.6627	.6678	.6729	.6780	.6831	.6881
2.0	.6932	.6981	.7031	.7080	.7130	.7178	.7227	.7276	.7324	.7372
2.1	.7419	.7467	.7514	.7561	.7608	.7655	.7701	.7747	.7793	.7839
2.2	.7885	.7930	.7975	.8020	.8065	.8109	.8154	.8198	.8242	.8286
2.3	.8329	.8373	.8416	.8459	.8502	.8544	.8587	.8629	.8671	.8713
2.4	.8755	.8796	.8838	.8879	.8920	.8961	.9002	.9042	.9083	.9123
2.5	.9163	.9203	.9243	.9282	.9322	.9361	.9400	.9439	.9478	.9517
2.6	.9555	.9594	0.9632	0.9670	0.9708	0.9746	0.9783	0.9821	0.9858	0.9895
2.7	0.9933	0.9970	1.0006	1.0043	1.0080	1.0116	1.0152	1.0189	1.0225	1.0260
2.8	1.0296	1.0332	.0367	.0403	.0438	.0473	.0508	1.0543	.0578	.0613
2.9	.0647	.0682	.0716	.0750	.0784	.0818	.0852	1.0886	.0919	.0953
3.0	.0986	.1019	.1053	.1086	.1119	.1151	.1184	.1217	.1249	.1282
3.1	.1314	.1346	.1378	.1410	.1442	.1474	.1506	.1537	.1569	.1600
3.2	.1632	.1663	.1694	.1725	.1756	.1787	.1817	.1848	.1878	.1909
3.3	.1939	.1970	.2000	.2030	.2060	.2090	.2119	.2149	.2179	.2208
3.4	.2238	.2267	.2296	.2326	.2355	.2384	.2413	.2442	.2470	.2499
3.5	.2528	.2556	.2585	.2613	.2641	.2670	.2698	.2726	.2754	.2782
3.6	.2809	.2837	.2865	.2892	.2920	.2947	.2975	.3002	.3029	.3056
3.7	.3083	.3110	.3137	.3164	.3191	.3218	.3244	.3271	.3297	.3324
3.8	.3350	.3376	.3403	.3429	.3455	.3481	.3507	.3533	.3558	.3584
3.9	.3610	.3635	.3661	.3686	.3712	.3737	.3762	.3788	.3813	.3838
4.0	.3863	.3888	.3913	.3938	.3962	.3987	.4012	.4036	.4061	.4085
4.1	.4110	.4134	.4159	.4183	.4207	.4231	.4255	.4279	.4303	.4327
4.2	.4351	.4375	.4398	.4422	.4446	.4469	.4493	.4516	.4540	.4563
4.3	.4586	.4609	.4633	.4656	.4679	.4702	.4725	.4748	.4771	.4793
4.4	.4816	.4839	.4861	.4884	.4907	.4929	.4952	.4974	.4996	.5019
4.5	.5041	.5063	.5085	.5107	.5129	.5151	.5173	.5195	.5217	.5239
4.6	.5261	.5282	.5304	.5326	.5347	.5369	.5390	.5412	.5433	.5454
4.7	.5476	.5497	.5518	.5539	.5560	.5581	.5603	.5624	.5644	.5665
4.8	.5686	.5707	.5728	.5749	.5769	.5790	.5810	.5831	.5852	.5872
4.9	.5892	.5913	.5933	.5953	.5974	.5994	.6014	.6034	.6054	.6074
5.0	.6094	.6114	.6134	.6154	.6174	.6194	.6214	.6233	.6253	.6273
5.1	.6292	.6312	.6332	.6351	.6371	.6390	.6409	.6429	.6448	.6467
5.2	.6487	.6506	.6525	.6544	.6563	.6582	.6601	.6620	.6639	.6658
5.3	.6677	.6696	.6715	.6734	.6752	.6771	.6790	.6808	.6827	.6846
5.4	1.6864	1.6883	1.6901	1.6919	1.6938	1.6956	1.6975	1.6993	1.7011	1.7029

$$\log_e N = 2.30259 \log_{10} N$$
$$\log_{10} N = 0.43429 \log_e N$$

Table V Natural Logarithms

	.00	.01	.02	.03	.04	.05	.06	.07	.08	.09
5.5	1.7048	1.7066	1.7084	1.7102	1.7120	1.7138	1.7156	1.7174	1.7192	1.7210
5.6	.7228	.7246	.7263	.7281	.7299	.7317	.7334	.7352	.7370	.7387
5.7	.7405	.7422	.7440	.7457	.7475	.7492	.7509	.7527	.7544	.7561
5.8	.7579	.7596	.7613	.7630	.7647	.7664	.7682	.7699	.7716	.7733
5.9	.7750	.7767	.7783	.7800	.7817	.7834	.7851	.7868	.7884	.7901
6.0	.7918	.7934	.7951	.7968	.7984	.8001	.8017	.8034	.8050	.8067
6.1	.8083	.8099	.8116	.8132	.8148	.8165	.8181	.8197	.8213	.8229
6.2	.8246	.8262	.8278	.8294	.8310	.8326	.8342	.8358	.8374	.8390
6.3	.8406	.8421	.8437	.8453	.8469	.8485	.8500	.8516	.8532	.8547
6.4	.8563	.8579	.8594	.8610	.8625	.8641	.8656	.8672	.8687	.8703
6.5	.8718	.8733	.8749	.8764	.8779	.8795	.8810	.8825	.8840	.8856
6.6	.8871	.8886	.8901	.8916	.8931	.8946	.8961	.8976	.8991	.9006
6.7	.9021	.9036	.9051	.9066	.9081	.9095	.9110	.9125	.9140	.9155
6.8	.9169	.9184	.9199	.9213	.9228	.9243	.9257	.9272	.9286	.9301
6.9	.9315	.9330	.9344	.9359	.9373	.9387	.9402	.9416	.9431	.9445
7.0	.9459	.9473	.9488	.9502	.9516	.9530	.9545	.9559	.9573	.9587
7.1	.9601	.9615	.9629	.9643	.9657	.9671	.9685	.9699	.9713	.9727
7.2	.9741	.9755	.9769	.9782	.9796	.9810	.9824	.9838	.9851	1.9865
7.3	1.9879	1.9892	1.9906	1.9920	1.9933	1.9947	1.9961	1.9974	1.9988	2.0001
7.4	2.0015	2.0028	2.0042	2.0055	2.0069	2.0082	2.0096	2.0109	2.0122	.0136
7.5	.0149	.0162	.0176	.0189	.0202	.0216	.0229	.0242	.0255	.0268
7.6	.0282	.0295	.0308	.0321	.0334	.0347	.0360	.0373	.0386	.0399
7.7	.0412	.0425	.0438	.0451	.0464	.0477	.0490	.0503	.0516	.0528
7.8	.0541	.0554	.0567	.0580	.0592	.0605	.0618	.0631	.0643	.0656
7.9	.0669	.0681	.0694	.0707	.0719	.0732	.0744	.0757	.0769	.0782
8.0	.0794	.0807	.0819	.0832	.0844	.0857	.0869	.0882	.0894	.0906
8.1	.0919	.0931	.0943	.0956	.0968	.0980	.0992	.1005	.1017	.1029
8.2	.1041	.1054	.1066	.1078	.1090	.1102	.1114	.1126	.1138	.1151
8.3	.1163	.1175	.1187	.1199	.1211	.1223	.1235	.1247	.1259	.1270
8.4	.1282	.1294	.1306	.1318	.1330	.1342	.1354	.1365	.1377	.1389
8.5	.1401	.1412	.1424	.1436	.1448	.1459	.1471	.1483	.1494	.1506
8.6	.1518	.1529	.1541	.1552	.1564	.1576	.1587	.1599	.1610	.1622
8.7	.1633	.1645	.1656	.1668	.1679	.1691	.1702	.1713	.1725	.1736
8.8	.1748	.1759	.1770	.1782	.1793	.1804	.1816	.1827	.1838	.1849
8.9	.1861	.1872	.1883	.1894	.1905	.1917	.1928	.1939	.1950	.1961
9.0	.1972	.1983	.1994	.2006	.2017	.2028	.2039	.2050	.2061	.2072
9.1	.2083	.2094	.2105	.2116	.2127	.2138	.2149	.2159	.2170	.2181
9.2	.2192	.2203	.2214	.2225	.2235	.2246	.2257	.2268	.2279	.2289
9.3	.2300	.2311	.2322	.2332	.2343	.2354	.2365	.2375	.2386	.2397
9.4	.2407	.2418	.2428	.2439	.2450	.2460	.2471	.2481	.2492	.2502
9.5	.2513	.2523	.2534	.2544	.2555	.2565	.2576	.2586	.2597	.2607
9.6	.2618	.2628	.2638	.2649	.2659	.2670	.2680	.2690	.2701	.2711
9.7	.2721	.2732	.2742	.2752	.2762	.2773	.2783	.2793	.2803	.2814
9.8	.2824	.2834	.2844	.2854	.2865	.2875	.2885	.2895	.2905	.2915
9.9	2.2925	2.2935	2.2946	2.2956	2.2966	2.2976	2.2986	2.2996	2.3006	2.3016

$$\log_e 10 = 2.30259$$
$$\log_e (10^k N) = \log_e N + 2.30259 k$$

Table VI   Amount of an Annuity of 1 per Annum

$$\frac{(1+r)^n - 1}{r}$$

n	$\frac{5}{12}\%$	$\frac{1}{2}\%$	$\frac{7}{12}\%$	$\frac{3}{4}\%$	1%
1	1.000 000	1.000 000	1.000 000	1.000 000	1.000 000
2	2.004 167	2.005 000	2.005 833	2.007 500	2.010 000
3	3.012 517	3.015 025	3.017 534	3.022 556	3.030 100
4	4.025 070	4.030 100	4.035 136	4.045 225	4.060 401
5	5.041 841	5.050 251	5.058 675	5.075 565	5.101 005
6	6.062 848	6.075 502	6.088 184	6.113 631	6.152 015
7	7.088 110	7.105 879	7.123 698	7.159 484	7.213 535
8	8.117 644	8.141 409	8.165 253	8.213 180	8.285 671
9	9.151 467	9.182 116	9.212 883	9.274 779	9.368 527
10	10.189 599	10.228 026	10.266 625	10.344 339	10.462 213
11	11.232 055	11.279 167	11.326 514	11.421 922	11.566 835
12	12.278 855	12.335 562	12.392 585	12.507 586	12.682 503
13	13.330 017	13.397 240	13.464 875	13.601 393	13.809 328
14	14.385 559	14.464 226	14.543 420	14.703 404	14.947 421
15	15.445 499	15.536 548	15.628 257	15.813 679	16.096 896
16	16.509 855	16.614 230	16.719 422	16.932 282	17.257 864
17	17.578 646	17.697 301	17.816 952	18.059 274	18.430 443
18	18.651 891	18.785 788	18.920 884	19.194 718	19.614 748
19	19.729 607	19.879 717	20.031 256	20.338 679	20.810 895
20	20.811 814	20.979 115	21.148 105	21.491 219	22.019 004
21	21.898 529	22.084 011	22.271 469	22.652 403	23.239 194
22	22.989 773	23.194 431	23.401 386	23.822 296	24.471 586
23	24.085 564	24.310 403	24.537 894	25.000 963	25.716 302
24	25.185 921	25.431 955	25.681 032	26.188 471	26.973 465
25	26.290 862	26.559 115	26.830 838	27.384 884	28.243 200
26	27.400 407	27.691 911	27.987 351	28.590 271	29.525 631
27	28.514 575	28.830 370	29.150 610	29.804 698	30.820 888
28	29.633 386	29.974 522	30.320 656	31.028 233	32.129 097
29	30.756 859	31.124 395	31.497 526	32.260 945	33.450 388
30	31.885 012	32.280 017	32.681 262	33.502 902	34.784 892
31	33.017 866	33.441 417	33.871 902	34.754 174	36.132 740
32	34.155 441	34.608 624	35.069 488	36.014 830	37.494 068
33	35.297 755	35.781 667	36.274 060	37.284 941	38.869 009
34	36.444 829	36.960 575	37.485 659	38.564 578	40.257 699
35	37.596 683	38.145 378	38.704 325	39.853 813	41.660 276
36	38.753 336	39.336 105	39.930 101	41.152 716	43.076 878
37	39.914 808	40.532 785	41.163 026	42.461 361	44.507 647
38	41.081 119	41.735 449	42.403 144	43.779 822	45.952 724
39	42.252 291	42.944 127	43.650 496	45.108 170	47.412 251
40	43.428 342	44.158 847	44.905 124	46.446 482	48.886 373
41	44.609 293	45.379 642	46.167 070	47.794 830	50.375 237
42	45.795 165	46.606 540	47.436 378	49.153 291	51.878 989
43	46.985 979	47.839 572	48.713 090	50.521 941	53.397 779
44	48.181 754	49.078 770	49.997 250	51.900 856	54.931 759
45	49.382 511	50.324 164	51.288 900	53.290 112	56.481 075
46	50.588 271	51.575 785	52.588 086	54.689 788	58.045 885
47	51.799 056	52.833 664	53.894 850	56.099 961	59.626 344
48	53.014 885	54.097 832	55.209 236	57.520 711	61.222 608
49	54.235 781	55.368 321	56.531 290	58.952 116	62.834 834
50	55.461 763	56.645 163	57.861 056	60.394 257	64.463 182

Table VI   Amount of an Annuity of 1 per Annum

$$\frac{(1+r)^n - 1}{r}$$

n	$\frac{5}{12}\%$	$\frac{1}{2}\%$	$\frac{7}{12}\%$	$\frac{3}{4}\%$	1%
51	56.692 854	57.928 389	59.198 579	61.847 214	66.107 814
52	57.929 074	59.218 031	60.543 904	63.311 068	67.768 892
53	59.170 445	60.514 121	61.897 077	64.785 901	69.446 581
54	60.416 989	61.816 692	63.258 143	66.271 796	71.141 047
55	61.668 726	63.125 775	64.627 149	67.768 834	72.852 457
56	62.925 679	64.441 404	66.004 140	69.277 100	74.580 982
57	64.187 869	65.763 611	67.389 165	70.796 679	76.326 792
58	65.455 319	67.092 429	68.782 268	72.327 654	78.090 060
59	66.728 049	68.427 891	70.183 498	73.870 111	79.870 960
60	68.006 083	69.770 031	71.592 902	75.424 137	81.669 670
61	69.289 442	71.118 881	73.010 527	76.989 818	83.486 367
62	70.578 148	72.474 475	74.436 422	78.567 242	85.321 230
63	71.872 223	73.836 847	75.870 634	80.156 496	87.174 443
64	73.171 691	75.206 032	77.313 213	81.757 670	89.046 187
65	74.476 573	76.582 062	78.764 207	83.370 852	90.936 649
66	75.786 892	77.964 972	80.223 664	84.996 134	92.846 015
67	77.102 671	79.354 797	81.691 636	86.633 605	94.774 475
68	78.423 932	80.751 571	83.168 170	88.283 357	96.722 220
69	79.750 698	82.155 329	84.653 318	89.945 482	98.689 442
70	81.082 993	83.566 105	86.147 129	91.620 073	100.676 337
71	82.420 838	84.983 936	87.649 654	93.307 223	102.683 100
72	83.764 259	86.408 856	89.160 944	95.007 028	104.709 931
73	85.113 276	87.840 900	90.681 049	96.719 580	106.757 031
74	86.467 915	89.280 104	92.210 022	98.444 977	108.824 601
75	87.828 198	90.726 505	93.747 914	100.183 314	110.912 847
76	89.194 149	92.180 138	95.294 777	101.934 689	113.021 975
77	90.565 791	93.641 038	96.850 663	103.699 199	115.152 195
78	91.943 149	95.109 243	98.415 625	105.476 943	117.303 717
79	93.326 245	96.584 790	99.989 716	107.268 021	119.476 754
80	94.715 104	98.067 714	101.572 989	109.072 531	121.671 522
81	96.109 751	99.558 052	103.165 498	110.890 575	123.888 237
82	97.510 208	101.055 842	104.767 297	112.722 254	126.127 119
83	98.916 500	102.561 122	106.378 440	114.567 671	128.388 390
84	100.328 653	104.073 927	107.998 981	116.426 928	130.672 274
85	101.746 689	105.594 297	109.628 975	118.300 130	132.978 997
86	103.170 633	107.122 268	111.268 477	120.187 381	135.308 787
87	104.600 511	108.657 880	112.917 543	122.088 787	137.661 875
88	106.036 346	110.201 169	114.576 229	124.004 453	140.038 494
89	107.478 164	111.752 175	116.244 590	125.934 486	142.438 879
90	108.925 990	113.310 936	117.922 684	127.878 995	144.863 267
91	110.379 848	114.877 490	119.610 566	129.838 087	147.311 900
92	111.839 764	116.451 878	121.308 294	131.811 873	149.785 019
93	113.305 763	118.034 137	123.015 926	133.800 462	152.282 869
94	114.777 871	119.624 308	124.733 519	135.803 965	154.805 698
95	116.256 112	121.222 430	126.461 131	137.822 495	157.353 755
96	117.740 512	122.828 542	128.198 821	139.856 164	159.927 293
97	119.231 098	124.442 684	129.946 647	141.905 085	162.526 565
98	120.727 894	126.064 898	131.704 670	143.969 373	165.151 831
99	122.230 927	127.695 222	133.472 947	146.049 143	167.803 349
100	123.740 222	129.333 698	135.251 539	148.144 512	170.481 383

Table VI   Amount of an Annuity of 1 per Annum

$$\frac{(1+r)^n - 1}{r}$$

n	$1\frac{1}{8}\%$	$1\frac{1}{4}\%$	$1\frac{1}{2}\%$	$1\frac{3}{4}\%$	2%
1	1.000 000	1.000 000	1.000 000	1.000 000	1.000 000
2	2.011 250	2.012 500	2.015 000	2.017 500	2.020 000
3	3.033 877	3.037 656	3.045 225	3.052 806	3.060 400
4	4.068 008	4.075 627	4.090 903	4.106 230	4.121 608
5	5.113 773	5.126 572	5.152 267	5.178 089	5.204 040
6	6.171 303	6.190 654	6.229 551	6.268 706	6.308 121
7	7.240 730	7.268 038	7.322 994	7.378 408	7.434 283
8	8.322 188	8.358 888	8.432 839	8.507 530	8.582 969
9	9.415 813	9.463 374	9.559 332	9.656 412	9.754 628
10	10.521 741	10.581 666	10.702 722	10.825 399	10.949 721
11	11.640 110	11.713 937	11.863 262	12.014 844	12.168 715
12	12.771 061	12.860 361	13.041 211	13.225 104	13.412 090
13	13.914 736	14.021 116	14.236 830	14.456 543	14.680 332
14	15.071 277	15.196 380	15.450 382	15.709 533	15.973 938
15	16.240 828	16.386 335	16.682 138	16.984 449	17.293 417
16	17.423 538	17.591 164	17.932 370	18.281 677	18.639 285
17	18.619 553	18.811 053	19.201 355	19.601 607	20.012 071
18	19.829 023	20.046 192	20.489 376	20.944 635	21.412 312
19	21.052 099	21.296 769	21.796 716	22.311 166	22.840 559
20	22.288 935	22.562 979	23.123 667	23.701 611	24.297 370
21	23.539 686	23.845 016	24.470 522	25.116 389	25.783 317
22	24.804 507	25.143 078	25.837 580	26.555 926	27.298 984
23	26.083 558	26.457 367	27.225 144	28.020 655	28.844 963
24	27.376 998	27.788 084	28.633 521	29.511 016	30.421 862
25	28.684 989	29.135 435	30.063 024	31.027 459	32.030 300
26	30.007 695	30.499 628	31.513 969	32.570 440	33.670 906
27	31.345 282	31.880 873	32.986 678	34.140 422	35.344 324
28	32.697 916	33.279 384	34.481 479	35.737 880	37.051 210
29	34.065 768	34.695 377	35.998 701	37.363 293	38.792 235
30	35.449 008	36.129 069	37.538 681	39.017 150	40.568 079
31	36.847 809	37.580 682	39.101 762	40.699 950	42.379 441
32	38.262 347	39.050 441	40.688 288	42.412 200	44.227 030
33	39.692 798	40.538 571	42.298 612	44.154 413	46.111 570
34	41.139 342	42.045 303	43.933 092	45.927 115	48.033 802
35	42.602 160	43.570 870	45.592 088	47.730 840	49.994 478
36	44.081 434	45.115 505	47.275 969	49.566 129	51.994 367
37	45.577 350	46.679 449	48.985 109	51.433 537	54.034 255
38	47.090 095	48.262 942	50.719 885	53.333 624	56.114 940
39	48.619 859	49.866 229	52.480 684	55.266 962	58.237 238
40	50.166 832	51.489 557	54.267 894	57.234 134	60.401 983
41	51.731 209	53.133 177	56.081 912	59.235 731	62.610 023
42	53.313 185	54.797 341	57.923 141	61.272 357	64.862 223
43	54.912 959	56.482 308	59.791 988	63.344 623	67.159 468
44	56.530 730	58.188 337	61.688 868	65.453 154	69.502 657
45	58.166 700	59.915 691	63.614 201	67.598 584	71.892 710
46	59.821 076	61.664 637	65.568 414	69.781 559	74.330 564
47	61.494 063	63.435 445	67.551 940	72.002 736	76.817 176
48	63.185 871	65.228 388	69.565 219	74.262 784	79.353 519
49	64.896 712	67.043 743	71.608 698	76.562 383	81.940 590
50	66.626 800	68.881 790	73.682 828	78.902 225	84.579 401

Table VI Amount of an Annuity of 1 per Annum

$$\frac{(1+r)^n - 1}{r}$$

n	$1\frac{1}{8}\%$	$1\frac{1}{4}\%$	$1\frac{1}{2}\%$	$1\frac{3}{4}\%$	2%
51	68.376 352	70.742 812	75.788 070	81.283 014	87.270 989
52	70.145 585	72.627 097	77.924 892	83.705 466	90.016 409
53	71.934 723	74.534 936	80.093 765	86.170 312	92.816 737
54	73.743 989	76.466 623	82.295 171	88.678 292	95.673 072
55	75.573 609	78.422 456	84.529 599	91.230 163	98.586 534
56	77.423 812	80.402 736	86.797 543	93.826 690	101.558 264
57	79.294 830	82.407 771	89.099 506	96.468 658	104.589 430
58	81.186 897	84.437 868	91.435 999	99.156 859	107.681 218
59	83.100 249	86.493 341	93.807 539	101.892 104	110.834 843
60	85.035 127	88.574 508	96.214 652	104.675 216	114.051 539
61	86.991 772	90.681 689	98.657.871	107.507 032	117.332 570
62	88.970 430	92.815 210	101.137 740	110.388 405	120.679 222
63	90.971 347	94.975 400	103.654 806	113.320 202	124.092 806
64	92.994 775	97.162 593	106.209 628	116.303 306	127.574 662
65	95.040 966	99.377 125	108.802 772	119.338 614	131.126 155
66	97.110 177	101.619 339	111.434 814	122.427 039	134.748 679
67	99.202 666	103.889 581	114.106 336	125.569 513	138.443 652
68	101.318 696	106.188 201	116.817 931	128.766 979	142.212 525
69	103.458 532	108.515 553	119.570 200	132.020 401	146.056 776
70	105.622 440	110.871 998	122.363 753	135.330 758	149.977 911
71	107.810 692	113.257 898	125.199 209	138.699 047	153.977 469
72	110.023 563	115.673 621	128.077 197	142.126 280	158.057 019
73	112.261 328	118.119 542	130.998 355	145.613 490	162.218 159
74	114.524 268	120.596 036	133.963 331	149.161 726	166.462 522
75	116.812 666	123.103 486	136.972 781	152.772 056	170.791 773
76	119.126 808	125.642 280	140.027 372	156.445 567	175.207 608
77	121.466 985	128.212 809	143.127 783	160.183 364	179.711 760
78	123.833 488	130.815 469	146.274 700	163.986 573	184.305 996
79	126.226 615	133.450 662	149.468 820	167.856 338	188.992 115
80	128.646 665	136.118 795	152.710 852	171.793 824	193.771 958
81	131.093 940	138.820 280	156.001 515	175.800 216	198.647 397
82	133.568 746	141.555 534	159.341 538	179.876 720	203.620 345
83	136.071 395	144.324 978	162.731 661	184.024 563	208.692 752
84	138.602 198	147.129 040	166.172 636	188.244 992	213.866 607
85	141.161 473	149.968 153	169.665 226	192.539 280	219.143 939
86	143.749 539	152.842 755	173.210 204	196.908 717	224.526 818
87	146.366 722	155.753 289	176.808 357	201.354 620	230.017 354
88	149.013 347	158.700 206	180.460 482	205.878 326	235.617 701
89	151.689 747	161.683 958	184.167 390	210.481 196	241.330 055
90	154.396 257	164.705 008	187.929 900	215.164 617	247.156 656
91	157.133 215	167.763 820	191.748 849	219.929 998	253.099 789
92	159.900 964	170.860 868	195.625 082	224.778 773	259.161 785
93	162.699 849	173.996 629	199.559 458	229.712 401	265.345 021
94	165.530 223	177.171 587	203.552 850	234.732 369	271.651 921
95	168.392 438	180.386 232	207.606 142	239.840 185	278.084 960
96	171.286 853	183.641 059	211.720 235	245.037 388	284.646 659
97	174.213 830	186.936 573	215.896 038	250.325 542	291.339 592
98	177.173 735	190.273 280	220.134 479	255.706 239	298.166 384
99	180.166 940	193.651 696	224.436 496	261.181 099	305.129 712
100	183.193 818	197.072 342	228.803 043	266.751 768	312.232 306

Table VI   Amount of an Annuity of 1 per Annum

$$\frac{(1+r)^n - 1}{r}$$

n	3%	4%	5%	6%	7%
1	1.000 000	1.000 000	1.000 000	1.000 000	1.000 000
2	2.030 000	2.040 000	2.050 000	2.060 000	2.070 000
3	3.090 900	3.121 600	3.152 500	3.183 600	3.214 900
4	4.183 627	4.246 464	4.310 125	4.374 616	4.439 943
5	5.309 136	5.416 323	5.525 631	5.637 093	5.750 739
6	6.468 410	6.632 975	6.801 913	6.975 319	7.153 291
7	7.662 462	7.898 294	8.142 008	8.393 838	8.654 021
8	8.892 336	9.214 226	9.549 109	9.897 468	10.259 803
9	10.159 106	10.582 795	11.026 564	11.491 316	11.977 989
10	11.463 879	12.006 107	12.577 893	13.180 795	13.816 448
11	12.807 796	13.486 351	14.206 787	14.971 643	15.783 599
12	14.192 030	15.025 805	15.917 127	16.869 941	17.888 451
13	15.617 790	16.626 838	17.712 983	18.882 138	20.140 643
14	17.086 324	18.291 911	19.598 632	21.015 066	22.550 488
15	18.598 914	20.023 588	21.578 564	23.275 970	25.129 022
16	20.156 881	21.824 531	23.657 492	25.672 528	27.888 054
17	21.761 588	23.697 512	25.840 366	28.212 880	30.840 217
18	23.414 435	25.645 413	28.132 385	30.905 653	33.999 033
19	25.116 868	27.671 229	30.539 004	33.759 992	37.378 965
20	26.870 374	29.778 079	33.065 954	36.785 591	40.995 492
21	28.676 486	31.969 202	35.719 252	39.992 727	44.865 177
22	30.536 780	34.247 970	38.505 214	43.392 290	49.005 739
23	32.452 884	36.617 889	41.430 475	46.995 828	53.436 141
24	34.426 470	39.082 604	44.501 999	50.815 577	58.176 671
25	36.459 264	41.645 908	47.727 099	54.864 512	63.249 038
26	38.553 042	44.311 745	51.113 454	59.156 383	68.676 470
27	40.709 634	47.084 214	54.669 127	63.705 766	74.483 823
28	42.930 923	49.967 583	58.402 583	68.528 112	80.697 691
29	45.218 850	52.966 286	62.322 712	73.639 798	87.346 529
30	47.575 416	56.084 938	66.438 848	79.058 186	94.460 786
31	50.002 678	59.328 335	70.760 790	84.801 677	102.073 041
32	52.502 759	62.701 469	75.298 829	90.889 778	110.218 154
33	55.077 841	66.209 527	80.063 771	97.343 165	118.933 425
34	57.730 177	69.857 909	85.066 959	104.183 755	128.258 765
35	60.462 082	73.652 225	90.320 307	111.434 780	138.236 878
36	63.275 944	77.598 314	95.836 323	119.120 867	148.913 460
37	66.174 223	81.702 246	101.628 139	127.268 119	160.337 402
38	69.159 449	85.970 336	107.709 546	135.904 206	172.561 020
39	72.234 233	90.409 150	114.095 023	145.058 458	185.640 292
40	75.401 260	95.025 516	120.799 774	154.761 966	199.635 112
41	78.663 298	99.826 536	127.839 763	165.047 684	214.609 570
42	82.023 196	104.819 598	135.231 751	175.950 545	230.632 240
43	85.483 892	110.012 382	142.993 339	187.507 577	247.776 496
44	89.048 409	115.412 877	151.143 006	199.758 032	266.120 851
45	92.719 861	121.029 392	159.700 156	212.743 514	285.749 311
46	96.501 457	126.870 568	168.685 164	226.508 125	306.751 763
47	100.396 501	132.945 390	178.119 422	241.098 612	329.224 386
48	104.408 396	139.263 206	188.025 393	256.564 529	353.270 093
49	108.540 648	145.833 734	198.426 663	272.958 401	378.999 000
50	112.796 867	152.667 084	209.347 996	290.335 905	406.528 929

Table VI   Amount of an Annuity of 1 per Annum

$$\frac{(1+r)^n - 1}{r}$$

n	3%	4%	5%	6%	7%
51	117.180 773	159.773 767	220.815 396	308.756 059	435.985 955
52	121.696 197	167.164 718	232.856 165	328.281 422	467.504 971
53	126.347 082	174.851 306	245.498 974	348.978 308	501.230 319
54	131.137 495	182.845 359	258.773 922	370.917 006	537.316 442
55	136.071 620	191.159 173	272.712 618	394.172 027	575.928 593
56	141.153 768	199.805 540	287.348 249	418.822 348	617.243 594
57	146.388 381	208.797 762	302.715 662	444.951 689	661.450 646
58	151.780 033	218.149 672	318.851 445	472.648 790	708.752 191
59	157.333 434	227.875 659	335.794 017	502.007 718	759.364 844
60	163.053 437	237.990 685	353.583 718	533.128 181	813.520 383
61	168.945 040	248.510 313	372.262 904	566.115 872	871.466 810
62	175.013 391	259.450 725	391.876 049	601.082 824	933.469 487
63	181.263 793	270.828 754	412.469 851	638.147 793	999.812 351
64	187.701 707	282.661 904	434.093 344	677.436 661	1070.799 216
65	194.332 758	294.968 380	456.798 011	719.082 861	1146.755 161
66	201.162 741	307.767 116	480.637 912	763.227 832	1228.028 022
67	208.197 623	321.077 800	505.669 807	810.021 502	1314.989 983
68	215.443 551	334.920 912	531.953 298	859.622 792	1408.039 282
69	222.906 858	349.317 749	559.550 963	912.200 160	1507.602 032
70	230.594 064	364.290 459	588.528 511	967.932 170	1614.134 174
71	238.511 886	379.862 077	618.954 936	1027.008 100	1728.123 566
72	246.667 242	396.056 560	650.902 683	1089.628 586	1850.092 216
73	255.067 259	412.898 823	684.447 817	1156.006 301	1980.598 671
74	263.719 277	430.414 776	719.670 208	1226.366 679	2120.240 578
75	272.630 856	448.631 367	756.653 718	1300.948 680	2269.657 419
76	281.809 781	467.576 621	795.486 404	1380.005 601	2429.533 438
77	291.264 075	487.279 686	836.260 725	1463.805 937	2600.600 779
78	301.001 997	507.770 873	879.073 761	1552.634 293	2783.642 833
79	311.032 057	529.081 708	924.027 449	1646.792 350	2979.497 831
80	321.363 019	551.244 977	971.228 821	1746.599 891	3189.062 680
81	332.003 909	574.294 776	1020.790 262	1852.395 885	3413.297 067
82	342.964 026	598.266 567	1072.829 776	1964.539 638	3653.227 862
83	354.252 947	623.197 230	1127.471 264	2083.412 016	3909.953 812
84	365.880 536	649.125 119	1184.844 828	2209.416 737	4184.650 579
85	377.856 952	676.090 123	1245.087 069	2342.981 741	4478.576 120
86	390.192 660	704.133 728	1308.341 422	2484.560 646	4793.076 448
87	402.898 440	733.299 078	1374.758 493	2634.634 285	5129.591 799
88	415.985 393	763.631 041	1444.496 418	2793.712 342	5489.663 225
89	429.464 955	795.176 282	1517.721 239	2962.335 082	5874.939 651
90	443.348 904	827.983 334	1594.607 301	3141.075 187	6287.185 427
91	457.649 371	862.102 667	1675.337 666	3330.539 698	6728.288 407
92	472.378 852	897.586 774	1760.104 549	3531.372 080	7200.268 595
93	487.550 217	934.490 245	1849.109 777	3744.254 405	7705.287 397
94	503.176 724	972.869 854	1942.565 266	3969.909 669	8245.657 515
95	519.272 026	1012.784 648	2040.693 529	4209.104 250	8823.853 541
96	535.850 186	1054.296 034	2143.728 205	4462.650 505	9442.523 288
97	552.925 692	1097.467 876	2251.914 616	4731.409 535	10104.499 919
98	570.513 463	1142.366 591	2365.510 346	5016.294 107	10812.814 913
99	588.628 867	1189.061 254	2484.785 864	5318.271 753	11570.711 957
100	607.287 733	1237.623 705	2610.025 157	5638.368 059	12381.661 794

Table VII   Present Value of 1 per Annum

$$\frac{1-(1+r)^{-n}}{r}$$

n	$\frac{5}{12}\%$	$\frac{1}{2}\%$	$\frac{7}{12}\%$	$\frac{3}{4}\%$	1%
1	0.995 851	0.995 025	0.994 200	0.992 556	0.990 099
2	1.987 569	1.985 099	1.982 635	1.977 723	1.970 395
3	2.975 173	2.970 248	2.965 337	2.955 556	2.940 985
4	3.958 678	3.950 496	3.942 340	3.926 110	3.901 966
5	4.938 103	4.925 866	4.913 677	4.889 440	4.853 431
6	5.913 463	5.896 384	5.879 381	5.845 598	5.795 476
7	6.884 777	6.862 074	6.839 484	6.794 638	6.728 195
8	7.852 060	7.822 959	7.794 019	7.736 613	7.651 678
9	8.815 329	8.779 064	8.743 018	8.671 576	8.566 018
10	9.774 602	9.730 412	9.686 513	9.599 580	9.471 305
11	10.729 894	10.677 027	10.624 537	10.520 675	10.367 628
12	11.681 222	11.618 932	11.557 120	11.434 913	11.255 077
13	12.628 603	12.556 151	12.484 295	12.342 345	12.133 740
14	13.572 053	13.488 708	13.406 093	13.243 022	13.003 703
15	14.511 588	14.416 625	14.322 545	14.136 995	13.865 053
16	15.447 224	15.339 925	15.233 682	15.024 313	14.717 874
17	16.378 978	16.258 631	16.139 534	15.905 025	15.562 251
18	17.306 867	17.172 768	17.040 133	16.779 181	16.398 269
19	18.230 904	18.082 356	17.935 510	17.646 830	17.226 008
20	19.151 108	18.987 419	18.825 693	18.508 020	18.045 553
21	20.067 494	19.887 979	19.710 714	19.362 799	18.856 983
22	20.980 077	20.784 059	20.590 602	20.211 215	19.660 379
23	21.888 873	21.675 681	21.465 387	21.053 315	20.455 821
24	22.793 898	22.562 866	22.335 099	21.889 146	21.243 387
25	23.695 169	23.445 638	23.199 767	22.718 755	22.023 156
26	24.592 699	24.324 018	24.059 421	23.542 189	22.795 204
27	25.486 505	25.198 028	24.914 089	24.359 493	23.559 608
28	26.376 603	26.067 689	25.763 800	25.170 713	24.316 443
29	27.263 007	26.933 024	26.608 583	25.975 893	25.065 785
30	28.145 733	27.794 054	27.448 467	26.775 080	25.807 708
31	29.024 796	28.650 800	28.283 480	27.568 318	26.542 285
32	29.900 212	29.503 284	29.113 650	28.355 650	27.269 589
33	30.771 995	30.351 526	29.939 006	29.137 122	27.989 693
34	31.640 161	31.195 548	30.759 575	29.912 776	28.702 666
35	32.504 725	32.035 371	31.575 385	30.682 656	29.408 580
36	33.365 701	32.871 016	32.386 464	31.446 805	30.107 505
37	34.223 105	33.702 504	33.192 840	32.205 266	30.799 510
38	35.076 951	34.529 854	33.994 538	32.958 080	31.484 663
39	35.927 254	35.353 089	34.791 587	33.705 290	32.163 033
40	36.774 029	36.172 228	35.584 014	34.446 938	32.834 686
41	37.617 290	36.987 291	36.371 845	35.183 065	33.499 689
42	38.457 053	37.798 300	37.155 107	35.913 713	34.158 108
43	39.293 330	38.605 274	37.933 826	36.638 921	34.810 008
44	40.126 138	39.408 232	38.708 029	37.358 730	35.455 454
45	40.955 490	40.207 196	39.477 742	38.073 181	36.094 508
46	41.781 401	41.002 185	40.242 991	38.782 314	36.727 236
47	42.603 885	41.793 219	41.003 803	39.486 168	37.353 699
48	43.422 956	42.580 318	41.760 201	40.184 782	37.973 959
49	44.238 628	43.363 500	42.512 213	40.878 195	38.588 079
50	45.050 916	44.142 786	43.259 864	41.566 447	39.196 118

Table VII Present Value of 1 per Annum

$$\frac{1-(1+r)^{-n}}{r}$$

n	$\frac{5}{12}\%$	$\frac{1}{2}\%$	$\frac{7}{12}\%$	$\frac{3}{4}\%$	1%
51	45.859 834	44.918 195	44.003 179	42.249 575	39.798 136
52	46.665 394	45.689 747	44.742 183	42.927 618	40.394 194
53	47.467 613	46.457 459	45.476 901	43.600 614	40.984 351
54	48.266 502	47.221 353	46.207 358	44.268 599	41.568 664
55	49.062 077	47.981 445	46.933 579	44.931 612	42.147 192
56	49.854 350	48.737 757	47.655 588	45.589 689	42.719 992
57	50.643 337	49.490 305	48.373 410	46.242 868	43.287 121
58	51.429 049	50.239 109	49.087 069	46.891 184	43.848 635
59	52.211 501	50.984 189	49.796 588	47.534 674	44.404 589
60	52.990 706	51.725 561	50.501 994	48.173 374	44.955 038
61	53.766 678	52.463 245	51.203 308	48.807 319	45.500 038
62	54.539 431	53.197 258	51.900 554	49.436 545	46.039 642
63	55.308 977	53.927 620	52.593 757	50.061 086	46.573 903
64	56.075 330	54.654 348	53.282 940	50.680 979	47.102 874
65	56.838 502	55.377 461	53.968 126	51.296 257	47.626 608
66	57.598 509	56.096 976	54.649 338	51.906 955	48.145 156
67	58.355 361	56.812 912	55.326 600	52.513 107	48.658 571
68	59.109 074	57.525 285	55.999 934	53.114 746	49.166 901
69	59.859 658	58.234 115	56.669 362	53.711 907	49.670 199
70	60.607 129	58.939 418	57.334 909	54.304 622	50.168 514
71	61.351 497	59.641 212	57.996 595	54.892 925	50.661 895
72	62.092 777	60.339 514	58.654 444	55.476 849	51.150 391
73	62.830 982	61.034 342	59.308 478	56.056 426	51.634 051
74	63.566 123	61.725 714	59.958 719	56.631 688	52.112 922
75	64.298 214	62.413 645	60.605 189	57.202 668	52.587 051
76	65.027 267	63.098 155	61.247 909	57.769 397	53.056 486
77	65.753 295	63.779 258	61.886 902	58.331 908	53.521 274
78	66.476 310	64.456 973	62.522 190	58.890 231	53.981 459
79	67.196 325	65.131 317	63.153 792	59.444 398	54.437 088
80	67.913 353	65.802 305	63.781 732	59.994 440	54.888 206
81	68.627 406	66.469 956	64.406 030	60.540 387	55.334 858
82	69.338 495	67.134 284	65.026 708	61.082 270	55.777 087
83	70.046 634	67.795 308	65.643 786	61.620 119	56.214 937
84	70.751 835	68.453 042	66.257 285	62.153 965	56.648 453
85	71.454 109	69.107 505	66.867 226	62.683 836	57.077 676
86	72.153 470	69.758 711	67.473 630	63.209 763	57.502 650
87	72.849 929	70.406 678	68.076 517	63.731 774	57.923 415
88	73.543 497	71.051 421	68.675 908	64.249 900	58.340 015
89	74.234 188	71.692 956	69.271 822	64.764 169	58.752 490
90	74.922 013	72.331 300	69.864 280	65.274 609	59.160 881
91	75.606 984	72.966 467	70.453 303	65.781 250	59.565 229
92	76.289 113	73.598 475	71.038 909	66.284 119	59.965 573
93	76.968 411	74.227 338	71.621 119	66.783 245	60.361 954
94	77.644 891	74.853 073	72.199 953	67.278 655	60.754 410
95	78.318 563	75.475 694	72.775 430	67.770 377	61.142 980
96	78.989 441	76.095 218	73.347 569	68.258 439	61.527 703
97	79.657 534	76.711 660	73.916 390	68.742 867	61.908 617
98	80.322 856	77.325 035	74.481 912	69.223 689	62.285 759
99	80.985 416	77.935 358	75.044 154	69.700 932	62.659 168
100	81.645 228	78.542 645	75.603 136	70.174 623	63.028 879

Table VII — Present Value of 1 per Annum

$$\frac{1 - (1 + r)^{-n}}{r}$$

n	$1\frac{1}{8}\%$	$1\frac{1}{4}\%$	$1\frac{1}{2}\%$	$1\frac{3}{4}\%$	2%
1	0.988 875	0.987 654	0.985 222	0.982 801	0.980 392
2	1.966 749	1.963 115	1.955 883	1.948 699	1.941 561
3	2.933 745	2.926 534	2.912 200	2.897 984	2.883 883
4	3.889 982	3.878 058	3.854 385	3.830 943	3.807 729
5	4.835 582	4.817 835	4.782 645	4.747 855	4.713 460
6	5.770 662	5.746 010	5.697 187	5.648 998	5.601 431
7	6.695 339	6.662 726	6.598 214	6.534 641	6.471 991
8	7.609 730	7.568 124	7.485 925	7.405 053	7.325 481
9	8.513 948	8.462 345	8.360 517	8.260 494	8.162 237
10	9.408 107	9.345 526	9.222 185	9.101 223	8.982 585
11	10.292 318	10.217 803	10.071 118	9.927 492	9.786 848
12	11.166 693	11.079 312	10.907 505	10.739 550	10.575 341
13	12.031 340	11.930 185	11.731 532	11.537 641	11.348 374
14	12.886 369	12.770 553	12.543 382	12.322 006	12.106 249
15	13.731 885	13.600 546	13.343 233	13.092 880	12.849 264
16	14.567 995	14.420 292	14.131 264	13.850 497	13.577 709
17	15.394 804	15.229 918	14.907 649	14.595 083	14.291 872
18	16.212 414	16.029 549	15.672 561	15.326 863	14.992 031
19	17.020 928	16.819 308	16.426 168	16.046 057	15.678 462
20	17.820 448	17.599 316	17.168 639	16.752 881	16.351 433
21	18.611 074	18.369 695	17.900 137	17.447 549	17.011 209
22	19.392 904	19.130 563	18.620 824	18.130 269	17.658 048
23	20.166 036	19.882 037	19.330 861	18.801 248	18.292 204
24	20.930 567	20.624 235	20.030 405	19.460 686	18.913 926
25	21.686 593	21.357 269	20.719 611	20.108 782	19.523 456
26	22.434 208	22.081 253	21.398 632	20.745 732	20.121 036
27	23.173 506	22.796 299	22.067 617	21.371 726	20.706 898
28	23.904 579	23.502 518	22.726 717	21.986 955	21.281 272
29	24.627 520	24.200 018	23.376 076	22.591 602	21.844 385
30	25.342 418	24.888 906	24.015 838	23.185 849	22.396 456
31	26.049 362	25.569 290	24.646 146	23.769 877	22.937 702
32	26.748 442	26.241 274	25.267 139	24.343 859	23.468 335
33	27.439 745	26.904 962	25.878 954	24.907 970	23.988 564
34	28.123 357	27.560 456	26.481 728	25.462 378	24.498 592
35	28.799 365	28.207 858	27.075 595	26.007 251	24.998 619
36	29.467 851	28.847 267	27.660 684	26.542 753	25.488 842
37	30.128 901	29.478 783	28.237 127	27.069 045	25.969 453
38	30.782 597	30.102 501	28.805 052	27.586 285	26.440 641
39	31.429 020	30.718 520	29.364 583	28.094 629	26.902 589
40	32.068 253	31.326 933	29.915 845	28.594 230	27.355 479
41	32.700 373	31.927 835	30.458 961	29.085 238	27.799 489
42	33.325 462	32.521 319	30.994 050	29.567 801	28.234 794
43	33.943 596	33.107 475	31.521 232	30.042 065	28.661 562
44	34.554 854	33.686 395	32.040 622	30.508 172	29.079 963
45	35.159 312	34.258 168	32.552 337	30.966 263	29.490 160
46	35.757 045	34.822 882	33.056 490	31.416 474	29.892 314
47	36.348 129	35.380 624	33.553 192	31.858 943	30.286 582
48	36.932 637	35.931 481	34.042 554	32.293 801	30.673 120
49	37.510 642	36.475 537	34.524 683	32.721 181	31.052 078
50	38.082 217	37.012 876	34.999 688	33.141 209	31.423 606

Table VII   Present Value of 1 per Annum

$$\frac{1-(1+r)^{-n}}{r}$$

n	$1\tfrac{1}{8}\%$	$1\tfrac{1}{4}\%$	$1\tfrac{1}{2}\%$	$1\tfrac{3}{4}\%$	2%
51	38.647 433	37.543 581	35.467 673	33.554 014	31.787 849
52	39.206 362	38.067 734	35.928 742	33.959 719	32.144 950
53	39.759 072	38.585 417	36.382 997	34.358 446	32.495 049
54	40.305 634	39.096 708	36.830 539	34.750 316	32.838 283
55	40.846 115	39.601 687	37.271 467	35.135 445	33.174 788
56	41.380 584	40.100 431	37.705 879	35.513 951	33.504 694
57	41.909 106	40.593 019	38.133 871	35.885 947	33.828 131
58	42.431 749	41.079 524	38.555 538	36.251 545	34.145 226
59	42.948 577	41.560 024	38.970 973	36.610 855	34.456 104
60	43.459 656	42.034 592	39.380 269	36.963 986	34.760 887
61	43.965 050	42.503 301	39.783 516	37.311 042	35.059 693
62	44.464 820	42.966 223	40.180 804	37.652 130	35.352 640
63	44.959 031	43.423 430	40.572 221	37.987 351	35.639 843
64	45.447 744	43.874 992	40.957 853	38.316 807	35.921 415
65	45.931 020	44.320 980	41.337 786	38.640 597	36.197 466
66	46.408 920	44.761 462	41.712 105	38.958 817	36.468 103
67	46.881 503	45.196 506	42.080 891	39.271 565	36.733 435
68	47.348 829	45.626 178	42.444 228	39.578 934	36.993 564
69	47.810 955	46.050 547	42.802 195	39.881 016	37.248 592
70	48.267 941	46.469 676	43.154 872	40.177 903	37.498 619
71	48.719 843	46.883 630	43.502 337	40.469 683	37.743 744
72	49.166 717	47.292 474	43.844 667	40.756 445	37.984 063
73	49.608 620	47.696 271	44.181 938	41.038 276	38.219 670
74	50.045 607	48.095 082	44.514 224	41.315 259	38.450 657
75	50.477 733	48.488 970	44.841 600	41.587 478	38.677 114
76	50.905 051	48.877 995	45.164 138	41.855 015	38.899 132
77	51.327 615	49.262 218	45.481 910	42.117 951	39.116 796
78	51.745 478	49.641 696	45.794 985	42.376 364	39.330 192
79	52.158 693	50.016 490	46.103 433	42.630 334	39.539 404
80	52.567 311	50.386 657	46.407 323	42.879 935	39.744 514
81	52.971 383	50.752 254	46.706 723	43.125 243	39.945 602
82	53.370 960	51.113 337	47.001 697	43.366 332	40.142 747
83	53.766 091	51.469 963	47.292 313	43.603 275	40.336 026
84	54.156 827	51.822 185	47.578 633	43.836 142	40.525 516
85	54.543 216	52.170 060	47.860 722	44.065 005	40.711 290
86	54.925 306	52.513 639	48.138 643	44.289 931	40.893 422
87	55.303 145	52.852 977	48.412 456	44.510 989	41.071 982
88	55.676 782	53.188 125	48.682 222	44.728 244	41.247 041
89	56.046 261	53.519 136	48.948 002	44.941 764	41.418 668
90	56.411 630	53.846 060	49.209 855	45.151 610	41.586 929
91	56.772 935	54.168 948	49.467 837	45.357 848	41.751 891
92	57.130 220	54.487 850	49.722 007	45.560 539	41.913 619
93	57.483 530	54.802 815	49.972 421	45.759 743	42.072 175
94	57.832 910	55.113 892	50.219 134	45.955 521	42.227 623
95	58.178 403	55.421 127	50.462 201	46.147 933	42.380 023
96	58.520 052	55.724 570	50.701 675	46.337 035	42.529 434
97	58.857 901	56.024 267	50.937 611	46.522 884	42.675 916
98	59.191 991	56.320 264	51.170 060	46.705 537	42.819 525
99	59.522 364	56.612 606	51.399 074	46.885 049	42.960 319
100	59.849 063	56.901 339	51.624 704	47.061 473	43.098 352

Table VII   Present Value of 1 per Annum

$$\frac{1 - (1 + r)^{-n}}{r}$$

n	3%	4%	5%	6%	7%
1	0.970 874	0.961 538	0.952 381	0.943 396	0.934 579
2	1.913 470	1.886 095	1.859 410	1.833 393	1.808 018
3	2.828 611	2.775 091	2.723 248	2.673 012	2.624 316
4	3.717 098	3.629 895	3.545 951	3.465 106	3.387 211
5	4.579 707	4.451 822	4.329 477	4.212 364	4.100 197
6	5.417 191	5.242 137	5.075 692	4.917 324	4.766 540
7	6.230 283	6.002 055	5.786 373	5.582 381	5.389 289
8	7.019 692	6.732 745	6.463 213	6.209 794	5.971 299
9	7.786 109	7.435 332	7.107 822	6.801 692	6.515 232
10	8.530 203	8.110 896	7.721 735	7.360 087	7.023 582
11	9.252 624	8.760 477	8.306 414	7.886 875	7.498 674
12	9.954 004	9.385 074	8.863 252	8.383 844	7.942 686
13	10.634 955	9.985 648	9.393 573	8.852 683	8.357 651
14	11.296 073	10.563 123	9.898 641	9.294 984	8.745 468
15	11.937 935	11.118 387	10.379 658	9.712 249	9.107 914
16	12.561 102	11.652 296	10.837 770	10.105 895	9.446 649
17	13.166 118	12.165 669	11.274 066	10.477 260	9.763 223
18	13.753 513	12.659 297	11.689 587	10.827 603	10.059 087
19	14.323 799	13.133 939	12.085 321	11.158 116	10.335 595
20	14.877 475	13.590 326	12.462 210	11.469 921	10.594 014
21	15.415 024	14.029 160	12.821 153	11.764 077	10.835 527
22	15.936 917	14.451 115	13.163 003	12.041 582	11.061 240
23	16.443 608	14.856 842	13.488 574	12.303 379	11.272 187
24	16.935 542	15.246 963	13.798 642	12.550 358	11.469 334
25	17.413 148	15.622 080	14.093 945	12.783 356	11.653 583
26	17.876 842	15.982 769	14.375 185	13.003 166	11.825 779
27	18.327 031	16.329 586	14.643 034	13.210 534	11.986 709
28	18.764 108	16.663 063	14.898 127	13.406 164	12.137 111
29	19.188 455	16.983 715	15.141 074	13.590 721	12.277 674
30	19.600 441	17.292 033	15.372 451	13.764 831	12.409 041
31	20.000 428	17.588 494	15.592 811	13.929 086	12.531 814
32	20.388 766	17.873 551	15.802 677	14.084 043	12.646 555
33	20.765 792	18.147 646	16.002 549	14.230 230	12.753 790
34	21.131 837	18.411 198	16.192 904	14.368 141	12.854 009
35	21.487 220	18.664 613	16.374 194	14.498 246	12.947 672
36	21.832 252	18.908 282	16.546 852	14.620 987	13.035 208
37	22.167 235	19.142 579	16.711 287	14.736 780	13.117 017
38	22.492 462	19.367 864	16.867 893	14.846 019	13.193 473
39	22.808 215	19.584 485	17.017 041	14.949 075	13.264 928
40	23.114 772	19.792 774	17.159 086	15.046 297	13.331 709
41	23.412 400	19.993 052	17.294 368	15.138 016	13.394 120
42	23.701 359	20.185 627	17.423 208	15.224 543	13.452 449
43	23.981 902	20.370 795	17.545 912	15.306 173	13.506 962
44	24.254 274	20.548 841	17.662 773	15.383 182	13.557 908
45	24.518 713	20.720 040	17.774 070	15.455 832	13.605 522
46	24.775 449	20.884 654	17.880 066	15.524 370	13.650 020
47	25.024 708	21.042 936	17.981 016	15.589 028	13.691 608
48	25.266 707	21.195 131	18.077 158	15.650 027	13.730 474
49	25.501 657	21.341 472	18.168 722	15.707 572	13.766 799
50	25.729 764	21.482 185	18.255 925	15.761 861	13.800 746

Table VII   Present Value of 1 per Annum

$$\frac{1-(1+r)^{-n}}{r}$$

n	3%	4%	5%	6%	7%
51	25.951 227	21.617 485	18.338 977	15.813 076	13.832 473
52	26.166 240	21.747 582	18.418 073	15.861 393	13.862 124
53	26.374 990	21.872 675	18.493 403	15.906 974	13.889 836
54	26.577 660	21.992 957	18.565 146	15.949 976	13.915 735
55	26.774 428	22.108 612	18.633 472	15.990 543	13.939 939
56	26.965 464	22.219 819	18.698 545	16.028 814	13.962 560
57	27.150 936	22.326 749	18.760 519	16.064 919	13.983 701
58	27.331 005	22.429 567	18.819 542	16.098 980	14.003 458
59	27.505 831	22.528 430	18.875 754	16.131 113	14.021 924
60	27.675 564	22.623 490	18.929 290	16.161 428	14.039 181
61	27.840 353	22.714 894	18.980 276	16.190 026	14.055 309
62	28.000 343	22.802 783	19.028 834	16.217 006	14.070 383
63	28.155 673	22.887 291	19.075 080	16.242 458	14.084 470
64	28.306 478	22.968 549	19.119 124	16.266 470	14.097 635
65	28.452 892	23.046 682	19.161 070	16.289 123	14.109 940
66	28.595 040	23.121 810	19.201 019	16.310 493	14.121 439
67	28.733 049	23.194 048	19.239 066	16.330 654	14.132 186
68	28.867 038	23.263 507	19.275 301	16.349 673	14.142 230
69	28.997 124	23.330 296	19.309 810	16.367 617	14.151 617
70	29.123 421	23.394 515	19.342 677	16.384 544	14.160 389
71	29.246 040	23.456 264	19.373 978	16.400 513	14.168 588
72	29.365 088	23.515 639	19.403 788	16.415 578	14.176 251
73	29.480 667	23.572 730	19.432 179	16.429 791	14.183 412
74	29.592 881	23.627 625	19.459 218	16.443 199	14.190 104
75	29.701 826	23.680 408	19.484 970	16.455 848	14.196 359
76	29.807 598	23.731 162	19.509 495	16.467 781	14.202 205
77	29.910 290	23.779 963	19.532 853	16.479 039	14.207 668
78	30.009 990	23.826 888	19.555 098	16.489 659	14.212 774
79	30.106 786	23.872 008	19.576 284	16.499 679	14.217 546
80	30.200 763	23.915 392	19.596 460	16.509 131	14.222 005
81	30.292 003	23.957 108	19.615 677	16.518 048	14.226 173
82	30.380 586	23.997 219	19.633 978	16.526 460	14.230 069
83	30.466 588	24.035 787	19.651 407	16.534 396	14.233 709
84	30.550 086	24.072 872	19.668 007	16.541 883	14.237 111
85	30.631 151	24.108 531	19.683 816	16.548 947	14.240 291
86	30.709 855	24.142 818	19.698 873	16.555 610	14.243 262
87	30.786 267	24.175 787	19.713 212	16.561 896	14.246 040
88	30.860 454	24.207 487	19.726 869	16.567 827	14.248 635
89	30.932 479	24.237 969	19.739 875	16.573 421	14.251 061
90	31.002 407	24.267 278	19.752 262	16.578 699	14.253 328
91	31.070 298	24.295 459	19.764 059	16.583 679	14.255 447
92	31.136 212	24.322 557	19.775 294	16.588 376	14.257 427
93	31.200 206	24.348 612	19.785 994	16.592 808	14.259 277
94	31.262 336	24.373 666	19.796 185	16.596 988	14.261 007
95	31.322 656	24.397 756	19.805 891	16.600 932	14.262 623
96	31.381 219	24.420 919	19.815 134	16.604 653	14.264 134
97	31.438 077	24.443 191	19.823 937	16.608 163	14.265 546
98	31.493 279	24.464 607	19.832 321	16.611 475	14.266 865
99	31.546 872	24.485 199	19.840 306	16.614 599	14.268 098
100	31.598 905	24.504 999	19.847 910	16.617 546	14.269 251

Answers to Even-Numbered Exercises

CHAPTER 1

Exercises 1.1

2. **a.** (2)(3)(7)
   b. (3)(19)
   c. (2)(7)(11)
   d. (7)(17)(23)

4. **a.** $\pm 1, \pm 2, \pm 2,$ and ± 6
   b. $\pm 1, \pm 2, \pm 3, \pm 6,$ and ± 12
   c. $\pm 1, \pm 3, \pm 5, \pm 15, \pm 25,$ and ± 75
   d. $\pm 1, \pm 2, \pm 3, \pm 4, \pm 6, \pm 8, \pm 12,$ and ± 24
   e. $\pm 1, \pm 3, \pm 9,$ and ± 27

6. **a.** Rational, since $1.24 = \dfrac{124}{100} = \dfrac{31}{25}$
   b. Rational, since $1.414 = \dfrac{1414}{1000} = \dfrac{707}{500}$
   c. Rational, since $0 = \dfrac{0}{1}$
   d. Rational, since $3.1416 = \dfrac{31{,}416}{10{,}000} = \dfrac{3927}{1250}$
   e. $3.142857142857142857\ldots$ is rational, since it is a repeating decimal.

f. 1.3201001000100001... is *not* rational, since its infinite decimal expansion is neither terminating nor repeating.

8.

Positive Integers	Integers	Rational Numbers	Irrational Numbers
7	7	7	
	−4	−4	
	−378	−378	
$\sqrt{9}$	$\sqrt{9}$	$\sqrt{9}$	
			$\dfrac{\pi}{3}$
		−2.46	
	0	0	
		$\dfrac{3}{7}$	
$\sqrt[3]{8}$	$\sqrt[3]{8}$	$\sqrt[3]{8}$	
$\dfrac{6 \cdot 4}{0 \cdot 32}$	$\dfrac{6 \cdot 4}{0 \cdot 32}$	$\dfrac{6 \cdot 4}{0 \cdot 32}$	
		$-\dfrac{5}{3}$	
			$\sqrt[3]{2}$
34,796	34,796	34,796	
	$\dfrac{0}{3}$	$\dfrac{0}{3}$	
$\sqrt{4}$	$\sqrt{4}$	$\sqrt{4}$	
			$4 + \sqrt{3}$

10. Proof by contradiction: Suppose $2 + \sqrt{3}$ is rational. Then $2 + \sqrt{3} = \dfrac{a}{b}$, where a and b are integers and $b \neq 0$. Then $2 + \sqrt{3} = \dfrac{a}{b}$ implies $\sqrt{3} = \dfrac{a}{b} - 2 = \dfrac{a - 2b}{b}$. Since a and b are integers, then $2b$ is an integer and $a - 2b$ is an integer, say c. Thus we can write $\sqrt{3} = \dfrac{c}{b}$, a ratio of two integers. But by Theorem 1.3, $\sqrt{3}$ is not rational, so the proof is done by contradiction.

Exercises 1.2

2. $(a + b + c) + d = (a + b) + (c + d)$
Proof:
$(a + b + c) + d = [(a + b) + c] + d$ (Definition 1.10)
$\qquad\qquad\qquad = (a + b) + (c + d)$ (Axiom 5)

Answers to Even-Numbered Exercises

4. **a.** -8
   $-\sqrt{2}$
   4
   $1/2$
   $-3/8$
   0
   -42.5

   b. $3/2$
   $5/2$
   $-8/7$
   $-3/4$
   $3/\pi$
   $0/2$ has no reciprocal.
   -4
   $1/\sqrt{5}$ or $\dfrac{\sqrt{5}}{5}$
   4
   3

6. **a.** $(2 \cdot 3)5 = 2(3 \cdot 5)$ by Axiom 6

   b. $abc = cba$

   Proof: $abc = (ab)c$ Definition 1.11
   $ = a(bc)$ by Axiom 6
   $ = (bc)a$ by Axiom 4
   $ = (cb)a$ by Axiom 4
   $ = cba$ Definition 1.11

8. Proof:
   Given a and b are even positive integers, then $a = 2k$ and $b = 2l$ for some positive integers k and l. Then $a + b = 2k + 2l = 2(k + l)$, which is by definition an even positive integer. Next $ab = (2k)(2l) = 4kl = 2(2kl)$, which is also an even positive integer. $\therefore$ by Definition 1.8, the set of even positive integers is closed under both addition and multiplication.

10. Let a and b be any two elements in S. Then $a = 5k$ and $b = 5l$ for some positive integers k and l. Hence

    $$a + b = 5k + 5l = 5(k + l)$$

    Thus $a + b = 5(k + l)$ is in S by definition of S. Therefore, S is closed under addition. Similarly,

    $$ab = (5k)(5l) = 5(k5)l = 5(5k)l = 5[5(kl)]$$

    is in S by definition of S. Therefore, S is also closed under multiplication.

12. **a.** $(7 \cdot 27) + (7 \cdot 13) = 7(27 + 13) = 7(40) = 280$
    b. $(17 \cdot 63) + (17 \cdot 37) = 17(63 + 37) = 17(100) = 1700$
    c. $(92 \cdot 58) + (92 \cdot 42) = 92(58 + 42) = 92(100) = 9200$
    d. $(375 \cdot 850) + (375 \cdot 150) = 375(850 + 150) = 375(1000) = 375{,}000$

14. **a.** $0 + 7 = 7$ Axiom 8
    b. $(-a) + (a + b) = [(-a) + (a)] + b$ Axiom 5
    $ = 0 + b$ Axiom 10
    $ = b$ Axiom 8

c. $1 \cdot (a + b) + 0 = 1 \cdot (a + b)$ Axiom 8
$= a + b$ Axiom 9
d. $-(ab) + a(a + b) = -(ab) + a(b + a)$ Axiom 3
$= -(ab) + ab + a^2$ Axiom 7
$= [-(ab) + ab] + a^2$ Definition 1.10
$= 0 + a^2$ Axiom 10
$= a^2$ Axiom 8

Exercises 1.3

2. a. 3   **b.** 17   **c.** -72 **d.** 0   **e.** 30
   f. 0   **g.** None   **h.** 0   **i.** 0   **j.** 12

4. a. 7   **b.** -3 **c.** 6   **d.** 17   **e.** -11
   f. $7a - 1$ **g.** $2 - 2a + 10b$

6. a. $a - (-b + c)$ **b.** $a - (b - c)$ **c.** $a - b - (c - d)$
   d. $a + b - (c + d)$

8. Approximately 5142.86 gallons
10. No
12. Yes
14. $(a + b)(c + d) = (a + b)c + (a + b)d$ Rule 1.4
$= c(a + b) + d(a + b)$ Axiom 4
$= ca + cb + da + db$ Rule 1.4
$= ac + bc + ad + bd$ Axiom 4

Exercises 1.4

2. a. $-\dfrac{3}{7} = \dfrac{-3}{7} = \dfrac{3}{-7} = -\dfrac{-3}{-7}$ **b.** $\dfrac{-4}{5} = -\dfrac{4}{5} = \dfrac{4}{-5} = -\dfrac{-4}{-5}$

   c. $\dfrac{2}{-9} = -\dfrac{2}{9} = \dfrac{-2}{9} = -\dfrac{-2}{-9}$ **d.** $\dfrac{5}{6} = \dfrac{-5}{-6} = -\dfrac{-5}{6} = -\dfrac{5}{-6}$

   e. $\dfrac{-8}{-15} = \dfrac{8}{15} = -\dfrac{-8}{15} = -\dfrac{8}{-15}$ **f.** $-\dfrac{-3}{-4} = -\dfrac{3}{4} = \dfrac{-3}{4} = \dfrac{3}{-4}$

4. a. 1/5   **b.** 1/7   **c.** 5/17   **d.** 1/3   **e.** 3/7   **f.** 2/847

6. a. 3/8   **b.** $-6/11$ **c.** 4   **d.** -2 **e.** $\dfrac{a + b}{a - b}$ **f.** $\dfrac{c - a}{a + b}$

8. a. 5/6   **b.** 19/15   **c.** $-3/4$ **d.** 55/24   **e.** 5/12
   f. 1/8   **g.** $\dfrac{143}{105}$ **h.** $\dfrac{17}{72}$

10. a. 5/4   **b.** $-3/10$ **c.** 6   **d.** $-4/3$ **e.** 0
   f. $\dfrac{8}{-27}$ **g.** $-12/7$ **h.** 6/5   **i.** 2/3

Answers to Even-Numbered Exercises

Exercises 1.5

2. $-6\frac{1}{4}$, -6, -4, -1, $-1/2$, 0, $\sqrt{3}$, 2, $\sqrt{7}$, 3, $\sqrt{100}$, 13, 21

4. a. $\frac{2}{3} > \frac{4}{7}$ **b.** $\frac{11}{12} > \frac{20}{31}$ **c.** $\frac{371}{586} < \frac{728}{1107}$ **d.** $\frac{40{,}783}{31{,}426} < \frac{81{,}565}{62{,}752}$

6. 5, 8, $3/4$, $\sqrt{2}$, 189, 17, 2.5

8. a. T **b.** F **c.** T **d.** T **e.** T
   f. T **g.** T **h.** F

10. $-2 < a < 2$

12. $a \geq 5$ or $a \leq -5$

14. $-3 \leq a \leq 3$

16. b.
$a < 0$ Hypothesis
$a + b < 0 + b$ Rule 1.31
$a + b < b$ Additive identity
$b < 0$ Hypothesis
$\therefore a + b < 0$ Rule 1.30

18. Case (1) $a > 0$ and $b > 0$
     $|a| = a$ and $|b| = b$, by Definition 1.21
     So $|a||b| = ab$, which is positive
     $\therefore |a||b| = |ab|$, by Definition 1.21

Case (2) $a < 0$, $b < 0$
     $|a| = -a$ and $|b| = -b$, by Definition 1.21
     So $|a||b| = (-a)(-b) = ab$, again positive
     $\therefore |a||b| = |ab|$, by Definition 1.21

Case (3) $a > 0$, $b < 0$
     $|a| = a$ and $|b| = -b$, by Definition 1.21
     So $|a||b| = a(-b) = -ab$, which is positive
     $|-ab| = |ab|$, by Definition 1.21
     $|a||b| = |ab|$, Transitive property

Case (4) $a = 0$, b any real number
     $|a| = |0| = 0$, by Definition 1.21
     $|a||b| = 0 \cdot |b| = 0$
     Since $a = 0$, $|ab| = |0| = 0$
     $|a||b| = |ab|$, Transitive property

Exercises 1.6

2.

A number line marked at -3, -2, -1, 0, 1, $\frac{3}{2}$, 2, $\sqrt{5}$, 2.4, $\sqrt{8}$, 3.

4.

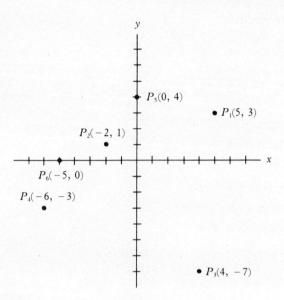

6. On a line parallel to y axis and 3 units to right of y axis
On a line parallel to y axis and 3 units to left of y axis
On a line parallel to x axis and 5 units above x axis
On a line parallel to x axis and 5 units below x axis

8.

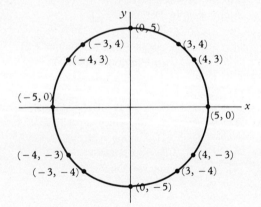

These points lie on a circle of radius 5, centered at the origin.

Exercises 1.7

2. a. $3 + 3i$ **b.** $5 - 3i$ **c.** $-8 - 2i$ **d.** $-2 + 2i$ $-2 - 5i$ **e.** $4 + 2i$
f. $6 + 17i$

4. a. i **b.** $\dfrac{6}{5} - \dfrac{2}{5}i$ **c.** $\dfrac{23}{29} - \dfrac{14}{29}i$ **d.** $-2 - 5i$ **e.** $-3 + 4i$
f. $\dfrac{2}{13} - \dfrac{16}{13}i$

Answers to Even-Numbered Exercises

6. a. $x = 4, y = -2$ b. $x = 3, y = 2$ c. $y = 7, x = 4$
   d. $x = 1, y = -3$

CHAPTER 2

Exercises 2.1

2. a. $3x + y + 5$ b. $-2x + 3$ c. $-11x + 9y$ d. $-x + 3xy - 10y$
   e. $(a - 2)x + (b + 5)y - 2z$

4. $-\dfrac{1}{3}x - \dfrac{1}{2}xy - \dfrac{7}{8}y$

6. a. $xy - 3x + y - 3$ b. $12xy + 21x - 8y - 14$
   c. $-\dfrac{3}{8}x + \dfrac{a}{7}x - \dfrac{1}{2}y + \dfrac{4}{21}ay - \dfrac{2}{5}xz - \dfrac{8}{15}yz + \dfrac{4}{5}z - \dfrac{2}{7}a + \dfrac{3}{4}$
   d. $-2x^2 + 9xy - 10y^2 - 6x + 15y$
   e. $15acx + \dfrac{9}{2}awx - 30ax - 20bcy - 6bwy + 40by + 10cz + 3wz - 20z - 40c - 12w + 80$

Exercises 2.2

2. a. $3^6 = 729$ b. x^8 c. $2^4 = 16$ d. a^5 e. x^3 f. $7^0 = 1$
   g. $2^6 = 64$ h. $\dfrac{a^3}{b^6}$ i. $\dfrac{x^2}{y^3}$ j. $12^3 = 1728$ k. 3 l. $\dfrac{1}{32xy^2z}$

Exercises 2.3

2. a. $3\sqrt{2}$ b. $4\sqrt{2}$ c. $2\sqrt[3]{2}$ d. $3\sqrt[3]{2}$ e. $10\sqrt{2}$ f. $5\sqrt{3}$
   g. $4\sqrt{10}$ h. $4\sqrt[3]{2}$ i. $2\sqrt[3]{3}$ j. $7\sqrt{2}$ k. $2\sqrt{17}$ l. $20\sqrt{2}$

4. a. 2 b. 3 c. $\dfrac{\sqrt[3]{4}}{2}$ d. $\dfrac{\sqrt{7}}{4}$ e. $\dfrac{3}{5}\sqrt{2}$ f. $\dfrac{\sqrt{6}}{4}$ g. x^2
   h. $\dfrac{\sqrt[3]{4}}{2}$ i. $3\sqrt{3}$

6. a. $x^{3/2}$ b. $\dfrac{4}{x^{1/2}}$ c. $x^{3/4}$ d. $\dfrac{4}{x^{1/3}}$ e. $(4 + x)^{2/3}$ f. $\dfrac{1}{(x + v^2)^{1/5}}$

Exercises 2.4

2. a. $5x - 5$ b. $-5x + 5$ c. $x^2 - 16x - 8$ d. $6x^9 - 5x^7 - 7$
   e. $-14x + 6$ f. $-\dfrac{1}{2}x^2 - \dfrac{3}{2}x + 3$ g. $-\dfrac{2}{21}x^3 + \dfrac{11}{23}x^2 - \dfrac{2}{45}x - \dfrac{100}{133}$
   h. $(\sqrt{2} - \sqrt{7})x^4 + \dfrac{68}{63}x^3 - 2\sqrt{3}x + \dfrac{16}{3}$

4. a. $x - 5$ b. $2x - 1$ c. $x + 5$ d. $2x + 10 + \dfrac{5}{x - 1}$
e. $x^2 - x - 6$ f. $x^3 - 9x^2 + 23x - 15$ g. $3x + 2$ h. $x^2 + 2x + 2$
i. $2x^2 + 7x + 12 + \dfrac{21}{x - 2}$ j. $x^3 + 11x^2 + 39x + 45$
k. $x^2 + 2x - \dfrac{5}{2} + \dfrac{-\tfrac{3}{2}}{2x + 1}$ l. $x^3 + 2x - 5$ m. $4x^2 + 4x + 13$
n. $x^4 + 2x^3 + 4x^2 + 8x + 16$ o. $\dfrac{3}{2}x^2 - \dfrac{5}{4}x + \dfrac{3}{8} + \dfrac{-\tfrac{5}{8}}{2x + 1}$
p. $x^3 - x^2 + x - 1 + \dfrac{2}{x + 1}$

Exercises 2.5

2. a. $-7x - 3y$ b. $-3x^2 + 5xy - y^2$ c. $8yz - 6xz - 3xy + 3xyz$
d. $\dfrac{3 + 2\sqrt{3}}{2}y^2 - \dfrac{11}{3}x^2 - 80$

4. a. $6x^5y - 15x^3y^3 + 9x^2y^3$ b. $x^3 + 4x^2y - 9xy^2 + 4y^3$
c. $x^3 - x^2y - 5xy^2 + 2y^3$ d. $20x^2 + 7xy - 6y^2$
e. $6x^2 + 5xy - 6y^2$ f. $8x^2 - 6xy - 35y^2$
g. $x^3 + y^3$ h. $x^3 + 3x^2y + 3xy^2 + y^3$
i. $x^2 + 2xy + y^2 - 1$ j. $8x^3 - y^3$
k. $25x^2 - 49y^2$ l. $a^2 + b^2 + c^2 + 2ab + 2ac + 2bc$
m. $3x^3 + 5x^2y - 4xy^2 - 4y^3$ n. $6x^3 - 7x^2y - xy^2 + 2y^3$
o. $36x^2 + 84xy + 49y^2$ p. $2x^2 + 2y^2 - 5xy + xz - 2yz + 6x - 3y + 3z$

Exercises 2.6

2. a. $5ax(ax^2 - 2y)$ b. $6x^2y^2(4x + y)$
c. $2b^2(3a^3 + 4a^2b - 2z^4)$ d. $5(3 + 5x - 6y)$
e. $6x^2y^2z^2(xyz - 2)$ f. $(2)(2)(x - 3)(x + 1)(x - 1)$
g. $(2x - y)(4x - 7)$ h. $2(2x + 1)^2(x^2 - 1)^3(11x^2 + 4x - 3)$

4. a. $(x + 3)^2$ b. $(x - 4)^2$
c. $(2x - 5)^2$ d. $(3x - 4y)^2$
e. $(6x - 7y)^2$ f. $(8x + 5y)^2$
g. $(\sqrt{2}x + 3y)^2$ h. $(\sqrt{x} + 2y)^2$
i. $(3\sqrt{x} - \sqrt{y})^2$ j. $(0.4x + 1.2)^2$

6. a. $(A - B)^2 = (A - B)(A - B)$
$= A(A - B) + (-B)(A - B)$
$= A^2 - AB - AB + B^2$
$= A^2 - 2AB + B^2$

Answers to Even-Numbered Exercises

b. $(A + B)(A - B) = A(A - B) + B(A - B)$
$= A^2 - AB + AB - B^2$
$= A^2 - B^2$

c. $(A + B)(A^2 - AB + B^2) = A(A^2 - AB + B^2) + B(A^2 - AB + B^2)$
$= A^3 - A^2B + AB^2 + A^2B - AB^2 + B^3$
$= A^3 + B^3$

d. $(A - B)(A^2 + AB + B^2) = A(A^2 + AB + B^2) + (-B)(A^2 + AB + B^2)$
$= A^3 + A^2B + AB^2 - A^2B - AB^2 - B^3$
$= A^3 - B^3$

Exercises 2.7

2. $\dfrac{8}{5}$ **4.** 14   **6.** $\dfrac{x}{x+5}$ **8.** $\dfrac{3x+5}{2}$ **10.** $\dfrac{2x+y}{2x}$ **12.** $\dfrac{x}{x+y}$

14. $\dfrac{x+4y}{x-3y}$ **16.** $\dfrac{y-x}{y+x}$ **18.** $\dfrac{x^2+y^2}{x^2-y^2}$ **20.** $\dfrac{51}{100}$ **22.** $\dfrac{1}{10}$ **24.** $\dfrac{11x+41}{8}$

26. $\dfrac{-x-16}{12}$ **28.** $\dfrac{2x+45}{4x^2-25}$ **30.** $\dfrac{1}{6x}$ **32.** $\dfrac{3(x+3)}{(x-2)(x+2)}$

34. $\dfrac{10}{(x+3)(x-3)}$ **36.** $\dfrac{2x^2-x+1}{x(x-1)(x-1)}$ **38.** $\dfrac{(x-1)(x-1)}{x(x+5)(x-2)}$

40. $\dfrac{x}{3} - \dfrac{2}{x}$ or $\dfrac{x^2-6}{3x}$ **42.** $\dfrac{4+x^2}{4-x^2}$ **44.** $-\dfrac{m^2+1}{2m-1}$ **46.** $\dfrac{-5x-4}{x^2(x+2)}$

48. $\dfrac{-1}{(x+k+3)(x+3)}$ **50.** $\dfrac{x+3}{x}$

Exercises 2.8

2. $x^4 + 8x^3 + 24x^2 + 32x + 16$

4. $x^5 + 10x^4y + 40x^3y^2 + 80x^2y^3 + 80xy^4 + 32y^5$

6. $x^5 - 10x^4y + 40x^3y^2 - 80x^2y^3 + 80xy^4 - 32y^5$

8. $81x^4 + 216x^3y + 216x^2y^2 + 96xy^3 + 16y^4$

10. $x^5 - 5x^3 + 10x - \dfrac{10}{x} + \dfrac{5}{x^3}$

12. $x^8 + 8x^{13/2} + 28x^5 + 56x^{7/2} + 70x^2 + 56x^{1/2} + 28\left(\dfrac{1}{x}\right) + 8\left(\dfrac{1}{x^{5/2}}\right) + \dfrac{1}{x^4}$

14. $a^{40} - 40a^{39}b + 780a^{38}b^2 - 9880a^{37}b^3 + \cdots$

16. $1 - 12x^2 + 66x^4 - 220x^6 + \cdots$

18. $\doteq 1.05101$

CHAPTER 3

Exercises 3.1

2. $3, -31, -9, a^3 - 4a^2 + 5a + 3$

4. a. $3x^7 - 13x^5 + 4x^3 + 6x^2 - 2$ b. $9x^4 - 6x^2 + 1$

Exercises 3.2

2. $10x^2 + x - 21$; degree $n = 2$

4. $2x - 10$; degree $n = 1$

6. $x + 5 = 0$; degree $n = 1$

8. $3x^3 + 2x^2 - 6x + 3 = 0$; degree $n = 3$

10. $\frac{17}{30}x^2 + (\sqrt{7} - 1)x - (\sqrt{3} + \sqrt{2}) = 0$; degree $n = 2$

Exercises 3.3

2. -2 4. 5 6. -2 8. $\frac{5}{2}$ 10. $\frac{384}{67}$ 12. 10 14. $\frac{4}{7}$ 16. $\frac{3}{4}$

18. $\cong 2.48$ 20. $\cong -1.30$

Exercises 3.4

2. $\{-3, 4, 5, 7\}$

4. $\{\sqrt{17}, 3/5\}$

6. $\left\{-9/14, \frac{2}{\sqrt{3}}, \frac{\sqrt{7}}{4}\right\}$

8. $\left\{\frac{\frac{1}{5}}{\frac{3}{2}}, \frac{2}{-\frac{3}{8}}, \frac{\sqrt{27}}{3}\right\}$ or $\{0.133, -5.33, 1.73\}$

10. $\left\{\frac{\sqrt{3}}{\sqrt{5}}, \frac{7}{13}\right\}$ or $\{0.775, 4.04\}$

Exercises 3.5

2. $\{-2, 5\}$ 4. $\{-1, 3\}$ 6. $\{-\sqrt{5}, \sqrt{5}\}$ 8. $\{-4, 1/2\}$ 10. $\{-5, 3/2\}$
12. $\{-2, -1\}$ 14. $\{1, 4\}$ 16. $\{0, 6\}$ 18. $\{-2, -2\}$ 20. $\{-1, -5/2\}$
22. $\{1/5, 4/3\}$ 24. $\{7/2, 2\}$ 26. $\{-3/4, 5/2\}$ 28. $\{a, -b\}$ 30. $\{-3b, 2a\}$

Answers to Even-Numbered Exercises

32. a. $x^2 - x - 6 = 0$ **b.** $4x^2 - 8x + 3 = 0$
 c. $x^2 + 6x + 9 = 0$ **d.** $35x^2 - x - 12 = 0$

Exercises 3.6

2. $\{\pm 4i\}$ **4.** $\{-2 \pm 3i\}$ **6.** $\{-1, -9\}$ **8.** $\left\{\dfrac{-2 \pm i\sqrt{5}}{3}\right\}$

10. $\left\{-\dfrac{21}{17} \pm \dfrac{7i\sqrt{5}}{2}\right\}$ **12.** $\{-1.134, 2.866\}$ **14.** $\{-0.649 \pm 5.1969i\}$

Exercises 3.7

2. $t^2 - 12t + 36 = (t - 6)^2$ **4.** $x^2 - \dfrac{3}{2}x + \dfrac{9}{16} = \left(x - \dfrac{3}{4}\right)^2$

6. $y^2 - \dfrac{7}{5}y + \dfrac{49}{100} = \left(y - \dfrac{7}{10}\right)^2$ **8.** $x^2 + \left(\dfrac{m}{n}\right)x + \dfrac{m^2}{4n^2} = \left(x + \dfrac{m}{2n}\right)^2$

10. $\{6 \pm 2\sqrt{6}\}$ **12.** $\left\{\dfrac{3 \pm \sqrt{65}}{4}\right\}$ **14.** $\left\{\dfrac{7 \pm \sqrt{129}}{10}\right\}$ **16.** $\left\{-\dfrac{12}{5}, 2\right\}$

18. $\left\{1 \pm \dfrac{\sqrt{14}}{2}\right\}$ **20.** $\left\{\dfrac{-m \pm \sqrt{m^2 - 4np}}{2n}\right\}$ **22.** $\left\{-\dfrac{1}{2}, -2\right\}$

24. $\left\{\dfrac{3 \pm i\sqrt{7}}{2}\right\}$ **26.** $\dfrac{3}{2}$

Exercises 3.8

2. $\{-2 - \sqrt{7}, -2 + \sqrt{7}\}$ **4.** $\{5/2, 7/3\}$ **6.** $\left\{\dfrac{4 \pm \sqrt{6}}{5}\right\}$ **8.** $\left\{\dfrac{5 \pm \sqrt{3}}{11}\right\}$

10. $\left\{\dfrac{3}{10}, -1\right\}$ **12.** $\left\{-\dfrac{27}{40}, \dfrac{32}{25}\right\}$ **14.** $\left\{\dfrac{5 \pm i\sqrt{71}}{4}\right\}$ **16.** $\{-1, 4\}$

18. $\{-3, 1/3\}$ **20.** $\left\{m, \dfrac{9}{5}m\right\}$ **22.** $\{3 \pm 2i\}$ **24.** $\left\{\pm\dfrac{5}{2}i\right\}$ **26.** $\{x \pm 5\}$

28. $\{-2x - 2, -2x + 1\}$ **30.** $\left\{x + 3, \dfrac{3x - 1}{2}\right\}$ **32.** $\{x + 6 \pm \sqrt{4x - x^2}\}$

Exercises 3.9

2. 109; real, unequal, irrational   **4.** 2820; real, unequal, irrational
6. -44; complex, unequal   **8.** 76; real, unequal, irrational
10. 21; real, unequal, irrational   **12.** 73; real, unequal, irrational   **14.** 0; real, equal
16. -32; complex, unequal   **18.** 0, 8   **20.** $k = 3$ **22.** $-2, 3$

24. $r_1 + r_2 = 3/4; r_1 r_2 = 2$ **26.** $r_1 + r_2 = 9/7; r_1 r_2 = -\dfrac{6}{7}$

28. $r_1 + r_2 = 3/8; r_1 r_2 = \dfrac{5}{8}$ **30.** $r_1 + r_2 = -\dfrac{3}{8}; r_1 r_2 = -\dfrac{1}{2}$

Exercises 3.10

2. $\{-1, 1, 2, 4\}$ **4.** $\{\pm 1, \pm 3\}$ **6.** $\{\pm 2, \pm 3\}$ **8.** $\{\pm i, \pm\sqrt{8}\}$

Exercises 3.11

2. -4 **4.** 30   **6.** 12   **8.** Yes   **10.** No   **12.** Yes

14. $(x - 2)(x + 1)(x + 3)$ **16.** $q(x) = 4x^2 - 6x + 18; r = -61$

18. $q(x) = 6x^4 + 3x^3 - \dfrac{13}{2}x^2 + \dfrac{3}{4}x - \dfrac{13}{8}; R = -\dfrac{13}{16}$

20. $q(x) = x^4 + 4x^3 + 6x^2 + 15x + 31; r = 57$ **22.** $9x^2 - 3x - 3; R = 7$

24. $q(x) = -2x^2 + 12x - 42; R = 160$ **26.** $k = 0$ **28.** No

Exercises 3.12

2. $2, -3$ **4.** $1, -2$ **6.** $2, -1$

10. 2 or no positive real roots
2 or no negative real roots
4, 2 or no imaginary roots
Any real root must lie between -3 and 3.

12. 4, 2 or no positive real roots; no negative real roots; 0, 2, or 4 complex roots
Any real root must lie between -1 and 2.

14. No positive real roots
1 negative real root between -1 and 1
4 complex roots

16. 3 or one positive real root
One negative real root
2 or no complex roots
Any real roots lie between -1 and 2.

Exercises 3.13

2. $\{-2, 1, 3\}$ **4.** $\{1, 1, 3\}$ **6.** $\{2, 1 \pm 2i\}$ **8.** $\left\{-4/3, -2, 1, \dfrac{5}{2}\right\}$

10. $\{0, 2, -1 \pm \sqrt{3}\}$ **12.** $\{-3, -3, 3, 3\}$ **14.** $\{3, -2, \pm 2i\}$ **16.** $\left\{-1, \dfrac{3 \pm 4i}{2}\right\}$

Answers to Even-Numbered Exercises

18. {2, only rational root} **20.** {−1/2, 2, 3} **22.** $\left\{3, 1, -2, \dfrac{3 \pm \sqrt{11}\,i}{2}\right\}$

24. {3/2, 3/2, 3/2} **26.** {−3, −1, 1, 3} **28.** {3, 1, −2, −3}

30. $\left\{5/2, \dfrac{3 \pm \sqrt{37}}{2}\right\}$ **32.** $(x-1)^2(x+2)(x+3)$ **34.** $x(x-1)(4x^2+4x+5)$

36. $(x+2)^3(x+3)$ **38.** $(x-2)^2(x-3)(x^2+6x+10)$

40. $(2x-5)(3x-1)(2x+1)$ **42.** $x^3 - 3x - 2 = 0$

Exercises 3.14

2. {2.2} **4.** {1, 2, 0.4, −2.4} **6.** {−2.1, 0.3, 1.9} **8.** {0.347}

10. {2.154} **12.** {1.516}

Exercises 3.15

2. {−1, 3} **4.** {0, −3/2} **6.** {2} **8.** {−2} **10.** ∅ **12.** {3} **14.** ∅

16. {5/2} **18.** {4} **20.** {−1} **22.** {0, 1/2}

Exercises 3.16

2. {4} **4.** {4} **6.** {5} **8.** {6} **10.** {3, 7} **12.** {1} **14.** {1} **16.** {4}

18. $(\infty, 1/2) \cup (2, \infty)$
$[-1, 4]$
$(0.99, 1.01)$
$x < 1/2$ or $x > 2$
$(-\infty, 2/3)$
$(-3, 5/3)$

Exercises 3.17

2. $7 \leq x \leq \infty$ **4.** $-8 < x < 3/5$ **6.** $-\infty < x \leq 2$ **8.** $-7 < x < \infty$

10. $-\infty < x < 5$ or $5 < x < \infty$ **12.** $-1 < x < 5$ **14.** $-7 < x < 1$

Exercises 3.18

2. $-\infty < x \leq 0$ or $1 \leq x \leq 4$ **4.** $0 \leq x \leq 1$ or $4 \leq x < +\infty$

6. $-1 \leq x \leq 1$ or $3 \leq x < +\infty$ **8.** $-\infty < x < -\dfrac{1}{2}$ or $\dfrac{1}{3} < x < 2$

10. $-\dfrac{1}{2} < x < 2$ or $3 < x < +\infty$ **12.** $1 < x < 3$ **14.** $-3 < x < 3$

16. $-3 < x < 1$ **18.** $-2 < x < -\dfrac{1}{2}$ or $2 < x < 3$

Exercises 3.19

2. $-2 < x < 0$ or $4 < x < +\infty$
4. $-3 < x < 2$ or $4 < x < +\infty$
6. $-\infty < x < -3/2$ or $-2/3 \le x \le 2$
8. $-\infty < x < -1/2$ or $5/3 < x < +\infty$
10. $-\infty < x < 11/5$ or $5/2 < x < 3$
12. $7/4 < x < 2$ or $5/2 < x < 3$
14. $-2 < x < 3$ or $13/2 < x < +\infty$

CHAPTER 4

Exercises 4.1

2. **a.** A is a function; dom $A = \{-2, -1, 0, 1, 2, 3\}$; rng $A = \{-9, -5, 2, 3, 5, 7\}$.
   b. B is not a function.
   c. C is not a function.
   d. D is a function; dom $D = \{-1, 0, 1, 2, 3, 4\}$; rng $D = \{-6, -5, -2, 1, 4, 9\}$.

4. $f(0) = 5 \quad f(4) = 3$
   $f(5) = 4 \quad f(5) = 0$
   $f(-3) = 4$

6. **a.** $f(2) = 9$ **b.** $f(2+h) = 3h^2 + 7h + 9$
   c. $\dfrac{f(2+h) - f(2)}{h} = 3h + 7$ **d.** $f(x+h) = 3x^2 + 6xh + 3h^2 - 5x - 5h + 7$
   e. $\dfrac{f(x+h) - f(x)}{h} = 6x - 5 + 3h$

8. **a.** $3x^2 + 3xh + h^2$ **b.** $4x^3 + 6x^2h + 4xh^2 + h^3$
   c. $-\dfrac{10}{x(x+h)}$ **d.** $-\dfrac{4}{(x+h+1)(x+1)}$ **e.** $\dfrac{1}{(x+h+2)(x+2)}$

10. $F(x^2) = \dfrac{1 - x^2}{1 + x^2}$ $\qquad \dfrac{1}{F(x)} = \dfrac{1+x}{1-x}$
    $[F(x)]^2 = \dfrac{1 - 2x + x^2}{1 + 2x + x^2}$ $\qquad F\left(\dfrac{1}{x}\right) = \dfrac{x-1}{x+1}$
    $F(2x + 3) = -\dfrac{x+1}{x+2}$ $\qquad F[F(x)] = x$

12. $a, b, c, e,$ and h are functions.

14. **a.** s^2 **b.** x^3 **c.** $\dfrac{\sqrt{3}}{4}s^2$ **d.** $x\sqrt{20^2 - x^2}$ **e.** $\pi\left(r^2 + \dfrac{200}{r}\right)$

Answers to Even-Numbered Exercises 743

16. **a.** ≈ 2.1540 **b.** ≈ 2.1817

18. Example 8: $\text{Rng } f = \{y \mid -\infty < y < \infty\}$
    Example 9: $\text{Rng } f = \{y \mid y \geq 0\}$
    Example 10: $\text{Rng } f = \{y \mid y \neq 0\}$
    Example 11: $\text{Rng } f = \{y \mid y \geq -1\}$

Exercises 4.2

2. Not a function graph

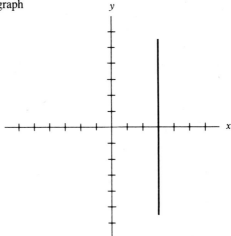

4. Function graph

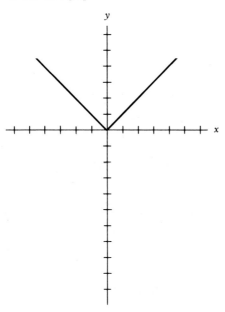

6. Not a function graph

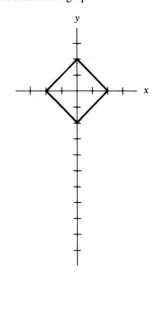

8. Not a function graph

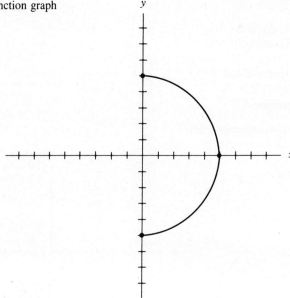

10. Not a function graph

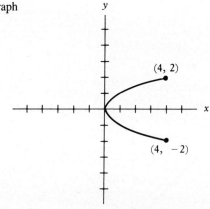

12.

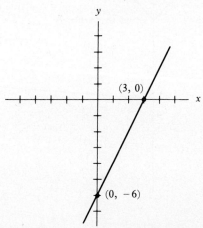

Answers to Even-Numbered Exercises

14.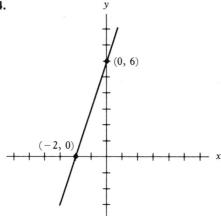

16.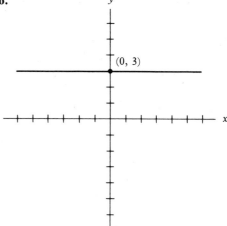

18.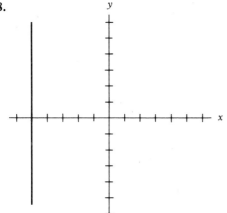

20.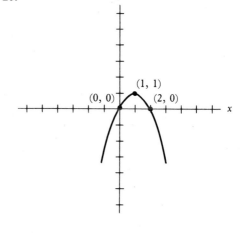

22.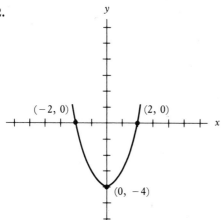

24.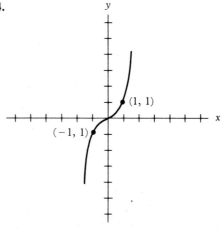

26.

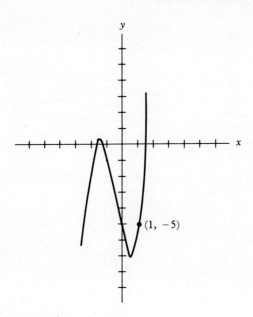

28. a. 26 units **b.** $400.00
30. a. Yes **b.** 88¢ **c.** $136.03

Exercises 4.3

2. a. $(\frac{3}{2}, 2)$ **b.** $(-\frac{1}{2}, 0)$ **c.** $(-1, -5)$ **d.** $(-3, \frac{1}{2})$
e. $(-\frac{5}{2}, 2)$ **f.** $(-1, \frac{5}{2})$

4. Let $A(8, 0)$, $B(6, 6)$, $C(-3, 3)$, $D(-1, -3)$.
$$d(A, C) = \sqrt{[8 - (-3)]^2 + (0 - 3)^2} = \sqrt{130}$$
$$d(B, D) = \sqrt{[6 - (-1)]^2 + [6 - (-3)]^2} = \sqrt{130}$$
∴ Parallelogram $ABCD$ is a rectangle.

6.

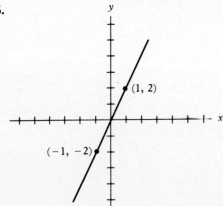 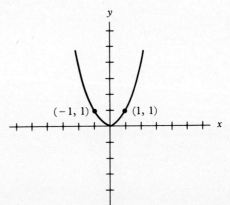

Answers to Even-Numbered Exercises

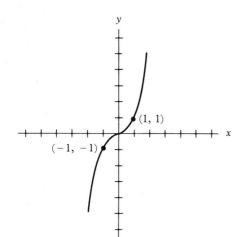

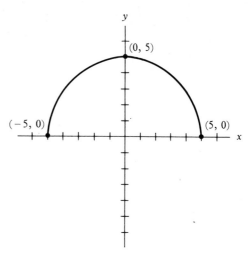

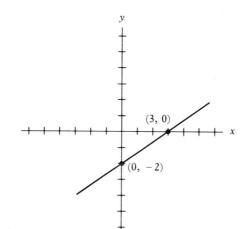

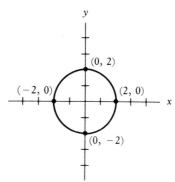

8. 1, 10, $\sqrt{85}$

12. $(3, 6), (0, 0), (12, -12), (\frac{4}{3}, 4), (48, 24)$

14. $10x + 8y = 51$

Exercises 4.4

2. a. $x^2 + y^2 = 25$ **b.**

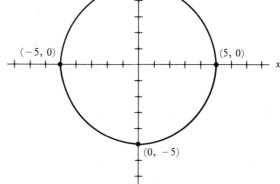

4. Absolute value of y is equal to absolute value of x.

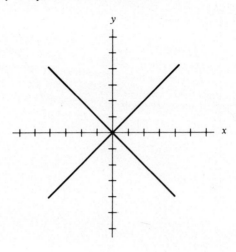

Exercises 4.5

2. a. $L = 2\sqrt{10}$, $m = -\frac{1}{3}$, $(5, 5)$
  b. $L = 2\sqrt{74}$, $m = -\frac{5}{7}$, $(-3, 2)$
  c. $L = 10\sqrt{2}$, $m = 7$, $(4, 2)$

4. a. $P_2(9, 4)$
  b.

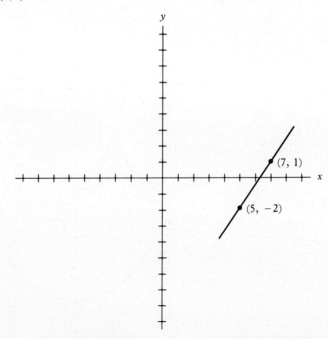

Answers to Even-Numbered Exercises

6. a.
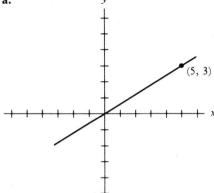

b.
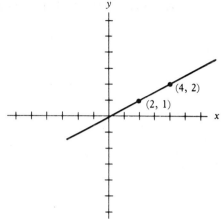

c.
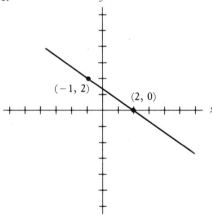

d.
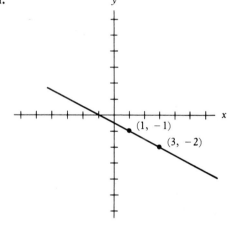

e.
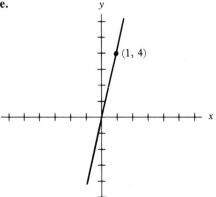

f.
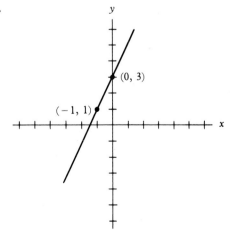

8. Let $A_1(-5, 2)$, $A_2(2, 1)$, $A_3(9, 4)$. Then

$$\text{Slope}(A_1, A_2) = \frac{1 - (-2)}{2 - (-5)}$$

$$= \frac{3}{7}$$

$$\text{Slope}(A_1, A_3) = \frac{4 - (-2)}{9 - (-5)} = \frac{6}{14} = \frac{3}{7}$$

∴ The 3 points lie on a straight line. Let $B_1(1, -2)$, $B_2(2, 0)$, $B_3(3, 2)$. Then

$$\text{Slope}(B_1, B_2) = \frac{0 - (-2)}{2 - 1} = \frac{2}{1} = 2$$

$$\text{Slope}(B_1, B_3) = \frac{2 - (-2)}{3 - 1} = \frac{4}{2} = 2$$

$$\text{Slope}(B_2, B_3) = \frac{2 - 0}{3 - 2} = \frac{2}{1} = 2$$

The 3 points lie in a straight line.

10. $(8, 6)$

12. Let the midpoint of $\overline{AB}$ be M_1 and the midpoint of $\overline{BC}$ be M_2. Then

$$M_1\left(\frac{4 + (-2)}{2}, \frac{-5 + 3}{2}\right) = M_1(1, -1)$$

and

$$M_2\left(\frac{8 + 4}{2}, \frac{1 + (-5)}{2}\right) = M_2(6, -2)$$

Hence

$$\text{Slope}(M_1, M_2) = \frac{-2 - (-1)}{6 - 1} = -\frac{1}{5}$$

and

$$\text{Slope}(A, C) = \frac{1 - 3}{8 - (-2)} = -\frac{1}{5}$$

∴ $\overline{AC} \parallel \overline{M_1 M_2}$
Also

$$d(M_1, M_2) = \sqrt{(6 - 1)^2 + [-2 - (-1)]^2} = \sqrt{26}$$

and

$$d(A, C) = \sqrt{[8 - (-2)]^2 + (1 - 3)^2} = \sqrt{104} = 2\sqrt{26}$$

$$d(M_1, M_2) = \tfrac{1}{2} d(A, C)$$

Answers to Even-Numbered Exercises

14. $3x^2 + 3y^2 - 46x - 36y + 203 = 0$

16. Let quadrilateral $ABCD$ be a parallelogram. Without loss of generality, let $A(0,0)$ and $B(b,0)$ and $C(a,c)$. Then vertex D has coordinates $(a+b, c)$. Prove that diagonals AD and BC bisect each other.

Proof:

$$\text{Midpoint of } AD \text{ is } \left(\frac{a+b}{2}, \frac{c}{2}\right)$$

$$\text{Midpoint of } BC \text{ is } \left(\frac{a+b}{2}, \frac{c}{2}\right)$$

Hence midpoints of diagonals AD and BC coincide and thus bisect each other.

Exercises 4.6

2. a. $y = \frac{1}{2}x + 3$ **b.** $y = -\frac{4}{3}x + 5$

4. a. $x = -4$ **b.** $x = 7$ **c.** $y = -5$

6. a. $y = 3/2\,x + 1$ **b.** $y = -1/2\,x + 7/2$ **c.** $y = 3/4\,x + 5/4$

8. a. $y = -x + 7$ **b.** $y = 3/2\,x - 1/2$ **c.** $y = -5/2\,x - 5/2$

10. $y = x + 1$

12. a. 2 **b.** $-3/2$ **c.** $-1/3$ **d.** $-4/7$ **e.** $3/4$ **f.** $5/7$

14. Median through vertex A: $y = \frac{1}{2}x$
Median through vertex B: $y = 3x - 10$
Median through vertex C: $y = \frac{4}{3}x - \frac{10}{3}$
Solving simultaneously the three equations yields the common point $(4, 2)$.

16. Altitude through vertex A: $y = \frac{1}{4}x + 1$
Altitude through vertex B: $y = \frac{3}{2}x - 1$
Altitude through vertex C: $y = -x + 3$
Common point: $D(\frac{8}{5}, \frac{7}{5})$

18. $x - 2y = 9 \Rightarrow m_1 = \frac{1}{2}$
$6x + 3y = 14 \Rightarrow m_2 = -2$
$4x - 8y = 15 \Rightarrow m_3 = \frac{1}{2}$
$2x + y + 5 = 0 \Rightarrow m_4 = -2$

20. a. $y = x + 4$ **b.** $(-1, 3)$ **c.** $d = 2\sqrt{2}$

Exercises 4.7

2. $(x - 2)^2 + (y - 3)^2 = 36$ **4.** $(x + 4)^2 + (y + 2)^2 = 25$

6. $(x + 4)^2 - (y - \frac{2}{3})^2 = 8$ 8. $(x - 2)^2 + (y - 1)^2 = 25$
10. $(x - 1)^2 + (y - 2)^2 = 25$ 12. $(x + 5)^2 + (y + 3)^2 = 169$
14. a. $C(2, -3)$ and $r = 4$ b. $C(-5, 4)$ and $r = 6$
c. $C\left(\dfrac{3}{2}, \dfrac{5}{2}\right)$ and $r = \dfrac{5}{2}$ d. $C(3, 0)$ and $r = 3$
e. $C(0, -2)$ and $r = 2$ f. $C\left(\dfrac{3}{2}, \dfrac{7}{2}\right)$ and $r = \dfrac{\sqrt{58}}{2}$

Exercises 4.8

2. a. $y^2 = 8x$

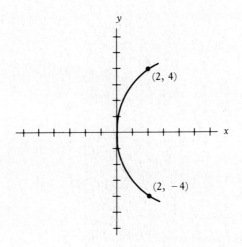

b. $x^2 = -12y$

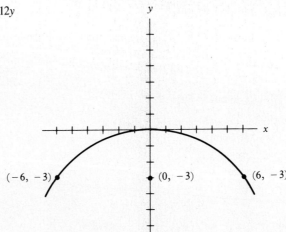

Answers to Even-Numbered Exercises

c. $y^2 = -12x$

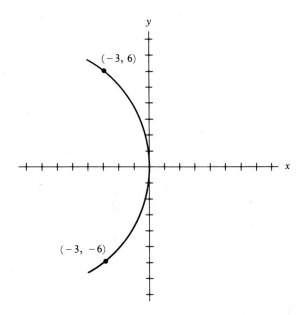

d. $(y - 1)^2 = 12(x - 1)$ **e.** $(x + 3)^2 = 8(y - 2)$

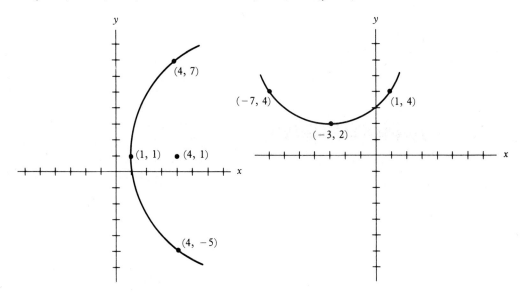

4. a. $V(-2,-2)$, $F(-2,\frac{1}{2})$
Directrix: $y = -\frac{9}{2}$
Ends of latus rectum: $(-7, 1/2)$ and $(3, 1/2)$

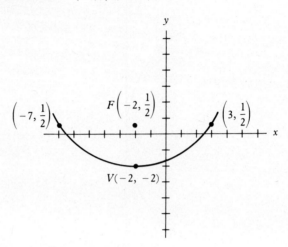

b. $V(-2,-2)$, $F(-1,-2)$
Directrix: $x = -2$
Ends of latus rectum: $(-1, 0)$ and $(-1, 4)$

c. $V(2,4)$, $F(2, 15/4)$
Directrix: $y = 17/4$
Ends of latus rectum: $(3/2, 15/4)$ and $(5/12, 15/4)$

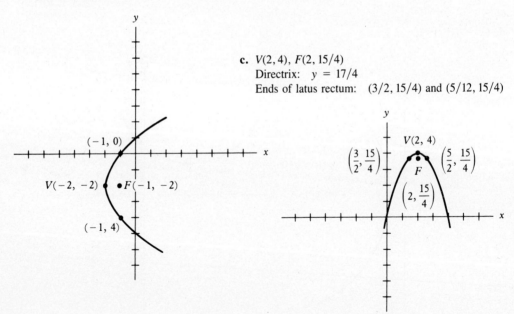

Answers to Even-Numbered Exercises

d. $V(3/2, -3)$, $F(0, -3)$
Directrix: $x = 3$
Ends of latus rectum: $(0, 0)$ and $(0, -6)$

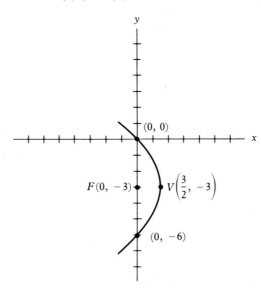

Exercises 4.9

2. a.

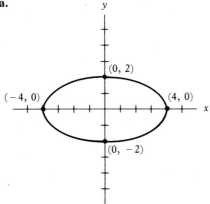

b.

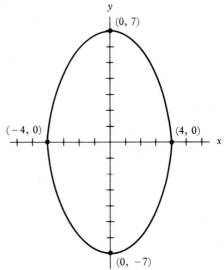

c.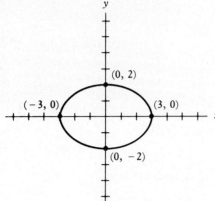

d.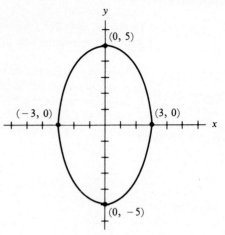

e.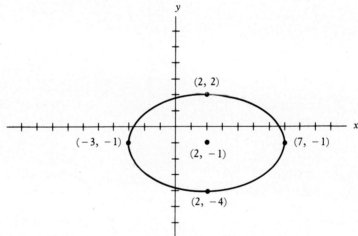

f.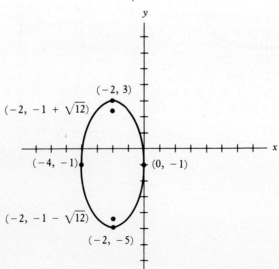

Answers to Even-Numbered Exercises

4. $y = 24$ ft

6. For the equation $9x^2 + 25y^2 = 225 \Leftrightarrow \dfrac{x^2}{25} + \dfrac{y^2}{9} = 1$ of the ellipse, its foci are $(4, 0)$ and $(-4, 0)$. The equation of the ellipse gives a locus that can be traced as follows: **Place two pins, one in each focus and with a loop of string of length $2a = 2(5) = 10$ units tied at each end of the pins. Insert a pencil in the loop, and if the string is kept taut, the pencil will trace an ellipse** with $C(0, 0)$ and foci $(4, 0)$ and $(-4, 0)$.

8. Let $C(0, 0)$ with foci along the x-axis. The coordinates of the foci are $(-12, 0)$ and $(12, 0)$, and $a = 200$ and $b = 199.64$. Thus

$$\frac{x^2}{(200)^2} + \frac{y^2}{39{,}856} = 1$$

10. a. $\dfrac{(x - 1)^2}{16} + \dfrac{(y - 3)^2}{4} = 1$

b. $\dfrac{(x + 3)^2}{16} + \dfrac{(y - 4)^2}{36} = 1$

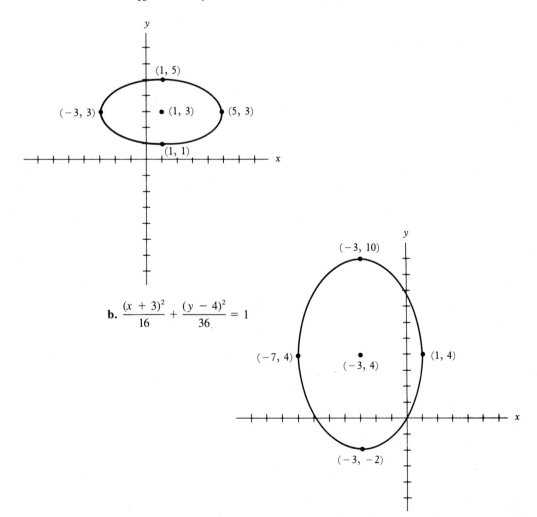

c. $\dfrac{x^2}{(14)^2} + \dfrac{(y-6)^2}{40} = 1$

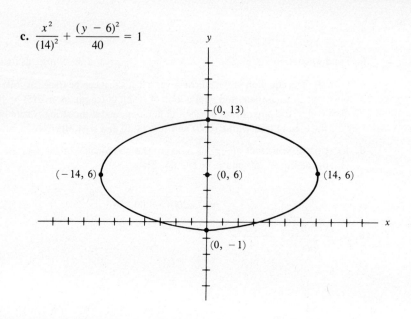

d. $\dfrac{x^2}{\frac{336}{16}} + \dfrac{(y+1)^2}{\frac{336}{25}} = 1$

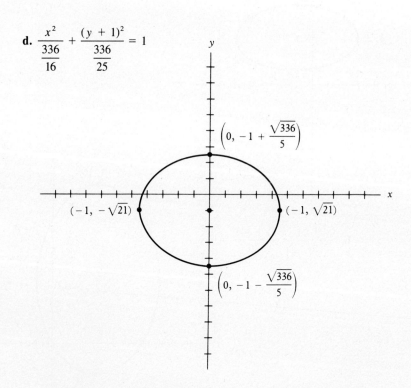

Answers to Even-Numbered Exercises

Exercises 4.10

2. a. $x^2 - y^2 = 4$
Asymptotes: $y = x$
$y = -x$

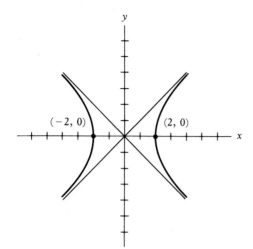

b. $y^2 - x^2 = 1$
Asymptotes: $y = x$
$y = -x$

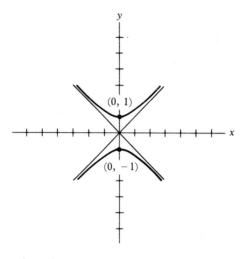

c. $\dfrac{x^2}{4} - \dfrac{y^2}{9} = 1$
Asymptotes: $y = \dfrac{3}{2}x$
$y = -\dfrac{3}{2}x$

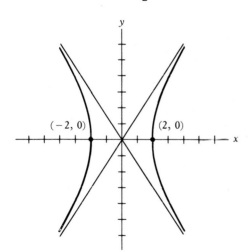

d. $\dfrac{y^2}{16} - \dfrac{x^2}{9} = 1$
Asymptotes: $y = \dfrac{4}{3}x$
$y = -\dfrac{4}{3}x$

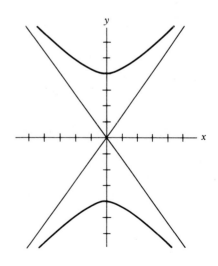

Exercises 4.11

2.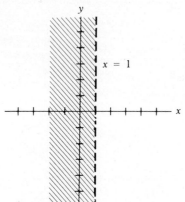

4.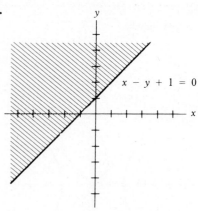

6.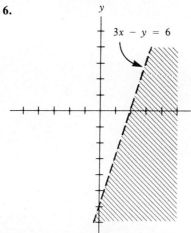

8.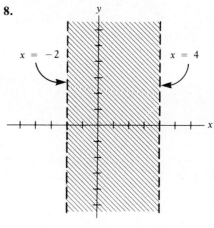

10.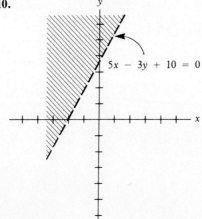

CHAPTER 5

Exercises 5.1

2. 3.75 bushels per acre
4. a. -1 unit/unit   b. 1 unit/unit   c. -1 unit/unit   d. $3(x_2 + x_1) - 7$
   e. $\dfrac{1}{(x_2 + 1)(x_1 + 1)}$
6. $56/unit
8. -1 unit/unit
10. 60 liters/minute
12. 60.002 units/unit
14. a. 3   b. 3   c. 3
    The graph of the function is a line.
16. 0.58 units/unit

Exercises 5.2

2. a. $12 + 6h + h^2, h \neq 0$ b. 12

Exercises 5.3

2. -21 4. -8 6. 4   8. $3x_1$ 10. -3 12. 66
14. 3   16. $4x_1 - 5$

Exercises 5.4

2. 3   4. 2   6. 3   8. 0   10. 12   12. $4x - 7$ 14. 12   16. 0
18. -2 20. $\dfrac{1}{2\sqrt{5}}$

Exercises 5.6

2. $6x - 8$ 4. $6 - 4x$ 6. $21x^2 - 5$ 8. $\dfrac{-4}{x^2}$ 10. $20x + 13$
12. $-4 + 8x$ 14. $\dfrac{-1}{(x + 2)^2}$

Exercises 5.7

2. $20x^3 + 14x$
4. $30t^2 - 4t^3$
6. $6u^2 + 6u - 2$
8. $D_x y = 2ax$, where z is considered a constant
10. $3v^4 + v^2$
12. $8t^3 + 15t^2 - 6t$
14. $\dfrac{2at + b}{c}, c \neq 0$
16. $1 + 0.6x - 1.5x^2 - 1.6x^3$
18. $30x - 11$
20. $12x^3 - 4x$
22. a. $-8x^{-9} - 24x^{-5}$ b. $-9x^{-4} - 16x^{-3} + 2x^{-2}$
    c. $-8x^{-3} + 10x^{-6}$ d. $-2x^{-5} + 2x^{-4}$

Exercises 5.8

2. $14x^{-3}$ 4. $5x^4 - 12x^2 + 4x$ 6. $\dfrac{-19}{(2x - 5)^2}$ 8. $8x^3 + 9x^2 - 2$

10. $\dfrac{20}{(1 - x)^3}$ 12. $-80x(1 - x^2)^4$ 14. $\dfrac{14}{(4 - 3x)^2}$ 16. $\dfrac{-16x}{(4 + x^2)^2}$

18. $\dfrac{20}{(1 - x)^3}$ 20. $6x - 2 + 8x^{-3}$ 22. $\dfrac{-10}{x^2}$ 24. $\dfrac{16}{(2 - 5x)^2}$

26. a. $2(5t^2 + 3t)(10t + 3)$ b. $\dfrac{-24t^2 - 18}{(4t^2 - 3)^2}$ c. $5t^4 + 3t^2$ d. $4(t^3 - t^{-5})$

Exercises 5.9

2. $-3x^{-3/2}$ 4. $\dfrac{5}{8}x^{-3/8}$ 6. $\dfrac{-x}{\sqrt{9 - x^2}}$ 8. $2x(x + 1)(2x^3 + 3x^2)^{-2/3}$

10. $x^{-1/2} + x^{-2/3} + x^{-3/4}$ 12. $x^2(x^3 + 2)^{-2/3}$ 14. $-8x^{-5/3}$

16. $-2x^3(4 - x^4)^{-1/2}$ 18. $-\dfrac{x}{y}$ 20. $\dfrac{-y - 3x}{x}$ 22. $\dfrac{3 - x}{y - 2}$ 24. $-\left(\dfrac{y}{x}\right)^{1/3}$

26. $2x\sqrt{1 - x^2} - \dfrac{x^3}{\sqrt{1 - x^2}}$ 28. $5x^{3/2} - \dfrac{3}{2}x^{1/2} - 2x^{-3/2}$

30. $-x^2(4 - x^2)^{-1/2} + (4 - x^2)^{1/2}$ 32. $-\dfrac{y}{x}$

Answers to Even-Numbered Exercises

CHAPTER 6

Exercises 6.1

2. **a.** $v(t) = 60 - 10t$ kilometers/hr
   $a(t) = -10$ kilometers/hr^2
   b. $v(t) = 60 - 10t$ kilometers/hr
   $a(t) = -10$ kilometers/hr^2
   Second car started 12 kilometers ahead of first car.

4. $v(0) = 0$ mph $v(1) = 2$ mph
   $a(0) = 0$ mi/hr^2 $a(1) = 3$ mi/hr^2
   $v(2) = 4$ mph $v(3) = 0$
   $a(2) = 0$ mi/hr^2 $a(3) = -9$ mi/hr^2

6. $m_c = \$425/$unit

8. 1

10. 1

12. 11

14. -1

16. 3/4

18. 0

20. 3/5

22. **a.** $3x - 5y + 11 = 0$ **b.** $3x - 4y = 7$

24. $(3, -26), (-1, 18)$

Exercises 6.2

2. **a.** $8/x^3$ **b.** $\dfrac{10}{(x-3)^2}$ **c.** $6 - \dfrac{4}{x^3}$ **d.** $-600/x^4$
   e. $\dfrac{-4y^2 - 9 + 12x - 4x^2}{4y^3}$

4. 12

Exercises 6.4

2. $m = -1$; $m' = +6$; increasing

4. $v(3) = 12$ units/unit; $a(1) = 12$ units/units^2; $a(5) = -12$ units/units^2

6. $x(0) = 4$, $v(0) = 9$, moving to the right
   $v(t) = 0$ when $t = 1$ and when $t = 3$
   $x(1) = 8$; $x(3) = 4$

8. $f(x) = 3 + 4x - x^2 \Rightarrow f'(x) = 4 - 2x = 2(2 - x)$
$\Rightarrow f''(x) = -2$

Increasing for $x < 2$; decreasing for $x > 2$; concave downward for all x; $f'(x) = 0$ when $x = 2$

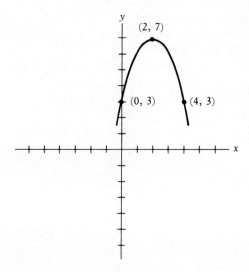

10. $f(x) = x^4 \Rightarrow f'(x) = 4x^3$
$\Rightarrow f''(x) = 12x^2$

Increasing for $x > 0$; decreasing for $x < 0$; concave upward for all x; $f'(x) = 0$ at $x = 0$

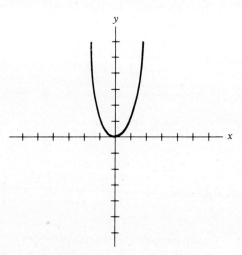

12. $f(x) = x^3 - 3x^2 + 3x + 5 \Rightarrow f'(x) = 3x^2 - 6x + 3 = 3(x - 1)(x - 1)$
$\Rightarrow f''(x) = 6x - 6 = 6(x - 1)$

Answers to Even-Numbered Exercises

Increasing for $x > 1$; increasing for $x < 1$; concave upward in $(1, \infty)$; concave downward in $(-\infty, 1)$; $f'(x) = 0$ at $x = 1$

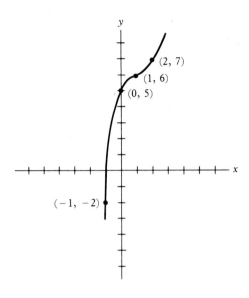

14. $f(x) = 3x^4 - 4x^3 - 5 \Rightarrow f'(x) = 12x^3 - 12x^2 = 12x^2(x - 1)$
$\Rightarrow f''(x) = 36x^2 - 24x = 12x(3x - 2)$

Increasing for $x > 1$; decreasing for $x < 1$; concave upward in $(-\infty, 0) \cup (\frac{2}{3}, \infty)$; concave downward in $(0, \frac{2}{3})$; $f'(x) = 0$ at $x = 0$ and at $x = 1$

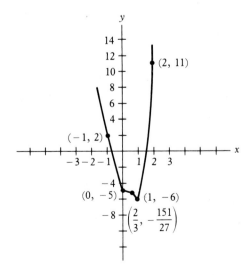

16. $f(x) = 4 - x^2 \Rightarrow f'(x) = \dfrac{-x}{(4-x^2)^{1/2}}$

$\Rightarrow f''(x) = \dfrac{-4}{(4-x^2)^{3/2}}$

Increasing in $(-2, 0)$; decreasing in $(0, 2)$; concave downward in $(-2, 2)$; $f'(x) = 0$ when $x = 0$

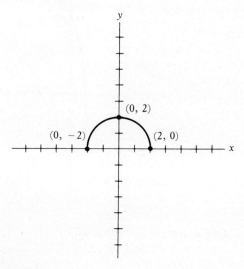

18. a. No point of inflection
b. Point of inflection at $(0, 0)$
c. Point of inflection at $(1, 6)$
d. Point of inflection at $(1/2, -3/2)$
e. Points of inflection at $(0, -5)$ and $(2/3, -5.72)$
f. Point of inflection at $(0, -10)$
g. No point of inflection
h. No point of inflection

Exercises 6.5

2. $x = 2$ is a critical value and $(2, -5)$ is a relative minimum.

4. $x = 2$ and $x = 4$ are critical values.
$(2, 13)$ is a relative maximum.
$(4, 9)$ is a relative minimum.

6. $x = 0$ and $x = 4$ are critical values.
$(0, -15)$ is a relative maximum.
$(4, -79)$ is a relative minimum.

8. Critical value is $x = 2$ and the point $(2, 0)$ is a relative minimum.

Answers to Even-Numbered Exercises

10. Critical values are $x = -1$
$x = 0$
$x = 1$
Point $(-1, -1)$ is a relative minimum.
Point $(0, 1)$ is a point of inflection.
Point $(1, 3)$ is a relative maximum.

12. Critical values are $x = -1$
$x = 0$
$x = 2$
Point $(-1, 18)$ is a relative maximum.
Point $(0, 5)$ is a point of inflection.
Point $(2, -171)$ is a relative minimum.

14. Critical values are $x = 0$ and $x = 4$.
Point $(0, -2)$ is a relative maximum.
Point $(4, 6)$ is a relative minimum.

16. Critical values are $x = -1$
$x = 0$
$x = 1$
Point $(-1, 0)$ is a relative minimum.
Point $(0, 1)$ is a relative maximum.
Point $(1, 0)$ is a relative minimum.

18. Critical values are $x = -1$ and $x = 3$.
$(-1, 15)$ is a relative maximum; also ABSOLUTE MAX in $[-2, 4]$.
$(3, -17)$ is a relative minimum; also ABSOLUTE MIN in $[-2, 4]$.

20. Hottest at $x = 10$ and $t = 20(10) - 10^2 = 100°$

Exercises 6.6

2. $x = 10$ and $y = 20$

4. $x = 20$ yards and $y = 40$ yards

6. $x = 20$ yards and $y = 20$ yards

8. $x = 2$ gives the maximum volume.

10. $x = 10$, $y = 10$, and $z = 5$

12. $x = 4$, $y = 4$, and $z = 2$

14. a. $M_c = 3x^2 - 24x + 60$ **b.** $x = 10$

16. $x = \dfrac{50}{3}$ inches; $V = \dfrac{250{,}000}{27}$ cubic inches

18. $x = 3$, $y = 3$, and $z = 2$

20. $x = 20$ ft and $y = 15$ ft

22. $\frac{56}{15}$ hours or 3.7333 hours

24. $t = 2$

Exercises 6.7

2. a. 96 ft/sec **b.** $60\sqrt{2}$ ft/sec

4. 1.6π ft²/sec

6. -22 km/hr; decreasing

8. 5 ft/sec

10. 6.4 ft/sec

CHAPTER 7

Exercises 7.1

2. $y = 2x^{5/2} + C$

4. $y = 3/5\ x^{5/3} + C$

6. $y = x^4 - \ln x + C$

8. $y = 2x^3 + x^2 - 20x + C$

10. $y = 3x^2 + 3x^{7/3} + C$

12. $y = x^{3/2} + 3/4\ x^{5/3} + C$

14. $y = \frac{x^4}{4} + \frac{4x^3}{3} - x^2 + 3x - \frac{38}{3}$

16. $U = v^4 + v^3 + v^2 - 8v + C$

18. $z = y^5 + \frac{1}{2}y^4 - 2y^2 + C$

20. $s = 96t - 16t^2 + 256$

Exercises 7.2

2. $s = 2t^2 - 20t + 18$; reverses direction at $t = 5$ sec at positive $S = -32$ ft

4. $v = -96$ ft/sec; speed is 96 ft/sec

6. $t = 128$ seconds and $s = 262{,}144$ ft

8. 1820 ft

10. $t = \frac{1}{20}$ hr or 3 minutes and $s = \frac{3}{2}$ miles

12. $C = 560$ ft/sec

Answers to Even-Numbered Exercises

Exercises 7.3

2. $y = \dfrac{3}{2}x - 6$ **4.** $y = -\dfrac{2}{3}x + \dfrac{1}{3}$ **6.** $y = -x - 2$

8. $y = 2x^2 - 2x - 1$ **10.** $y = \dfrac{2x^3}{3} - \dfrac{3x^2}{2} + x + \dfrac{65}{2}$ **12.** $y = \dfrac{2}{x} - 1$

Exercises 7.4

2. $5x + C$ **4.** $7t - 2t^2 + C$ **6.** $-\dfrac{4}{x} + C$ **8.** $4x^{3/2} + C$

10. $\dfrac{2}{5}x^{5/2} + C$ **12.** $2x^{1/2} + C$ **14.** $-\dfrac{ax^3}{3} + \dfrac{bx^2}{2} + C$ **16.** $\dfrac{3}{2}x^4 - 2x^2 + C$

18. $\dfrac{2}{9}x^3 - \dfrac{5}{14}x^2 + 16x + C$ **20.** $\dfrac{1}{3}(2x - 5)^{3/2} + C$ **22.** $\dfrac{1}{15}(4 + x^3)^5 + C$

24. $2\sqrt{3x^2 + 5x - 7} + C$ **26.** $\sqrt{2x + 3} + C$ **28.** $-\dfrac{1}{4(4 + x^2)^2} + C$

30. $2x^3 - \dfrac{5x^2}{2} - 4x + C$ **32.** $-\dfrac{1}{3}(4 - x^2)^{3/2} + C$ **34.** $-5\sqrt{2 - t^2} + C$

36. $\sqrt{x^2 + 4x + 7} + C$ **38.** $\dfrac{4}{5}x^{5/2} + 3x + C$ **40.** $-\dfrac{3}{2}(6 - x^2)^{2/3} + C$

42. $y^2 = 4x + 25$ **44.** 0.19

Exercises 7.5

2. 50 **4.** 132 **6.** 20 **8.** 9 **10.** 1

12. 20 sq. units **14.** $\dfrac{32}{3}$ sq. units

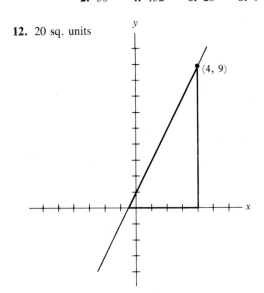

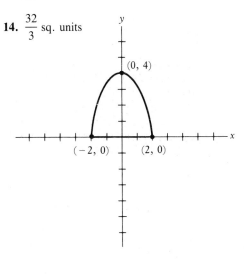

16. $\frac{9}{2}$ sq. units

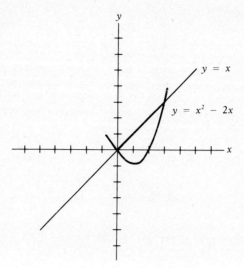

18. $\frac{32}{3}$

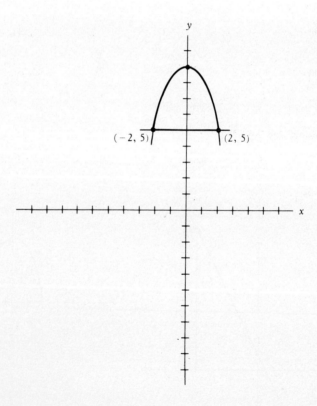

Answers to Even-Numbered Exercises

20. $\dfrac{45}{4}$ sq. units

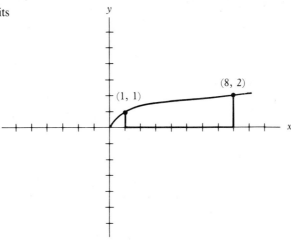

22. $\dfrac{9}{2}$ sq. units

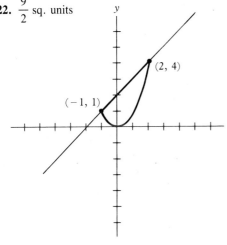

24. $\dfrac{9}{2}$ sq. units

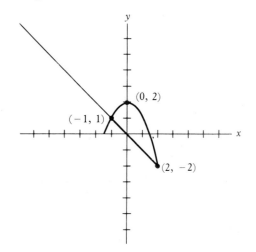

Exercises 7.6

2. $\dfrac{16{,}000}{3}\pi$ cubic meters   **4.** $\dfrac{215}{3}\pi$ cubic units   **6.** $\dfrac{\pi r^2 h}{3}$ cubic units

Answers to Even-Numbered Exercises

Exercises 7.7

2. 36 sq. units

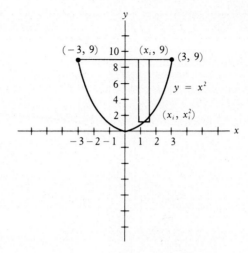

4. 12 sq. units

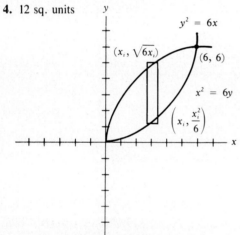

6. $\dfrac{9}{2}$ sq. units

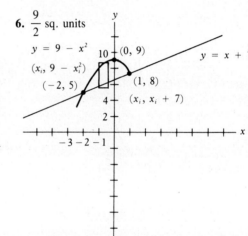

8. $\dfrac{32}{3}$ sq. units

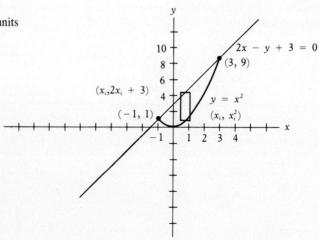

CHAPTER 8

Exercises 8.1

2. $\{(-2, 3)\}$ **4.** $\{(2, 3)\}$ **6.** $\{(3, -1)\}$ **8.** $\{(1, 2)\}$ **10.** $\{(-3, -4)\}$
12. $\left\{\left(-\dfrac{12}{163}, -\dfrac{267}{163}\right)\right\}$ **14.** $\{(-5, -2)\}$ **16.** $\left\{\left(\dfrac{6}{5}, -3\right)\right\}$ **18.** $\{(1, 2, 3)\}$
20. $\{(2, 3, 4)\}$ **22.** $\{(-4, 3, 2)\}$ **24.** \$3600 @ 4% and \$2400 @ 6%
26. \$125/horse and \$105/cow **28.** 125 lb of 79¢ and 50 lb of \$1.00 candy
30. $\Rightarrow \{(-0.69970 - 0.86248)\}$ **32.** $\Rightarrow \{(-0.44046, -0.07719, -1.06478)\}$

Exercises 8.2

2. 3 **4.** 1 **6.** -4 **8.** 0 **10.** 0 **12.** 1 **14.** $\{(0, 2)\}$
16. $\left\{\left(-\dfrac{19}{5}, -\dfrac{49}{5}\right)\right\}$ **18.** $\left\{\left(\dfrac{1}{3}, \dfrac{1}{6}, \dfrac{1}{2}\right)\right\}$ **20.** $\left\{\left(2, -1, -\dfrac{1}{3}\right)\right\}$
22. $\left(\dfrac{9}{4}, -\dfrac{1}{2}, 1\right)$

24. 0, since by multiplying the first column by a factor k, $k \neq 0$, and adding to the second column gives us all zero entries on the second column

26. $2x + 3y = 13$

28. $\begin{vmatrix} a & d & g \\ c & f & j \\ b & e & h \end{vmatrix} = -\begin{vmatrix} a & d & g \\ c & f & j \\ b & e & h \end{vmatrix}$

30. a. $\{(-0.69969, -0.86248)\}$ **b.** $\{(-0.44046, -0.07719, -1.06478)\}$

Exercises 8.3

2. $\{(1, 2)\}$ **4.** $\{(-3, -4)\}$ **6.** $\{(3, 2)\}$ **8.** $\left\{\left(1, -\dfrac{1}{2}\right)\right\}$ **10.** $\{(1, 2, 3)\}$
12. $\{(2, 3, -1)\}$ **14.** $\{(1, 2, 3)\}$ **16.** $\{(3, 1, 2, -1)\}$
18. $\{(-0.69969, -0.86248)\}$ **20.** $\{(-0.44046, -0.07719, -1.06478)\}$

Exercises 8.4

2. $\begin{bmatrix} 8 & 7 \\ -5 & 8 \end{bmatrix}$

4. $\begin{bmatrix} 3 & -3 & 6 \\ -6 & 0 & 3 \end{bmatrix}$

6. $\begin{bmatrix} 2 & 3 \\ -2 & -2 \end{bmatrix}$

8. $\begin{bmatrix} 1 & -2 & -4 \\ 4 & -4 & -1 \\ -3 & 13 & -1 \end{bmatrix}$

10. $\begin{bmatrix} \sqrt{2}+\sqrt{3} & \dfrac{15}{2} \\ -\dfrac{2}{55} & -7+\sqrt{7} \end{bmatrix}$

12. Not defined

14. $\begin{bmatrix} 14 & 0 \\ 7 & 7 \end{bmatrix}$

16. $\begin{bmatrix} 21 & 0 \\ -1 & -5 \end{bmatrix}$

18. $\begin{bmatrix} 4 & 13 & 13 \\ 1 & -2 & -6 \\ 3 & 1 & 1 \end{bmatrix}$

20. $\begin{bmatrix} 26 \\ 10 \end{bmatrix}$

22. $\begin{bmatrix} -1 & 8 \\ 0 & 25 \end{bmatrix}$

24. $\begin{bmatrix} 1 & -5 & -6 & 6 \\ -8 & 5 & 3 & -13 \\ 11 & 8 & 15 & 3 \\ 15 & 2 & 9 & 13 \end{bmatrix}$

26. $[-1495.3689]$

28. $\begin{bmatrix} 564.66569 & 4875.9735 & -14,648.79436 \\ -33.12978 & -424.1706 & -1266.9557 \\ 180.34283 & 555.3413 & -1721.8841 \end{bmatrix}$

Exercises 8.5

2. $\begin{bmatrix} 1 & -3 \\ -1 & 4 \end{bmatrix}$

6. Has no inverse

4. $\begin{bmatrix} -1 & -\dfrac{3}{2} \\ -2 & -\dfrac{5}{2} \end{bmatrix}$

8. Has no inverse

10. $\begin{bmatrix} 0.02320 & 0.02821 \\ 0.07709 & -0.09514 \end{bmatrix}$

Exercises 8.6

2. $\{(3, -2)\}$

4. $\{(2, -1)\}$

6. $\{(-2, 7)\}$

8. $\begin{bmatrix} 2 \\ -1 \\ 1 \end{bmatrix}$

Exercises 8.7

2.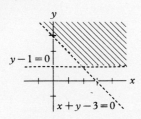

4.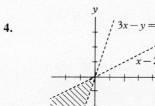

Answers to Even-Numbered Exercises

6.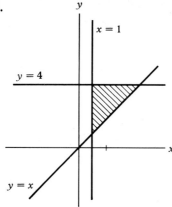

8.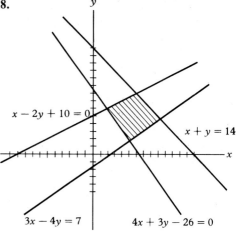

Exercises 8.8

2. Max is $\dfrac{73}{3}$ and min is 1.

4. Max is 26 and min is 17.

6. Yes

8. a. $\begin{cases} y + x \geq -3 \\ y - x \leq 3 \end{cases}$ b. $\begin{cases} x + y - 5 \geq 0 \\ 5x - 3y + 15 \geq 0 \\ 3x - 5y - 15 \geq 0 \end{cases}$

10. a. For minimum cost ($217,500) the developer should build either 30 Type I units and 3 Type II units or 5 Type I units and 15 Type II units.

    b. For minimum construction time (900 hours) the developer should build 5 Type I units and 15 Type II units.

12. The director should administer 42 aptitude tests and 5 performance tests.

CHAPTER 9

Exercises 9.1

2. **a.** 0.075   **b.** 0.14   **c.** 0.23   **d.** 1.35   **e.** 2.50

4. **a.** 7%   **b.** 23%   **c.** 124%   **d.** 62.4%   **e.** 105.6%

6. **a.** 8.33%   **b.** 75%   **c.** 125%   **d.** 2%

8. 34

10. $3840

12. $6175

14. $323.08

16. Lot A cost $1200 and resold @ 9% profit.
Lot B cost $1600 and resold @ 2% loss.

18. 44.11486

20. 145,627.09

Exercises 9.2

2. $156; $3,756   **4.** $280   **6.** 12%   **8.** 5%   **10.** $380

12. $I_e = I_0 - \frac{1}{73}I_0; I_0 = I_e + \frac{1}{72}I_e$

14. $I_0 = \$86.63$
$I_e = \$85.45$

16. $97.82   **18.** 23.39%   **20.** 9.82%   **22.** 180 days; $743.79   **24.** $268.24

Exercises 9.3

2. $943.40 present value; $56.60 discount   **4.** $809.03; $836   **6.** $1556.14
8. $6000 down and $4000 in a year   **10.** $808.06   **12.** $274.64

Exercises 9.4

2. $1345.11   **4.** $3794.32   **6.** $337.97   **8.** 2.39278   **10.** $1247.97
12. $1159.27   **14.** $28,758.70

Exercises 9.5

2. $1420.84   **4.** $1602.56   **6.** $1064.44   **8.** 9.38%   **10.** $1416.36
12. 7.71%   **14.** $6314.31

Exercises 9.6

2. $338.18   **4.** $224.31   **6.** $197.84   **8.** $26,070.40   **10.** $540.61

Exercises 9.7

2. a. $47,434.91   **b.** $48,916.03   **c.** $49,693.22
4. a. $11,041.85   **b.** $25,639.77
6. $4948.73

8. $\frac{(1 + r)^{k+m} - 1}{r} \stackrel{?}{=} \frac{(1 + r)^k - 1}{r} + (1 + r)^k \left[\frac{(1 + r)^m - 1}{r}\right]$

$\stackrel{?}{=} \frac{(1 + r)^k - 1}{r} + \frac{(1 + r)^k(1 + r)^m - (1 + r)^k}{r}$

Answers to Even-Numbered Exercises

$$\stackrel{?}{=} \frac{(1+r)^k - 1 + (1+r)^{k+m} - (1+r)^k}{r}$$

$$= \frac{(1+r)^{k+m} - 1}{r}$$

10. $S_{180} = \$13{,}364.44$
$I = \$55.68$, interest for one month

12. $6788.87

Exercises 9.8

2. $14,669.52 **4.** $110.37 **6.** $105.10 **8.** $n = 24$; $408.92

10. $n = 108$; $38.11 **12.** 7.3%; too low

Exercises 9.9

2. $408.96
$12.09
$130.37

4. a. $1040.81 **b.** $2858.77

6. $2571.77

8. $341.66

10. $n = 12$; $73.75

12. $n = 8$; $115.77

14.

AMORTIZATION SCHEDULE

Annual Interest Rate 0.0925	Amount of Loan $52,000	Number of Payments 240	Periodic Payment $476.25
Payment Number	Interest	Principle	Remaining Balance
1	400.83	75.42	51,924.58
2	400.25	76.00	51,848.58
3	399.66	76.59	51,771.99
4	399.07	77.18	51,694.81
5	398.48	77.77	51,617.04
6	397.88	78.37	51,538.67
7	397.28	78.97	51,459.70
8	396.67	79.58	51,380.12
9	396.05	80.20	51,299.92
10	395.44	80.81	51,219.11
⋮	⋮	⋮	⋮
240	3.64	472.61	0

Exercises 9.10

2. $3173.99 **4.** $5707.65 **6.** $695.93 **8.** $152.28 **10.** $1143.36
12. a. $8920.70 **b.** $9100.01 **c.** $9191.01 **d.** $10,356.66 **e.** $16,697.27
14. $166.58

CHAPTER 10

Exercises 10.1

2. $N = 100e^{0.20t}$; 7389.1; 54598.15

4. 40.91 mg

6. 448,169; 738,906; 1,218,269; 2,008,554; 3,311,545

8. $i = 20e^{-100t}$

Exercises 10.2

2. a. 3.23 **b.** 4.89 **c.** 8.92 **d.** 4.645 **e.** 8.725 **f.** 6.3942
4. a. 0.1679 **b.** 1.445 **c.** 48.826 **d.** -0.8945

Exercises 10.3

2. a. 376 **b.** 17600.81 **c.** 967.58 **d.** 0.783127 **e.** 0.09237
f. 0.36788 **g.** 148.41

4. 196.38

6. 4.62 hours

8. 4.225

10. 30

Exercises 10.4

2. $D_x y = -40e^{-4x}$ **4.** $D_x y = 30x^2 e^{x^3}$ **6.** $D_x y = (2 - x)xe^{-x}$

8. $D_x y = 7x^{-1}$ **10.** $D_x y = 6x^{-1}$ **12.** $D_x y = \dfrac{-19}{6x^2 + x - 15}$

14. $D_x y = \dfrac{2x^2 - 1}{x^3 - x}$ **16.** $D_x y = \dfrac{e^x}{x}(1 + x \ln x)$ **18.** $D_x y = \dfrac{2}{x(3x + 2)}$

20. $D_x y = \dfrac{1}{x(x^2 + 1)}$ **22.** $D_x y = 1 + \ln x$ **24.** $D_x y = 2 - e^{2x}$

26. $D_x y = \ln x$ **28.** $D_x y = \dfrac{4}{(e^x + e^{-x})^2}$ **30.** $D_x y = \dfrac{x + 1}{x}$

Answers to Even-Numbered Exercises

Exercises 10.5

2. $e^{x^2} + C$ **4.** $-\frac{1}{2}e^{2-x^2} + C$ **6.** $\ln|x + 1| + C$ **8.** $\ln(4 + e^x) + C$
10. $-\frac{1}{3}\ln|5 - 3x| + C$ **12.** $2x - 5\ln|x| + C$ **14.** $\frac{1}{5}(\ln x)^5 + C$
16. 19.086   **18.** 1.3863   **20.** 2.33   **22.** 0.2311

Exercises 10.6

2. 1/2   **4.** 4   **6.** 0   **8.** 64   **10.** 2   **12.** 10,000   **14.** 1   **16.** 0.7124
18. 1.7784   **20.** 1.5350   **22.** $2x5^{x^2}\ln 5$ **24.** $-\dfrac{31}{(6x^2 - 11x - 35)\ln 10}$
26. $-\dfrac{8\ln 3}{3^{2x}}$ **28.** $\left(\dfrac{1}{x}\ln 8\right)8^{\ln x}$ **30.** $\dfrac{4(\log_5 x)^3}{x \ln 5}$ **32.** $-\dfrac{2^{-x^2}}{2\ln 2} + C$
34. $\log PQR$ **36.** $\log \dfrac{x^2}{y}$ **38.** $\log \dfrac{x\sqrt{x^2+1}}{(2x+3)^2}$ **40.** $\log \sqrt[3]{\dfrac{xy}{z^2}}$

CHAPTER 11

Exercises 11.1

2. a. 0.62832   **b.** 1.19031   **c.** 1.91986   **d.** 2.45393   **e.** 0.85172
   f. 4.01426   **g.** 2.48376
4. a. 80° 12′ 51″   **b.** 34° 22′ 39″   **c.** 134° 4′ 20″   **d.** 24° 3′ 51″
   e. 183° 20′ 47″   **f.** 95° 41′ 2″   **g.** 90° 0′ 1″   **h.** 272° 43′ 40″

Exercises 11.2

2. a. -12.4 **b.** 8   **c.** -5 **d.** 24   **e.** 0   **f.** 3
4. See illustrations on pages 780-781.
6. 5 mph
8. $\dfrac{3}{10}$ radians $= \dfrac{54}{\pi}$ degrees

Exercises 11.3

2. a. $\sin 0° = 0$
      $\cos 0° = 1$
      $\tan 0° = 0$
      ctn 0° does not exist
      $\sec 0° = 1$
      csc 0° does not exist
   b. $\sin \pi = 0$
      $\cos \pi = -1$
      $\tan \pi = 0$
      ctn π does not exist
      $\sec \pi = -1$
      csc π does not exist

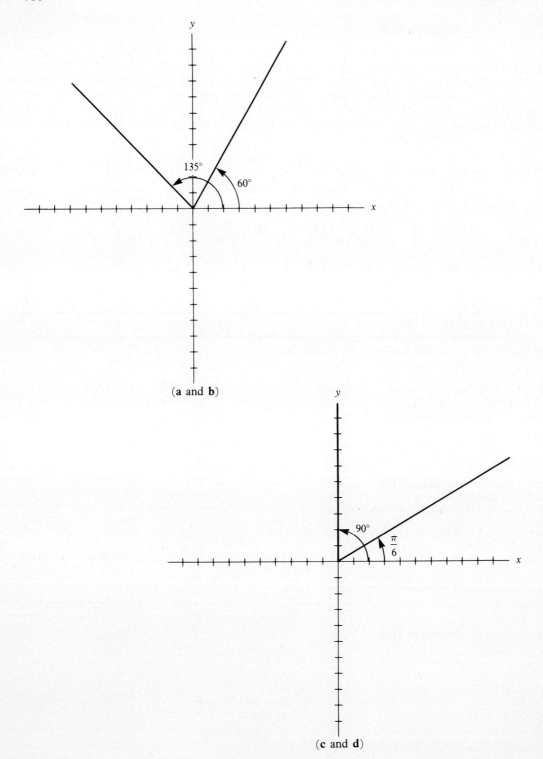

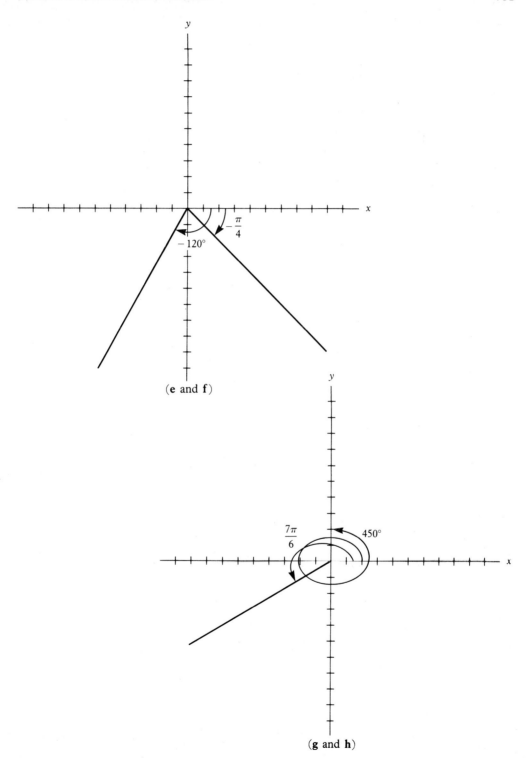

(e and f)

(g and h)

c. $\sin 270° = -1$
$\csc 270° = 0$
tan 270° does not exist
$\ctn 270° = 0$
sec 270° does not exist
$\csc 270° = -1$

d. $\sin \dfrac{\pi}{2} = 1$
$\cos \dfrac{\pi}{2} = 0$
$\tan \dfrac{\pi}{2}$ does not exist
$\ctn \dfrac{\pi}{2} = 0$
$\sec \dfrac{\pi}{2}$ does not exist
$\csc \dfrac{\pi}{2} = 1$

4. a. $\theta = 45° = \dfrac{\pi}{4}$ rad

b. $\sin 45° = \dfrac{1}{2} = \dfrac{\sqrt{2}}{2}$
$\cos 45° = \dfrac{1}{2} = \dfrac{\sqrt{2}}{2}$
$\tan 45° = 1$
$\ctn 45° = 1$
$\sec 45° = \sqrt{2}$
$\csc 45° = \sqrt{2}$

c. $\theta = -315°$

d. $\sin -315° = \sin 45° = \dfrac{\sqrt{2}}{2}$
$\cos -315° = \cos 45° = \dfrac{\sqrt{2}}{2}$
$\tan -315° = \tan 45° = 1$
$\ctn -315° = \ctn 45° = 1$
$\sec -315° = \sec 45° = \sqrt{2}$
$\csc -315° = \csc 45° = \sqrt{2}$

6. a. First case: $x = -8$; $y = 15$
$\sin \theta = \dfrac{15}{17}$
$\cos \theta = -\dfrac{8}{17}$
$\tan \theta = -\dfrac{15}{8}$
$\ctn \theta = \dfrac{-8}{15}$
$\sec \theta = -\dfrac{17}{8}$
$\csc \theta = +\dfrac{17}{15}$

Second case: $x = -8$; $y = -15$
$\sin \theta = -\dfrac{15}{17}$
$\cos \theta = -\dfrac{8}{17}$
$\tan \theta = \dfrac{15}{8}$
$\ctn \theta = \dfrac{8}{15}$
$\sec \theta = -\dfrac{17}{8}$
$\csc \theta = -\dfrac{17}{15}$

b. $y = 24$ and $r = 25$
First case: $x = 7$
$\sin \theta = \dfrac{24}{25}$

Second case: $x = -7$
$\sin \theta = \dfrac{24}{25}$

Answers to Even-Numbered Exercises

$$\cos\theta = \frac{7}{25} \qquad \cos\theta = \frac{-7}{25}$$

$$\tan\theta = \frac{24}{7} \qquad \tan\theta = \frac{24}{-7}$$

$$\text{ctn}\,\theta = \frac{7}{24} \qquad \text{ctn}\,\theta = \frac{-7}{24}$$

$$\sec\theta = \frac{25}{7} \qquad \sec\theta = \frac{25}{-7}$$

$$\csc\theta = \frac{25}{24} \qquad \csc\theta = \frac{25}{24}$$

8. **a.** Impossible, since $-1 \le \cos\theta \le 1$
   b. Impossible, since $\csc\theta \ge 1$ or $\csc\theta \le -1$
   c. Possible
   d. Possible
   e. Impossible, since $-1 \le \sin\theta \le 1$
   f. Impossible, since $\sec\theta \ge 1$ or $\sec\theta \le -1$

Exercises 11.4

2. Second quadrant: tan
   ctn
   sec
   cos
   Third quadrant: sin
   csc
   cos
   sec
   Fourth quadrant: sin
   csc
   tan
   ctn
   Each function occurs in exactly two lists.

4. **a.** 150°; −40   **b.** 126°; −54   **c.** 107° 20′; (−72° 40′)   **d.** 95°; −85°

6. **a.** $\sin 120° = \sin 60° = \dfrac{\sqrt{3}}{2}$

   $\cos 120° = -\cos 60° = -\dfrac{1}{2}$

   $\tan 120° = -\tan 60° = -\sqrt{3}$

   b. $\sin 210° = -\sin 30° = -\dfrac{1}{2}$

   $\cos 210° = -\cos 30° = -\dfrac{\sqrt{3}}{2}$

   $\tan 210° = \tan 30° = \dfrac{1}{\sqrt{3}}$

c. $\sin 135° = \sin 45° = \dfrac{\sqrt{2}}{2}$

$\cos 135° = -\cos 45° = -\dfrac{\sqrt{2}}{2}$

$\tan 135° = -\tan 45° = -1$

d. $\sin 240° = -\sin 60° = -\dfrac{\sqrt{3}}{2}$

$\cos 240° = -\cos 60° = -\dfrac{1}{2}$

$\tan 240° = \tan 60° = \sqrt{3}$

e. $\sin 405° = \sin 45° = \dfrac{\sqrt{2}}{2}$

$\cos 405° = \cos 45° = \dfrac{\sqrt{2}}{2}$

$\tan 405° = \tan 45° = 1$

Exercises 11.5

2. $\sin 45° = \dfrac{1}{\sqrt{2}} = \dfrac{\sqrt{2}}{2}$ $\quad \operatorname{ctn} 45° = 1$

$\cos 45° = \dfrac{1}{\sqrt{2}} = \dfrac{\sqrt{2}}{2}$ $\quad \sec 45° = \sqrt{2}$

$\tan 45° = 1$ $\quad \csc 45° = \sqrt{2}$

4. $\sin \theta = \dfrac{3}{5}$ $\quad \operatorname{ctn} \theta = \dfrac{4}{3}$

$\cos \theta = \dfrac{4}{5}$ $\quad \sec \theta = \dfrac{5}{4}$

$\tan \theta = \dfrac{3}{4}$ $\quad \csc \theta = \dfrac{5}{3}$

6. $\sin 135° = \sin 45° = \dfrac{1}{\sqrt{2}} = \dfrac{\sqrt{2}}{2}$ $\quad \tan 135° = -\tan 45° = -1$

$\cos 135° = -\cos 45° = -\dfrac{1}{\sqrt{2}}$ $\quad \operatorname{ctn} 135° = -\operatorname{ctn} 45° = -1$

8. $\sin 150° = \sin 30° = \dfrac{1}{2}$ $\quad \tan 150° = -\tan 30° = -\dfrac{1}{\sqrt{3}}$

$\cos 150° = -\cos 30° = -\dfrac{\sqrt{3}}{2}$ $\quad \operatorname{ctn} 150° = -\operatorname{ctn} 30° = -\sqrt{3}$

10. $\sin 300° = -\sin 60° = -\dfrac{\sqrt{3}}{2}$ $\quad \tan 300° = -\tan 60° = -\sqrt{3}$

$\cos 300° = \cos 60° = \dfrac{1}{2}$ $\quad \operatorname{ctn} 300° = -\operatorname{ctn} 60° = -\dfrac{1}{\sqrt{3}} = -\dfrac{\sqrt{3}}{3}$

Answers to Even-Numbered Exercises

12. $\sin\left(-\dfrac{5\pi}{6}\right) = -\sin\dfrac{\pi}{6} = -\dfrac{1}{2}$ $\tan\left(-\dfrac{5\pi}{6}\right) = \tan\dfrac{\pi}{6} = \dfrac{1}{\sqrt{3}} = \dfrac{\sqrt{3}}{3}$

 $\cos\left(-\dfrac{5\pi}{6}\right) = -\cos\dfrac{\pi}{6} = -\dfrac{\sqrt{3}}{2}$ $\text{ctn}\left(-\dfrac{5\pi}{6}\right) = \text{ctn}\dfrac{\pi}{6} = \sqrt{3}$

14. $\sin 390° = \sin 30° = \dfrac{1}{2}$ $\tan 390° = \tan 30° = \dfrac{1}{\sqrt{3}}$

 $\cos 390° = \cos 30° = \dfrac{\sqrt{3}}{2}$ $\text{ctn } 390° = \text{ctn } 30° = \sqrt{3}$

16. $\sin 780° = \sin 60° = \dfrac{\sqrt{3}}{2}$ $\tan 780° = \tan 60° = \sqrt{3}$

 $\cos 780° = \cos 60° = \dfrac{1}{2}$ $\text{ctn } 780° = \text{ctn } 60° = \dfrac{1}{\sqrt{3}}$

18. $\sin(-315)° = \sin 45° = \dfrac{1}{\sqrt{2}} = \dfrac{\sqrt{2}}{2}$ $\tan(-315)° = \tan 45° = 1$

 $\cos(-315)° = \cos 45° = \dfrac{1}{\sqrt{2}} = \dfrac{\sqrt{2}}{2}$ $\text{ctn}(-315)° = \text{ctn } 45° = 1$

20. **a.** $\sin^2 30° = 1 - \cos^2 30°$ **b.** $\sin 240° = 2(\sin 120°)(\cos 120°)$

 $\left(\dfrac{1}{2}\right)^2 = 1 - \left(\dfrac{\sqrt{3}}{2}\right)^2$ $-\sin 60° = 2(\sin 60°)(-\cos 60°)$

 $\dfrac{1}{4} = 1 - \dfrac{3}{4}$ $-\dfrac{3}{2} = 2\left(\dfrac{\sqrt{3}}{2}\right)\left(-\dfrac{1}{2}\right)$

 $\dfrac{1}{4} = \dfrac{1}{4}$ $-\dfrac{\sqrt{3}}{2} = -\dfrac{\sqrt{3}}{2}$

Exercises 11.6

2. 1.0724 **4.** 0.68241 **6.** 0.90996 **8.** 4.70463 **10.** 1.33673

12. 0.53395 **14.** −1.00000 **16.** −3.21458 **18.** 17° 46′ **20.** 30° 24′

22. 56° 26′ **24.** 52° 6′ **26.** 77° 23′ **28.** 45° 0′ **30.** 36° 11′

Exercises 11.7

2. $A = 35° 42'; b = 104.37; C = 128.53$

4. $A = 48°; b = 133.86; B = 42°$

6. $B = 48° 40'; a = 92.46; b = 105.12$

8. $b = 0.9702; A = 3° 49' 16''; B = 86° 10' 44''$

10. $b = 1.8922; A = 18° 53' 56''; B = 71° 6' 4''$

12. $B = 35° 48'; a = 55.46; c = 68.38$

14. $A = 45°; b = 28.74; c = 40.64$

16. $c = 772.14$; $A = 39° 0' 27''$; $B = 50° 59' 63''$

18. $c = 10$; $A = 36° 52' 12''$; $B = 53° 7' 78''$

20. $A = 30°$; $a = 64.66$; $c = 129.33$

Exercises 11.8

2. $\theta = 56° 19'$ **4.** $h = 5.4$ ft **6.** $d = 389.71$ ft **8.** $S = 30.72$ ft

10. 10.73 ft **12.** 223.97 ft **14.** 21.40 ft

16. $A = 66° 25' 19''$; $C = 66° 25' 19''$; $B = 47° 9' 22''$

Exercises 11.9

2. 40 ft **4.** 10 mph **6.** 694.59 lb; 3939.23 lb **8.** 33.92 lb; 51.2 lb

Exercises 11.10

2. $C = 45° 15'$; $a = 86.39$; $c = 36.06$

4. $B = 43° 51' 14''$; $C = 76° 8' 46''$; $c = 11.21$

6. $B = 35° 15' 52''$; $C = 24° 44' 8''$; $c = 14.49$

8. $A = 46° 34'$; $B = 57° 54' 36''$; $c = 75° 31' 21''$

10. $b = 90.29$; $A = 38° 38' 31''$; $C = 31° 21' 29''$

12. $c = 234.87$ yd

14. Let h be the altitude from the vertex B. Then $b = c \sin A$. Thus

$$\text{Area} = \frac{1}{2}bh$$
$$= \frac{1}{2}b(c \sin A)$$
$$= \frac{1}{2}bc \sin A$$

Similarly,

$$\text{Area} = \frac{1}{2}ab \sin C$$
$$= \frac{1}{2}ac \sin B$$

Exercises 11.11

2. Since $\cot \theta = \dfrac{x}{y} = \dfrac{\cos \theta}{\sin \theta}$ and $\tan \theta = \dfrac{y}{x} = \dfrac{\sin \theta}{\cos \theta}$, then

$$\cot \theta = \frac{\cos \theta}{\sin \theta} = \frac{1}{\sin \theta / \cos \theta} = \frac{1}{\tan \theta}$$

Answers to Even-Numbered Exercises

4. $\cot\theta = \dfrac{x}{y} = \dfrac{x/r}{y/r} = \dfrac{\cos\theta}{\sin\theta}$

6. $\cos\theta = \pm\sqrt{1 - \sin^2\theta} \qquad \sec\theta = \pm\dfrac{1}{\sqrt{1 - \sin^2\theta}}$

$\tan\theta = \pm\dfrac{\sin\theta}{\sqrt{1 - \sin^2\theta}} \qquad \csc\theta = \dfrac{1}{\sin\theta}$

$\cot\theta = \pm\dfrac{\sqrt{1 - \sin^2\theta}}{\sin\theta}$

8. $\sin\theta$ **10.** $1 + \sin\theta$ **12.** $\cos\theta + 1$ **14.** $\tan\theta$ **16.** $\sin\theta$ **18.** $\sin\theta$
20. 1   **22.** 1   **24.** $\sin\theta\tan\theta$ **26.** $\sin\theta\cos\theta + 1$ **28.** 0.601815

Exercises 11.12

2. $\tan(A + B) = \dfrac{\tan A + \tan B}{1 - \tan A \tan B}$

$= \dfrac{\dfrac{1}{2} + \dfrac{1}{3}}{1 - \dfrac{1}{2}\left(\dfrac{1}{3}\right)}$

$= \dfrac{\dfrac{3 + 2}{6}}{1 - \dfrac{1}{6}}$

$= \dfrac{\dfrac{5}{6}}{\dfrac{5}{6}}$

$= 1$

$\therefore A + B = 45°$ since $\tan 45° = 1$

4. $\sin 3\theta = \sin(2\theta + \theta)$
$= \sin 2\theta \cos\theta + \cos 2\theta \sin\theta$
$= (2\sin\theta\cos\theta)\cos\theta + (\cos^2\theta - \sin^2\theta)\sin\theta$
$= 2\sin\theta\cos^2\theta + \cos^2\theta\sin\theta - \sin^3\theta$
$= 3\sin\theta\cos^2\theta - \sin^3\theta$
$= 3\sin\theta(1 - \sin^2\theta) - \sin^3\theta$
$= 3\sin\theta - 3\sin^3\theta - \sin^3\theta$
$= 3\sin\theta - 4\sin^3\theta$

6. a. $-\sin\theta$ **b.** $\sin\theta$ **c.** $-\cos\theta$

8. $\dfrac{3}{4}$

10. a. True   **b.** False   **c.** True   **d.** True

Exercises 11.13

2. $y = 3 \sin 2x$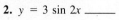

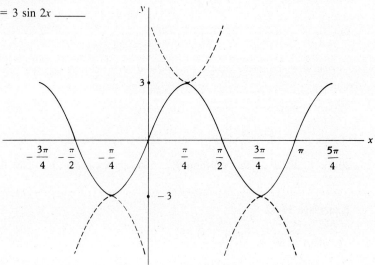

4. $y = 3 \csc 2x$ - - - -

6. $y = \cot 2x$

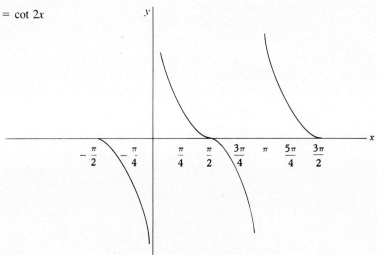

8. $y = \sin\left(x - \dfrac{\pi}{4}\right)$

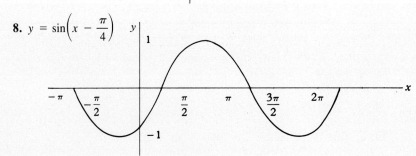

Answers to Even-Numbered Exercises

Exercises 11.14

2. $D_x y = 14 \cos 7x$ **4.** $D_x y = 3x^2 \cos(x^3 + 5)$ **6.** $\cos x$
8. $-10 \cos^4 2x \sin 2x$ **10.** $2(\sec 2x)(\tan 2x)$ **12.** 0 **14.** $-3(\csc 3x)(\cot 3x)$
16. $-\dfrac{1}{5} \cos 5x + C$ **18.** $\dfrac{7}{4} \sin 4x + C$ **20.** $\tan x - x + C$
22. $-3 \cot x + C$ **24.** $-\dfrac{\csc^4 x}{4} + C$

CHAPTER 12

Exercises 12.1

2. $\dfrac{1}{4}; \dfrac{4}{13}; \dfrac{3}{4}$

4. $\dfrac{1}{2} = P$ (at least two beads)

$\dfrac{7}{8} = P$ (at most two beads)

6. a. $\dfrac{5}{18}; \dfrac{25}{36}; \dfrac{11}{36}$ **b.** $\dfrac{1}{6}$

8. a. 0.80521 **b.** 0.87245

Exercises 12.2

2. 210 **4.** 34650 **6.** 10,140; 7740 **8.** $\dfrac{1}{9}$ **10.** 0.20508

12. $C(n, n-4) = \dfrac{n!}{[n-(n-r)]!(n-r)!} = \dfrac{n!}{(n-n+r)!(n-r)!} = \dfrac{n!}{r!(n-r)!}$
$C(n, r) = C(n, n-r)$

14. 1,712,304 **16.** 0.245 **18.** 0.043956; 0.31648 **20.** 0.31056

22. 9 ways if there are no restrictions on how to travel
6 ways if the return trip is not to go back on the same route
3 ways if the return trip is back on the same route

Exercises 12.3

2. 0.21875 **4.** $\dfrac{3}{28}; \dfrac{1}{196}$ **6.** 0.512 **8.** 0.1937 **10.** 0.30199

12. 0.5574 **14.** 0.3407407

Exercises 12.4

2. 7¢   **4.** 25¢   **6.** 1.25   **8.** 0.50

Exercises 12.5

2. 0.9534   **4.** 0.0007

Exercises 12.6

2. 5.3; 5.41; 2.32594

4. 12; 51.6; 7.183314

6. $\{-1.25, 0.00, 0.75, -0.75, 1.75, -0.50\}$; 0; 1

Exercises 12.7

2. $S^2 = \dfrac{(x_1 - \bar{x})^2 f_1 + (x_2 - \bar{x})^2 f_2 + \cdots + (x_k - \bar{x})^2 f_k}{n}$

where $\bar{x} = \dfrac{\sum_{i=1}^{n} x_i}{n}$. Thus

$S^2 = \dfrac{(x_1 - \bar{x})^2 f_1 + (x_2 - \bar{x})^2 f_2 + \cdots + (x_k - \bar{x})^2 f_k}{n}$

$\Rightarrow S^2 = \dfrac{(x_1 - \bar{x})^2 + (x_2 - \bar{x})^2 + \cdots + (x_n - \bar{x})^2}{n}$

$= \dfrac{(x_1^2 - 2x_1\bar{x} + \bar{x}^2) + (x_2^2 - 2x_2\bar{x} + \bar{x}^2) + \cdots + (x_n^2 - 2x_n\bar{x} + \bar{x}^2)}{n}$

$= \dfrac{(x_1^2 + x_2^2 + \cdots + x_n^2) - 2\bar{x}(x_1 + x_2 + \cdots + x_n) + n\bar{x}^2}{n}$

$= \dfrac{x_1^2 + x_2 + \cdots + x_n^2}{n} - 2\bar{x}\left(\dfrac{\bar{x}_1 + \bar{x}_2 + \cdots + \bar{x}_n}{n}\right) + \bar{x}^2$

$= \dfrac{x_1^2 + x_2^2 + \cdots + x_n^2}{n} - \bar{x}^2$ Q.E.D.

Exercises 12.8

2. a. Between 50 and 350: 95.44%
   Greater than 275: 15.87%
 b. 84% of 600 $\doteq$ 504
 c. 219

4. $P = 0.2776$

6. $N = 2009$

APPENDIX A

Exercises A.1

2. a. $\sim p \wedge q$ **b.** $\sim p \wedge \sim q$ **c.** $p \wedge \sim q$ **d.** $\sim p \vee q$
   e. $\sim p \wedge \sim q$ **f.** $\sim (\sim p \wedge \sim q)$

4. a. $\sim p$ **b.** $p \wedge q$ **c.** $q \vee p$ **d.** $p \to q$

6. a.

p	q	$p \vee q$	$p \wedge q$	$p \to q$	$q \to p$
T	T	T	T	T	T
T	F	T	F	F	T
F	T	T	F	T	F
F	F	F	F	T	T

If we assume both p and q to be true, then all four statements, $p \vee q$, $p \wedge q$, $p \to q$, and $q \to p$ are T.

b. All four statements are T if p: Five is a counting number.
q: $3 + 2 = 5$.

c. p: Five is a counting number.
q: $3 + 2 = 6$.

$p \vee q$	$p \wedge q$	$p \to q$	$q \to p$
T	F	F	T

d. Assume both are T. p: Bill does not do his homework.
q: Bill calls Mary.
All four statements are T.

e. p: Six is not a prime number.
q: Three times 5 does not equal 15.

8. a. Converse: If I am not his brother, then he is my father.
Contrapositive: If I am his brother, then he is not my father.
Inverse: If he is not my father, then I am his brother.

b. Converse: If 2 is greater than 4, then 2 is not greater than 5.
Contrapositive: If 2 is not greater than 4, then 2 is not greater than 5.
Inverse: If 2 is not greater than 5, then 2 is not greater than 4.

c. Converse: If Joan will marry Tom, then he is rich.
Contrapositive: If Joan will not marry Tom, then he is not rich.
Inverse: If Tom is not rich, then Joan will not marry him.

d. Converse: If $a^2 = b^2$, then $a = b$.
Contrapositive: If $a^2 \neq b^2$, then $a \neq b$.
Inverse: If $a \neq b$, then $a^2 \neq b^2$.

e. Converse: If $a + b$ is positive, then $a^2 + b^2$ is positive.
   Contrapositive: If $a + b$ is not positive, then $a^2 + b^2$ is positive.
   Inverse: If $a^2 + b^2$ is not positive, then $a + b$ is not positive.

f. Converse: If Bill or Joe's wives are not sisters, then they are not brothers.
   Contrapositive: If Bill and Joe's wives are sisters, then they are brothers.
   Inverse: If Bill and Joe are brothers, then their wives are sisters.

10.

p	$\sim p$	q	$\sim q$	$p \to q$	$q \to p$	$\sim p \to \sim q$
T	F	T	F	T	ⓣ	F Ⓣ F
T	F	F	T	F	ⓣ	F Ⓣ T
F	T	T	F	T	Ⓕ	T Ⓕ F
F	T	F	T	T	ⓣ	T Ⓣ T

$\therefore (q \Rightarrow p) \Leftrightarrow (\sim p \Rightarrow \sim q)$

12.

p	$\sim p$	q	$\sim q$	$p \wedge q$	$\sim(\sim p) \vee (\sim q)$	$(p \wedge q)$
T	F	T	F	T	F	F
T	F	F	T	F	T	T
F	T	T	F	F	T	F
F	T	F	T	F	T	F

$\therefore \sim(p \wedge q) \leftrightarrow (\sim p) \vee (\sim q)$

p	$\sim p$	q	$\sim q$	$p \vee q$	$\sim(p \vee q)$	$(\sim p) \wedge (\sim q)$
T	F	T	F	T	F	F
T	F	F	T	T	F	F
F	T	T	F	T	F	F
F	T	F	T	F	T	T

$\therefore \sim(p \vee q) \leftrightarrow (\sim p) \wedge (\sim q)$

p	$\sim p$	q	$\sim q$	$p \Rightarrow q$	$\sim(p \to q)$	$p \wedge (\sim q)$
T	F	T	F	T	F	F
T	F	F	T	F	T	T
F	T	T	F	T	F	F
F	T	F	T	T	F	F

$\therefore \sim(p \to q) \leftrightarrow p \wedge (\sim q)$

14. a. Disjunction
    b. Implication
    c. Negation
    d. Conjunction

Answers to Even-Numbered Exercises

e. Conjunction

16. a.

$[(p \to q) \land (q \to r)] \to (p \to r)$

							T			
T	T	T	T	T	T	T	**T**	T	T	T
T	T	T	F	T	F	F	**T**	T	F	F
T	F	F	F	F	T	T	**T**	T	T	T
T	F	F	F	F	T	F	**T**	T	F	F
F	T	T	T	T	T	T	**T**	F	T	T
F	T	T	F	T	F	F	**T**	F	T	F
F	T	F	T	F	T	T	**T**	F	T	T
F	T	F	T	F	T	F	**T**	F	T	F

Tautology

b.

$\sim(\sim p) \to p$

T	FT	**T**	T
F	TF	**T**	F

Tautology

c.

$[(p \land q) \to r] \to [p \to (q \to r)]$

T	T	T	T	T	**T**	T	T	T	T	T
T	T	T	F	F	**T**	T	F	T	F	F
T	F	F	T	T	**T**	T	T	F	T	T
T	F	F	T	F	**T**	T	T	F	T	F
F	F	T	T	T	**T**	F	T	T	T	T
F	F	T	T	F	**T**	F	T	T	F	F
F	F	F	T	T	**T**	F	T	F	T	T
F	F	F	T	F	**T**	F	T	F	T	F

Tautology

d.

$(p \land q) \to p$

T	T	T	**T**	T
T	F	F	**T**	T
F	F	T	**T**	F
F	F	F	**T**	F

Tautology

e.

$[p \land (q \lor r)] \to [(p \land q) \lor (p \land r)]$

T	T	T	T	T	**T**	T	T	T	T	T	T	T
T	T	T	T	F	**T**	T	T	T	T	T	F	F
T	T	F	T	T	**T**	T	F	F	T	T	T	T
T	F	F	F	F	**T**	T	F	F	F	T	F	F
F	F	T	T	T	**T**	F	F	T	F	F	F	T
F	F	T	T	F	**T**	F	F	T	F	F	F	F
F	F	F	T	T	**T**	F	F	F	F	F	F	T
F	F	F	F	F	**T**	F	F	F	F	F	F	F

Tautology

f.

$[p \lor (q \land r)] \to [(p \lor q) \land (p \lor r)]$

T	T	T	T	T	**T**	T	T	T	T	T	T	T
T	T	T	F	F	**T**	T	T	T	T	T	T	F
T	T	F	F	T	**T**	T	T	F	T	T	T	T
T	T	F	F	F	**T**	T	T	F	T	T	T	F
F	T	T	T	T	**T**	F	T	T	T	F	T	T
F	F	T	F	F	**T**	F	T	T	F	F	F	F
F	F	F	F	T	**T**	F	F	F	F	F	T	T
F	F	F	F	F	**T**	F	F	F	F	F	F	F

Tautology

18. a.

$p \to (q \lor r)$

T	**T**	T	T	T
T	**T**	T	T	F
T	**T**	F	T	T
T	**F**	F	F	F
F	**T**	T	T	T
F	**T**	T	T	F
F	**T**	F	T	T
F	**T**	F	F	F

b.

$(p \lor \sim q) \land r$

T	T	FT	**T**	T
T	T	FT	**F**	F
T	T	TF	**T**	T
T	T	TF	**F**	F
F	F	FT	**F**	T
F	F	FT	**F**	F
F	T	TF	**T**	T
F	T	TF	**F**	F

Exercises A.2

2.

$p \leftrightarrow q$	$q \leftrightarrow q$
T T T	T T T
T F F	F F T
F F T	T F F
F T F	F T F

Equivalent

4. a. A parallelogram is not a rectangle if and only if its diagonals are equal.
  b. A parallelogram is not a rectangle if and only if a vertex angle is a right angle.
  c. A necessary and sufficient condition that $x^2 - 3x \neq 0$ is that $x = 0$ or $x = 3$.
  d. x^2 is greater than 4 and x is not greater than 2.
  e. x is a multiple of 4 and x is not an even integer.

6. a. $q \rightarrow p$, i.e., if the weather is fair, then the softball game on Friday will be played (using contrapositive).
  b. The softball game on Friday will be played.
  c. No definite information, since an inverse is not logically equivalent to the original statement.

8. a.

$(p \wedge q) \rightarrow r$		
T T T	T	T
T T T	F	F
T F F	T	T
T F F	T	F
F F T	T	T
F F T	T	F
F F F	T	T
F F F	T	F

b.

$r \leftrightarrow (p \vee q)$		
T T	T T T
T T	T T F
T T	F T T
T F	F F F
F F	T T T
F F	T T F
F F	F T T
F T	F F F

Exercises A.3

2. a. $\{1, 2, 3\}$ **b.** $\{1, 2, 3, 4, 5, \ldots\}$
  c. $\{\ \} = \{x | x = x + 3\}$ **d.** $\{x | x \text{ is a wealthy person}\}$

4. a. The set of all states in the Union starting with capital letter A
  b. The set of counting numbers less than 31
  c. The solution set of the equation $x + 9 = 5$ on the number line
  d. The set of all even counting numbers
  e. The set of all even counting numbers less than 20
  f. The set of all counting numbers exactly divisible by 3
  g. The set of all baseball teams in the American League

Answers to Even-Numbered Exercises

6. **a.** The set of all cities in Texas
   b. The set of letters in our alphabet
   c. The set of all women
   d. The set of all teachers
   e. The set of all students

Exercises A.4

2.

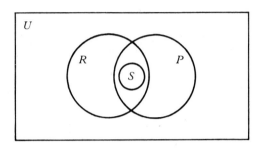

4.
$$\{\,\}$$
$$\{1\}\quad\{2\}\quad\{3\}\quad\{4\}$$
$$\{1,2\}\quad\{1,3\}\quad\{1,4\}\quad\{2,3\}\quad\{2,4\}\quad\{3,4\}$$
$$\{1,2,3\}\quad\{1,2,4\}\quad\{2,3,4\}\quad\{1,3,4\}$$
$$\{1,2,3,4\}$$

There are a total of sixteen subsets.

6. For a fixed element a of X, we know that $a \in Y$ since $X \subset Y$. Now, since $Y \subset Z$, we also known that $a \in Z$. Hence, every element of X is also an element of Z. Since Y is a proper subset of Z, there is at least one element of Z (say c) which is not an element of Y. But c cannot be an element of X since each element of X is also an element of Y and c is not an element of Y. Hence, according to Definition A.11, $X \subset Z$.

8. **a.** 4 **b.** 7 **c.** 14 **d.** 10 **e.** 13

10.

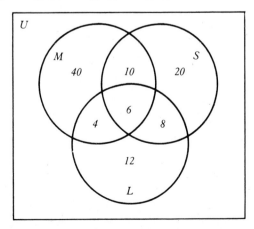

Exercises A.5

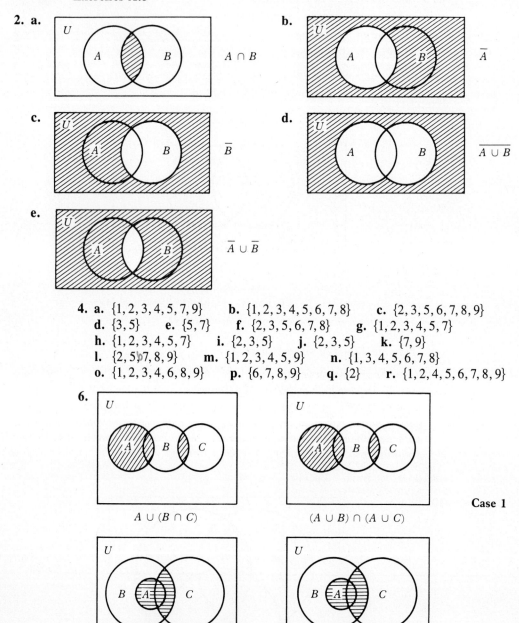

4. a. {1, 2, 3, 4, 5, 7, 9}   **b.** {1, 2, 3, 4, 5, 6, 7, 8}   **c.** {2, 3, 5, 6, 7, 8, 9}
d. {3, 5}   **e.** {5, 7}   **f.** {2, 3, 5, 6, 7, 8}   **g.** {1, 2, 3, 4, 5, 7}
h. {1, 2, 3, 4, 5, 7}   **i.** {2, 3, 5}   **j.** {2, 3, 5}   **k.** {7, 9}
l. {2, 5, 7, 8, 9}   **m.** {1, 2, 3, 4, 5, 9}   **n.** {1, 3, 4, 5, 6, 7, 8}
o. {1, 2, 3, 4, 6, 8, 9}   **p.** {6, 7, 8, 9}   **q.** {2}   **r.** {1, 2, 4, 5, 6, 7, 8, 9}

These Venn diagrams suggest that the sets $A \cup (B \cap C)$ and $(A \cup B) \wedge (A \cup C)$ are equal.

8. Proof: $x \in A \cap (B \cup C) \leftrightarrow x \in A$ and $x \in (B \cup C)$
$\leftrightarrow x \in A$ and $(x \in B$ or $x \in C)$
$\leftrightarrow (x \in A$ and $x \in B)$ or $(x \in A$ and $x \in C)$
$\leftrightarrow (x \in A \cap B)$ or $(x \in A \cap C)$
$\leftrightarrow x \in [(A \cap B) \cup (A \cap C)]$

Hence $A \cap (B \cup C) \subseteq (A \cap B) \cup (A \cap C)$
and $(A \cap B) \cup (A \cap C) \subseteq A \cap (B \cup C)$
$\therefore A \cap (B \cup C) = (A \cap B) \cup (A \cap C)$ Q.E.D.

10. a. 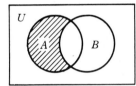   **b.** 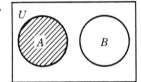   **c.**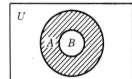

d. 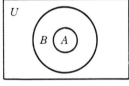   **e.**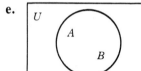

$A - B = \emptyset$ $A - B = \emptyset$

12. a. $A = \{1, 2\}, \quad n(A) = 2$
$B = \{3, 4, 5\}, \quad n(B) = 3$
$A \cup B = \{1, 2, 3, 4, 5\}, \quad n(A \cup B) = 5$

b. $A = \{1, 2\}, \quad n(A) = 2$
$B = \{2, 3, 4\}, \quad n(B) = 3$
$A \cup B = \{1, 2, 3, 4\}, \quad n(A \cup B) = 4$

c. $A = \{1, 2, 3, 4\}, \quad n(A) = 4$
$B = \{2, 4\}, \quad n(B) = 2$
$A - B = \{1, 3\}, \quad n(A - B) = 2$

14. 45 families

16. 20 with raincoat or umbrella: 25
                                          -20
                                            5   neither raincoat nor umbrella

18. 70

20. a. 31   **b.** 13   **c.** 11   **d.** 22

Exercises A.6

2. pq

4.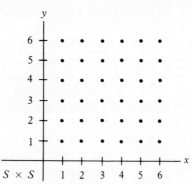

6.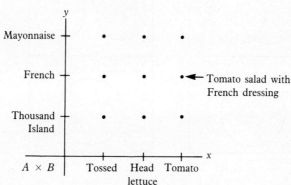

8. **a.** $M = \{(1,6), (2,5), (3,4), (4,3), (5,2), (6,1)\}$
   b. $N = \{(1,6), (2,5), (3,4), (4,3), (5,2), (6,1), (2,6), (3,5), (4,4), (5,3),$
   $(6,2), (3,6), (4,5), (5,4), (6,3), (4,6), (5,5), (6,4), (5,6), (6,5),$
   $(6,6)\}$
   c. $Q = 5 \times 5 - \{(2,6), (3,5), (4,4), (5,3), (6,2)\}$
   d. $K = 5 \times 5 - \{(1,6), (2,6), (3,6), (4,6), (5,6), (6,6), (6,5), (6,4),$
   $(6,3), (6,2), (6,1)\}$
   e. $L = \{(6,1), (6,2), (6,3), (6,4), (6,5)\}$
   f. $J = \{(1,6), (2,6), (3,6), (4,6), (5,6), (6,5), (6,4), (6,3), (6,2), (6,1)\}$

Exercises A.7

2. $\{1,2\}, \{1,3\}, \{2,4\}, \{3,5\}$ *(Answers may vary)*

4. $\{6, 12, 18, 24, \ldots\}$ *(Answers may vary)*

6. *Answers may vary;* e.g., a lamp directly overhead will cast the same shadow, etc.

Exercises A.9

2. *Answers may vary:*
   a. p_x: x is an even number, $U = \{1, 2, 3, 4, 5\}$.

Answers to Even-Numbered Exercises

   b. p_x: x is a vowel; U is the set of all letters in our alphabet.
   c. p_x: x is a nonnegative number; $U = I$, the set of integers.
   d. p_x: x is a prime number, $U = N$.

4. a. $\forall x$, x is a prime number, $x > 2$, $x + 1$ is even.
   Negation: $\exists x$, x is a prime number, $x > 2$, $x + 1$ is odd.
   b. $\forall x$, $x^2 - 4x + 7$ is positive.
   Negation: $\exists x$, $x^2 - 4x + 7$ is nonpositive.
   c. $\forall x$, $x > 2 \rightarrow x^2 > 4$
   Negation: $\exists x$, $x > 2 \wedge x^2 \leq 4$
   d. $\forall x$, x is a man; x's height is 8 feet tall.
   Negation: $\exists x$, x is a man; x's height is not 8 feet tall.
   e. $\forall x$, x is odd; x^2 is odd.
   Negation: $\exists x$, x is odd; x^2 is even.

6. a. $\exists x$, x is prime; $x + 1$ is a prime number.
   b. $\exists x$, x is a man; x is rich.
   c. $\exists x$, x is a number; $x^2 = 1$.
   d. $\exists x$, x is a plane triangle; x is isosceles.
   e. $\exists x$, x is a number; $1 - 17x - 5x^2$ is positive.

8. a. $N = \{1, 2, 3, 4, 5, 6, \ldots, n, \ldots\}$
   $A = \{7, 14, 21, 28, 35, 42, \ldots, 7n, \ldots\}$
   b. $N = \{1, 2, 3, 4, 5, 6, \ldots, n, \ldots\}$
   $B = \left\{\dfrac{1}{1}, \dfrac{1}{2}, \dfrac{1}{3}, \dfrac{1}{4}, \dfrac{1}{5}, \dfrac{1}{6}, \ldots, \dfrac{1}{n}, \ldots\right\}$
   c. $N = \{1, 2, 3, 4, 5, 6, \ldots, n, \ldots\}$
   $C = \left\{\dfrac{1}{2}, \dfrac{2}{3}, \dfrac{3}{4}, \dfrac{4}{5}, \dfrac{5}{6}, \dfrac{6}{7}, \ldots, \dfrac{n}{n+1}, \ldots\right\}$
   d. $N = \{1, 2, 3, 4, 5, 6, \ldots, n, \ldots\}$
   $D = \{2, 1, 0, -1, -2, -3, \ldots, 3 - n, \ldots\}$
   e. $N = \{1, 2, 3, 4, 5, 6, \ldots, n, \ldots\}$
   $E = \{4, 9, 14, 19, 24, 29, \ldots, 5n - 1, \ldots\}$
   f. $N = \{1, 2, 3, 4, 5, 6, \ldots, n, \ldots\}$
   $F = \left\{1, \dfrac{1}{2}, \dfrac{1}{4}, \dfrac{1}{8}, \dfrac{1}{16}, \dfrac{1}{32}, \ldots, \left(\dfrac{1}{2}\right)^{n-1}, \ldots\right\}$

Index

Abscissa of point, 53
Absolute inequalities, definition of, 184
Absolute maxima, 327
Absolute maximum value, 324, 330
Absolute minima, 327
Absolute minimum value, 324
Absolute value
   inequalities involving, 188–189
   maximum, 324, 330
   minimum, 324
   of real numbers, 44–45
   rules of operation for, 45–46
Absolute-value symbol, using, 274–275
Acceleration, 308
   angular, 529
   derivation of, 315
   of freely falling body, 349–350
   instantaneous, 284–286
Accumulation factor, 461
Accumulation of principal, 461
Acute angles
   definition of, 526
   positive, 541, 542
   trigonometric functions of, 536, 541–547
Addition
   of algebraic fractions, 107–109
   basic properties of, 15–16
   of complex numbers, 57
   of fractions, 30–37
   of matrices, 408–410
   method of elimination by, 234–235
   of polynomials, 83–84, 92–93
Additive identity for matrices, 410
Additive inverse
   of matrix, 410
   of real numbers, 11, 16
Algebra, definition of elementary, 61. *See also* Linear algebra; Matrix algebra
Algebraic expressions, 61–68
   definition of, 62
   exponents for, 76
   multiplying two, 93–95
   rules for operation for, 63
   terms of, 64
Algebraic fractions, 106–113
   addition of, 107–109
   division of, 110
   multiplication of, 110
   simplest form of, 106
   subtraction of, 107–109
Amortization of debt, 483–488
Amortization schedule, 485, 486
Amount, compound, 458, 459–463

Amplitude, 578
Analytic geometry
  definition of, 217
  introduction to, 217–222
Angles
  acute, 536, 541–547
  complementary, 542, 543
  coterminal, 532–533
  of depression, 552–555
  double-formula, 576
  of elevation, 552–555
  magnitude of, 525–526
  measures of, 525–529
  obtuse, 526
  quadrantal, 532
  related, 537–541
  right, 526
  straight, 526
Angular acceleration, 529
Angular velocity, 528
Annual rate
  effective, 463
  nominal, 459, 462, 463
Annuities, 473–493
  amount of, 476
  annuities certain, 473–476
  contingent, 473
  formulas for, 481–482
  perpetuity, 473
  present value of, 478–483
  simple, 473–476
  types of, 473
  value of, 489–493
Antiderivatives, 343–346, 517
  symbol for, 355
Antilogarithms, 506
Arbitrary constants, 61, 119, 120
Arbitrary exponents, 76
Area
  between two curves, 369–371
  found by integration, 363–372
  under a curve, 364–369
Associative lives, 9–10, 652
Asymptotes of hyperbola, 253–254
Average marginal cost, 262, 263, 264
Average rates, 259–265
  definition of, 260–261
  over flexible interval, 226–269
Average slope of curve, 262, 264
Average velocity, 261, 264

Axioms, 9–11, 638. *See also* Propositions
Axis, 281
  of ellipse, 244
  of parabola, 244
  x and y, 51–53

Bi-implications, and theorems, 637–640
Binomial, definition of, 113–117
Binomial expansion, 601–605
Binomial probability distribution, 607–608
Binomial theorem, 113–117
Bodies, freely falling, 349–351
Brackets, use of, 14–15
Broken-line graph, 212

Calculators, hand
  exponential function keys on, 501
  interest computations and, 446
  logarithms and, 504, 505
  programmable, 175–176
  quadratic equations and, 146
  trigonometric functions and, 447, 547–548
Calculus, elementary, 259–307
  average rates and, 259–269
  continuous functions in, 276–281
  definition of, 197
  derivatives of, 286–289, 295–303
  fundamental theorem of, 377–380
  implicit differentiation and, 303–307
  instantaneous rates and, 281–286
  limit concept in, 269–271
  limit theorem and, 272–276
  rates for differentiation in, 289–294
Cartesian product of two sets, 51, 658–662
Chain rule of differentiation, 337–341
Circles
  center of, 238, 241, 242
  definition of, 238
  determined by geometric conditions, 240–242
  equations of, 238–243
  formulas for length of, 528
  radius of, 238
Closed intervals, 50, 277, 327
Closure laws, 9
Closure property of set, 8
Coefficients of polynomials, 82–83
Cofactors, 394–395
Column index, 402

Index

Column matrix, 403
Column vector, 403
Combinations, 598–599
Common denominators, 28–30
   adding fractions with, 30–31
   of algebraic fractions, 107
   least, 29–30
Common factors, 3, 27
Common logarithms, 521
Commutative laws, 9, 652
Commuting into an equivalent set, 470
Complementary angles, 542, 543, 548
Complement of set, 653
Completing the square, method of, 139–145
Complex fractions, 110–111
Complex numbers, 55–60, 162–163
   addition and multiplication of, 57–59
   conjugate, 162–163
   definition of, 57
   imaginary part of, 57
   properties of, 56–57
Complex number system, 57
Components of vectors, 558–560
Composite numbers, 2
Compound amount, 458, 459–463
Compound discount, 466
Compounding continuously
   definition of, 494
   and exponential functions, 495–501
Compound interest, 457–469
   definition of, 437
   formulas for, 467–468
   present value at, 464–469
   terminology for, 457–458, 459
Compound interest law, 500, 501
Compound propositions, 627, 628–629
   logically equivalent, 632
Concavity of curve, 321–322
Condition, initial, 348
Conditional connectives, 631
Conditional equations
   definitions of, 120
   solutions for, 121–126
Conditional inequalities, definition of, 184
Conjugate complex numbers, 162–163
Conjugate of complex numbers, 56, 57
Conjunctions, 628
Connectives, 627
   conditional, 631

   truth tables for, 633
Constant function, 204
Constants
   arbitrary, 61, 62, 119, 120
   definition of, 643
   derivatives of, 290
   of integration, 343
Constant term, 82
Constraints, 433
Contingent annuity, 473
Continuous functions, 276–281
Contradictory inequalities, definition of, 184
Contrapositive of implications, 632
Converse of implications, 632
Conversion date, value of annuity on, 489–493
Conversion period, 457, 459
   average rate of growth of, 498
Converted interest, 457
Convex polygon, 427
Convex sets, 426
   polygonal, 426, 427
Coordinates of a point, 53, 54
Coordinate systems
   one-dimensional, 48–50
   rectangular, 50–55
Corollary, definition of, 18
Correspondence, one-to-one, 663
Cosines, law of, 563–565
Cost, average marginal, 262, 263, 264, 311
Coterminal angles, 532–533
Countable sets, 665–667
Counting numbers, set of, 2
Cramer's rule, for three linear equations, 396–399
Critical value for function, 327
Cubic polynomial, 83
Curves
   area between, 369–371
   area under, 364–369
   concavity of, 321–322
   equations of, 353–354
   as graph of function, 210
   normal, 610–612
   point of inflection of, 321
   slope of a, 309–312

Debt
   amortization of, 483–488

Debt (cont'd.)
  consolidation of, 470
  interest-bearing, 450, 451
Decimals
  infinite, 3, 4
  repeating, 3
Decreasing function, 318–324
Definite integrals, 367, 376
Degree(s)
  of an angle, 526
  of a polynomial equation, 126, 157–158
Delta, use of symbol, 287
Delta process, 288
Denominators, 28–30
  common, 28–30
  of fractions, 17, 106, 107
  least common, 29–30
  positive, 26
  sign of, 25–26
Depressed equation, 171–172
Depression, angle of, 552–555
Derivatives
  anti-, 343–346, 355, 517
  applications of, 308–341
  calculus and, 197
  of a constant, 290
  of a constant times a function, 291–292
  finding (see Differentiation)
  first test, 327–329
  of function, 286–289
  of function raised to power, 299–303
  of independent variable, 513–514
  interpretation of, 308–313
  natural logarithms, 514–516
  and positive integers, 290–291
  of products, quotients, and powers, 295–303
  second, 314
  second test, 329–330
  symbol for taking, 355
  theorems for application of, 315–318
  of trigonometric functions, 591
  of variable with respect to itself, 290
Derived equations, 124–126
Descartes' rule of signs, 167–168
Determinants
  cofactors of, 394–395
  element of, 393–396
  method of, 390–401

  second order, 390–391
  of square matrix, 418–419
  of third order, 392–394
  value of, 394–395
Diagrams, Venn, 647–652
Differential equations, 361
Differential of function, 357–361
Differentiation
  chain rule for, 337–341
  definition of, 287, 342
  of exponential functions, 517–520
  formula for, 519
  implicit, 303–307
  of logarithmic functions, 517–520
  to other bases, 520–524
  of polynomial, 293
  process of, 197
  rules for, 289–294, 301–303
  successive, 314–315
  of trigonometric functions, 582–588
Directrix of parabola, 243
Discontinuous function, 277
Discount, compound, 466
Discount factor, 465
Discount short-term loans, 455
Discriminant, 151
Disjoint sets, 647
Disjunction of two equations, 131
Disjunctions, 628–629
Distance between points, 54–55
Distance formula, 218–220
Distributions
  normal probability, 620–624
  probability, 607–613
Distributive law, 10–11
Dividend, 88
Division
  of algebraic fractions, 110
  of fractions, 38–41
  of polynomials, 87–89, 94–95
  process of, 17
  synthetic, 159–160
  by zero, 19–20
Divisor, the, 2, 88
Double-angle formulas, 576

Echelon form, of linear equations, 403–404
Effective rate equivalent, 463

Index

Elementary row operations of matrix, 405–406
Elements, 640–645
Elements of determinants, 393
   cofactor of, 394–395
Elements of a matrix, 402
Elements of ordered pairs, 658–659
Elevation, angle of, 552–555
Elimination method
   for linear equations and matrices, 403–408
   for solving linear equations, 383–389
Ellipses
   definition of, 248
   equations for, 248–252
Empty sets, 644–645
Endpoint of ray, 525
Entries of determinants, 393
Equalities
   definition of, 7
   of fractions, 24–25
   recurring, 97
Equal subsets, 646
Equals symbol, cautions in using, 124
Equations, 119–126
   of circles, 238–243
   conditional, 120
   containing fractions, 177–182
   containing radicals, 182–184
   of curves, 353–354
   depressed, 171–172
   derived, 124–126
   differential, 361
   of ellipses, 248–252
   equivalent, 123–124
   as formulas (*see* Determinants)
   functions defined by, 203
   of hyperbola, 252–255
   of line, 231, 232
   linear in one variable, 127–130
   numerical, 120
   polynomial (*see* Polynomial equations)
   in quadratic form, 153–154
   quadratic in one variable, 130–133
   of parabolas, 243–247
   rational roots of, 170–174
   of sets of points in plane, 222–224
   single matrix, 420–423
   solutions of, 121–126
   solution sets of (*see* Solution sets)
   of straight lines, 230–238
   of value, 469–472
   See also Linear equations; Polynomial equations; Quadratic equations
Equilateral triangles, definition of, 530
Equipollent sets, 662–665
Equivalence
   definition of, 638
   symbol for, 131
Equivalent equations, 123–124
Equivalent interest rates, 463
Exact simple interest, 444
Exclusion disjunction, 628, 629
Existential quantifiers, 670
Expectation, 605–607
Expected value of set, 614
Exponential functions, 501–504
   and logarithmic functions, 494–524
Exponents, 68–78
   for algebraic expression, 76
   arbitrary, 76
   different bases and same, 71–72
   five properties of, 68–72
   laws of, 502–503, 512
   logarithms and, 510–511
   negative and zero, 72–73
   rules of operation for, 79
   same base and different, 68–70
Extraneous roots, 182

Face value of debt, 451
Factorial, 597–598
Factoring
   of polynomials, 97–105
   of quadratic equations, 133–136
   as trial and error process, 102–104
Factoring expressions, 101–104
Factorization theorem, unique, 3
Factors
   accumulation, 461
   common, 3, 27
   definition of, 2
   integers as, 26–27
   simple, 106
Factor theorem, 158–159
Feasible point, 433
Finite set, 643
First derivative test, 327–329

Flexion of curve, 311
Focal date, 470
Focus
   of hyperbola, 252
   of parabola, 243
Formulas
   annuity, 481–482
   for compound amount, 459–463
   compound interest, 467–468
   for distance between points, 218–220
   double-angle, 576
   integration, 343–346
   for length of circle, 528
   midpoint, 220–221
   simple interest, 455–456
   for slope of line, 224, 225
Fractional exponents, 73–75
Fractions, 2, 23–42
   addition of, 30–36
   algebraic, 106–113
   clearing equation of, 128–129
   with common denominators, 28–30
   complex, 110–111
   division of, 37–40
   equality of, 24–25
   equations containing, 177–182
   improper, 35–36
   least common denominator (lcd) of, 29–30
   limit of, 273
   lowest form of, 3
   method of clearing, 179–182
   mixed, 36
   multiplying, 37–40
   natural logarithms of, 507
   notation for, 11
   as number pair, 658
   rules of operation for, 40
   signs associated with, 25–26
   simplest form of, 26–28
   subtraction of, 30–36
Freely falling bodies, 349–351
Frequency of conversion, 459
Functional notation, 200–202
Function graph, 211–212
Functions
   antiderivatives of, 343
   average rate of change of (*see* Average rates)
   concept of, 198–200
   constant, 204
   continuous, 276–281
   defined by equation, 203
   defined by graph, 263
   defined by table of values, 204–208
   derivative of, 286–289
   derivative of product of, 295–297
   differential of, 357–361
   equivalent definitions for, 199
   exponential and logarithmic, 494–524
   finding limit of, 272–276
   graphs of relations and, 209–217
   increasing and decreasing, 318–324
   notations for, 200–202
   periodic, 578
   power, 501
   raised to power, 299–303
   range of, 200
   and relations, 197–208
   relative value of, 324–331
   rules in defining, 202–203
   trigonometric, 525–592
   of trigonometric identities, 570–577
Fundamental theorem of integral calculus, 377–380

Geometry. *See* Analytic geometry
Going rate, 450
Graphs
   broken line, 212
   of Cartesian product, 660–661
   of equations with two variables, 212–215
   of functions, 212, 322
   histogram, 621
   interpolation, 210–212, 263
   and linear equations, 382–383, 429–433
   and linear inequalities, 256–258, 423–426
   plotting, 217
   of relations and functions, 209–217
   of set of real numbers, 49
   of trigonometric functions, 578–582
Gross profit, 439
Growth
   law of, 500, 501
   rate of, 494, 498

Half-line, definition of, 525

Index

Half-planes, 257
Higher-degree polynomial inequalities, 189–193
Histogram, 621
Hyperbolas
  asymptotes of, 253–254
  center of, 252
  equations for, 252–255
  foci of, 252
  transverse axis of, 253
  vertices of, 253

Identities
  additive, 11
  algebraic, 139–140
  basic trigonometric, 565–570
  common trigonometric, 570–577
  uses of, 120–121
Imaginary numbers, 55–57
Imaginary units, 56
Implications, 631–634
Implicit differentiation, 303–307
Improper fractions, 35–36
Inclusive disjunction, 628
Increasing function, 318–324
Indefinite integral, 367, 376
Independent variables, derivatives of, 513–514
Index, 78
  column, 402
  row, 402
Inequalities
  absolute, 184
  direction of, 43, 44
  graphs of linear, 256–258
  higher-degree polynomial, 189–193
  involving absolute values, 188–189
  linear, 187–189, 191–192, 256–258
  in one variable, 184–189
  procedure for solving polynomial, 192
  properties of, 45–46
  quotients of polynomials, 193–195
  rules of operation for, 186–187
  solution set of, 184–185
  solving types of, 185–186
  symbols for, 43
  systems of linear, 423–426
  types of, 184
Infinite decimals, 3, 4, 5

Infinite set, 643
  countably, 666, 667
  properties of, 665
Initial conditions, 348
Initial point of ray, 525
Instantaneous acceleration, 284–286
Instantaneous rates, 259, 266–269, 281–286. *See also* Derivatives
Instantaneous velocities, 266, 281–284. *See also* Acceleration
Integers
  as factors, 26–27
  positive, 2
Integral calculus, fundamental theorem of, 377
Integrals, 354–357
  definite vs indefinite, 367
  definition of, 517
  of trigonometric functions, 591
Integral sign, 355
Integrand, 355
Integration, 342–380
  antiderivatives and, 343–346
  applications of, 346–352
  areas found by, 363–372
  definition of, 343
  differential equations and, 361
  differential functions and, 357–361
  equations of curves and, 353–354
  exponential and logarithmic functions and, 517–520
  notation for, 354–357
  other bases and, 520–524
  process of, 197
  as process of summation, 375–380
  of trigonometric functions, 588–592
  upper and lower limits of, 367, 376
  and volume of solids, 373–375
Interest. *See* Compound interest; Simple interest
Interest-bearing debt, 450, 451
Interest rate
  definition of, 437, 442
  equivalent, 463
Interior points of ray, 525
Intermediate value theorem, 316
Interpolation
  calculators and, 505
  graphical, 210–212, 263

Interpolation (*cont'd.*)
   by proportional parts, 205–206
Intersection
   of two lines, 234–236
   of two sets, 653
Intervals on number line, 49–50. *See also* Open interval
Inverse elements, axioms defining, 11–13
Inverse of implication, 633
Inverse of square matrix, 416–420
Investments
   average rate of growth of, 498–499
   rate of growth of (*see* Compounding continuously)
   total rate of growth of, 494
Irrational numbers, definition of, 4
Irrational roots of polynomial equation, 174–177
Isosceles triangles, 218
   definition of, 530
   trigonometric functions of, 545–546

Latus rectum, 244
Law(s)
   common interest, 500, 501
   of cosines, 563–565
   of exponents, 502–503, 512
   of growth, 500, 501
   of natural logarithms, 507–513
   of sines, 561–563, 565
   for union of sets, 652
Least common denominators (LCD), 29–30
Length, units of, 363–364
Limits, 269
   concept of, 269–271
   of fraction, 273
   theorems concerning, 272–276
Limit theorems, 275–276
Linear algebra, introduction to, 381–436
Linear equations
   in echelon form, 403–404
   and matrices, 403–408
   and method of determinants, 390–401
   in one variable, 127–130
   as single matrix equation, 420–423
   standard form of, 129
   for straight lines, 232–234
   in three variables, 381–394, 396–399
   in two variables (*see* Linear equations in two variables)
Linear equations in two variables, 381–389
   graphical method for solving, 382–383
   method of elimination and, 383–389
Linear inequalities
   in one variable, 187–189
   procedure for solving, 191–192
   systems of, 423–426
   in two variables, 256–258
Linear polynomial, value of, 429–433
Linear programming, 427–436
Linear velocity of point, 528, 529
Lines
   parallel, 226
   perpendicular, 226–227
   point of intersection of, 234–236
   slope of, 224–226
   straight, 224–229
Line segments
   definition of, 525
   formulas to calculate midpoint of, 220–221
   projection of, 556
Logarithms
   base of natural, 520
   common, 521
   defining natural, 504–513
   and exponents, 510–511
   laws of, 512
   to other bases, 520–524
Logic, and sets, 625–672
Logical equivalents, 632
Lower limit of integration, 367

Magnitude of angle, 525
Marginal cost
   average, 262, 263, 264
   definition of, 311
Marginal revenue, 311
Markup, 439
Mathematical expectation, 605–607
Mathematic system, definition of, 7
Matrices
   additive identity for, 410
   additive inverse of, 410
   concept of, 401–408

Index

Matrices (*cont'd.*)
  definition of, 419
  elementary row operations on, 405–406
  elimination method and, 403–408
  equality of, 408
  inverse square of, 416–420
  multiplication of, 410–413
  sum of two, 408–410
Matrix algebra, 408–416
Matrix equation, linear equation as single, 420–423
Maturity value of note, 451
Maxima, 324–331
  applied problems in, 331–337
  steps for solving problems in, 334–335
Mean, sample, 618
Mean value, 613–614
Mean value theorem, 317–318
Midpoint formulas, 220–221
Minima, 324–331
  applied problems in. 331–337
  steps for solving problems in, 334–335
Minor of element of a determinant, 393, 394
Minus sign, ways of using, 16
Minute, definition of, 566
Mixed fractions, 36
Money, going rate for, 450
Multiplication
  of algebraic fraction, 110
  basic properties of, 17–20
  of complex numbers, 58–59
  as distributive, 10–11
  of fractions, 37–38
  of a matrix, 410–413
  of polynomials, 85–87
  of two algebraic expressions, 93–94
Multiplicative identity matrix, 413–414, 416

Natural logarithms
  base of, 504, 520
  definition of, 504–513
  derivative of, 514–516
  laws of, 507–513
  table for, 505–506
Natural numbers, 2
Negations, 629–631

  of proposition, 627
  symbol for, 631
Negative direction, 453
Negative exponents, 72–73
Negative real numbers, 74
Nominal annual rate, 459, 462, 463
Nondifferentiability, 326–327
Normal distribution, standard, 610–612
Normal probability distributions, 620–624
Notation
  for fractions, 11, 23
  scientific, 508
Null set, 644–645
Number lines, 48–50
  intervals on, 49–50
Number of roots of polynomial equation, 163–164
Numbers
  additive inverse of real, 11
  complex, 55–60
  conjugal complex, 162–163
  counting, 2
  decimal representation of rational, 3
  imaginary, 55–57
  irrational, 41
  natural, 2
  prime, 2
  rational, 2, 3
  real (*see* Real numbers)
  reciprocal of real, 11–12
  sets of, 2
Number system, real, 7–13
Numerator
  of fractions, 17, 106
  sign of, 25–26
Numerical coefficient of term, 64, 65
Numerical equations, definitions of, 120
Numerical identity, 120

Oblique triangles, 560–565
  definition of, 530
Obtuse angles, definition of, 526
One-dimensional coordinate system, 48–50
One-to-one correspondence, 663
Open interval, 277
  on number line, 50
  of real numbers, 270

Open statements, 668–669, 671–672
  truth set of, 668–669
Optimal point, 433
Order axioms, for comparing real numbers, 43–46
Ordered pairs, 658–662
  definition of, 51
  graph of set of, 209
Order relations, 42–48
Ordinary simple interest, 444
Ordinate of point, 53
Origin on number line, 48
Overlapping sets, 646–647, 649

Pairs, ordered. *See* Ordered pairs
Parabola
  axis of, 243
  equations for, 243–247
  focus of, 243
  vertex of, 244
Parallel lines, 226
Parallelogram, definition of, 218
Parentheses, use of, 14–15
Percentage, 437–441
Periodic functions, definition of, 578
Permutations, 596–598
Perpendicular lines, 227–228
Perpetuity, annuity, 473
Physical objects, measuring. *See* Functions; Relations
Planes
  equations for sets of points in, 222–224
  half-, 257
  points in, 51–55
Point(s)
  abscissa of, 53
  convex set of, 426
  coordinates of, 53, 54
  distance between, 54–55
  feasible, 433
  of inflection, 321
  initial, 525
  linear velocity, 528, 529
  optimal, 433
  ordinate of, 53
  in planes, 51–55, 222–224
  plotting, 54
  projection of, 555
  slope of curve at, 309–312

zero, 48
Point-slope form of equation, 231
Polygon, convex, 427
Polygonal convex set, 426, 427
Polynomial equations
  complex roots of, 162–163
  degree of, 126
  Descartes' rule of signs and, 167–168
  irrational roots of, 174–177
  number of roots of, 163
  in one variable, 121–126
  rational roots of, 170–174
  roots of, 151, 157–161
  solutions and solution sets for, 121–126
  solving form of, 125–126
  theorem on roots of, 163–164
  upper and lower limits of real roots of, 165–167
  See also Linear equations; Quadratic equations
Polynomial functions, continuity of, 278–281
Polynomial inequalities, higher-degree, 189–193
Polynomials
  addition of, 83–85
  coefficients of, 82–83
  cubic, 83
  differentiating any, 293
  division of, 87–89
  in equations containing fractions, 177–182
  factoring, 97–105
  factor theorem and, 158–159
  inequalities involving, 193–195
  multiplication of, 85–87
  nth degree, 82
  in one variable, 82–91, 117–119
  quadratic, 83
  subtraction of, 83–85
  two major problems with, 154–161
  in two or more variables, 92–96
  value of linear, 429–433
  zeros of, 157–161
Populations, and samples, 617–619
Positive acute angle, 541, 542
  trigonometric function of, 543
Positive denominators, 26
Positive direction, 453

Positive integers, 2
　closure on set of, 8–9
　expression of, 3
Positive real numbers, 43
　fractional powers of, 74–75
Power function, 501
Powers, and exponents, 70
Present values
　of annuity, 478–483
　at compound interest, 464–469
　at simple interest, 449–457
Prime numbers, 2, 28
Principal
　accumulation of, 461
　changing (*see* Compound interest)
　definition of, 437, 442
Probability
　definition of, 593–594
　and statistics, 593–624
Probability distributions, 607–613
　normal, 620–624
Probability rules, 601–605
Product
　involving zero, 19–20
　natural logarithm of, 507
　special, 97–105, 139
　of two functions, 295–297
　See also Multiplication
Profit, gross, 439
Programmable calculator, and irrational roots of equations, 175–176
Programming linear, 427–436
Projection
　of line segment, 556
　of point, 555
Promissory note, 451
Propositions, 626–627, 670
　compound, 628–629, 632
　negation of, 627
　simple, 627
　true, 639
　truth values of, 626–628
Pure imaginary numbers, 56
Pythagorean theorem, 218–219
　and right triangle, 544

Quadrantal angle, 532
Quadratic equations
　in one variable, 130–133
　roots of, 134
　solving factored, 131–132
Quadratic equation solution methods, 131–151
　completing the square as, 139–145
　factoring as, 133–136
　facts about roots of, 151–153
　modified, 136–137
　quadratic formula method as, 145–151
Quadratic form, equations in, 153–154
Quadratic formula, 157
　derivation of, 147–148
　and roots of quadratic equation, 151–153
Quadratic formula method, 145–151
Quadratic polynomial, 83
Quadrilateral, as parallelogram, 221
Quadrilateral triangle, 218
Quantifiers, 670–672
Quotients, 17, 88
　and exponents, 69–70, 71
　of functions derivatives, 297–298
　involving zero, 19–20
　polynomial inequalities involving, 193–195

Radian of angle, 526–529
　definition of, 527
Radical notations, 78
Radicals, 78–82
　equations containing, 182–184
　properties of, 78–79
　rules of operations for, 80
Radicand, 78
Radius of circle, 238
Range of function, 200
Range of relation, 198
Rate of growth
　over conversion period, 498
　total investment, 494
Rate of interest, 437, 442, 463
Rate problem in calculus, 197
Rates. *See* Average rates; Instantaneous rates
Rates of change, types of, 259
Rational numbers, 170
　decimal representation of, 3
　definition of, 2
Rational roots, 170–174

Ray, definition of, 525
Real number(s), 1–5
  absolute value of, 44–45
  adding, 5
  additive inverse of, 11, 16
  axioms for, 9–11
  basic properties of, 14–22
  comparing two, 42–44
  concept of, 1–5
  coordinate systems and, 48–59
  definition of, 4
  fractional powers of positive, 74–75
  negative, 74
  and number lines, 48–50
  open interval of, 270
  order axioms and, 43–46
  ordered pairs of, 51 (*see also* Relations)
  positive, 43
  reciprocal of, 11–12
  roots of (*see* Radicals)
  rules of operation for, 20–21
  symbols of grouping, 14–15
Real number system, 7–13
Reciprocal relations, 566
Rectangle, definition of, 218
Rectangular coordinate system, 50–55
  and analytic geometry, 217
  and circles, 238–239, 240–242
  and ordered pairs, 209–217
Rectangular plot, finding dimensions of, 333–334
Recurring equalities, 97
Reflexive property of equality, 7
Related angles, 537–541
Related-angle theorem, 538–539, 545, 546
Related rates, 338–340
Relations
  domain of definition of, 198
  and functions, 197–208
  graphs of functions and, 209–217
  order, 42–48
  range of, 198
  reciprocal, 566
  squared, 567
Relative maximum and minimum values. *See* Maxima; Minima
Remainder theorem, 158
Repeating decimals, 3
Revolution of angle, 526

Right angles, 526
Right triangles, 218
  definition of, 530
  solution of, 550–552
Rolle's theorem, 316–317
Roots
  of equations, 121, 123
  Descartes' rule of signs, 167–168
  extraneous, 182
  irrational, 174–177
  number of, 163–164
  of polynomial equations, 157–161
  of quadratic equations, 134, 151–153
Rotation of angle, 526
Row index of matrix, 402
Row matrix, 403
Row vector, 403
Rule of signs, Descartes, 167–168
Rules
  for defining functions, 202–203
  for differentiation, 293, 301–303
Rules of operation
  for absolute values and inequalities, 45–46
  for exponents, 77
  for fractions, 40
  for inequalities, 186–187
  for radicals, 80
  for real numbers, 20–21

Samples from populations, 617–619
Scalar matrices, 410
Scientific notation, 508
Secant, slope of, 310
Secant line, 262
Second, definition of, 566
Second derivative, 314
Second derivative test, 329–330
Sets
  Cartesian product of two, 658–662
  closure property of, 8–9
  complement of, 653
  convex, 426, 427
  countable, 665–667
  elements of, 640–641, 642, 643
  empty, 644–645
  equipollent, 662–665
  finite, 643
  infinite, 665, 666, 667

Index

Sets (cont'd.)
   intersection of, 653
   and logic, 625–672
   method of constructing, 641–643
   null, 644–645
   of numbers, 2 (see also Mean; Standard deviation; Variance)
   operations forming, 652–658
   ordered pair, 209
   overlapping, 646–647, 649
   solution (see Solution sets)
   standard, 616
   truth, 668–669
   union of, 652
Simple annuity, 473–476
Simple factor, definition of, 106
Simple fraction, 111
Simple interest, 441–457
   definition of, 437
   formula for, 455–456
   ordinary, 444
   present value at, 449–457
Simple proposition, 627
Sines, laws of, 561–563, 565. See also Trigonometric functions
Slope-intercept form of equation of line, 232
Slope of curve
   average, 262, 264
   and integration, 353–354
   at a point, 309–312
Slope of a line, 224–226
Solids, volumes of, 373–375
Solution of an equation
   definition of, 123
   involving one variable, 121
Solutions
   for factored polynomial equations, 131–132
   for polynomial equations, 121–126
Solution sets, 213, 669
   definition of, 123
   of inequalities, 184–185, 256
   for polynomial equations, 121–126
   union of, 131–132
Special products, 97–105, 139
Speed of object, 284
Square
   definition of, 218

   method of completing, 139–145
Squared relations, 567
Square matrices, 403
   inverse of, 416–420
   of same size, 412–413
Square root radicals. See Radicals
Standard deviation
   sample, 618
   of set, 616–617
Standard normal curve, 610–612
Statements
   definition of, 626
   open, 668–669, 671–672
Statistics, and probability, 593–624
Straight angle, definition of, 526
Straight line, 224–229
   equations of, 230–238
Subsets, 646–652
   equal, 646
   overlapping, 646–647
   Venn diagrams and, 647–650
Substitution, synthetic, 155–157, 159
Substitution principle, 8
Subtraction
   of algebraic fractions, 107–109
   of fractions, 34–35
   method of elimination by, 234–235
   of polynomials, 84–85, 92–93
   process of, 16
Successive differentiation, 314–315
Sum
   value of, 451
   of vectors, 557
Summation, integration as process of, 375–380
Symbols
   for determinant of second order, 390
   for determinant of third order, 392–394
   of grouping, 14–15
   for inequalities, 43
Symmetry property of equality, 7
Synthetic division, 159–160
Synthetic substitution, 155–157, 159

Tables
   of trigonometric functions, 547–550
   use of natural-logarithm, 505–506
   of value, 204–208
   See also Appendix B

Tangent line, 321
  definition of, 310
Tautology, 634
Test for limits of roots of polynomial equations, 166–167
Theorems
  and bi-implications, 637–640
  for functions, 315–318
  intermediate value, 316
  mean value, 317–318
  Rolle's, 316–317
Time-line diagram, and interest, 452, 453, 456, 460
Triangles, 529–532
  equilateral, 530
  isosceles, 530, 545–546
  oblique, 560–565
  right, 550–552
  trigonometric functions of, 529–532, 544–545
  types of, 218
  vertices of, 529, 541
Trigonometric functions, 525–592
  acute angles and, 536, 541–547
  coterminal angles and, 532–533
  differentiation and, 582–588
  graphs of, 578–582
  integration of, 588–592
  of isosceles triangles, 545–546
  oblique triangles and, 560–565
  of positive acute angle, 543
  related angles and, 537–541
  right triangles and, 550–552
  six, 534–535
  table of, 547–550
  triangles and, 529–532, 544–545
  vectors and, 556–560
Trigonometric identities
  basic, 565–570
  list of, 569
Trigonometry
  calculator for, 447
  definition of, 535
Truth in Lending Act, 455
Truth set of open statement, 668–669
Truth tables
  for basic connectives, 633
  construction, 633–634

Unique factorization theorem, 3
Union of two sets, 131–132, 652
Units per unit, 261
Universal quantifiers, 670
Universal sets, 643–645
Upper limit of real roots of polynomial equation, 165–166

Value, equations of, 469–472
Value of annuity, 474–476
  on conversion date, 489–493
Variable(s), 120
  definition of, 61, 643
  derivative of, 290
  graphs of equations involving, 212–215
  graphs of linear inequalities in, 256–258
  independent, 513–514
  inequalities in one, 184–189
  linear equations in one, 127–130
  polynomial functions in one, 279
  polynomials in one, 82–91, 117–119
  polynomials in two or more, 92–96
Variance
  sample, 618
  of set of numbers, 614–616
Vector quantity, 284
Vectors, 556–560
  column and row, 403
  components of, 558–560
Velocity
  angular, 528
  average, 264
  and distance, 346–349
  instantaneous, 266, 281–284
  linear, 528, 529
Venn diagrams, 647–652
Vertex
  of ellipse, 249
  of hyperbola, 253
  of parabola, 244
  of triangle, 529, 541
Volume of box, maximum, 331–333
Volume of solids, 373–375

X axis, constructing, 51–52

Y axis, constructing, 51–52

Zero
- division by, 19–20
- of polynomial, 157–161
- products and quotients involving, 19–20
- rules concerning, 11–12
- special property of, 11

Zero exponents, 72–73
Zero matrix, 409–410
Zero point on number line, 48
Zero polynomial, 82
Zeros of polynomial, 157–161